Geomorphology and Global Environmental Change

How will global environmental change affect our landscape and the way we interact with it? The next 50 years will determine the future of the environment in which we live, whether catastrophe or reorganisation. Global climate change will potentially have a profound effect on our landscape, but there are other important drivers of landscape change, including relief, hydroclimate and runoff, sea level change and human activity. This volume summarises the state of the art concerning the landscape-scale geomorphic implications of global environmental change.
It analyses the potential effects of environmental change on a range of landscapes, including mountains, lakes, rivers, coasts, reefs, rainforests, savannas, deserts, permafrost, and ice sheets and ice caps.

Geomorphology and Global Environmental Change provides a benchmark statement from some of the world's leading geomorphologists on the state of the environment and its likely near-future change. It is invaluable as required reading in graduate advanced courses on geomorphology and environmental science, and as a reference for research scientists. It is highly interdisciplinary in scope, with a primary audience of earth and environmental scientists, geographers, geomorphologists and ecologists, both practitioners and professionals. It will also have a wider reach to those concerned with the social, economic and political issues raised by global environmental change and be of value to policy-makers and environmental managers.

OLAV SLAYMAKER is Professor Emeritus in the Department of Geography, University of British Columbia. He is a Senior Associate of the Peter Wall Institute for Advanced Studies and Senior Fellow of St John's College, University of British Columbia. He is a Former President of the Canadian Association of Geographers and the International Association of Geomorphologists, and a Linton Medallist. He has held visiting professorships at the universities of Vienna, Canterbury, Oslo, Southern Illinois, Taiwan, and Nanjing. He has authored 120 refereed journal articles and authored and edited 20 books. He is a Co-Editor-in-Chief of *Catena* and member of nine international editorial boards.

THOMAS SPENCER is University Senior Lecturer in the Department of Geography, Director of the Cambridge Coastal Research Unit, University of Cambridge, and Official Fellow, Magdalene College, Cambridge. His research interests in wetland hydrodynamics and sedimentation, coral reef geomorphology, sea level rise and coastal management have taken him to the Caribbean Sea, the Pacific and Indian oceans, Venice and its lagoon and the coastline of eastern England. He has authored and co-edited numerous books on coastal problems, environmental challenges and global environmental change.

CHRISTINE EMBLETON-HAMANN is a Professor in the Department of Geography and Regional Research at the University of Vienna. Her main interest is in alpine environments. Within this field she focusses on the history of ideas concerning the evolution of alpine environments, genesis and development of specific landforms and human impact on alpine environments, and has written extensively on geomorphological hazards and risks and the assessment of scenic quality of alpine landscapes. She is Past President of the Austrian Commission on Geomorphology and Secretary-General of the International Association of Geomorphologists Working Group.

Praise for *Geomorphology and Global Environmental Change:*

'Global change, whether due to global warming or other human impacts, is one of the great issues of the day. In this volume some of the world's most distinguished geomorphologists give an expert and wide-ranging analysis of its significance for the movement.'

ANDREW GOUDIE, *University of Oxford and President of the International Association of Geomorphologists*

'*Geomorphology and Global Environmental Change*, with chapters by a truly global group of distinguished geomorphologists, redresses the imbalance that has seen an overemphasis on climate as the prime driver of landscape change. This comprehensive book summarises the deepening complexity of multiple drivers of change, recognising the role that relief plays in influencing hydrological processes, that sea level exerts on coastal environments, and the far-reaching impacts of human activity in all the major biomes, in addition to climate. The lags and thresholds, the changing supply to the sediment cascade, and the influence of fire on vegetation ensure that uncertain near-future process regimes will result in unforeseen landscape responses. The potential collapse and reorganisation of landscapes provide fertile research fields for a new generation of geomorphologists and this book provides an authoritative synthesis of where we are today and a basis for embarking on a more risk-based effort to forecast how the landforms around us are likely to change in the future.'

COLON D. WOODROFFE, *University of Wollongong*

'A robust future for geomorphology will inevitably have to be founded on greater consideration of human impacts on the landscape. An intellectual framework for this will necessarily have environmental change as a central component. This volume represents an important starting point. Coverage is comprehensive, and a set of authoritative voices provide individual chapters serving as both benchmarks and signposts for critical disciplinary topics.'

COLIN E. THORN, *University of Illinois at Urbana-Champaign*

'According to the World Resources Institute, 21 metric tons of material, including materials not actually used in production (soil erosion, over-burden, construction debris, etc.) are processed and discharged as waste every year to provide the average Japanese with goods and services. The figure for the US is an astonishing 86 tonnes per capita. The OECD says that in 2002, 50 billion tonnes of resources were extracted from the ecosphere to satisfy human needs and the number is headed toward 80 billion tonnes per year by 2020. Most of this is associated with consumption by just the richest 20% of humanity who take home 76% of global income, so the human role in global mass movement and landscape alteration may only be beginning. These data show unequivocally that the human enterprise in an integral and growing component of the ecosphere and one of the greatest geological forces affecting the face of the earth. Remarkably, however, techno-industrial society still thinks of itself as separate from "the environment". Certainly geomorphologists have historically considered human activities as external to geosystems. This is about to change. In *Geomorphology and Global Environmental Change*, Slaymaker, Spencer and Embleton-Hamann provide a comprehensive treatment of landscape degradation in geosystems ranging from coral reefs to icecaps that considers humans as a major endogenous forcing mechanism. This long-overdue integration of geomorphology and human ecology greatly enriches the global change debate. It should be a primary reference for all serious students of contemporary geomorphology and the full range of environmental sciences.'

WILLIAM E. REES, *University of British Columbia; co-author of* Our Ecological Footprint; *Founding Fellow of the One Earth Initiative*

Geomorphology and Global Environmental Change

EDITED BY

Olav Slaymaker

The University of British Columbia

Thomas Spencer

University of Cambridge

Christine Embleton-Hamann

Universität Wien

CAMBRIDGE UNIVERSITY PRESS
Cambridge, New York, Melbourne, Madrid, Cape Town, Singapore, São Paulo, Delhi

Cambridge University Press
The Edinburgh Building, Cambridge CB2 8RU, UK

Published in the United States of America by Cambridge University Press, New York

www.cambridge.org
Information on this title: www.cambridge.org/9780521878128

First published 2009

Printed in the United Kingdom at the University Press, Cambridge

A catalogue record for this publication is available from the British Library

ISBN 978-0-521-87812-8 hardback

Contents

List of contributors — *page* x

Preface — xiii

Acknowledgements — xiv

List of acronyms and abbreviations — xv

1 Landscape and landscape-scale processes as the unfilled niche in the global environmental change debate: an introduction — 1

OLAV SLAYMAKER, THOMAS SPENCER AND SIMON DADSON

1.1 The context — 1
1.2 Climatic geomorphology — 4
1.3 Process geomorphology — 5
1.4 Identification of disturbance regimes — 6
1.5 Landscape change — 8
1.6 Systemic drivers of global environmental change (I): hydroclimate and runoff — 10
1.7 Systemic drivers of global environmental change (II): sea level — 14
1.8 Cumulative drivers of global environmental change (I): topographic relief — 17
1.9 Cumulative drivers of global environmental change (II): human activity — 19
1.10 Broader issues for geomorphology in the global environmental change debate — 22
1.11 Landscape change models in geomorphology — 25
1.12 Organisation of the book — 28

2 Mountains — 37

OLAV SLAYMAKER AND CHRISTINE EMBLETON-HAMANN

2.1 Introduction — 37
2.2 Direct driver I: relief — 42
2.3 Direct driver II: hydroclimate and runoff — 44
2.4 Direct driver III: human activity, population and land use — 45
2.5 Twenty-first-century mountain landscapes under the influence of hydroclimate change — 49

2.6 Twenty-first-century mountain landscapes under the influence of land use and land cover change 55
2.7 Vulnerability of mountain landscapes and relation to adaptive capacity 61

3 Lakes and lake catchments 71
KENJI KASHIWAYA, OLAV SLAYMAKER AND MICHAEL CHURCH
3.1 Introduction 71
3.2 Lakes and wetlands 72
3.3 The lake catchment as geomorphic system 74
3.4 Internal lake processes 78
3.5 Hydroclimate changes and proxy data 80
3.6 Effects of human activity 86
3.7 Scenarios of future wetland and lake catchment change 92

4 Rivers 98
MICHAEL CHURCH, TIM P. BURT, VICTOR J. GALAY AND G. MATHIAS KONDOLF
4.1 Introduction 98
4.2 Land surface: runoff production 98
4.3 River channels: function and management 103
4.4 Fluvial sediment transport and sedimentation 109
4.5 Water control: dams and diversions 114
4.6 River restoration in the context of global change 121
4.7 Conclusions 125

5 Estuaries, coastal marshes, tidal flats and coastal dunes 130
DENISE J. REED, ROBIN DAVIDSON-ARNOTT AND GERARDO M. E. PERILLO
5.1 Introduction 130
5.2 Estuaries 133
5.3 Coastal marshes and tidal flats 136
5.4 Coastal sand dune systems 142
5.5 Managing coastal geomorphic systems for the twenty-first century 150

6 Beaches, cliffs and deltas 158
MARCEL J. F. STIVE, PETER J. COWELL AND ROBERT J. NICHOLLS
6.1 Introduction 158
6.2 Coastal classification 159
6.3 The coastal-tract cascade 162
6.4 Applications of the quantitative coastal tract 167
6.5 Risk-based prediction and adaptation 174
6.6 Conclusions 176

180
THOMAS SPENCER
180
vironments: the reef carbonate
182
t geomorphology 188
195
y landforms 202
205

214
H. BLAKE
d morphoclimatic zone 214
f the rainforest zone: a synthesis 217
rest zone 231
ting geomorphological change:
empirical approaches 234
nd geomorphological responses to
rainforest areas 235
8.6 Research gaps and ... provement to geomorphological predictions in the humid tropics 243
8.7 Summary and conclusions 243

9 Tropical savannas 248
MICHAEL E. MEADOWS AND DAVID S. G. THOMAS
9.1 Introduction 248
9.2 Key landforms and processes 255
9.3 Landscape sensitivity, thresholds and 'hotspots' 262
9.4 A case study in geomorphic impacts of climate change: the Kalahari of southern Africa 265
9.5 Concluding remarks 269

10 Deserts 276
NICHOLAS LANCASTER
10.1 Introduction 276
10.2 Drivers of change and variability in desert geomorphic systems 278
10.3 Fluvial geomorphic systems in deserts 283
10.4 Aeolian systems 286
10.5 Discussion 291

11 Mediterranean landscapes 297
MARIA SALA
11.1 Introduction 297
11.2 Geology, topography and soils 297
11.3 Climate, hydrology, vegetation and geomorphological processes 299
11.4 Long-term environmental change in Mediterranean landscapes 303

11.5 Traditional human impacts in Mediterranean landscapes and nineteenth- and twentieth century change 307
11.6 Contemporary and expected near-future land use changes 310
11.7 Global environmental change in Mediterranean environments and its interaction with land use change 312
11.8 Concluding remarks 315

12 Temperate forests and rangelands 321
ROY C. SIDLE AND TIM P. BURT
12.1 Introduction 321
12.2 Global distribution of mid-latitude temperate forests and rangelands 323
12.3 Potential climate change scenarios and geomorphic consequences 325
12.4 Types, trajectories and vulnerabilities associated with anticipated mass wasting responses to climate change 325
12.5 Anthropogenic effects on geomorphic processes 328
12.6 Techniques for assessing effects of anthropogenic and climate-induced mass wasting 334
12.7 Summary and conclusions 337

13 Tundra and permafrost-dominated taiga 344
MARIE-FRANÇOISE ANDRÉ AND OLEG ANISIMOV
13.1 Permafrost regions: a global change 'hotspot' 344
13.2 Permafrost indicators: current trends and projections 348
13.3 Permafrost thaw as a driving force of landscape change in tundra/taiga areas 350
13.4 Impact of landscape change on greenhouse gas release 354
13.5 Socioeconomic impact and hazard implications of thermokarst activity 356
13.6 Vulnerability of arctic coastal regions exposed to accelerated erosion 358
13.7 Discriminating the climate, sea level and land use components of global change 360
13.8 Lessons from the past 361
13.9 Geomorphological services and recommendations for future management of permafrost regions 362

14 Ice sheets and ice caps 368
DAVID SUGDEN
14.1 Introduction 368
14.2 Distribution of ice sheets and ice caps 369
14.3 Ice sheet and ice cap landscapes 374
14.4 Ice sheets and ice caps: mass balance 378
14.5 Ice flow and ice temperature 380
14.6 External controls and feedbacks 381
14.7 Landscapes of glacial erosion and deposition 384
14.8 How will ice sheets and ice caps respond to global warming? 389
14.9 Conclusion and summary 399

15 Landscape, landscape-scale processes and global environmental change: synthesis and new agendas for the twenty-first century 403

THOMAS SPENCER, OLAV SLAYMAKER AND CHRISTINE EMBLETON-HAMANN

15.1 Introduction: beyond the IPCC Fourth Assessment Report 403
15.2 Geomorphological processes and global environmental change 405
15.3 Landscapes and global environmental change 407
15.4 Conclusions: new geomorphological agendas for the twenty-first century 416

Index 424

The colour plates are situated between pages 80 and 81

Contributors

Professor Marie-Françoise André
University of Clermont-Ferrand
4 rue Ledru, 63057 Clermont-Ferrand Cedex 1, France

Professor Oleg Anisimov
State Hydrological Institute
23, second Line V.O.,
St Petersburg, Russia

Dr Will H. Blake
School of Geography, University of Plymouth
Drake Circus, Plymouth PL4 8AA, UK

Professor Tim P. Burt
Department of Geography, Durham University
South Road, Durham DH1 3LE, UK

Professor Michael Church
Department of Geography,
The University of British Columbia
1984 West Mall, Vancouver, British Columbia,
Canada V6T 1Z2

Dr Peter J. Cowell
University of Sydney, Institute of Marine Science
Building H01, Sydney, Australia

Dr Simon Dadson
Centre for Ecology and Hydrology
Crowmarsh Gifford, Wallingford OX10 8BB, UK

Professor Robin Davidson-Arnott
Department of Geography, University of Guelph
Guelph, Ontario, Canada, N1G 2WI

Professor Christine Embleton-Hamann
Institut für Geographie und Regionalforschung,
Universität Wien
Universitätsstr. 7, A-1010 Wien, Austria

Dr Victor J. Galay
Northwest Hydraulic Consultants Ltd
30 Gostick Place, North Vancouver, British Columbia,
Canada V7M 3G2

Professor Kenji Kashiwaya
Institute of Nature and Environmental Technology,
Kanazawa University
Kakuma, Kanazawa 920–1192, Japan

Dr Paul Kench
School of Geography and Environmental Science,
University of Auckland
Private Bag 92019, Auckland, New Zealand

Professor G. Mathias Kondolf
Department of Landscape Architecture
and Environmental Planning,
University of California
202 Wurster Hall, Berkeley, CA 94720, USA

Professor Nicholas Lancaster
Center for Arid Lands Environmental Management,
Desert Research Institute
2215 Raggio Parkway, Reno, NV 89512, USA

Professor Michael E. Meadows
Department of Environmental and Geographical Sciences,
University of Cape Town
Rondebosch 7701, South Africa

Professor Robert J. Nicholls
School of Civil Engineering and the Environment,
University of Southampton
Lanchester Building, Southampton SO17 1BJ, UK

Professor Gerardo M. E. Perillo
Departamento de Geología, Universidad Nacional del Sur
San Juan 670, (8000) Bahía Blanca, Argentina

Professor Chris Perry
Department of Environmental and Geographical Sciences,
Manchester Metropolitan University
John Dalton Building, Chester St.,
Manchester M1 5GD, UK

Professor Denise J. Reed
Department of Earth and Environmental Sciences,
University of New Orleans
New Orleans, LA 70148, USA

Professor Maria Sala
Departament de Geografia Física,
Facultat de Geografia i Història, Universitat de Barcelona
Montalegre 6, 08001 Barcelona, Spain

Professor Roy C. Sidle
Director, Environmental Sciences Program,
Department of Geology, Appalachian State University
P.O. Box 32067, Boone, NC 28608, USA

Professor Olav Slaymaker
Department of Geography,
The University of British Columbia
1984 West Mall, Vancouver, British Columbia, Canada
V6T 1Z2

Dr Thomas Spencer
Cambridge Coastal Research Unit,
Department of Geography,
University of Cambridge
Downing Place, Cambridge CB2 3EN, UK

Professor Marcel J. F. Stive
Section of Hydraulic Engineering,
Faculty of Civil Engineering and Geosciences,
Delft University of Technology
P.O. Box 5048, 2600 GA Delft, The Netherlands

Professor David Sugden
Institute of Geography, School of GeoSciences,
University of Edinburgh
Edinburgh EH9 3JW, UK

Professor David S. G. Thomas
School of Geography,
Oxford University Centre for the Environment
South Parks Road, Oxford OX1 3QY, UK

Professor Rory P. D. Walsh
Department of Geography,
University of Wales Swansea
Singleton Park, Swansea SA2 8PP, UK

Preface

The catalyst for this book was the Presidential Address delivered by Professor Andrew Goudie, Master of St Cross College, Oxford, to the Sixth International Conference on Geomorphology held in Zaragoza, Spain in September 2005. He identified the question of landform and landscape response to global environmental change as one of the five central challenges for geomorphology (the science of landform and landscape systems). He called for the establishment of an international Working Group to address this question and the chapters of this volume constitute the first product of that process. We applaud Professor Goudie's vision and trust that this first modest effort to respond to his call to arms will be reinforced by further research contributions on the topic. The book was written under the editorial guidance of Professor Olav Slaymaker (The University of British Columbia), Dr Thomas Spencer (University of Cambridge) and Professor Christine Embleton-Hamann (University of Vienna). The editors wish to pay tribute to three mentors, Clifford Embleton, Ian Douglas and Denys Brunsden, who, in different ways, have been instrumental in stimulating their enthusiasm for a global geomorphological perspective.

The editors and authors share a common professional interest in landforms, landform systems and terrestrial landscapes. Love of landscape and anxiety over many of the contemporary changes that are being imposed on landscape by society are also driving emotions that unite the authors. All authors perceive a heightened awareness of the critical issue of global climate change in contemporary public debate, but at the same time see a worrying neglect of the role of landscape in that environmental problematique. The two topics (climate change and landscape change) are closely intertwined. This book certainly has no intention of downplaying the importance of climate change but it does attempt to counterbalance an overemphasis on climate as the single driver of environmental change.

All the contributors strongly believe that a greater understanding of geomorphology will contribute to the sustainability of our planet. It is our hope that this understanding will be turned into practical policy. It is our gratitude for the beauty and integrity of landscape that motivates us to present this perspective on the crucial global environmental debate that involves us all.

Acknowledgements

This book has been entirely dependent on the expertise, hard work and commitment to excellence shown throughout by the Lead Authors of the 15 chapters:

Olav Slaymaker (Chapters 1 and 2), Kenji Kashiwaya (3), Michael Church (4), Denise Reed (5), Marcel Stive (6), Paul Kench (7), Rory Walsh (8), Michael Meadows (9), Nick Lancaster (10), Maria Sala (11), Roy Sidle (12), Marie-Françoise André (13), David Sugden (14) and Thomas Spencer (15) with important assistance from the many Contributing Authors:

Thomas Spencer (Chapters 1 and 7), Simon Dadson (1), Christine Embleton-Hamann (2 and 15), Olav Slaymaker (3 and 15), Michael Church (3), Tim Burt (4 and 12), Vic Galay (4), Matt Kondolf (4), Robin Davidson-Arnott (5), Gerardo Perillo (5), Peter Cowell (6), Robert Nicholls (6), Chris Perry (7), William Blake (8), David Thomas (9) and Oleg Anisimov (13).

The external reviewers brought a balance and perspective which modified some of the enthusiasms of the authors and have produced a better product:

John Andrews, Peter Ashmore, James Bathurst, Hanna Bremer, Denys Brunsden, Bob Buddemeier, Nel Caine, Celeste Coelho, Arthur Conacher, John Dearing, Richard Dikau, Ian Douglas, Charlie Finkl, Hugh French, Thomas Glade, Andrew Goudie, Dick Grove, Pat Hesp, David Hopley, Philippe Huybrechts, Johan Kleman, Gerd Masselink, Ulf Molau, David Nash, Jan Nyssen, Frank Oldfield, Phil Owens, Volker Rachold, John Schmidt, Ashok Singhvi, John Smol, Marino Sorriso-Valvo, David Thomas, Michael Thomas, Colin Thorn, Ian Townend, Sandy Tudhope, Theo Van Asch, Heather Viles, Andrew Warren and Colin Woodroffe.

A critical element in the gestation of this volume were two intensive meetings between authors held in Cambridge, England and Obergurgl, Austria. We are grateful to The Master and Fellows of Magdalene College, Cambridge for the use of their superb facilities. The Austrian meeting was part of a joint meeting with the Austrian Commission on Geomorphology (now known as the Austrian Research Association on Geomorphology and Environmental Change). We are indebted to Margreth Keiler, Andreas Kellerer-Pirklbauer and Hans Stötter for making the Obergurgl meeting a successful international exchange of ideas. We gratefully acknowledge a grant of €2000 from the International Association of Geomorphologists to assist with the costs of these meetings.

We thank Matt Lloyd, Senior Editor, Earth and Life Sciences at Cambridge University Press for encouragement and advice. We also thank Annie Lovett and Anna Hodson at the Press for their support throughout the production process. The technical expertise of Eric Leinberger, Senior Computer Cartographer, Department of Geography, UBC, is responsible for the excellence of the illustrative material. Dr Dori Kovanen, Research Associate, Department of Geography, UBC, provided critical comments, many hours of editorial assistance and technical expertise in data handling (Plates 9–12). Dr Pamela Green, Research Scientist, Complex Systems Research Center, University of New Hampshire, provided Plate 7.

Acronyms and abbreviations

ACD Arctic Coastal Dynamics
ACIA Arctic Climate Impact Assessment
AMIP Atmospheric Model Intercomparison Project
AO Arctic Oscillation
AOGCM Atmosphere–Ocean General Circulation Model
AVHRR Advanced Very High-Resolution Radiometer
BP Before Present
CAESAR Cellular Automaton Evolutionary Slope And River Model
CALM Circumpolar Active Layer Monitoring
CAP Common Agricultural Policy (EU)
CCC Canadian Centre for Climate
CCIAV Climate Change Impacts, Adaptation and Vulnerability (IPCC)
CHILD Channel–Hillslope Integrated Landscape Development Model
CIESIN Centre for International Earth Science Information Network
CLIMAP Climate Long-Range Investigation, Mapping and Prediction
CORINE Coordination of Information on the Environment Programme (EC)
CORONA First operational space photo reconnaissance satellite (USA), 1959–72
CPC Climate Prediction Centre (NOAA)
CSIRO Commonwealth Scientific and Industrial Research Organization
DEM Digital Elevation Model
dSLAM Distributed Shallow Landslide Analysis Model
EC European Commission
ECHAM European Centre Hamburg Model
ECMWF European Centre for Medium-Range Weather Forecasts
ELA Equilibrium Line Altitude
EMDW Eastern Mediterranean Deep Water
ENSO El Niño–Southern Oscillation
ENVISAT Environmental Satellite
EOSDIS Earth Observing System Data and Information System
EPA Environmental Protection Agency (USA)
EPICA European Programme for Ice Coring in Antarctica
EROS Earth Resources Observation System
ERS-1 European Remote-Sensing Satellite-1
ERS-2 European Remote-Sensing Satellite-2
ESA European Space Agency
ESF European Science Foundation
ETM Enhanced Thematic Mapper (Landsat)
EU European Union
FAO Food and Agriculture Organization of the United Nations
FAR Fourth Assessment Report (IPCC)
GATT General Agreement on Tariffs and Trade
GCES Glen Canyon Environmental Study
GCM General Circulation Model
GEO Global Environmental Outlook
GFDL Geophysical Fluid Dynamics Laboratory
GIS Geographic Information System
GLASOD Global Assessment of Soil Degradation
GLIMMER Ice Sheet Model
GLOF Glacier Lake Outburst Flood
GLP Global Land Project (IGBP–IHDP)
GOES Geostationary Operational Environmental Satellite
GPR Ground Penetrating Radar
GPS Global Positioning System
GTN-P Global Terrestrial Network for Permafrost
HadCM Hadley Centre Model (UK Met Office)

HOP	Holocene Optimum
IAG	International Association of Geomorphologists
IASC	International Arctic Science Committee
IDSSM	Integrated Dynamic Slope Stability Model
IGBP	International Geosphere–Biosphere Programme
IGU	International Geographical Union
IHDP	International Human Dimensions Programme on Global Environmental Change
IPA	International Permafrost Association
IPCC	Intergovernmental Panel on Climate Change
IPO	Interdecadal Pacific Oscillation
ITCZ	Inter-Tropical Convergence Zone
IUCN	International Union for Conservation of Nature
IUGS	International Union of Geological Sciences
Landsat	Land Remote Sensing Satellite
LER	Local Elevation Range
LGM	Last Glacial Maximum
LIA	Little Ice Age
LIDAR	Light Detection and Ranging Instrument
LUCC	Land Use Cover Change project (IGBP – IHDP)
MA	Millennium Ecosystem Assessment
METEOSAT	Geosynchronous Meteorology Satellite (ESA)
MMD	Multi-model Data Set
MODIS	Moderate Resolution Imaging Spectroradiometer
MOHSST	Met Office (UK) Historical Sea Surface Temperature
MSLP	Monthly Sea Level Pressure
NAO	North Atlantic Oscillation
NASA	National Aeronautics and Space Administration (USA)
NCAR	National Centre for Atmospheric Research
NDVI	Normalised Difference Vegetation Index
NESDIS	National Environmental Satellite, Data, and Information Service (NOAA)
NGO	Non-governmental Organisation
NOAA	National Oceanic and Atmospheric Administration (USA)
NRCS	Natural Resources Conservation Service (USA)
NSIDC	National Snow and Ice Data Center (USA)
OECD	Organization for Economic Cooperation and Development (Paris)
OSL	Optically Stimulated Luminescence
PAGES	Past Global Changes (IGBP)
PDO	Pacific Decadal Oscillation
PDSI	Palmer Drought Severity Index
PESERA	Pan-European Soil Erosion Risk Assessment
QBO	Quasi-Biennial Oscillation
RCP	Representative Concentration Pathway
RIL	Reduced Impact Logging
SHALSTAB	Shallow landslide model
SIR	Spaceborne Imaging Radar
SOC	Soil Organic Carbon
SOI	Southern Oscillation Index
SPOT	Système pour l'Observation de la Terre (France)
SRES	Special Report on Emissions Scenarios (IPCC)
SST	Sea Surface Temperature
STM	Shoreface Translation Model
SUDS	Sustainable Urban Drainage Systems
TAR	Third Assessment Report (IPCC)
TGD	Three Gorges Dam (China)
TOC	Total Organic Carbon
TOPEX/Poseidon	Ocean Topography Experiment (USA and France)
TOPOG	Catchment Hydrological Model (CSIRO)
UN	United Nations
UNCCD	United Nations Convention to Combat Desertification
UNCHS	United Nations Centre for Human Settlements
UNEP	United Nations Environment Programme
UNESCO	United Nations Educational, Scientific and Cultural Organization
UNFCCC	United Nations Framework Convention on Climate Change
UNPD	United Nations Population Division
USDA	United States Department of Agriculture
USFS	United States Forest Service
USGS	United States Geological Survey
USLE	Universal Soil Loss Equation
WCMC	World Conservation Monitoring Centre
WHO	World Health Organization
WMDW	Western Mediterranean Deep Water
WMO	World Meteorological Organization
WWF	World Wide Fund for Nature

1 Landscape and landscape-scale processes as the unfilled niche in the global environmental change debate: an introduction

Olav Slaymaker, Thomas Spencer and Simon Dadson

1.1 The context

Whatever one's views, it cannot be doubted that there is a pressing need to respond to the social, economic and intellectual challenges of global environmental change. Much of the debate on these issues has been crystallised around the activities of the IPCC (Intergovernmental Panel on Climate Change). The IPCC process was set up in 1988, a joint initiative between the World Meteorological Organization and the United Nations Environment Programme. The IPCC's First Assessment Report was published in 1990 and thereafter, the Second (1996), the Third (2001) and the Fourth Assessment Report (2007) have appeared at regular intervals. Each succeeding assessment has become more confident in its conclusions.

The conclusions of the Fourth Assessment can be summarised as follows:

(a) warming of the climate system is unequivocal;
(b) the globally averaged net effect of human activities since AD 1750 has been one of warming (with high level of confidence);
(c) palaeoclimate information supports the interpretation that the warmth of the last half century is unusual in at least the previous 1300 years;
(d) most of the observed increase in globally averaged temperature since the mid twentieth century is very likely due to the observed increase in anthropogenic greenhouse gas concentrations; and
(e) continued greenhouse gas emissions at or above current rates will cause further warming and induce many changes in the global climate system during the twenty-first century that would very likely be larger than those observed in the twentieth century. Details of the methodology used to reach these conclusions can be found in Appendix 1.1.

The IPCC assessments have been complemented by a number of comparable large-scale exercises, such as the UNEP GEO-4 Assessment (Appendix 1.2) and the Millennium Ecosystem Assessment (Appendix 1.3) and, for example, at a more focussed level, the Land Use and Land Cover Change (LUCC) Project (Appendix 1.4) and the World Heritage List (Appendix 1.5). There is no doubting the effort, value and significance of these enormous research programmes into global environmental change (Millennium Ecosystem Assessment, 2005; Lambin and Geist, 2006).

1.1.1 Defining landscape and appropriate temporal and spatial scales for the analysis of landscape

It is important to establish an appropriate unit of study against which to assess the impacts of global environmental change in the twenty-first century and to identify those scales, both temporal and spatial, over which meaningful, measurable change takes place within such a unit. The unit of study chosen here is that of the landscape. There are strong historical precedents for such a choice. Alexander von Humboldt's definition of 'Landschaft' is the 'Totalcharakter einer Erdgegend' (Humboldt, 1845–1862). Literally this means the total character of a region of the Earth which includes landforms, vegetation, fields and buildings. Consistent with Humboldt's discussion, we propose a definition of landscape as 'an intermediate scale region, comprising landforms and landform assemblages, ecosystems and anthropogenically modified land'.

The preferred range of spatial scales is 1–100 000 km^2 (Fig. 1.1). Such a range, of six orders of magnitude, is valuable in two main ways:

(a) individual landforms are thereby excluded from consideration; and

Geomorphology and Global Environmental Change, eds. Olav Slaymaker, Thomas Spencer and Christine Embleton-Hamann. Published by Cambridge University Press.

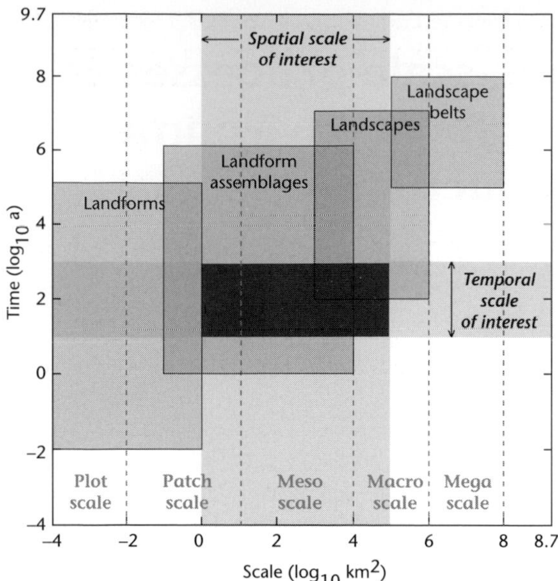

FIGURE 1.1. Spatial and temporal scales in geomorphology. On the *x*-axis, the area of the surface of the Earth in km² is expressed as 8.7 logarithmic units; on the *y*-axis, time since the origin of the Earth in years is expressed as 9.7 logarithmic units.

(b) landscape belts (*Landschaftgürtel*) and biomes, which provide an organising framework for this volume, are nevertheless so large that their response to environmental disturbance is impossible to characterise at century or shorter timescales.

The preferred range of timescales is decades–centuries (Fig. 1.1). These are intermediate temporal scales that are relevant to human life and livelihoods (and define timescales required for mitigation and adaptive strategies in response to environmental change). The determination of the future trajectory of landscape change is unthinkable for projections into a more distant future. Nevertheless, as is argued below, an understanding of changes in landscapes and biomes over the past 20 000 years (i.e. since the time of maximum continental ice sheet development over North America and Eurasia) provides essential context for a proper understanding of current and near-future landscape dynamics.

1.1.2 The global human footprint and landscape vulnerability

The human imprint on the landscape has become global (Turner *et al.*, 1990a; Messerli *et al.*, 2000) and positive feedbacks between climate, relief, sea level and human activity are leading in the direction of critical system state 'tipping points'. This is both the threat and the opportunity of global environmental change. Some of the implications of arriving at such a tipping point are that gradual change may be overtaken by rapid change or there may even be a reversal of previously ascertained trends. A few examples of the most vulnerable landscapes, in which small environmental changes, whether of relief, sea level, climate or land use, can produce dramatic and even catastrophic response, are listed here:

(a) Low-lying deltas in subsiding, cyclone-prone coasts are highly vulnerable to changes in tropical storm magnitude and/or frequency. It is clear that societal infrastructure is poorly attuned to disaster response in such heavily populated landscapes, in both developed (e.g. Hurricane Katrina, Mississippi Delta, August 2005) and developing (Cyclone Nargis, Irrawaddy Delta, May 2008) countries;
(b) Shifting sand dunes respond rapidly to changing temperature and rainfall patterns. Dunes migrate rapidly when vegetation is absent; the vast areas of central North America, central Europe and northern China underlain by loess (a mixture of fine sand and silt) are highly vulnerable to erosion when poorly managed, but are also an opportunity for continuing intensive agricultural activity guided by the priority of the ecosystem;
(c) Glacier extent and behaviour are highly sensitive to changing temperatures and rising sea level. In most parts of the world, glaciers are receding; in tropical regions, glaciers are disappearing altogether, with serious implications for late summer water supply; in Alaska, British Columbia, Iceland, Svalbard and the Antarctic Peninsula glaciers are surging, leading to catastrophic drainage of marginal lakes and downstream flooding. Transportation corridors and settlements downstream from surging glaciers are highly vulnerable to such dynamics;
(d) Permafrost is responding to rising temperatures in both polar and alpine regions. In polar regions, landscape impacts include collapse of terrain underlain by massive ice and a general expansion of wetlands. Human settlements, such as Salluit in northern Quebec, Canada, are highly vulnerable to such terrain instability and adaptation strategies are required now to deal with such changes; and
(e) In earthquake-prone, high-relief landscapes, the damming of streams in deeply dissected valleys by landslides has become a matter of intense concern. The 12 May 2008 disaster in Szechwan Province, China saw the creation of over 30 'quake lakes', one of which reached a depth of 750 m before being successfully drained via overspill channels. If one of these dams had

been catastrophically breached, the lives of 1.5 million downstream residents would have been endangered. Although one example does not make a global environmental concern, the quake lakes phenomenon is representative of the natural hazards associated with densely populated, tectonically active, high-relief landscapes.

1.1.3 Multiple drivers of environmental change

There is an imbalance in the contemporary debate on global environmental change in that the main emphasis is on only one driver of environmental change, namely climate (Dowlatabadi, 2002; Adger *et al.*, 2005). In fact, environmental change necessarily includes climate, relief, sea level and the effects of land management/anthropogenic factors *and* the interactions between them. It is important that a rebalancing takes place now, to incorporate all these drivers. Furthermore, the focus needs to be directed towards the landscape scale, such that global environmental changes can be assessed more realistically. Human safety and well-being and the maintenance of Earth's geodiversity will depend on improved understanding of the reciprocal relations between landscapes and the drivers of change.

In his book *Catastrophe*, for example, Diamond (2005) has described a number of ways in which cultures and civilisations have disappeared because, at least in part, those civilisations have not understood their vulnerability to one or more of the drivers of environmental change. Montgomery (2007) has developed a similar thesis with a stronger focus on the mismanagement of soils.

1.1.4 Systemic and cumulative global environmental change

Global environmental change is here defined as environmental change that consists of two components, namely systemic and cumulative change (Turner *et al.*, 1990b). Systemic change refers to occurrences of global scale, physically interconnected phenomena, whereas cumulative change refers to unconnected, local- to intermediate-scale processes which have a significant net effect on the global system.

In this volume, hydroclimate and sea level change are viewed as drivers of systemic change (see Sections 1.6 and 1.7 of this chapter below). The atmosphere and ocean systems are interconnected across the face of the globe and the modelling of the coupled atmosphere–ocean system (AOGCM) has become a standard procedure in application of general circulation models (or GCMs). A GCM is a mathematical representation of the processes that govern global climate. At its core is the solution to a set of physical equations that govern the transfer of mass, energy and momentum in three spatial dimensions through time. The horizontal atmospheric resolution of most global models is between 1°–3° (~100–300 km). Processes operating at spatial scales finer than this grid (such as cloud microphysics and convection) are parameterised in the model. In the vertical direction, global models typically divide the atmosphere into between 20 and 40 layers.

Topographic relief, and land cover and land use changes, by contrast, are viewed as drivers of cumulative change (see Sections 1.8 and 1.9). The patchiness of relief and land use and difficulties of both definition and spatial resolution make the incorporation of their effects into GCMs a continuing challenge. Nevertheless, developments in global climate modelling over the past decade have seen the improvement in land-surface modelling schemes in which an explicit representation of soil moisture, runoff and river flow routing has been incorporated into the modelling framework (Milly *et al.*, 2002). This trend, coupled with the widespread implementation of dynamic vegetation models (in which vegetation of different plant functional types is allowed to grow according to prevailing environmental conditions) has resulted in a generation of models into which such a range of complex interacting processes are embedded that they have become termed global *environmental* models instead (Johns *et al.*, 2006).

1.1.5 The role of geomorphology

In these contexts, geomorphology (from the Greek *geo* Earth and *morphos* form) has an important role to play; it involves the description, classification and analysis of the Earth's landforms and landscapes and the forces that have shaped them, over a wide range of time and space scales (Fairbridge, 1968). In particular, geomorphology has the obligation to inform society as to what level of disturbance the Earth's landforms and landscapes can absorb and over what time periods the landscape will respond to and recover from disturbance.

In this book, we have chosen to view geomorphology (changing landforms, landform systems, landscapes and landscape systems) as dependent on the four drivers of environmental change, namely climate, relief, sea level and human activity, but also as an independent variable that has a strong effect on each of the drivers at different time and space scales. The relationship in effect is a reflexive one and it is important to avoid the implication of unique deterministic relations.

Two important intellectual strands in geomorphology have been so-called 'climatic' and 'process' geomorphology; they have tended to focus on different spatio-temporal scales of inquiry.

1.2 Climatic geomorphology

Climate's role in landscape change has long been of interest to geomorphology. Indeed in the continental European literature this was a theme that was already well developed by the end of the nineteenth century (Beckinsale and Chorley, 1991). The greatest impetus to climatic geomorphology came from the global climatic classification scheme of Köppen (1901). A clear statement of the concept of climatic geomorphology was made by de Martonne (1913) in which he expressed the belief that significantly different landscapes could be developed under at least six present climatic regimes and drew particular attention to the fact that humidity and aridity were, in general, more important as differentiators of landscape than temperature. The identification of morphoclimatic/morphogenetic regions and attempts to identify global erosion patterns (Büdel in Germany, Tricart in France and Strakhov in Russia) were also important global-scale contributions. Strakhov's map of global-scale erosion patterns is reproduced here (Fig. 1.2) to illustrate the style and scale of this research. He attempted to estimate world denudation rates by extrapolating from sediment yields for 60 river basins. His main conclusions were:

(a) arid regions of the world have distinctive landforms and landscapes;

(b) the humid areas of the tropics and subtropics, which lie between the +10 °C mean annual isotherm of each hemisphere, are characterised by high rates of denudation, reaching maximum values in southeastern Asia;

(c) the temperate moist belt, lying largely north of the +10 °C mean annual isotherm, experiences modest denudation rates;

(d) the glaciated shield areas of the northern hemisphere, largely dominated by tundra and taiga on permafrost and lying north of the 0 °C mean annual isotherm, have the lowest recorded rates of denudation; and

(e) mountain regions, which experience the highest rates of denudation, are sufficiently variable that he was forced to plot mountain denudation data separately in graphical form.

The map is an example of climatic geomorphology in so far as it demonstrates broad climatic controls but perhaps the most important contribution of twentieth-century climatic geomorphology was that it maintained a firm focus on the landscape scale, the scale to which this volume is primarily directed. The weakness of the approach is that regional and zonal generalisations were made primarily on the basis of form (in the case of arid regions) and an inadequate sampling of river basin data. There was a lack of field measurements

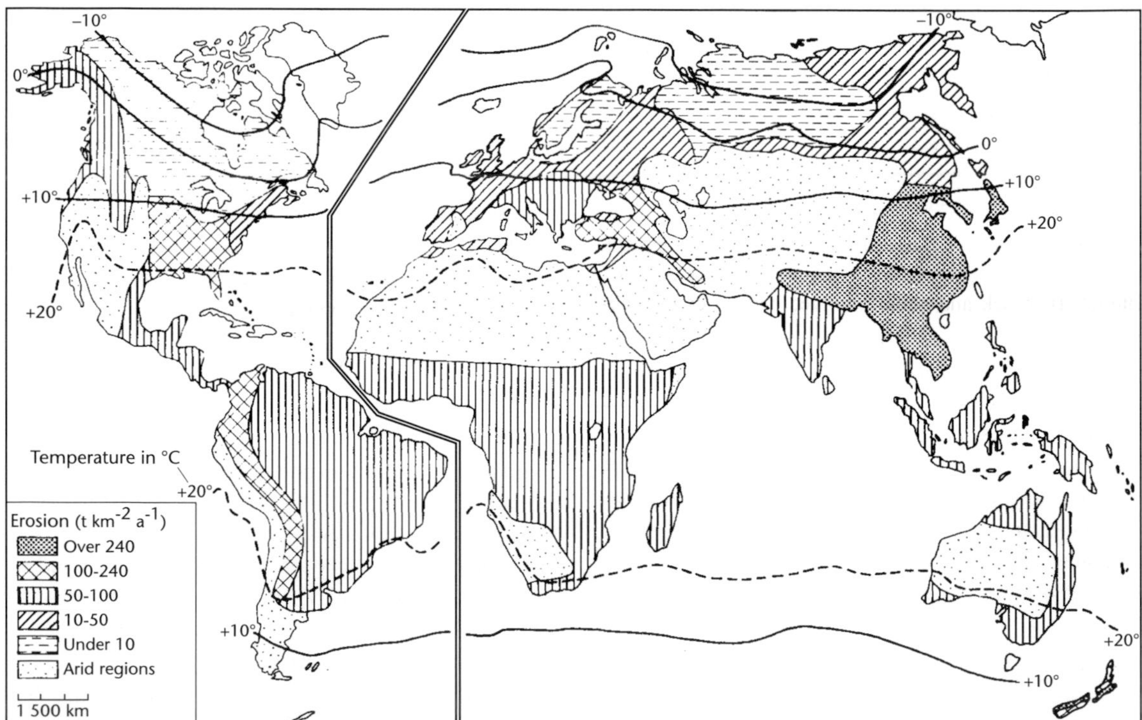

FIGURE 1.2. Climatic geomorphology (modified from Strakhov, 1967).

of contemporary process and no discussion of the scale dependency of key rainfall, runoff and sediment relations.

Whilst one may be critical of these earlier attempts to deal with landscape-scale geomorphology, now is a good time to revisit the landscape scale, with a firmer grasp of the relief, sea level and human activity drivers, for the following reasons:

(a) the development of plate tectonic theory and its geomorphological ramifications has given the study of earth surface processes and landforms a firmer geological and topographic context;
(b) a better understanding of the magnitudes and rates of geomorphological processes has been achieved not only from contemporary process measurements but also from the determination of more precise and detailed records of global environmental change over the last 20 000 years utilising improved chronologies (largely ocean rather than terrestrially based) and benefiting from the development of whole suites of radiometric dating techniques, covering a wide range of half-lives and thus timescales; and
(c) the ability to provide, at a range of scales, quantitative measurements of land surface topography and vegetation characteristics from satellite and airborne remote sensing.

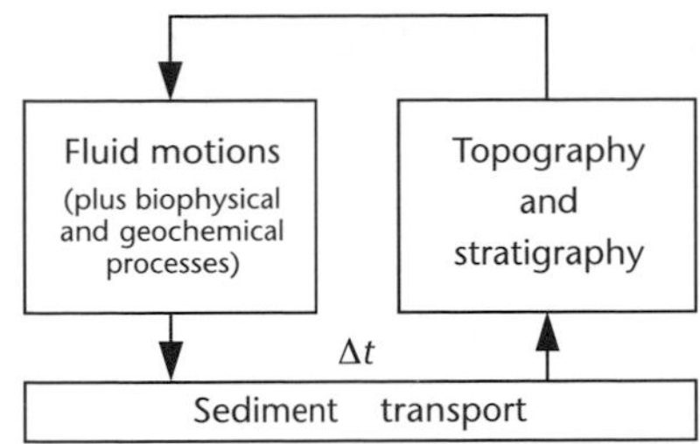

FIGURE 1.3. A simplified conceptual model of a process–response system.

1.3 Process geomorphology

From the 1950s onwards an Anglo-American geomorphology came to be reorientated towards quantitative research on the functional relations between form, materials and earth surface processes. These 'process studies', generally at the scale of the small drainage basin or below, began to determine local and regional rates of surface lowering, or denudation, material transport and deposition and their spatial differentiation. The rates at which these processes take place are dependent upon local relief and topography, the materials (bedrock and soils) involved and, of course, climate, both directly and indirectly through the relations between climate, vegetation characteristics and surface processes. The emphasis on rates of operation of processes led to a greater interest in the role of hydroclimate, runoff and sediment transport both in fluvial and in coastal systems. The role of vegetation in landscape change also assumed a new importance for its role in protecting the soil surface, in moderating the soil moisture and climate and in transforming weathered bedrock into soil (Kennedy, 1991).

1.3.1 Process–response systems

One of the most influential papers in modern geomorphology concerned the introduction of general systems thinking into geomorphology (Chorley, 1962). General systems thinking provided the tool for geomorphologists to analyse the critical impacts of changes in the environmental system on the land surface, impacts of great importance for human society and security. One kind of general system that has proved to be most fruitful in providing explanations of the land surface–environment interaction is the so-called process–response system (Fig. 1.3). Such systems are defined as comparatively small-scale geomorphic systems in which deterministic relations between 'process' (mass and energy flows) and 'response' (changes in elements of landscape form) are analysed with mathematical precision and attempted accuracy. There is a mutual co-adjustment of form and process which is mediated through sediment transport, a set of relations which has been termed 'morphodynamics' and which has been found to be particularly useful in coastal studies (e.g. Woodroffe, 2002).

Morphodynamics explains why, on the one hand, physically based models perform well at small spatial scales and over a limited number of time steps but, on the other hand, why model predictions often break down at 'event' and particularly 'engineering' space-timescales. Unfortunately, these are exactly the scales that are of greatest significance in the context of predicting landscape responses to global environmental change and the policy and management decisions that flow from such responses.

1.3.2 The scale linkage problem

The issue of transferring knowledge between systems of different magnitude is one of the most intransigent problems in geomorphology, both in terms of temporal scale and spatial scale (Church, 1996).The problem of scale linkage can be summarised by the observation that landscapes are characterised by different properties at different scales of investigation. Each level of the hierarchy includes the cumulative effects of lower levels in addition to some new considerations (called emergent properties in the technical literature) (Fig. 1.4).

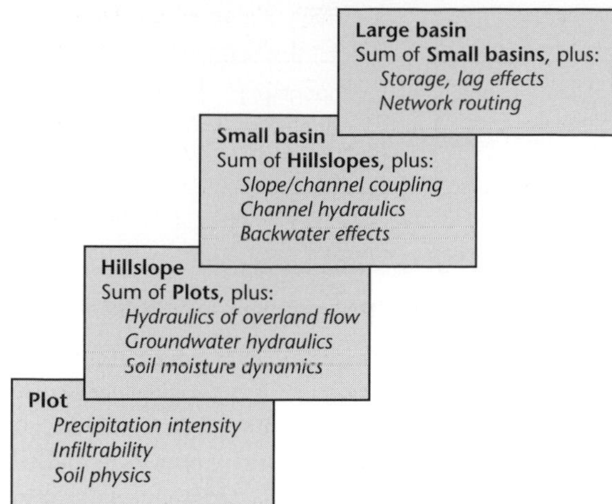

FIGURE 1.4. The scale linkage problem (modified from Phillips, 1999) illustrated in terms of a spatial hierarchy which contains new and emergent properties at each successive spatial scale.

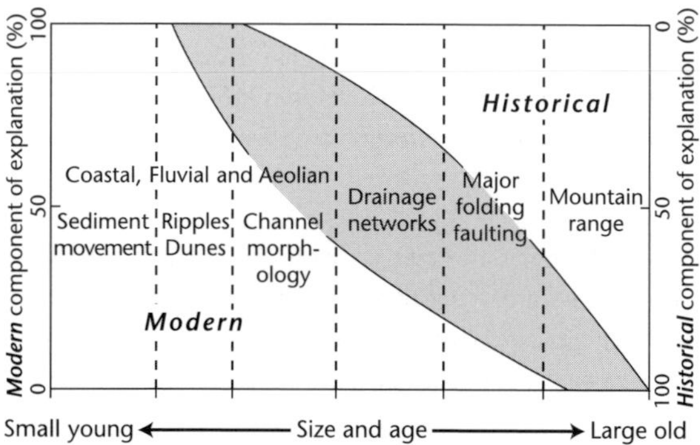

FIGURE 1.5. The relative importance of historical vs. modern explanation as a function of size and age of landforms and landscapes (modified from Schumm, 1985). Note the assumption that size and age are directly correlated, an assumption that is most appropriate for coastal, fluvial and aeolian landscapes, but does not easily fit volcanic and tectonic landscapes.

At the landscape scale, here taken to be larger than the large basin scale in Fig. 1.4, there are further emergent properties which have to be considered such as regional land use and hydrology.

Figure 1.5 combines a consideration of both temporal and spatial scales. At one extreme of very small spatial scale, such as the movement of individual sand grains over very short timescales, the process–response model works well. At the other extreme, large landscapes that have evolved over millions of years owe their configuration almost exclusively to past processes. Discontinuous sediment disturbances have a history of variable magnitude and frequency of occurrence. The practical implication is that, in general, the larger the landscape we wish to consider the more we have to take into account past processes and the slower will be the response of that landscape in its entirety to sediment disturbance regimes. Coastal morphology and drainage networks, which occupy the central part of Fig. 1.5, exemplify the scales of interest in this volume.

1.4 Identification of disturbance regimes

Global environmental change has become a major concern in geomorphology because it poses questions about the magnitude, frequency and kinds of disturbance to which geomorphic systems are exposed. What then are the major drivers of that change? Discussions about the rhythm and periodicity of geological change have spilled over into geomorphology. In his discussion of rhythmicity in terrestrial landforms and deposits, Starkel (1985) directed attention to the fact that the largest disturbance in the geologically recent past is that of continental-scale glaciation (see Plates 1 and 2). Periods of glaciation alternating with warmer episodes define a disturbance regime characterised by varying rates of soil formation and erosional and depositional geomorphological processes during interglacial and glacial stades (Fig. 1.6).

Some of the excitement in the current debate over global environmental change concerns precisely the question of the rate at which whole landscapes have responded to past climate changes and disturbances introduced by tectonism (e.g. volcanism, earthquakes and tsunamis) or human activity.

1.4.1 Landscape response to disturbance

The periodicity of landscape response to disturbance in Fig. 1.6 is controlled by the alternation of glacial and interglacial stades. The magnitude and duration of this response is a measure of the sensitivity and resilience of the landscape. In the ecological and geomorphic literature, this response is commonly called the system vulnerability. Conventionally, human activity has been analysed outside

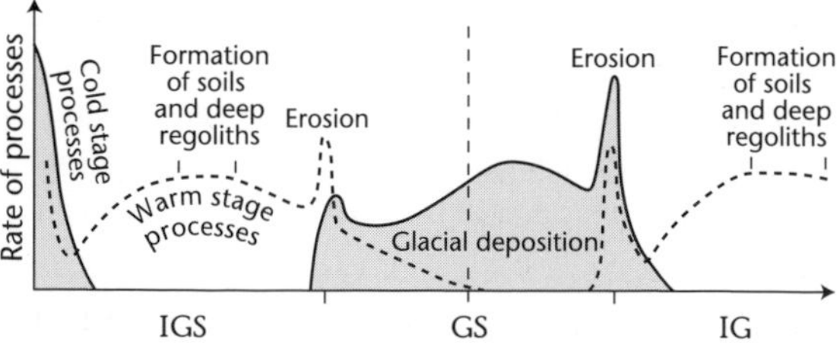

FIGURE 1.6. Periodicity of erosion and sedimentation (modified from Starkel, 1985). IGS is interglacial stade; GS is glacial stade; and IG is the present interglacial.

the geosystem (and Fig. 1.6 contains no human imprint) but the weakness of this approach is that it fails to recognise the accelerating interdependence of humankind and the geosystem. The IPCC usage of the term 'vulnerability', by contrast, addresses the ability of society to adjust to disturbances caused by environmental change. We therefore follow, broadly, the IPCC approach in defining sensitivity, adaptive capacity and vulnerability as follows. 'Sensitivity' is the degree to which a system is affected, either adversely or beneficially, by environment-related stimuli; 'adaptive capacity' is the ability of a system to adjust to environmental change, to moderate potential damages, to take advantage of new opportunities or to cope with the consequences; and 'vulnerability' is the degree to which a system is susceptible to, or unable to cope with, adverse effects of environmental change. In sum, 'vulnerability' is a function of the character, magnitude and rate of environmental change and variation to which a system is exposed, its sensitivity and its adaptive capacity (Box SPM-1 in IPCC, 2001b, p. 6.).

In general, those systems that have the least capacity to adapt are the most vulnerable. Geomorphology delivers a serious and often unrecognised constraint to the feasible ways of dealing with the environment in so far as it controls vulnerability both in the ecological sense (in the absence of direct human agency) *and* in the IPCC sense. A number of unique landscapes and elements of landscapes are thought to be more likely to experience harm than others following a perturbation. There are seven criteria that have been used to identify key vulnerabilities:

(a) magnitude of impacts;
(b) timing of impacts;
(c) persistence and reversibility of impacts;
(d) estimates of uncertainty of impacts;
(e) potential for adaptation;
(f) distributional aspects of impacts; and
(g) importance of the system at risk.

In the present context, such landscapes are recognised as hotspots with respect to their vulnerability to changes in climate, relief, sea level and human activities. We think immediately for example of glaciers, permafrost, coral reefs and atolls, boreal and tropical forests, wetlands, desert margins and agricultural lands as being highly vulnerable. Some landscapes will be especially sensitive because they are located in zones where it is forecast that climate will change to an above average degree. This is the case for instance in the high arctic where the degree of warming may be three to four times greater than the global mean. It may also be the case with respect to some critical areas where particularly substantial changes in precipitation may occur. For example, the High Plains of the USA may become markedly drier. Other landscapes will be especially sensitive because certain landscape forming processes are particularly closely controlled by thresholds, whether climatic, hydrologic, relief, sea level or land use related. In such cases, modest amounts of environmental change can switch systems from one state to another (Goudie, 1996).

1.4.2 Azonal and zonal landscape change

The overarching problem of assessing probable landscape change in the twenty-first century is approached here in two main ways. A group of chapters which are 'azonal' in character concern themselves with ways in which geomorphic processes are influenced by variations in mass, energy and information flows, and this self-evidently includes human activity. These azonal chapters deal with land systems that are larger than individual slopes, stream reaches and pocket beaches, but generally smaller than continental-scale regions. By comparison, the zonal chapters use whole biomes as their organising principle, similar to those used in the Millennium Ecosystem Assessment (2003) (Plate 3). In these chapters also, environmental change is driven, not only by hydroclimate, relief and sea level but also by human activity.

In addition to understanding the terrestrial distribution of biomes, it is also important to recognise the broad limits to coral reef and associated shallow water ecosystems, such that the upper ocean's vulnerability to global environmental change can also be assessed (Fig. 1.7).

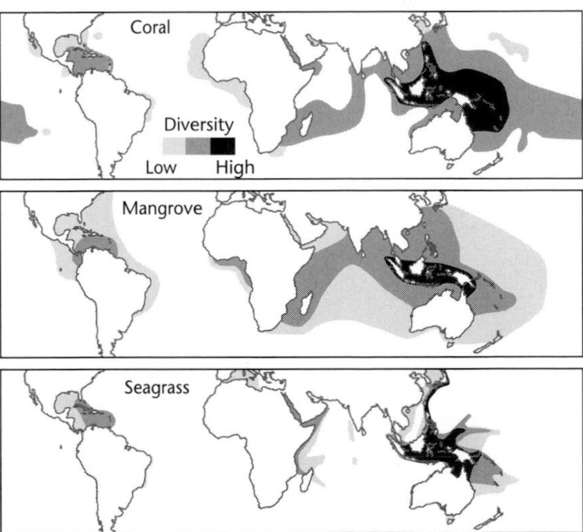

FIGURE 1.7. Global distribution of coral reefs, mangroves and seagrass. Scale of diversity ranges from 0–10 genera (low); 10–25 genera (medium); and >50 (high) (modified from Veron, 1995).

The decision to structure the book chapters using a bottom–up (azonal) and a top–down (zonal) approach reflects the fact that both approaches have complementary strengths.

1.5 Landscape change

Geomorphology emphasises landscape change under the influence of climate, relief, sea level change and human activity (Chorley *et al.*, 1984) and does so at a range of space and timescales. With respect to temporal scales, attention is confined in this volume to the last complete glacial–interglacial cycle and forward towards the end of the twenty-first century (Fig. 1.8). The reasons for the selection of these end points are that they include one complete glacial–interglacial cycle (see Chapter 14), and thus the widest range of climates and sea levels in recent Earth history. This period includes the rise of *Homo sapiens sapiens*; and extends forward to a time when future landscapes can be modelled with some confidence and for which credible scenarios of landscape change can be constructed.

Included in this timescale are the closing stages of the Pleistocene Epoch (150 000 to 10 000 years ago); the Holocene Epoch (10 000 years BP until the present) and a recent, more informally defined, Anthropocene, extending from about 300 years ago when human impact on the landscape became more evident, and into the near future. The comprehensive ice core records from Greenland (GISP and GRIP) and from Antarctica (Vostok and EPICA) (Petit *et al.*, 1999; EPICA, 2004) (Fig. 1.8); lake sediments from southern Germany (Ammersee) (Burroughs, 2005) (Fig. 1.9) and elsewhere; and a number of major reconstructions of the climate of the last 20 000 years using past scenarios (Plates 1 and 2) provide a well-authenticated record of the Earth's recent climatic history.

The record of changing ice cover and biomes since the Last Glacial Maximum (LGM) has been reconstructed by an international team of scientists working under the general direction of the Commission for the Geological Map of the World (Petit-Maire and Bouysse, 1999; Plates 1 and 2). The authors stress that the maps are tentative but contain the best information that was available in 1999. The maps depict the state of the globe during the two most

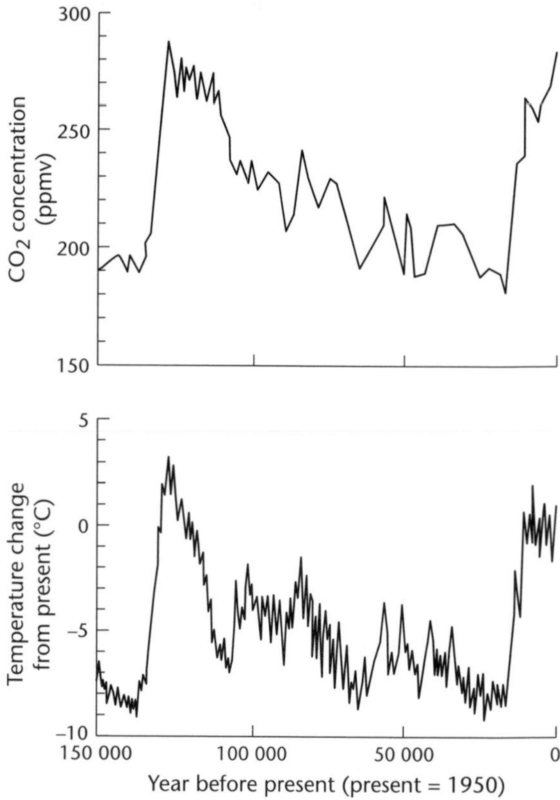

FIGURE 1.8. Climate records from East Antarctica (Vostok ice core) covering the last glacial–interglacial cycle (modified from Petit *et al.*, 1999). Note the rapid warming followed by a gentler, stepped cooling process and also the close correlation of temperature and CO_2.

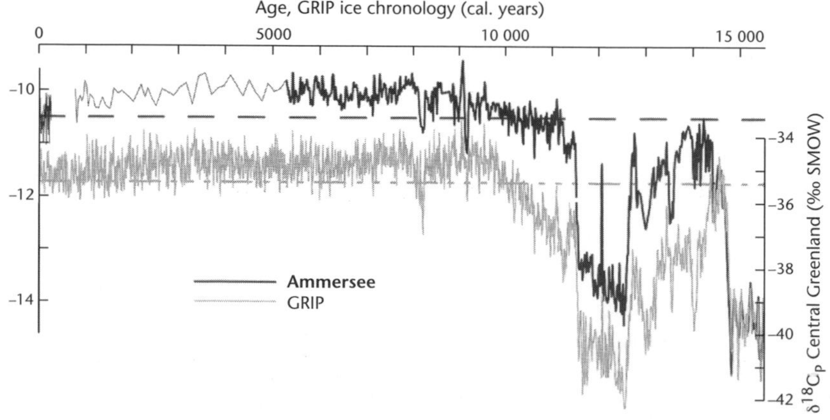

FIGURE 1.9. A comparison of the record from Ammersee, in southern Germany, and the GRIP ice core from Greenland showing the close correlation between the Younger Dryas cold event from 12.9 to 11.6 ka BP at the two sites (from von Grafenstein *et al.*, 1999).

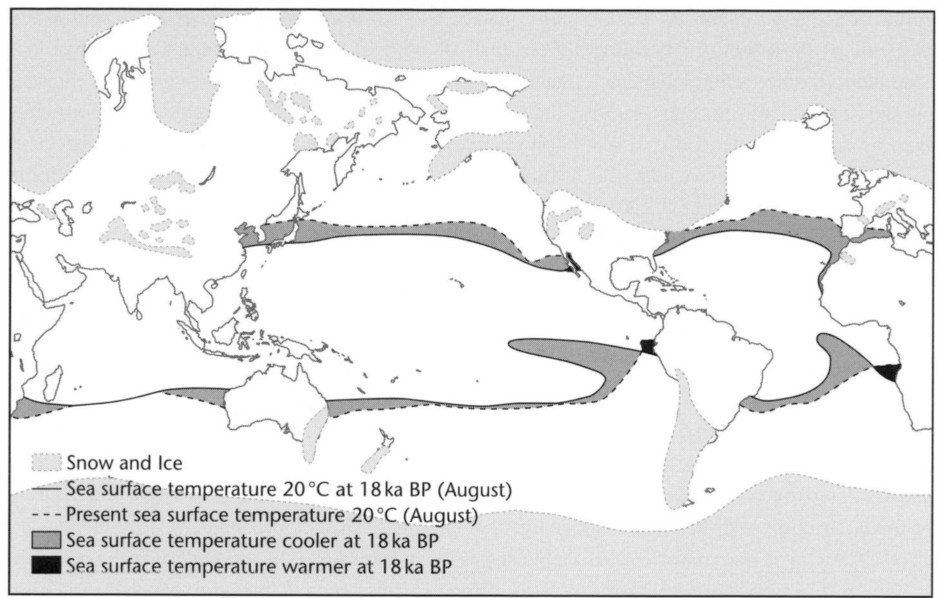

FIGURE 1.10. Changing tropical ocean temperatures, LGM to present (modified from CLIMAP, 1976 and Spencer, 1990).

contrasted periods of the last 20 ka. The LGM was the coldest (*c.* 18 ka ± 2 ka BP) and the Holocene Optimum (HOP) was the warmest (*c.* 8 ka ± 1 ka BP) period. These periods were only 10 ka apart and yet there was a dramatic reorganisation of the shorelines, ice cover, permafrost, arid zones, surface hydrology and vegetation at the Earth's surface over that interval. Thus within a 10-ka time-span (in many places less) the two vast ice sheets of Canada and Eurasia, which reached a height of 4 km and covered about 25 million km², disappeared; 20 million km² of continental platform were submerged by the sea; biomes of continental scale were transformed and replaced by new ones; and humans could no longer walk from Asia to America nor from New Guinea to Australia nor from France to England.

It is also interesting to compare these shifts in the terrestrial landscape with change in sea surface temperatures over the same period of time. In particular, in the tropical oceans, these changes were relatively small – as illustrated by the change in the 20 ºC isotherm (which provides a broad limit to coral growth) – with the greatest changes being in the variable strength of the equatorial upwelling systems on the eastern margins of the ocean basins (Fig. 1.10).

1.5.1 The Last Glacial Maximum

First of all, there needs to be a caveat with respect to the timing of the LGM (Plate 1). There is strong evidence that the maximum extent of ice was reached in different places at different times. The ice distribution that is mapped corresponds to the maximum extent during the time interval 22 ka to 14 ka years BP, which covers the global range within which the maximum is believed to have occurred. During the LGM, mean global temperature was at least 4.5 °C colder than present. Permafrost extended southwards to latitudes of 40–44º N in the northern hemisphere (although in the south, only Patagonia and the South Island of New Zealand experienced permafrost). Mean sea level was approximately 125 m lower than at present. Large areas of continental shelf were above sea level and colonised by terrestrial vegetation, particularly off eastern Siberia and Alaska, Argentina, and eastern and southern Asia. New Guinea was connected to Australia, the Persian Gulf dried up and the Black Sea, cut off from the Mediterranean Sea, became a lake.

There was a general decrease in rainfall near the tropics. Loess was widespread in periglacial areas and dunes in semi-arid and arid regions. All desert areas were larger than today but in the Sahara there was the greatest southward extension of about 300–400 km. Surface hydrology reflected this global aridity except in areas that received meltwaters from major ice caps, such as the Caspian and Aral seas. Grasslands, steppes and savannas expanded at the expense of forests.

1.5.2 The record from the ice caps and lake sediments

The transition between the LGM and the Holocene was marked by a partial collapse of the Laurentide/Eurasian ice sheets. This led to a surge of icebergs, recorded in the sediments of the North Atlantic by the last of the so-called Heinrich events (thick accumulations of ice-rafted sediments) around 16.5 ka. There followed a profound warming around 14.5 ka (Fig. 1.9) which coincided with a rapid rise in sea level (see Section 1.7), presumably associated with the break-up of part of the Antarctic ice sheet (Burroughs, 2005).

Between 14.5 and 12 ka BP the mean annual temperature oscillated violently and between 12.9 and 11.6 ka the last great cooling of the ice age (known as the Younger Dryas stade) occurred. Rapid warming continued until around 10 ka but thereafter, the climate seems to have settled into what looks like an extraordinarily quiet phase when compared with the earlier upheavals. The Holocene Epoch is conventionally said to start around 10 ka because the bulk of the ice sheet melt had occurred by that time, but the Laurentide ice sheet, for example, did not disappear until 6 ka BP.

Although climatic fluctuations during the Holocene have been much more modest than those which occurred during the previous 10 ka, there have been fluctuations which have affected glacier distribution in the mountains, treeline limits in the mountains and in the polar regions, and desiccation of the Sahara. The CASTINE project (Climatic Assessment of Transient Instabilities in the Natural Environment) has identified at least four periods of rapid climate change during the Holocene, namely 9–8 ka; 6–5 ka; 3.5–2.5 ka and since 0.6 ka. In terms of landscape history, it is also important to recognise that the mean global temperature may not be the most significant factor in landscape change. Precipitation amounts and soil moisture availability and their variability of occurrence and intensity over space and through time have had a strong influence on regional and local landscape evolution.

1.5.3 The Holocene Optimum

A caveat also needs to be applied with respect to the timing of the HOP (Plate 2). The maximum values of the signals for each of the various indicators of environmental change are far from being coeval. During the HOP, the mean global temperature was about 2 °C warmer than today. By 6 ka BP, mean relative land and sea level was close to that of the present day except in two kinds of environments:

(a) the Canadian Arctic and the Baltic Sea where isostatic (land level rebound after ice sheet load removal) adjustments were at a maximum;
(b) deltas of large rivers, such as the Mississippi, Amazon, Euphrates–Tigris and Yangtze, had not reached their present extent.

The glacier and ice sheet cover cannot be distinguished from that of today at this global scale. Permafrost, both continuous and discontinuous, was within the present boundary of continuous permafrost in the northern hemisphere. Significantly wetter conditions were experienced in the Sahara, the Arabian Peninsula, Rajasthan, Natal, China and Australia, where many lakes that have subsequently disappeared were formed. In Canada the Great Lakes were formed following the melting of the ice sheet and the isostatic readjustment of the land. Rainforest had recolonised extensive areas and the taiga and boreal forest had replaced a large part of the tundra and areas previously covered by ice sheets (Petit-Maire, 1999).

This time-span of 20 000 years has been selected in order to encapsulate the extremes of mean global cold and warmth experienced between the LGM and the HOP, a range that one might expect to contain most of the reasonable scenarios of environmental change over the next 100 years. Certainly, this range defines the 'natural' variability of Earth's landscapes but, notably, little distinctive human impact was discernible at this global scale of analysis.

Recently, however, Ruddiman (2005) has claimed to recognise the effects of human activity in reversing the trends of CO_2 and methane concentrations around 8–5 ka BP. His hypothesis is that clearing of the land for agriculture and intensification of land use during the Holocene has so altered the climate as to delay the arrival of the next glacial episode. This is a controversial hypothesis which requires further testing. If the hypothesis is supported, it emphasises the importance of the warning issued by Steffen *et al.* (2004) against the use of Pleistocene and Holocene analogues to interpret the Anthropocene, the contemporary epoch which is increasingly dominated by human activity and is therefore a 'no analogue' situation.

1.6 Systemic drivers of global environmental change (I): hydroclimate and runoff

1.6.1 Introduction

Water plays a key role in the transfer of mass and energy within the Earth system. Incoming solar radiation drives the evaporation of approximately $425 \times 10^3\ km^3\ a^{-1}$ of water from the ocean surface and approximately $71 \times 10^3\ km^3\ a^{-1}$ from the land surface; precipitation delivers about $385 \times 10^3\ km^3\ a^{-1}$ of water to the ocean and $111 \times 10^3\ km^3\ a^{-1}$ to the land surface. The balance is redressed through the flow of $40 \times 10^3\ km^3\ a^{-1}$ of water from the land to the oceans in rivers (Berner and Berner, 1996). Global environmental change affecting any one of these water transfers will lead to changes in runoff and river flows. However, the prediction of changes may not be simple because the role of hydrological processes in the land surface system is complex and involves interactions and feedbacks between the atmosphere, lithosphere and vegetation.

The hydrological cycle is affected by changes in global climate, but also by changes that typically occur on a smaller, regional scale, such as changes in vegetation type and land use (for example the change from forest to agricultural pasture land) and changes in land management. These latter changes may also include reservoir construction, abstractions of water for human use, and discharges of water into river courses and the ocean.

Increasing atmospheric carbon dioxide levels and temperature are intensifying the global hydrological cycle, leading to a net increase in rainfall, runoff and evapotranspiration (Huntingdon, 2006). Changes are projected to occur not only to mean precipitation and runoff, but also to their spatial patterns. Within the tropics, precipitation rates increased between 1900 and 1950 but have declined since 1970. In contrast, mid-latitude regions have seen a more consistent increase in precipitation since 1900 (IPCC, 2007a).

The intensification of the hydrological cycle is likely to mean an increase in hydrological extremes (IPCC, 2001a). Changes to the frequency distribution of rainfalls and flows of different magnitude can have a disproportionately large effect on environmental systems such as river basins, vegetation and aquatic habitats. The reason for this disproportionality is because extreme flows provoke changes when certain thresholds in magnitude or in the duration of runoff are exceeded. There are suggestions that interannual variability will increase, with an intensification of the natural El Niño and North Atlantic Oscillation (NAO) cycles, leading to more droughts and large-scale flooding events. Key questions that this section will address include:

(a) what changes in precipitation, evaporation and consequent runoff have been observed over the historical period; and
(b) what changes are projected under future climate and land use scenarios.

1.6.2 Observed changes in precipitation, evaporation, runoff and streamflow

Surface temperatures

Global mean surface temperatures have increased by 0.74 °C ± 0.18 °C over the period 1906–2005, although the rate of warming in the last 50 years of that period has been almost double that over the last 100 years (IPCC, 2007a). With this change in surface temperature comes the theoretical projection that warming will stimulate evaporation and in turn precipitation, leading to an intensified hydrological cycle. In the earliest known theoretical work on the subject, Arrhenius (1896) showed that specific humidity would increase roughly exponentially with air temperature according to the Clausius–Clapeyron relation. Numerical modelling studies have since indicated that changes in the overall intensity of the hydrological cycle are controlled not only by the availability of moisture but also by the ability of the troposphere to radiate away latent heat released by precipitation. An increase in temperature of 1 Kelvin would lead to an increase in the moisture-holding capacity of the atmosphere by approximately 3.4% (Allen and Ingram, 2002). The convergence of increased moisture in weather systems leads to more intense precipitation; however the frequency or duration of intense precipitation events must decrease because the overall amount of water does not change a great deal (IPCC, 2007a).

Pollutant aerosols

In addition to the effects of temperature increases, the effects of pollutant aerosols can be significant. Increases in sulphate, mineral dust and black carbon can suppress rainfall in polluted areas by increasing the number of cloud condensation nuclei. This leads to a reduction in the mean size of cloud droplets and reduces the efficiency of the process whereby cloud droplets coalesce into raindrops (Ramanathan *et al.*, 2001). An additional effect of increased atmospheric aerosol loading is a reduction in the amount of solar radiation that reaches the land surface; this may affect the amount of evaporation and therefore precipitation. The effects of aerosol loading are expected to be highest in highly polluted areas such as China and India and evidence is emerging that aerosol loading may explain the recent tendency towards increased summer flooding in southern China and increased drought in northern China (Menon *et al.*, 2002). Recent analyses indicate that aerosol loading led to a reduction in solar radiation reaching the land surface of 6–9 W m^{-2} (4–6%) between the start of measurements in 1960 and 1990 (Wild *et al.*, 2005). However, between 1991 and 2002, the same authors estimate that the amount of solar radiation reaching the Earth's surface increased by approximately 6 W m^{-2}. This observed shift from dimming to brightening, has prompted Andreae *et al.* (2005) to suggest that the decline of aerosol forcing relative to greenhouse forcing may lead to atmospheric warming at a much higher rate than previously predicted.

Precipitation

Precipitation trends are harder to detect than temperature trends, because the processes that cause precipitation are more highly variable in time and space. Nevertheless, global mean precipitation rates have increased by about 2% over the twentieth century (Hulme *et al.*, 1998). The spatial pattern of trends has been uneven, with much of the increase focussed between 30° N and 85° N. Evidence for a change in the nature of precipitation is reported by Brown (2000), who finds a shift in the amount of snowfall in western North America between 1915 and 1997. Zhang *et al.* (2007) used an ensemble of fourteen climate models to show that the observed changes in the spatial patterns of precipitation between 1925 and 1999 cannot be explained by internal climate variability or natural forcing, but instead are consistent with model projections in which anthropogenic forcing was included.

Runoff

Rates of surface and subsurface runoff depend not just on trends in precipitation, because the water balance for any soil column dictates that runoff is the difference between precipitation and evaporation. When an unlimited amount of water is available at the surface, the rate of evaporation is controlled by the amount of energy available and the water vapour pressure deficit (i.e. the difference between the actual vapour pressure and the vapour pressure at saturation) in the overlying air (Penman, 1948). The amount of energy available depends on the net solar radiation received at the surface. The water vapour pressure deficit depends on the temperature of the air and its specific humidity. The rate of transport of air across the surface exerts a key control on the potential evaporation rate because it determines how readily saturated air is refreshed so that further evaporation can occur. In practice, an unlimited amount of water is available only over persistent open water in oceans, lakes and rivers. Over the land surface, the rate of actual evaporation depends not only on meteorological variables but also on the nature of the vegetation and on the amount of available soil moisture.

Evapotranspiration

Only limited observations of evaporation are available. Sparse records of potential evaporation from evaporation pans show a generally decreasing trend in many regions, including Australia, China, India and the USA (IPCC, 2007a). Roderick and Farquar (2002) demonstrate that the decreasing trends in pan evaporation are likely to be a result of a decrease in surface solar radiation that may be related to increases in atmospheric aerosol pollution. On the other hand, Brutsaert and Parlange (1998) have pointed out that pan evaporation does not represent actual evaporation and that in regions where soil moisture exerts a strong control on actual evaporation any increase in precipitation which drives an increase in soil moisture will also lead to higher rates of actual evaporation.

In any case, at the scale of a river basin, vegetation controls the amount of evaporation from the land surface through its effects on interception and transpiration. A significant fraction of precipitation falling on land can be intercepted by vegetation and subsequently evaporated. The rate of evaporation from intercepted water depends on the vegetation type and structure of the plant canopy; it is normally higher for forest canopies than for grassland. In contrast, transpiration occurs through the evaporation of water through plant stomata. The rate of transpiration depends on available energy and vapour pressure deficit but also on stomatal conductance: the ease with which the stomata of a particular plant species permit evaporation under given environmental conditions. Stomatal conductance depends on light intensity, CO_2 concentration, the difference in vapour pressure between leaf and air, leaf temperature and leaf water content. All of these properties change over timescales relevant to global change, but the most significant variant may be CO_2 concentration (Arnell, 2002).

Increased CO_2 concentration stimulates photosynthesis and may encourage plant growth in some plant species that use the C3 photosynthesis pathway, which includes all trees and most temperate and high-latitude grasses (Arnell, 2002). Carbon dioxide enrichment has the additional effect of reducing stomatal conductance by approximately 20–30% for a doubling of the CO_2 concentration. Thus, for a given set of meteorological conditions, the water use efficiency of plants increases. Considerable uncertainty exists over whether this leaf-scale process can be extrapolated to the catchment scale; in many cases the decreased transpiration caused by CO_2-induced stomatal closure is likely to be offset by additional plant growth (Arnell, 2002).

Trends in streamflow

Only patchy historical data are available to assess global patterns of streamflow. Dai and Trenberth (2002) estimate that only about two-thirds of the land surface has ever been gauged and the length and availability of observed records are highly variable. Detecting trends in streamflow is problematic too, because runoff is a spatially integrated variable which does not easily permit discrimination between changes caused by any of its driving factors. Streamflow and groundwater recharge exhibit a wide range of natural variability and are open to a host of other human or natural influences.

In an analysis of world trends in continental runoff, Probst and Tardy (1987) found an increase of approximately 3% between 1910 and 1975. This trend has been confirmed in a reanalysis of data between 1920 and 1995 by Labat *et al.* (2004). In areas where precipitation has increased over the latter half of the twentieth century, runoff has also increased. This is particularly true over many parts of the USA (Groisman *et al.*, 2004). Streamflow records exhibit a wide range of variability on timescales ranging from inter-annual to multi-decadal and for most rivers, secular trends are often small. There is evidence that flood peaks have increased in the USA because increases in surface air temperature have hastened the onset of snowmelt (Hodgkins *et al.*, 2003). In many rivers in the Canadian Arctic, earlier break-up of river ice has been observed (Zhang *et al.*, 2001).

Caution must be exercised in the interpretation of long-term hydrological trends. Using flow observations made during the last 80–150 years, Mudelsee *et al.* (2003) found a decrease in winter flooding in the Elbe and Oder rivers in Eastern Europe, but no trend in summer flooding. They concluded that the construction of reservoirs and deforestation may have had minor effects on flood frequency. Svensson *et al.* (2006) showed that, in a study of long time series of annual maximum river flows at 195 gauging stations worldwide, there is no statistically significant trend at over 70% of sites. They attributed the lack of a clear signal to the wide ranging natural variability of river flow across multiple time and space scales. Hannaford and Marsh (2006) described a set of benchmark UK catchments defined to represent flow regimes that are relatively undisturbed by anthropogenic influences. Over the past 30–40 years they found a significant trend towards more protracted periods of high flow in the north and west of the UK, although trends in flood magnitude were weaker. However, they pointed out that much of the trend is a result of a shift towards a more positive NAO index since the 1960s. The NAO is a climatic phenomenon in the North Atlantic Ocean and is measured by the difference in sea level pressure between the Icelandic Low and the Azores High. This difference controls the strength and direction of westerly winds and storm tracks across the North Atlantic.

Some component of many observed runoff trends can be explained by changes to environmental properties other than climate. Increases in runoff have resulted from land use change, particularly from the conversion of forest to grassland or agricultural land (Vörösmarty and Sahagian, 2000). The most significant effect of removing trees is the reduction in canopy evaporation; that is, evaporation from water intercepted by the tree canopy and stored on leaves. In the Plynlimon experimental catchments in upland Wales, United Kingdom, Roberts and Crane (1997) found that clearcutting of ~30% of the coniferous forest cover led to an increase in runoff of 6–8%. In snow-dominated catchments, the effects of deforestation are more significant because a large amount of snow is permitted to accumulate if forest is cleared (Troendle and Reuss, 1997). Conversely, abandonment of agricultural land and upland afforestation can reduce the volume of runoff. In the Plynlimon catchments experiment, annual evaporation losses (estimated as the difference between annual precipitation and annual runoff) in the forested Severn catchment were ~200 mm greater than in the grassland Wye catchment. This represents a 15% reduction in the flow (Robinson *et al.*, 2000). Most studies of the effects of land cover on the water balance have involved catchment-scale measurements and it is, at present, uncertain whether these findings can be extrapolated to regional and planetary scales.

1.6.3 Projections for future changes

An assessment of the potential impacts of future climate change on precipitation, evaporation, and therefore runoff is a fundamental influence on the strategies adopted by land and river managers. It is impossible to make a reliable prediction of the weather more than about a week in advance, but projections of the statistical properties of future climate can be obtained by using general circulation models (GCMs) to construct climate scenarios (see Section 1.1.4).

Models that encompass an ever-wider range of environmental processes have led to a shift in the goals of climate modelling. Instead of simply providing projections of future average weather, current environmental models provide a powerful tool to examine numerically the complex feedbacks that exist within the Earth system. They have also fuelled studies that fall under the broad title of 'detection and attribution', in which historical observed changes in measured environmental variables are partly explained through understanding gained using climate simulations. For example, Gedney *et al.* (2006) attempt to show that statistically significant continent-wide increases in twentieth-century streamflow can be explained by the effect of CO_2-driven stomatal closure on continental-scale transpiration.

In contrast to numerical weather prediction, which is an initial-value problem where governing equations are solved to find the time-evolution of the system given a set of initial conditions, a typical GCM experiment corresponds to a boundary-value problem in mathematics. In a boundary-value problem, boundary conditions for the problem are used to constrain solutions to the governing equations that are consistent with the imposed conditions. For example, the frequency distribution of rainfall magnitudes may be required under different scenarios of atmospheric CO_2 concentration. The detailed trajectory of changes in CO_2 is highly dependent on socioeconomic factors which govern the behaviour of human societies. The IPCC has published a range of plausible alternatives in its Special Report on Emissions Scenarios (SRES) in which the scenarios are defined (Appendix 1.1).

Temperature and precipitation

Globally averaged mean water vapour, evaporation and precipitation are projected to increase with global environmental warming (IPCC, 2007a). Current models indicate that future warming of 1.8–4.0 °C by 2100 (depending on the scenario chosen) will drive increases in precipitation in the tropics and at high latitudes; decreases in precipitation are expected in the subtropics (Plate 4). The intensity of individual precipitation events is predicted to increase, especially in areas seeing greatest increases in mean precipitation, but also in areas where mean precipitation is projected to fall. Summer drying of continental interiors is a consistent feature of model projections (IPCC, 2007a).

Hydroclimate and runoff

Despite some consistent patterns, the spatial response of precipitation to climate change is much more highly variable than that of temperature. Plate 5 shows the spatial pattern of precipitation and other hydrologically related changes projected for the A1B scenario (Appendix 1.1). Runoff is expected to fall in southern Europe and increase in Southeast Asia and at high latitudes. The impact of these changes has been assessed by Nohara *et al.* (2006), who find that in high latitudes river discharge is predicted to increase but in much of Central America, Europe and the Middle East, decreases in river flow are expected.

1.7 Systemic drivers of global environmental change (II): sea level

1.7.1 Introduction

Variations in sea level form part of a complex set of relations between atmosphere–ocean dynamics, ice sheets and the solid Earth, all of which have different response timescales. Thus changes in sea level resulting from global environmental change are, on the one hand, masked by shorter-term variations in the elevation of the sea surface. These fluctuations can be considerable: El Niño–Southern Oscillation (ENSO)-related inter-annual changes in ocean level in the western Pacific Ocean are ~45 cm (Philander, 1990), of comparable magnitude to many estimates of the magnitude of sea level rise to 2100. Furthermore, near-future sea level changes will take place against a backdrop of ongoing geological processes. The geographical variability in Holocene sea level histories has been well established (e.g. Pirazzoli, 1991) and geophysical models have identified large-scale sea level zones, with typical sea level curves, for the near- (Zone I), intermediate- (Zone II) and far-fields (Zones III–VI) in relation to former ice sheets (Clark *et al.*, 1978) (Fig. 1.11).

These differences have implications for future coastal vulnerabilities. Some coasts (in Zones IV and V for example) will have adjusted to coastal processes operating at present, or close to present, sea level for at least 5000 years whereas other regions (Zone II for example) will only have experienced present sea level being reached within the last 1000 years.

At the inter-ocean basin scale, coral reef systems in the Indo-Pacific Reef Province lie in the far-field, with near present sea level being reached at ~6 ka BP. Thereafter, hydro-isostatic adjustments and meltwater migration back to former ice margins, and in some cases local tectonics, resulted in a mid-Holocene sea level high of ~ +1 m, followed by a gradual fall to present mean sea level (Fig. 1.11). In the western Indian Ocean, sea level rose rapidly (6 mm a^{-1}) until 7.5 ka BP then dramatically slowed (to ~1 mm a^{-1}), with near present sea level being attained at 3.0–2.5 ka BP (Camoin *et al.*, 2004); however, in the central and eastern Indian Ocean a mid-Holocene high

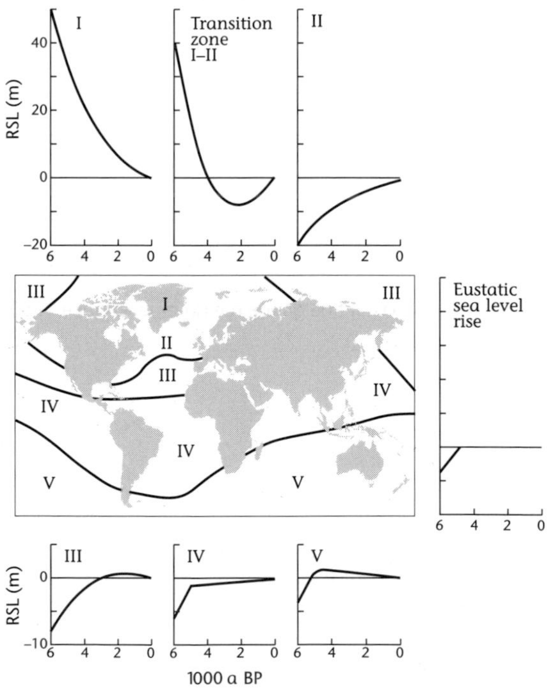

FIGURE 1.11. Sea level regions and associated sea level curves, 6–0 ka BP, assuming no eustatic component after 5 ka BP. RSL, relative sea level (modified from Clark *et al.*, 1978).

stand of ~ +0.5 m has been reported (Woodroffe, 2005; P. Kench, personal communication, 2007). By contrast, the reefs of the Atlantic Reef Province lie within the intermediate field and experienced a decelerating rate of sea level rise through the Holocene, with present sea level being reached only within the last 1000 years (Toscano and Macintyre, 2003) (Fig. 1.11). These different sea level histories go some way to explaining the gross morphological differences between reef provinces. Thus Indo-Pacific reef margins are generally characterised by wide, low-tide-emergent reef flats and, in some locations, by supratidal conglomerates and raised reef deposits. By contrast, in the Atlantic reef province, emergent reef features are lacking and relatively narrow reef crests are backed by shallow backreef environments and lagoons. These differences are of importance: as sea level rises, near-future reef responses will take place over these different topographies.

Finally, at the within-region scale, continuing adjustments to the unloading of ice, reflooding of shallow shelf seas and sedimentation in coastal lowlands in the British Isles since the Last Glacial Maximum have resulted in maximum rates of land uplift of ~1.5 mm a^{-1} on coasts in western Scotland but corresponding maximum subsidence of –0.85 mm a^{-1} on the Essex coast of eastern England (Shennan and Horton, 2002). In Scotland, therefore, sea level rise will be partially offset by uplift whereas in south-east England sea level rise will be additive to existing rates of relative rise.

Long-term trends in coastal erosion – such as along the eastern seaboard of the USA where 75% of the shoreline removed from the influence of spits, tidal inlets, and engineering structures is eroding (Zhang, 2004) and the eastern coastline of the UK where 67% of the shoreline length has retreated landward of the low-water mark (Taylor *et al.*, 2004) – can be reasonably associated with sea level rise over the last 100–150 years. The exact linkage, however, is most probably more event-based. Komar and Allan (2007) have identified increased ocean wave heights along the eastern seaboard of the USA and related them to an intensification of hurricane activity from the late 1990s; they have also argued that increased erosion along the US west coast since the 1970s has been associated with increasing wave heights due to higher water levels and storm intensities (Allan and Komar, 2006). In addition to sea level change and its variability, coastal erosion is also driven by other natural factors, including sediment supply and local land subsidence. Anthropogenic activities can intensify these controls on coastal change. Finally, the impact of such processes depends upon the ability of coastal landforms to migrate to new locations in the coastal zone and to occupy the accommodation space that becomes available (or not) to them (Fitzgerald *et al.*, 2008). These issues, and many more, are dealt with in Chapters 5, 6 and 7.

1.7.2 Recent sea level rise

Global sea level has been rising at a rate of ~1.7–1.8 mm a^{-1} over the last century, with an acceleration of ~3 mm a^{-1} during the last decade (Church *et al.*, 2004; Church and White, 2006) (Fig. 1.12). The total twentieth-century rise was 0.17 ± 0.05 m (IPCC, 2007a).

Observations since 1961 indicate that the oceans have been absorbing 80% of the heat added to the climate system, causing the expansion of seawater and perhaps

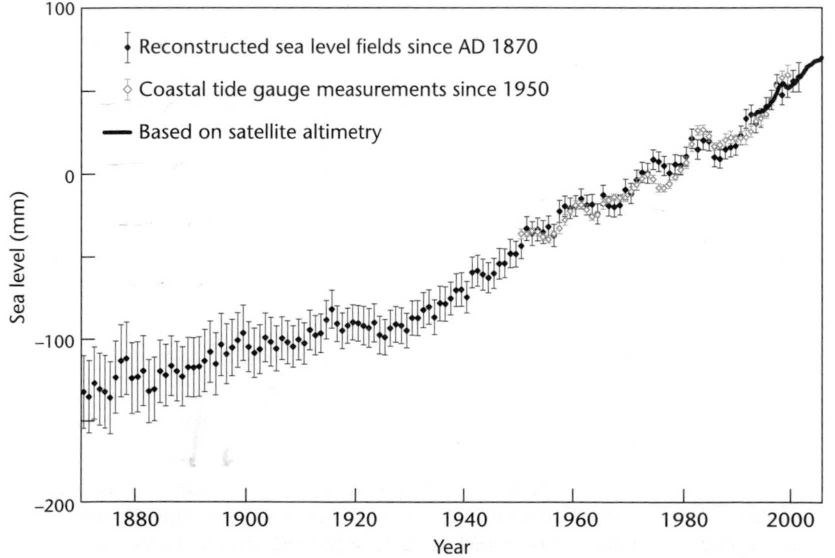

FIGURE 1.12. Annual averages of global mean sea level (mm), 1870–2003. Sea level fields have been reconstructed since 1870; coastal tide gauge measurements are available since 1950; and altimetry estimates date from 1993.

TABLE 1.1. *Observed rate of sea level rise and estimated contributions from different sources, 1961–2003 and 1993–2003*

Source of sea level rise	Rate of sea level rise ($mm\ a^{-1}$)[a] 1961–2003	1993–2003
Thermal expansion	0.42 ± 0.12	1.6 ± 0.5
Glaciers and ice caps	0.50 ± 0.18	0.77 ± 0.22
Greenland ice sheet	0.05 ± 0.12	0.21 ± 0.35
Antarctic ice sheet	0.14 ± 0.41	0.21 ± 0.35
Sum of individual climate contributions to sea level rise	1.1 ± 0.5	2.8 ± 0.7
Observed total sea level rise	1.8 ± 0.5	3.1 ± 0.7
Difference (observed minus sum of estimated climate contributions)	0.7 ± 0.7	0.3 ± 1.0

[a] Data prior to 1993 are from tide gauges and after 1993 are from satellite altimetry.
Source: From IPCC (2007a).

TABLE 1.2. *Projected globally averaged surface warming and sea level rise at the end of the twenty-first century*

Case	Temperature change (°C at 2090–2099 relative to 1980–1999)[a] Best estimate	Likely range	Sea level rise (m at 2090–2099 relative to 1980–1999) Model-based range excluding future rapid dynamical changes in ice flow
Constant Year 2000 concentrations[b]	0.6	0.3–0.9	NA
B1 scenario	1.8	1.1–2.9	0.18–0.38
A1T scenario	2.4	1.4–3.8	0.20–0.45
B2 scenario	2.4	1.4–3.8	0.20–0.43
A1B scenario	2.8	1.7–4.4	0.21–0.48
A2 scenario	3.4	2.0–5.4	0.23–0.51
A1FI scenario	4.0	2.4–6.4	0.26–0.59

[a] These estimates are assessed from a hierarchy of models that encompass a simple climate model, several Earth Models of Intermediate Complexity (EMICs) and a large number of Atmosphere–Ocean General Circulation Models (AOGCMs).
[b] Year 2000 constant composition is derived from AOGCMs only.
Source: From IPCC (2007a).

explaining half the observed global sea level rise since 1993 (Table 1.1), although with considerable decadal variability (IPCC, 2007a). Smaller component contributions are estimated to have come from glacier and ice cap shrinkage and, with less certainty, from the Greenland, and particularly, the Antarctic ice sheets. There is, however, no consensus on Antarctica. The balance of opinion appears to be for a shrinking West Antarctic ice sheet and a growing East Antarctic ice sheet, leading to a near-neutral effect (possibly perturbed by ice shelf losses on the Antarctic Peninsula). It should be noted, however, that the fit between estimated and actual sea level rise over the much longer period 1961–2003 is poor (Table 1.1).

This may partly be a function of the change in the measurement base after 1993, to satellite altimetry and away from tide gauge records. In the latter case, water level records are complicated by local vertical land movements, driven by glacial isostatic adjustments, neotectonics and/or subsurface fluid withdrawal, which need to be subtracted from the water level record. An additional uncertainty is that knowledge of changes in the storage of water on land (from extraction and increases in dams and reservoirs) remains poor. However, it is also not clear to what extent the sea level record reflects secular rise and to what extent it is a measure of regional inter-annual to inter-decadal climate variability from phenomena such as ENSO, the NAO and Pacific Decadal Oscillation (PDO) (Woodworth *et al.*, 2005) the effects of which are also strongly spatially variable (Church *et al.*, 2004). Thus, for example, elevated sea levels and anomalously high sea surface temperatures in the Pacific Ocean were closely correlated during the major 1997–1998 El Niño event (Allan and Komar, 2006). It appears likely that this variability also underpins the relations between coastal and global sea level rise, with faster coastal rise typical of the 1990s and around 1970 and faster ocean rise during the late 1970s and late 1980s (White *et al.*, 2005).

1.7.3 Future sea level rise

The average rate of sea level rise throughout the twenty-first century is likely to exceed the rate of 1.8 $mm\ a^{-1}$ recorded over the period 1961–2003. For the period 2090–2099, the central estimate of the rate of sea level rise is predicted to be 3.8 $mm\ a^{-1}$ under scenario A1B (Appendix 1.1) (the scenario spread is in any case small: Table 1.2), comparable to the observed rate of sea level rise 1993–2003, a period thought to contain strong positive

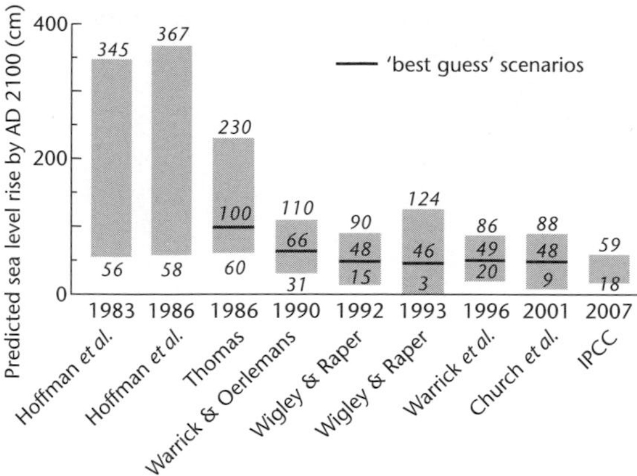

FIGURE 1.13. Changing estimates of the range of potential sea level rise to 2100 (predictions in the period 1983–2001) or 2099 (data from IPCC 2007a).

residuals within the long-term pattern of sea level variability. Under the A1B scenario, by the mid-2090s sea level rises by +0.22 to +0.44 m above 1990 levels, at a rate of 4 mm a^{-1}. The thermal expansion term accounts for 70–75% of the total rise, being equivalent to a sea level rise of 2.9 ± 1.4 mm a^{-1} in the period 2080–2100 (IPCC, 2007a).

In terms of the history of sea level rise projections, the IPCC Fourth Assessment predictions appear to suggest a further reduction in the expected rate of sea level rise (Fig. 1.13). However, these new projections do not allow for a contribution from ice flow from the Greenland and Antarctic ice sheets. Conservative estimates suggest that accelerated ice flow rates might add +0.1 to +0.2 m to the upper range of sea level rise (i.e. a sea level rise of up to 0.79 m under scenario A1FI: Appendix 1.1).

Non-model-based estimates of future sea level rise, based upon a semi-empirical relationship connecting temperature change and sea level rise through a proportionality constant of 3.4 mm a^{-1} per °C, suggest a potential rise at 2100 of 0.5 to 1.4 m (Rahmstorf, 2007), although this methodology has been highly contested.

1.8 Cumulative drivers of global environmental change (I): topographic relief

1.8.1 Introduction

By contrast with hydroclimate and sea level changes, relief and human activity (Section 1.9) are discontinuous both over space and through time. The implications of this simple fact are profound. Spatial discontinuities dictate that certain parts of the landscape are more sensitive to the drivers of change than others; temporal discontinuities mean that very old and very young landscape elements can exist side by side (Fig. 1.5). Global impact then becomes the net effect of change at a large number of disparate sites. A further implication is that the geomorphologist is best suited to identify those aspects of the landscape which are particularly sensitive to disturbance.

1.8.2 The sediment cascade

Continental-scale relief defines a pathway from the highest mountains to the ocean; local-scale relief, with associated sinks, defines a complex of pathways which can be best described by the term 'sediment cascade'. The type and rate of weathering, erosion and denudation are profoundly controlled by these pathways, which ultimately owe their variety to relief at a wide range of scales.

Weathering is the alteration by atmospheric and biological agents of rocks and minerals. The physical, mineralogical and chemical characteristics of the materials are modified so that these weathering products can be removed by either mechanical or biochemical erosion. Mechanical and biochemical erosion involves mobilisation, transport and export of rock and soil materials (Fig. 1.14a). Surface and subsurface water is critical to these processes. How much of these so-called 'clastic' sediments are broken up into their constituent chemical elements, the extent to which these elements have become combined with nutrients and the amount of comparatively unchanged rocks and minerals depends on the type and rate of weathering. The sediment which is exported is engaged to a greater or lesser extent in exchange processes between minerals, solutes and nutrients.

Primary sediment mobilisation can usefully be divided into 'normal regime' processes that are more or less pervasive and occur regularly, and episodic or 'catastrophic' events which occur less frequently. The former include soil creep, tree throw, surface disturbance by animals and surface erosion from exposed soils; the latter include rockfalls, rockslides, earthflows, landslides, debris avalanches and debris flows.

Sediment transport requires both a supply of transportable sediment and runoff competent to entrain the sediment. Sediments in suspension and bedload have the greatest influence on river channel form, whereas solutes, nutrients and pollutants are important indices of biogeochemical cycling processes.

The sediment cascade involves inputs and outputs with delays caused by: rates of rock breakdown, intermittent sediment mobilising events, the limited capacity of the flows to entrain sediment, and trapping points that occur downstream, so that sediment is intercepted and

(a)

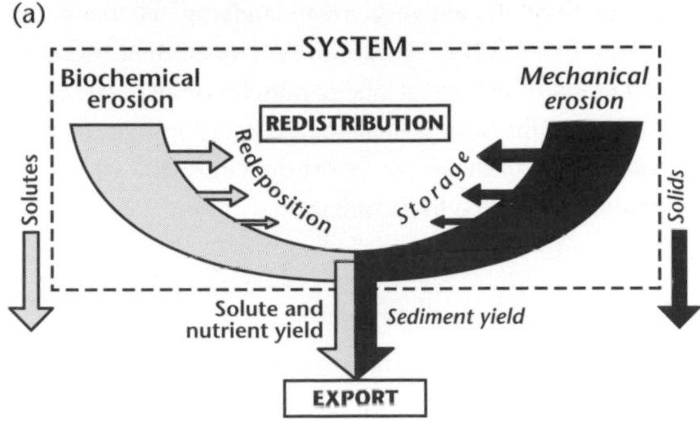

(b)

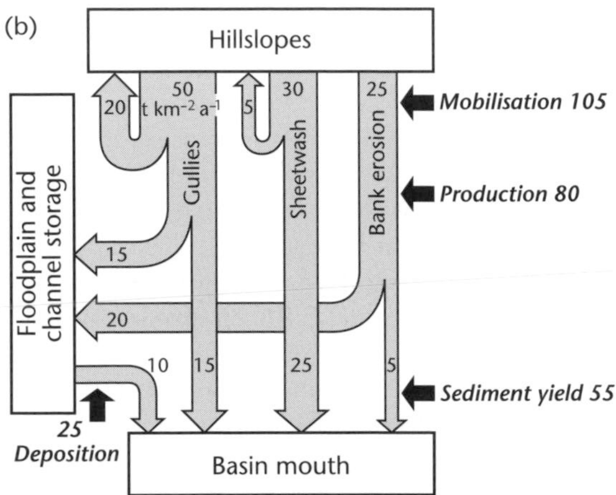

FIGURE 1.14. (a) The sediment cascade system from biochemical and mechanical weathering sources to export of sediment, solutes and nutrients. (b) Relations between sediment mobilisation, production, deposition and yield. Line widths are proportional to the amount of sediment transferred and values are shown in t $km^{-2} a^{-1}$. An average of 105 t $km^{-2} a^{-1}$ is mobilised on hillslopes and only 55 t $km^{-2} a^{-1}$ is exported from the basin (from Reid and Dunne, 1996).

remobilised at a later time. Sediment in a drainage system thus moves through a cascade of reservoirs (Fig. 1.14b).

This simple picture is distorted by the fact that sediment transport is also a sorting process. Mixtures of sedimentary particles of differing size and density segregate in the cascade into subpopulations with different transit times through the landscape. In the case of disturbance, there is a spatial complication in that the sediments have diffuse sources. The sedimentary signal originating in the uppermost part of the basin may move through a reach subject to similar disturbance and be reinforced. But at downstream points, if the disturbance is of varying intensity, the reinforcement will vary spatially. Sediment on its way from source to sink gets sidetracked in a number of ways, not only into channel and floodplain storage but also into other hillslope locations (Fig. 1.14b).

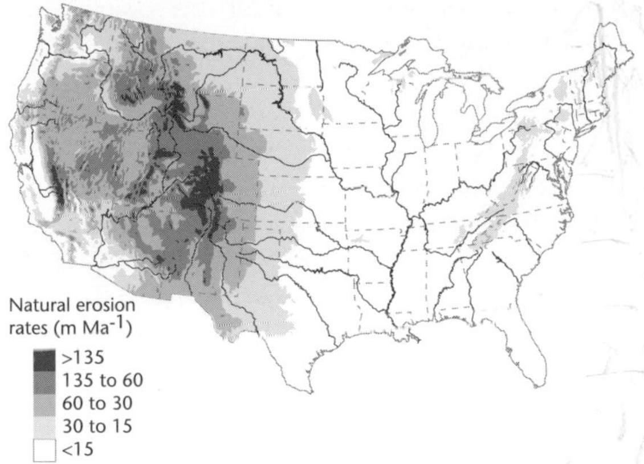

FIGURE 1.15. Estimates of average natural erosion (denudation) rates inferred from GTOPO30 area–elevation data and global fluvial erosion–elevation relations from Summerfield and Hulton (1994) (from Wilkinson and McElroy, 2007). Units are in metres of denudation per million years.

1.8.3 Topographic relief and denudation

In spite of the variety of ways in which sediment is delayed on its way to the ocean, there are extensive data sets that demonstrate a close relationship between topographic relief and rates of fluvial erosion. With the increasing availability of digital elevation models, it has also become possible to infer average natural erosion rates, as illustrated in Fig. 1.15 for the conterminous USA. The units used may be unfamiliar as they are given in metres of denudation per million years, but the main message of the map is that relief is a strong driver of landscape change (see Chapter 2 for an expansion of this theme and discussion of shortcomings of this generalisation).

1.8.4 The sediment budget

In order to account for the sources, pathways of movement and rates of delivery of sediments along coasts and through river basins, a method of budgeting of sediments has been developed. A river basin sediment budget is 'an accounting of the sources and disposition of sediment as it travels from its point of origin to its eventual exit from a drainage basin' (Reid and Dunne, 1996).

If the sediment budget is balanced, the residual will be zero. Thus the sediment budget can be expressed as

$$I - O = \Delta S \tag{1.1}$$

where I is input, O is output and ΔS is change of storage over the time period of measurement. This equation will hold for any landform or landscape as long as the budget cell can be unambiguously defined. In the case of the river basin, the budget cell defines itself. But Cowell *et al.* (2003), for example, in working with the problem of determining an objective budget cell in coastal studies, have coined the term 'coastal tract'. They define the coastal tract as the morphological composite comprising the lower shoreface, upper shoreface and back barrier. They use this framework in defining boundary conditions and internal dynamics to separate low-order from higher-order coastal behaviour (see Chapter 6 for further details).

The sediment budget is a mass-balance-based approach, which necessarily, in principle, includes a consideration of water, sediments, chemical elements and nutrients. This accounting of sources, sinks and redistribution pathways of sediments in a unit region over unit time is a complex exercise (Dietrich and Dunne, 1978). The primary application of sediment budgets to landscapes is in the accounting of clastic sediment transfers because clastic sediments provide the most direct indication of changing surface form. The US Corps of Engineers working on the coasts of southern California and New Jersey and geomorphologists working in the Swiss Alps and northern Scandinavia were the first to employ this methodology in coastal and river basin geomorphology in the 1950s.

Sediment budget studies have become a fundamental element of coastal, drainage basin and regional management, but only when somewhat simplified (Reid and Dunne, 1996). They can be formulated to aid in the design of a project, characterise sediment transport patterns and magnitudes and determine a project's erosion or accretionary impacts on adjacent beaches and inlets (Komar, 1998). Although Trimble (1995) and Inman and Jenkins (1999) have developed flexible models to discriminate the effects of climate change and land use activities on sediment budgets, their application is at an early stage.

In conclusion, the growing interest in global environmental change has spurred interest in a variety of mass-balance calculations that permit quantitative comparisons from one region to another and from one time period to another, with appropriate scaling. The biggest challenge in the use of sediment budgets is in making the link between intensively studied smaller-scale systems to the global scale and in extrapolating from the short term to the longer term and future changes.

1.8.5 Limitations of the sediment budget approach in determining the role of relief

1. Summerfield and Hulton (1994) demonstrated that 60% (but no more) of the spatial variation of global sediment yield can be explained by relief and runoff.
2. The sediment budget approach often ignores the role of tectonics. This is unfortunate because the uplift of mountain ranges involves the input of new mass which directly influences the mass balance of the geosystem. When longer term studies of sediment flux are undertaken, it is essential to incorporate the style and rate of tectonic processes (see Chapter 15).
3. Global riverine changes such as chemical contamination, acidification, and eutrophication which are of great interest in global environmental change are rarely included in sediment budgets (see Chapter 3).
4. Order of magnitude effects produced by global-scale river damming (Nilsson *et al.*, 2005) and general decrease of river flow due to irrigation (Meybeck and Vörösmarty, 2005) scarcely require sediment budgeting (see Chapter 4).
5. Recent studies of the Ob and Yenisei rivers in Russia (Bobrovitskaya *et al.*, 2003) and of the Huanghe River in China (Wang *et al.*, 2007) suggest that the decline in sediment load delivered to the sea in recent years is caused primarily by soil conservation practices and reservoir construction; relief and runoff have relatively little influence.
6. These observations lead logically to an analysis of the critical role of human activity.

1.9 Cumulative drivers of global environmental change (II): human activity

Global population growth and the attendant land cover and land use changes pose serious challenges for management and planning in the face of global environmental change (Lambin and Geist, 2006). Wasson (1994) has emphasised the value of mapping past land use and land cover changes in attempting to interpret contemporary landscape change and degradation. De Moor *et al.* (2008) have demonstrated that in the past 2000 years, the impact of climate on river systems can be neglected when compared with the human impact. They link the considerable changes in hillslope processes induced by humans with equivalent sediment supply to the river valleys of the Netherlands. Factors that influence land cover and land use change can be divided into indirect and direct factors.

1.9.1 Indirect factors

Indirect factors include such broadly contextual factors as demographic change, socioeconomic, cultural and religious practices and global trade, which enhance the importance of governance *sensu lato*. The indirect drivers are of critical importance when we come to discuss scenario building (Appendix 1.1) in the sense that population level and density, socioeconomic context, societal values and the institutions available to implement change must all be incorporated into any forward-looking thinking.

Population growth

The history of humans as geomorphic agents is now becoming clear. Virtually all of Earth's biomes have been significantly transformed by human activity because of the enormous growth of the world's population during the second half of the twentieth century and the even more rapid growth in energy use (Turner *et al.*, 1990a). Some 80% of humankind lives in developing countries in Asia, Africa and South and Middle America and those countries account for more than 90% of the more than 100 million births each year. These populations are becoming increasingly urbanised; thus whilst only 3% of the population of less developed regions lived in cities larger than 100 000 inhabitants in 1920, this figure is set to rise to 56% by 2030. By 2015, Asia is predicted to have 1.02 billion people in urban centres of over 500 000 inhabitants. Many of these populations are being squeezed into narrow coastal and estuarine margins. Several of the fastest-growing cities, with rates of increase of up to +4% per year and all projected to have populations in excess of 15 million by 2015, have long histories of exposure to coastal hazards; they include Shanghai, Kolkata, Dhaka and Jakarta (United Nations Population Fund, 2007).

As a result, the biomes shown in Plate 3 are actually an abstract depiction of the climax vegetation in the absence of human intervention. Only in the areas unmodified by agriculture, forestry and urbanisation do these biomes correspond to the terrestrial ecosystems of today.

Socioeconomic context of soil degradation

If land loss continues at current rates, an additional 750 000–1.8 million km² will go out of production because of soil degradation between 2005 and 2020. There are 95 developing countries, each with less than 100 000 km² of arable land, for which the loss of their most vulnerable lands will mean either loss of economic growth potential or famine or both (Scherr, 1999). The development of long-term programmes to protect and enhance the quality of soils in these countries would seem to be a priority. Countries with large areas of high quality agricultural land (Brazil, China, India, Indonesia and Nigeria, for example) will probably focus on the more immediate economic effects of soil degradation. Because the poor are particularly dependent on agriculture, on annual crops and on common property lands, the poor tend to suffer more than the non-poor from soil degradation. Countries or sub-regions that depend upon agriculture as the engine of economic growth will probably suffer the most. Furthermore, as growing populations in the developing world become increasingly urbanised, the difficulties in maintaining food supply chains between cities and their rural hinterlands are likely to increase (Steel, 2008).

1.9.2 Direct factors

Direct factors include deliberate habitat change; physical modification of rivers, water withdrawal from rivers, pollution, urban growth and suburban sprawl (Goudie, 1997).

Cultivated systems

The most significant change in the structure of biomes has been the transformation of approximately one-quarter (24%) of Earth's terrestrial surface to cultivated systems. More land was converted to cropland in the 30 years after 1950 than in the 150 years between 1700 and 1850. There is a direct connection between soil erosion on the land (net loss of agriculturally usable soil to reservoirs) and coastal erosion resulting from a reduction in sediment delivery to the coast. It is estimated that human activity affects directly 8.7 million km² of land globally; about 3.2 million km² are potentially arable, of which a little less than a half is used to grow crops. Soil quality on three-quarters of the world's agricultural land has been relatively stable since the middle of the twentieth century, but the remaining 400 000 km² are highly degraded and the overall rate of degradation has been accelerating over the past 50 years. Almost 75% of Central America's agricultural land has been seriously degraded, as has 20% of Africa's and 11% of Asia's.

In the United States, cropland is being eroded at 30 times the rate of natural denudation (Figs. 1.15 and 1.16) and this is creating two problems in addition to the obvious loss of usable land *in situ*. First of all, there are large sediment sinks being created on land, where the sink consists mostly of clastic sediments and secondly, nutrient sinks are being created offshore. Trimble and Crosson (2000) have commented helpfully on the implications of the build-up of new sediment stores in low-order basins.

It has become increasingly difficult to assess the future impact of environmental changes on the sediment flux to the coastal zone because of the complex impacts of human activities (Walling, 2006). Globally, soil erosion is accelerating as a result of deforestation and some agricultural practices; but at the same time, sediment flux to the coastal zone is globally decelerating, because of dams and water diversion schemes. A summary statement by Syvitski *et al.* (2005) notes that human activities have simultaneously increased the sediment transport by global rivers through soil erosion

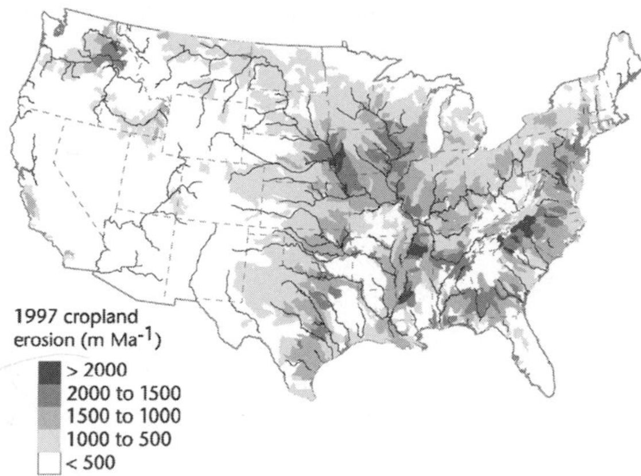

FIGURE 1.16. Rates of cropland erosion derived from estimates by the Natural Resources Conservation Service using the Universal Soil Loss Equation, and scaled to a farmland mean of 600 m Ma^{-1}. (from Wilkinson and McElroy, 2007).

by 2.3 ± 0.6 Gt a^{-1}, yet reduced the flux of sediment reaching the world's coasts by 1.4 ± Gt a^{-1} because of retention within reservoirs. Over 100 Gt of sediment and 1–3 Gt of carbon are now sequestered in reservoirs constructed largely within the last 50 years. Furthermore, the impact of dams on large rivers is likely to continue to reduce the global sediment load and will alter the patterns of sediment flux in many coastal zones (Dearing and Jones, 2003).

Reduced sediment loads delivered to the coastal zone result in accelerated coastal erosion and a decrease in habitat. The reduction of the seasonal flood wave also means that the sediment is dispersed over smaller areas of the continental margin. Although this is a correct aggregate picture, it should be borne in mind that tropical deforestation continues unabated while temperate forests are being enlarged. There are therefore numerous individual unregulated basins within the tropical world (e.g. Indonesia and Malaysia) where sedimentation at the coast continues to be considerable.

Milliman and Syvitski (1992) concluded that mountainous rivers, particularly in the island nations of Southeast Asia, contribute the largest proportion of global sediment flux. These regions have recent histories of severe deforestation and are dominated by low-order streams. It is therefore probable that the most rapid increase in future sediment flux to the ocean may well come from disturbed, small–medium-size basins feeding directly to the coast. This sediment is likely to contain enhanced loads of adsorbed nutrients and surface pollutants, a phenomenon which has already been documented in some detail from the west coast of India (Naqvi *et al.*, 2000). The primary factor responsible for increased nitrous oxide production is the intensification of agriculture and the increasingly widespread application of anthropogenic nitrate and its subsequent denitrification. The western Indian continental shelf and the Gulf of Mexico immediately to the south and west of the Mississippi delta are well-documented hypoxic zones; that is, the concentration of oxygen in the water is less than 2 ppm. In the case of the Mississippi–Missouri drainage basin both phosphates (industrial) and nitrates (agricultural) constitute diffuse sediment sources which follow both surface and subsurface hydrological pathways (Fig. 1.17) (Goolsby, 2000; Alexander *et al.*, 2008). The average extent of the hypoxic zone was 16 700 km^2 between 2000 and 2007. The goal of reducing the extent of the hypoxic zone to an average of 5000 km^2 by 2015 looks increasingly difficult as there has been no significant reduction in nutrient loading (Turner *et al.*, 2008).

Desertification

Desertification is defined by the UN Convention to Combat Desertification as 'land degradation in arid, semi-arid and dry sub-humid areas resulting from various factors, including climatic variation and human activities'. Land degradation in turn is defined in that context as the reduction or loss of biological or economic productivity. From the perspective of the global hydrological cycle the role of vegetation in land surface response is critically important, especially through vegetation–albedo–evaporation feedbacks. Desertification is a global phenomenon in drylands, which occupy 41% of Earth's land area and are home to more than 2 billion people. Some 10–20% of drylands are already degraded. This estimate from the Desertification Synthesis of the Millennium Assessment is consistent with the 15% global estimate of soils permanently degraded. Excessive loss of soil, change in vegetation composition, reduction in vegetative cover, deterioration of water quality, reduction in available water quantity and changes in the regional climate system are implicated. Desertified areas are likely to increase and proactive land and water management policies are needed (Reynolds *et al.*, 2007).

1.9.3 Conclusion

The fourth driver of global environmental change, namely human activity, is the most rapidly changing driver. Land use and land cover change, especially the transformation of forest and grasslands to logged forests, agricultural lands and urbanisation, have the most profound effect. The intensification of human activity in the temperate and tropical zones is especially effective in landscape change; by contrast, in polar latitudes, where population densities are low, climate in particular, but also relief and sea level, continue to be the more important drivers.

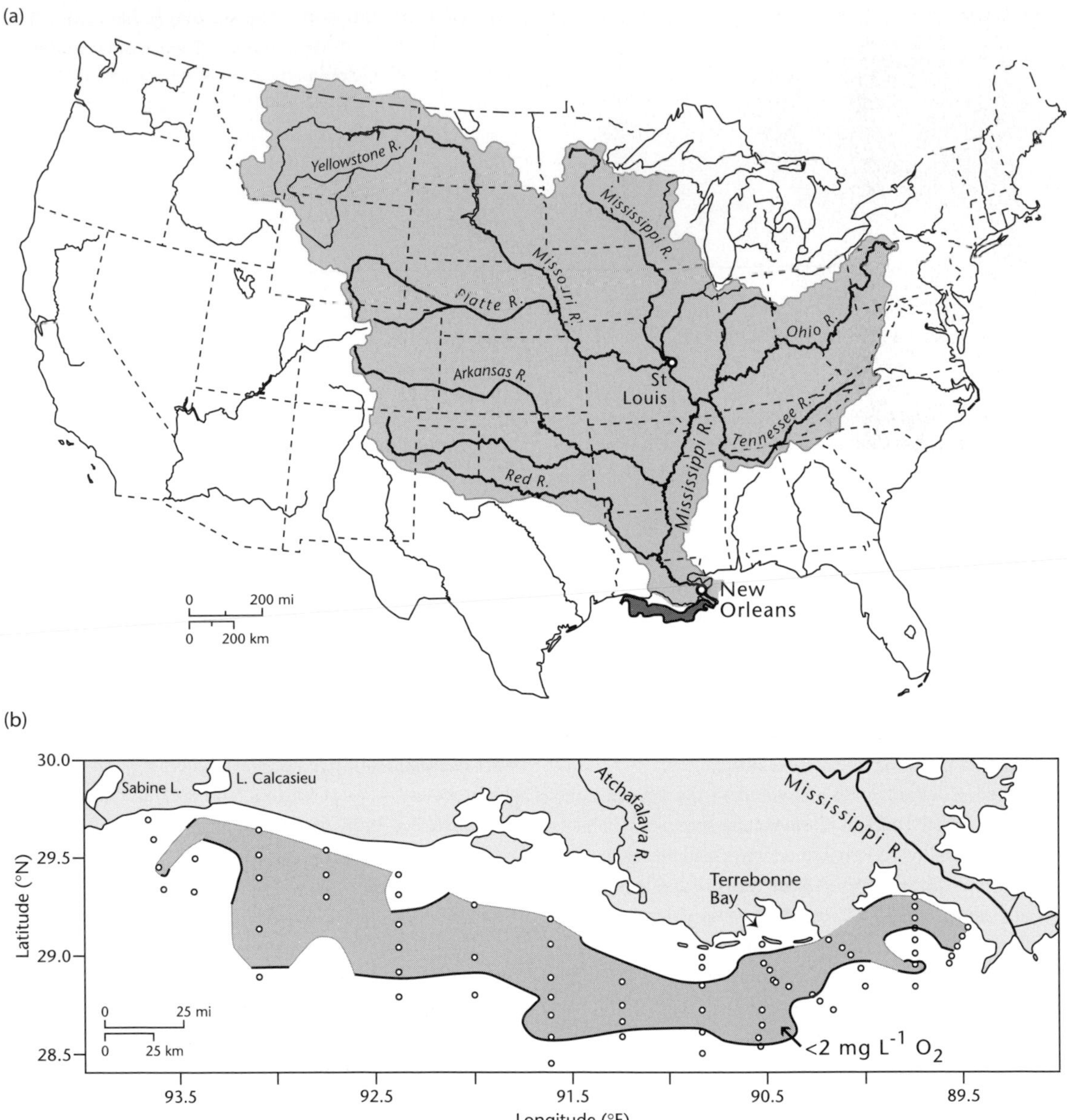

FIGURE 1.17. (a) Diffuse sediment sources of phosphates and nitrates from the Mississippi–Missouri drainage basin. (b) The resultant hypoxia in the Gulf of Mexico (modified from Goolsby, 2000).

1.10 Broader issues for geomorphology in the global environmental change debate

1.10.1 Putting the 'geo' into the 'bio' debates

Geodiversity is defined as a measure of the variety and uniqueness of landforms, landscapes and geological formations (Goudie, 1990) in geosystems at all scales; biodiversity is defined as the variation of life forms within a given ecosystem, biome or the entire Earth. There are thus strong parallels between the two concepts. The term natural heritage is more easily understood by the general public and sums up the totality of geodiversity and biodiversity. Geodiversity has its own intrinsic value, independent of any role in sustaining living things. The World Heritage Convention came into force in 1972 and after 35 years the World Heritage List now identifies, as of 2007, 851 sites of 'outstanding universal value' (http://whc.unesco.org/en/list/). Of these, 191 are

FIGURE 1.18. Bright Angel Canyon, Grand Canyon National Park: a landscape of 'outstanding universal value' (from Strahler and Strahler, 1994).

justified in terms of at least one component of geodiversity (Appendix 1.4). It is useful to note that biodiversity cannot exist without geodiversity, the natural diversity of our non-living environment. There is a tendency to think of biodiversity as being fragile, with geodiversity having greater robustness. However, this is not necessarily the case, but in those sites where geodiversity is relatively robust the loss of any one element of that diversity becomes doubly serious. Geomorphology has an important role to play in identifying physically sensitive landscapes that are highly responsive to environmental change. Geoconservation aims to preserve geodiversity (O'Halloran *et al.*, 1994). This means protecting significant examples of landforms and soils, as well as a range of distinctive Earth surface processes. It is proposed that we have an ethical obligation to retain landforms and landscapes of unique natural beauty and diversity for future generations. There is a very good reason for active geoconservation management because, when a landform is lost, it can only be replaced over geological timescales. Glen Canyon in Colorado has disappeared and, in a significant sense, the Three Gorges of the Yangtze (Li-Jiang) River have disappeared. Although it is not anticipated that the Grand Canyon will disappear sometime soon, it is only through active geoconservation management that its pristine status can be maintained (Fig. 1.18).

But there are other reasons why biodiversity concerns cannot be divorced from geodiversity. There are linkages, for example, between above-sediment and sediment biota in freshwater ecosystems, which suggest that the monitoring of the bio- and the geodiversities should be collaborative (Lake *et al.*, 2000). The standard view in ecology is that biodiversity confers stability and resilience and, to a more limited extent, natural floodplains conserve the resilience of landforms to changing environmental drivers.

Nevertheless, the analogy between biodiversity and geodiversity should not be overemphasised given the detailed functional differences between eco- and geosystems.

1.10.2 Geomorphology, natural hazards and risks

The dynamic nature of landscapes and landforms over space and through time has generated interest in defining geomorphological 'hotspots'. Biodiversity hotspots are defined as regions that contain at least 1500 species of vascular plants and which have lost at least 70% of their original habitat (Myers *et al.*, 2000). In other words they are regions of special value and at the same time highly vulnerable to disturbance. By analogy, geomorphological hotspots are sites or regions of special value in terms of geodiversity and highly vulnerable to environmental change. 'Geoindicators' have been defined as one way of identifying such hotspots (Berger and Iams, 1996). They are measures of geomorphological processes and phenomena that vary significantly over periods of less than 100 years and are thus high-resolution measures of short-term changes in the landscape. Geoindicators have been designed as an aid to state-of-the-environment reporting and long-term ecological monitoring. They are also useful in the identification of potential natural hazards, where irreversible damage to property and loss of life occur. A wide range of slope failure, river, coastal and wind erosion related problems which have been exacerbated by human activity are recognised as natural hazards. Where the geoindicator detects unusual movement or instability, a geomorphological 'hotspot' is identified. If this instability is located close to human settlements or transport corridors, a geomorphological hazard

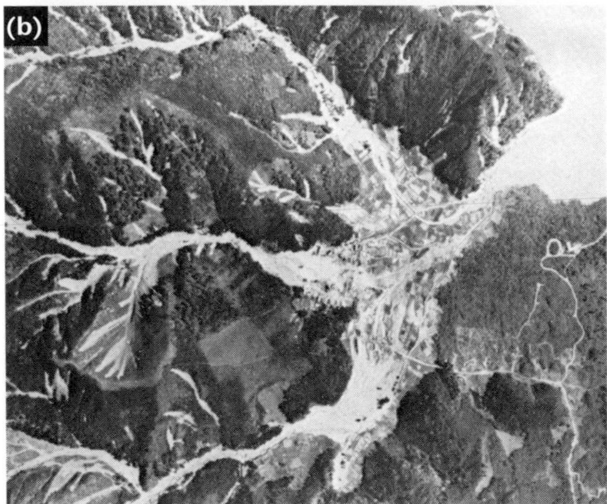

FIGURE 1.19. Debris flows as geomorphological hazard, illustrated from the results of an intense rainstorm in 1966 at Ashiwada in Yamanashi Prefecture, Japan. (a) Pre-1966; (b) post-typhoon 26, 1966 (from Akagi, 1973).

is declared and risk studies may be undertaken. It is the point of intersection of the human use system with the extreme geomorphological event which defines a geomorphological hazard (Glade, 2003).

Because of the high population density, the high-relief volcanic slopes, the intense rainstorms produced by multiple annual typhoons, the vulnerable coastline and the tendency to build settlements in the most vulnerable locations, there is a tragic history of loss of lives and infrastructure in the Japanese islands (Akagi, 1973). In the example illustrated (Fig. 1.19) a village, located on a debris flow fan on the western flanks of Mount Fujiyama, Japan is the human use system. The village was overwhelmed by debris flows (an extreme geomorphological event) following intense rainstorms accompanying Typhoon 26, in 1966. The unfortunate coincidence of a human use system interacting with a geomorphological event system (or sediment cascade) defines a geomorphological (or natural) hazard. Surrounding the question of natural hazards is the question of natural risk (Hufschmidt *et al.*, 2005). Natural risk is a measure of the probability of adverse effects on health, property and society, resulting from the exposure to a natural hazard of a given type and magnitude within a certain time and area. If risk studies had been undertaken before the event shown in Fig. 1.19, it might have been possible to develop a plan for mitigation of the disaster and to save many lives and much property as a result.

1.10.3 Geomorphology and unsustainable development

The concept of sustainable development has infused the policy world since the publication of Gro Harlem Brundtland's *Our Common Future* (WCED, 1987). The concept of sustainability is highly contested in a field like geomorphology because the drivers of change are themselves constantly changing and landscapes and their soils are, over century timescales, frequently collapsing, due to overexploitation. While the value of sustainability has achieved wide currency in principle, the implementation has proved difficult. A distinction should be made between 'strong sustainability' in the case of a landscape and a 'weak sustainability' in the case of a society. Strong sustainability concerns preserving the structure and function of an ecosystem and its supporting geophysical substrate. Weak sustainability admits that landscapes may be changed by deliberate human agency so long as the total resource value is not devalued.

It would perhaps be more realistic in geomorphology, and in related sciences, to seek evidence of 'unsustainability' such that adaptation and/or mitigation can be planned well in advance of the system collapse. Four key concerns have been identified by the policy community and two promising approaches have emerged. Concerns are: the need to integrate natural and physical science with social science, health science and humanities research; a focus on the future; the need to involve stakeholders; and a sensitivity to the appropriate temporal and spatial scale of analysis required. The term 'back-casting' was coined by Robinson (2004) for an approach that involves working backward from a desired future end point to the present and finding the appropriate policy measures required to reach that point. Scenario building, discussed in Chapter 3, is one specific application of the back-casting approach. 'Integrated assessment' is another interdisciplinary process of 'combining, interpreting and communicating knowledge from diverse scientific disciplines in such a way that the whole set of cause–effect interactions of a problem can be evaluated from a synoptic perspective' (Rotmans and Dowlatabadi, 1998).

Whilst it must be admitted that these are promising policy approaches, especially with respect to their holistic emphasis, geomorphologists have to ask the question 'in what sense can landscapes be sustained over century or millennial timescales in the face of constantly changing human activities, sea level changes and climate change?'

Not only is there a problem of constant change and the possibility of achieving sustainability in the future, but many would claim that we have already exceeded the Earth's capacity to support our consuming lifestyle. Rapid increase in interest in the ecological footprint idea (Rees, 1992) confirms a move of public opinion towards greater concern about our degrading environment.

Ecological footprint analysis attempts to measure the human demand on nature and compares human consumption of natural resources with the Earth's ecological capacity to regenerate them (Plate 6). Ecological footprinting is now widely used as an indicator of environmental sustainability. The only prudent policy would seem to be the absolute minimisation of land cover change especially where urban sprawl impacts landforms of exceptional interest or soils whose optimal use is an agricultural one.

1.10.4 Geomorphology and the land ethic

Aldo Leopold's land ethic (Leopold, 1949) offered this general principle of land management: 'A thing is right when it tends to preserve the integrity, stability and beauty of the biotic community. It is wrong when it tends otherwise.' Land to Leopold had a broad meaning. It included soils, waters, plants and animals. The boundaries included people as members of nature. Leopold's lifelong land ethics journey led from a stewardship resource management mentality to stewardship entwined with ecological conscience.

Geomorphology provides certain constraints that limit how people should act towards the land and, at another level, it challenges the unspoken assumptions of an unwisely managed neoliberal society, which accepts uncritically the rule of the market. Only recently has the neoliberal view been willing to attempt proper valuation of environmental resources (e.g. Millennium Ecosystem Assessment (2005) and the Stern Review (2007) on the economics of climate change).

It is clear that the removal of soil cover will reduce livelihood options for people and agriculture. The careful management of land and its biogeochemical and aesthetic properties enhances long-term human security.

1.11 Landscape change models in geomorphology

1.11.1 Landscape change over long time periods

Brunsden (1980) has developed the idea of 'characteristic form time' (Fig. 1.20). This is a variant on the 'punctuated equilibrium' idea, but its importance is based on the fact that it is possible for a landscape to remain sufficiently unperturbed by environmental change such that it develops a form that is representative of the balance of resistance and driving forces, a balance which remains relatively unchanged over long periods of time. Whereas this was the prevailing assumption in geomorphology during pre-systems thinking, such has been the swing towards emphasis on change that some have questioned the possibility of a characteristic form ever being achieved. But Brunsden (1980) has pointed out that there is a need for the concept of characteristic form because the geological record gives evidence of past landscapes that appear to have retained constant characteristics over millions of years (e.g. much of the interior landscapes of Australia and the extremely low relief extensive erosion surfaces exposed in the stratigraphy of the Grand Canyon; Fig. 1.18).

The idea of the landscape as a palimpsest, with evidence preserved in layers which have simply to be unpeeled in

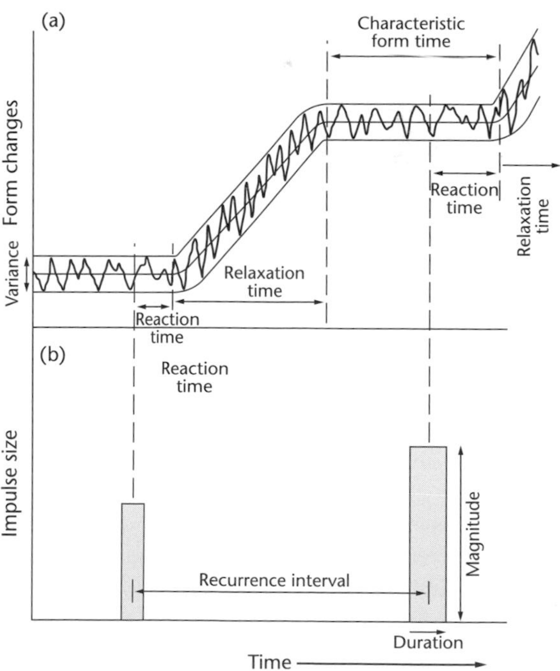

FIGURE 1.20. Definitions of geomorphological time, impulse recurrence, reaction, relaxation and characteristic form (modified from Brunsden, 1980).

order to discover the evidence of past landscape forming processes, is well enshrined in the literature (Wasson, 1994).

Frequency and magnitude of geomorphic events

However, in rapidly changing global environments, it is no longer possible to maintain the fiction of statistical stationarity in historical environmental data. Even in a landscape as quiescent as that of the UK, Higgitt and Lee (2001) demonstrated the difficulty of the concept of frequency and magnitude over the past 1000 years and Steffen *et al.* (2004), by invoking the 'no analogue' argument, have made it doubly difficult to sustain the traditional hydrological approach to the topic. It seems probable that the coupling of temporal and spatial dimensions through the analysis of lake and marine sediments (Dearing and Jones, 2003) and ice cores will provide the best estimates of changing frequencies and magnitudes of geomorphic events.

1.11.2 Thresholds and complex response

Individual slopes and river channels lie at variable distances from 'thresholds' (points at which major changes occur in response to perturbation) (Schumm, 1973; Brunsden and Thornes, 1979). A simple example of this effect is that a perturbation may generate a landslide or debris flow at a location in the headwaters of basin A; and the same perturbation may generate a similar slope failure close to the mouth of basin B. The response at the mouth of basin B will differ significantly from that at the mouth of basin A. In effect the communication of threshold exceedance through the river basin system is far more complex and has given rise to the complex response theory (Schumm, 1973).

Schumm (1973, 1977, 1985, 1991) has demonstrated the importance of discriminating between intrinsic and extrinsic variables in the evolution of landscapes. If we consider a river basin, we see a system which has its own organisation of river networks and slopes, adjusted to a greater or lesser extent to the prevailing internal properties of the basin, such as lithology, hydrology and biomass. These are intrinsic variables. At the same time, the system is subject to land use and land cover change and climate change. These are extrinsic variables. The point at issue is that the intrinsic and extrinsic variables operate independently and that systems with apparently identical intrinsic properties will respond differently to different extrinsic variables Even more subtly, side-by-side systems with apparently identical intrinsic properties may respond quite differently to apparently identical extrinsic variables. This is a result of the fact that the internal organisation of a geomorphic system contains historical elements that have not yet reached 'characteristic form' as defined above.

1.11.3 Landscapes of transition

Hewitt *et al.* (2002) have developed the idea of landscapes of transition. Nearly all landscapes bear the imprint of past conditions. Some of these past conditions no longer apply. Biomes have expanded and contracted geographically over large areas over geological timescales. There are at least two ways in which regional landscapes must be considered as transitional:

(a) some landforms reflect different past conditions. They are not merely relict forms; they operate as constraints upon present-day developments and
(b) there are also parts of landscapes and patterns of sedimentation that are in incomplete transition from past conditions.

The notion of temporal transition is intended to cover ways in which environmental changes are uniquely expressed. Although processes and landforms are driven by contemporary heat and moisture conditions, available relief and lithologies, they are not directly responding to them. They involve mechanisms or patterns of adjustment that are not readily obvious in climate change or tectonics. When geomorphic processes are modifying past landscapes, the adjustments do not travel directly from one equilibrium to another. There are combinations of intervening constraints, self-adjusting mechanisms or 'epicycles' peculiar to the earth surface processes affected.

In Canada, Church *et al.* (1999) have suggested that, at all scales above the order of 1 km^2, the landscape is still adjusting to the perturbation of continental glaciation. The fact that geomorphic systems have variable relaxation times following disturbance (i.e. time taken to return to the same conditions as those which prevailed prior to disturbance) has been well understood, but few careful quantitative studies have been available until recently.

The response of the cryosphere to global environmental change, which demonstrates high sensitivity to temperature change through the threshold 0 °C, is a more obvious example than the response of such an extensive and resistant landscape as the whole of Canada.

There are radical differences in the role of geomorphology in global environmental change whether one takes the characteristic form approach or the transitional landscapes approach. At one level, it is a question of different punctuated equilibrium models of the landscape, where one is dominated by quiescence and the other by change. But it is not necessary to choose between the two approaches. There are elements of the landscape that have remained unchanged for long periods of time and others that are highly sensitive. The challenge and the opportunity of global environmental

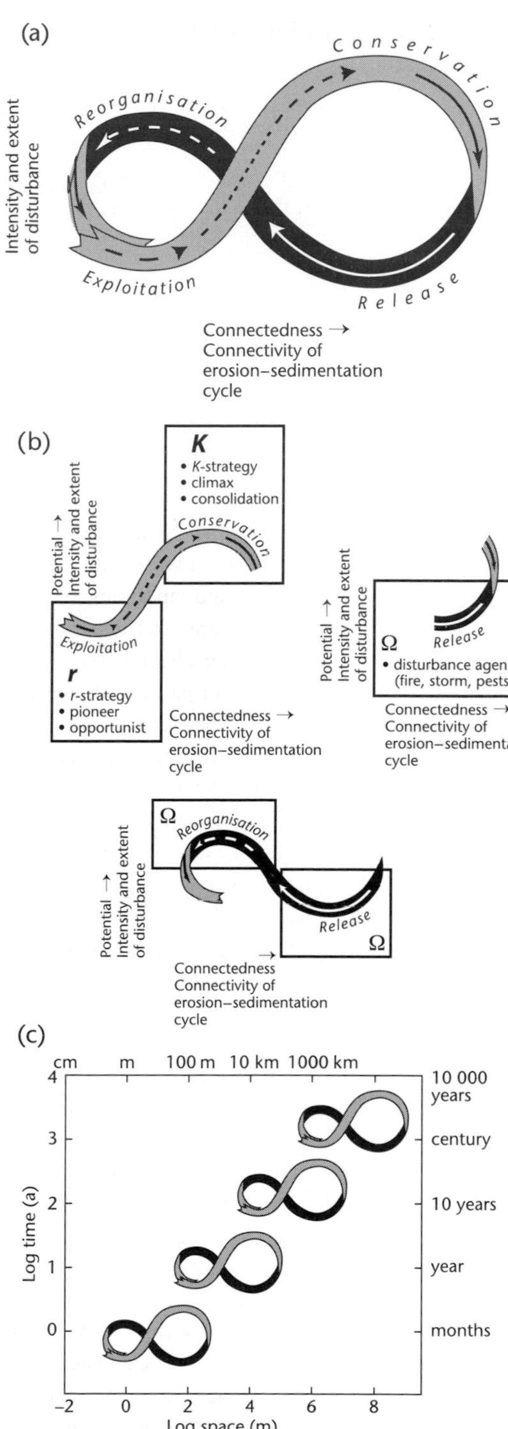

FIGURE 1.21. The panarchy framework (modified from Gunderson and Holling, 2002). (a) the adaptive cycle; (b) components of the adaptive cycle; and (c) panarchy in time and space.

change is to distinguish those parts of landscapes that are close to threshold and which, because of their vulnerability, deserve priority attention. The short-term challenge of global environmental change makes the transitional landscapes and complex response models more relevant; but this in no way invalidates the characteristic form time model for interpreting longer-timescale changes.

1.11.4 Adaptive systems

The transience of components of the landscape and the idea of transitional landscapes constantly undergoing change are interesting concepts that are directly relevant to contemporary global environmental change. If the IPCC definition of vulnerability is seriously engaged, then geomorphologists also have to investigate adaptive systems. Reliance on reactive, autonomous adaptation to the cumulative effects of environmental change is ecologically and socioeconomically costly. Planned and anticipatory adaptation strategies can provide multiple benefits. But there are limits on their implementation and effectiveness. Enhancement of adaptive capacity reduces the vulnerability of landscapes to environmental change, but adaptive capacity varies considerably among regions, cultures and socioeconomic groups.

It is possible to learn from ecology where recent models have placed environmental change and system collapse as central to an understanding of contemporary change and where there are similar complexity problems. Panarchy is a metaphor designed to describe systems of ecosystems at varying spatial and temporal scales. The terminology developed for panarchy (Holling, 2001) is entirely ecological and needs to be translated for the needs of geomorphology. Holling suggests that complex systems are driven through adaptive cycles which exist at a range of spatial scales. The term adaptive is self-evident in ecological systems; in geomorphic systems we often speak of self-regulating systems (Phillips, 2003). Adaptive cycles are defined as consisting of four phases, namely exploitation (the environmental disturbance regime), conservation (the response), collapse (threshold exceedance and unpredictable behaviour of the system) and reorganisation (recovery) (Fig. 1.21). A geomorphic analogue would be a glacial–interglacial cycle characterised by both orderly evolution and system collapse. The duration of these phases of adaptation in ecological systems depends on three factors: intrinsic wealth, connectivity and resilience of the system. A geomorphic analogue would be the intensity and extent of a glaciation (intrinsic wealth), the connectivity of the erosion–sedimentation cycles (connectivity) and the time required for the recovery of the system during interglacials (resilience).

The panarchy metaphor has a fascinating flexibility in dealing with complex systems and geomorphologists are working with this concept in an effort to introduce more flexible theory into the discussion of geomorphic complexity (Cammeraat, 2002). Although there are evident differences between geophysical systems and ecosystems, there are many parallels in the behaviour of self-organising systems which can assist in improving attempts to understand and manage the environment sensitively. The concepts and terminology of the panarchy model are consistent with geomorphic concepts such as complex response (Schumm, 1973), threshold exceedance, landscape sensitivity and barriers to change. Complex adaptive cycles throw new light on complex response in landscape systems. Resistances in the panarchy model of ecosystem behaviour perform a similar function to that of geomorphic barriers to change in geosystems.

In the context of rapid and accelerating global environmental change, geomorphologists are well placed to assess the potential for collapse and/or reorganisation of many of our desired landscapes. A scenario is a coherent, internally consistent and plausible description of a possible future state of the world (IPCC, 2007b.). Scenarios are not predictions or forecasts but are alternative images without ascribed likelihoods of how the future might unfold. The development of scenarios of landscape trajectories is a worthy aim. With such an approach, it may be possible to forestall the most serious negative outcomes and make wise use of the new opportunities created by global environmental change.

1.12 Organisation of the book

Following this introductory chapter, the volume is divided into two substantive sections, consisting respectively of six chapters (2–7) and seven chapters (8–14). The concluding Chapter 15 contains the highlights of conclusions from the substantive chapters, focussing on the identification of geomorphic hotspots, major uncertainties and matters of special concern that ought to be communicated to policy-makers, environmental managers and the scientifically literate public.

The two substantive sections of the book concern respectively azonal and zonal topics. In Chapters 2–7 (azonal topics) we attempt to follow the energy, mass and information cascade from mountain tops to the sea and focus attention on the ways in which environmental change is governed by a limited number of geophysical processes and their interactions with human activity. The hydrological cycle and the sediment budget constitute the central themes of this section and environmental change is discussed under the influence of this primarily geophysical framework. The chapters are sequenced along the topographic gradient: mountain environments; lake systems; river systems; estuaries, sedimentary coasts and coastal wetlands; beaches, cliffs and deltas; and coral reefs.

In Chapters 8–14 (zonal topics) we have attempted to identify the critical landscapes thought to be the most sensitive to potential environmental change although it is accepted that this list may not be exhaustive. The biome is used as the unit of zonality as vegetation is a good integrator of average climate conditions. The biomes selected for treatment in depth are: tropical rainforest; tropical wet and dry savanna; deserts; Mediterranean landscapes; mid-latitude temperate forest and rangelands (with particular reference to landslides and other slope processes in this biome); tundra and permafrost-dominated taiga; and ice sheets and ice caps. In some cases the whole zone is composed of critical landscapes (e.g. savanna and permafrost-dominated taiga), but in other cases the hotspots are located around the boundaries of the zone (zones of transition).

The mid-latitude temperate forest lands of Europe have been deliberately excluded for two reasons: firstly because the biome has been so intensively modified by human activity that climate change will probably produce only marginal geomorphological change and secondly because the azonal chapters include good coverage of this biome.

The final Chapter 15 sums up the key conclusions on landscape change trajectories, both positive and negative; provides warnings about special concerns; lists opportunities for landscape enhancement; and, where appropriate, makes recommendations for action by policy-makers and concerned citizens. It also considers the relation between this book and the IPCC Fourth Assessment Reports and recommends a closer integration of future landscape change discussions with the anticipated IPCC Fifth Assessment.

APPENDIX 1.1 The IPCC scenarios

Assumptions contained within the IPCC scenarios (Table 1.3) are as follows:

A1. The A1 storyline and scenario family describes a future world of very rapid economic growth, global population that peaks in mid-century and declines thereafter, and the rapid introduction of new and more efficient technologies. Major underlying themes are convergence among regions, capacity building and increased cultural and social interactions, with a substantial reduction in regional differences in per capita income. The A1 scenario family develops into three

TABLE 1.3. *Description of SRES emissions scenarios*

Scenario	Description
A1	Globally integrated world with rapid economic growth A1FI – Continued dependence on fossil fuels A1B – Balance of fossil and renewable energy sources A1T – Shift to renewable energy sources
A2	A world of independently operating self-reliant nations
B1	Globally integrated but with ecologically friendly growth
B2	Emphasis on local solutions to environmental problems

Source: After IPCC (1996b).

groups that describe alternative directions of technological change in the energy system. The three A1 groups are distinguished by their technological emphasis: fossil intensive (A1FI), non-fossil energy sources (A1T), or a balance across all sources (A1B) (where balanced is defined as not relying too heavily on one particular energy source, on the assumption that similar improvement rates apply to all energy supply and end use technologies).

A2. The A2 storyline and scenario family describes a very heterogeneous world. The underlying theme is self-reliance and preservation of local identities. Fertility patterns across regions converge very slowly, which results in continuously increasing population. Economic development is primarily regionally oriented and per capita economic growth and technological change more fragmented and slower than other storylines.

B1. The B1 storyline and scenario family describes a convergent world with the same global population, that peaks in mid-century and declines thereafter, as in the A1 storyline, but with rapid change in economic structures toward a service and information economy, with reductions in material intensity and the introduction of clean and resource-efficient technologies. The emphasis is on global solutions to economic, social and environmental sustainability, including improved equity, but without additional climate initiatives.

B2. The B2 storyline and scenario family describes a world in which the emphasis is on local solutions to economic, social and environmental sustainability. It is a world with continuously increasing global population, at a rate lower than A2, intermediate levels of economic development, and less rapid and more diverse technological change than in the B1 and A1 storylines. While the scenario is also oriented towards environmental protection and social equity, it focusses on local and regional levels.

How the IPCC process deals with uncertainty

The question of uncertainty in prediction is discussed explicitly in each of the IPCC reports (IPCC, 2001b, Box 1–1, p. 79) and is reported in two different ways:

(a) a five-point confidence scale is used to assign confidence levels to many of the conclusions. The confidence levels are stated as probabilities, meaning that they represent the degree of belief among the authors of the report in the validity of a conclusion, based on their collective expert judgement of all observational evidence, modelling results and theory currently available. The confidence levels are: very high (>95%); high (67–95%); medium (33–67%); low (5–33%); and very low (<5%).

(b) Qualitative 'state of knowledge' descriptors, labelled: well-established; established but incomplete; competing explanations; speculative descriptors are used when a quantitative scale seems inappropriate.

All the following initiatives (Appendices 1.2–1.5) have started with the same 'storylines' by envisaging six different future world contexts as identified in Appendix 1.1.

APPENDIX 1.2
Global Environmental Outlook scenarios to 2032 (GEO-3: see UNEP, 2002) and the fourth Global Environmental Outlook: environment for development (GEO-4)

The Global Environmental Outlook process has over the past 10 years (1997–2007) produced a series of global integrated environmental assessment reports aimed at providing comprehensive, reliable, scientifically credible and policy relevant assessments on the interaction between environment and society. GEO-1 (1997), GEO-2 (1999), GEO-3 (2002) were UNEP's three previous global assessments and GEO-4 appeared in 2007. Special attention is paid to the role and impact of the environment on human well-being and vulnerability. There are five themes:

(a) Present global and regional issues in the context of the development of international environmental governance;
(b) State and trends of the global environment analysing human drivers and the impact of natural phenomena on

the environment, the consequences of environmental change for ecosystem services and human well-being;

(c) Interlinkages between major environmental challenges and their consequences for policy and technology response options;

(d) Cross-cutting issues on how environment can contribute to sustainable development goals and how environmental degradation impedes progress; and

(e) Global and sub-global outlooks, including short-term (to 2015) and medium-term (up to 2050) scenarios for the major environmental issues.

APPENDIX 1.3 The Millennium Ecosystem Assessment scenarios to 2100

The Millennium Ecosystem Assessment (2005) was carried out between 2002 and 2005 to assess the consequences of ecosystem change for human well-being and to establish the scientific basis for actions needed to enhance the conservation and sustainable use of ecosystems and their contributions to human well-being. The Assessment responds to government requests for information received through four global conventions: the Convention on Biological Diversity, the UN Convention to Combat Desertification, the Ramsar Convention on Wetlands and the Convention on Migratory Species.

The underlying assumption of this Assessment is that everyone in the world depends completely on Earth's ecosystems and the services they provide, such as food, water, disease management, climate regulation, spiritual fulfilment and aesthetic enjoyment. Over the past 50 years, humans have changed these ecosystems more rapidly and extensively than in any comparable period of time in human history, largely to meet rapidly growing demands for food, fresh water, timber, fibre and fuel. This transformation of the planet has contributed substantial net gains in human well-being and economic development. But not all regions have benefited from this process. In fact, most have been harmed or at least rendered more vulnerable to further change and to reduction in material productivity. Moreover, the full costs associated with these gains are only now becoming apparent.

The Assessment focusses on the linkages between ecosystems and human well-being and in particular on 'ecosystem services'. Ecosystem services are the benefits people obtain from ecosystems. In sum, these services are quite similar to those provided by geomorphology as discussed in the previous section. In showing how ecosystems influence human well-being, human well-being is defined to include basic material for a good life, such as: secure and adequate livelihoods, enough food at all times, shelter, clothing and access to goods; health, clean air and access to clean water; good social relations; security, including secure access to natural and other resources; and freedom of choice and action, including the opportunity to achieve what an individual values doing and being.

Five overarching questions guided the issues that were assessed: What are the current condition and trends of ecosystems, ecosystem services and human well-being? What are future plausible changes in ecosystems and their ecosystem services and the consequent changes in human well-being? What can be done to enhance well-being and conserve ecosystems? What are the strengths and weaknesses of response options that can be considered to realise or avoid specific futures? What are the key uncertainties that hinder effective decision-making concerning ecosystems? What tools and methodologies developed and used in the Assessment can strengthen capacity to assess ecosystems, the services they provide, their impacts on human well-being and the strengths and weaknesses of response options?

APPENDIX 1.4 The Land Use and Land Cover Change (LUCC)Project

Land use and land cover change is an important component of global environmental change. The International Geosphere–Biosphere Programme (IGBP) and the International Human Dimensions Programme on Global Environmental Change (IHDP) commissioned a Core Project Planning Committee/Research Programme Planning Committee for Land Use and Land Cover Change (CPPC/RPPC/LUCC), working from 2000 to 2005.

The main topics covered areas of deforestation or forest degradation over the last 20 years (1980–2000); the main areas of land degradation in the drylands and hyper-arid zones (1980–2000); the main areas of change in cropland extent (1980–90); the main areas of change in urban extent (1990–2000); and the fire events with long-term impact on land cover. There were three research foci: land use dynamics; land cover dynamics; and regional and global models.

In response to the question 'How has land cover been changed by human use over the last 300 years?' it was noted that human activities have transformed Earth's landscapes for a long time. The pace and intensity of land cover change increased rapidly over the last three centuries and accelerated over the last three decades. Rapid land cover changes are clustered on forest edges and along transportation networks, mostly in humid forests. Decreases in croplands in temperate regions and expansion of forest lands display opposite trends from those seen in the tropics, where croplands are expanding and deforestation is marked. Land cover change data sets are inadequate for many parts of the world, but dryland degradation, soil degradation in croplands, wetland drainage and urban expansion are some of the most common changes.

In response to the question 'What are the major human causes of land cover change?' the most important finding was that the mix of driving forces of land use change varies in time and space, according to specific human–environment conditions. A distinction is made between decadal timescales, influenced by individual and social responses to changing economic conditions, mediated by institutions and a centennial timescale, dominated by demographic trends.

In response to the question 'How will changes in land use affect land cover in the next 50–100 years?' a number of predictive models were developed and tested against past land use patterns, but on the whole a regional approach was preferred to global modelling. Urbanisation was seen to be the dominant factor in future land use patterns.

In response to the question 'How do human and biophysical dynamics affect the coupled human–environment system?' attention was directed to historical and contemporary examples of land use transitions associated with societal and biophysical changes. Finally, in response to the question 'How might changes in climate and biogeochemistry affect both land cover and land use and vice versa?' it was noted that slow land cover conversion takes place against a background of high-frequency regional-scale fluctuations in land cover conditions caused by harvest cycles and climatic variability. Abrupt short-term ecosystem changes are often caused by the interaction of climatic and land use factors.

APPENDIX 1.5
World Heritage Sites, the World Conservation Union (IUCN) and UNEP's Global Programme of Action

Global Scenarios Group (GSG) Scenarios to 2050 (Raskin *et al.*, 2002) have incorporated ecological and land use variables. World Heritage sites have been established under the Convention concerning natural and cultural sites (1972). Roughly 20% of the 851 World Heritage Sites have been selected on the basis of 'natural' criteria. From a world landscapes perspective, the most interesting part of this process is the identification of 'natural sites' that are threatened or are particularly unique and worthy of preservation (Fig. 1.18).

The IUCN, operating within the Convention on Biodiversity, has established a Red List of threatened species and ecosystems. As always, improved governance mechanisms, capacity-building and awareness raising and improved financing are the broad policy recommendations. But within those broad parameters, basin-wide pollution treatment and basin management of land use and better allocation of resources, with particular focus on the effects on coastal areas are emphasised (in cooperation with UNEP's Global Programme of Action). In summary, the most critical factors that influence land and land use change can be divided into indirect and direct drivers.

Indirect drivers include such broadly contextual factors as:

(a) demographic change;
(b) economic, sociopolitical, cultural and religious drivers;
(c) science and technology;
(d) global trade which magnifies the importance of governance and management practices; and
(e) urban growth, and more particularly suburban sprawl.

Direct drivers include:

(a) deliberate habitat change resulting from land use change;
(b) physical modification of rivers;
(c) water withdrawal from rivers;
(d) pollution;
(e) sea level change; and
(f) climate change.

References

ACIA (2005). *Arctic Climate Impact Assessment*. Cambridge: Cambridge University Press.

Adger, W. N., Brown, K. and Hulme, M. (2005). Redefining global environmental change. *Global Environmental Change A*, **15**, 1–4.

Akagi, M. (1973) *Sabo Works in Japan*. Tokyo: The National River Conservation-Sabo Society.

Alexander, R. B. *et al.* (2008). Differences in phosphorus and nitrogen delivery to the Gulf of Mexico from the Mississippi River basin. *Environmental Science and Technology*, **42**, 822–830.

Allan, J. C. and Komar, P. D. (2006). Climate controls on US West Coast erosion processes. *Journal of Coastal Research*, **22**, 511–529.

Allen, M. R. and Ingram, W. J. (2002). Constraints on future changes in climate and the hydrological cycle. *Nature*, **419**, 224–232.

Andreae, M. O., Jones, C. D. and Cox, P. M. (2005). Strong present-day aerosol cooling implies a hot future. *Nature*, **435**, 1187–1190.

Arnell, N. (2002). *Hydrology and Global Environmental Change*. Harlow: Addison-Wesley Longman.

Arrhenius, S. (1896). On the influence of carbonic acid in the air upon the temperature of the ground. *Philosophical Magazine*, **41**, 237–275.

Ballantyne, C. K. (2002). Paraglacial geomorphology. *Quaternary Science Reviews*, **21**, 1935–2017.

Beckinsale, R. P. and Chorley, R. J. (1991). *The History of the Study of Landforms*, vol. 3. London: Routledge.

Berger, A. R. and Iams, W. J., eds. (1996). *Geoindicators: Assessing Rapid Environmental Change in Earth Systems*. Rotterdam: Balkema.

Berner, E. K. and Berner, R. A. (1996). *Global Environment, Water, Air and Geochemical Cycles*. Upper Saddle River: Prentice-Hall.

Betts, R. A., Cox, P. M. and Woodward, F. I. (2000). Simulated responses of potential vegetation to doubled-CO_2 climate change and feedbacks on near-surface temperature. *Global Ecology and Biogeography*, **9**, 171–180.

Bobrovitskaya, N. N., Kokorev, A. V. and Lemeshko, N. A. (2003). Regional patterns in recent trends in sediment yields of Eurasian and Siberian rivers. *Global and Planetary Change*, **39**, 127–146.

Brown, R. D. (2000). Northern hemisphere snow cover variability and change, 1915–1997. *Journal of Climate*, **13**, 2339–2355.

Brunsden, D. (1980). Applicable models of long term landform evolution. *Zeitschrift für Geomorphologie, Supplementband*, **36**, 16–26.

Brunsden, D. (1993). The persistence of landforms. *Zeitschrift für Geomorphologie, Supplementband*, **93**, 13–28.

Brunsden, D. (2001). A critical assessment of the sensitivity concept in geomorphology. *Catena*, **42**, 99–123.

Brunsden, D. and Thornes, J. B. (1979). Landscape sensitivity and change. *Transactions of the Institute of British Geographers (NS)*, **4**, 463–484.

Brutsaert, W. and Parlange, M. B. (1998). Hydrological cycle explains the evaporation paradox. *Nature*, **396**, 30.

Burroughs, W. J. (2005). *Climate Change in Prehistory*. Cambridge: Cambridge University Press.

Cammeraat, L. H. (2002). A review of two strongly contrasting geomorphological systems within the context of scale. *Earth Surface Processes and Landforms*, **27**, 1201–1222.

Camoin, G. F., Montaggioni, L. F. and Braithwaite, C. J. R. (2004). Late glacial to post glacial sea levels in the Western Indian Ocean. *Marine Geology*, **206**, 119–146.

Cayan, D. R. *et al.* (2001). Changes in the onset of spring in the western United States. *Bulletin of the American Meteorological Society*, **82**, 399–415.

Chorley, R. J. (1962). *Geomorphology and General Systems Theory*, Professional Paper 500-B. Washington DC: US Geological Survey.

Chorley, R. J., Schumm, S. A. and Sugden, D. E. (1984). *Geomorphology*. London: Methuen.

Church, J. A. and White, N. J. (2006). A 20th century acceleration in global sea-level rise. *Geophysical Research Letters*, **33**, L01602.

Church, J. A. *et al.* (2004). Estimates of the regional distribution of sea level rise over the 1950–2000 period. *Journal of Climate*, **17**, 2609–2625.

Church, M. (1996). Space, time and the mountain: how do we order what we see? In B. L. Rhoads and C. E. Thorn, eds., *The Scientific Nature of Geomorphology*. Chichester: John Wiley, pp. 147–170.

Church, M. *et al.* (1999). Fluvial clastic sediment yield in Canada: a scaled analysis. *Canadian Journal of Earth Sciences*, **36**, 1267–1280.

Clark, J. A., Farrell, W. E. and Peltier, W. R. (1978). Global change in post glacial sea level: a numerical calculation. *Quaternary Research*, **9**, 265–287.

CLIMAP Project Members (1976). The surface of the ice-age Earth. *Science*, **191**, 1131–1144.

Cowell, P. J. *et al.* (2003). The coastal tract: 1. A conceptual approach to aggregated modelling of low order coastal change. *Journal of Coastal Research*, **19**, 812–827.

Dai, A. and Trenberth, K. E. (2002). Estimates of freshwater discharge from continents: latitudinal and seasonal variations. *Journal of Hydrometeorology*, **3**, 660–687.

Dai, A., Fung, I. Y. and Del Genio, A. D. (1997), Surface observed global land precipitation variations during 1900–88. *Journal of Climate*, **10**, 2943–2962.

Dearing, J. A. and Jones, R. T. (2003). Coupling temporal and spatial dimensions of global sediment flux through lake and marine sediment records. *Global and Planetary Change*, **39**, 147–168.

Diamond, J. (2005). *Collapse: How Societies Choose to Fail or Succeed*. London: Penguin.

Dietrich, W. E. and Dunne, T. (1978). Sediment budget for a small catchment in mountainous terrain. *Zeitschrift für Geomorphologie, Supplementband*, **29**, 191–206.

Dowlatabadi, H. (2002). Global change: much more than a matter of degrees. *Meridian (Canadian Polar Commission, Ottawa)*, **Spring/Summer**, 8–12.

EPICA (European Project for Ice Coring in Antarctica) (2004). Eight glacial cycles from an Antarctic ice core. *Nature*, **429**, 623–628.

Fairbridge, R. W. (1968). *The Encyclopedia of Geomorphology*. New York: Reinhold.

Falloon, P. D. and Betts, R. A. (2006). The impact of climate change on global river flow in HadGEM1 simulations. *Atmospheric Science Letters*, **7**, 62–68.

Fitzgerald, D. M. *et al.* (2008). Coastal impacts due to sea level rise. *Annual Review of Earth and Planetary Sciences*, **36**, 601–647.

Freeze, R. A. (1974). Streamflow generation. *Reviews of Geophysics and Space Physics*, **12**, 627–647.

Gedney, N. *et al.* (2006). Detection of a direct carbon dioxide effect in continental river runoff records. *Nature*, **439**, 835–838.

Glade, T. (2003). Landslide occurrence as a response to land use change: a review of evidence from New Zealand. *Catena*, **51**, 297–314.

Goolsby, D. A. (2000). Mississippi Basin nitrogen flux believed to cause gulf hypoxia. *EOS, Transactions of the American Geophysical Union*, **321**, 326–327.

Goudie, A. S., ed. (1990). *Geomorphological Techniques*. London: Routledge.

Goudie, A. S. (1996). Geomorphological 'hotspots' and global warming. *Interdisciplinary Science Reviews*, **21**, 253–259.

Goudie, A., ed. (1997). *The Human Impact Reader*. Oxford: Blackwell.

Grafenstein, U. von *et al.* (1999). A mid-European decadal isotope climate record from 15 500–5000 years BP. *Science*, **284**, 1654–1657.

Groisman, P. Y. *et al.* (2004). Contemporary changes of the hydrological cycle over the contiguous United States: trends derived from *in situ* observations. *Journal of Hydrometeorology*, **5**, 64–85.

Gunderson, L. H. and Holling, C. S., eds. (2002). *Panarchy: Understanding Transformations in Human and Natural Systems*. Washington, DC: Island Press.

Hannaford, J. and Marsh, T. (2006). High flow and flood trends in a network of undisturbed catchments in the UK. *International Journal of Climatology*, **28**, 1325–1338.

Hewitt, K. *et al.*, eds. (2002). *Landscapes of Transition: Landform Assemblages and Transformations in Cold Regions*. Dordrecht: Kluwer.

Higgitt, D. L. and Lee, E. M., eds. (2001). *Geomorphological Processes and Landscape Change: Britain in the Last 1000 Years*. Oxford: Blackwell.

Hodgkins, G. A., Dudley, R. W. and Huntington, T. G. (2003). Changes in the timing of high river flows in New England over the 20th century. *Journal of Hydrology*, **278**, 244–252.

Holling, C. S. (1986). The resilience of terrestrial systems: local surprise and global change. In W. C. Clark and R. Munn, eds., *Sustainable Development of the Biosphere*. Cambridge: Cambridge University Press, pp. 292–317.

Holling, C. S. (2001). Understanding the complexity of economic, ecologic and social systems. *Ecosystems*, **4**, 390–405.

Hufschmidt, G., Crozier, M. and Glade, T. (2005). Evolution of natural risk: research framework and perspectives. *Natural Hazards and Earth Systems Sciences*, **5**, 375–382.

Hulme, M., Osborn, T. J. and Johns, T. C. (1998). Precipitation sensitivity to global warming: comparisons of observations with HadCM2 simulations. *Geophysical Research Letters*, **25**, 3379–3382.

Humboldt, A. von (1845–1862). *Kosmos. Entwurf einer physischen Weltbeschreibung*. Bd. 1–5. Stuttgart.

Huntington, T. (2006). Evidence for intensification of the global water cycle: review and synthesis. *Journal of Hydrology*, **319**, 83–95.

Inman, D. L. and Jenkins, S. A. (1999). Climate change and the episodicity of sediment flux of small California rivers. *Journal of Geology*, **107**, 251–270.

IPCC (1996a). *Climate Change 1995: The Science of Climate Change. Contribution of Working Group I to the Second Assessment Report of the Intergovernmental Panel on Climate Change*. Houghton, J. T. *et al.*, eds. Cambridge: Cambridge University Press.

IPCC (1996b). *Climate Change 1995: Impacts, Adaptations and Mitigation of Climate Change. Contribution of Working Group II to the Second Assessment Report of the Intergovernmental Panel on Climate Change*. Watson, R. T. *et al.*, eds. Cambridge: Cambridge University Press. (pp. 75–92, Nakicenovic, N., ed.).

IPCC (2000). *Special Report on Emissions Scenarios. A Special Report of Working Group III of the Intergovernmental Panel on Climate Change*. Nakicenovic, N. *et al.*, eds. Cambridge and New York: Cambridge University Press.

IPCC (2001a). *Climate Change 2001: The Scientific Basis. Contribution of Working Group I to the Third Assessment Report of the Intergovernmental Panel on Climate Change*. Houghton, J. T. *et al.*, eds. Cambridge: Cambridge University Press.

IPCC (2001b). *Climate Change 2001: Impacts, Adaptations and Vulnerability. Contribution of Working Group II to the Third Assessment Report of the Intergovernmental Panel on Climate Change*. McCarthy, J. J. *et al.*, eds. Cambridge: Cambridge University Press. (pp. 75–103, Schneider, S. H. *et al.*, eds.; pp. 191–233, Arnell, N. *et al.*, eds.)

IPCC (2007a). *Climate Change 2007: The Physical Science Basis. Contribution of Working Group I to the Fourth Assessment Report of the Intergovernmental Panel on Climate Change*. Solomon, S. *et al.*, eds. Cambridge: Cambridge University Press. (pp. 1–18, Alley, R. B. *et al.*, eds.; pp. 237–336, Trenberth, K. E. *et al.*, eds.; pp. 339–383, Lemke, P. *et al.*, eds.; pp. 385–432, Bindoff, N. L. *et al.*, eds.; pp. 747–846, Meehl, G. A., *et al.*, eds.)

IPCC (2007b). *Climate Change 2007: Impacts, Adaptations and Vulnerability. Contribution of Working Group II to the Fourth Assessment Report of the Intergovernmental Panel on Climate Change*. Parry, M. L. *et al.*, eds. Cambridge: Cambridge University Press. (pp. 175–210, Kunzewicz, Z. W. *et al.*, eds.; pp. 317–356, Nicholls, R. J. *et al.*, eds.; pp. 543–580, Alcamo, J. *et al.*, eds.; pp. 655–685, Anisimov, O. A. *et al.*, eds.; pp. 689–716, Mimura, N. *et al.*, eds.; pp. 781–810, Schneider, S. H. *et al.*, eds.)

Jäckli, H. (1957). Gegenwartsgeologie des bündnerischen Rheingebietes: Ein Beitrag zur exogenen Dynamik alpiner Gebirgslandschaften (Exogene dynamics of an alpine landscape). *Beiträge zur Geologie der Schweiz, Geotechnische Serie*, **36**.

Johns, T. C. *et al.* (2006). The new Hadley Centre climate model HadGEM1: evaluation of coupled simulations. *Journal of Climate*, **19**, 1327–1353.

Kareiva, P. *et al.* (2007). Domesticated nature: shaping landscapes and ecosystems for human welfare. *Science*, **316**, 1866–1869.

Kennedy, B. A. (1991). Trompe l'oeil. *Journal of Biogeography*, **27**, 37–38.

Komar, P. D. (1998). *Beach Processes and Sedimentation*, 2nd edn. Upper Saddle River: Prentice-Hall.

Komar, P. D. and Allan, J. C. (2007). Higher waves along US coast linked to hurricanes. *EOS, Transactions of the American Geophysical Union*, **88**, 301.

Köppen, W. (1901). Versuch einer Klassifikation der Klimate vorzugsweise nach ihren Beziehungen zur Pflanzenwelt. *Geographische Zeitschrift*, **6**, 593–611, 657–679.

Kundzewicz, Z. W. *et al.* (2005). Trend detection in river flow series: 1. Annual maximum flow. *Hydrological Sciences Journal*, **50**, 797–810.

Labat, D. *et al.* (2004). Evidence for global runoff increase related to climate warming. *Advances in Water Resources*, **27**, 631–642.

Lake, P. S. *et al.* (2000). Global change and the biodiversity of freshwater ecosystems: impacts on linkages between above-sediment and sediment biota. *BioScience*, **50**, 1099–1107.

Lambin, E. F. and Geist, H. J., eds. (2006). *Land-Use and Land-Cover Change*. Berlin: Springer-Verlag.

Leopold, A. (1949). *A Sand County Almanac*. Oxford: Oxford University Press.

Lins, H. F. and Slack, J. R. (1999). Streamflow trends in the United States. *Geophysical Research Letters*, **26**, 227–230.

Lyman, J. M., Willis, J. K. and Johnson, G. C. (2006). Recent cooling of the upper ocean. *Geophysical Research Letters*, **33**, L18604.

Martonne, E. de (1913). Le climat facteur du relief. *Scientia*, **13**, 339–355.

Meehl, G. A. *et al.*, eds. (2007) Global climate projections. In *Climate Change 2007: The Physical Science Basis. Contribution of Working Group I to the Fourth Assessment Report of the Intergovernmental Panel on Climate Change*. Solomon, S. *et al.*, eds. Cambridge: Cambridge University Press, pp. 747–846.

Menon, S. *et al.* (2002). Climate effects of black carbon aerosols in China and India. *Science*, **297**, 2250–2253.

Messerli, B. *et al.* (2000). From nature-dominated to human-dominated environmental changes. In K. D. Alverson, F. Oldfield and R. S. Bradley, eds., *Past Global Changes and their Significance for the Future*. Amsterdam: Elsevier, pp. 459–479.

Meybeck, M. and Vörösmarty, C. (2005). Fluvial filtering of land-to-ocean fluxes: from natural Holocene variations to Anthropocene. *Comptes Rendus Geosciences*, **337**, 107–123.

Millennium Ecosystem Assessment (2003). *Ecosystems and Human Well-Being: A Framework for Assessment*. Washington, DC: Island Press.

Milliman, J. D. and Syvitski, J. P. M. (1992). Geomorphic/tectonic control of sediment discharge to the ocean: the importance of small, mountainous rivers. *Journal of Geology*, **100**, 525–544.

Milly, P. C. D. *et al.* (2002). Increasing risk of great floods in a changing climate. *Nature*, **415**, 514–517.

Montgomery, D. R. (2007). *Dirt: The Erosion of Civilizations*. Berkeley: University of California Press.

Moor, J. J. W. de *et al.* (2008). Human and climate impact on catchment development during the Holocene: the Geul River, the Netherlands. *Geomorphology*, **98**, 316–339.

Mudelsee, M. *et al.* (2003). No upward trend in the occurrence of extreme floods in central Europe. *Nature*, **425**, 166–169.

Myers, N. *et al.* (2000). Biodiversity hotspots for conservation priorities. *Nature*, **403**, 853–858.

Naqvi, S. W. A. *et al.* (2000). Increased marine production of nitrous oxide due to intensifying anoxia on the Indian continental shelf. *Nature*, **408**, 346–349.

National Geographic Society (2005). *National Geographic Atlas of the World*, 8th edn. Washington, DC: National Geographic Society.

Nilsson, C. *et al.* (2005). Fragmentation and flow regulation of the world's large river systems. *Science*, **308**, 405–408.

Nohara, D. *et al.* (2006). Impact of climate change on river discharge projected by multimodel ensemble. *Journal of Hydrometeorology*, **7**, 1076–1089.

O'Halloran, D. *et al.* (1994). *Geological and Landscape Conservation*. London: The Geological Society.

Penman, H. L. (1948). Natural evaporation from open water, bare soil and grass. *Proceedings of the Royal Society of London A*, **192**, 120–145.

Petit, J. R. *et al.* (1999). Climate and atmospheric history of the past 420 000 years from the Vostok ice core, Antarctica. *Nature*, **399**, 429–436.

Petit-Maire, N. (1999). Natural variability of the Earth's environments: the last two climatic extremes (18 000 ± 2000 and 8000 ± 1000 years BP). *Earth and Planetary Sciences*, **328**, 273–279.

Petit-Maire, N. and Bouysse, Ph. (1999). *Maps of World Environments during the Last Two Climatic Extremes*. Paris: Commission for the Geological Map of the World (CGMW) and the Agence Nationale pour la Gestion des Dechets Radioactifs (ANDRA).

Philander, S. G. H. (1990). *El Niño, La Niña, and the Southern Oscillation*. San Diego: Academic Press.

Phillips, J. D. (1999). *Earth Surface Systems*. Oxford: Blackwell.

Phillips, J. D. (2003). Sources of nonlinearity and complexity in geomorphic systems. *Progress in Physical Geography*, **27**, 1–23.

Pielke, R. A. (2005). Land use and climate change. *Science*, **310**, 1625–1626.

Pirazzoli, P. A. (1991). *World Atlas of Holocene Sea-Level Changes*. Amsterdam: Elsevier.

Probst, J. L. and Tardy, Y. (1987). Long range streamflow and world continental runoff fluctuations since the beginning of this century. *Journal of Hydrology*, **94**, 289–311.

Rahmstorf, S. (2007). A semi-empirical approach to projecting future sea-level rise. *Science*, **315**, 368–370.

Ramanathan, V. *et al.* (2001). Aerosols, climate, and the hydrological cycle. *Science*, **294**, 2119–2124.

Rapp, A. (1960). Recent development of mountain slopes in Karkevagge and surroundings, northern Scandinavia. *Geografiska Annaler*, **42**A, 73–200.

Raskin, P. *et al.* (2002). *Great Transition: The Promise and Lure of the Times Ahead*, a report of the Global Scenario Group. Boston: Stockholm Environment Institute.

Rees, W. E. (1992). Ecological footprints and appropriated carrying capacity: what urban economics leaves out. *Environment and Urbanisation*, **4**, 121–130.

Reid, L. M. and Dunne, T. (1996). *Rapid Evaluation of Sediment Budgets*. Reiskirchen: Catena.

Reynolds, J. F. *et al.* (2007). Global desertification: building a science for dryland development. *Science*, **316**, 847–851.

Roberts, G. and Crane, S. B. (1997). The effects of clear-felling established forestry on stream-flow losses from the Hore sub-catchment. *Hydrology and Earth System Sciences*, **1**, 477–482.

Robinson, J. (2004). Squaring the circle? Some thoughts on the idea of sustainable development. *Ecological Economics*, **48**, 369–384.

Robinson, M. *et al.* (2000). Land use change. In M. Acreman, ed., *The Hydrology of the UK: A Study of Change*. London: Routledge, pp. 30–54.

Roderick, M. L. and Farquhar, G. D. (2002). The cause of decreased pan evaporation over the past 50 years. *Science*, **298**, 1410–1411.

Rotmans, J. and Dowlatabadi, H. (1998). Integrated assessment modelling. In S. Rayner and E. Malone, eds., *Human Choice and Climate Change*, vol. 3. Columbus: Battelle, pp. 291–378.

Ruddiman, W. F. (2005). How did humans first alter global climate? *Scientific American*, **292**(3), 46–53.

Scherr, S. J. (1999). *Soil Degradation: A Threat to Developing-Country Food Security by 2020?* Food, Agriculture and the Environment Discussion Paper No. 27. Washington, DC: International Food Policy Research Institute.

Schumm, S. A. (1973). Geomorphic thresholds and complex response of drainage systems. In M. Morisawa, ed., *Fluvial Geomorphology*. New York: Binghamton, pp. 299–310.

Schumm, S. A. (1977). *The Fluvial System*. New York: John Wiley.

Schumm, S. A. (1979). Geomorphic thresholds: the concept and its applications. *Transactions of the Institute of British Geographers (NS)*, **4**, 485–515.

Schumm, S. A. (1985). Explanation and extrapolation in geomorphology: seven reasons for geologic uncertainty. *Transactions of the Japanese Geomorphological Union*, **6**, 1–18.

Schumm, S. A. (1991). *To Interpret the Earth: Ten Ways to be Wrong*. Cambridge: Cambridge University Press.

Schumm, S. A. and Lichty, R. W. (1965). Time, space and causality in geomorphology. *American Journal of Science*, **273**, 110–119.

Shennan, I. and Horton, B. (2002). Holocene land- and sea-level changes in Great Britain. *Journal of Quaternary Science*, **17**, 511–526.

Slaymaker, O. and Spencer, T. (1998). *Physical Geography and Global Environmental Change*. Harlow: Addison-Wesley Longman.

Smil, V. (2001). *Enriching the Earth: Fritz Haber, Carl Bosch and the Transformation of World Food Production*. Cambridge: MIT Press.

Smith, L. C. (2000). Trends in Russian Arctic river ice formation and breakup, 1917–1994. *Physical Geography*, **21**, 46–56.

Spencer, T. (1990). Tectonic and environmental histories in the Pitcairn Group, Palaeogene to present: reconstructions and speculations. *Atoll Research Bulletin*, **322**, 1–41.

Starkel, L. A. (1985). The reflection of Holocene climatic variations in the slope and fluvial deposits and forms in the European mountains. *Ecologia Mediterranea*, **11**, 91–98.

Steel, C. (2008). *Hungry City: How Food Changes Our Lives*. London: Chatto & Windus.

Steffen, W. *et al.*, eds. (2004). *Global Change and the Earth System: A Planet under Pressure*. Berlin: Springer-Verlag.

Stern, N. (2007). *The Economics of Climate Change: The Stern Review*. Cambridge: Cambridge University Press.

Strahler, A. H. and Strahler, A. N. (1994). *Introducing Physical Geography*. New York: John Wiley.

Strakhov, N. M. (1967). *Principles of Lithogenesis*, vol. 1. New York: Consultants Bureau.

Summerfield, M. A. and Hulton, N. J. (1994). Natural controls of fluvial denudation rates in major world drainage basins. *Journal of Geophysical Research*, **99**, 13 871–13 883.

Svensson, C. *et al.* (2006). Trends in river floods: why is there no clear signal? In I. Tchiguirinskaia, K. N. N. Thein and Hubert, P., eds., *Frontiers in Flood Research*, IAHS Publication No. **305**. Wallingford: IAHS Press, pp. 1–18.

Syvitski, J. P. M. *et al.* (2005). Impact of humans on the flux of sediment to the global coastal ocean. *Science*, **308**, 376–380.

Taylor, J. A., Murdock, A. P. and Pontee, N. I. (2004). A macro-scale analysis of coastal steepening around the coast of England and Wales. *Geographical Journal*, **170**, 179–188.

Toscano, M. A. and Macintyre, I. G. (2003). Corrected western Atlantic sea-level curve for the last 11 000 years based on calibrated ^{14}C dates from *Acropora palmata* and mangrove intertidal peat. *Coral Reefs*, **22**, 257–270.

Trimble, S. W. (1995). Catchment sediment budgets and change. In A. Gurnell and G. Petts, eds., *Changing River Channels*. Chichester: John Wiley, pp. 201–215.

Trimble, S. W. and Crosson, P. (2000). US soil erosion rates: myth and reality. *Science*, **289**, 248–250.

Troendle, C. A. and Reuss, J. O. (1997). Effect of clearcutting on snow accumulation and water outflow at Fraser, Colorado, *Hydrology and Earth System Sciences*, **1**, 325–332.

Turner, B. L. II *et al.* (1990a). *The Earth as Transformed by Human Action*. Cambridge: Cambridge University Press.

Turner, B. L. II *et al.* (1990b). Two types of global environmental change: definitional and spatial scale issues in their human dimensions. *Global Environmental Change*, **1**, 14–22.

Turner, R. E., Rabalais, N. N. and Justic, D. (2008). Gulf of Mexico hypoxia: alternate states and a legacy. *Environmental Science and Technology*, **42**, 2323–2327.

United Nations Population Fund (2007). *State of World Population 2007: Unleashing the Potential of Urban Growth*. New York: UN Population Fund.

Veron, J. E. N. (1995). *Corals in Space and Time: The Biogeography and Evolution of the Scleractinia*. Ithaca: Cornell University Press.

Vörösmarty, C. J. and Sahagian, D. 2000 Anthropogenic disturbance of the terrestrial water cycle. *BioScience*, **50**, 753–765.

Walling, D. E. (2006). Human impact on land–ocean sediment transfer by the world's rivers. *Geomorphology*, **79**, 192–216.

Wang, H. *et al.* (2007). Stepwise decreases of the Huanghe (Yellow River) sediment load (1950–2005): impacts of climate change and human activities. *Global and Planetary Change*, **57**, 331–354.

Wasson, R. J. (1994). Living with the past: uses of history for understanding landscape change and degradation. *Land Degradation and Rehabilitation*, **5**, 79–87.

White, N. J., Church, J. A. and Gregory, J. M. (2005). Coastal and global averaged sea level rise for 1950 to 2000. *Geophysical Research Letters*, **32**, L01601.

Wild, M. *et al.* (2005). From dimming to brightening: decadal changes in solar radiation at Earth's surface. *Science*, **308**, 847–850.

Wilkinson, B. H. and McElroy, B. J. (2007). The impact of humans on continental erosion and sedimentation. *Bulletin of the Geological Society of America*, **119**, 140–156.

Woodroffe, C. D. (2002). *Coasts: Form, Process and Evolution*. Cambridge: Cambridge University Press.

Woodroffe, C. D. (2005). Late Quaternary sea-level highstands in the central and eastern Indian Ocean: a review. *Global and Planetary Change*, **49**, 121–138.

Woodworth, P. L., Gregory, J. M. and Nicholls, R. J. (2005). Long-term sea level changes and their impacts. In A. R. Robinson and K. H. Brink, eds., *The Global Coastal Ocean: Multiscale Interdisciplinary Processes*. Cambridge: Harvard University Press, pp. 715–753.

World Commission on Economic Development (WCED) (1987). *Our Common Future*. Oxford: Oxford University Press.

Zhang, K. Q., Douglas, B. C. and Leatherman, S. P. (2004). Global warming and coastal erosion. *Climatic Change*, **64**, 41–58.

Zhang, X. *et al.* (2001). Trends in Canadian streamflow. *Water Resources Research*, **37**, 987–999.

Zhang, X. *et al.* (2007). Detection of human influence on twentieth-century precipitation trends. *Nature*, **448**, 461–465.

2 Mountains

Olav Slaymaker and Christine Embleton-Hamann

2.1 Introduction

There is greater geodiversity in mountains than in most other landscapes (Barsch and Caine, 1984). Mountain geosystems are not exceptionally fragile but they show a greater range of vulnerability to disturbance than many landscapes. Forested slopes give place to alpine tundra over short vertical distances; resistant bedrock slopes alternate with intensively cultivated soils and erodible unconsolidated sediments over short horizontal distances. Mountain systems account for roughly 20% of the terrestrial surface area of the globe.

Mountains are high and steep so that when natural hazards occur, whether seismic, volcanic, mass movements or floods, the disturbance is transmitted readily through the geosystem. When inappropriate land use changes are made, vegetation and soils are rapidly removed. Because of the steep terrain, low temperatures and the relatively thin soils, the recovery of mountain geosystems from disturbance is often slow and sometimes fails completely. Mountains provide the direct life support base for 10–20% of humankind (statistics differ on this point; see Appendix 2.1) and indirectly affect the lives of more than 50%. Because of significant elevation differences, mountains such as the Himalayas, the Andes, the Rocky Mountains and the Alps show, within short horizontal and vertical distances, climatic regimes similar to those of widely separate latitudinal belts. Because of the compressed life zones with elevation and small-scale biodiversity caused by different topoclimates, mountain systems are of prime conservation value. Körner and Ohsawa (2005) estimate that 32% of protected areas are in mountains (9345 protected areas covering about 1.7 Mkm^2).

Human well-being also depends on mountain geodiversity and biodiversity. Mountain systems are especially important for the provision of clean water and the safety of settlements and transport routes depends directly on ability to cope with natural hazards. Slope stability and erosion control are also closely interdependent with a healthy and continuous vegetation cover. Key mountain resources and services include water for hydroelectricity and irrigation, flood control, agriculture, mineral resources, timber, tourism and medicinal plants. Geographically fragmented mountains also support a high ethnocultural diversity (Körner and Ohsawa, 2005). For many societies, mountains have spiritual significance, and scenic landscapes and clean air make mountains target regions for recreation and tourism.

During the past three decades, the world's population has doubled, the mountain regions' population has more than tripled and stresses on the physical and biological systems of the Earth have intensified many fold. The implications of the emergence of the human factor and the ramifications in terms of environmental degradation and enhancement in mountains have still to be fully explored. Ninety percent of the global mountain population (between 600 million and 1.2 billion people) lives in developing countries and countries in transition. Some 90–180 million mountain people, and almost everyone living above 2500 m above sea level, live in poverty and are considered to be especially vulnerable to environmental change (Huddleston, 2003). Some claim that most, if not all, of the major mountain problems and their solutions are triggered and shaped by developments outside the mountains (Price, 1999). Deforestation, accelerated erosion, overpopulation and depopulation are processes that are heavily influenced by 'indirect' drivers, such as outside socioeconomic forces. But our chief concern here is to determine the triggers of global environmental change in mountain regions. We make no claims in this volume to do more than to document the direct drivers as we understand them and to urge that the implications of these changes, both positive and negative, be incorporated into informed policy recommendations affecting twenty-first-century mountain landscapes.

Geomorphology and Global Environmental Change, eds. Olav Slaymaker, Thomas Spencer and Christine Embleton-Hamann. Published by Cambridge University Press.

The major direct drivers of environmental change in mountains are relief, hydroclimate and land use. Not only are they important in themselves but they are commonly so closely interrelated that it becomes difficult to rank their relative importance and, indeed, their status, whether dependent or independent. Precisely which of these drivers is most important in any specific mountain setting and how they should be ranked individually and in combination is a matter for research. One of the greatest challenges facing mountain scientists is to separate environmental change caused by human activities from change that would have occurred without human interference (i.e. relief and hydroclimate) (Marston, 2008). Linking cause and effect is especially difficult in mountain regions where physical processes alone can operate at exceptionally high rates. Some of the major issues in mountain landscapes are the measurement and modelling of geomorphic change; the role of mountain land use and land cover change; and the assessment of mountain landform/landscape vulnerability and sustainability.

Perhaps the most distinctive characteristic of the three drivers of change is the temporal scale at which they operate and make their impacts on landforms and landscapes. This is especially important to note in the context of making projections for the twenty-first century and in assessing the possibilities of or necessity for remedial action. Mountain relief evolves over millions of years but may generate natural hazards within minutes that may take decades to mitigate; mountain climate has evolved over the Quaternary Period, may participate in almost instantaneous extreme weather and flood events and may also take years to mitigate. Population and land use have evolved over the Holocene Epoch; their collective impact was first seen between 5 and 8 ka BP, but this impact has dramatically increased during the past century to the point that in 1990 it was noted for the first time that human activities had now impacted more than 50% of the terrestrial globe (Turner *et al.*, 1990). Change, however, is not always negative: there is commonly a balance sheet of both positive and negative effects: deltas may be devastated by floods but will benefit in the following years from the addition of fertile soil; arid landscapes may incur costs and benefits from extra moisture; cold landscapes both lose and gain from warming; and landscapes undergoing change of land use may be either enhanced or ruined, depending on the way in which that land use change is implemented. The combination of extreme geophysical events with exceptional population growth and land use modifications underlines the urgency of better understanding of these interactions and working out the implications for adaptation to and mitigation of the effects of these drivers of change on landforms and landscapes.

2.1.1 Definition

There are two necessary conditions to define 'mountains': high gradient and high absolute elevation above sea level. Relief roughness, or local elevation range, is a useful surrogate for gradient (Plate 7). We therefore adopted a definition of mountains as land systems with both high gradient (a local elevation range (LER) greater than 300 m 5 km^{-1}) and elevation (greater than 500 m above sea level). This definition excludes large plateaus, such as Tibet. Using similar criteria (Appendix 2.1), Meybeck *et al.* (2001) generated nine global relief types, only two of which are mountain regions: (a) low to mid-elevation mountains and (b) high and very high mountains. In this chapter, we have expanded these mountain regions to four, all four of which have a LER which exceeds 300 m 5 km^{-1}. These four mountain regions are identified as:

- Class 1: Low mountains having a zonal elevation range from 500 to 1000 m above sea level. They occupy an estimated 6.3 Mkm^2 and have between 175 and 350 million inhabitants; population estimates vary by a factor of 2 (Appendix 2.1);
- Class 2: Mid-elevation mountains having a zonal elevation range from 1000 to 2500 m above sea level (*c.* 11 Mkm^2 and 290–580 million inhabitants);
- Class 3: High mountains having a zonal elevation range from 2500 to 4500 m above sea level (*c.* 3.9 Mkm^2 and 60–120 million inhabitants); and
- Class 4: Very high mountains more than 4500 m above sea level (*c.* 1.8 Mkm^2 and 4–8 million inhabitants).

(N.B. the upper estimate of population, from Meybeck *et al.* (2001), is 1420 million, which includes approximately 360 million in 'mountains' between 300 and 500 m above sea level.)

High and very high mountains (Classes 3 and 4; greater than 2500 m above sea level) are closely associated with the most recent alpine orogenesis in Europe, Asia, Australasia and the Americas, rifting and active volcanism and isostatically rebounded glaciated regions. Included are the North and South American cordilleras, the European Alps–Zagros, the Caucasus–Elburz, the Pamir-Alai–Tien-Shan and the Karakoram–Himalaya, East Africa, Hawaii, the western Pacific Rim and the north and northeast Siberian ranges.

2.1.2 Holocene climate change in mountains

It is important to recognise that mountain landscapes have a history. They have a tectonic history and a denudational history, and the relative effectiveness of these processes determines the absolute scale and the rate of change of

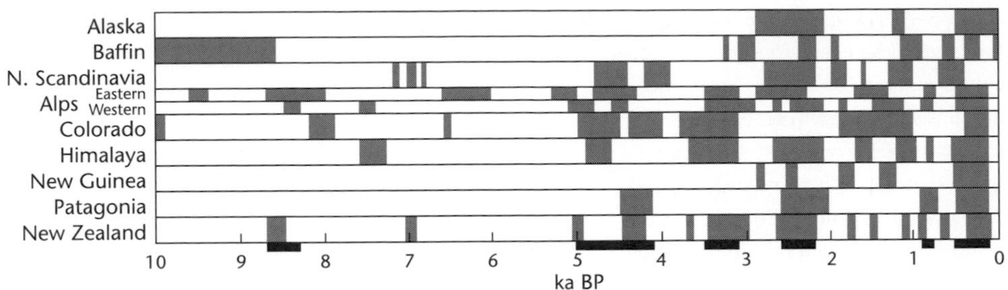

FIGURE 2.1. Holocene climate changes in the mountains. Summary of glacier expansion phases in different areas of the world during the Holocene (possible times of widespread advances are indicated by black bars on the horizontal axis) (modified from Bradley, 1999).

mountains (Schumm, 1963). Mountain landscapes also have a climatic history, and past climates leave traces of their past impacts on the contemporary landscape. This effect is most obvious in the case of alternations of glacial and interglacial climates during the Quaternary Period (see Chapter 1).

Glacier advances and historical records

Alpine glacier extent has varied over the past millennium within a range defined by the extremes of the Little Ice Age and today's reduced glacier stage. We appear to be evolving towards or even beyond the warmest phases of Holocene variability. Such a conclusion is reinforced by the discovery of the Oetztal ice man, who had been buried by snow/ice over 5000 years ago and melted out in 1991, and the discovery of well-preserved wooden bows, dated around 4 ka BP (Haeberli, 1994). Artefacts melting from glacier and ice patches have been documented from a number of mountain areas in North America (Dixon *et al.*, 2005).

There is little support for the notion that Holocene glacier fluctuations were exactly synchronous throughout the world, though attempts have been made to define episodes in which glacier advance has occurred in many regions (Fig. 2.1). Such episodes are called neoglacials. Reasons for the complexity of the mountain glacier record include regional climatic fluctuations (which are neither hemispheric nor global), poor dating control and discontinuous or incomplete data sets (Bradley, 1999). Information on cooling in mountain and upland areas during the Little Ice Age (AD 1550–1850) compared with the Medieval Warm Period are available from European countries, China and Japan (Grove, 1988). Barry (1992, 1994) and Diaz and Bradley (1997) give the best summary of data sources for high-elevation sites in Europe and globally respectively. Auer (2007) provides the richest source of historical instrumental climatological surface data for the European alpine region.

Lake sediments

Because the information from moraines and trim lines is often partial, attention has moved towards the interpretation of glacier-fed lakes, the changes in organic matter and the increase in sediment input to lakes (e.g. Batterbee, 2002). Von Grafenstein (1999) used oxygen isotope ratios of precipitation inferred from deep lake ostracods from the Ammersee (in the foothills of the Alps, southern Germany) to provide a climate record with decadal resolution. The correlation with central Greenland ice cores between 15 ka and 5 ka BP is impressive.

Palaeoecology

Pollen records from lake sediments and peat bogs have a typical resolution of 50–200 years over time intervals of 1 ka–10 ka respectively (Bradley, 1999, pp. 357–96). The method involves the examination of the relative frequency of pollen grains from various plant species in long cores taken from marshes and peat bogs. Dendrochronology, which involves converting tree ring width indices to proxy climate data, depends on the availability of instrumental records long enough to allow correlation between temperature (or precipitation, where moisture availability is the critical environmental control) and ring widths. Annual or even seasonal resolution over hundreds to a few thousand years is possible (Bradley, 1999, pp. 397–438). Evidence of past changes in tree line position is generally interpreted in terms of variations in summer temperature and/or summer moisture. Tree stumps or wood fragments from above the modern tree line suggest warmer conditions in the past and this has been documented widely in western North America, the Urals and Scandinavia. The broad conclusion is that tree lines were higher from 8 to 6 ka BP and that tree lines declined after 5 ka BP (Bradley, 1999, pp. 337–56).

Ice cores

Ice cores provide evidence of changing climates over the Holocene Epoch through measurement of ^{18}O ratios which vary systematically with depth within the glacier. At least eight high-altitude sites have provided ice cores to bedrock: Quelccaya and Huascarán in Peru, Sajama in Bolivia, Colle

Gnifetti and Mont Blanc in Switzerland, Dunde and Guliya ice caps in western China and Mt Kilimanjaro in East Africa (Cullen, 2006). At Quelccaya, Huascarán and Dunde there is evidence of dramatic climatic change in recent decades; melting is occurring at such an accelerated rate that there is a danger of permanent loss of ice. Where records extend back to the last glacial period (Sajama, Huascarán and Dunde) ice is thin, making a detailed interpretation difficult (Thompson, 1998).

2.1.3 Ecological zonation

Ecological zonation in polar, temperate and tropical mountains compared

Halpin (1994) has investigated the differing sensitivity of mountain ecosystems to changing climatic conditions at tropical, temperate and polar sites. A 3900 m hypothetical mountain with 100 m elevation intervals was digitised into a raster GIS and used to represent a typical mountain at each site. A single +3.5 °C temperature and +10% precipitation change was imposed for all sites. The conceptual model implied linear shift of all vegetation belts upslope and the progressive loss of the coolest climatic zones at the peaks of the mountains. The resulting ecological zonation of the sites can be seen in Fig. 2.2. According to the simulation, at the wet tropical Costa Rican site (a), the five ecological zones would be reduced to four under a 3.5 °C temperature and 10% precipitation increase: the subalpine *paramo* would probably disappear. At the dry temperate Californian site (b), eight ecological zones would be reduced to six, but low and mid-elevation ecological zones would have expanded ranges; and at the boreal/arctic mountain site (c), in Alaska, the only ecosystem loss would occur near the base of the mountain. It should be borne in mind that assumptions of symmetrical change for mountain systems under global environmental change can be misleading. Predictions of vegetation shifts are complicated by uncertainties in species-specific responses to changing atmospheric CO_2, in addition to projected regional temperature, precipitation and soil moisture changes. There are photoperiod constraints in cold climates and the duration and depth of snow cover have major ecophysiological impacts.

Geoecological zonation

In order to express the landscape implications of ecological zoning, the upper timber line or tree line; the modern snow line and the lower limit of solifluction have been identified as being especially sensitive to environmental change. Messerli (1973) noted that, in the case of African mountain systems at least, thermal criteria for geoecological zonation were less important than availability of moisture. He also

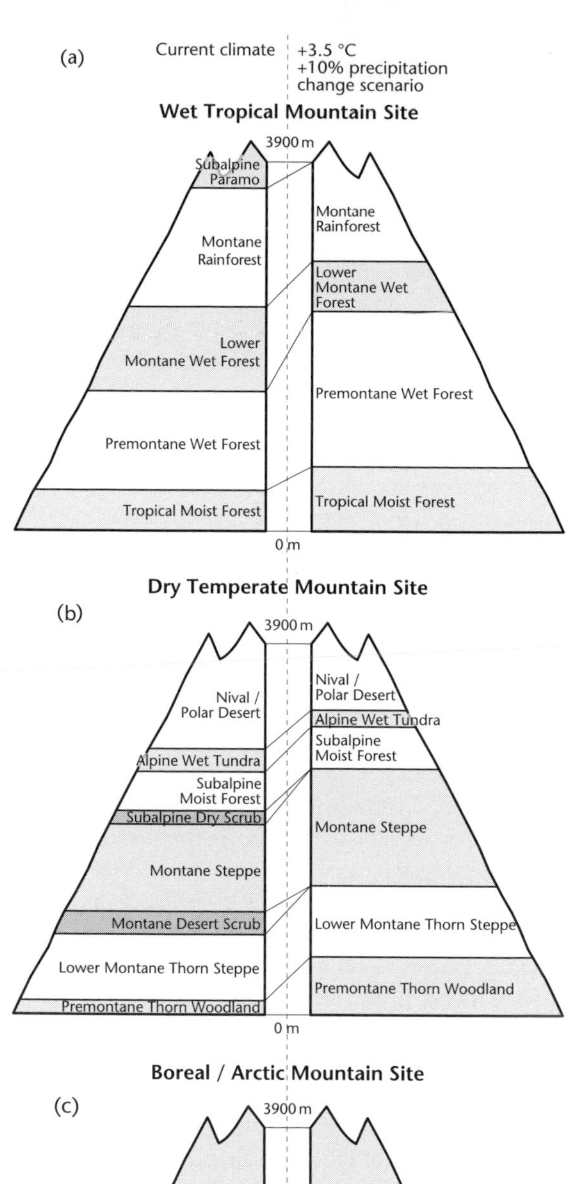

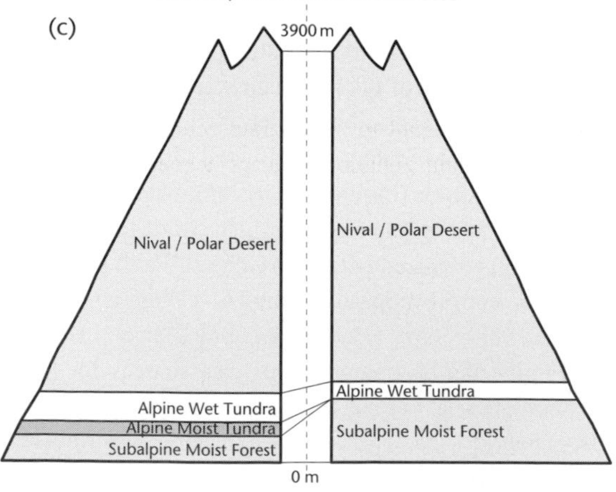

FIGURE 2.2. Current ecological zonation for a tropical, temperate and polar site and hypothetical zonation after climate change (modified from Halpin, 1994).

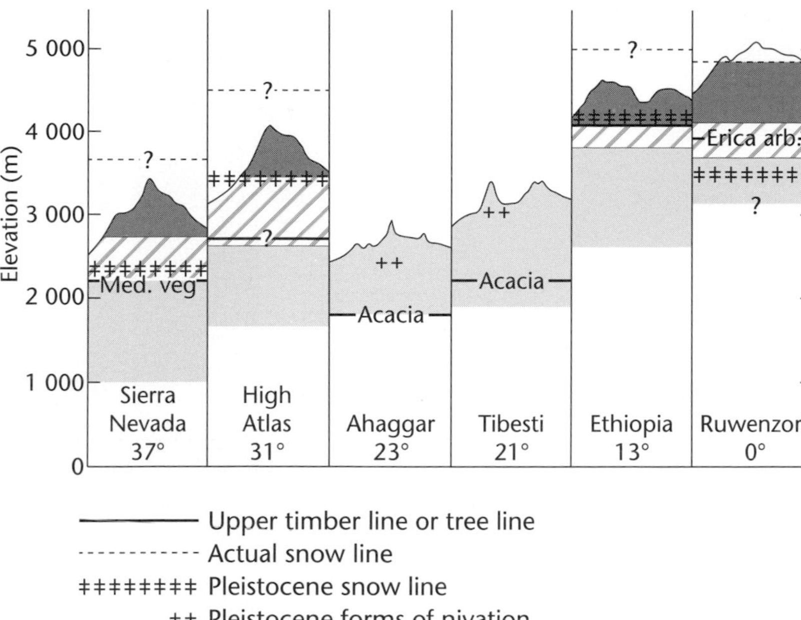

FIGURE 2.3. Horizontal zonation of geomorphological and climatic elements from Mediterranean to tropical high mountains of Africa (modified from Messerli, 1973).

emphasised the importance of latitudinal zonation over continental scales of mountain systems (Fig. 2.3). The idea of geoecological zonation of mountains is fruitful as it provides a framework within which the extreme heterogeneity of mountain landscapes, both past and present, can be considered. The changing magnitude and frequency of operation of geomorphic processes over Quaternary time is indexed by the LGM snow line and the lower limit of periglacial activity.

Geomorphic process zones and sediment cascades

In an explicit attempt to incorporate geomorphic activity into the mountain zonation concept a general model of geomorphic process variation with elevation has been developed (Fig. 2.4a). In temperate mountain systems high-elevation slopes are characterised by solifluction, nivation, talus development and glaciation; mid-elevation slopes, below the tree line, have landslide, avalanche and debris flow features; and low-elevation slopes are commonly sediment storage zones. Local-scale lateral zonation of mountains is caused by gravity-driven processes of slope erosion, mass movement and river action as well as by other influences, such as aspect. These processes intersect the vertical zonation pattern and subdivide the landscape into a series of slope facets bounded on the top and bottom by timber line, snow line, permafrost or periglacial activity and bounded at the sides by a channel way or preferred pathway for sediment, snow, ice or water movement. Aspect, slope erosion, mass movement and river action give rise to a lateral zonation which cuts across the vertical zonation described above (Slaymaker, 1993). The net effect is one of a lateral zonation which emphasises varying sensitivity to disturbance (Fig. 2.4b).

2.1.4 Summary

A working definition of mountains is offered and mountain landscapes are shown to have high geodiversity, a geodiversity which is threatened by the rapid rate of growth of mountain populations. The three drivers of environmental change in mountains are relief, as a proxy for tectonics (Tucker and Slingerland, 1994), hydroclimate and runoff (Vandenberghe, 2002) and human activity (Coulthard and Macklin, 2001). The first two of these drivers can be interpreted through proxy records contained in glacier ice cores, soils and lake sediments. The net effect of the operation of these drivers over long time periods is a variety of ecological and geomorphic zones, which divide mountain landscapes into clearly delimited facets. The third driver, human activity, results in land use and land cover patterns that often cut across the geoecological zones and generate accelerated landscape disturbance.

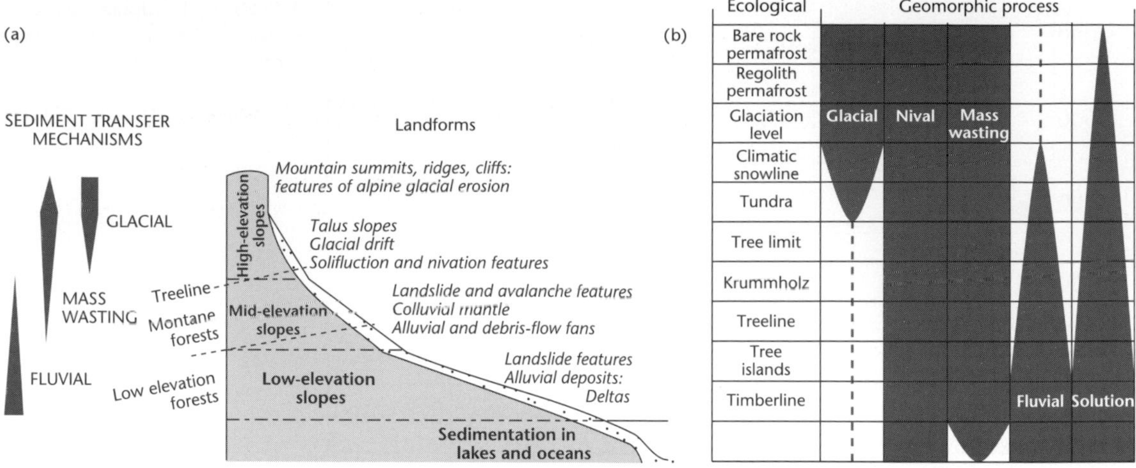

FIGURE 2.4. (a) Vertical and (b) lateral zonation of major Holocene landforms and processes in Coast Mountains, British Columbia (from Slaymaker, 1993).

2.2 Direct driver I: relief

Over geological timescales, relief is controlled by tectonic plate movements and climate, via rates of denudation (Schumm, 1963). Over contemporary timescales, however, relief controls climate. At the larger mountain system and global scales, elevation and gradient are the most important relief elements in so far as they influence temperature and precipitation. Elevation controls the incidence and intensity of freeze–thaw events as well as orographic precipitation, and many associated climatic effects. Gradient defines the gravitational driving force ($g \sin \alpha$) and influences radiation and precipitation receipt, wind regimes and snow. Erosion rates reported for the Nanga Parbat massif are among the highest measured (22 ± 11 m ka^{-1}) and reported rates of uplift for the Himalayas vary from 0.5 to 20 m ka^{-1} (Owen, 2004). Ahnert (1970) developed an equation relating denudation and local relief:

$$D = 0.1535h \qquad (2.1)$$

where D is denudation in mm ka^{-1} and h is local relief in m km^{-1}.

Summerfield and Hulton (1994) analysed 33 basins with areas greater than 500 000 km^2 from every continent except Antarctica. Total denudation (suspended plus dissolved load) varied from 4 mm ka^{-1} (Kolyma in the Russian Far East) to 688 mm ka^{-1} (Brahmaputra). They found that more than 60% of the variance in total denudation was accounted for by basin relief ratio and runoff.

A contentious issue is the relation between drainage basin area and specific sediment yield established for basins in British Columbia (Slaymaker, 1987) and for Canada as a whole (Church *et al.*, 1999). They demonstrated that in most of Canada basins of intermediate size and relief have the highest specific sediment yields (Caine, 2004). They contended that the presence in the contemporary landscape of sediment storage areas, formed in the early Holocene, is the most important control of specific sediment yield. They insisted that it is critical to examine not only relative relief and elevation, but also the sediment cascade. In so doing it becomes apparent that there are numerous complications that modify the direct relation between relief, absolute elevation and denudation rate, the most important of which seems to be the presence of a whole variety of sediment sinks. Sediment sinks include lakes, alluvial fans, proglacial zones and floodplains, inter alia, and are storage zones in which sediments may be stored for shorter or longer periods.

2.2.1 The sediment cascade in mountains

Elevation, downslope gradient, across-slope gradient, aspect, vertical convexity and horizontal convexity are the six fundamental components of relief (Evans, 1972). Vertical and horizontal convexity, which are the rates of change of gradient downslope and across-slope, and aspect, which is the preferred orientation of the slope, are important influences on environmental change at local scale. Concave slopes tend to accumulate water and sediment whereas convex slopes tend to shed water and sediment. Aspect controls the amount of solar radiant energy received at the surface and, especially in mountain systems, leads to highly contrasted slope climates on, for example, north- and south-facing slopes (known in the French literature as *ubac* (shady) and *adret* (sunny) slopes). The potential energy of

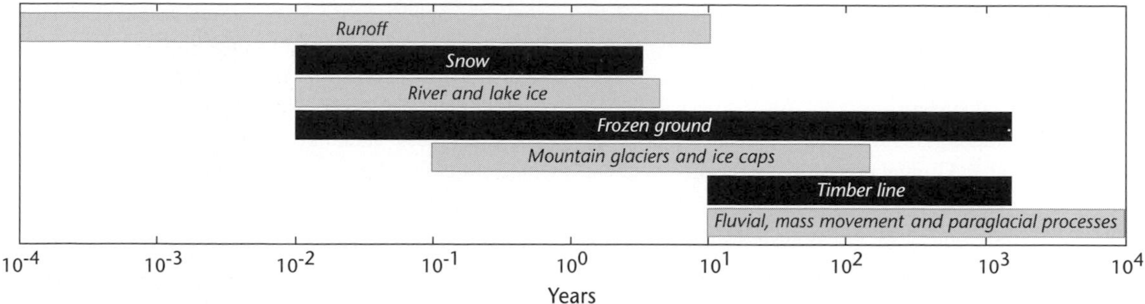

FIGURE 2.5. Components of the mountain cryosphere, ecosphere and sediment cascade and their response times following disturbance by hydroclimate or by human activity.

position (indexed by elevation above sea level) and the kinetic energy of motion (indexed by relief roughness) combine to form preferred erosional and transportational pathways from the highest summits to the ocean or nearest base level. En route from the summit, sediments are moved downslope and tend to be sorted as they go. The sediment cascade that results can be characterised in terms of four environments (after Caine, 1974) which are differentiated by dominant processes and forms as:

- the mountain cryosphere system;
- the coarse debris system;
- the fine-grained sediment system; and
- the geochemical system.

Note that the categories overlap and are identified only in terms of their dominant characteristics.

The mountain cryosphere system

The mountain cryosphere includes snow, seasonally frozen ground, mountain permafrost, glaciers, and lake and river ice. Each of the elements of the cryosphere is sensitive to environmental change, whether in response to relief, temperature, precipitation, runoff, sediment transport or land use changes. The cryosphere stores water and changes the timing and magnitude of runoff which erodes and transports sediment. Snow responds to environmental changes on a daily timescale; lake and river ice on an annual timescale; permafrost and glaciers on annual to century timescales; associated ecosystem responses are measured in decades to centuries; and sediment systems may take decades to millennia to respond (Fig. 2.5). In glacierised mountains (with present glacier coverage), the highest rates of change are found in the proglacial environment and hence at the higher elevations (Hallet *et al.*, 1996); in past glaciated mountains, the highest rates of change occur at intermediate elevations (Church and Slaymaker, 1989); and in never-glaciated mountains the global-scale relation of monotonic decline of denudation from high to low elevation would seem to hold (Milliman and Syvitski, 1992).

The coarse debris system

The coarse debris system involves the transfer of coarse detritus between cliffs and the talus and associated deposits. Rockfalls, landslides, avalanches and debris flows are processes that drive the coarse-grained sediment system. This system may

(a) be strongly controlled by the cryosphere;
(b) supply sediment inputs into the fine-grained sediment system; or
(c) be a closed system that remains uncoupled with neighbouring systems over long periods of time (Barsch and Caine, 1984).

But where the hydroclimate supplies sufficient runoff, gravel bed rivers are formed, and they are commonly coupled with adjacent systems. Where hillslopes, river channels and glaciers combine to move sediment, basin denudation rates are high. Where the coarse-grained sediment system is uncoupled there may nevertheless be high local rates of denudation, as demonstrated in the earliest sediment budget study published from the Swiss Alps in 1957 (Embleton-Hamann and Slaymaker, 2006).

The fine-grained sediment system

The fine-grained sediment system is commonly coupled with adjacent systems, located at progressively lower elevations of the sediment cascade. According to Church (1998), it is convenient (for reasons that have to do with mode of transport) to define the upper size limit of fine sediments at 1 mm. Weathering, surface erosion and fluvial processes, moving sediment within basins and aeolian processes, moving sediment from sources external to the basin, drive the fine-grained sediment system. Fluvial processes also transport sediment in suspension, in solution and as bedload from mountain summits (Reid and Dunne, 1996).

The geochemical system

Solutional weathering, nivation or snow-related processes, hydrogeological and fluvial processes are the drivers of the geochemical system. The contact time between water and material sources, and thus the residence time of water within the basin, temperature changes, vegetation cover and thickness of unconsolidated sediments all influence the importance of the geochemical system (Drever and Zobrist, 1992). As the sediment cascade enters more vegetation-dominated zones, and soils are also more maturely developed at these lower elevations, the breakdown of minerals into solutes and nutrients becomes more rapid and complex. Rapp (1960) was apparently the first to report the importance of solutes in the sediment cascade in non-calcareous polar environments.

2.2.2 Summary

The details of the sediment cascade need to be understood in order to predict the role of relief in landscape change.

- at the global scale, there is a direct relation between elevation plus runoff and rate of landscape change, with 60–70% variance explanation;
- at the regional scale, there is a difference between glacierised (present glacier coverage), glaciated (where glaciers were formerly present) and never-glaciated mountains;
- at the regional longer-term scale, it is important to consider the rate of tectonic uplift and the extent to which the uplift is balanced by denudational processes; and
- at the local scale, much depends on slope concavity/convexity and aspect and the degree of coupling of slope and channel processes.

2.3 Direct driver II: hydroclimate and runoff

2.3.1 Hydroclimatic variables of interest

Precipitation, snow storage, glacier storage, available soil moisture, groundwater storage, actual evapotranspiration and surface runoff are the components of the hydrological cycle which influence and respond to environmental change. The magnitude, frequency and duration of storm events is vital information which often does not exist because the usual presentation of precipitation data is in the form of daily totals. In terms of impact on the landscape, it is the extreme events (both high and low) and the freshet flows which mobilise most of the sediments and, to a lesser extent, the solutes (see Chapter 4). In mountain environments, the primary input of precipitation is regulated by the regional thermal climate, which determines whether snow or ice storage may occur. The local thermal climate determines the precise nature of the snow and/or ice storage. If temperatures are above freezing, the infiltration capacity of the surface materials regulates water movement either to soil moisture or groundwater or as direct runoff. In turn, soil moisture storage capacity regulates the contribution of soil moisture to delayed runoff. At the same time the ground thermal climate regulates ground ice occurrence and storage and the local thermal climate regulates evaporation and transpiration. Runoff may be stored in lakes, bogs and channels at which point regional thermal climate and thermal regime of water regulates the surface ice storage. Finally, water above the freezing point runs off. The effects of elevation, relief and surface materials in mountain environments ensure that a full range of water pathways and storage zones can be found. Nearing (2005) discusses the relations between precipitation, land cover changes, soil erosion and runoff.

2.3.2 The water balance model as an integrator of hydroclimate

The complexity of the hydrological cycle as described above precludes the possibility of exhaustive analysis at landscape scale. Therefore, a method of analysis which integrates the components of the cycle is sought. Solving the water balance is commonly the best way to assess regional impacts of potential environmental change. Some variant of the Thornthwaite–Mather water balance method can be used down to the monthly timescale (Thornthwaite and Mather, 1955). Water balance models are compatible with the scale of output from general circulation models (GCMs) and therefore are useful in predicting the hydrologic effects of environmental change. They are seen to be accurate, flexible and easy to use with respect both to hydrologic consequences and their ability to incorporate month to month and seasonal variations. If the range of hypothetical temperature and precipitation data is presented independently for winter and summer months, hydrologic scenarios can be developed with the aid of water balance approaches (Woo, 1996).

2.3.3 Runoff and sediment transport

The absence of long-term stable monitoring stations in mountains has led to an increasing reliance on lake sediments as proxies of environmental change. Hinderer (2001) has shown a direct exponential relation ($r = 0.97$) between basin area and annual rate of delta growth in Swiss alpine lakes. Cores of annually laminated sediments (varves) from five lakes in the southern Coast Mountains of British Columbia, for example, have provided information on past annual flood magnitudes, the effects of glacier recession and the influence of reorganisation of the North Pacific climate system on the magnitude of autumn flooding (Menounos, 2005). In tectonically active mountain regions, such as Taiwan, there are direct links between seismicity, runoff variability and erosion rates (Dadson, 2003).

2.3.4 Summary

The general conclusion from this brief overview of the role of hydroclimate and runoff in mountains is as follows:

- at the global scale, there is a direct relation between elevation plus runoff and rates of landscape change (see Section 2.2 above);
- at the regional scale, the water balance permits prediction of hydrologic consequences of environmental change;
- at the regional scale, runoff intensity commonly reaches a maximum within basins whose dimensions approximate those of the extreme event-producing storm cells. At the local scale, it is necessary to consider the relative importance of weathering, sediment transport and sediment depositional processes in response to hydroclimate to determine rates of change; and
- systematic analysis of the relative magnitude of change in sediment flux in basins of different size shows that relative changes diminish with increasing basin size. This is because smaller basins are more sensitive and vulnerable to environmental change.

2.4 Direct driver III: human activity, population and land use

Human activity, in the form of population density and land use, is a direct driver of environmental change in mountains. It is not, however, the sheer numbers of people but aspects of population composition and distribution, especially the level of urbanisation and household size, which exercise the greatest demands on the land (Lambin and Geist, 2006). It is 'population in context' (Rindfuss *et al.*, 2004) that matters. High population densities in the developing world, for example, may lead to better management, such as in Kenya and Bolivia, described by Tiffen *et al.* (1994) as cases where the presence of more people has led to less erosion. The creation of infrastructure, especially roads, is a crucial step in triggering land use intensification. The largest mountain populations are found in the mid-elevation mountain category (Class 2; Table 2.1) in developing countries (where developing countries are defined as having a relatively low standard of living, an undeveloped industrial base and a moderate to low human development index). In developed and transitional countries, by contrast, low mountains (Class 1) are most heavily populated.

TABLE 2.1. *Population in millions by mountain class (1, 2, 3 or 4)*

	Class 1[a]	Class 2	Class 3	Class 4
Asia and Pacific	101	113	14	3
Latin America	21	42	27	1
Near East/North Africa	19	55	5	0.03
Sub-Saharan Africa	9	58	12	Nil
Transition countries[b]	12	7	0.5	0.01
Developed countries[c]	22	12	0.3	Nil
Total	184	287	59	4.1

[a] 50% reduction of Class 1 population because of omission of the 300–500 m elevation band.
[b] A transition country is one which has recently changed from a centrally planned economy to a free market economy.
[c] A developed country is one which has a developed economy in which the tertiary and quaternary sectors of industry dominate and there is a relatively high human development index.
Source: Huddleston *et al.* (2003).

2.4.1 A typology of mountain systems with respect to human influence

Mountain systems can be differentiated not only in terms of relief and hydroclimate but in ways that reflect demography and land use. Relief has already been defined by the fourfold classification introduced in Section 2.1.1; and the descriptors polar, temperate and tropical have been introduced in Section 2.1.3 as surrogates for hydroclimate. In this typology we incorporate population density as a proxy for the intensity of the human signature on the landscape. It is fully understood that technology and farming practices are also critical factors. Higher population densities lead to a higher pressure on land resources and intensified land use and therefore the human signature (though not necessarily a negative impact) will be higher. In Table 2.2, a representative group of mountainous countries/regions is compared in terms of the population densities within their mountain regions.

TABLE 2.2. *Mountainous countries/territory/province, their mountain areas, the estimated total population and the population density in those mountains (as of AD 2000)*

	Mountain area (1000s of km ^{2})	Population in the mountains (millions)	Population density in mountains (persons km $^{-2}$)
Svalbard	48	0.001	0.02
Japan	185	15	81
Ethiopia	471	35.2	77
Tajikistan	131	2.9	22
Ecuador	108	5.3	49
Austria	55	3.3	60
British Columbia	750	0.5	0.7

Source: Huddleston *et al.* (2003).

At the most simplistic level, four categories of mountain region can be defined:

Polar mountains (e.g. Svalbard)
Svalbard has few permanent residents (around 2300 as of AD 2000) and a few isolated mining activities but increasing numbers of ecotourists and scientific researchers. It is now estimated that 40 000 tourists visit each year (as of 2007). Sixty-five percent of the surface of Svalbard consists of protected areas. Of all the drivers of environmental change the effects of hydroclimate are most readily observable.

Low population density temperate mountains (e.g. Canadian Cordillera, Tajikistan)
These mountains are providers of services such as water resources, tourism, agriculture, forestry, mining. Mountains with a history of less dense settlement retain more of their traditional agriculture and forestry. Relief and hydroclimate are the most important drivers of environmental change.

High population density temperate mountains (e.g. Austria, Japan)
Mountains having a history of relatively dense settlement, such as those in central Europe and Japan, are facing a decline of traditional agriculture and forestry. In western Europe and Japan, mountain regions are experiencing depopulation but in general land use pressures are increasing because of competition between conservation use, mineral extraction and processing, recreation development and market-oriented agriculture, forestry and livestock grazing. The human impact (both positive and negative) on these mountains far exceeds the documented effects of relief and hydroclimate.

Tropical mountains (e.g. Ecuador, Ethiopia)
In tropical and semi-arid environments, mountain areas are usually wet and/or cool and hospitable for living and commercial exploitation. They also have deeper soils, lower temperatures and fewer diseases. Human encroachment has reduced vegetation cover, increasing erosion and siltation, thereby adversely affecting water quality and other resources. Direct anthropogenic influence on these mountain regions appears to greatly surpass hydroclimate effects.

2.4.2 Land use in mountain areas

Anthropogenic impacts may be created by agricultural use, forestry, extractive industries and public utilities, dam construction, tourism or Mountain Protected Areas. It is important to make the assertion that not all anthropogenic impacts are negative and it is useful to recall that, under certain circumstances, more people generate less erosion (Tiffen *et al.*, 1994). It is also the case that some forms of land use change can enhance the environment (e.g. Fig. 2.6). Afforestation conducted between 1932 and 1960 transformed this Swiss landscape, which had been severely damaged by landslides and debris flows following a 33-hour rainstorm in June 1910.

Agriculture has been the greatest force of land transformation on this planet (Lambin and Geist, 2006). Nearly a third of the Earth's land surface is currently being used for growing crops or grazing cattle (FAO, 2007). Much of this agricultural land has been created at the expense of natural forests, grasslands and wetlands. In Africa, for example, most of the Class 2 and 3 mountains are under pressure from commercial and subsistence farming activities. In unprotected areas, mountain forests are cleared for cultivation of high altitude adapted cash crops like tea, pyrethrum and coffee.

Grazing and forestry are the predominant uses of mountain land in all regions. Areas above the timber line and dry mountain areas are commonly used for grazing (Table 2.3). Grazing land continues to predominate in Class 1 and 2 mountains. Extensive grazing has little impact on slope processes, but overgrazing can have severe impacts (see below). Huddleston (2003) notes that all mountain areas below 3500 m have exceeded the population density that is thought to be able to support extensive grazing methods (25 persons km $^{-2}$). Forestry is important in Class 1 and 2 mountains. Timber harvesting is a major land management practice whose precise influence on slope stability depends on the method of harvesting, density of residual trees and understorey vegetation, rate and type of regeneration, site characteristics and patterns of water inflow after harvesting (Sidle, 2002). As is demonstrated clearly in Chapter 12,

FIGURE 2.6. The beneficial effects of afforestation illustrated from the source area of the Gangbach River, Canton Uri, Switzerland. (a) The extensive debris flow activity that followed an extreme 33-hour rainfall event, 13–14 June 1910; (b) shows the same landscape in 1981, after the completion of afforestation between 1932 and 1960 (R. Kellerhals, personal communication, 2008).

roads can indeed be the focus of the highest rates of denudation in the landscape. Humans accelerate slope failures through road building, especially when roads are situated in mid-slope locations instead of along ridge tops (Marston *et al.* 1998). Roads in eastern Sikkim and western Garhwal have caused an average of two major landslides for every kilometre constructed. Road building in Nepal has produced up to 9000 m^3 km^{-1} of landslide material, and it has been estimated that, on average, each kilometre of road constructed will eventually trigger 1000 tons of land lost from slope failures (Zurick and Karan, 1999).

In the developed world, winter tourism has expanded significantly during the past three decades and the winter tourism industry has a strong impact on the natural system. Extensive infrastructure and many technological measures,

TABLE 2.3. *Rural mountain population (in thousands) in developing and transition countries, by land use category and mountain area class*

Mountain class	1[a]	2	3	4	Total
Wildland	2 293	10 009	5 021	2 284	19 607
Protected	4 224	7 895	4 017	426	16 562
Grazing mixed[b]	53 345	117 054	28 204	1 387	199 990
Grazing	21 241	17 666	705		39 612
Closed forest	14 724	19 088	1 712		35 524
Mixed[c]	10 500	19 907	1 304		31 711
Cropland	14 483	11 227	727		26 437

[a] Class 1 population total has been reduced by 50%.
[b] Grazing land with some cropland, closed forest and wildland.
[c] Mixed use, closed forest, grazing land and cropland.
Source: Huddleston *et al.* (2003).

like grooming of ski slopes or artificial snow-making are required (Section 2.6.4).

One of the most dramatic effects of human activity on landscape processes is the construction of large dams (see Chapters 1 and 4). Many of the world's dams are located in mountains, because of the presence of stable, bedrock controlled river cross-sections, deeply incised channels and glacially modified valley morphology (Plate 8). Natural peak flows are reduced and distributed over time and the discharge of streams in adjacent watersheds is altered through water diversion. Water and sediment flushing of reservoirs disturbs natural discharge and sediment movement in the fluvial system below the reservoir. There are massive changes in the sediment transport of major rivers pre- and post-dam construction, as illustrated by the Nile, Ebro and Orange rivers (Fig. 2.7).

In the case of temperate mountain areas, the only space for urban agglomerations and transport routes is on the flat floor of broader valleys and mountain basins (Bätzing, 2003). These used to be flood-prone wetland areas and thus all rivers in these areas have needed to be turned into artificial channels. The consequence of this action has been the enhancing of natural hazards as the concreting of riverbanks and the waterproofing of urban areas represent aggravating factors for floods. By contrast in the case of tropical mountains, the towns are generally in the uplands and roads are often in ridge top locations. This has a significant impact on runoff coefficients (Nyssen, 2002). The growth of cities in the mountain world places further stress on mountain ecology. In Latin America, the Caribbean and countries in transition, nearly half of the mountain population lives in urban areas (Appendix 2.2).

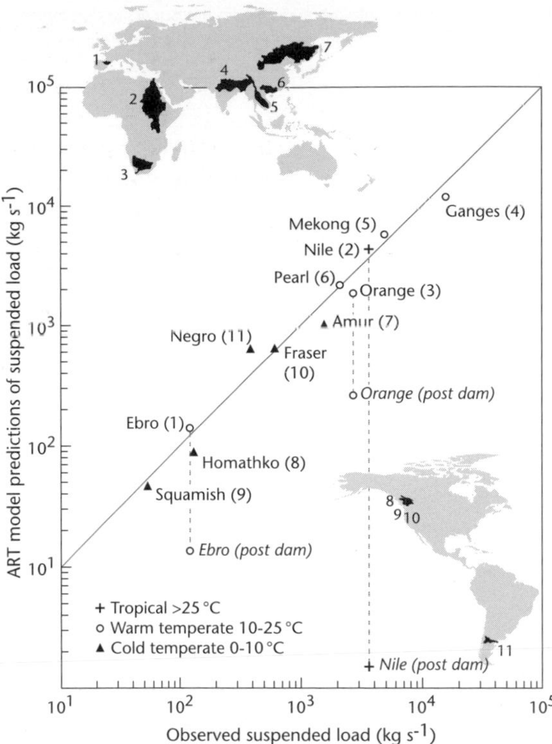

FIGURE 2.7. A plot of observed sediment load versus predicted values for 11 major rivers, eight of which are undammed and three of which (Nile, Orange and Ebro) have records of sediment transport before and after damming. The dashed vertical lines indicate the effect of the dams on downstream sediment transport (modified from Syvitski *et al.*, 2005).

FIGURE 2.8. Machapuchare, in the Annapurna Range, Nepal. The government of Nepal has declared the sacred mountain off limits to climbers (from Bernbaum, 1998).

Protection and enhancement of mountain systems

The same land uses that lead to land degradation can also enhance mountain environments, when ecosystem needs are taken into account. Wu and Thornes (1995) note that in the Middle Mountains of Nepal *khet* terraces constructed with a berm can protect slopes with a local relief of thousands of metres; at higher elevations rain-fed *bari* terraces, without a berm and sloping outwards, experience erosion rates two orders of magnitude higher than khet terraces.

Mountain Protected Areas both protect and enhance the mountain environment. There are six categories of Mountain Protected Areas according to the International Union for the Conservation of Nature (IUCN, 2000), ranging from strict nature reserves to managed resource protected areas. The World Heritage List is a notable example of an attempt to conserve natural and cultural landscapes; as of 2002, Thorsell and Hamilton (2002) identified 55 natural and mixed (natural/cultural) mountain sites on the List. This amounts to 33% of the total of all sites in these two categories (167) and demonstrates that a substantial portion of the natural World Heritage sites is 'mountain land'. This finding is even more interesting when it is noted that their criterion for identifying mountain sites is that they contain a minimum of 1500 m relief within the borders of the protected area, a criterion that is more restrictive than that of the World Conservation Monitoring Centre (WCMC). Bernbaum (1998) has discussed the cultural significance and religious symbology of mountains as a strong reason for mountain conservation (Fig. 2.8).

Similarly, cultural landscapes of outstanding universal value also consist disproportionately of mountain landscapes because of their aesthetic appeal and also because so many of the world's great religions grew up in mountainous regions: Buddhism, Taoism, Hinduism, Islam, Judaism and Christianity (Bernbaum, 1998).

2.4.3 Summary

From this brief overview of population and land use the implications for landscape change are as follows:

- at the global scale, demography and land use effects have become more important than relief and hydroclimate. The highest population densities are often found at intermediate elevations (Class I and II mountains) and landscape changes are greater there than in high and very high mountains;

- at the regional scale, the picture is more complex: in polar regions, the impact of humankind on the landscape is still minor and the geoecological and geomorphic process zonation is dominant; in wild lands (uncultivated and unprotected land), such as the North American Cordillera, the human impact remains comparatively low; in the European Alps, land use effects are dominant; and in tropical mountains, demography and land use are overwhelmingly important;
- also at the regional scale, the potential effects of depopulation and changed land use on traditional agricultural landscapes may well reduce tourism appeal; and
- at the local scale, it is in general clear that land use effects are dominant wherever road systems and land clearance are extended.

2.5 Twenty-first century mountain landscapes under the influence of hydroclimate change

Predictions from GCMs are for continental-scale regions and there is no guarantee that these trends will hold for mountain systems. The precipitation data which are unreliable for larger regions are even less reliable for individual mountain systems. Computing limitations also necessitate relatively coarse spatial resolution (equivalent to about 5° latitude typically) and therefore mountain terrain is greatly smoothed. Simulated wind fields and the distribution of clouds and precipitation are also in error. Moreover, model climate outputs are conventionally presented for sea level and ignore conditions at the surface over mountain landscapes. Scenarios of climate change in mountains are therefore highly uncertain; they are poorly resolved even in the highest resolution GCMs.

2.5.1 Snow

Mountain snowpacks are important as: (a) primary water resources for adjacent lowlands or for power generation in the spring months; (b) essential elements of winter sports development; (c) determinants of potential snow avalanche hazard; and (d) creators of a mosaic of microenvironments for flora and small animals. But the data on snow cover area and snow water equivalent provide contradictory evidence of directions of change.

In western North American mountains, the date of maximum snow water equivalent has shifted earlier, by about 2 weeks, in the period 1950 to 2005. Trends in Europe and Eurasia are conflicting. Satellite data for the lowlands surrounding the European Alps show 3–4 weeks less snow cover during the 1980s and 1990s than earlier. But for higher elevations, total snow accumulation is unchanged in spite of warmer temperatures. No long-term trend from 1979 to 2002 in mid-latitude alpine regions of the Andes has been detected. Snow depths in Australia show significant decline since 1962 (Kosciusko Mountains) but no trend has been found in New Zealand from 1930 to 1985 (IPCC, 2007a.). If snow cover area is reduced, there will be a number of implications: early season runoff may increase but this will be followed by drier soil and vegetation in summer and greater fire risk. In alpine areas, the snow line could rise by 100–400 m, depending on precipitation. Less snow will accumulate at low elevations but there may be more above the freezing level.

2.5.2 Snowmelt

Since most precipitation at high elevation falls as snow, the timing of runoff is primarily determined by snowmelt and therefore spring temperatures. On the assumption of a 4 °C increase in mean annual temperature and no precipitation change in the high mountains of China, Liu and Woo (1996) have shown that over a 3000 m range in elevation, evaporation will increase at all elevations and this will be accompanied by a decrease in runoff volume. In his work on runoff from the mountain tributaries of Mackenzie River, Woo (1996) speculated that seasonal snow accumulation might increase in high-elevation zones and increased summer storminess may reduce snowmelt at intermediate elevations as a result of increased cloudiness and summer snowfall. At lower elevations, however, rainfall and rain-on-snow melt events will probably increase. Effectively, there will be a shift from a nival to a pluvial runoff.

2.5.3 River and lake ice

The timing and extent of river and lake water freeze-up and break-up is of importance in that it controls the magnitude and timing of annual water flux to the oceans. However, such data for mountain regions are difficult to access (Prowse, 2000). Caine (2002) attributes declining ice thickness on an alpine lake in the Colorado Front Range, USA to increased winter precipitation.

2.5.4 Frozen ground

This includes near-surface soil affected by short-term freeze–thaw cycles, seasonally frozen ground and permafrost. In

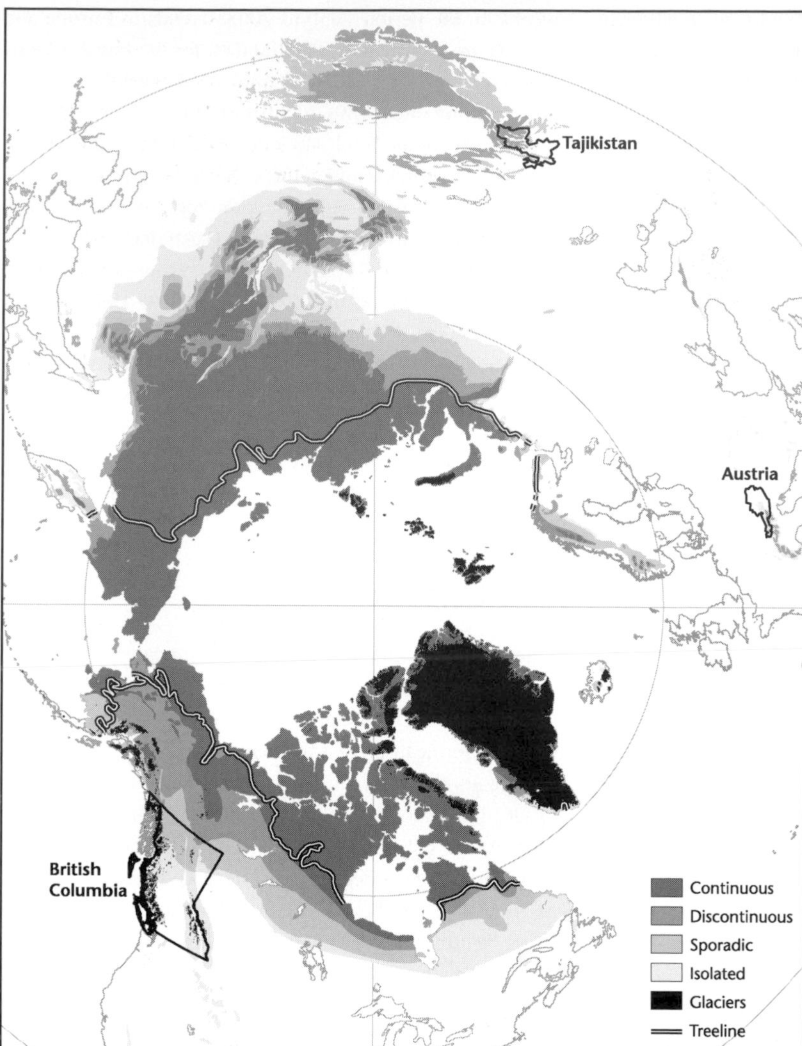

FIGURE 2.9. Permafrost distribution in the northern hemisphere (20–90° N). The data classes have been grouped into spatially continuous (90–100% cover), discontinuous (50–90%), sporadic (10–50%) and isolated (<10%) permafrost zones (modified from Brown *et al.*, 2001). British Columbia, Tajikistan and Austria are outlined to provide permafrost context for Plates 9, 10 and 11.

terms of areal extent, frozen ground is the single largest component of the cryosphere (Zhang, 2003; see Chapter 13). Seasonally frozen ground has decreased since 1900 by a maximum of 7%. Stations in southern Norway, Austria, Mongolia, Tibet and the Tien Shan Mountains indicate increased active layer thickness of up to 1 m since the 1990s (Zhang, 2003).

Permafrost

Permafrost regions represent approximately 25% of the exposed land area of the northern hemisphere (Fig. 2.9) but taking into account the extensive discontinuous permafrost zone, only 13–18% is actually underlain by permafrost (Zhang, 2003). Gradual warming of ground temperature has been reported over the past decade in southern Norway and Svalbard. In China, along the Qinghai–Xizang Highway and in the Tien Shan, Kunlun and Da Hinggan mountains, noticeable warming of the permafrost has occurred over the past three decades. The southern limit of permafrost in 2002 was 2 km further north than in 1975 along the Qinghai–Xizang Highway (IPCC, 2007a). The lower limit of permafrost has moved upward by *c.* 25 m between 1975 and 2002 on the north-facing slopes of the Kunlun Mountains and the areal extent of *taliks* (unfrozen zones within the permafrost) has increased.

Janke (2005) has used the distribution of active and inactive rock glaciers to model current permafrost distribution. He found that in an area of the Colorado Front Range which currently has 12% permafrost, a 4 °C warmer climate would cause its complete disappearance.

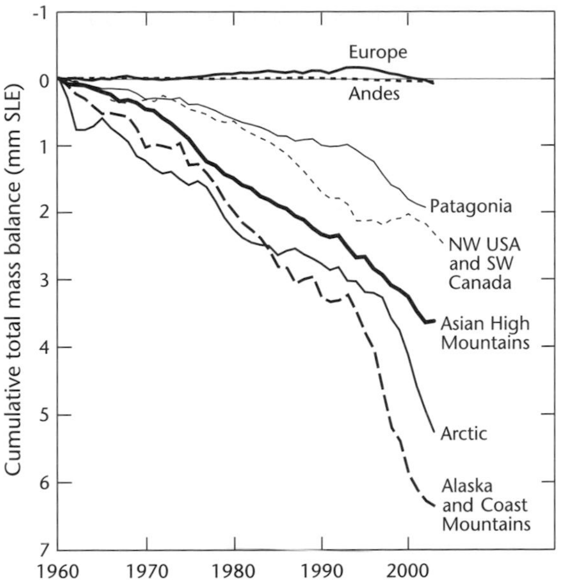

FIGURE 2.10. Cumulative total mass balances of glaciers and ice caps, calculated for seven large regions. Total mass balance is a measure of the contribution from each region to sea level rise. SLE, sea level equivalent (from Lemke *et al.*, 2007).

2.5.5 Glaciers and ice caps

The effects of temperature and precipitation changes on glaciers are complex. Measurements of mass balances outside the polar regions are unambiguously negative, by contrast with the approximately balanced mass balances of the major ice sheets (see Chapter 14). In cold-based glaciers at polar latitudes and high mid-latitude elevations, atmospheric warming simply leads to ice warming. In areas of temperate ice, atmospheric warming can directly impact the mass and geometry of glaciers. Alpine glacier and permafrost signals therefore constitute some of the clearest evidence available for climate change. During the twentieth century there was an obvious thinning, mass loss and retreat of mountain glaciers (Fig. 2.10). Oerlemans (2001) has shown the consistent reduction of glacier length since the end of the Little Ice Age (i.e. the period 1850–1990) from 48 glaciers located in nine different regions. Mean rates of retreat of individual glaciers varied from 1.3 to 86 m a^{-1}. By introducing a scaling factor, adjusting for the greater sensitivity of maritime glaciers, the mean scaled rates were shown to vary from 6.3 to 14.9 m a^{-1}. Alpine and subpolar glacier melting has contributed *c.* 60% of sea level rise in the last decade (Meier, 2007). It is estimated that by AD 2050 a quarter of mountain glacier mass will have melted. Glaciers are likely to shrink even where mountains become wetter. An upward shift of the equilibrium line by 200–300 m and annual ice thickness losses of 1–2 m are expected for temperate glaciers. Zemp (2006) estimated that a 3 °C warming of summer air temperature would reduce the currently existing alpine glacier cover by some 80%. Over the last half century, both global mean winter accumulation and summer melting have increased steadily (IPCC, 2007a). The strongest mean negative specific mass balances have occurred in Patagonia, the northwest USA, southwest Canada and Alaska. Only Europe has shown a mean value close to zero, with strong losses in the Alps compensated by strong gains in maritime Scandinavia. Glaciers in the high mountains of Asia have generally retreated but the central Karakorams have seen recent advances. In East Africa, there is also a debate as to whether glacier shrinkage is due to local decreased rainfall resulting from deforestation. Tropical glaciers in Africa (Hastenrath, 2005), South America (Georges, 2004) and Irian Jaya (Klein and Kincaid, 2006) have shrunk from maximum positions in the mid nineteenth century, with rapid shrinkage in the 1940s and since the 1970s. The smallest glaciers have been the most impacted (Fig. 2.11).

2.5.6 Glacier–runoff–sediment transport relations

One of the most significant impacts of climate change in glacierised basins may be the changing pattern of glacier melt runoff (Walsh, 2005). Glaciers will provide extra runoff as the ice disappears. In most mountain regions, this will happen for a few decades and then cease. For those regions with very large glaciers, the effect may last for a century or more. Kotlyakov *et al.* (1991) have provided estimates of change for Central Asia which give a threefold increase of runoff from glaciers by AD 2050 and a reduction to two-thirds of present runoff by AD 2100. In the short term, a significant increase in the number of flood events in Norway is projected (Bogen, 2006). Bogen found that suspended sediment concentrations and volumes were dependent on the availability of sediments, the type and character of the erosion processes and the temporal development of the flood. In the Norwegian case, it appears that the glacier-controlled rivers are unlikely to respond dramatically in terms of sediment transport because of limited sediment availability. Nevertheless, in global sediment yield terms, Hallet *et al.* (1996) have conclusively demonstrated the importance of glacier meltwaters. Both the global data and the regional Norwegian data demonstrate the difficulty of making generalisations about the probable effects of climate change on sediment transport in glacierised basins.

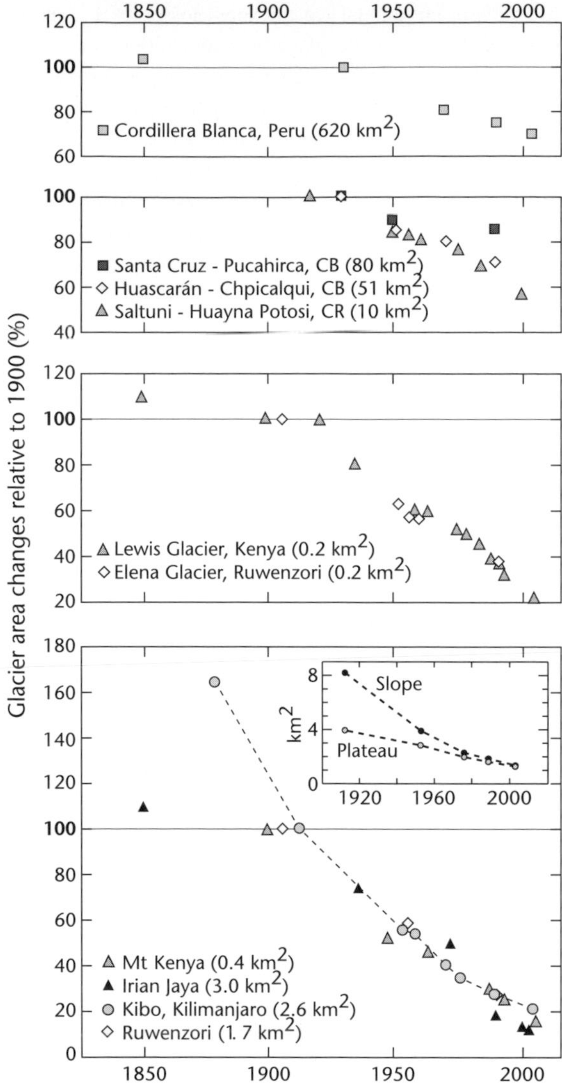

FIGURE 2.11. Changes in the surface area of tropical glaciers relative to their extent around 1900, grouped according to different glacier sizes. The sizes are given for 1990 or the closest available date to 1990. The insert shows the area change of the Kilimanjaro Plateau and slope glaciers as separated by the 5700 m contour line. CB, Cordillera Blanca; CR, Cordillera Real (from Lemke *et al.*, 2007).

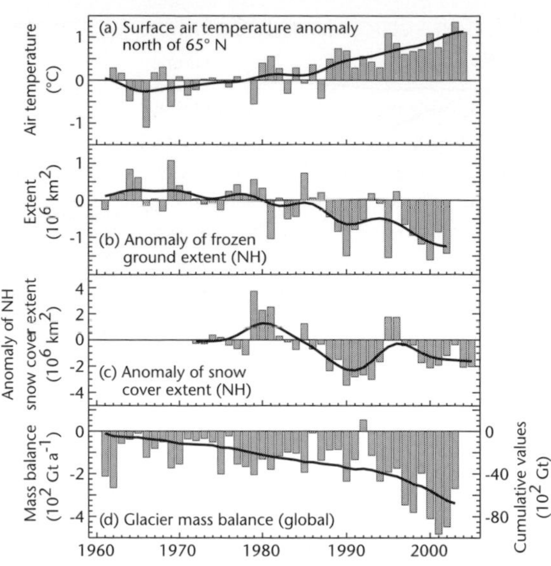

FIGURE 2.12. Time series of departure from the long-term mean of (a) arctic surface air temperature; (b) northern hemisphere (NH) frozen ground extent; (c) northern hemisphere snow cover extent; and (d) global glacier mass balance (from Lemke *et al.* (2007).

2.5.7 Extreme events

If climate warming is accompanied by increasing storminess (IPCC, 2007b) and intense rainfall, then peak stream discharges will increase and erosion, sediment transport and sediment deposition downstream will presumably also increase. Hazards will include flooding, possibly increasing both magnitude and frequency of floods and increased lateral instability of stream banks. A prominent indication of a change in extremes is the observed evidence of increases in heavy precipitation events over the mid-latitudes in the last 50 years, even in places where mean precipitation amounts are not increasing (Kunkel, 2003). In central and southwest Asia, the 1998–2003 drought provides an example of the unanticipated effects of extreme events. Flash flooding occurred over hardened ground desiccated by prolonged drought in Tajikistan, central and southern Iran and northern Afghanistan, leading to accelerated erosion in early 2002 (IPCC, 2007a).

2.5.8 Summary of global implications of hydroclimate change in mountains

Trends in the mountain cryospheric elements (Fig. 2.12) are summarised from IPCC (2007a). It can be seen that glacier mass balances, frozen ground extent and mean surface air temperatures provide trends that are consistent with global warming. The snow data are less clear. Indeed, it should be stated that the evidence for snowpack reduction is less strong than has frequently been asserted (Fig. 2.12c). Increasing glacier melt runoff and high-magnitude, low-frequency rainfall events will produce variable sediment transport changes depending on proportion of the basin occupied by the cryosphere and the nature of sediment availability.

2.5.9 Case study: British Columbia's mountains and hydroclimate

The Canadian province of British Columbia (BC) has an area of approximately 950 000 km² and a population of 4.4 million inhabitants (2007 estimate). Perhaps as many as 500 000 persons live in the mountains which cover about 80% of the province and are almost entirely Class 1 and 2 mountains (Plate 9). There are isolated pockets of Class 3 mountains in the Rocky, Columbia, St Elias and Coast mountains and a single Class 4 mountain in Mount Fairweather (4663 m above sea level) in the St Elias Mountains. Those same mountain areas contain over 29 000 km² of glaciers and ice caps. Warming since the Little Ice Age has been accompanied by marked glacier retreat. Schiefer *et al.* (2007) estimate total ice loss of 22.5 km^3 a^{-1} over the previous 15-year period.

The regional climate projection of the Fourth Assessment Report (IPCC, 2007a) for BC anticipates increases in both mean annual temperature and mean annual precipitation, the latter incorporating a summer decrease and a winter increase in precipitation. A decrease in snow season length and snow depth is also projected. The decrease in snow and summer rain combined with the increased mean annual temperature has potentially devastating effects on the semi-arid Bunchgrass and Ponderosa pine zones of the BC interior. The increase in absolute winter precipitation is unlikely to influence the perhumid Mountain hemlock and Coastal Western hemlock zones, unless this larger volume of moisture comes in the form of extreme events.

The postglacial landscape of BC

The postglacial landscape of BC is in transition between the LGM and present (see Chapter 1). The Cordilleran ice sheet covered almost the whole of the province but rapid melting of the ice sheet commenced as early as 14.5 ka BP and since that time the process of transition towards a fluvially dominated landscape has been ongoing (Church *et al.*, 1999). During the millennia immediately following the melting of the ice sheet, a condition described as 'paraglacial' prevailed (Ryder, 1971; Church and Ryder, 1972). A distinctive pattern of sediment yield (Fig. 2.13) can still be seen as evidence of a transitional landscape with a relaxation time of the order of 10 ka (Church and Slaymaker, 1989). This transitional landscape lasts until the glacially conditioned sediment stores are either removed or attain stability (Schumm and Rea, 1995). The landscape of BC is, therefore, a disturbance regime landscape (Hewitt, 2006), in so far as the postglacial landscape has had insufficient time to recover from the effects of the last major disturbance, namely the LGM (and see also Brardinoni and Hassan, 2006). The conclusion is that many landforms in the rugged BC mountain landscape are conditionally unstable and it can be anticipated that small changes in hydroclimate and/or land use may cause landscape change.

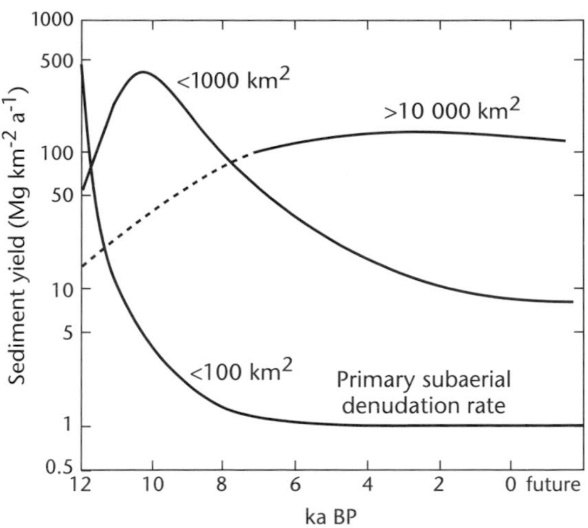

FIGURE 2.13. Temporal pattern of paraglacial sediment yield in formerly glaciated upland and valley sites in coastal British Columbia and Alaska. Values are for basins of order 100, 1000 and 10 000 km² based on the contemporary spatial pattern of sediment yield (from Church, 1998).

Geomorphic hazards associated with glacier retreat include rock avalanches, deep-seated slope sagging (sackung), debris flows, debris avalanches, debris slides, rockfall, moraine dam failures and glacier outburst floods (Geertsema, 2006). The effects of glacier retreat on sediment transport are controversial. Schiefer *et al.* (2006) identified three periods of accelerated sedimentation in a montane lake in the Coast Mountains over the past 70 years: a period of intense rainstorms; a year of massive slope failure; and a period of rapid glacier retreat between 1930 and 1946. The present period of rapid glacier retreat does not seem to be generating exceptional sedimentation events.

The general tendency under climate warming will be an upslope shifting of hazard zones and widespread reduction in stability of formerly glaciated or perennially frozen slopes (Barsch, 1993). Climate change can be expected to alter the magnitude and frequency of a wide variety of geomorphic processes (Holm *et al.*, 2004). Increased triggering of rockfalls and landslides could result from increased groundwater seepage and pressure. Large landslides are propagated by increasing long-term rainfall whereas small landslides are triggered by high-intensity

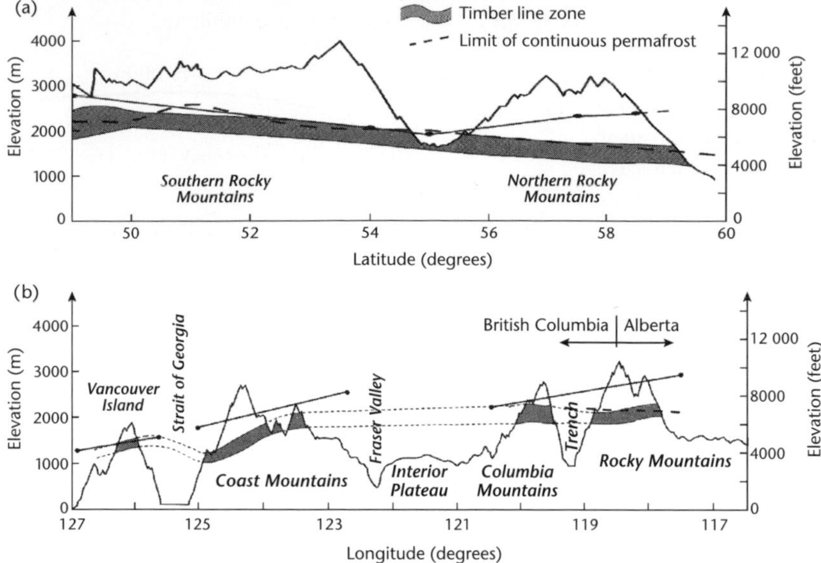

FIGURE 2.14. S–N and W–E profiles of British Columbia, showing the timber line zone, the lower limit of continuous permafrost and the annual snow line (modified from Ryder, 1998). (a) S–N transect along the crest of the Rocky Mountains (49–60° N); (b) W–E transect from Vancouver Island to the Rocky Mountains.

rainfall. These tendencies will probably lead to enhanced sediment transport. Increased sediment input to glacier-fed rivers may lead to increased channel instability, erosion and flooding. The hazard zones related to most of these fluvial processes will extend a long way beyond the limits of the mountain area (Ashmore and Church, 2001).

The timber line

The timber line is a zone of transitional vegetation that lies between continuous forest cover and the treeless alpine tundra. This zone, which is included in the definition of alpine areas, includes forest with openings (forest tundra), park-like areas with tree clusters (tree islands and ribbon forests), stunted and wind distorted trees, and the krummholz (low, horizontally spreading bushy conifers) zone. The transition takes place over a range in elevation of about 300 m, depending on local edaphic conditions. If temperature change proceeds as predicted, then one of the most apparent results will be a rise in the timber line (Fig. 2.14). This will result in a decrease in the extent of the alpine tundra zone, thus altering the appearance of many mountain areas and their potential for recreation and tourism. The impact of a reduction in the extent of treeless alpine areas is hard to assess because it will occur slowly and irregularly. Alpine meadows will be transformed into parkland or forest and present subalpine parkland will develop a continuous tree cover (Ryder, 1998). In the long run, a rise in the timber line will result in an increase in the land base for the timber industry and a decrease in the zone where most snow avalanches are initiated.

Anticipated changes in the cryosphere

Anticipated reduction in snowfall at lower elevations (Moore and McKendry, 1996) will have a detrimental effect on low-lying ski resorts but will improve winter access for other activities. However, increased snowfall at higher elevations will benefit alpine resorts. In relatively cold mountain regions, such as the interior and north of BC, a more general increase in snowfall is expected. Warmer summer temperatures will probably lead to a rise in the summer snow line and significant loss of glaciers, thereby changing the scenery in southern mountain areas. In regions where snowfall will increase, snow avalanche activity is also likely to increase but there may also be counterbalancing effects from the anticipated tree colonisation of avalanche initiation zones and from the increased stability of snowpacks under warmer temperatures. Seasonally frozen ground is most widespread in northern and more continental mountain systems, where winters are relatively cold and snowpacks are thin. Climate warming will have the greatest effects where permafrost is relatively warm and thin, as is the case at the southernmost limits of discontinuous, sporadic and isolated permafrost or where massive ice is present. In BC, most of the permafrost is either sporadic or isolated, as can be seen in Fig. 2.12, and massive ice permafrost is not a factor. Thin permafrost may disappear. Daily freezing and thawing is most effective in continental and high-latitude mountains where seasonal snow cover is thin and contributes to the shattering of bedrock and the heaving and churning of soil. Periglacial activity is associated with frequent freeze–thaw events covering a wider range of environments. Periglacial activity is thus substantially more intense in the continental Rocky Mountains than in the maritime Coast Mountains of BC (Slaymaker, 1990). Heavy snowfall in all the coastal and insular mountains protects the ground and inhibits permafrost development, even where intense freeze–thaw activity has been monitored above the snow.

2.5.10 Summary of anticipated hydroclimatic effects on British Columbia's mountains

Changes envisaged in BC will have both positive and negative effects. More extreme storms, with intense rain, wind and snowstorms, could fell trees and increase the incidence of flooding and landslides. The interaction of more people with less stable land could well generate higher hazard ratings. Land cover changes, such as the current infestation by pine beetle of the Lodgepole pine over 300 000 km^2 of the interior of the province, can be anticipated. This infestation has been blamed, in part, on higher mean temperatures. Higher temperatures will lead to an upward shift of biotic and cryospheric zones and perturb the hydrological cycle. Warming creates new opportunities for enhanced forest growth but hotter and drier summers may eliminate some tree species. Warmer weather in itself will reduce the extent of the alpine tundra zone and will have unpredictable effects on the integrity of ecosystems that are close to their environmental limits.

2.6 Twenty-first century mountain landscapes under the influence of land use and land cover change

2.6.1 The distinction between land cover and land use

'Land change is a forcing function in global environmental change' (Turner, 2006). This land change may be either land cover change (with relatively small human involvement) or land use change (resulting from a change of purpose for which people exploit the land cover). Land cover change involves material and energy flows that sustain the biosphere, including trace gas emissions and the hydrological cycle; such changes are visible in remotely sensed data and require interpretation and ground referencing. Land use change, by contrast, involves both the manner in which the biophysical attributes of the land are manipulated and the purpose for which the land is used. Such data are commonly derived from detailed ground-based analysis. The conceptual model showing links between environmental change and socioeconomic activity in mountains (Fig. 2.15) is highly generalised.

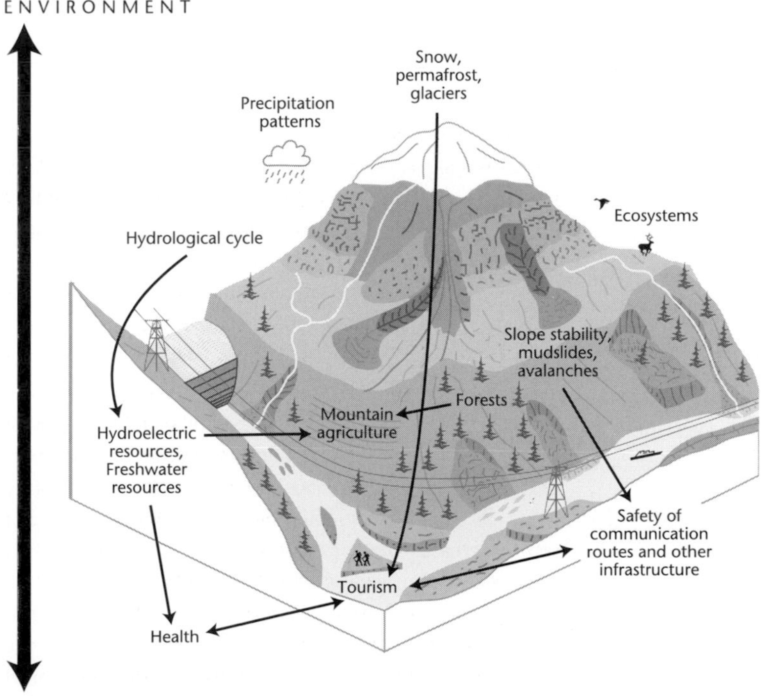

FIGURE 2.15. Suggested links between socioeconomic activity, landscape and global environmental change in mountains.

2.6.2 Population change

Bearing in mind that it is 'population in context' (Rindfuss *et al.*, 2004) which is the driver of environmental change, it is nevertheless important to recognise that global population has doubled in the past 40 years and increased by 2 billion people between 1975 and 2000. Except for parts of Europe, Japan and the Russian Federation, most regions are expected to experience population growth. The rate of growth will vary between just below or above replacement rate to 3% or more in Africa. Urbanisation will continue with the most rapid urban growth rates occurring in the developing world at 2.7% per annum (UN, 2006). The number of cities with more than 1 million inhabitants, which was 80 in 1950, is expected to exceed 500 by 2010 (UN Centre for Human Settlements, 1996).

2.6.3 Agriculture and forestry

Erosional effects of increased settlement and the expansion of livestock grazing in the western United States have been documented in mountain lake sediments in Colorado. The dust load in those sediments increased during the nineteenth and early twentieth centuries by more than 500% above the late Holocene average (Neff, 2008). The degree of soil erosion associated with crop agriculture depends on slope angle and farming system applied. Soil erosion can be minimised by terracing but decay of terraces when not maintained is associated with erosion and gullying. In the Chinese Loess plateau programmes of re-terracing, supported by the World Bank, have converted over 3000 km² of degraded land between 1971 and 2000. Under projected levels of climate change, agricultural yields will increase for most crops but increasing temperatures will shorten growth duration and lead to severe water shortages in southern and eastern Europe. Northern Europe should experience overall positive effects.

Forest management in the developing world is chiefly characterised by deforestation and global decline in forest cover is expected to continue. By contrast, in North America and Europe afforestation and reafforestation will likely continue to expand forest cover. Under climate change timber harvest will probably increase in commercial forests in northern Europe but in the Mediterranean, increased drought and fire risk may compromise forest management (see Chapter 11). Upward expansion of forests would also lead to decreased erosion and potentially reduce flood peaks and sediment transport.

2.6.4 Recreation

Recreational preferences are likely to change with higher temperatures. Outdoor activities will be stimulated in northern Europe but heat waves may reduce peak summer demand in the Mediterranean region. Less reliable snow conditions will certainly impact adversely on winter recreation (IPCC, 2007b). Winter recreation involves substantial infrastructure development and environmental modification, such as the following two examples.

Grooming of ski slopes

This strategy involves the landscaping of large ski areas with the aim of reducing the snow depth required for skiing. Additional snow management measures to support this strategy are the erection of snow fences to capture moving snow, the planting or retention of trees to partially shade the ski runs, and the drainage of wetland areas in order to avoid delayed snow accumulation and premature snowmelt (Agrawala, 2007). Landscaping, in particular bulldozing, has a huge impact on alpine vegetation and slope denudation. Wipf (2005) found that the proportion of bare ground was almost five times higher on graded than on non-graded plots and that revegetation is therefore difficult to achieve, especially at high elevations. Machine grading of ski slopes in Bavaria appears to have influenced the area's sensitivity to erosion, as 63% of all damage caused by erosion occurred on modified ski slopes (Dietmann and Kohler, 2005). Bulldozing ski runs is also having a negative effect on the attractiveness of the alpine environment. This may negatively impact summer tourism. Another measure is snow compaction on ski runs with heavy snow-grooming vehicles. This increases soil freezing and the formation of ice layers and leads to mechanical damage and a delay in plant development.

Artificial snow-making

Snow-making is the most widespread adaptation strategy used by ski area operators. It is used to extend the operating season and to increase ability to cope with climate variability. In order to secure the water supply for snow-making, many mountain reservoirs or artificial lakes have been built in recent years. However, not only does the construction of such reservoirs involve high costs, but it can also be destructive on the environment and lead to 'scars' on the alpine landscape, as new roads have to be built to facilitate access (Agrawala, 2007). Artificial snow also has an impact on alpine vegetation. Later snowmelt leads to a further delay in plant development and, in addition, artificial snow increases the input of water and nutrients to ski runs, which can have a fertilising effect and change the composition of plant species. On the other hand, some of the impacts of ski slope preparation can be mitigated by the deeper snow cover, e.g. soil freezing and the mechanical impacts of snow-grooming vehicles.

2.6.5 Natural hazards

In mountain landscapes, extreme geophysical events interact with social systems in dramatic ways. Thus, for example, the 12 May 2008 earthquake in Szechwan, China generated hundreds of large landslides, blocked more than 30 large lakes, killed 75 000 people and caused millions of dollars in damages. Earthquakes and floods, which are not exclusively mountain hazards, account for more than 50% of the damages caused by natural hazards globally; and the highest damages in recent years in Switzerland, Austria and France, for example, were due to floods and windstorms (Agrawala, 2007).

Seismic hazards: the case of Tajikistan

The country of Tajikistan is part of the Pamir–Alai mountain system. Ninety-two percent of its land area of *c.* 140 000 km^2 is mountainous; nearly 50% lies above 3000 m and is classified as dry, cold desert (Plate 10). It is drained by the headwaters of the Amu Darya and Syr Darya rivers, the major feeders of the Aral Sea. This is a zone of intensive seismic activity and steeplands which are geomorphically highly active. Earthquakes with magnitudes exceeding 5 on the Richter scale have a recurrence interval of 75 days and the region is well known for the earthquake-triggered natural dams which have blocked large lakes, holding as much as 17 km^3 of water in the case of Sarez Lake, and presenting a permanent risk of catastrophic draining (Alford *et al.*, 2000). Rockfalls and massive rockslides have accounted for the deaths of more than 100 000 victims during the twentieth century, and this in a country of barely 6 million people. Thirty-one per cent of the country is said to be agricultural land and 13% is under forest, but overgrazing of the rangelands by sheep and rapid deforestation has encouraged widespread erosion on slopes and more frequent occurrence of mudflows and landslides.

Glacier hazards

Surging glaciers are concentrated in a few mountain ranges, notably subpolar maritime and subtropical regions: Alaska–Yukon, Iceland and Svalbard (Jiskoot *et al.*, 1998) are examples of the former and the Karakoram and Pamir ranges and the Argentinian Andes are examples of the latter. Surging glaciers commonly surge once or twice a century. They are often accompanied by ice margin mass movement, ponding and sudden releases of meltwater (*jökulhlaups*) and heavy proglacial sediment loads. Glacial retreat also favours the formation of glacial lakes and ice avalanches, and disastrous events such as glacial lake outburst floods (GLOFs); these are the most destructive hazards originating from glaciers due to the large water volume and large areas covered. Luckily, glacial lakes, from which GLOFs originate, usually form slowly and can be monitored. McKillop and Clague (2007) have estimated the probability of the occurrence of GLOFs in southern British Columbia.

Mass movement hazards

Of special concern are debris avalanches, debris torrents, large debris flows, catastrophic rock wall failures and rock avalanches. Rock avalanches (or *Bergstürze*) are among the most catastrophic of subaerial processes. In mountain regions debris torrents are triggered either by high discharges associated with thunderstorms, snowmelt and prolonged rainfall or by debris avalanches and are sudden and violent phenomena which can carry large amounts of mud, stones and timber. Timber presents a particular danger, because of its ability to block streams in narrow sections, followed by the catastrophic breakthrough of water and debris.

The debris flow hazard in Austria has been intensively studied (Fig. 2.16). The zone boundaries in Fig. 2.16 were drawn according to the debris loads involved in about 2000 torrent disasters from catchments up to 80 km^2 in size (Table 2.4).

Permafrost-related hazards

Permafrost degradation is likely to contribute to rockfall activity and hiking and climbing could become more dangerous (Behm *et al.*, 2006). In this context, it is quite often postulated that permafrost degradation will trigger higher debris flow activity but research into this link so far remains inconclusive. A recent study showed that frequency of debris flows originating from permafrost areas in Ritigraben (Swiss Alps) has been decreasing, although a lower frequency may also be associated with more intense events due to larger accumulation of materials between events (Stoffel and Beniston, 2006). Fischer *et al.* (2006) have identified permafrost degradation as a cause of slope instabilities on Monte Rosa; and Gruber and Haeberli (2007) provide a comprehensive review of permafrost and slope instability.

Snow avalanches

Steep slopes and heavy snowfall at high elevations are the main factors affecting avalanche incidence. Large avalanches can run for kilometres, and create massive destruction of the lower forest and inhabited structures. The real costs are perhaps best expressed through the costs of avalanche defences. In countries such as Canada, Switzerland, Austria, France, Italy and Norway expenditures involve tens of millions of dollars annually. No clear trends have been identified in the frequency and number of avalanches in the European Alps in the past century (Agrawala, 2007) but in many avalanche-prone areas the density of buildings and other investments has increased.

2.6.6 Case study: Austria's mountains under the influence of land use and land cover changes

Two-thirds of Austria is occupied by the mountainous terrain of the Eastern Alps (Plate 11). The westernmost half of the Eastern Alps is deeply dissected and consists of 10% Class 3 mountains, 75% Class 2 and 5% Class 1. They culminate in Grossglockner at 3800 m above sea level and have summit heights which range from 3000 to 3800 m above sea level and a relative relief of 1500–2000 m 5 km^{-1}. The eastern half of the Eastern Alps consists of 60% Class 2, 30% Class 1 and 10% non-mountains. Over the last 80 km of the Eastern Alps to their termination at Vienna, the mountains are all reduced to Class 1. Areas suitable for habitation are scarce in the Eastern Alps. Some lesser inclined surfaces can be found on the shoulders of glacial troughs, but the main part of the living space is concentrated in the lower valley slopes and the valley floors. Mountain people in the Eastern Alps have always lived in the main valley bottoms which are below 1000 m above sea level (see Table 2.5).

TABLE 2.4. *Criteria for debris flow hazard boundaries*

Zone	Observed maximum debris loads
Zone A	>100 000 m^3, exceptionally < 1 000 000 m^3
Zone B	100 000–200 000 m^3
Zone C	<60 000 m^3
Zone D	<20 000 m^3
Zone E	Loess gullies up to several thousand m^3; other torrents up to several hundred m^3

Source: Kronfellner-Kraus (1989).

The agricultural era: population development and landscape change AD 1100–1880

Rapid expansion of settlement and cultivation in the Eastern Alps commenced around AD 1100, favoured by the political consolidation of Europe and by significant climatic amelioration. Three significant landscape changes took place during the next eight centuries: (a) strong depression of the timber line, expansion of the alpine meadows and alteration of the vegetation composition of the high-altitude meadows through grazing; (b) fragmentation of the mountain forest through clear-cutting of all lesser inclined surfaces; and (c) nineteenth-century draining and reclamation of the flat valley floors (Bätzing, 2003). Thus 800 years of intense agricultural use converted the former dense and monotonous forest cover into an ecologically stable cultural landscape of high floristic biodiversity. The aesthetic value of this unique cultural landscape with its small-scale mosaic of different land uses is high. Severe disruptions of the environment were limited in time and space, and were usually connected with mining or warfare. Regulation of rivers and draining of the valley floors at the end of the era were precursors of the industrial and service economy era.

The industrial and service economy era: population development and landscape change since AD 1880

After the onset of the industrial revolution a new industrial economy, which culminated in the 1970s, led to long ribbons of densely built-up areas on the floors of the main

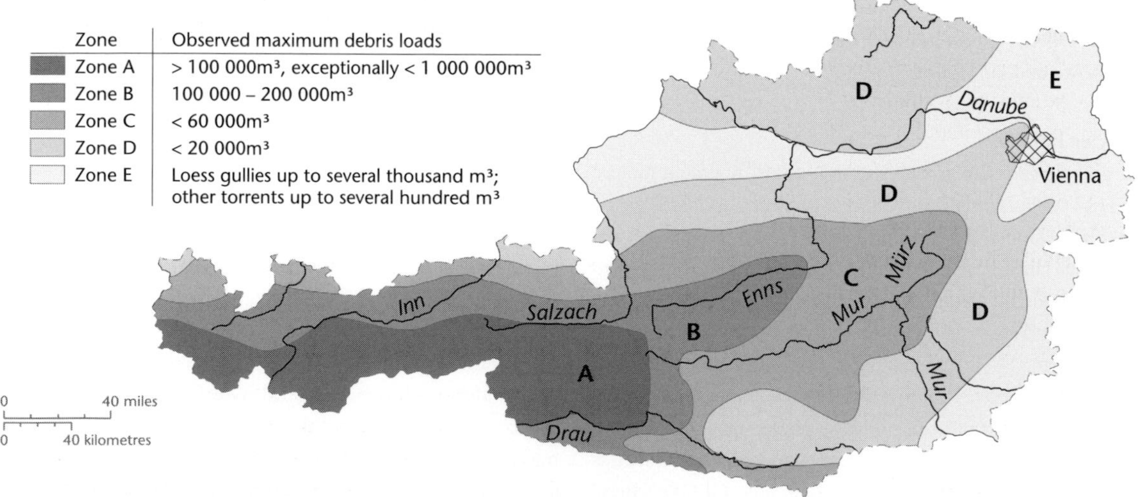

FIGURE 2.16. Regional distribution of torrent and debris flow hazard in Austria, based on observed maximum debris load (from Embleton-Hamann, 2007) (see Table 2.4).

TABLE 2.5. *Vertical population distribution in the Austrian Alps and population changes 1871–2000*

	Lower than 500 m	500–999 m	1000–1499 m	1500–2042 m	Total for Eastern Alps area in Austria	Total for Austrian territory
Area in km^2	11 590 (21%)	33 151 (60%)	9 882 (18%)	323 (1%)	54 946 (100%)	83 871
Population in 1871, total and per km^2	644 157 56	817 194 25	107 377 11	900 3	1 569 628 29	4 498 000 54
Population in 2000, total and per km^2	1 460 053 126	1 672 678 50	165 836 17	1 213 4	3 299 780 60	8 144 000 97
Population changes 1871–2000	227%	205%	154%	135%	210%	181%

Source: After Bätzing (2003) and additional data kindly provided by W. Bätzing from his personal alpine databank.

valleys. New typical 'mountain' economic sectors that evolved in the twentieth century were tourism and hydroelectric power generation. Mass tourism started in the late 1950s and the most rapid growth rates were soon associated with winter sports while summer tourism stagnated. Between 1931 and 1990, 31 reservoirs for hydroelectric power generation were built in Austria (Fig. 2.17; also Plate 8), with large-scale projects starting in the 1950s. Finally, starting around 1970, conservation units, such as parks and natural reserves, were established; they stopped further anthropogenic landscape change in the high-altitude zone. As of 2006 the protected area in the Austrian Alps covers about 10 550 km^2 or 19% (N. Weixlbaumer, personal communication, 2008).

FIGURE 2.17. Two dam sites in Austrian Alps (from Beckel and Zwittkovits, 1981). Upper Kapruner valley, Salzburg Province, with two reservoirs, Moserboden and Wasserfallboden. (See also Plate 8 for colour version.)

Current population development

The demographic development given in Table 2.5 depicts the massive changes that started with the industrial revolution, namely strong population growth in the bottom of the main valleys and lesser growth in the higher zones. In 1870 vertical population distribution already shows a concentration of people below 1000 m above sea level but in the twentieth century the disparity between low and high elevation sites became much greater (see last line of Table 2.5). Population densities in Table 2.5 are based on population figures divided by the area of the administrative unit and are thus comparable with other mountain areas of the world. But they are unable to depict actual population densities on the main valley floors of the Eastern Alps which are currently on the order of 400 to 800 inhabitants km^{-2} and higher around the urban centres marked in Plate 11.

Land cover changes

The Stubai Valley is a microcosm of the massive land cover changes which have occurred in the western part of the Austrian Alps since 1865 (Fig. 2.18). Over the past 150 years all former cropland on the floor of the Stubai Valley has been converted to intensively used hay meadows and built-up areas, accompanied by a high degree of soil sealing. At the next elevation level of the lower valley slopes, former cropland has changed to either settlement or planted forest. Finally there have been marked land use shifts at the high-altitude level. Seventy per cent of the subalpine meadows, which were cleared in the agricultural area and in 1865 were still used for grazing and hay-making, are now abandoned. Forest had been left to regenerate and take over these areas. Figure 2.18 shows that a very strong pulse of

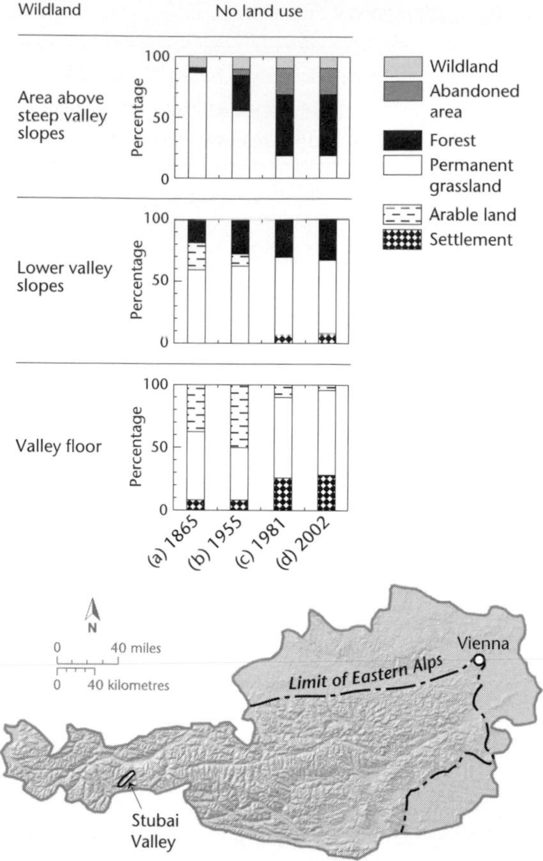

FIGURE 2.18. Land cover changes in the Stubai Valley: (a) 1865; (b) 1955; (c) 1981; and (d) 2002 (modified from Tappeiner *et al.*, 2006); location of the Stubai Valley; and delimitation of the Eastern Alps following the Alpine Convention, which determines the population database.

land use change took place between 1955 and 1981. The trend towards increasing forest cover in the higher elevation zone continues at the present time, but at a slower rate, as indicated by data which refer to total forest area in the Austrian Alps (Borsdorf and Bender, 2007). They note that between 1991 and 2000, forest cover increased by 5.5% (1450 km^2) and subalpine pastures declined by 5.5% (501 km^2).

Impact of changing land cover

The winter tourism industry provides a significant contribution to the economy of Austria. In the 1970s and 1980s the environmental impacts through machine grading or bulldozing of ski runs were greatest (Bätzing, 2003). Since the 1980s, a heightened conservation ethic, coupled with an improving economic situation, has led to the more careful management of ski slopes. New techniques of bioremediation, together with efficient methods of controlling runoff and erosion, have further improved the situation. However, snow-making as a back-up strategy to compensate for climate variability began at the same time. In order to secure the water supply for snow-making, many mountain reservoirs and artificial ponds were built in high-elevation sites associated with considerable impact on the alpine landscape, especially during the construction phase.

On abandoned hay meadows and pastures below the timber line, vegetation succession leads to a dense shrub cover and finally to forest regeneration. The transitional stage has however potential implications for snow gliding and hillslope erosion through the development of small landslides in topsoil. Newesely (2000) showed that the long-stemmed grasses and low-growing dwarf shrub communities of fallow hay meadows promote snow gliding and the formation of avalanches. At a later succession stage, when young trees take over, the snow gliding risk decreases again, as the trees serve as obstacles. At the same time abandoned land with shrub encroachment is significantly more prone to landslide development than meadows and pastures that are still in use (Tasser *et al.*, 2005). Snow gliding seems to play a role in this dynamic: the pull on aerial parts of plants frozen into the base of the snow cover opens tension cracks that become starting points for further mass wasting. Boulders and woody debris transported within the sliding snow slabs can also cause initial erosion fissures. A further factor favouring the break-off of grass and soil clods is that the soil at depths of 25–35 cm under bush or dwarf shrub communities is less densely rooted than under grass communities (Tasser *et al.*, 2005). These shallow landslides developing on abandoned pastures are not very large, but the process becomes important because of the large land areas of the subalpine belt in the Austrian Alps affected by them.

Finally, reservoirs change hydrological processes and fluvial sediment transport. The disturbance of natural processes is not limited to those valleys with reservoirs, as large-scale water diversions also involve surrounding valleys. Since the completion of the last large reservoir in 1990, all further plans of power plant operators to develop new reservoir sites have met with strong opposition from the mountain population.

Future scenarios: implications of climate warming on land cover in the Eastern Alps

Natural snow reliability for a financially viable winter tourism industry is currently provided by areas above 1200 m in western Austria and above 1050 m in the colder continental eastern part of Austria. With a +2 °C scenario by AD 2050, the baseline of snow reliability would rise by 300 m and the number of snow-reliable ski areas would drop to between 8% (eastern Austria) and 62 % (western Austria) of the present level (Agrawala, 2007). Artificial snow-making remains the dominant adaptation strategy. Other measures include moving ski areas to higher altitudes and glaciers, protecting against glacier melt with white plastic sheets and diversification of tourism revenues by promoting summer tourism. Rising temperatures will favour summer tourism. Nevertheless, three assets of the alpine landscape for summer tourism are endangered. These are: the reduced presence of glaciers through climate warming; the loss of amenity of a unique cultural landscape through the current trend of reforestation (as completely overgrown areas are perceived to be less scenic) (Hunziker and Kienast, 1999); and the unspoiled aesthetic quality of the alpine landscape (Hamann, 1994). Hikers are critical of intrusive man-made elements, such as prepared ski slopes, pylons visible on the skyline or cuttings for ski lift tracks. Positive effects of current climate and land use trends, especially afforestation, include a reduced risk of erosion, avalanche and flooding for the residential areas in the valleys and a reduction in costs currently spent on torrent and avalanche control in Austria.

2.6.7 Summary

Population pressure, expressed especially through agriculture, forestry, recreation and urbanisation processes, is expected to exacerbate the effects of climate change on mountain lands. In particular, it is projected that natural hazards, a product of the interaction of society and extreme geophysical events, will become more widespread and more economically damaging. There are, however, compensating trends, illustrated by examples drawn from the Eastern Alps, where mountain landscapes are being enhanced. The expansion of protected areas, net afforestation and reduced intensity of forest harvesting are human impacts which increase the sustainability of mountain landscapes in the face of anticipated hydroclimatic changes.

2.7 Vulnerability of mountain landscapes and relation to adaptive capacity

2.7.1 Mountain landscape disturbance regimes

There is an increasing sense that almost all mountain landscapes are transitional from one landscape-forming regime to another and almost all mountain landscapes contain evidences of past processes, which operated at different rates under different relief, climate and, in many cases, sea level conditions (Section 1.11.1). Landscapes in transition from one regime to the next would seem to retain, in principle, an almost continuous series of transient solutions to a given set of relief, hydroclimate and sea level conditions. This condition has been described by Hewitt (2006) as a disturbance regime landscape; by this he meant that disturbances occur so frequently that the landscape gets no chance to equilibrate with contemporary processes (Section 1.11.3). Mega-landslides in the Himalayas are the regional example which Hewitt has described. This idea of a landscape which is in a constant state of disturbance is also consistent with the case where there is an approximate balance between tectonic, hydroclimatic and land use related processes under effectively constant sea level.

Sensitive mountain environments

One of the implications of the fact that many mountain landscapes are disturbance regime landscapes is that they are exceptionally sensitive to environmental change and are in this sense geomorphically vulnerable. More than one-sixth of the world's population, which lives in glacier- or snowmelt-fed river basins, is expected to be affected by a decrease in water volume stored in glaciers and snowpack, an increase in the ratio of winter to annual flows, increased flood hazard and possibly a reduction in low flows caused by decreased glacier extent or melt season snow water storage by 2050 (IPCC, 2007b).

Geomorphic vulnerability is not only expressed by the frequent incidence of earthquake-triggered landslides, but, for example, by the fate of Himalayan glaciers which cover 17% of the mountain area, and are predicted to shrink from their present area of 500 000 km^2 to 100 000 km^2 by AD 2035 (WWF, 2005). The glaciers on Mt Kilimanjaro are likely to disappear by AD 2020 (Thompson, 2002). Destabilising mountain walls, increasing frequency of rockfall, incidence of rock avalanches, increase and enlargement of glacial lakes and destabilisation of moraines damming these lakes are expected to accompany increased risk of outburst floods.

Rainfall amounts and intensities are the most important factors in water erosion (Nearing, 2005) and they affect slope stability, channel change and sediment transport. Increased precipitation intensity and variability is projected to increase the risk of floods and droughts in many areas. Changes in permafrost will affect river morphology through destabilising of banks and slopes, increased erosion and sediment supply (Vandenberghe, 2002).

2.7.2 Uncertainties surrounding adaptive capacity in mountain landscapes

Mountain societies contain a higher incidence of poverty than elsewhere and therefore have a lower adaptive capacity (for definition see Chapter 1) and a higher vulnerability to environmental change. Mountain peoples have been made more vulnerable to natural extreme events by vast numbers being uprooted and resettled in unfamiliar and more dangerous settings (Hewitt, 1997). There are at least two interesting conclusions from the 2007 IPCC Assessment:

- there is a high degree of confidence (8 out of 10 chance) that climate change will disrupt mountain resources needed for subsistence and that competition between alternative mountain land uses is likely to increase under projected climate change and population increase scenarios; but
- there is only a medium degree of confidence (5 out of 10 chance) in being able to predict the response of mountain populations to such changes.

These conclusions go some way towards explaining the inadequacy of an exclusively geosystem-based definition of landscape vulnerability and why the IPCC has incorporated adaptive capacity into the definition.

Environmental disasters are a function of both climatic and non-climatic drivers and that is why greater attention to systemic thresholds of relief and land use, and their interactions with climate, is needed. Unfortunately, there has been inadequate representation of the interactive coupling between relief, land use and climate in the climate change discussions to date (Osmond, 2004). Based on a number of criteria in the literature such as magnitude, timing, persistence/reversibility, potential for adaptation, distributional aspects of the impacts, and likelihood and importance of the impacts, some of these vulnerabilities have been identified as 'key' (Fussel and Klein, 2006). It is the point of exceedance of thresholds, where non-linear processes cause a system to shift from one major state to another, which expresses this key vulnerability. The example quoted in IPCC (2007b) is exclusively concerned with climatic thresholds. For example, where global mean temperature increases by 2 °C above the 1990–2000 levels some current key vulnerabilities will be exacerbated, but some systems such as global agricultural activity at mid and high latitudes and elevation could benefit. However, global mean temperature changes greater than 4 °C above 1990–2000 levels would lead to major increase in vulnerability, exceeding the adaptive capacity of many systems and having no discernible benefits.

Development of improved scenarios of future mountain landscapes

The spatial variability of mountain landscapes makes scenario building difficult. Baseline conditions are simply too variable to allow easy generalisation. Results achieved by the IPCC Special Report on Emissions Scenarios (SRES) and the approaches and methods being used in climate change impacts, adaptation and vulnerability (CCIAV) can be viewed in IPCC (2007b).

Global studies suggest that at least until AD 2050 land use change will be the dominant driver of change in human-dominated regions (UNEP, 2002). Not only are there geosystem disturbance regimes, such as those discussed by Hewitt (1997) but land use change, fire and insect outbreaks can also be analysed as disturbance regimes, using a shorter response timescale (Sala, 2005). Similarly, overgrazing, trampling and vegetation destabilisation have been analysed in this way in the Caucasus and Himalayas (IPCC, 2001b). There is growing evidence since the IPCC Third Assessment Report (IPCC, 2001b) that adaptations that deal with non-climatic drivers are being implemented in both developed and developing countries. Examples of adaptations to land use change, such as construction, decommissioning and management of reservoirs, and adaptations to extreme sediment cascades and relief, as well as their interactions with climate include the following:

- partial drainage of the Tsho Rolpa glacial lake in Nepal designed to relieve the threat of GLOFs (Shrestha and Shrestha, 2004);
- Sarez Lake in Tajikistan and the so-called 'quake lakes' in Szechwan Province, China are commanding national and international funding;
- increased use of artificial snow-making by the alpine ski industry in Europe, Australasia and North America); and
- more frequent/intense occurrences of extreme weather events will exceed the capacity of many developing countries to cope.

2.7.3 Case study: the Ethiopian Highlands

Ethiopia has an area of 1.2 billion km^2, with 43% over 1500 m above sea level. Much of the Ethiopian landmass is part of the East African Rift Plateau. Its general elevation ranges from 1500 to 3000 m above sea level and reaches a maximum elevation of 4620 m. It is divided into four physiographic regions: the Northwest Highlands, the Great Rift Valley, the Southeast Highlands and the Peripheral Lowlands (Plate 12). The terms plateau and highlands are somewhat misleading as the country is genuinely mountainous, especially the Northwest Highlands, which are deeply dissected by the Blue Nile and Tekeze rivers and their tributaries. The Northwest Highlands consist of 15% Class 3, 70% Class 2 and 5% Class 1 mountains; the Southeast Highlands include 10% Class 3, 70% Class 2 and 5% Class 1 mountains. Only a small proportion (10% and 15% respectively) is not mountainous on the basis of the definition used here. Elevation-induced climatic conditions form the basis for three traditional environmental zones: cool, temperate and hot. The cool zone consists of the central parts of the Northwest Highlands. The terrain there is generally above 2400 m and average daily highs range from near freezing to 16 °C. The temperate zone is found between 1500 and 2400 m where daily highs range from 16 to 30 °C. The hot zone consists of areas where the elevation is below 1500 m and where the average annual daytime temperature is 27 °C with daily highs reaching 50 °C in the northernmost part of the Great Rift Valley.

Population and land cover

The population of the country was 77 million in 2007 and is increasing by 2 million per annum. The Population Division of the UN Department of Economic and Social Affairs (2006) predicts the population will reach 100 million by AD 2015. About 80% of the population and 70% of the livestock live in the ecologically fragile highlands where population growth and poverty are higher than in urban areas. The mountain area is responsible for nearly all the food crops and most of the cash crops produced in the country. The intermediate slopes (1500–1800 m) and upper (>3500 m) belts in the northern massifs are most sensitive to disturbance (Getahun, 1984). Out of the 540 000 km^2 of mountain lands, 70% is grassland used as grazing and browsing land, 15% is cultivated land, 12% is degraded badlands and 3% is forest. The FAO (2007) estimated a deforestation rate of 1410 $km^2\ a^{-1}$ over the past three decades. The present natural forest area is around 43 000 km^2 (Earth Trends, 2003). The Afromontane rainforests of southwestern Ethiopia are the world's birthplace of *Coffea arabica* and contain their last wild populations. If current deforestation rates were to continue there would be no coffee-bearing forests left in about 27 years. The main reasons for the deforestation are population pressure, insecurity of land tenure and governance issues, reinforced by one extended drought period (Whiteman, 1998).

Contemporary erosion and sediment yield in the highlands

Over the past several decades, Ethiopia's high population growth, unsustainable land use and ambiguous land ownership policies have led to a rapid loss of biomass cover, increased soil erosion and desertification (Tekle and Hedlund, 2000). Past and present social relations are responsible for poverty leading to the present land degradation. A reversal of the present desertification and land degradation would seem eminently possible under improved socioeconomic conditions, but the second of the conclusions from the 2007 IPCC Assessment noted under Section 2.7.2 must be borne in mind.

Soil and water conservation structures have been widely implemented since the 1970s. Local knowledge and farmers' initiatives have been integrated with conservation measures and impact assessments have shown clear benefits in controlling runoff and soil erosion. In high-rainfall areas runoff management is expensive and beyond the means of private farms (Nyssen, 2004). Temporal rain patterns, apart from the catastrophic impact of dry years on the degraded environment (1978–1988) cannot explain the current desertification in the driest parts of the country and the accompanying land degradation elsewhere (Nyssen, 2004). Hurni *et al.* (2005) show that surface runoff and sediment yield from the Northwest Highlands into the upper Nile basin have most probably increased due to intensified land use and land degradation induced by population increase. Climate change has intensified these problems by altering the region's rainfall patterns. Although relief and human activity are the root causes of the environmental decline, hydroclimate is also important in the form of the incidence of drought, such as that which occurred in 2003 and affected 13 million people and the decade of the 1980s (the Sahel crisis).The most important present-day geomorphic processes are sheet and rill erosion and gullying (Fig. 2.19). Based on existing data from the central and northern Ethiopian highlands:

$$SY = 2595A^{0.29} \qquad (2.2)$$

where SY is specific sediment yield in tonnes $km^{-2}\ a^{-1}$, and A is drainage area (km^2), $n=20$; $r^2=0.59$ (Nyssen, 2004).

Environmental rehabilitation in the Tigray Highlands

Nyssen *et al.* (2007) and Munro *et al.* (2008) have made a multi-scale assessment over a period of 30 years of environmental rehabilitation in the Tigray Highlands of northern Ethiopia (1975–2006). Their study shows that sheet and rill erosion rates have decreased, infiltration and spring discharge are enhanced and vegetation and crop cover have improved. The environmental changes are assessed on the basis of 10 years' field research (on-farm and catchment scale) and comparison of 51 historical photographs taken in 1975 with the current status. One example of the kinds of evidence used in these studies can be seen in Plate 13. Six experts interpreted ten visible indicators of soil erosion on a five-point scale from −2 to +2. Whereas the population of Ethiopia has increased from 34 to 77 million between 1975 and 2006 land management and vegetation cover have improved on average. Universal Soil Loss Equation (USLE) estimates of average soil loss by sheet and rill erosion and suspended sediment yield to local reservoirs confirm the unanimous conclusions from the time lapse photography. Five of the elements interpreted from the time lapse photography (vegetation, sheet and rill erosion,

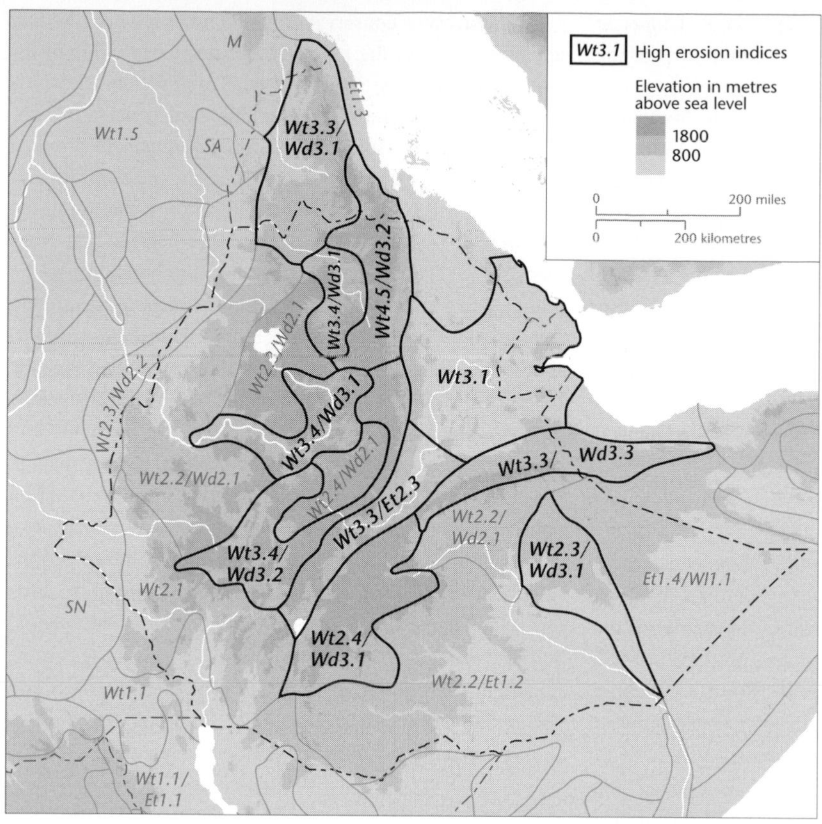

FIGURE 2.19. Extract from the Global Assessment of Soil Degradation (GLASOD) (Oldeman *et al.*, 1990) map of Ethiopia (modified from Nyssen *et al.*, 2004). Letters indicate the type of erosional process that is dominant; the first digit after the letter indicates the degree of seriousness on a scale of 1–4 and the second digit indicates the spatial extent of the problem on a scale of 1–4. Wt, loss of topsoil due to water erosion; Wd, terrain deformation by rill and gully erosion and/or mass wasting; Et, loss of topsoil due to wind erosion.

management of farm and other land and overall assessment) indicated significant improvement over the 30-year period (Plate 13). Land cover dynamics in some small experimental watersheds have confirmed improvements over the past three or four decades but there is much regional variability in the Ethiopian highlands. Dessie and Kleman (2007) note the radical reduction in forest cover in the south–central Rift Valley over the past three decades. At the same time it should be noted that local examples of exceptional degradation and expansion of gully systems require priority attention. The observed changes are not climate-driven and cast doubt on the hypothesis of the irreversibility of land degradation in marginal semi-arid mountain systems.

2.7.4 Summary

Not only are the mountains themselves changing through high rates of denudation as a function of steep slopes and active tectonism and volcanism, but hydroclimate and runoff and land use patterns are changing simultaneously. Future landscape scenarios can only be tentatively described and the full range of options presented in the metaphor 'panarchy' (Holling, 2001) provides valuable interpretations (see Chapter 1). Panarchy, in so far as it deals with adaptive systems, is superbly suited to the analysis of disturbance regimes within geosystems, ecosystems and socioeconomic systems as illustrated in the following paragraph.

Landscapes in polar mountains (such as Svalbard) are in the exploitation phase of an adaptive cycle; landscapes in low population density temperate mountains (such as British Columbia) are in the conservation phase; landscapes in high population density temperate mountains (such as Austria) are in a reorganisation phase and high population density tropical environments (such as Ethiopia) may be close to collapse. If the panarchy metaphor is applied, it indicates possibilities of adaptation where system connectivity, resilience and internal strength are adequate.

With respect to location within these mountain systems, the evidence available suggests that in polar mountains, the highest elevations and the steepest slopes are the sites of most rapid change; in low population density temperate environments, the sites of most rapid change vary as a function of glacierisation, past glaciation and never-glaciated conditions; in high population density temperate mountains, the valley floors and intermediate slopes have experienced the greatest land use and land cover change and in high population density tropical mountains intermediate elevation zones experience the most damaging natural hazards, due to the combination of infrastructure investment and high relief.

The three drivers of environmental change in mountains are relief, hydroclimate and runoff, and human activity. Relief controls the instability of the landscape and ensures that the impacts of changes in hydroclimate and runoff and human activity are transferred rapidly through the system. Not only are mountains unstable in themselves because of their high relief but the pace and character of development and the magnitude and frequency of hydroclimate and run-off all militate against sustainability. It might be more practical to speak of avoiding unsustainability in mountain systems, rather than attempting to achieve the impossible goal of sustainability in a disturbance regime dominated landscape.

APPENDIX 2.1
The components of topographic relief

The primary driver of geomorphic change is the downslope force of $g \sin \alpha$. Land at elevations greater than 2500 m above sea level is commonly classified as mountain, but the major exception to this generalisation is the Tibetan Plateau. In this respect, the Meybeck *et al.* (2001) typology which includes a relief roughness criterion at all altitudes is a significant improvement. The UNEP-WCMC (2000) classification defines 29.4 Mkm^2 of mountain terrain (even though it includes 'mountains' as low as 300 m above sea level and also includes the Tibetan Plateau); Meybeck *et al.* (2001) define 33.3 Mkm^2 (excluding the Tibetan Plateau and excluding the 300–500 m above sea level range); Kapos (2000) defines 35.8 Mkm^2 (including the Tibetan Plateau and the 300–500 m range). Estimates would seem to be in the range of 28–33 Mkm^2 for all mountains above 500 m. UNEP-WCMC counted 719 million people in their mountain regions; Meybeck *et al.* (2001) found as many as 1420 million people, using Vorosmarty's (2003) estimates based on both national census and nocturnal light emission data. Huddleston (2003) uses population estimates from the Oak Ridge National Laboratory (Appendix 2.2).

APPENDIX 2.2
Methodology for classification of mountain regions and their population

Because the focus of this global classification scheme is on how to characterise mountain ranges rather than individual mountains or hills, the resolution is limited to 1 km within a 30 × 30 minute cell of land.

(1) Mountain region type and extent
Sources: United Nations Environment Programme-World Conservation Monitoring Centre (UNEP-WCMC) (2000); *Mountains of the World* (2000) Cambridge, UK.
Note: the UNEP-WCMC map, published in *Mountains of the World* (2000) has been used as the reference map for this analysis. To produce this map, UNEP-WCMC used topographical data from the 1996 USGS EROS Data Center (USGS-EDC) global digital elevation model (GTOPO30) to generate slope and local elevation range (LER) on a 30 arc-second grid of the world. These parameters were combined with elevation to define the six mountain area classes defined by UNEP-WCMC (2000) (see Appendix 2.3).

(2) Global population database
Source: Oak Ridge National Laboratory (2000). LandScan 2000 global population database. Oak Ridge, USA.
Note: the LandScan data set is a worldwide population data base compiled on a 30 arc-second grid. Census counts (mainly at sub-national level) were apportioned to each grid cell based on probability coefficients, which are based on proximity to roads, slope land cover and night-time lights.

(3) Global cities database
Sources: United Nations Environment Programme (UNEP)-GRID-Arendal, (1990). The World Cities Population Data Base (WCPD). Environmental Systems Research Institute (ESRI) 2002. Redlands, USA.

APPENDIX 2.3
Processing SRTM (Shuttle Radar Topographic Mission) 90 m Digital Elevation Data (DEM) Version 3.0

Source: Jarvis *et al.* (2006).
Database: http://srtm.csi.cgiar.org
Note: with the use of hole-filling algorithms Version 3 provides seamless, complete coverage of elevation for the globe.

References

Agrawala, S., ed. (2007). *Climate Change in the European Alps: Adapting Winter Tourism and Natural Hazard Management*. Paris: OECD.

Ahnert, F. (1970). Functional relationships between denudation, relief and uplift in large mid-latitude drainage basins. *American Journal of Science*, **268**, 243–263.

Alford, D., Cunha, S. F. and Ives, J. D. (2000). Mountain hazards and development assistance: Lake Sarez, Pamir Mountains, Tajikistan. *Mountain Research and Development*, **20**, 20–23.

Ashmore, P. and Church, M. (2001). The impact of climate change on rivers and river processes in Canada. *Geological Survey of Canada Bulletin*, **555**.

Auer, I. (2007). HISTALP: Historical Instrumental Climatological Surface Time Series of the Greater Alpine Region. *International Journal of Climatology*, **27**, 17–46.

Barry, R. G. (1992). *Mountain Weather and Climate*. London: Routledge.

Barry, R. G. (1994). Past and potential future changes in mountain environments: a review. In M. Beniston, ed., *Mountain Environments in Changing Climates*. London: Routledge, pp. 3–33.

Barsch, D. 1993. Periglacial geomorphology in the twenty-first century. *Geomorphology*, **7**, 141–163.

Barsch, D. and Caine, N. (1984). The nature of mountain geomorphology. *Mountain Research and Development*, **4**, 287–298.

Battarbee, R. W. (2002). Comparing paleolimnological and instrumental evidence of climate change for remote mountain lakes over the last 200 years. *Journal of Paleolimnology*, **28**, 161–179.

Bätzing, W. (2003). *Die Alpen: Geschichte und Zukunft einer europäischen Kulturlandschaft*. München: Ch. Beck.

Beckel, L. and Zwittkovits, F., eds. (1981). *Landeskundlicher Flugbildatlas*. Salzburg: Otto Müller Verlag.

Behm, M., Raffeiner, G. and Schöner, W. (2006). *Auswirkungen der Klima- und Gletscheränderung auf den Alpinismus*. Wien: Umweltdachverband.

Beniston, M. (2000). *Environmental Change in Mountains and Uplands*. London: Arnold.

Bernbaum, E. (1998). *Sacred Mountains of the World*. Berkeley: University of California Press.

Bogen, J. (2006). Sediment transport rates of major floods in glacial and non-glacial rivers in Norway in the present and future climate. In J. S. Rowan, R. W. Duck and A. Werritty, eds., *Sediment Dynamics and the Hydromorphology of Fluvial Systems*. Wallingford: IAHS Press, pp. 148–158.

Borsdorf, A. and Bender, O. (2007). Kulturlandschaftsverlust durch Verbuschung und Verwaldung im subalpinen und hochmontanen Höhenstockwerk: Die Folgen des klimatischen und sozioökonomischen Wandels. In *Alpine Kulturlandschaft im Wandel*. Innsbruck: Innsbrucker Geographische Gesellschaft, pp. 29–50.

Bradley, R. S. (1999). *Paleoclimatology: Reconstructing Climates of the Quaternary.* San Diego: Academic Press.

Brardinoni, F. and Hassan, M. A. (2006). Glacial erosion, evolution of river long profiles and the organization of process domains in mountain drainage basins of coastal B.C. *Journal of Geophysical Research*, **11**, doi:10.1029/2005JF000358.

Brown, J. (2001). *Circum-Arctic Map of Permafrost and Ground Ice Conditions*. Boulder: National Snow and Ice Center Data/World Data Center for Glaciology.

Brunsden, D. (1980). Applicable models of long term landform evolution. *Zeitschrift für Geomorphologie, Supplementband*, **36**, 16–26.

Brunsden, D. (1993). The persistence of landforms. *Zeitschrift für Geomorphologie, Supplementband*, **93**, 13–28.

Caine, N. (1974). The geomorphic processes of the alpine environment. In J. D. Ives and R. G. Barry, eds., *Arctic and Alpine Environments*. London: Methuen, pp. 721–748.

Caine, N. (2002). Declining ice thickness on an alpine lake is generated by increased winter precipitation. *Climatic Change*, **54**, 463–470.

Caine, N. (2004). Mechanical and chemical denudation in mountain systems. In P. N. Owens and O. Slaymaker, eds., *Mountain Geomorphology.* London: Edward Arnold, pp. 132–152.

Church, M. (1998). The landscape of the Pacific Northwest. In D. L. Hogan, P. J. Tschaplinski, and S. Chatwin, eds., *Carnation Creek and Queen Charlotte Islands Fish/Forestry Workshop: Applying Twenty Years of Coast Research to Management Solutions*. Victoria: BC Forestry Research Branch, pp. 13–22.

Church, M. (1999). Fluvial clastic sediment yield in Canada: scaled analysis. *Canadian Journal of Earth Sciences*, **36**, 1267–80.

Church, M. and Ryder, J. M. (1972). Paraglacial sedimentation: a consideration of fluvial processes conditioned by glaciation. *Bulletin of the Geological Society of America*, **83**, 3059–3072.

Church, M. and Slaymaker, O. (1989). Disequilibrium of Holocene sediment yield in glaciated British Columbia. *Nature*, **337**, 452–454.

Coulthard, T. J. and Macklin, M. G. (2001). How sensitive are river systems to climate and land use changes? A model based evaluation. *Journal of Quaternary Science*, **16**, 347–351.

Cullen, N. J. (2006). Kilimanjaro glaciers: recent areal extent from satellite data and new interpretation of observed twentieth century retreat rates. *Geophysical Research Letters*, **33**, L16502, doi:10.1029/2006GL027084.

Dadson, S. J. (2003). Links between erosion, runoff variability and seismicity in the Taiwan orogen. *Nature*, **426**, 648–651.

Dessie, G. and Kleman, J. (2007). Pattern and magnitude of deforestation in the south central Rift Valley region of Ethiopia. *Mountain Research and Development*, **27**, 162–168.

Diaz, H. F. and Bradley, R. S. (1997). Temperature variations during the last century at high elevation sites. *Climatic Change*, **36**, 253–279.

Dietmann, T. and Kohler, U. (2005). *Die Skipistenuntersuchung Bayern: Landschaftsökologische Untersuchung in den bayerischen Skigebieten*. Augsburg: Bayerisches Landesamt für Umweltschutz.

Dixon, E. J., Manley, W. F. and Lee, C. M. (2005). The emerging archeology of glaciers and ice patches: examples from Alaska's Wrangell–St.Elias National Park and Preserve. *American Antiquity*, **70**, 129–143.

Drever, J. J. and Zobrist, J. (1992). Chemical weathering of silicate rocks as a function of elevation in the southern Swiss Alps. *Geochimica et Cosmochimica Acta*, **56**, 3209–3216.

Earth Trends (2003). http://earthtrends.wri.org/pdf_library/country_profiles/for_cou_231.pdf

Embleton-Hamann, C. (2007). Geomorphological hazards in Austria. In A. Kellerer-Pirklbauer, ed., *Geomorphology for the Future*. Innsbruck: Innsbruck University Press, pp. 33–56.

Embleton-Hamann, C. and Slaymaker, O. (2006). Classics in physical geography revisited: Jäckli, H. (1957). *Progress in Physical Geography*, **30**, 779–783.

Evans, I. S. (1972). General geomorphometry, derivatives of altitude and descriptive statistics. In R. J. Chorley, ed., *Spatial Analysis in Geomorphology*. London: Harper & Row, pp. 17–90.

FAO (2007). *State of the World's Forests Report 2007*. Rome: Food and Agriculture Organization of the United Nations.

Fischer, L. *et al*. (2006). Geology, glacier retreat and permafrost degradation as controlling factors of slope instabilities in a high mountain rock wall: the Monte Rosa east face. *Natural Hazards and Earth System Sciences*, **6**, 761–772.

Füssel, H.-M. and Klein, R. J. T. (2006). Climate change vulnerability assessments: an evolution of conceptual thinking. *Climatic Change*, **75**, 301–329.

Geertsema, M. (2006). An overview of recent large catastrophic landslides in northern British Columbia. *Engineering Geology*, **83**, 120–143.

Georges, C. (2004). The twentieth-century glacier fluctuations in the Cordillera Blanca, Peru. *Arctic, Antarctic and Alpine Research*, **36**, 100–107.

Getahun, A. (1984). Stability and instability of mountain ecosystems in Ethiopia. *Mountain Research and Development*, **4**, 39–44.

Grove, J. M. (1988). *The Little Ice Age*. London: Methuen.

Gruber, S. and Haeberli, W. (2007). Permafrost in steep bedrock slopes and its temperature-related destabilization following climate change. *Journal of Geophysical Research*, **112** (F2), doi:10.1029/2006JF000547.

Haeberli, W. (1994). Accelerated glacier and permafrost changes in the Alps. In M. Beniston, ed., *Mountain Environments in Changing Climates*. London: Routledge, pp. 91–107.

Hallet, B., Hunter, L. and Bogen, J. (1996). Rates of erosion and sediment evacuation by glaciers: a review of field data and their implications. *Global and Planetary Change*, **12**, 213–235.

Halpin, P. N. (1994). Latitudinal variation in the potential response of mountain ecosystems to climatic change. In M. Beniston, ed., *Mountain Environments in Changing Climates*. London: Routledge, pp. 180–203.

Hamann, C. (1994). The role of geomorphological mapping in scenery appraisal. *Proceedings of the National Science Council of Taiwan, Part C, Humanities and Social Sciences*, **4**, 231–245.

Hastenrath, S. (2005). The glaciers of Mount Kenya 1899–2004. *Erdkunde*, **59**, 120–125.

Hewitt, K. (1997). Risks and disasters in mountain lands. In B. Messerli and J. D. Ives, eds., *Mountains of the World: A Global Priority*. London: Parthenon, pp. 371–408.

Hewitt, K. (2006). Disturbance regime landscapes: mountain drainage systems interrupted by large rockslides. *Progress in Physical Geography*, **30**, 365–393.

Hinderer, M. (2001). Late Quaternary denudation of the Alps, valley and lake fillings and modern river loads. *Geodinamica Acta*, **14**, 231–263.

Holling, C. S. (2001). Understanding the complexity of economic, ecological and social systems. *Ecosystems*, **4**, 390–405.

Holm, H., Bovis, M. and Jakob, M. (2004). The landslide response of alpine basins to post-Little Ice Age glacial thinning and retreat in southwestern British Columbia. *Geomorphology*, **57**, 201–216.

Huddleston, B. A. *et al*. (2003). *Towards a GIS-Based Analysis of Mountain Environments and Populations*, Working Paper No. 10, Environment and Natural Resources. Rome: Food and Agriculture Organization of the United Nations.

Hunziker, M. and Kienast, F. (1999). Potential impacts of changing agricultural activities on scenic beauty: a prototypical technique for automated rapid assessment. *Landscape Ecology*, **14**, 161–176.

Hurni, H., Tato, K. and Zeleke, G. (2005). The implications of changes in population, land use and land management for surface runoff in the upper Nile basin of Ethiopia. *Mountain Research and Development*, **25**, 147–154.

IPCC (2001a). *Climate Change 2001: The Scientific Basis. Contribution of Working Group I to the Third Assessment Report of the Intergovernmental Panel on Climate Change*. Houghton, J. T. *et al*., eds. Cambridge: Cambridge University Press.

IPCC (2001b). *Climate Change 2001: Impacts, Adaptation and Vulnerability. Contribution of Working Group II to the Third Assessment Report of the Intergovernmental Panel on Climate Change*. McCarthy, J. J. *et al*., eds. Cambridge: Cambridge University Press.

IPCC (2007a). *Climate Change 2007: The Physical Science Basis. Contribution of Working Group 1 to the Fourth Assessment Report of the Intergovernmental Panel on Climate Change*. Solomon, S. *et al*., eds. Cambridge: Cambridge University Press.

IPCC (2007b). *Climate Change 2007: Impacts, Adaptation and Vulnerability. Contribution of Working Group II to the Fourth Assessment Report of the Intergovernmental Panel on Climate Change*. Parry, M. L. *et al*., eds. Cambridge: Cambridge University Press.

IUCN (2000). *Guidelines for Protected Areas Management Categories*. Grafenau: IUCN, EUROPARC Federation, IUCN World Commission on Protected Areas and WCMC.

Janke, J. R. (2005). Modelling past and future alpine permafrost distribution in the Colorado Front Range. *Earth Surface Processes and Landforms*, **30**, 1495–1508.

Jarvis, A. *et al*. (2006). Hole-filled SRTM for the global Version 3. CGIAR-CSI SRTM 90 m. Available at http://srtm.csi.cgiar.org/

Jiskoot, H., Boyle, P. and Murray, T. (1998). The incidence of glacier surging in Svalbard. *Computers and Geoscience*, **24**, 387–399.

Kapos, V. (2000). Developing a map of the world's mountain forests. In M. F. Price and N. Butt, eds., *Forests in Sustainable Mountain Development*. Wallingford: CAB International, pp. 4–9.

Klein, A. G. and Kincaid, J. L. (2006). Retreat of glaciers on Puncak Jaya, Irian Jaya, determined from 2000 and 2002 IKONOS satellite images. *Journal of Glaciology*, **52**, 65–79.

Körner, C. and Ohsawa, M. (2005). Mountain systems. In *Millennium Ecosystem Assessment: Ecosystems and Human Well-Being*, vol.1. Washington, DC: Island Press, pp. 681–716.

Kotlyakov, V. M. *et al.* (1991). The reaction of glaciers to impending climate change. *Polar Geography and Ecology*, **15**, 203–217.

Kronfellner-Kraus, G. (1989). Die Änderung der Feststofffrachten von Wildbächen. *Informationsbericht des bayerischen Landesamtes für Wasserwirtschaft*, **1989/4**, 101–115.

Kunkel, K. E. (2003). North American trends in extreme precipitation. *Natural Hazards*, **29**, 291–305.

Lambin, E. F. and Geist, H. J., eds. (2006). *Land Use and Land Cover Change: Local Processes and Global Impacts*. Berlin: Springer-Verlag.

Lemke, P. *et al.* eds. (2007). Observations: changes in snow, ice and frozen ground. In S. Solomon *et al.*, eds., *Climate Change 2007: the Physical Science Basis. Contribution of Working Group I to the Fourth Assessment Report of the Intergovernmental Panel on Climate*. Cambridge: Cambridge University Press, pp. 337–383.

Liu, C. and Woo, M.-K. (1996). A method to assess the effects of climatic warming on the water balance of mountainous regions. In J. A. A. Jones, ed., *Regional Hydrological Response to Climate Change*. Dordrecht: Kluwer, pp. 301–315.

McKillop, R. J. and Clague, J. (2007). Statistical, remote sensing-based approach for estimating the probability of catastrophic drainage from moraine-dammed lakes in south-western British Columbia. *Global and Planetary Change*, **56**, 153–171.

Marston, R. A. (2008). Land, life and environmental change in mountains. *Annals of the Association of American Geographers*, **98**, 507–520.

Marston, R. A., Miller, M. M. and Devkota, L. (1998). Geoecology and mass movement in the Manaslu-Ganesh and Langtang-Jugal himals, Nepal. *Geomorphology*, **26**, 139–150.

Meier, M. F. (2007). Glaciers dominate eustatic sea-level rise in the twenty-first century. *Science*, **317**, 1064–1067.

Menounos, B. (2005). Environmental reconstruction from a varve network in the southern Coast Mountains, BC, Canada. *The Holocene*, **15**, 1163–1171.

Messerli, B. (1973). Problems of vertical and horizontal arrangements in the high mountains of the extreme arid zone (central Sahara). *Arctic and Alpine Research*, **5**, 139–147.

Messerli, B. and Ives, J. D., eds. (1997). *Mountains of the World: A Global Priority*. London: Parthenon.

Meybeck, M., Green, P. and Vorosmarty, C. J. (2001). A new typology for mountains and other relief classes: an application to global continental water resources and population distribution. *Mountain Research and Development*, **21**, 34–45.

Milliman, J. D. and Syvitski, J. P. M. (1992). Geomorphic/tectonic control of sediment discharge to the ocean: the importance of small, mountainous rivers. *Journal of Geology*, **100**, 525–544.

Moore, R. D. and McKendry, I. G. (1996). Spring snowpack anomaly patterns and winter climatic variability, British Columbia. *Water Resources Research*, **32**, 623–632.

Munro, R. N. *et al.* (2008). Soil and erosion features of the central Plateau region of Tigray: learning from photo monitoring with 30 years interval. *Catena*, **75**, 55–64.

Nearing, M. (2005). Modeling response of soil erosion and runoff to changes in precipitation and cover. *Catena*, **61**, 131–154.

Neff, J. C. (2008). Increasing eolian dust deposition in the western United States linked to human activity. *Nature Geoscience*, **1**, 189–195.

Newesely, Ch. (2000). Effects of land-use changes on snow gliding processes in alpine ecosystems. *Basic and Applied Ecology*, **1**, 61–67.

Nyssen, J. (2002). Impact of road building on gully erosion risk: a case study from the northern Ethiopian Highlands. *Earth Surface Processes and Landforms*, **27**, 1267–1283.

Nyssen, J. *et al.* (2004). Human impact on the environment in the Ethiopian and Eritrean Highlands: a state of the art. *Earth Science Reviews*, **64**, 273–320.

Nyssen, J. *et al.* (2007). *Understanding the Environmental Changes in Tigray: A Photographic Record over 30 years*, Tigray Livelihood Papers No. 3. Mekelle, Ethiopia: VLIR-Mekelle University IUC Programme and Zala-Daget Project.

Oerlemans, J. (2001). *Glaciers and Climate Change*. Lisse: Balkema.

Oldeman, L. R., Hakkeling, R. T. A. and Sombroek, W. G. (1990). *World Map on the Status of Human-Induced Soil Degradation*. Wagemingen and Nairobi: UNEP/ISRIC.

Osmond, B. (2004). Changing the way we think about global change research: scaling up in experimental ecosystem science. *Global Change Biology*, **10**, 393–407.

Owen, L. A. (2004). Cenozoic evolution of global mountain systems. In P. Owens and O. Slaymaker, eds., *Mountain Geomorphology*. London: Edward Arnold, pp. 33–58.

Price, M. (1999). *Global Change in the Mountains*. New York: Parthenon.

Prowse, T. D. (2000). *River Ice Ecology*. Saskatoon: National Water Research Institute, Environment Canada.

Rapp, A. (1960). Recent development of mountain slopes in Kärkevagge and surroundings, northern Scandinavia. *Geografiska Annaler*, **42**A, 71–200.

Reid, L. M. and Dunne, T. (1996). *Rapid Evaluation of Sediment Budgets*. Reiskirchen: Catena.

Rindfuss, R. R. *et al.* (2004) Developing a science of land change: challenges and methodological issues. *Proceedings of the National Academy of Sciences of the USA*, **101**, 13 976–13 981.

Ryder, J. M. (1971). The stratigraphy and morphology of para-glacial alluvial fans in south-central British Columbia. *Canadian Journal of Earth Sciences*, **8**, 279–298.

Ryder, J. M. (1998). Geomorphological processes in the alpine areas of Canada: the effects of climate change and their impacts on human activities. *Geological Survey of Canada Bulletin*, **524**.

Sala, O. E. (2005). Global biodiversity scenarios for the year 2100. *Science*, **287**, 1770–1774.

Schiefer, E., Menounos, B. and Slaymaker, O. (2006). Extreme sediment delivery events recorded in the contemporary sediment record of a montane lake, southern Coast Mountains, British Columbia. *Canadian Journal of Earth Science*, **43**, 1777–1790.

Schiefer, E., Menounos, B. and Wheate, R. (2007). Recent volume loss of British Columbia glaciers. *Geophysical Research Letters*, **34**, L16503, 1–6.

Schimel, D. S. (2002). Carbon sequestration studied in western US mountains. *EOS, Transactions of the American Geophysical Union*, **83**, 445–449.

Schumm, S. A. (1963). *The Disparity between Present Rates of Denudation and Orogeny*, US Geological Survey Professional Paper No. 454-H. Washington, DC: US Geological Survey.

Schumm, S. A. and Rea, D. K. (1995). Sediment yield from disturbed earth systems. *Geology*, **23**, 391–394.

Shrestha, M. L. and Shrestha, A. B. (2004). *Recent Trends and Potential Climate Change Impacts on Glacier Retreat/Glacier Lakes in Nepal and Potential Adaptation Measures*, ENV/EPOC/GF/SD/RD(2004)6/FINAL. Paris: OECD.

Sidle, R. C., ed. (2002). *Environmental Changes and Geomorphic Hazards in Forests*. Wallingford: CAB International.

Slaymaker, O. (1987). Sediment and solute yields in British Columbia and Yukon: their geomorphic significance re-examined. In V. Gardiner, ed., *International Geomorphology 1986*, vol. 1, Chichester: J. Wiley, pp. 925–945.

Slaymaker, O. (1990). Climate change and erosion processes in mountain regions of western Canada. *Mountain Research and Development*, **10**, 171–182.

Slaymaker, O. (1993). Cold mountains of western Canada. In H. French and O. Slaymaker, eds., *Canada's Cold Environments*. Montreal: McGill-Queens University Press, pp. 171–197.

Stoffel, M. and Beniston, M. (2006). On the incidence of debris flows from the early Little Ice Age to a future greenhouse climate: a case study from the Swiss Alps. *Geophysical Research Letters*, **33**, L16404, doi:10.1029/2006GL026805.

Summerfield, M. A. and Hulton, N. J. (1994). Natural controls of fluvial denudation rates in world drainage basins. *Journal of Geophysical Research*, **99** (B7), 13 871–13 883.

Syvitski, J. P. M. 2005. Impact of humans on the flux of terrestrial sediment to the global coastal ocean. *Science*, **308**, 376–380.

Tappeiner, U. (2006). Landnutzung in den Alpen: historische Entwicklung und zukünftige Szenarien. In R. Psenner and R. Lackner, eds., *Die Alpen im Jahr 2020*. Innsbruck: Innsbruck University Press, pp. 23–39.

Tasser, E., Mader, M. and Tappeiner, U. (2005). Auswirkungen von Bewirtschaftsänderungen auf die Blaikenbildung im Gebirge. *Mitteilungen der österreichischen Bodenkundlichen Gesellschaft*, **72**, 193–215.

Tekle, K. and Hedlund, L. (2000). Land cover changes between 1958 and 1986 in Kalu District, southern Wello, Ethiopia. *Mountain Research and Development*, **20**, 42–51.

Thompson, L. G. (1998). A 25,000 year tropical climate history from Bolivian ice cores. *Science*, **282**, 1858–1864.

Thompson, L. G. (2002). Kilimanjaro ice core records: evidence of Holocene change in tropical Africa. *Science*, **298**, 589–593.

Thornthwaite, C. W. and Mather, J. R. (1955). The water balance. *Publications in Climatology*, **8**, 1–86.

Thorsell, J. and Hamilton, L. (2002). *A Global Overview of Mountain Protected Areas on the World Heritage List: A Contribution to the Global Theme Study of World Heritage Natural Sites*. Paris: UNESCO.

Tiffen, M., Mortimore, M. and Gichuki, F. (1994). *More People Less Erosion: Environmental Recovery in Kenya*. Chichester: John Wiley.

Tucker, G. E. and Slingerland, R. L. (1994). Erosional dynamics, flexural isostasy and long-lived escarpments: a numerical modelling study. *Journal of Geophysical Research*, **99**, 12 229–12 243.

Turner, B. L. II (2006). Land change as a forcing function in global environmental change. In H. J. Geist, ed., *Our Earth's Changing Land: An Encyclopedia of Land-Use and Land-Cover Change*, vol. 1 (A–K). Westport: Greenwood Press, pp. xxv–xxxii.

Turner, B. L. II *et al.* (1990). *The Earth as Transformed by Human Action*. Cambridge: Cambridge University Press.

UN Centre for Human Settlements (1996). *An Urbanizing World: Global Report on Human Settlements*. Oxford: Oxford University Press.

United Nations Environment Programme – World Conservation Monitoring Centre (UNEP-WCMC) (2000). *Mountains of the World*. Cambridge: Cambridge University Press.

United Nations Environment Programme (UNEP) (2002). *Global Environmental Outlook 3: Past, Present and Future Perspectives*. Nairobi: UNEP.

UN Department of Economic and Social Affairs, Population Division (2006). *World Population Prospects: The 2006 Revision*. New York: UN.

Vandenberghe, J. (2002). The relation between climate and river processes, landforms and deposits during the Quaternary. *Quaternary International*, **91**, 17–23.

Velazquez, A. (2003). A landscape perspective on biodiversity conservation. *Mountain Research and Development*, **23**, 240–246.

Von Grafenstein U. (1999). A mid-European decadal isotope-climate record from 15 500 to 5000 years BP. *Science*, **284**, 1654–1657.

Vorosmarty, C. J. (2003). Anthropogenic sediment retention: major global impact from registered river impoundments. *Global and Planetary Change*, **39**, 169–190.

Walsh, J. (2005). Cryosphere and hydrology. In *Arctic Climate Impact Assessment*. Cambridge: Cambridge University Press, pp. 183–242.

Whiteman, P. T. S. (1998). Mountain agronomy in Ethiopia, Nepal and Pakistan. In N. J. R. Allan, G. W. Knapp and C. Stadel, eds., *Human Impacts on Mountains*. Totowa: Rowan & Littlefield, pp. 57–82.

Wipf, S. (2005). Effects of ski piste preparation on alpine vegetation. *Journal of Applied Ecology*, **42**, 306–316.

Woo, M.-K. (1996). Hydrology of northern North America under global warming. In J. A. A. Jones, ed., *Regional Hydrological Response to Climate Change*. Dordrecht: Kluwer, pp. 73–86.

World Wildlife Fund (WWF) (2005). *An Overview of Glaciers, Glacier Retreat and Subsequent Impacts in Nepal, India and China*, World Wildlife Fund, Nepal Programme. Available at http://assets.panda.org/downloads/himalayaglaciersretreat/

Wu, K. and Thornes, J. B. (1995). Terrace origination of mountainous hill slopes in the Middle Hills of Nepal: stability and instability. In G. P. Chapman and M. Thompson, eds., *Water and the Quest for Sustainable Development in the Ganges Valley*. London: Mansell, pp. 41–63.

Zemp, M. (2006). Alpine glaciers to disappear within decades. *Geophysical Research Letters*, **33**, L13504, doi:10.1029/2006GL026319.

Zhang, T. (2003). Distribution of seasonally and perennially frozen ground in the Northern Hemisphere. *Proceedings of the 8th International Conference on Permafrost*, 21–25 July 2003, Zurich, pp. 1289–1294.

Zurick, D. and Karan, P. P. (1999). *Himalaya: Life on the Edge of the World*. Baltimore: Johns Hopkins University Press.

3 Lakes and lake catchments

Kenji Kashiwaya, Olav Slaymaker and Michael Church

3.1 Introduction

Wetlands (including rivers – see Chapter 4 – and coastal wetlands – see Chapter 5), lakes and reservoirs are distinct elements of the hydrological cycle from the atmosphere to the ocean. In this chapter, the focus is on lakcs and lakc catchments but certain aspects of wetlands and reservoirs require comment. Saline lakes, such as the Caspian, Aral and Dead seas, also come under the purview of this chapter. Lakes, reservoirs and freshwater wetlands vary spatially as a function of relief, climate and human activity and respond variably over time to changes in hydroclimate and human activity. They act as integrators of processes that are taking place in their tributary catchments and are sensitive indicators of environmental change. At the smallest scales (less than 0.01 km^2) lakes merge with wetlands, which are among thc most ephemeral landscape elements (Table 3.1).

Wetlands represent the ecotone between terrestrial and aquatic environments. They occur in depressions in the landscape where the groundwater table intersects the surface for much or all of the year. Wetlands are variously classified as bogs (acidic peatlands maintained by incoming precipitation), fens (peatlands receiving groundwater drainage), marshes (seasonally or perennially flooded land with emergent herbaceous vegetation), swamps (wooded marshes), saline wetlands (pans or playas), permanent rivers (see also Chapter 4) and open water (characteristically less than 2 m deep). Wetlands are perennially or seasonally saturated or flooded ground. Seasonal wetlands are saturated for sufficient time to support vegetation that thrives under saturated conditions. Freshwater wetlands are of major hydrological importance, for they store substantial volumes of terrestrial runoff and are therefore an important flood-mitigating factor. Conversely, they contribute water to the downstream drainage basin far into the dry season and so are a significant source of base flow maintenance. Freshwater wetlands also accumulate biomass which has its own stratigraphy and provides a number of proxies for past environmental conditions. Saline wetlands such as pans and playas are closed topographic depressions which occur in areas of relatively low effective precipitation.

Lakes are also of hydrologic importance in that they reflect changes in precipitation, evaporation, water balance, and hydrochemical and hydrobiological regimes. Interpreting indications of past or future environmental change in lakes is complicated by the fact that they have their own internal processes. These physical, chemical and biological processes transform both lake waters and their contained sediments. It is therefore necessary to understand the basics of the physics, chemistry and biology of lake systems in order to interpret past lacustrine environments and to infer potential future ones. The term 'reservoir' is used explicitly for man-made lakes. Reservoirs may be especially useful as indicators of environmental change because the precise date of their formation is known and rates of sediment accumulation are regularly monitored. Reservoirs are in themselves indicators of environmental change caused by human activities, as discussed in Chapters 2 and 4.

3.1.1 Objective of the chapter

The objective of this chapter is to consider potential changes that may occur at the landscape scale in lake catchment systems. These changes will be the result of the changing global environmental context described in Chapter 1 and driven by relief, hydroclimate and human activities over the course of the next century or so. From a geological perspective, lakes are an ephemeral feature of the landscape, and wetlands are even more transient. Under past assumptions of climatic stationarity, little attention was paid to environmental changes at the scale of a human lifetime. Under the present assumption of large climatic perturbations and exponential

Geomorphology and Global Environmental Change, eds. Olav Slaymaker, Thomas Spencer and Christine Embleton-Hamann. Published by Cambridge University Press.

TABLE 3.1. *Classification and extent of lakes and wetlands*

	Extent (M km^2)	Per cent of total global land surface area (excluding large ice sheets)
(1) Lakes		
Lakes	2.428	1.8
Reservoirs	0.251	0.2
Total lakes and reservoirs	2.679	2.0
(2) Wetlands		
Rivers	0.360	0.3
Marshes/ floodplains	2.529	1.9
Swamp forest	1.165	0.9
Pans or saline wetland	0.435	0.3
Bog, fen or mire	0.708	0.5
Wetland complexes[a]	2.362	1.8
Coastal wetland	0.660	0.5
Total wetlands	8.225	6.2

[a] A wetland complex is defined as a composite of wetlands of different types, typically determined from remotely sensed imagery.
Source: Data as presented in a Global Lakes and Wetlands Database (GLWD) (from Lehner and Döll, 2004).

increases in energy consumption and land use by an increasing global population, geomorphologists and limnologists are thus intensifying their research into medium time and space scale changes.

Discussion of the record of lake catchment response is restricted to the changing environment of the last 20 ka, a period within which landscapes have been perturbed by continental-scale glaciation, wherein the hydroclimatic variations are relatively well known, and during which the influence of human activity on the landscape has rapidly accelerated. Although much is known about lake systems such as Lake Baikal, whose terrestrial sediments cover as long a time period as 10 Ma, this chapter does not consider evidence from periods before the LGM (and see Chapter 1).

3.2 Lakes and wetlands

3.2.1 Lake types by origin

Hutchinson (1957) defined 76 different lake types by mode of origin which are here reduced to six broad categories (Table 3.2). Lakes and wetlands of glacial and periglacial origin are the most numerous by far and are found in what Meybeck (1979; 1995) defined as the deglaciated zone. The limnic ratio (lake area divided by land area) of the deglaciated zone is one order of magnitude higher than that of any other zone discussed by Meybeck, but the lakes and their catchments are highly varied. Illustrated here are just two of the many different kinds of lakes in the deglaciated zone.

Lakes in deglaciated mountains are located in glacially excavated depressions (Fig. 3.1) and have clearly defined riverine inputs and outputs. The lake catchment tributary to the lake has relatively unambiguous water and sediment source areas. By contrast, lakes in deglaciated zones of low relief are shallow and ice-covered for the winter months and resemble wetlands (Fig. 3.2; Plate 14) more than the glacial lakes illustrated in Fig. 3.1. The most obvious contrast is that there is no surface stream tributary to many of these lakes nor is there a riverine output. The main water source is groundwater or, in regions underlain by permafrost, seepage from the seasonally thawed 'active layer', and temperature changes, both seasonal and their longer term, dominate the response regime.

Outbursts of dammed lakes, including landslide-related ones, have caused natural disasters and/or climatic shifts in some regions (e.g. Hermanns *et al*., 2004). Outbursts with demonstrated global effects are less common, although recently glacier lake outburst floods (GLOFs) associated with global warming have been a target of discussion.

The world's largest lakes are found in tectonic basins and include the Caspian and Aral seas, and lakes Victoria, Tanganyika and Baikal. Lakes linked to river erosion and sedimentation processes – so-called 'fluvial lakes' – are found in floodplains and deltas. They are evidently the most vulnerable of all lake systems and their inventories are least reliable.

3.2.2 Lake types by climatic zone and area

In order to address the issue of global environmental change, it is useful to discriminate between lakes of differing climatic regime and size. Meybeck (1995) identified five broad climatic zones: deglaciated (including tundra and taiga biomes); temperate (including wet taiga); dry and arid (subsuming dry temperate, arid and savanna); desert; and wet tropics.

Closed basins, described as endorheic catchments, are numerous in many parts of the world's arid zones. Pans are depressions associated with limited vegetation cover, localised accumulation of water, localised salt development and wind action (Goudie and Thomas, 1985). They differ from classic closed-basin lakes, such as Qinghai in Tibet or Turkana in East Africa (Yan *et al*., 2002), not only in terms

TABLE 3.2. *Lake by area and total area occupied by lakes of specific origin (Meybeck, 1995)*

	Area of individual lakes			
	<1 km^2	1–100 000 km^2	>100 000 km^2	Total (M km^2)
Glacial/periglacial	0.244	1.033	0	1.277
Fluvial	0.050	0.168	0	0.218
Volcanic	0.000	0.003	0	0.003
Tectonic	0.015	0.504 (inc. Aral)	0.374 (Caspian)	0.893
Coastal	0.004	0.056	0	0.060
Solution/aeolian/mass movement	0.013	0.045	0	0.058
Total	0.326	1.809	0.374	2.509

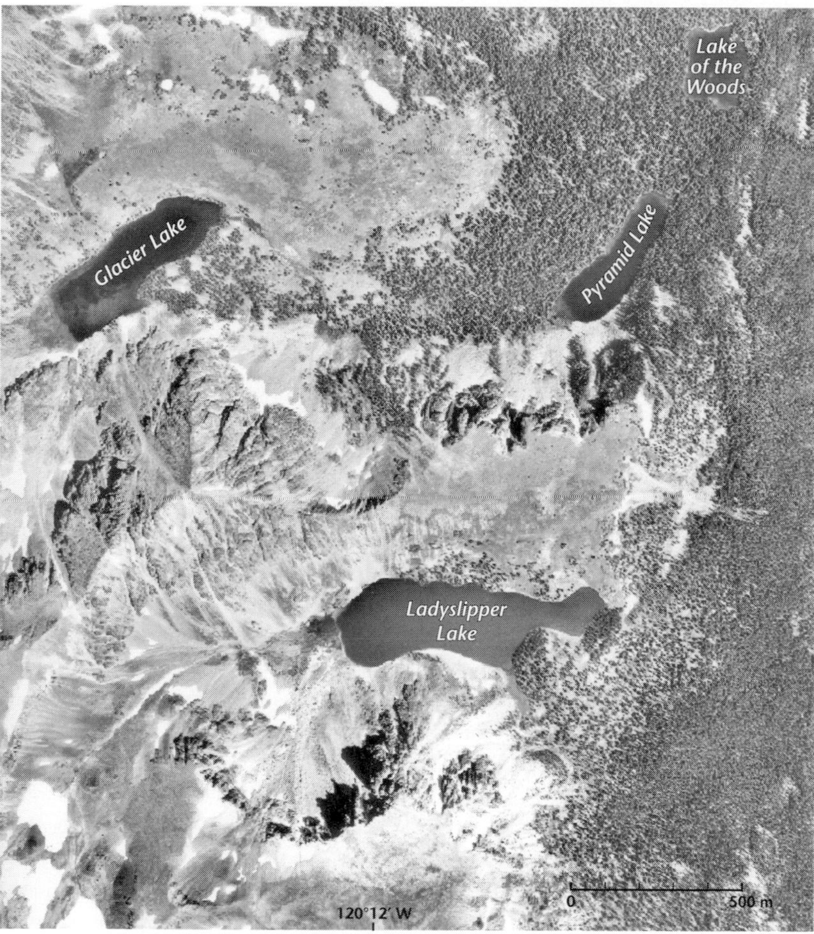

FIGURE 3.1. Glacial lakes in Cascade Mountains, British Columbia, illustrating glacially overdeepened lakes and high limnic ratio.

of size but, more importantly, in terms of mode of origin. As discussed in Chapter 1, it is necessary to restrict discussion to medium time and space scales of enquiry. Lakes between 1 and 100 000 km^2 in area, and timescales of tens to hundreds of years, have been chosen, consistent with Fig. 1.1 in Chapter 1. Lakes smaller than 1 km^2 are extremely variable in their behaviour and fail to integrate landform and landscape systems; as it happens, there is only one lake (the Caspian Sea) that is larger than 100 000 km^2 in area.

3.2.3 Summary

In the absence of human intervention, lakes and wetlands can be differentiated in terms of mode of origin, climatic

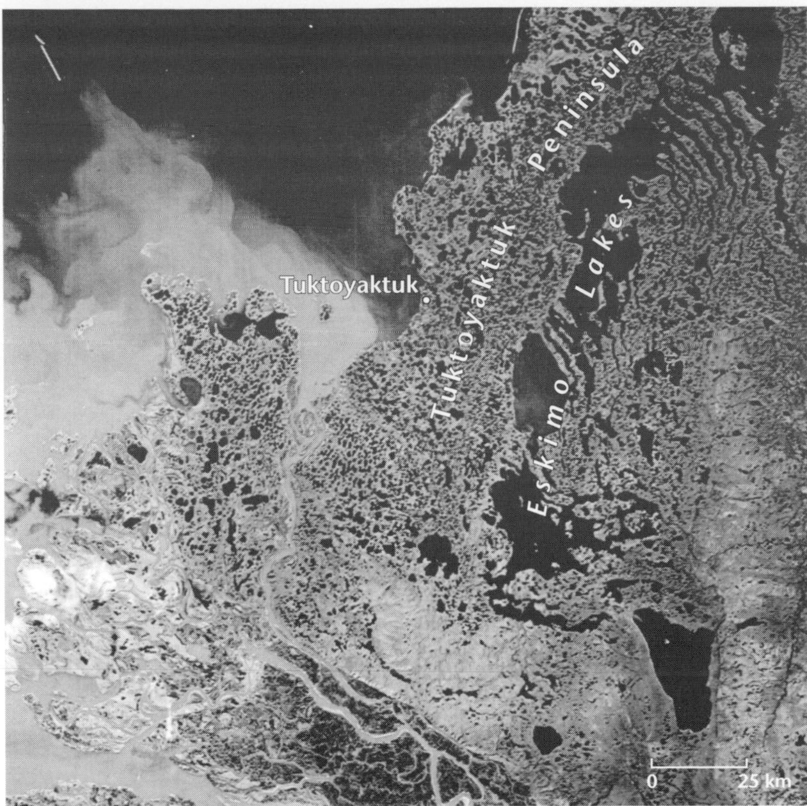

FIGURE 3.2. Periglacial lakes and ponds in Tuktoyaktuk Peninsula, District of Mackenzie, Northwest Territories, Canada, a region underlain by permanently frozen ground, illustrating shallow unstable lakes, highly sensitive to temperature changes. This deglaciated zone has an exceptionally high limnic ratio. (See also Plate 14 for colour version.) (Data available from USGS/EROS, Sioux Falls, SD.)

zone and surface area. The limnic ratio is highest, by one order of magnitude, in landscapes that have been recovering from continental glaciation during the last 20 ka. In other landscapes, the limnic ratio is controlled by the mean water balance, which is itself a function of the regional hydroclimate, and the presence of large tectonically controlled lakes, such as the Caspian and Aral seas. Differentiation by area is important for a number of reasons:

- shallow small lakes are transitional to wetlands and should be considered in the context of wetland landscapes, rather than as individual lakes;
- small lakes are highly variable in their response to hydroclimate and fail to integrate a consistent landscape signal;
- wetlands and lakes in arid regions are especially vulnerable to human activity and global warming; and
- lakes in high-relief regions are vulnerable to high rates of sedimentation and earthquake-induced slope failures which form new, but often short-lived lakes.

3.3 The lake catchment as geomorphic system

Lakes contain water, biota and sediment from a large variety of sources (Fig. 3.3). These sources can be broadly divided into allochthonous (from outside the lake, and including both terrestrial and atmospheric inputs) and autochthonous (from within the lake itself). Lake catchments have well-defined spatial limits; these spatial limits have remained relatively unchanged over Holocene time, and it is anticipated that they will remain essentially constant over the next century. Lake catchments exist at all spatial scales. The number of lake catchments in the landscape varies according to region and they are most numerous in deglaciated regions, such as tundra and taiga biomes (see Section 3.2 above and Chapter 13).

Oldfield (1977) proposed that the lake catchment system is a useful methodological unit for sediment-based ecological studies since the quantity and quality of sedimentary deposits contained within a lake are an integrated function of basin dynamics. This lake-sediment-based framework for reconstructing past sediment transfer processes and historical sediment yields for lake basins has subsequently been advocated by Dearing *et al.* (1987), Kashiwaya *et al.* (1997), Smol (2002) and many others. Last and Smol (2001) explicitly confirm the value of lake-sediment-based research in the context of whole catchments. Considerable progress has been achieved in understanding past environmental change from the evidence of lake sediments. But the extent to which

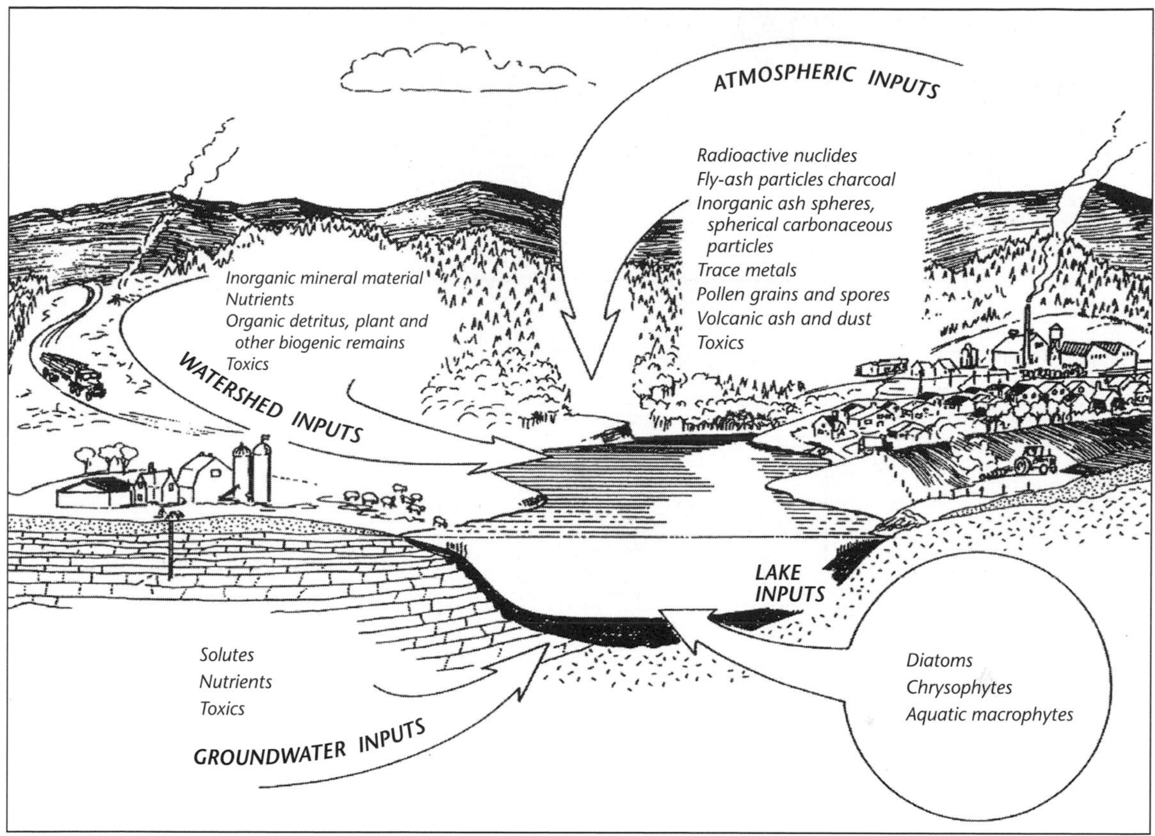

FIGURE 3.3. Allochthonous and autochthonous sources of materials in lake catchments (modified from Smol, 2002).

this information will provide a reliable guide for future environmental change is uncertain, especially because of the accelerating land use and land cover changes which are generating a 'no analogue' condition.

3.3.1 Catchment controls

Foster *et al.* (1988) developed a sediment yield and budget model within a lake catchment framework which describes the coupled lake and catchment systems by means of a flow diagram. The forcing variables climate and human activity act independently on the subsystems hillslope, channel and floodplain (the catchment system) and delta, open water and bottom sediments (the lake system). A third driver, relief, has been added in order to interpret spatial variability in lake catchment response (Fig. 3.4).

The spatial variability of lake catchments is controlled by regional climate and catchment area as well as relief, local hydroclimate and human activity. These variables are seen to affect hillslopes and collectively influence weathering rates, hydrological processes, mass movement, soil erosion, vegetation cover and land transformation. Sediment production, sediment transfer rates and sedimentation in channels and floodplains are distinctive to each lake catchment system. The lake system receives sediments from the catchment system and the sediments are distributed across deltas, through open water and deposited as bottom sediments. There is also throughput of water and of sediments, the amount of which depends on the trapping efficiency of the lake.

Temporal variability of lake catchments is controlled primarily by hydroclimate and human activity because, at the scale of 100 years or less, catchment relief and area are constant and the regional climatic zone is relatively constant. Variation through time of biogeochemical processes is also important in lake catchment systems. In Lake Baikal sediments, for example, high rates of chemical weathering in the catchment were found during interglacial periods; and crustal weathering and soil formation in the catchment were enhanced under warm climatic conditions (Sakai *et al.*, 2005). Under present conditions, human alteration of flow regimes; alteration of water stored in lakes, rivers and groundwater; and introduction of pollutants and

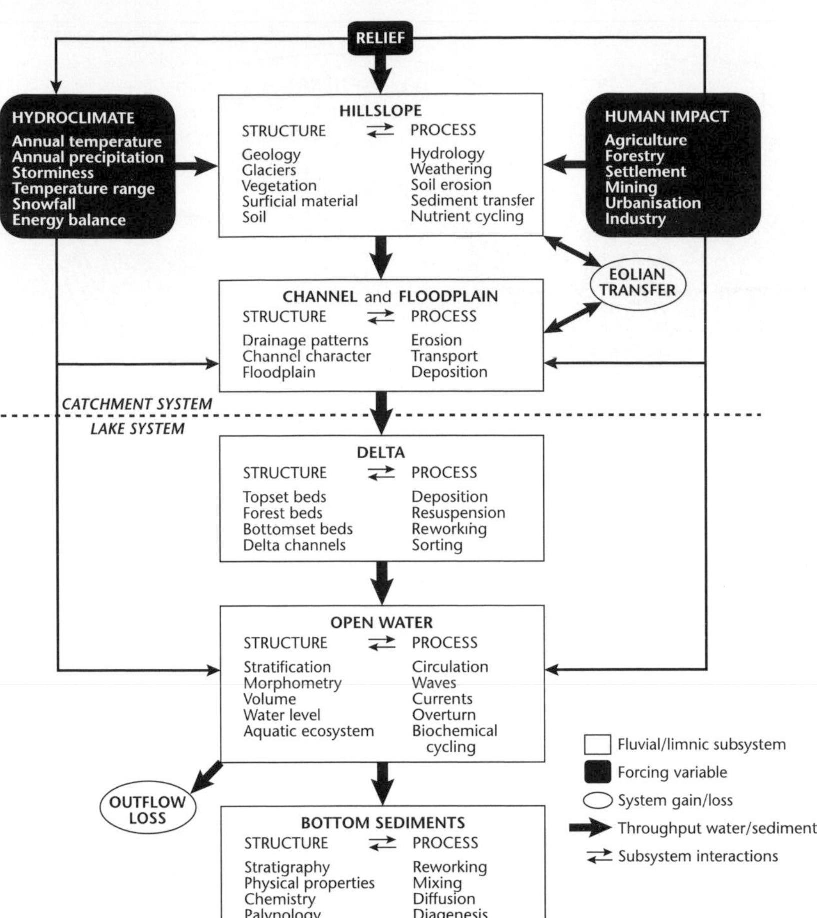

FIGURE 3.4. A geomorphological lake catchment model (modified from Foster *et al.*, 1988).

nutrients into waterways, land use change, and overexploitation of resources are thought to be the major catchment controls on lakes (Beeton, 2002).

3.3.2 Relief, catchment area and regional climate

The effect of relief on lake catchment spatial variability can be assessed by examining lake responses in regions of variable relief typology (see Plate 7). Meybeck *et al.* (2001) propose nine relief types from plains (1) to high and very high mountains (9). Lake catchments in high and very high mountains (elevations in excess of 2500 m above sea level and local relief ranges greater than 300 m 5 km^{-1}) are strongly influenced by the behaviour of snow, ice and permafrost, and experience rapid sedimentation in response to intense precipitation and rapid runoff. Meybeck's regional climatic zonation, which was introduced with respect to lake behaviour, is an equally important consideration with respect to total lake catchment response (Meybeck, 1995). The presence of glaciers has a strong influence on the rate of sedimentation in lake catchments (Hallet *et al.*, 1996).

Deltas and fans are distinct morphological features that provide information on the amount and rate of coarse clastic sediment delivered to a lake. In the example of Lillooet Lake delta, Coast Mountains of British Columbia, rates of advance of the delta front have been recorded over a 150-year period (Gilbert, 1975) (Fig. 3.5). Hinderer (2001) has analysed suspended load, bedload and rates of delta growth in a number of European alpine lake catchments. He demonstrated a close correlation between contemporary delta sediment growth and catchment area in 12 lake catchments and also showed that elevation and slope of catchment were the main drivers of rates of lake sedimentation in 20 lake catchments.

Lake catchments in plains (elevations below 500 m above sea level) and undulating relief are generally less active in terms of sediment delivery. In warm, tropical environments, however, geochemical processes of soil formation and within-lake sediment and solute transformations are significant.

FIGURE 3.5. Lillooet Lake delta, illustrating rapid delta growth in a glacierised mountain lake, Coast Mountains, British Columbia (modified from Gilbert, 1975).

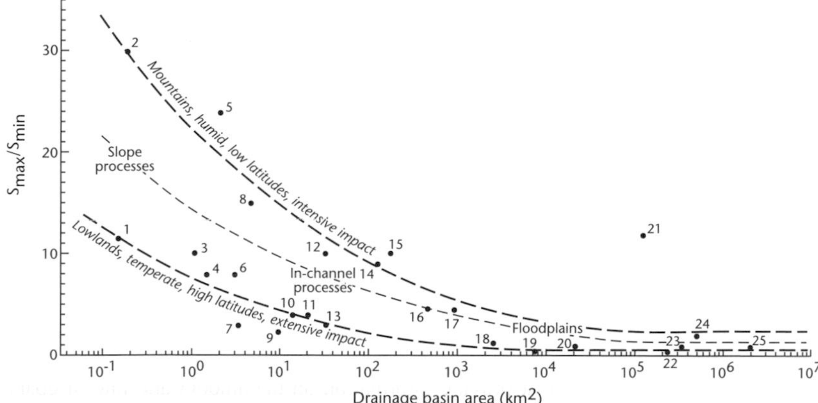

FIGURE 3.6. Relative magnitude of change between low or baseline sediment flux and maximum sediment flux (dimensionless ratio S_{max}/S_{min}) plotted against lake catchment area (km²) (from Dearing and Jones, 2003).

Foster *et al.* (1990) showed that specific sediment yield to lakes in northwestern European lake catchments is inversely related to catchment area. This finding for lakes confirms the conventional plot of decreasing specific sediment yield with increasing basin area in fluvial systems. The effect is commonly explained as a function of the fact that larger catchments have more storage areas in which to trap sediment than smaller catchments. By contrast, Schiefer *et al.* (2001) examined 70 lake catchments, varying in size from 0.9–190 km², in the Skeena region of northwestern British Columbia and showed a direct relation between specific sediment yield and catchment area in that region. Their study confirms for lake catchments the conclusions reached by Church and Slaymaker (1989) for fluvial suspended sediments in the postglacial environment of British Columbia (see Chapter 2 for further discussion).

3.3.3 Variability and lake catchment behaviour

Beyond the question of how much sediment moves through the lake catchment system, Dearing and Jones (2003) asked the question whether the relative magnitude of change in lake catchments is a function of basin area. They showed (Fig. 3.6) that indeed the parameter S_{max}/S_{min} achieved maximum values in small, highly disturbed catchments, where S_{max} is the maximum sediment accumulation rate in a record sustained over at least 10–100 years and S_{min} is the lowest, pre-human impacts flux in the same record. The data suggest that the upper limit of the distribution is associated with mountainous landscapes, humid climate (especially at low attitudes) and intense human impact. The important implication of this analysis is that the relative magnitude of sediment fluxes generated by disturbances in headwaters is expected to be greatly attenuated in downstream reaches.

3.3.4 Coupling of temporal and spatial scales

Dearing and Jones (2003) also take a different approach to the scale problem by examining the coupling of temporal and spatial scale in records of global lake sediment flux. Results from small–medium catchments show that, during the Holocene Epoch, climate has been largely subordinate to human impact in driving long-term sediment loads (though the evidence for short-term climate impacts is also clear). The rise in sediment delivery following major human impact is typically 5–10 times that under undisturbed conditions. For larger catchments (greater than 1000 km^2) modern fluxes lie close to long-term averages. Dearing and Jones make a systematic analysis of the relative magnitude of temporal change in sediment flux in different catchment sizes and demonstrate a global trend where relative changes diminish with increasing catchment size. Large catchments with effective storage zones and very slow reaction times lead to weaker levels of spatio-temporal coupling. The important practical conclusion from the work of Dearing and Jones (2003) is that it makes no sense to compare results from lake catchments of greatly differing spatial scale over different temporal scales.

3.3.5 Summary

As noted in Chapter 2, relief at all spatial scales promotes variability in hydroclimate and runoff. The most prominent effect of relief is to increase the absolute rate of landscape change. As a global generalisation, the combined effect of relief and runoff explains approximately 60% of the variance in denudation rates. Nevertheless, it must be borne in mind that sediment availability, especially in recently glaciated environments, can override the effects of relief in small- to medium-scale lake catchments.

The absolute rate of infill of lake depressions is controlled by the extent to which rivers and their adjacent slopes are actively interconnected and the availability of sediment, as well as the relief. It is therefore, in general, impossible to come to definitive global conclusions on denudation rates from small to medium lakes by the volume of sediment they contain. For this reason, some of the smallest mountain lakes, with almost no sediment input, can anticipate greater longevity than larger downstream lakes.

3.4 Internal lake processes

Sedimentation in lakes can be dominated by either physical, chemical or biological processes. Open lake systems in humid regions can be characterised as having four main types of sediments: clastic, carbonate, biogenic silica and organic matter.

3.4.1 Physical mixing

Lake waters tend to become thermally or chemically stratified, either permanently or seasonally. From a sedimentological perspective, the stratification has bearings on the fate and distribution of the material dissolved and suspended in the water mass. The stratification of the water and possibilities for the overturning of the stratification depend on the unique property of water to assume its maximum density at +4 °C.

Tropical lakes are temperature stratified with colder, heavier water near the bottom. Vertical mixing of the lake waters is therefore impossible (oligomictic). In temperate climates, surface lake water is warmer and less dense than deeper water during the summer, but during the winter, when surface water cools, it may reach and surpass the densities of water at depth. Hence vertical mixing of the whole water body will occur at least once and sometimes twice every year (monomictic or dimictic). The net effect of this stratification is illustrated in Fig. 3.7. Complications occur when bottom waters are more saline than surface waters (meromictic). In these cases the bottom waters become stagnant and oxygen and nutrients are depleted. If this stratification continues for several thousand years,

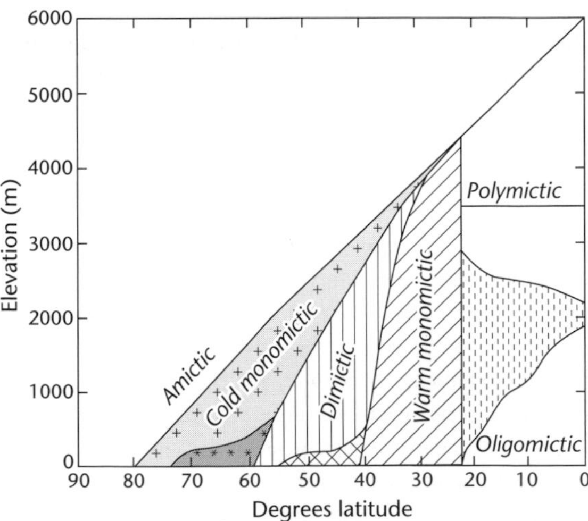

FIGURE 3.7. Classification of thermal lake types with latitude and altitude (modified from Wetzel, 1975).

laminated muds, rich in organic matter, will accumulate. By contrast, in those lakes with regular mixing, the water is well oxygenated and biogenic production of diatoms and algae occurs.

3.4.2 Biological activity

Biological activity in lakes is influenced by the degree of nutrient enrichment, a statistic which defines the trophic state of a lake. The degree of nutrient enrichment varies from being low in oligotrophic high-latitude or mountain lakes to being very high in shallow eutrophic lakes. In shallow eutrophic lakes the level of biological and chemical oxygen demand may exceed the capacity of the lake to supply dissolved oxygen and oxidise the organic material. Partly degraded organic material then accumulates on the lake floor. This organic material is supplied by lake plankton, bottom- and shoreline-dwelling organisms and by debris carried in by rivers from the drainage basin. Eutrophication affects water clarity and large plant growth. In addition to compounds containing carbon and hydrogen, sediments may accumulate organic sulphides, biogenic silica or calcium carbonate. The main natural factors influencing lake trophic state are: soils/geology, phosphorous concentration, lake depth and flushing. Human activity accelerates the process of eutrophication through wastewater disposal, agricultural practices and urban runoff.

3.4.3 Chemical activity

Lakes act as large reaction vessels for a variety of chemical processes. Hydrologically closed lakes (called endorheic systems) have no outflow. Water levels of such lakes may fluctuate considerably and provide a sensitive indicator of changing hydroclimate from the differing positions of their past shorelines. During longer time periods, precipitation of salts occurs and there is a predictable sequence of chemical pathways followed as a function of the nature and concentration of solutes entering the lake, the amount of evaporitic concentration and the removal of chemical species by burial, regradation and precipitation of minerals. The most common ions in lake waters are $Ca > Mg > Na > K$ and $HCO_3 > SO_4 > Cl$ (where > means 'is greater than'). In waters of this type, a typical concentration sequence might include precipitation of calcite followed by precipitation of gypsum and finally mirabilite precipitation. Ephemeral or playa lakes contain highly concentrated brines. Such endorheic lakes produce a variety of precipitates, whether carbonate, sulphate or sodic, depending on the lithological characteristics of their basins. Classic examples of such lakes include the Caspian and Dead seas, Great Salt Lake and lakes Chad and Eyre. Lakes with significant biogenic sediments are transitional between over-supplied and sediment starved conditions. They produce predominantly biogenic carbonates (calcium rich) and opaline silica (silica rich), such as Lake Baikal, in Russia.

3.4.4 Sedimentation processes

Sedimentation processes are closely related to the internal flow patterns and bathymetry of the lake basin, which influence the hydrodynamic regime of the lake (Håkanson and Jansson, 1983). River inflow, winds and atmospheric heating are the major energy inputs which are controlled by lake morphometry, relief of the surrounding terrain and the prevailing hydrologic regime of the lake. In the river mouth zone of lakes, the rapid change of hydraulic conditions results in deposition of the coarse sediment load in a delta and river plume dispersion of finer-calibre sediments beyond the delta. Studies of delta sedimentation indicate a high degree of spatial variability in sedimentary characteristics, reflecting the complex and dynamic hydraulic mechanisms occurring in the delta area (Gilbert, 1975).

There is commonly a transition zone between delta front and mid-lake which is characterised by subaqueous slumping and sediment surges; sampling of sediment in this zone may produce complex and unrepresentative sedimentary signals. Mid-lake spatial sedimentation patterns are controlled by river inflow dispersion processes which are more predictable. Density contrasts between the inflowing river water and the vertical density distribution of the lake water produce distinctive inflow plumes. Where river water density is less than lake water density, a vertically restricted 'overflow' occurs; where river water density is intermediate between lake surface and lake bottom waters, 'interflow ' occurs and where river water density is greater than lake water density, 'underflow' occurs.

The implications of these density contrasts for the nature of lake sedimentation are fundamental. River inflow mixing patterns usually fluctuate seasonally between overflow, interflow and underflow, leading to mixed depositional processes. River inflow patterns may also vary over time due to changing catchment conditions such as upland sediment supply, downstream storage capacity and lake inflow configuration. Overall sedimentation rates, mean rhythmite thickness and the grain size of inorganic materials generally decrease along a central transect from the lake inflow point. High spatial uniformity is observed in central, flat-lying lake basins, a phenomenon which is called 'sediment focussing' (Schiefer, 2006) and which allows sampling of representative sedimentation sequences beyond a critical distance from the delta front (Evans and Church, 2000). Continuous stratigraphic records are normally sought and it is therefore

TABLE 3.3. *Lakes by area and total area of lakes in each climatic zone (Meybeck, 1995)*

	Area of individual lakes				
	<1 km^2	1–100 000 km^2	>100 000 km^2	Total (Mkm^2)	Limnic ratio
Deglaciated	0.229	1.018	0	1.247	6.9
Dry temperate/savanna	0.015	0.213	0.374	0.602	0.54 (ex-Caspian)
Wet temperate	0.020	0.168	0	0.188	0.7
Wet tropics	0.039	0.184	0	0.223	0.9
Desert	0.000	0.127	0	0.127	0.27 (ex-Aral)
Total	0.303	1.710	0.374	2.387	

also necessary to differentiate between lake zones of accumulation, transportation and erosion. This distinction is based on a water depth threshold.

Lakes dominated by mineral sediments, which are the lakes that figure most prominently in the geomorphic literature, often reflect the influence of relief, rock type and hydroclimate. Proglacial and mountain lakes are largely characterised by high clastic sedimentation rates and short lifetimes. Lakes with marked seasonal changes of sedimentation tend to produce alternating coarse (summer or storm period) and fine (winter or quiet period) couplets, called 'varves'. Varves are commonly assumed to be annual but this is by no means always the case; it is safer to refer to such sediments as rhythmites unless the annual frequency of the events can be confirmed independently.

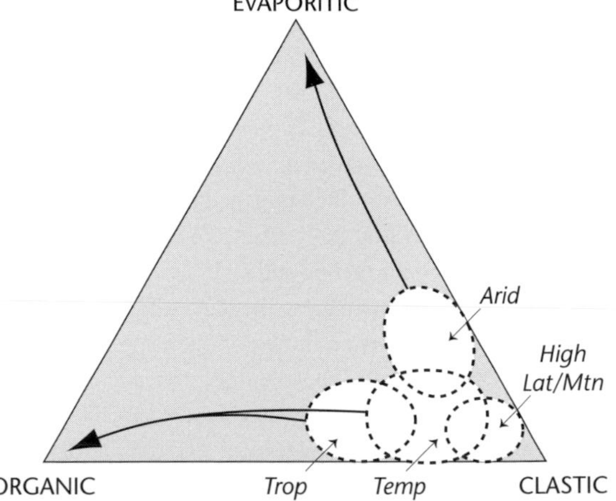

FIGURE 3.8. Model for lake sedimentation in relation to climatic environment and geomorphic stage (modified from Rust, 1982). Arrows indicate trends of sedimentation in arid, temperate and tropical lakes as they evolve.

3.4.5 Summary

Physical mixing of lake waters and biological, chemical and sedimentation processes in lakes create a complex within-lake environment. Physical or clastic sedimentation dominates in high-latitude and high-altitude lakes (where chemical and biological activity may be relatively low) and in lakes with high relief either of drainage basin or lake floor (Rust, 1982). Climatic zone (Table 3.3) and geomorphic stage of lake are also important factors in lacustrine deposition (Fig. 3.8). The end members of evaporitic precipitation relate to arid climates, organic precipitation to tropical and temperate climates, and clastic accumulation to low productivity non-arid climates, such as high latitudes and mountains. Secondary influences relate to the stage of evolution of the lake and its catchment. For example, arid or temperate lakes become (respectively) increasingly dominated by evaporitic or organic accumulation as their lake catchments are worn down.

3.5 Hydroclimate changes and proxy data

Although instrumental measurements and observations from the past are most important, they are rarely available or of sufficient quality or duration. Comparison of disturbed lake catchments with natural lake catchments in similar climatic and relief zones, and of comparable area, allows the researcher to infer a temporal sequence of events, from natural to disturbed lake catchment. The technique has been applied most satisfactorily in lake acidification experiments, comparing anthropogenically fertilised lakes with natural lakes (e.g. Schindler, 1974). In watershed hydrology, the paired watershed approach is a similar methodology, where, for example, harvested and pristine forested lake catchments are compared in terms of water and sediment balances (see Chapter 12 for further discussion). There are many problems with ensuring an acceptable level of similarity between two lake catchments. For these and other reasons, almost all hydrological processes inferred from the sedimentary record are determined from proxy evidence (Table 3.4).

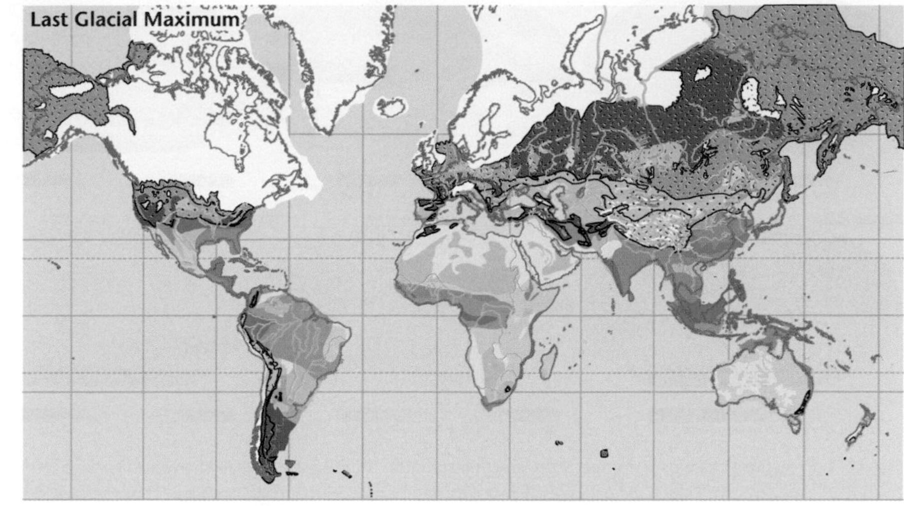

Vegetation cover
- Arctic or alpine tundra
- Tundra–steppe
- Taiga
- Boreal forest (southern taiga)
- Mixed temperate forest
- Deciduous temperate forest
- Xerophytic or Mediterranean woodland
- Steppe
- Savanna
- Tropical/equatorial forest
- Arid area
- Mangrove
- Wooded steppe/ wooded savanna
- Loess
- Sand
- Glacier/ice cap

Permafrost
- Continuous
- Discontinuous

PLATE 1. The Last Glacial Maximum (18 ± 2 ka BP) (from Petit-Maire and Bouysse, 1999).

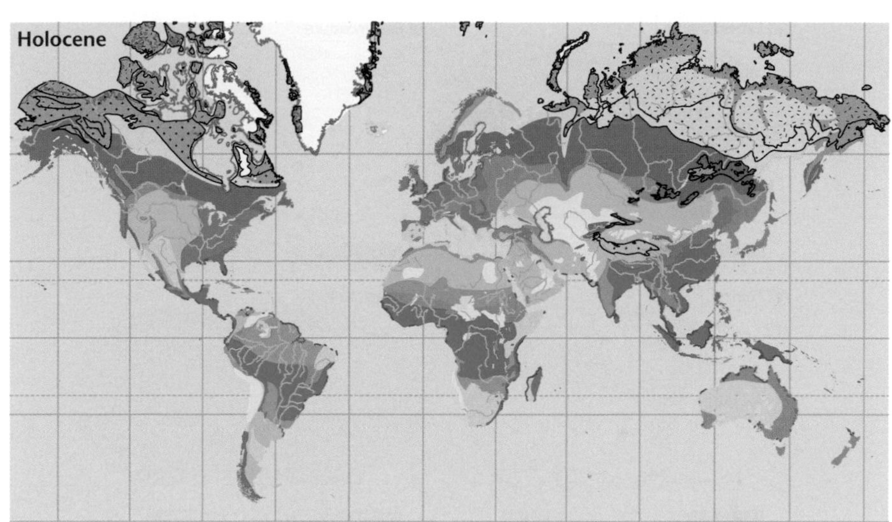

PLATE 2. The Holocene Optimum (8 ± 1 ka BP) (from Petit-Maire and Bouysse, 1999).

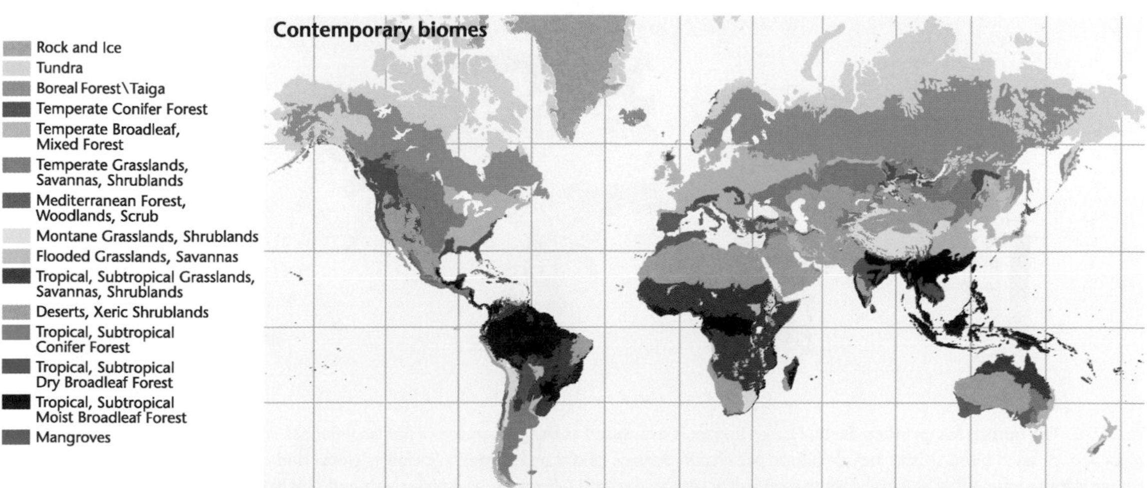

PLATE 3. Map of world biomes (Source: www.worldwildlife.org/science/ecoregions/item1267.html).

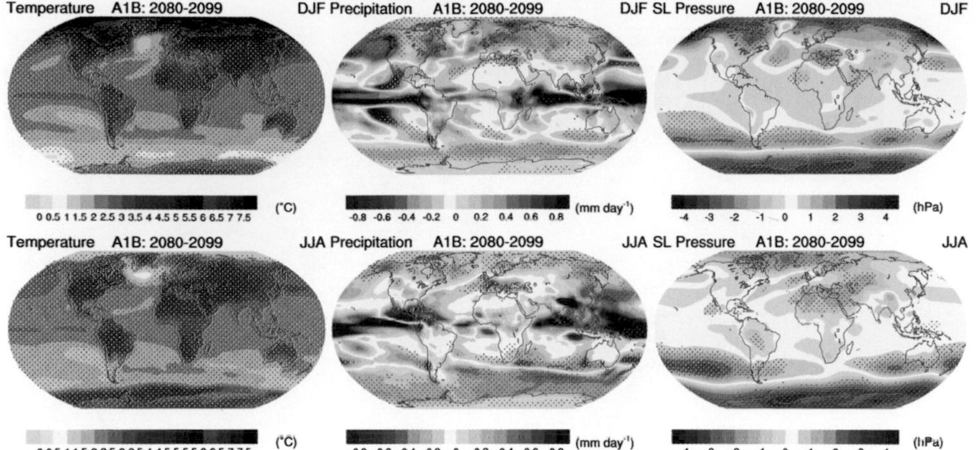

PLATE 4. Projected changes in temperature and precipitation. Multi-model mean changes in surface temperature (°C, left), precipitation (mm day^{-1}, middle) and sea level pressure (hPa, right) for boreal winter (DJF, top) and summer (JJA, bottom). Changes are given for the SRES A1B scenario (Appendix 1.1) for the period AD 2080–2099 relative to AD 1980–1999. Stippling denotes areas where the magnitude of the multi-model ensemble mean exceeds the inter-model standard deviation (from Meehl *et al.*, 2007).

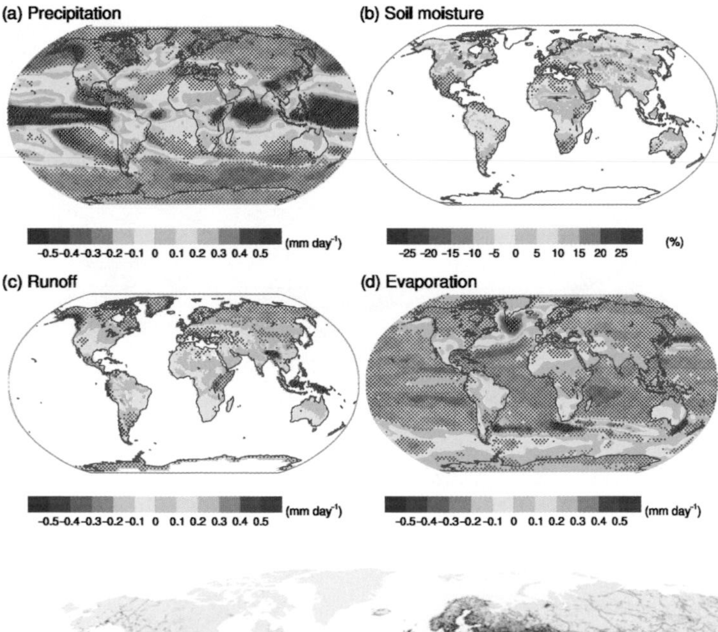

PLATE 5. Projected changes in global hydroclimate and runoff. Multi-model mean changes in (a) precipitation (mm day^{-1}); (b) soil moisture content (%); (c) runoff (mm day^{-1}); and (d) evaporation (mm day^{-1}). Regions are stippled where at least 80% of models agree on the sign of the mean change. Changes are annual means for the SRES A1B scenario (Appendix 1.1) for the period 2080–2099 relative to 1980–1999. Note that runoff from the melting of glaciers and ice sheets is not included in these calculations (from Meehl *et al.*, 2007).

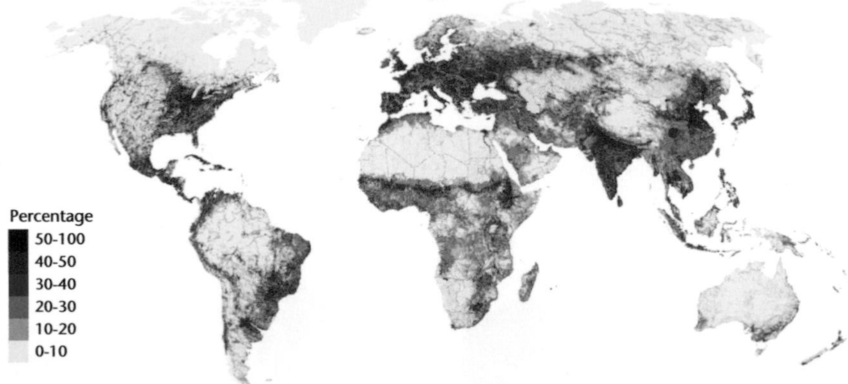

PLATE 6. The human footprint on Earth. Human impact is expressed as the percentage of human influence relative to the maximum influence recorded for each biome. Data include human population density, land transformation (including global land cover, roads and cities), electrical power infrastructure (NOAA night-lights data) and access to the land (via roads, navigable rivers and coastline). (Map created from data downloaded at www.ciesin.columbia.edu/wild_areas from the Human Footprint data set generated by the Center for International Earth Science Information Network (CIESIN) at Columbia University and The Wildlife Conservation Society (from Kareiva *et al.*, 2007).)

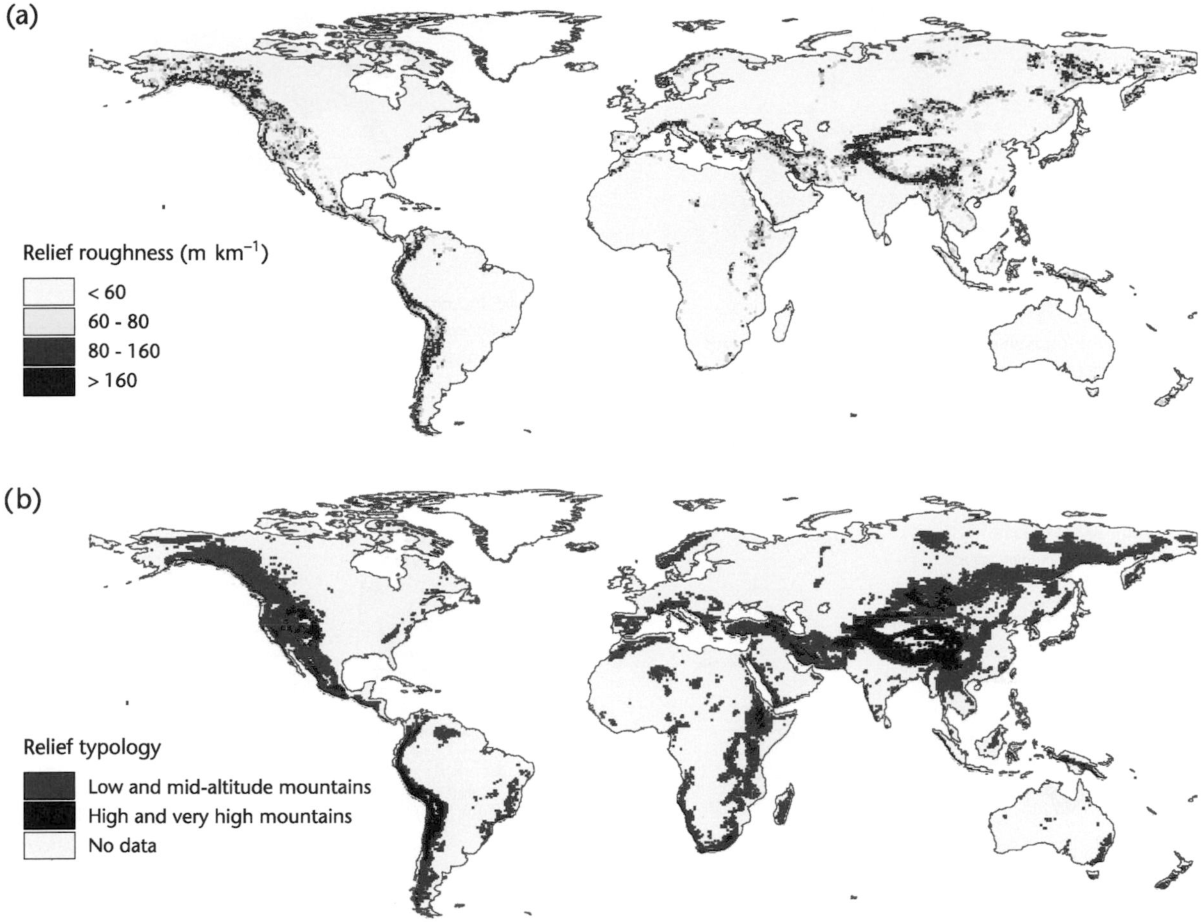

(a)

Relief roughness (m km^{-1})

< 60
60 - 80
80 - 160
> 160

(b)

Relief typology

Low and mid-altitude mountains
High and very high mountains
No data

PLATE 7. Mountains of the world. Global distribution of: (a) relief roughness (m km^{-1}) and (b) two aggregated relief types, defined in terms of elevation band and local elevation range (modified by Pamela Green from Meybeck *et al.*, 2001).

PLATE 8. Two dam sites in the Austrian Alps (from Beckel and Zwittkovits, 1981). Upper Kapruner valley, Salzburg Province, with two reservoirs, Moserboden and Wasserfallboden. (Also shown as Fig. 2.17 within the text.)

PLATE 9. Mountain classes and major mountain ranges in British Columbia (classes as defined in Section 2.1.1). (SRTM data analysis by Dori Kovanen, 2007; see Appendix 2.3.)

PLATE 10. Mountain classes and major mountain ranges in Tajikistan (classes as defined in Section 2.1.1). (SRTM data analysis by Dori Kovanen, 2007.)

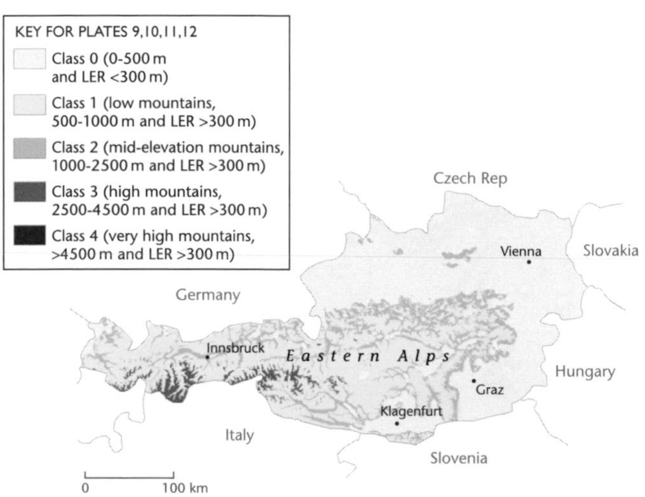

PLATE 11. Mountain classes and major urban centres within the Austrian Alps (classes as defined in Section 2.1.1). (SRTM data analysis by Dori Kovanen, 2007.)

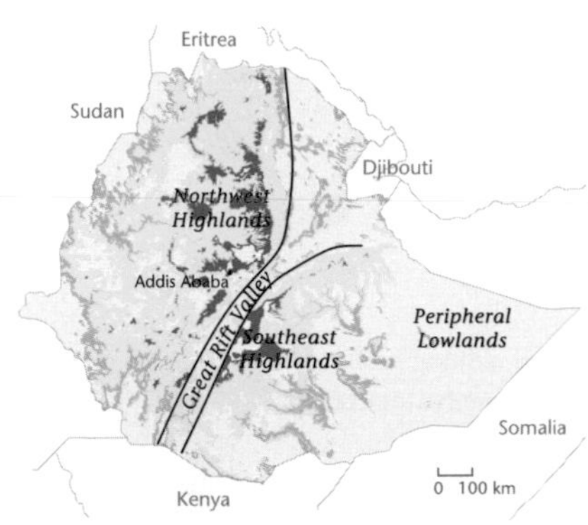

PLATE 12. Mountain classes and major mountain regions of Ethiopia (classes as defined in Section 2.1.1). (SRTM data analysis by Dori Kovanen, 2007.)

In 1975 slopes on this escarpment were severely degraded by livestock. Recently terraces have been built on the steep slopes, stone bunds on agricultural land and stock routes carefully delineated. In the background there is an increase in number of homesteads, noticeable improvement to vegetation around homesteads and a major forest is new on previously degraded land.

Landscape interpretation

Visible erosion
- \+ Overall assessment
- \+ Vegetation

Gully erosion
- \+ Sheet and rill erosion
- \+ Management of farmland
- \+ Management of other land

Soil loss by sheet and rill erosion

(Assessment of relative change, whereby the situation in 2006 is expressed as a percentage of the situation in 1975)

100
50
0
1975
2006

PLATE 13. Evidence of improved landscapes in Ethiopia (after Nyssen *et al*., 2007; Munro *et al*., 2008). Landscape interpretation of time lapse photography and estimates of soil loss by sheet and rill erosion.

PLATE 14. Periglacial lakes and ponds in Tuktoyaktuk Peninsula, District of Mackenzie, Northwest Territories, Canada, a region underlain by permanently frozen ground, illustrating shallow, unstable lakes, highly sensitive to temperature changes. This deglaciated zone has an exceptionally high limnic ratio. (Also shown as Fig. 3.2 within the text).

PLATE 15. False-colour image of Lake Dian-chi and its catchment, Kunming, Yunnan, illustrating the combined impact of urbanisation and intensive agriculture on lake water quality.

PLATE 16. Photograph of the beach and foredune system at Greenwich Dunes, Prince Edward Island, Canada taken from a fixed camera mounted on a mast on the foredune crest (photograph taken on 18 October 2007 with vegetation cover at a maximum). (Also shown as Fig. 5.12a within the text.)

PLATE 17. Development of a parabolic dune in a transgressive dunefield, New South Wales coast, Australia: photograph of the active slip face of the parabolic dune showing burial of vegetation. (Also shown as Fig. 5.15b within the text.)

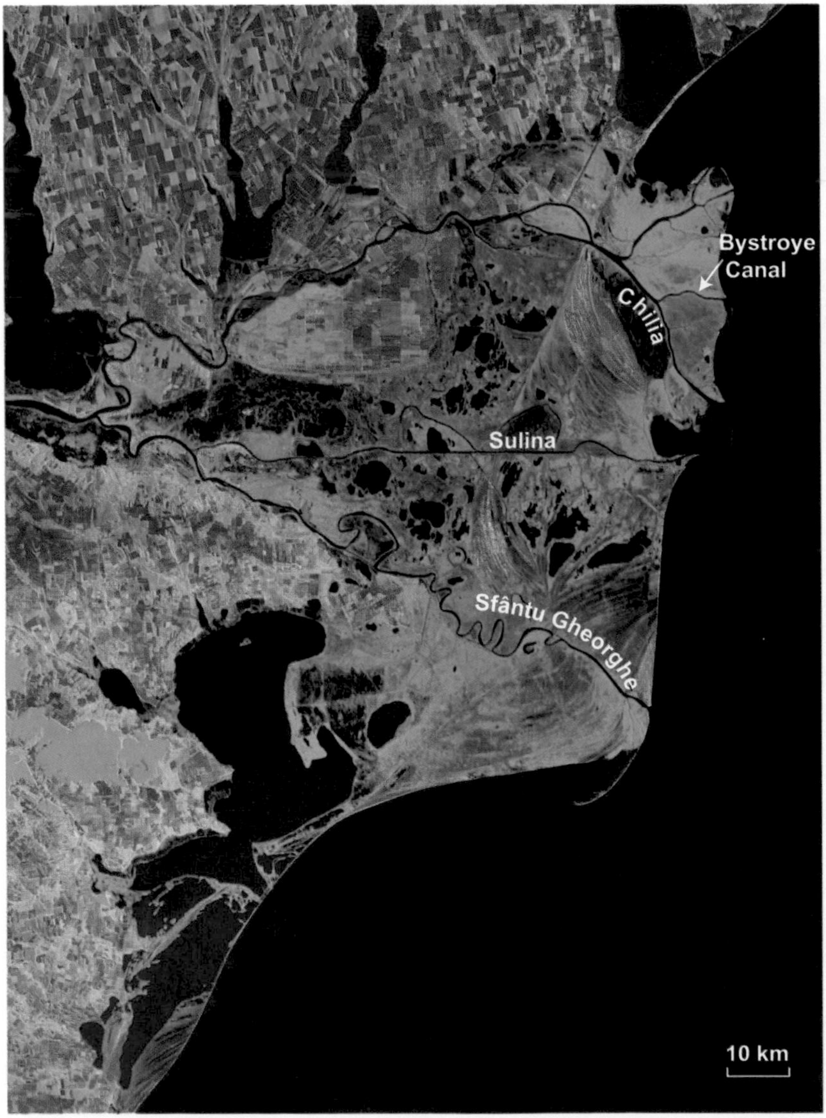

PLATE 18. The Danube Delta. (Also shown as Fig. 5.16 within the text.)

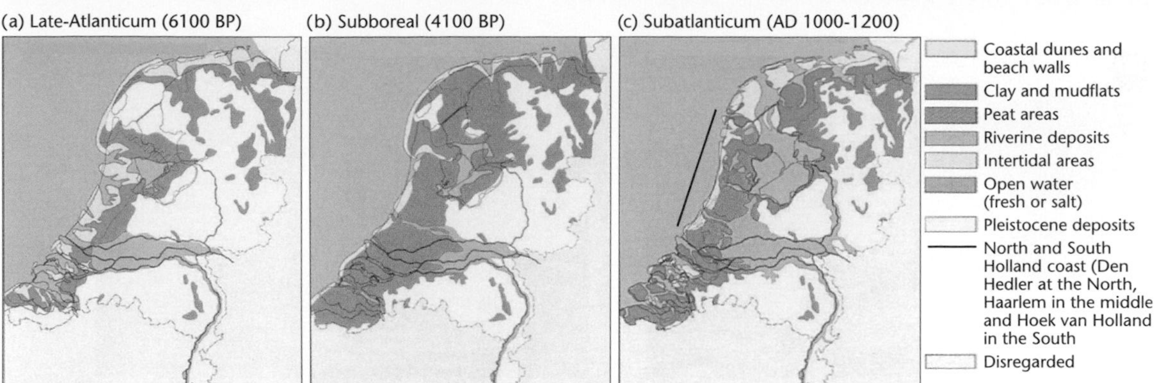

PLATE 19. Development of the Netherlands during the Holocene (modified from Zagwijn, 1986).

PLATE 20. Prograded barrier at Tuncurry, 220 km north of Sydney, southeastern Australia, where progradation has averaged 0.3 m a^{-1} over the past 6 ka. (a) Aerial view of the 2-km wide barrier extending 11.5 km alongshore between headlands (© NSW Department of Lands, 2008), showing transect for section featured in Fig. 6.6a; (b) view to south with barrier in foreground and prominent headlands that have blocked alongshore supply of sand to the barrier, ensuring that the lower shoreface has been the exclusive sand source for barrier growth during the past 6 ka under conditions of stable sea level (photograph by Andrew Short).

PLATE 21. A soft eroding cliff at Alum Bay, Isle of Wight, UK (photograph by Robert Nicholls).

Fringing reef, Lizard Island, Great Barrier Reef, Australia.

Barrier reef setting, Mauritius.

Platform reef, Lady Elliot Island, Great Barrier Reef, Australia.

Atoll reef rim and islands, Tarawa Atoll, Kiribati.

Emergent reef flat exposed at low tide, Majuro Atoll, Marshall Islands.

Coral cover on reef platform (2 m depth), Nadi Bay, Fiji.

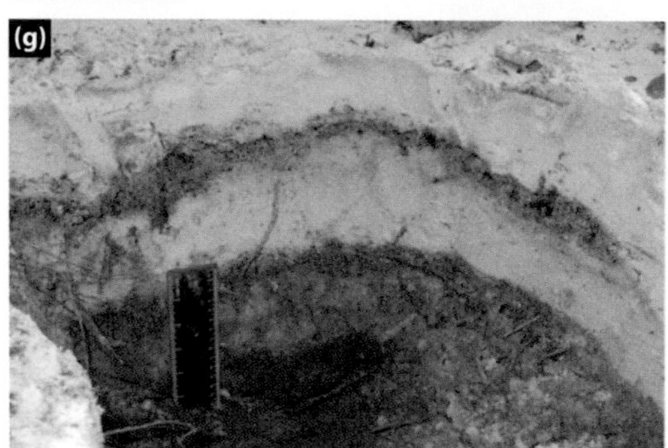

Multiple overwash layers contributing to island building on a Maldivian reef island since 2003.

Densely urbanised reef platform island, Male, Maldives, Indian Ocean.

PLATE 22. Images of coral reef landforms:
(a)–(d) differences in geomorphic state of reefs;
(e)–(g) sediment overwash deposition on island margin; and
(h) example of anthropogenic modification of reefs.
(Also shown as Fig. 7.11 within the text.)

PLATE 23. Erosional impacts of oil palm cultivation at Danum, Sabah.

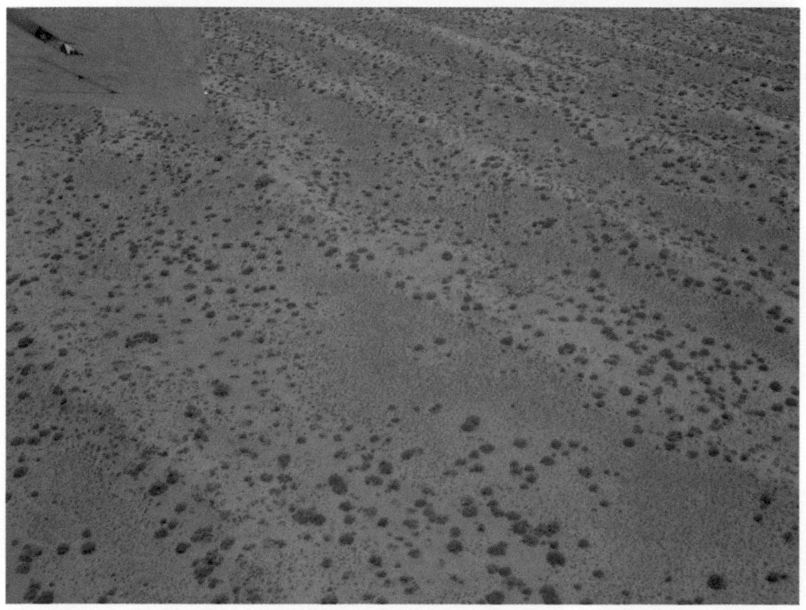

PLATE 24. Longitudinal dunes in the Kalahari near Upington, South Africa (photograph by Michael Meadows). These dunes are thought to be close to the threshold of reactivation.

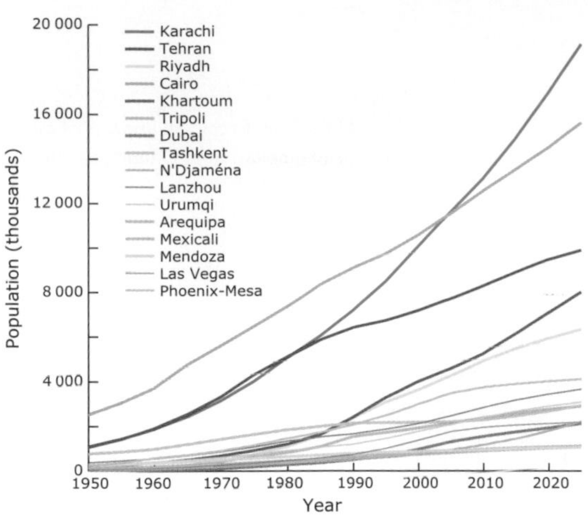

PLATE 25. Population growth in selected desert and desert margin cities (from UN Population Division of the Department of Economic and Social Affairs, 2006, 2007).

PLATE 26. Fluvial system response to climate change: Long Valley Creek, Great Basin Desert, Nevada. Note Holocene aggradation (dated by Mazama Ash) and historical (twentieth-century) incision.

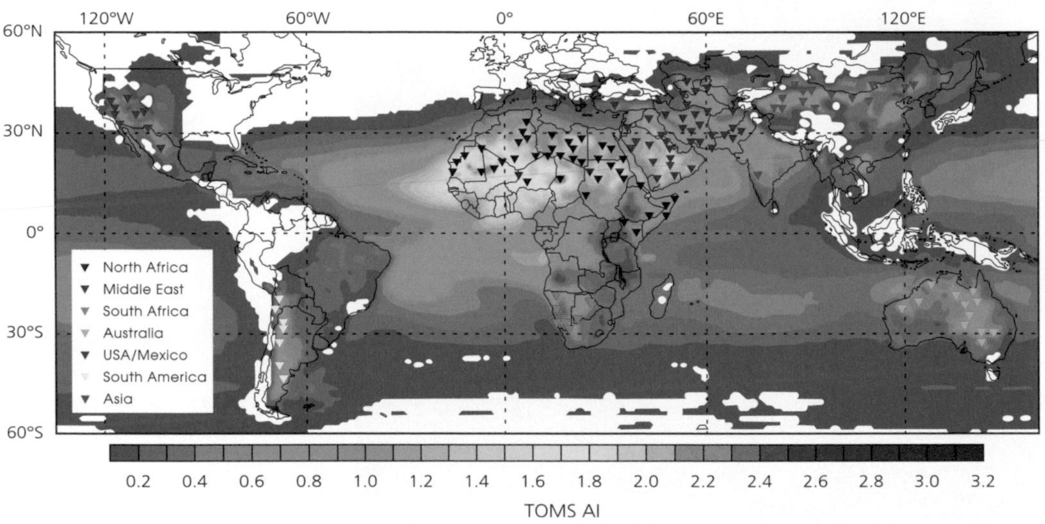

PLATE 27. Global dust hotspots (from Engelstaeder and Washington, 2007).

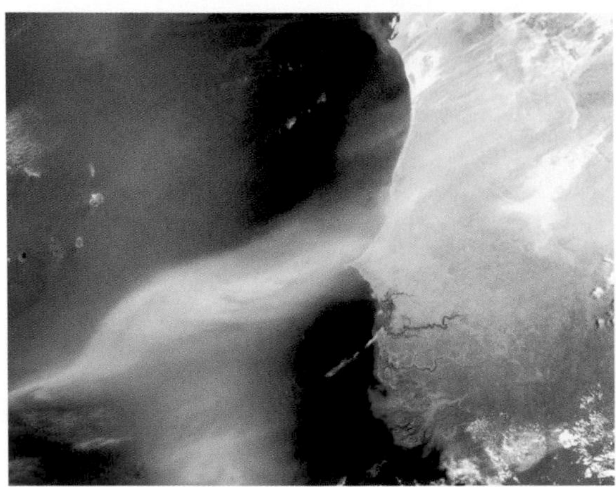
PLATE 28. Dust plume emanating from the western Sahara in the boreal winter, driven by intense trade winds.

PLATE 29. Examples of (a) relict, (b) dormant and (c) active dunes in the Gran Desierto, Mexico.

PLATE 30. Ridaura stream at Platja d'Aro (Girona, northeast Spain) during the dry season (photograph by Maria Sala).

PLATE 31. Ridaura stream at Platja d'Aro (Girona, northeast Spain) during a flash flood event in which flood waters extended along the coastal plain near the stream outlet (photograph by Water Authority, by kind permission).

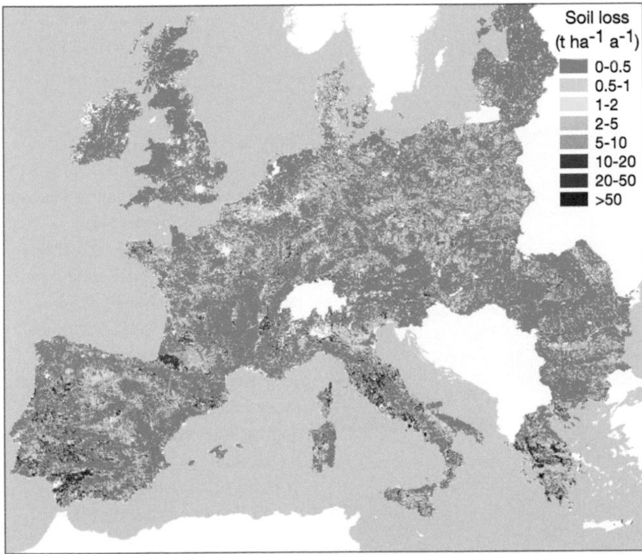

PLATE 32. Pan-European Soil Erosion Risk Assessment (PESERA) (from http://eusoils.jrc.it/ESDB_Archive/pesera/docs/EROSIONA4.pdf).

PLATE 33. A burnt slope in the Cadiretes Massif, Catalan Coastal Ranges, northeast Spain (photograph by Maria Sala).

PLATE 34. *Dehesa* landscape in Caceres, Extremadura, Spain (photograph by Maria Sala).

PLATE 35. Soil salinisation near Perth, Australia (photograph by Maria Sala).

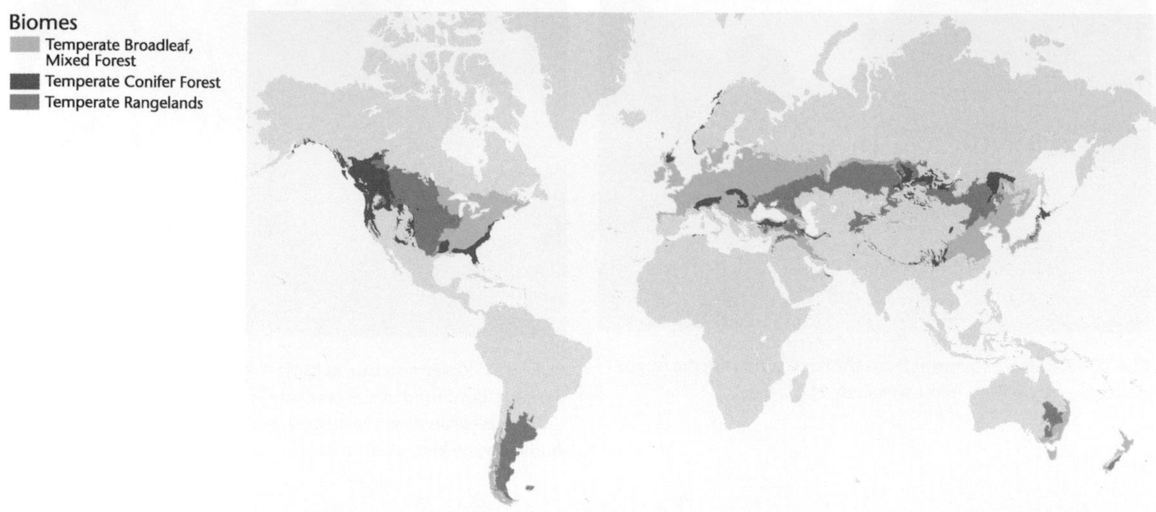

PLATE 36. Map of land cover categories in the mid-latitude zone.

PLATE 37. Formerly forested areas that have been converted to agriculture in the upper reaches of the Salween River basin, northwestern Yunnan Province, China. Note that little sediment appears to be coming to the river from dispersed agriculture, whereas the landslides in the lower right photograph all emanate from hillslope trails.

(a) Agricultural land use adjacent to river;
(b) and (c) agricultural land buffered by riparian forest;
(d) landslides initiated by trails supply sediment directly to river.

PLATE 38.

(a) Epic levels of landslide erosion along the newly constructed Weixi–Shangri-La road in Yunnan, China;
(b) High level of connectivity of landslide sediment with a tributary of the Mekong River.

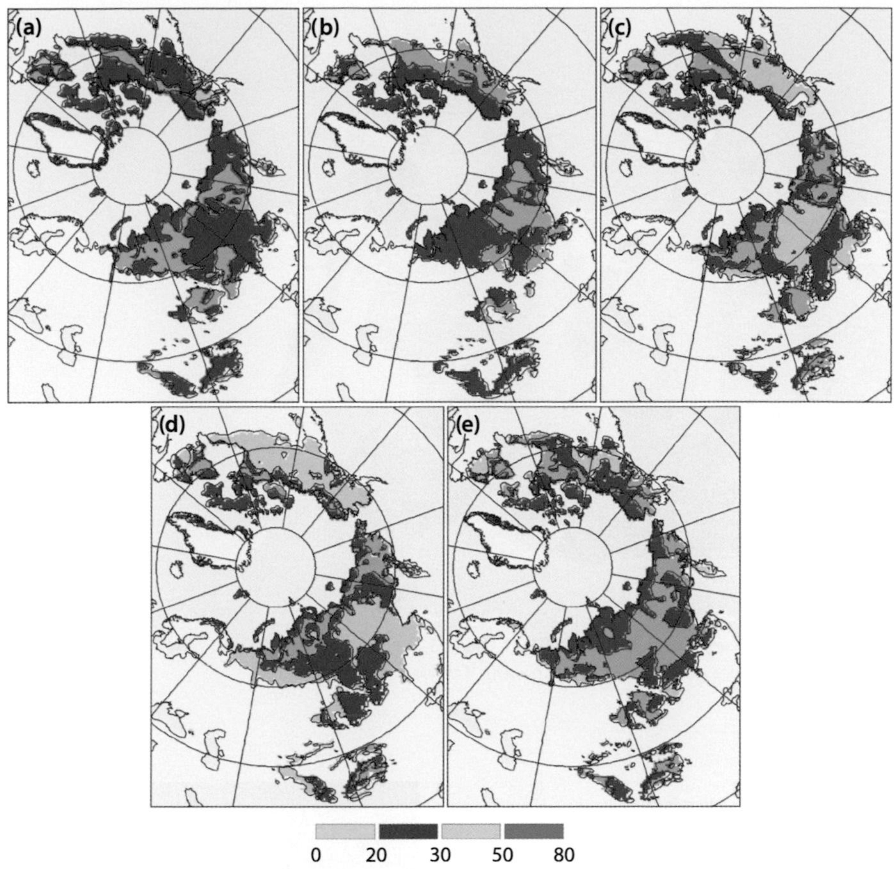

PLATE 39. Predictive tools for future land management of permafrost regions. Projected by 2050 changes in the depth of seasonal thawing (per cent from modern) under five climatic scenarios: (a) CCC, (b) ECHAM, (c) GFDL, (d) HadCM3 and (e) NCAR (from Walsh *et al.*, 2005).

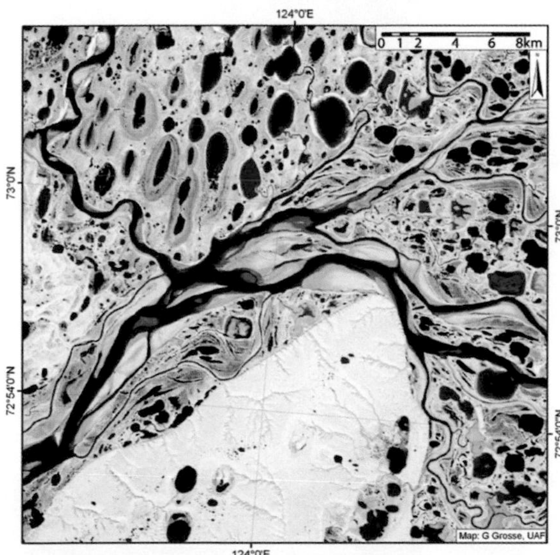

PLATE 40. Landsat-7 satellite image map (RGB colour bands 5–4–3) of thermokarst landscape on Lena Delta terraces, Siberia. First terrace (active floodplain) in the east, with channels and oxbow lakes; second terrace to the north, with nicely oriented thermokarst lakes; third terrace in the south, with circular thermokarst lakes that developed in very ice-rich permafrost deposits (from G. Grosse, Permafrost Laboratory, UAF, 2007–08).

PLATE 41. Collapsing palsas and thermokarst ponds in subarctic Québec (from F. Calmels, CEN, Université Laval).

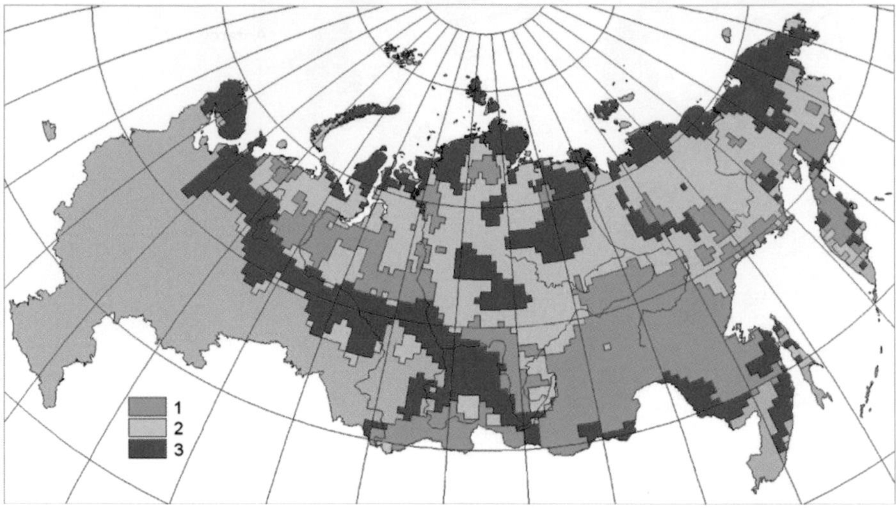

PLATE 42. Geocryological hazard map (from Anisimov and Reneva, 2006). 1, 2 and 3, regions with low, moderate and high potential threat to infrastructure due to permafrost thawing. The map was constructed using the GFDL climate scenario for 2050.

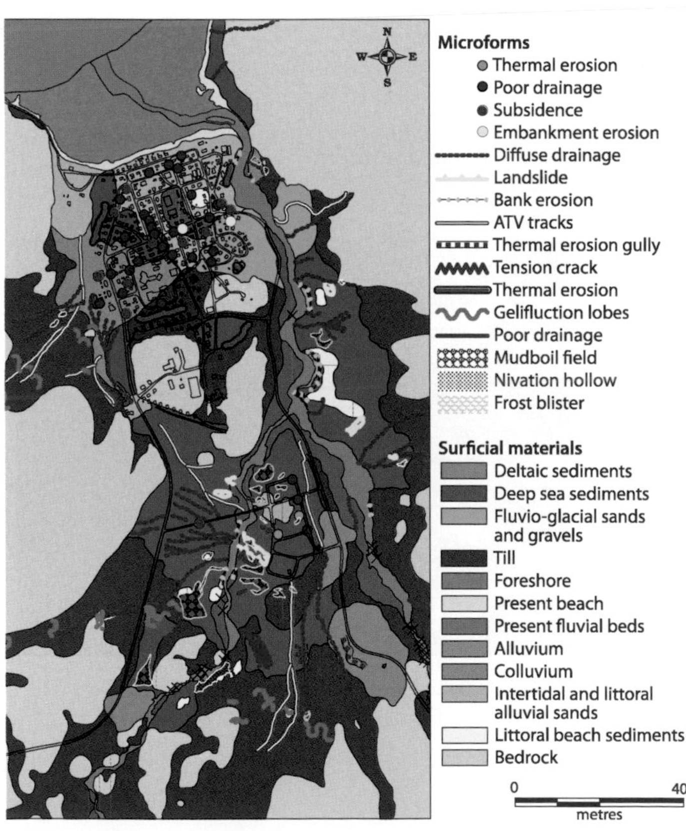

PLATE 43. Geomorphological services for future land management in sensitive permafrost terrain: the case of the Salluit Community, Nunavik, North Canada (Allard *et al.*, 2004). Detailed map of surficial deposits and microforms associated with permafrost instability.

PLATE 44. Map of recommended and not recommended zones for building purposes (same source and key as Plate 43).

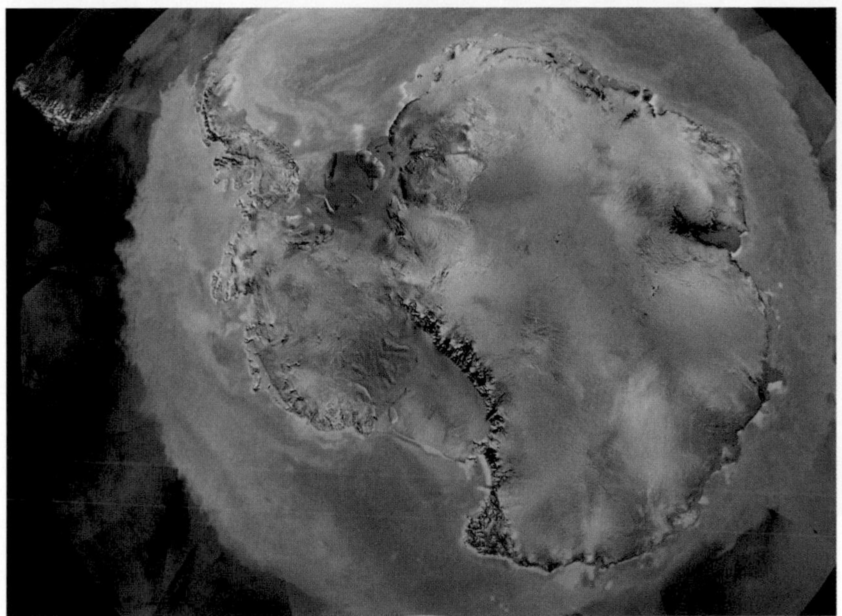

PLATE 45. Composite satellite image of Antarctica showing the main ice domes of East and West Antarctica, the ice shelves of the Ross Sea and Weddell Sea embayments and the Antarctic Peninsula extending towards southernmost South America. The dark areas are mountains protruding above the ice and coastal oases. The sea ice extent represents winter conditions. The land ice is from AVHRR imagery and sea ice from Seawinds imagery. (Courtesy NASA/Goddard Space Flight Center, Scientific Visualization Studio.) (Also shown as Fig. 14.1 within the text.)

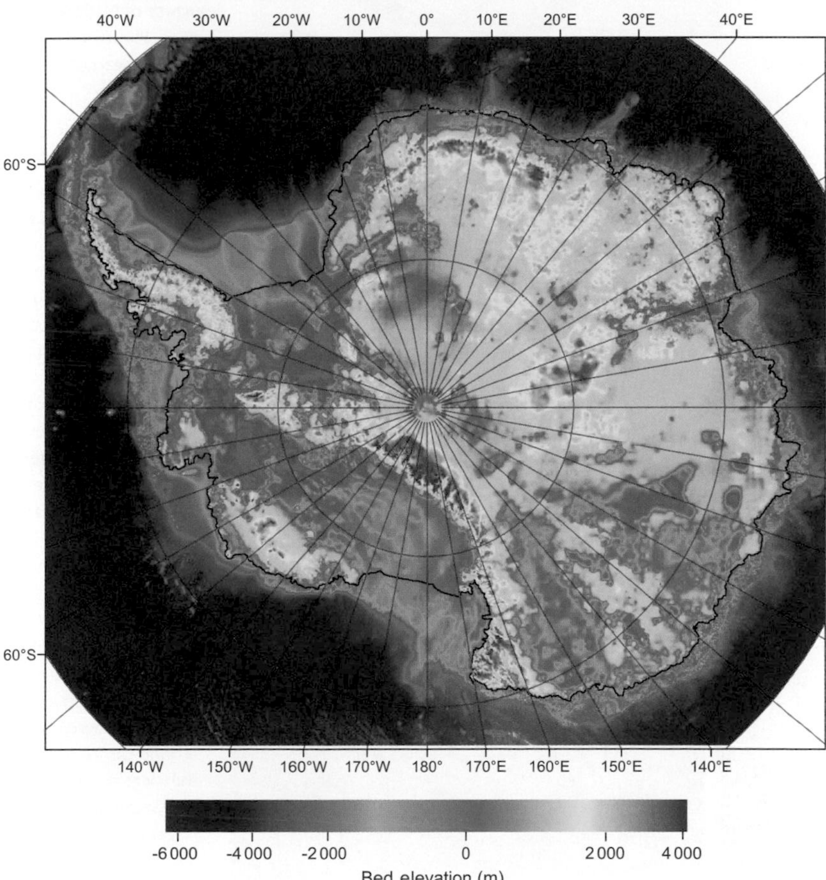

PLATE 46. The present subglacial topography of Antarctica showing the continent and mountains of East Antarctica and the archipelago of West Antarctica. This is the idealised view of Antarctica if you remove the ice, maintain present sea level and ignore any isostatic adjustment (from Lythe *et al.*, 2001). (Also shown as Fig. 14.3 within the text.)

PLATE 47. The fjord coast of southern Chile and the adjacent glacierised Andes Mountains contrast with the Patagonian lowlands of southern Argentina. The mountains were formed along the boundary where two plates have collided. The denser oceanic Nazca plate which underlies the Pacific Ocean is being pushed under the lighter continental South American plate. This results in a geologically active region with many geomorphic hazards (NASA Visible Earth imagery; credit Jeff Schmaltz; http://visibleearth.nasa.gov/view_rec.php?id=8245).

PLATE 48. The community of Gilchrist, Texas, on the Bolivar Peninsula between eastern Galveston Bay and the Gulf of Mexico, was almost totally destroyed by the US landfall of Hurricane Ike on 13 September 2008. The geomorphological damage has been described as landscape 'scarring', affecting over 400 km of the Gulf coastline (photograph: David J. Phillip – Pool/Getty Images).

PLATE 49. Hurricane Ike as seen by Houston/Galveston radar just before landfall at 06:07 UTC, 13 September 2008. At US landfall, Ike was a category 2 hurricane with winds of 175 km hr^{-1}. Total estimated property damage was US$27 billion (2008 prices), making Ike the third costliest hurricane in US history, after Katrina (2005) and Andrew (1992) (image courtesy of National Weather Service, Houston/Galveston, NOAA Research).

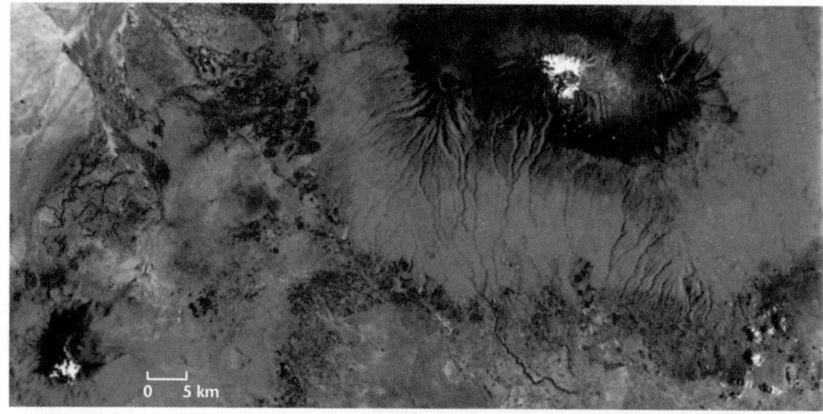

PLATE 50. Mt. Kilimanjaro, showing Kibo Peak (5985 m a.s.l.) with larger glacier area, and Mawenzi Peak (5148 m a.s.l.) to the west. Encircling the peaks is the reddish-brown of Kilimanjaro's cold desert. A second ring of blackish-brown is the zone of mountain grass and moorland. The transition from brown to green denotes the treeline ecotone and further changes in the shades of green mark the transition from mountain forest to cultivated areas. The savanna landscape at lower levels is inhabited by the pastoral tribe of the Masai (Landsat Multispectral Scanner, 24 January 1976; © NASA ID20367–06582.)

PLATE 51. (a) Las Vegas in AD 1972; (b) Las Vegas in AD 2000, showing the huge urban expansion into the desert of southern Nevada. The area around Las Vegas belongs to the Ridge and Valley Province of the Great Basin. The mountain ridges run north–south and the intervening valleys are *bajadas*, flanked by massive alluvial and colluvial fans. The vegetated areas result from artificial irrigation with water from Lake Mead behind the Hoover Dam on the Colorado River.

TABLE 3.4. *Sample palaeohydrological processes inferred from proxy evidence*

Proxy evidence	Hydroclimatic inference
Stratigraphy and past shoreline	Lake level change, water balance
Particle size, stratigraphy	Flood events
Sediment laminations, rhythmites	Sediment accumulation rates
Carbon, lead and caesium isotopes	Sediment accumulation rates
Geochemistry, particle size, magnetics	Minerogenic particulate changes
Organic chemistry, isotopes	Organic particulate changes
Inorganic/organic phosphorus	Changing weathering rates
Diatoms	pH changes, salinity changes
Enrichment of heavy isotopes	Water balance, lake water–groundwater interaction
Polychlorinated biphenyls, polycyclic aromatic hydrocarbons, lead	Long-distance transport pollution

3.5.1 Proxy data

Lake catchments are a tool for understanding past physical environments on the assumption (a) that current sedimentation can be related to measurements of current drivers of change and (b) that past sedimentation can be assumed to respond to the same drivers in the same way. Proxy data are physical, chemical or biological indicators of change (Table 3.4) incorporated into the properties of the sediments in the bottom of the lake. The objective of the use of proxy data is to make a causal link between environmental sedimentary responses, archived in lakes, with contemporary observational data on geomorphic, hydrological or climatic processes in the lake catchment. In this way past information recorded in sediments has a possibility to be properly interpreted (Burroughs, 2005). The relevance of palaeo-information for future prediction in lake catchment systems is however much debated: it depends on both the resolution of the data and the extent to which future changes of hydroclimate and land use can be known.

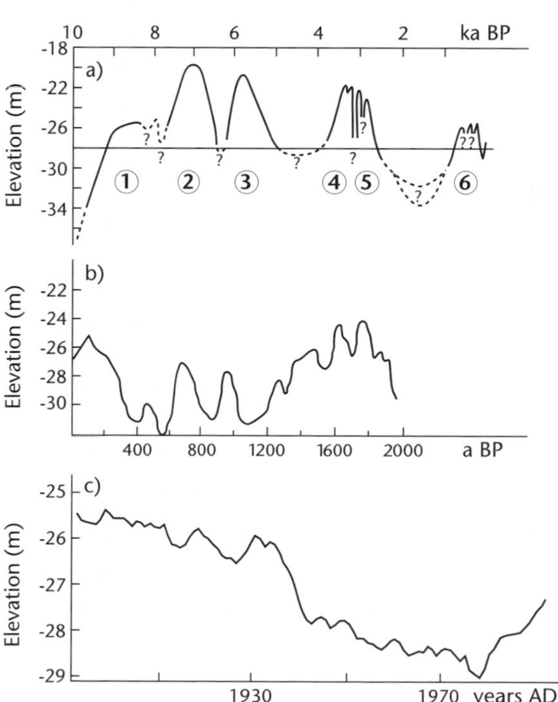

FIGURE 3.9. Caspian Sea level variation at different time scales: (a) Holocene; (b) the last 2 ka; and (c) modern changes (from Kislov and Surkova, 1998).

Past shorelines as proxy for palaeo-water balance

Dry, temperate and arid climate zones with endorheic drainage provide examples of lake catchment systems that respond most directly to water balance changes. The largest lake in the world, the Caspian Sea, is one such highly responsive lake. Lake level fluctuations at a variety of temporal scales, from 15 ka BP to the last century have been documented. This record demonstrates the rapidity of the response of an endorheic lake system to hydroclimate during the Holocene Epoch (e.g. Kislov and Surkova, 1998) (Fig. 3.9). The high sedimentation rate of 1–10 km Ma^{-1}, not to mention repeated dislocations of the ice sheet meltwater drainage, make lake levels difficult to interpret over longer timescales. Many endorheic lakes, such as Great Salt Lake, Utah, the Dead Sea in Israel/Palestine/Jordan and Lake Eyre in Australia, provide sensitive indicators of water balance changes from their past shorelines.

Particle size and stratigraphy as proxy for palaeo-precipitation

Particle size and its stratigraphic variations are often used as a proxy for palaeo-precipitation. The archive of sediments of a 1.74 km^2 lake in central Japan (Lake Yogo) has been

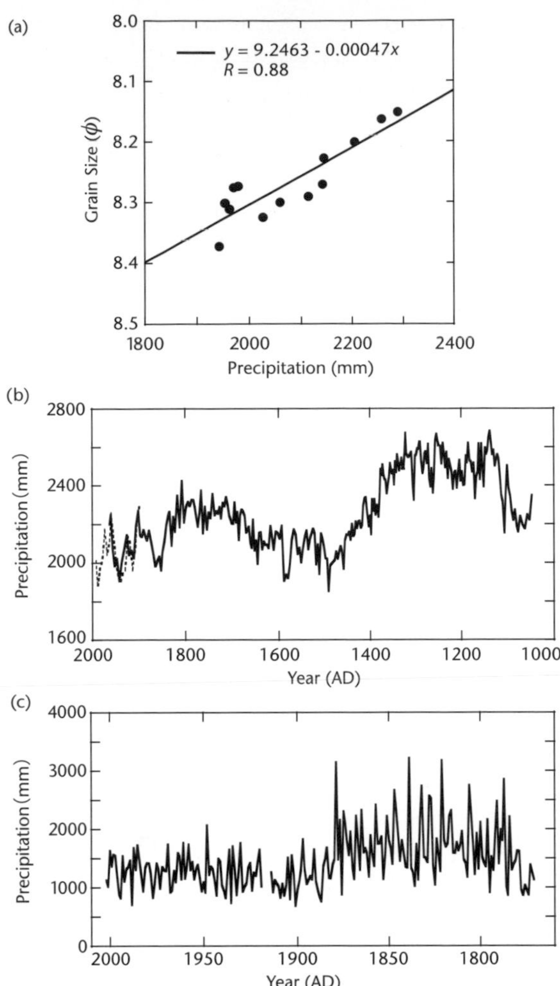

FIGURE 3.10. Estimation of past precipitation. (a) A relation between annual precipitation and mineral grain size; (b) estimated annual precipitation during the past 1000 years (modified from Shimada *et al.*, 2002); and (c) instrumental–observed precipitation in Korea (Wada, 1916; Korea Meteorological Administration: www.kma.go.kr/intro.html).

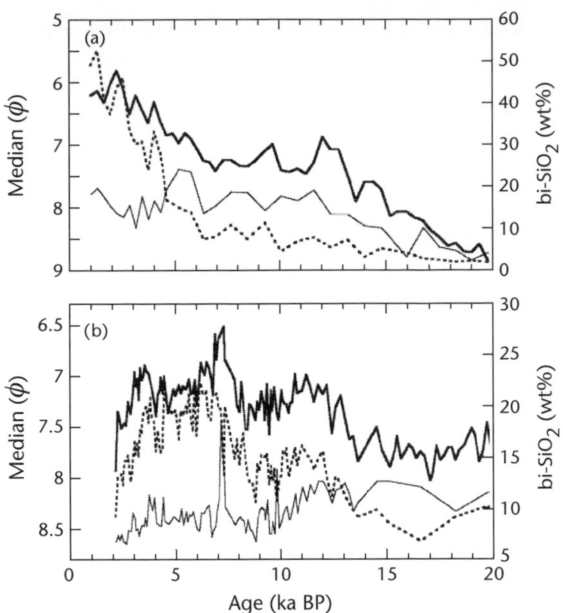

FIGURE 3.11. (a) Fluctuations in sediment grain size and biogenic silica content in Lake Baikal during the past 20 ka. Mineral particle size is highly correlated with sediment discharge in the catchment (H. Nakagawa, personal communication, 2000). (b) Fluctuation in sediment grain size and biogenic silica in Lake Biwa during the past 20 ka. Mineral particle size is highly correlated with sediment discharge in the catchment (K. Ishikawa, personal communication, 2004). Thick line, total sediment; thin line, inorganics; dotted line, biogenic silica.

used to link present and past environmental changes via sediment particle size (Shimada *et al.*, 2002). A linear relation between particle size and annual rainfall was first established using the past 100 years of rainfall instrumental observations (Fig. 3.10a). This relation was then used to reconstruct rainfall fluctuations over a 1000-year period from fluctuation in lake sediment particle size (Fig. 3.10b).

This estimated rainfall fluctuation was then checked against a 300-year instrumental observation record for precipitation in Seoul, Korea, on the reasonable assumption that there is little difference in annual rainfall trend between Japan and Korea (Fig. 3.10c). It was confirmed that the estimated rainfall fluctuation at Lake Yogo has a similar trend to that which obtained in Seoul over the period of record. This gives some confidence that the estimated rainfall during the 1000-year period is plausible. In a separate study, it was noted that the Yogo lake sediments became coarser (higher per cent coarse fraction and coarser mean particle size) around AD 700 (Shimada *et al.*, 2002). This was interpreted as a change from low water to high water inflow regime. More recently, acceleration in sedimentation rate occurred following the construction of an artificial channel which diverted water from an adjacent drainage. In this case, the sudden increase in lead in the sediments was an accurate proxy for the change of sedimentation rate that occurred in 1960.

In the cases of ancient lakes, such as Lake Baikal in Russia and Lake Biwa in Japan, robust relations between sediment discharge and bottom sediment particle size have been established for contemporary hydroclimatic conditions. Aota *et al.* (2006) have described the recording of sedimentation processes in Lake Biwa on a continuing basis. After the drilling of a 200-m long core in Lake Biwa in 1971 this first long sediment core sampling in ancient lakes established terrestrial long-term environmental changes (Horie, 1984; Kashiwaya *et al.*, 1991). Figures 3.11a and 3.11b are the long-term proxy records for Lake Baikal and Lake Biwa respectively, showing the close relations amongst particle size, organic sediment content and biogenic silica. These parameters are also correlated with the input of sediment discharge.

Sediment laminations and rhythmites as proxies

Sediment deposited in Lillooet Lake, British Columbia during a severe rainfall event in 2004, which produced the largest floods in almost a century of record, is clearly distinguished by changes in stratigraphy, colour, texture, magnetic properties and organic content. These characteristics can be explained by variations in lacustrine processes, especially turbid underflow. The 2004 flood layers are also clearly identified from 2004 non-flood layers and different year layers by their quite different thickness. This suggests that present observational records offer a means of assessing the changing nature of extreme hydroclimatic events, and their relation to more ubiquitous, lower-energy processes (Gilbert *et al.*, 2006).

Lacustrine varve deposits from 2 km^2 Green Lake, British Columbia, covering a period of 70 years (1930–2000) were analysed by Schiefer *et al.* (2006). They identified three circumstances under which extreme sediment delivery coincided with high discharge conditions and elevated sediment availability:

- rapid glacier recession of the early twentieth century;
- late summer and autumn rainstorm-generated floods; and
- freshet floods caused by unusual snowmelt conditions.

They also noted the complexity of disentangling the effects of hydroclimatic factors from the geomorphic effects of landslides and channel changes.

Recent core-sediment-based reconstruction in Cheakamus Lake, British Columbia reveals not only glacier advance and retreat in the past millennium, but also seasonal floods associated with present instrumental observation data (Menounos and Clague, 2008).

Geochemical data as proxies

Lake sediment records also provide information about changes in soil and catchment development via phosphorus geochemistry. Phosphorus is a limiting nutrient in terrestrial ecosystems and is made available to catchments mainly through *in situ* soil development processes. New phosphorus (P) is provided by the weathering of mineralised P forms in rocks, which in turn are converted to organic forms by plants and soil development processes. By examining the fraction of P in mineralised, organic and occluded forms in lake sediment records, soil development and catchment development in contributing catchments can be inferred. A number of lakes in the southern Coast Mountains have been examined and provide a consistent regional history of soil and catchment development. Lake sediments show a dominance of mineralised P forms in the early (11–10 ka BP) portion of the record, reflecting a regional landscape dominated by bedrock and unweathered glacial deposits just after ice sheet retreat. The landscape was unstable and the rate of sedimentation was high (paraglacial sedimentation) at this time. From 10 to 8.5 ka BP, an increase in organic and occluded P formed in the lake sediments, indicating increased soil formation and maturity. The interval between 8.5 and 1 ka BP was marked by relatively constant environmental conditions with progressive soil and ecosystem development. The last 1 ka, coinciding with the Little Ice Age, has seen a return to higher concentrations of mineralised P and higher relative sedimentation rates (Souch, 1994; Filippelli and Souch, 1999; Slaymaker *et al.*, 2003).

The use of geochemical data as proxies for environmental change can also be illustrated from sediments from lake catchments in the Highlands of Scotland (Dalton *et al.*, 2005). They reveal that catchment processes, including degradation of soils and post-industrial acidification of the lakes, have had a strong influence on lake biota during the late Holocene (Dalton *et al.*, 2005).

Diatoms as proxies

Diatoms are only one of many algal indicators in lake sediments, but historically they have been used extensively for environmental reconstruction. Diatoms are often the dominant algal group in most freshwater systems, frequently contributing more than half of the overall primary production (Smol, 2002). There are thousands of species of diatoms and different taxa have different environmental tolerances and optimal conditions. Assemblages change rapidly in response to environmental change. Also the silica (SiO_2) which forms the skeletons of diatoms is generally resistant to decomposition, dissolution and breakage. Certain diatom taxa are closely linked to the pH of lake waters, nutrient concentrations and salinity (Battarbee *et al.*, 2001) and many empirical indices have been developed to infer lake water quality from them. Ecologically sound and statistically robust inference models have become available to reconstruct past lake water pH.

3.5.2 Models and limitations for prediction

The use of models in observational science encounters a dilemma that was well stated by Oreskes *et al.* (1994), namely that model verification is only possible in closed systems, in which all the components of the system are established independently and known to be correct. Real-world systems are not closed, model predictions are not unique, model input parameters and assumptions are almost always poorly known, scaling issues for non-additive properties are poorly known and data themselves pose problems. In spite of these limitations, models help us to recognise, describe and measure the relation between proxy data from sedimentary archives and water and sediment fluxes in lakes (Ochiai and Kashiwaya, 2003; Endo *et al.*, 2005). Mathematical models can be helpful for the quantitative understanding of processes in lake catchment systems, but they do incorporate explicit assumptions about causal relations in the catchments. Accordingly, their usefulness for future prediction is vigorously debated, especially given the increasingly common notion that we are in a 'no analogue' situation since the rise of the human factor as the major driver of environmental change.

As illustrations of the application of models in lake catchment studies, models expressing the relation between sedimentation rate and erosion rate are discussed below. Firstly, an experimental model based on instrumental observation data is introduced; and secondly, a process model relating mass transport rate to precipitation and catchment surface properties is presented.

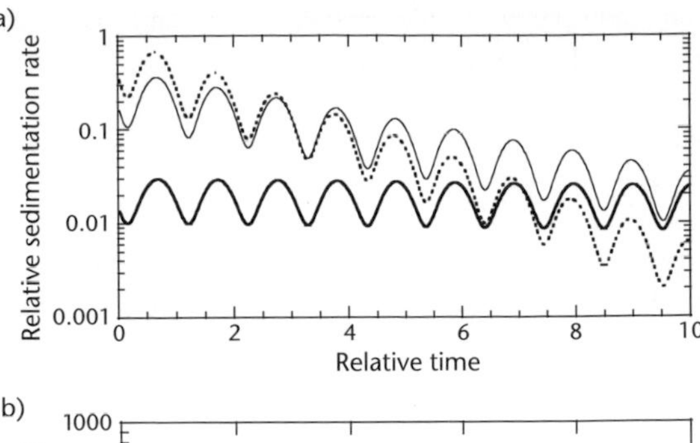

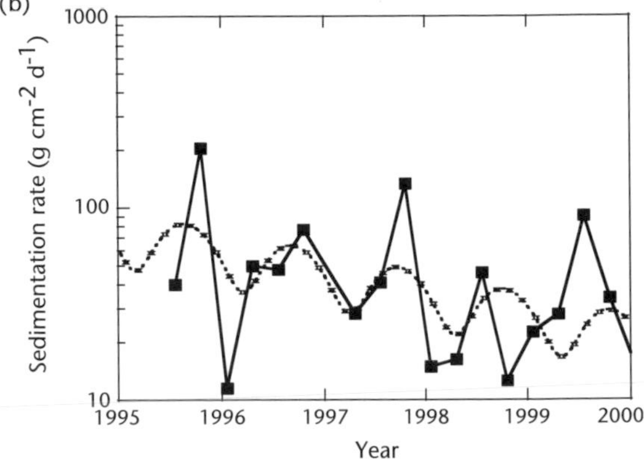

FIGURE 3.12. (a) Numerical calculations for a short-term conceptual model (thin solid line: $a = 0.3$, $c = 0.5$; dotted line: $a = 0.15$, $c = 0.25$; thick solid line: $a = 0.01$, $c = 0.02$), and (b) observed values (black square) and calculated results with the model (dotted curve).

A short-term experimental model for a lake catchment system

Following a study of sedimentation in the small lake catchment Kawauso-ike near Kobe, Japan, a short-term experimental model was introduced with the simple assumptions that sedimentation rate in a lake (SR(t)) is proportional to the size of external force ($\lambda(t)$) (in this context hydroclimatic conditions) and amount of erodible material in its catchment at time t ($V(t)$) (Kashiwaya *et al.*, 2004). A sine function is introduced in order to express the effects of seasonality on erosion and the exponential decay term describes the way in which the effect of a disturbance gradually dissipates. Then, sedimentation rate is expressed by Kashiwaya *et al.* (2004) as:

$$\mathrm{SR}(t) = \alpha\lambda(t)V_0 e^{-\int^{e}\lambda(\tau)d\tau} \tag{3.1}$$

where the erosion factor $\lambda(t)$ is a sine function of the form (a sin $bt + c$) and α is a constant proportionality factor which physically resembles a delivery ratio. Numerical calculations varying the values of a and c produce a plot of relative sedimentation rate versus relative time (Fig. 3.12a). This model has been applied to the small lake catchment mentioned above. The relation between seasonal rainfall and sedimentation rate fits reasonably well with the model (Fig. 3.12b). It would be possible to adapt the model to describe the effects of human activity on the lake catchment.

A process-oriented model for lake catchment systems

A number of small lake catchments in Japan and Korea have been studied in terms of the relations between sedimentation and seasonal rainfall. The model has been developed on the basis of two assumptions:

- that lake catchment conditions in the ith region of the catchment are approximately constant over a period of decades; and
- that rainfall erosivity controls erosion.

Average sedimentation rate in the ith region of a catchment (SR_i) is then expressed as:

$$\mathrm{SR}_i = R_i \cdot f(P_i)(P_i\Delta T) \tag{3.2}$$

where R_i is the erodibility of the ith region, $f(P_i)$ is a rainfall erosivity function and $P_i\Delta T$ is the average rainfall during the interval ΔT for the ith region. In general, $f(P_i)$ is experimentally determined for each climatic zone. Here, examples from Japan and Korea are introduced (Fig. 3.13). For the data obtained in this region, a power function is available of the form:

$$\mathrm{SR}_i = R_i f(P_i)(P_i\Delta T) \tag{3.3}$$

where $f(P_i\Delta T) = (P_i\Delta T)^b$ and where b is a runoff factor. The physical meaning of this equation is that a sediment yield is

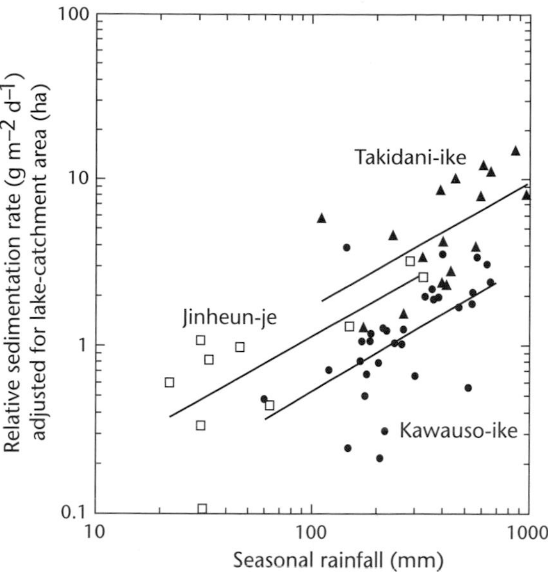

FIGURE 3.13. Relations between sedimentation rate and seasonal rainfall in small pond-catchment systems in Japan (black circles, black triangles) and Korea (open squares). Data were obtained by a team from Kanazawa University and Korea Institute of Geoscience and Mineral Resources (KIGAM).

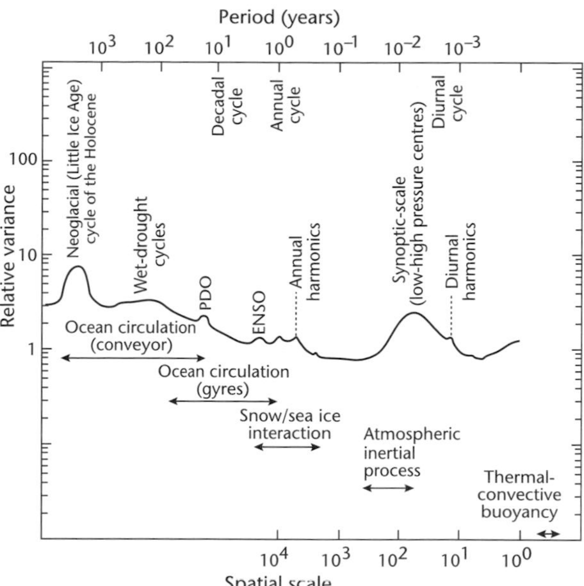

FIGURE 3.14. Spectrum of observed climate variability in instrumental and proxy climate records (modified from Mitchell, 1976).

the product of a basin erodibility factor and a forcing function raised to the power of b. This equation was applied to some small pond catchment systems in Japan and Korea (Fig. 3.13), showing that the model is appropriate for simple lake catchment processes. In this case R_i takes the values 0.015, 0.039 and 0.054, indicating a wide range of catchment erodibility conditions, but the fact that b has a small range from 0.7 to 0.8 suggests that the catchments respond similarly to hydroclimatic forcing.

The basic concept underlying this model is that it can be used to make explicit other environmental changes over time. From the same studies there have also emerged some clear relations amongst sedimentation rate, organic material and particle density. Large relative sedimentation rates are related to low loss on ignition and high particle density, suggesting that these factors are sensitive to land transformation. This implies that any land transformation in forested catchments requires careful management to prevent surface erosion, especially in the initial stage when the erodibility factor increases most rapidly. This conclusion is consistent with findings reported in Chapters 2, 4, 11 and 12 of this volume.

3.5.3 Hydroclimate changes interpreted from lake sediments

Observed climate variability in instrumental and proxy records has been summarised by Mitchell (1976) (Fig. 3.14). The plot expresses the relative variance of climate against temporal and spatial scale and shows the effect of each component of the hydroclimate on that variance. Spatial and temporal abrupt environmental changes (shift in regime, system or subsystem) are recorded in lake catchment systems both at regional and global scales as well as short-term and long-term scales. The fact that sediment transport can be linked to climatic variability over a range of timescales, and that the proportion of variance attributable to climatic forcing increases with decreasing frequency are recurring themes in studies from arctic and alpine lake studies. The El Niño-Southern Oscillation (ENSO) is the dominant year-to-year climate signal on Earth (McPhaden *et al.*, 2006). ENSO affects the frequency, intensity and spatial distribution of tropical storms and, with respect to lake catchments, alters patterns of rainfall, surface temperature and sunlight availability. Impacts may also be exaggerated by land use practices (Siegert *et al.*, 2001). As a result, interest in identifying past effects of ENSO in lake sediments has intensified.

Menounos *et al.* (2005) documented clastic sediment response to climate and geomorphic change over the past 120 years in five lakes in the southern Coast Mountains of British Columbia. The lakes varied in area from 2 to 30 km² and the lake catchment relief varied from 1580 to 2640 m. They confirmed substantial concordance among the records, with a notable break occurring around 1976. This break coincided with a major reorganisation of the North Pacific climate system (the PDO) leading to an increase in

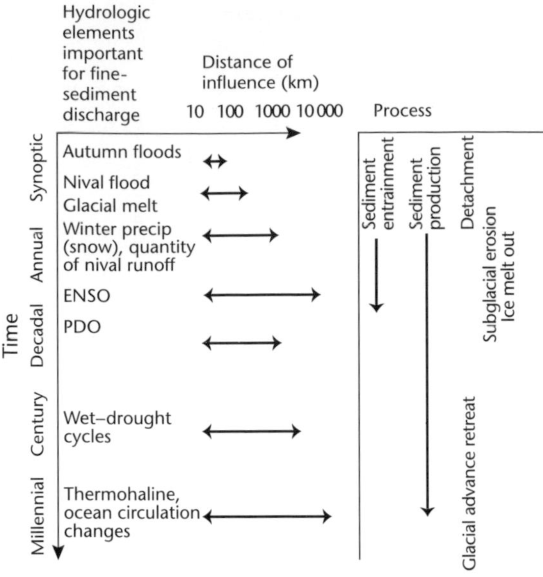

FIGURE 3.15. Time-space scaling of observed and inferred climate–sediment transport linkages in Coast Mountains of British Columbia (from Menounos, 2002).

the magnitude of autumn flooding. Both before and after 1980 inter-annual varve thickness correlated well with annual flood magnitude and inter-decadal trends were influenced by glacier fluctuations. Menounos (2002) has related sources of variance in the hydroclimate to sediment entrainment and production in the Coast Mountains (Fig. 3.15).

Factors controlling sediment entrainment and delivery at the event scale are primarily driven by meteorological events which occur locally and over periods of hours to days. At this scale, controls such as relief also influence the intensity of a given hydrologic event. In the above example, the divide of the Coast Mountains controls the magnitude of autumn runoff events. At the annual scale, variations in the intensity of nival and glacial melt are influenced by winter precipitation and temperature changes which are moderately linked to climate variability at the regional (PDO) and hemispheric (ENSO) spatial scales. Inter-annual to inter-decadal changes in the phase of the PDO or ENSO can be shown to be associated with minor changes in snow cover and, possibly, the frequency of extreme autumn runoff events. Century to millennial changes in sediment delivery are related to large-scale changes in ice cover. For at least the last six centuries, glacier fluctuations in the southern Coast Mountains appear to be driven by persistent anomalies in temperature. Over longer periods, variations in glacier cover were probably caused by large-amplitude changes in climate driven by processes operating at global spatial scales.

3.5.4 Summary

Proxy data incorporated into conceptual and mathematical models provide a link between climate and lake catchment behaviour. When several proxies provide convergent information, detailed reconstruction of past environments at the lake catchment scale can be generated with some confidence. It remains uncertain that these models can be used for predictive purposes (though Kashiwaya is more optimistic). It is however increasingly clear that hydrological variability, and hence lake level and lake catchment behaviour, can be interpreted in terms of large-scale climatic anomalies, such as those associated with ENSO, and that there are strong relations between hydrologic anomalies in different parts of the world. If the future frequency of ENSO could be predicted more accurately then at least it would be more feasible to predict the behaviour of undisturbed lake catchments over the next few decades. Until such a time, we suspect that it will be necessary to employ conceptual models, such as one which was proposed by Menounos (Fig. 3.16). This model is a valuable summary assessment of the major relations between climate and landscape change in undisturbed catchments in the Canadian Cordillera. Although the ENSO effect is well evidenced in catchments that have had minimal disturbance by human activity (e.g. Desloges and Gilbert, 1994; Lamoureux, 2002; Menounos *et al.*, 2005; Schiefer *et al.*, 2006), the effect is rarely documented in lake catchments that are heavily disturbed by human activity. It is therefore necessary to now move to the consideration of human impacts on lake catchment systems.

3.6. Effects of human activity

3.6.1 Overuse of water for irrigation: Aral Sea and Lake Chad

One of the most disastrous results caused by recent land use changes in a large lake basin is that of the Aral Sea, located in a desert region of central Asia (Austin *et al.*, 2007; Micklin, 2007; Sorrel *et al.*, 2007). Long-term changes in the area of the Aral Sea are closely related to changes in the Amu Darya drainage basin (Fig. 3.17a). Recent shrinkage of the water area of the Aral Sea is attributed to intensive irrigation from the 1960s to the present (Fig. 3.17b), which had a severe effect on the lake catchment hydrology and enhanced desiccation in the area (Small *et al.*, 2001). Ninety-four reservoirs and 24 000 km of channels on the Amu Darya and Syr Darya rivers have been constructed to support the irrigation of 7 Mha of agricultural land. The volume of water in the Aral Sea has been reduced by 75% since 1960. It is interesting to note that this shrinkage is not without precedent in the past as the result of entirely natural

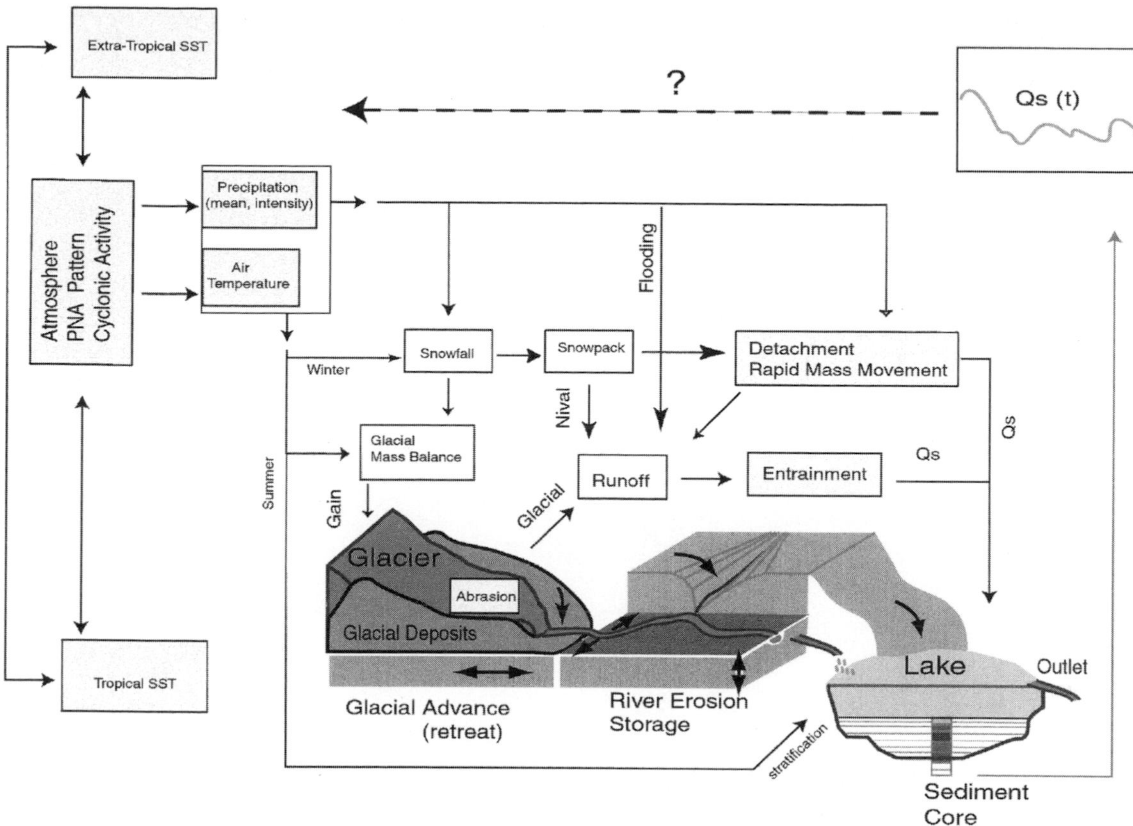

FIGURE 3.16. Proposed lake-sediment linkages with relief and hydroclimate (from Menounos, 2002).

drainage modifications under changing hydroclimate (Boomer *et al.*, 2000).

Lake Chad in central Africa is facing a similar situation due to land use and hydroclimate change. Lake Chad is especially sensitive to environmental change because of its shallowness (Coe and Foley, 2001). This lake has also experienced several hydrological shifts during the Holocene, including a so-called Megalake Chad stage (7.7–5.5 ^{14}C ka BP) (Sepulchre *et al.*, 2008). The most recent high water level has been dated to 3.7–3.0 ^{14}C ka BP; thereafter the level decreased (Leblanc *et al.*, 2006). These hydrological shifts have seriously influenced human activities and culture, as they are dependent on the availability and quality of accessible water (Gasse, 2002). During the severe drought periods of the 1970s, 1980s and the early 1990s, Lake Chad shrank significantly from approximately 23 000 km^2 in 1963 to less than 2000 km^2 in the mid-1980s. Overgrazing, deforestation contributing to a drier climate and large, unsustainable irrigation projects in Chad, Cameroon, Niger and Nigeria which have diverted water from the lake and the Chari and Logone rivers have all contributed to this shrinkage. Although there was partial recovery in the late 1990s, reduced rainfall and increased drought frequencies are predicted by the IPCC (2001, 2007a, 2007b) and continuing unsustainable irrigation projects will potentially eliminate Lake Chad altogether.

3.6.2 Accelerated erosion and sedimentation

Although erosion is a function of both erosivity of rainfall and erodibility of the land surface, erosivity is independent of human activity. Human activity, however, does affect the erodibility of lake catchment land surfaces. The erodibility of these surfaces, expressed in terms of resistance to water erosion, is estimated through particle size analysis of the surface materials, organic matter content, soil structure and permeability. Physical disturbance of the natural vegetation cover and chemical and biological changes in the structure of the near-surface earth materials have direct consequences for the mobilisation of sediment on a slope in the form of accelerated erosion. The term 'accelerated erosion' literally refers to the net effect of human disturbance of the landscape.

Recent land use changes have caused large environmental changes. Kashiwaya *et al.* (1997) have conducted a

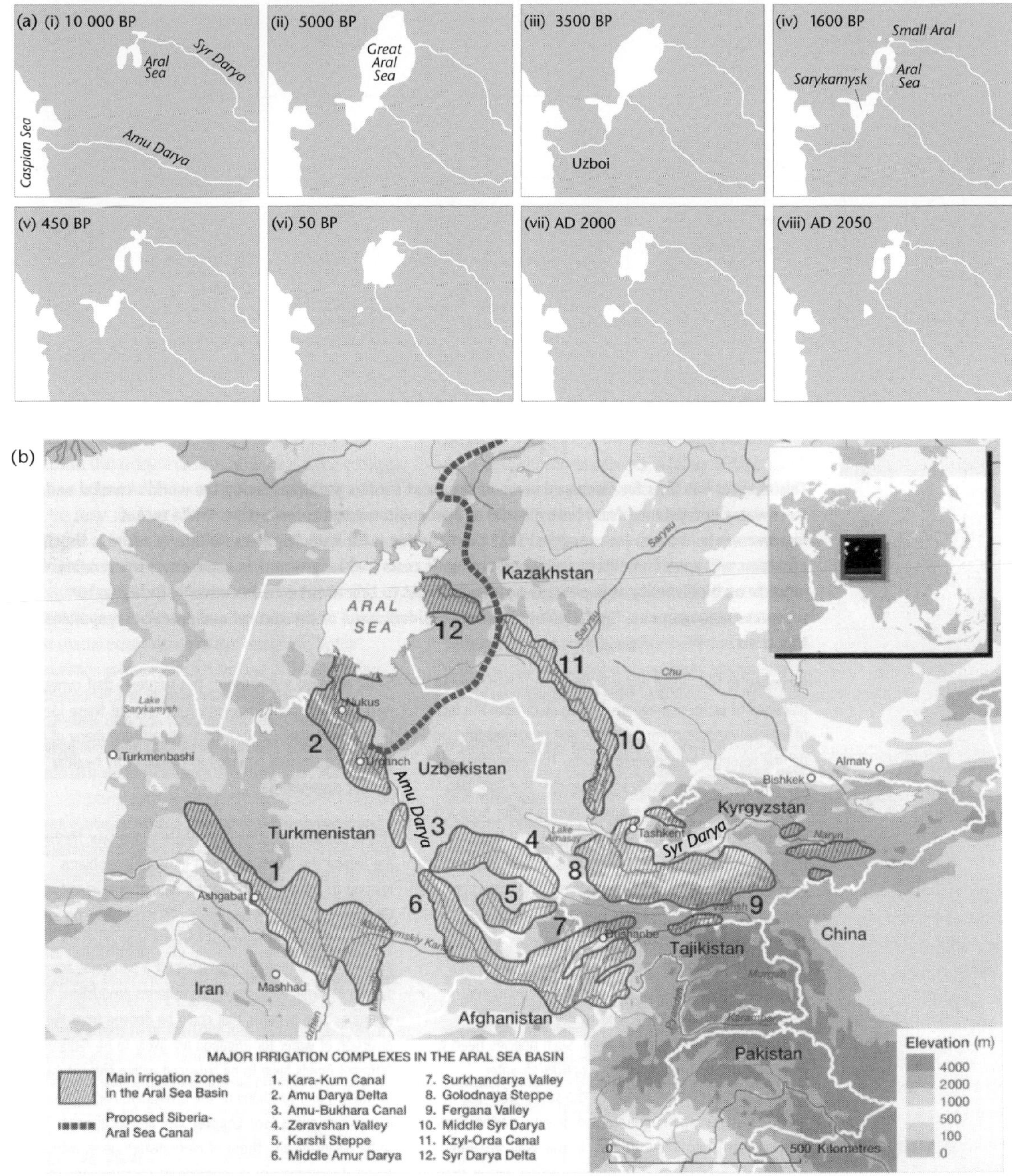

FIGURE 3.17. (a) Long-term changes in the Aral Sea system (from Boomer et al., 2000); (b) major irrigation areas in the Aral Sea basin (modified from Kreutzberg-Mukhina, 2004).

geomorphological experiment over several years using two pond catchments in the Kobe district, central Japan. These pond catchments were located side by side, one forested and the other clearcut, in order to show relations between sediment response and land transformation. Sedimentation rate in the undisturbed pond was roughly proportional to rainfall intensity, whereas in the pond with land transformation sedimentation rate was a direct function of area cleared, as well as rainfall intensity. A catchment erosion factor was here defined as accumulated material in the pond

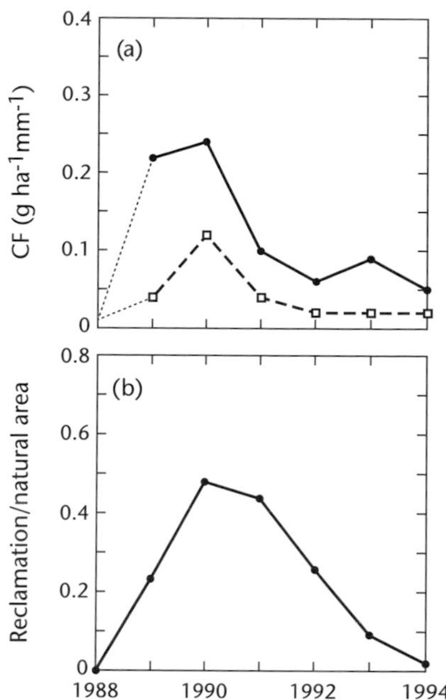

FIGURE 3.18. (a) Temporal change in catchment erosion factor (CF) (relative sedimentation rate) in two pond catchment systems (bold solid line, catchment under reclamation, KR; thin solid line, catchment without reclamation, KN); (b) change in reclaimed/natural area in the reclaimed catchment. (Kashiwaya *et al.*, 1997.)

divided by the catchment area. Rainfall intensity increased in the initial stage of land transformation and then decreased (Fig. 3.18).

Modification of lakes by human activity includes river diversion, water supply mismanagement, acidification, contamination with toxic chemicals, eutrophication and enhanced siltation due to accelerated land erosion (Lerman *et al.*, 1995).

3.6.3 Land clearance

Much of the pioneering work on relating land clearance to rates of sediment accumulation in lakes has been carried out in lake catchments of less than 1000 km^2 in area. Davis (1976) used pollen analysis to chronicle land clearance in the Frains Lake catchment in Michigan from the early nineteenth century. She showed a 30–80-fold increase in catchment erosion rates in the mid nineteenth century following forest clearance, based on sediment accumulation rates in the lake. Even though erosion rates levelled off after three to four decades, they remained about one order of magnitude higher than pre-disturbance rates.

Well-known examples of environmental reconstruction are described below to illustrate the importance of land clearance as recorded in lake sediments. Few better ways

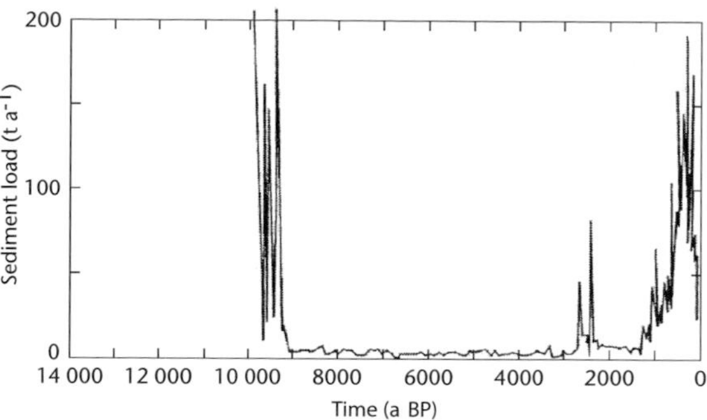

FIGURE 3.19. Holocene sediment flux in Bussjösjön, southern Sweden (from Dearing *et al.*, 1990).

are available for reconstructing the landscape changes of the Holocene Epoch.

Havgårdssjön and Bussjösjön, southern Sweden

These systems are typical of the lakes of previously glaciated parts of the northern hemisphere. Sediment accumulation rates in lakes in southern Sweden show high pre-Holocene rates declining to minimum values during the early to mid-Holocene and rising rapidly during the late Holocene or Anthropocene. Havgårdssjön sediments provided a record of sediment accumulation over the last 5000 years (Dearing *et al.*, 1987) and Bussjösjön sediments document *c.* 10 000 years of sedimentation, covering the whole of the Holocene Epoch (Dearing *et al.*, 1990) (Fig. 3.19). Four major periods of lake sediment responses to environmental conditions were identified. Periglacial or paraglacial times, when high levels of sediment movement were interpreted as being associated with glacial sediments left behind by the retreating Eurasian ice sheet, continued until *c.* 9250 BP. The next 6500 years were a period of relative catchment stability after forest cover became established. Human influence on catchment erosion became evident around 2500 years ago and this modest influence lasted until about 300 BP. Detailed interpretation of the higher lake sedimentation rates during this more recent period invokes the influence of varying surface drainage networks and subsoil erosion as a result of undersoil drainage associated with more intensive agriculture.

Lake Patzcuaro, Mexico

Lake Patzcuaro's catchment is a 126 km^2 closed lake basin at 2036 m above sea level in the Michoacan Highlands of Mexico (Bradbury, 2000). O'Hara *et al.* (1993) identified three major episodes of forest clearing before the period of European contact and also established that erosion rates apparently decreased following the arrival of the Spanish in AD 1521, doubtless due to depopulation and forest regeneration.

Schwarzsee and Seebergsee, Swiss Alps
Vegetation changes over the past 7000 years were reconstructed and past landslide events traced from sediment cores from the Schwarzsee in the Swiss Alps (Dapples *et al.*, 2002). Replacement of forested areas with pastures and meadows accelerated after 3650 BP, with a progressive lowering of slope stability in the catchment. Thirty-six turbidity events have been recorded in lake sediments over the last 2000 years, compared with only 16 in the previous 4300 years. These data suggest that human-induced landslide activity is more important than climate change in this area. Thus a multi-proxy, high-resolution study of the past 2600 years for Seebergsee, also in the Swiss Alps, has indicated the importance of land use changes (alpine pasturing) associated with the introduction of fertilisers in the seventeenth and eighteenth centuries (Hausmann *et al.*, 2002).

3.6.4 Pastoral land use

An important example of the late Holocene lake sediment record of erosional response to land use changes is provided by Page and Trustrum (1997). They recorded changes in lake sediment thickness in cores taken from lakes Tutira and Waikopiro in North Island, New Zealand as a proxy for the severity of catchment erosion. In doing so, they noted that such measures must take account of variations in sediment density, aquatic organic matter, biogenic silica, lake bank erosion, volcanic ash and wind-borne dust, trap efficiency of the lake and changes in sediment storage sites. Depending on the area of the lake catchment and specifically the length, gradient and nature of the sediment pathway from source to lake, a variable proportion of the sediment mobilised will arrive in the lake. The so-called 'sediment delivery ratio' expresses the proportion of mobilised sediment that reaches the lake. Having made all these adjustments, they came to two conclusions:

- The natural variability in sedimentation rates was high, even under indigenous forest because of sediment-mobilising storms, earthquakes, droughts and volcanic eruptions; and
- at no time since the lakes were formed (around 6500 BP) were sedimentation rates as high as the rate under pastoral land use, when rates accelerated 8- to 17-fold.

3.6.5 Eutrophication

The problem of nutrient enrichment of water bodies can really be summarised as the environmental consequence of overfertilisation. Nutrients such as phosphorus and nitrogen enter water bodies from non-point sources such as overland drainage, fertiliser runoff from agriculture and groundwater, or from point sources such as municipal and industrial effluents. Algal blooms, aquatic macrophytes, decaying organic matter, taste and odour problems, decreased deep-water oxygen and shifts in the structure of the food web (including loss of fish species) are the salient evidences of eutrophication.

Incipient eutrophication
Lake Baikal is by far the oldest lake known, as it exists in a tectonic basin that began to form some 30 million years ago. It contains the largest single source of fresh water on Earth, accounting for 20% of the total freshwater reserve. Diatom species analyses do not record any marked eutrophic signals (Mackay *et al.*, 1998). This is a most important indicator of global lake health and the fact that Lake Baikal remains relatively undisturbed is encouraging. Intensive monitoring of future changes is an international priority, particularly given the industrialisation of its southern shoreline.

Artificial eutrophication
In a paired catchment experiment in the Experimental Lakes, Ontario, one catchment was fertilised with phosphorus, nitrogen and carbon whilst a neighbouring catchment was fertilised, but with no phosphorus. Algal blooms dominated the first lake within 2 months (Schindler, 1974). Shortly after these results were reported, legislation was adopted limiting the phosphorus levels in detergents and other effluents.

Industrial eutrophication
As an example of industrial eutrophication, industrial pollution from mining, milling and smelting complexes in Sudbury, Ontario has led to the creation of a landscape denuded of most vegetation. Kelly Lake and its input tributary, Junction Creek, have retained a record of the environmental and erosional impacts, both through the growth of the Junction Creek delta and through the accumulation of polluted sediments at the bottom of the lake. During the period 1928–56, deforestation of the catchment coincided with the most accelerated period of erosion. Smol (2002) comments that, with anticipated warming, lake levels are expected to drop. One potential consequence will be that the Junction Creek delta, with its massive archive of metal pollutants, will be exposed and the pollutants will become remobilised and redistributed into the lake.

Urbanisation, sewering and phosphate detergents
The St Lawrence River–Great Lakes system in Canada and the USA is the largest freshwater system in the world,

containing over 25% of the world's stored fresh water. The Great Lakes catchment is currently highly developed and supports over 35 million inhabitants. It was not until the 1960s that the symptoms of eutrophication began to be clearly recognised. This was in response to the increased urbanisation, sewering and the use of phosphate detergents. Lake Erie is the most eutrophic lake in the Great Lakes system. Stoermer (1998) has summarised some of the findings:

- Lake Ontario was severely impacted as early as *c.* AD 1830, when substantial European populations were established;
- Nutrient loading seems most important;
- A significant source of nitrogen and trace metals is the deposition from airborne nitrates, sulphates and metal dust;
- Exotic invaders, like zebra mussel (*Dreissena polymorpha*), have had important impact; and
- Nutrient abatement was successful in reducing eutrophication in some systems but severely disturbed systems will probably never recover to their original states.

However, eutrophication is becoming less severe in developed countries and new issues such as global warming, and the influx of pharmaceutical products such as endocrine disruptors, will have to be significantly considered. In developing countries, eutrophication continues in many cases unabated.

Urbanisation and urban agriculture

Lake Dian-chi, located in Kunming, Yunnan, southwestern China, and on the eastern margin of the Tibetan Plateau (1800 m above sea level), is highly eutrophic and one of the most drastically polluted lakes (Plate 15). The lake area is 300 km² and the city sits on a series of alluvial fans sloping directly towards the lake. Between the city and the lake are irrigated lands and ponds and the eastern shore of the lake is a cultivated plain. Effects of economic reform, booming industrial development and population increase in the catchment are recorded in lake surface sediments (Gao *et al.*, 2004). The physical conditions of the lake, with a limited outflow at the east end, mean that the lake is effectively a closed system. However, this lake catchment has long records of human activity. Notable impact of human activity has been detected in lake sediments since 1700 BP mainly because of land use change, both agricultural activity and urbanisation, in the catchment (Wu *et al.*, 1998), and gradual increase in aridity (Chena *et al.*, 2008).

During the spring of 2007 a massive blue–green algae bloom broke out in Lake Taihu, one of the largest lakes in China. This freshwater lake is located in Jiangsu on the Yangtze River Delta, one of the world's most urbanised and heavily populated areas. The massive bloom event became an environmental crisis that prompted officials to cut tap water supply to several million residents in nearby Wuxi City. The Chinese government identified the outbreak as a major natural disaster (Wang and Shi, 2008).

3.6.6 Rainfall acidification

Acid rain, produced by the release of sulphur and nitrogen oxides into the atmosphere, transportation through the atmosphere, conversion to sulphuric and nitric acids and precipitated back to Earth, has been occurring for more than a century. However it was not until the 1960s that the scale of the problem became apparent (Oden, 1968). There are four major steps in the process of acidification which must be differentiated in order to obtain a clear understanding of the problem:

- emission, both natural and anthropogenic, with sulphur and nitrogen oxides being the chief anthropogenic sources (coal and oil combustion, smelting of ores and agricultural processes);
- transport through the atmosphere, with long-range transport being a key issue;
- transformation – the combination of water with sulphur and nitric oxides produces sulphuric and nitric acids;
- deposition or fallout, where local geology, depth and type of soils, slope and vegetation will determine the seriousness of the environmental effects of the acid deposition by determining the degree of acid buffering.

Major successes have been achieved in unravelling the problem, one of the most notable being the transformation of Sudbury, Ontario, Canada from the 'acid rain capital of the world' to the scene of a remarkable recovery of many lake systems, following a 90% reduction in sulphur emissions. Not only have lakewater pH levels recovered rapidly but also diatom populations. It is not entirely clear, however, why some lakes are still acidifying and not showing any signs of recovery. Smol *et al.* (1998) compared lakes in the Adirondack Park, New York State, with those in the Sudbury region. They concluded that, in general, there has been little recovery in the Adirondack lakes whereas the Sudbury lakes have shown much stronger recovery. They point to the higher background pHs and natural buffering capacities of the Sudbury lakes and suggest that the Sudbury lakes can absorb higher critical loads than the more naturally acidic lakes of the Adirondacks.

As with many environmental problems, while acidic deposition in Europe and North America is being overcome as an environmental problem, it is increasing in many developing nations, including many tropical and subtropical countries (Kuylenstierna *et al.*, 2001).

3.6.7 Reservoirs and impoundments

It is estimated that there are 260 000 km² of reservoirs and 77 000 km² of farm ponds in the world today, as compared with 4.2 M km² of lakes and between 8.2 and 10.1 M km² of wetlands. Chapter 4 considers the implications of such features for the riverine landscape, including the environmental impact of the establishment of very large reservoirs in the twentieth century.

The Three Gorges Dam

As an example of the problems raised by large reservoirs, it is instructive to consider the Three Gorges Dam (TGD) in China. The structure was partially opened in 2003 with full pool (normal maximum level of the reservoir lake) being reached in 2008; it is having major environmental impacts. Physical, chemical and biological conditions of the Yangtze (Changjiang) River have dramatically changed in the reservoir, deposition in the reservoir is occurring and an increase of erosion in downstream sections has already been detected (Yang *et al.*, 2007). These physical impacts have been complemented by the socioeconomic impacts of the loss of land and displacement of people. In his report to the US Export–Import Bank, warning of the unpredictable effects of constructing this dam and advising against providing financing and loan guarantees to US firms bidding on related contracts, Luna Leopold (1996) made the following comments:

> The TGD is designed to operate under conditions practically untested in the world and never before tested in such a large structure. Projections of controlling sedimentation within the reservoir are subject to significant uncertainties. China has about 83,000 reservoirs built for various purposes, of which 330 are major in size. In 230 of these sediment deposition has become a significant problem, resulting in a combined loss of 14% of the total storage capacity. In some, more than 50% of the storage capacity has been lost.

He also added his critique to the widely discredited Canadian Yangtze (Changjiang) Joint Venture study which concluded that a 175-m dam was feasible. Now (2008) that the dam is fully operational, it has been suggested that the climatic effect of the TGD will be on the regional (*c.* 100 km) scale rather than on the local scale (*c.* 10 km) as projected in previous studies (Wu *et al.*, 2006). Reports on risks for the ecosystem of TGD area have been published (e.g. Wu *et al.*, 2003; Shen and Xie, 2004; Nilsson *et al.*, 2005) and it is expected that the completion will also produce downstream effects offshore, by reducing the upwelling and thus productivity in the East China Sea (Chen, 2000).

TABLE 3.5 *Order of magnitude effects of human activities on erosion rates in lake catchments*

Land use	Relative erosion rate
Forest	1
Grassland	10
Cropland	200
Harvested forest	500
Construction sites	2000

Source: After Morris and Fan (1998).

3.6.8 Summary

River impoundment and regulation behind dams, anthropogenic activities that affect the land surface, such as urbanisation, agriculture, deforestation and afforestation and overgrazing within lake catchments, affect the quantity and quality of water and the rate of sediment production and delivery to lakes. The order of magnitude of impacts of human activity in accelerating erosion is summarised in Table 3.5.

3.7 Scenarios of future wetland and lake catchment change

Over the course of the next century, changes in hydroclimate and land use will have major impacts on lakes and wetlands. Erosion, transportation and sedimentation in lake catchment systems are closely related to increases and decreases of rainfall and/or runoff. Lake level changes following increased evaporation, and superimposed on lower runoff, will provoke concerns over water shortage, as is already the case in the Great Lakes of North America (Löfgren *et al.*, 2002) and the Caspian Sea (Elguindi and Giorgi, 2007). It can be anticipated that the knock-on effects of these hydrological trends will include sediment budget changes in lake catchments (Milly *et al.*, 2005).

No globally consistent trend in recent lake level fluctuations has been found (IPCC, 2007b). While some lake levels have risen in Mongolia and China in response to glacier and snowmelt, other lake levels in China, Australia, Africa, North America and Europe have declined due to the combined effects of drought, warming and human activities. Since the 1960s, surface water temperatures have warmed by 0.2–2 °C in lakes in Europe, North America and Asia. Deepwater temperatures of the large East African lakes have warmed by 0.2–0.7 °C since the early 1900s. Increased water temperature and longer ice-free seasons have influenced the thermal stratification and internal hydrodynamics of lakes. In several lakes in Europe and North America, lengthened periods of stratification have increased thermal stability. Increased stratification in European and East African lakes has inhibited the mixing which provides essential nutrients to the food web. These changes imply reduced nutrients in surface waters and concomitant increases in deep waters. By contrast, in alpine and subalpine lakes, the warming of lake waters has led to increased organic acid inputs. Within permafrost areas in the Arctic, formation of temporary lakes has been the result of the onset of permafrost degradation.

3.7.1 Terrestrial wetlands

Wetlands, because they accumulate organic matter and nutrients, are also unusually productive and diverse ecosystems (see also Chapter 5 for discussion of coastal wetlands). Many species of plants and animals are obligate wetland-dwellers, and many more rely on them seasonally or in passage, as recognised in the 1971 Convention on Wetlands of International Importance (commonly known as the Ramsar Convention). Whilst climate change will ineluctably alter the water balance of wetlands and marginally shrink or expand them, eliminating certain seasonal wetlands, it is direct human manipulation that has been vastly more important in effecting change in the world's wetlands. An important change has been their elimination by drainage. Wetlands provide habitats for many disease-carrying pests so that much early drainage – of which the drainage of the Roman marshes of the River Tiber is perhaps the archetypical example – was directed toward improving community health. But marshes, once drained, are also potentially attractive agricultural or forest land, and so massive modern efforts in wetland drainage have been conducted for agricultural improvements. More than half of the 100 million hectares of wetland originally located in the conterminous United States have been lost to drainage, most of it for land development. In Canada, about 14% of 127 million hectares has been lost, nearly all to agricultural land development. The contrast in fractional loss between the two neighbouring countries emphasises the overwhelming importance of population size and distribution in trends of wetland loss.

Wetlands are, in any case, one of the most ephemeral of landforms. They are former shallow water bodies being transformed by accumulation of organic material into terrestrial surfaces. In many places, eutrophication has accelerated this process dramatically and indeed ponds that were open water bodies in the nineteenth century are, today, terrestrial surfaces in highly industrialised areas or subjected to intensive modern agricultural practices. At the other extreme, humans maintain large areas of artificial wetland, often on hillsides, for paddy agriculture.

3.7.2 Lake catchments

Lake catchment systems deliver a wide range of ecosystem services such as water supply, flood regulation, recreational opportunities, tourism and fish and forest products. The degradation and loss of lakes and lake catchment integrity, though not as rapid as the loss of terrestrial wetlands, is a serious concern. The primary drivers of lake degradation are population growth, increasing economic development, lakeshore infrastructure development, land use changes in the catchment, water withdrawal, eutrophication and pollution, overharvesting, overexploitation and the introduction of invasive species. Accelerated erosion and sedimentation, and excessive nutrient loading are a growing threat to lake catchment functions. Abrupt changes in the behaviour of lake catchment systems are potentially impossible to reverse. Physical and economic water scarcity and reduced access to water are major challenges to society. Lake catchment management which is cross-sectoral and ecosystem-based can consider trade-offs between agricultural production and water quality, land use and biodiversity, water use and aquatic biodiversity and land use and geodiversity (Cohen *et al.*, 2003). Closed lakes serve as sediment and carbon sinks. Lake catchment management is a framework that supports the promotion and delivery of the Ramsar Convention's 'wise use' concept and, in the long term, contributes to human well-being.

3.7.3 Vulnerability of terrestrial wetlands and lake catchments

In assessing the global vulnerability of terrestrial wetlands and lakes, it is important to note both the biomes in which they occur and the intensity of socioeconomic activity within these zones. For example, the majority of natural freshwater lakes are located in the higher latitudes, most artificial lakes occur in mid- and lower latitudes and many saline lakes occur at elevations up to 5000 m, especially in the Andes, Himalaya and Tibet.

Freshwater lakes in higher latitudes are still subject to relatively modest disturbance and, as has been noted, undisturbed records of lake sedimentation in these biomes have provided some of the finest records of the influence of hydroclimate on landscape evolution. However, the lack of immediate threat to these high-latitude lake catchment systems and wetlands should not blind us to the fact of the highly variable vulnerability of these landscapes to future disturbance. Both physical vulnerability of permafrost to regional warming and ecological vulnerability to persistent organic pollutants transported from lower latitudes are uniquely characteristic of high-latitude environments (ACIA, 2005)

Wetlands in the tropics and subtropics are experiencing a kind of reverse development where paddy agriculture has been developed. Large areas have been artificially turned into seasonal wetlands for the cultivation of rice and semi-aquatic crops, such as lotus (*Nelumbo nucifera*). But these systems remain dramatically simplified wetlands that serve only part of the hydrological function of natural wetlands and little of the ecological function. Today, much of the remaining, and still extensive, tropical wetlands are under threat of development for paddy or raised field agriculture.

The phenomena of excessive nutrient and sediment loading are projected to increase substantially during the next few decades, leading to eutrophication, acidification and reductions in both geodiversity and biodiversity in terrestrial wetlands and lakes.

3.7.4 Conclusions

There are at least six probable outcomes from climate change and intensification of human activities in lake catchments over the next few decades:

- the absolute loss of water, both in lakes and, more urgently, in wetlands of many kinds can be anticipated if global warming continues and if intensification of land use change continues;
- the reduction in water storage capacity in lakes and wetlands will reduce flood control;
- increased sedimentation will exacerbate both the absolute loss of water and the reduction in water storage capacity;
- acidification will continue and more 'dead' lakes can be anticipated;
- eutrophication and more overly productive lakes will be associated with intensification of land use practices; and
- potential loss of geodiversity and biodiversity will affect both lake catchments and wetlands.

All these trends can be mitigated in the developed world at a cost, but in the developing world the indications are that all of these negative outcomes are accelerating in magnitude and frequency of occurrence.

In the last analysis, lake catchments and wetlands integrate and reflect both human socioeconomic values and geophysical events. Vulnerability to climate change is the most pressing issue in the least densely populated biomes. But in temperate and tropical biomes, vulnerability to climate change pales into insignificance by comparison with vulnerability to the disturbances provoked by *Homo sapiens sapiens*.

APPENDIX 3.1
Global extent of lakes and wetlands

The global extent and distribution of lakes and wetlands is a matter of active debate. Three respected sources are Meybeck (1995) Lehner and Döll (2004) and Downing *et al.* (2006). According to Meybeck, there are 0.8–1.3 million lakes larger than 0.1 km² occupying a total area of 2.3–2.6 Mkm²; at the same time there are 5.6–9.7 Mkm² of wetlands. Lehner and Döll report only 246 000 lakes occupying 2.4 Mkm² and 9.2 Mkm² of wetlands. Downing *et al.* (2006), using size–frequency relations, estimate that the global extent of natural lakes is twice as large as previously known, but they include lakes as small as 0.0001 km².

References

ACIA (2005). *Arctic Climate Impact Assessment*. Cambridge: Cambridge University Press.

Aota, Y., Kumagai, M. and Kashiwaya. K. (2006). Estimation of vertical mixing based on water current monitoring in the hypolimnion of Lake Biwa. *Japan Society of Mechanical Engineers International Journal Series B*, **49**, 621–625.

Austin, P. *et al.* (2007). A high-resolution diatom-inferred palaeoconductivity and lake level record of the Aral Sea for the last 1600 yr. *Quaternary Research*, **67**, 383–393.

Battarbee, R. W. *et al.* (2001). Diatoms. In J. P. Smol, J. B. Birks and W. M. Last, eds., *Tracking Environmental Change Using Lake Sediments*, vol. 3. Dordrecht: Kluwer, pp. 155–202.

Beeton, A. N. (2002). Large freshwater lakes: present state, trends, and future. *Environmental Conservation*, **29**, 21–38.

Boomer, I. *et al.* (2000). The palaeolimnology of the Aral Sea: a review. *Quaternary Science Reviews*, **19**, 1259–1278.

Bradbury, J. P. (2000). Limnologic history of Lago de Patzcuaro, Michoacan, Mexico for the past 48 000 years: impacts of climate and man. *Palaeogeography, Palaeoclimatology, Palaeoecology*, **163**, 69–95.

Burroughs, W. J. (2005). *Climate Change in Prehistory*. Cambridge: Cambridge University Press.

Chen, C. T. A. (2000). The Three Gorges Dam: reducing the upwelling and thus productivity in the East China Sea. *Geophysical Research Letters*, **27**, 381–383.

Chena, F. *et al.* (2008). Holocene moisture evolution in arid central Asia and its out-of-phase relationship with Asian monsoon history. *Quaternary Science Reviews*, **27**, 351–364.

Church, M. and Slaymaker, O. (1989). Disequilibrium of Holocene sediment yield in glaciated British Columbia. *Nature*, **337**, 452–454.

Coe, M. T. and Foley, J. A. (2001). Human and natural impacts on the water resources of the Lake Chad basin. *Journal of Geophysical Research*, **106**, D4, 3349–3356.

Cohen, W. B. *et al.* (2003). An improved strategy for regression of biophysical variables and Landsat ETM+ data. *Remote Sensing of Environment*, **84**, 561–571.

Dalton, C. *et al.* (2005). A multi-proxy study of lake-development in response to catchment changes during the Holocene at Lochnagar, north-east Scotland.*Palaeogeography, Palaeoclimatology, Palaeoecology*, **221**, 175–201.

Dapples, F. *et al.* (2002). Paleolimnological evidence for increased landslide activity due to forest clearing and land use since 3600 cal BP in the western Swiss Alps. *Journal of Paleolimnology*, **27**, 239–48.

Davis, M. B. (1976). Erosion rates and land use history in southern Michigan. *Environmental Conservation*, **3**, 139–148.

Dearing, J. A. and Jones, R. T. (2003). Coupling temporal and spatial dimensions of global sediment flux through lake and marine sediment records. *Global and Planetary Change*, **39**, 147–168.

Dearing, J. A. *et al.* (1987). Lake sediments used to quantify the erosional response to land use change in southern Sweden. *Oikos*, **50**, 60–78.

Dearing, J. A. *et al.* (1990). Past and present erosion in southern Sweden. In J. Boardman, I. D. L. Foster and J. A. Dearing, eds., *Soil Erosion on Agricultural Land*. Chichester: John Wiley, pp. 687–701.

Desloges, J. R. and Gilbert, R. (1994). Sediment source and hydroclimatic inferences from glacial lake sediments: the post-glacial sedimentary record of Lillooet Lake, British Columbia. *Journal of Hydrology*, **159**, 375–393.

Downing, J. A. *et al.* (2006). The global abundance and size distribution of lakes, ponds and impoundments. *Limnology and Oceanography*, **51**, 2388–2397.

Elguindi, N. and Giorgi, F. (2007). Simulating future Caspian Sea level changes using regional climate model outputs. *Climate Dynamics*, **28**, 365–379.

Endo, N., Sunamura, T. and Takimoto, H. (2005). Barchan ripples in the laboratory: formation and planar morphology. *Earth Surface Processes and Landforms*, **30**, 1675–1682.

Evans, M. and Church, M. (2000). A method for error analysis of sediment yields derived from estimates of lacustrine sediment accumulation. *Earth Surface Processes and Landforms*, **25**, 1257–1267.

Filippelli, G. M. and Souch, C. (1999). Effects of climate and landscape development on the terrestrial phosphorous cycle. *Geology*, **27**, 171–174.

Foster, I. D. L., Dearing, J. A. and Grew, R. (1988). Lake catchments: an evaluation of their contribution to studies of sediment yield and delivery processes. *International Association of Hydrological Sciences Special Publication*, **174**, 413–424.

Foster, I. D. L. *et al.* (1990). The lake sedimentary database: an appraisal of lake and reservoir-based studies of sediment yield. *International Association of Hydrological Sciences Special Publication*, **189**, 19–43.

Gao, L. *et al.* (2004). Lake sediments from Dianchi Lake: a phosphorus sink or source? *Pedosphere*, **14**, 483–490.

Gasse, F. (2002). Diatom-inferred salinity and carbonate oxygen isotopes in Holocene waterbodies of the western Sahara and Sahel (Africa). *Quaternary Science Reviews*, **21**, 737–767.

Gilbert, R., (1975). Sedimentation in Lillooet Lake, British Columbia. *Canadian Journal of Earth Science*, **12**, 1697–1711.

Gilbert, R. *et al.* (2006). The record of an extreme flood in the sediments of montane Lillooet Lake, British Columbia: implications for paleoenvironmental assessment. *Journal of Paleolimnology*, **35**, 737–745.

Goudie, A. S. and Thomas, D. S. G. (1985). Pans in southern Africa with particular reference to South Africa and Zimbabwe. *Zeitschrift für Geomorphologie*, **29**, 1–19.

Håkanson, L. and Jansson, M. (1983). *Principles of Lake Sedimentology*. Berlin: Springer-Verlag.

Hallet, B., Hunter, L. and Bogen, J., (1996). Rates of erosion and sediment evacuation by glaciers: a review of field data and their implications. *Global and Planetary Change*, **12**, 213–235.

Hausmann, S. *et al.* (2002). Climate alters grazing regimes: a quantitative multi-proxy, high resolution study of varved sediments from a small Swiss alpine lake. *The Holocene*, **12**, 279–289.

Hermanns, R. L. *et al.* (2004). Rock avalanching into a landslide-dammed lake causing multiple dam failure in Las Conchas valley (NW Argentina): evidence from surface exposure dating and stratigraphic analyses. *Landslides*, **2**, 113–122.

Hinderer, M. (2001). Late Quaternary denudation of the Alps, valley and lake filling and modern river loads. *Geodinamica Acta*, **14**, 231–263.

Horie, S. (1984). *Lake Biwa*. Dordrecht: Dr W. Junk.

Hutchinson, G. E. (1957). *A Treatise on Limnology*, vol. 1, *Geography, Physics and Chemistry*. New York: J. Wiley.

IPCC (2001). *Climate Change 2001: The Scientific Basis. Contribution of Working Group I to the Third Assessment Report of the Intergovernmental Panel on Climate Change*. Houghton, J. T. *et al.*, eds. Cambridge: Cambridge University Press.

IPCC (2007a). *Climate Change 2007: The Physical Science Basis. Contribution of Working Group I to the Fourth Assessment Report of the Intergovernmental Panel on Climate Change*. Solomon, S. *et al.*, eds. Cambridge: Cambridge University Press.

IPCC (2007b). *Climate Change 2007: Impacts, Adaptation and Vulnerability. Contribution of Working Group II to the Fourth Assessment Report of the Intergovernmental Panel on Climate Change*. Parry, M. L. *et al.*, eds. Cambridge: Cambridge University Press. (pp. 173–272, Kundzewicz, Z. W. *et al.* and Fischlin, A. *et al.*, eds.)

Kashiwaya, K., Fukuyama, K. and Yamamoto, A. (1991). Time variations in coarse materials from lake bottom sediments and secular paleoclimatic change. *Geophysical Research Letters*, **18**, 1245–1248.

Kashiwaya, K., Okimura, T. and Harada, T. (1997). Land transformation and pond sediment information. *Earth Surface Processes and Landforms*, **22**, 913–922.

Kashiwaya, K., Tsuya, Y. and Okimura, T. (2004). Earthquake-related geomorphic environment and pond sediment information. *Earth Surface Processes and Landforms*, **29**, 785–793.

Kislov, A. V. and Surkova, G. V. (1998). Simulation of Caspian Sea level changes during the last 20 000 Years. In G. Benito, V. R. Baker and K. J. Gregory, eds., *Paleohydrology and Environmental Change*. Chichester: J. Wiley, pp. 235–244.

Kreutzberg-Mukhina, E. (2004). Effect of drought on waterfowl in the Aral Sea region: Monitoring of anseriformes at the Sudochie Wetland. In *Management and Conservation of Waterfowl in Northern Eurasia*, Abstracts, Petrozavodsk, pp. 202–203.

Kuylenstierna, J. C. I. *et al.* (2001). Acidification in developing countries: ecosystem sensitivity and the critical load approach on a global scale. *Ambio*, **30**, 20–28.

Lamoureux, S., 2002. Temporal patterns of suspended sediment yield following moderate to extreme hydrological events recorded in varved lacustrine sediments. *Earth Surface Processes and Landforms*, **27**, 1107–1124.

Last, W. M. and Smol, J. P., eds. (2001). *Tracking Environmental Change Using Lake Sediments,* vol. 1, *Basin Analysis, Coring and Chronological Techniques*. Dordrecht: Kluwer.

Leblanc, M. *et al.* (2006). Reconstruction of Megalake Chad using shuttle radar topographic mission data. *Palaeogeography, Palaeoclimatology, Palaeoecology*, **239**, 16–27.

Lehner, B. and Döll, P., 2004. Development and validation of a global data base of lakes, reservoirs and wetlands. *Journal of Hydrology*, **296**, 1–22.

Leopold, L. B. (1996). *Sediment Problems at Three Gorges Dam*, Report to the US Export–Import Bank. Available at www.nextcity.com/probeinternational/threegorges/rdhcapdxb.html.

Lerman, A., Imboden, D. M. and Gat, J. R., eds. (1995). *Physics and Chemistry of Lakes*. Berlin: Springer-Verlag.

Löfgren, B. *et al.* (2002). Evaluation of potential impacts on Great Lakes water resources based on climate scenarios of two GCMs. *Journal of Great Lakes Research*, **28**, 537–554.

Mackay, A. *et al.* (1998). Diatom succession trends in recent sediments from Lake Baikal and their relation to atmospheric pollution and to climate change. *Philosophical Transactions of the Royal Society of London A*, **335**, 1011–1055.

McPhaden, M. J., Zebiak, S. E. and Glantz, M. H., (2006). ENSO as an integrating concept in Earth science. *Science*, **314**, 1740–1745.

Menounos, B. (2002). Climate and fine-sediment transport linkages, Coastal Mountains, BC. Ph.D. thesis, The University of British Columbia.

Menounos, B. and Clague, J. J. (2008). Reconstructing hydro-climatic events and glacier: fluctuations over the past millennium from annually laminated sediments of Cheakamus Lake, southern Coast Mountains, British Columbia, Canada. *Quaternary Science Reviews*, **27**, 701–713.

Menounos, B. *et al.* (2005). Environmental reconstruction from a varve network in the southern Coast Mountains, British Columbia. *The Holocene*, **15**, 1163–1171.

Meybeck, M. (1979). Concentrations des eaux fluviales en éléments majeurs et apports en solution aux oceans. *Revue de Géologie Dynamique et Géographie Physique*, **21**, 215–246.

Meybeck, M. (1995). Global distribution of lakes. In A. Lerman *et al.*, eds., *Physics and Chemistry of Lakes*. Berlin: Springer-Verlag, pp. 1–35.

Meybeck, M., Green, P. and Vörösmarty, C. (2001). A new typology for mountains and other relief classes: an application to global continental water resources and population distribution. *Mountain Research and Development*, **21**, 34–45.

Micklin, P. (2007). The Aral Sea disaster. *Annual Reviews in Earth and Planetary Science*, **35**, 47–72.

Milly, P. C. D., Dunne, K. A. and Vecchia, A. V. (2005). Global pattern of trends in streamflow and water availability in a changing climate. *Nature*, **438**, 347–350.

Mitchell, J. M. Jr, (1976). An overview of climatic variability and its causal mechanisms. *Quaternary Research*, **6**, 481–493.

Morris, G. L. and Fan, J. (1998). *Reservoir Sedimentation Handbook*. New York: McGraw-Hill.

Nilsson, C. *et al.* (2005). Fragmentation and flow regulation of the world's large river systems. *Science*, **308**, 405–408.

Ochiai, S. and Kashiwaya, K. (2003). A conceptual model on sedimentation processes for hydro-geomorphological study in Lake Baikal. In K. Kashiwaya, ed., *Long Continental Records from Lake Baikal*. Tokyo: Springer-Verlag, pp. 297–312.

Oden, S. (1968). *The Acidification of Air Precipitation and its Consequences in the Natural Environment*, Ecology Committee Bulletin No. 1. Stockholm: Swedish National Research Council.

O'Hara, S. L., Street-Perrott, F. A. and Burt, T. P. (1993). Accelerated soil erosion around a Mexican highland lake caused by prehispanic agriculture. *Nature*, **362**, 48–51.

Oldfield, F. (1977). Lakes and their drainage basins as units of sediment-based ecological study. *Progress in Physical Geography*, **1**, 460–504.

Oreskes, N., Shrader-Frechette, K. and Berlitz, K. (1994). Verification, validation and confirmation of numerical models in the earth sciences. *Science*, **263**, 641–646.

Page, M. J. and Trustrum, N. A. (1997). A late Holocene lake sediment record of the erosion response to land use change in a steepland catchment, New Zealand. *Zeitschrift für Geomorphologie*, **41**, 369–392.

Rust, B. R. (1982) Sedimentation in fluvial and lacustrine environments. In P. G. Sly, ed., *Sediment Freshwater Interaction*. Dordrecht: Dr W. Junk, pp. 71–84.

Sakai, T. *et al.* (2005). Influence of climate fluctuation on clay formation in the Baikal drainage basin. *Journal of Paleolimnology*, **33**, 105–121.

Schiefer, E. (2006). Predicting sediment physical properties within a montane lake basin, British Columbia. *Lake and Reservoir Management*, **22**, 69–78.

Schiefer, E., Slaymaker, O. and Klinkenberg, B. (2001). Physiographically controlled allometry of specific sediment yield in the Canadian Cordillera: a lake sediment based approach. *Geografiska Annaler*, **83**A, 55–65.

Schiefer, E., Menounos, B. and Slaymaker, O. (2006). Extreme sediment delivery events recorded in the contemporary sediment record of a montane lake, southern Coast Mountains. *Canadian Journal of Earth Science*, **43**, 1–14.

Schindler, D. W. (1974). Eutrophication and recovery in experimental lakes: implications for lake management. *Science*, **184**, 897–899.

Sepulchre, P. *et al.* (2008). Evolution of Lake Chad Basin hydrology during the mid-Holocene: a preliminary approach from lake to climate modeling. *Global and Planetary Change*, **61**, 41–48.

Shen, G. and Xie, Z. (2004). Three Gorges Project: chance and challenge. *Science*, **304**, 681.

Shimada, T. *et al.* (2002). Hydro-environmental fluctuation in a lake catchment system during the late Holocene inferred from Lake Yogo sediments. *Transactions of the Japanese Geomorphological Union*, **23**, 415–431 (in Japanese with English abstract and captions).

Siegert, F. *et al.* (2001). Increased damage from fires in logged forests during droughts caused by El Niño. *Nature*, **414**, 437–440.

Slaymaker, O. *et al.* (2003). Advances in Holocene mountain geomorphology inspired by sediment budget methodology. *Geomorphology*, **55**, 305–316.

Small, E. E. *et al.* (2001). The effects of desiccation and climatic change on the hydrology of the Aral Sea. *Journal of Climate*, **14**, 300–322.

Smol, J. P. (2002). *Pollution of Lakes and Rivers: A Palaeoenvironmental Perspective*. London: Arnold.

Smol, J. P. *et al.* (1998). Tracking recovery patterns in acidified lakes: a paleolimnological perspective. *Restoration Ecology*, **6**, 318–326.

Sorrel, P. *et al.* (2007). Climate variability in the Aral Sea basin (Central Asia) during the late Holocene based on vegetation changes. *Quaternary Research*, **67**, 357–370.

Souch, C. (1994). A methodology to interpret downvalley lake sediments as records of Neoglacial activity: Coast Mountains, B.C. *Geografiska Annaler*, **76**A, 169–185.

Stoermer, E. F. (1998). Thirty years of diatom studies on the Great Lakes at the University of Michigan. *Journal of Great Lakes Research*, **24**, 518–530.

Wada, Y. (1916). *Chosen Kodai Kansoku kiroku Chousa Houkoku* (A report on past observational records in Korea). Seoul: Observatory at Governor of Korea's Office (in Japanese).

Wang, M. and Shi, W. (2008). Satellite observed algae blooms in China's Lake Taihu. *EOS, Transactions of the American Geophysical Union*, **89**, 201–202.

Wetzel, R. G. (1975). *Limnology*. Philadelphia: W. B. Saunders.

Wu, J. *et al.* (2003). Three Gorges Dam: experiment in habitat fragmentation? *Science*, **300**, 1239–1240.

Wu, L., Zhang, Q. and Jiang, Z. (2006). Three Gorges Dam affects regional precipitation. *Geophysical Research Letters*, **33**, L13806, doi:10.1029/2006 GL026780.

Wu, Y. *et al.* (1998). Paleoenvironmental evolution in Dianchi Lake area since 13 ka BP. *Journal of Lake Sciences*, **10**, 5–9.

Yan, J. P., Hinderer, M. and Einsele, G. (2002). Geochemical evolution of closed-basin lakes: general model and application to lakes Qinghai and Turkana. *Sedimentary Geology*, **148**, 105–122.

Yang, S. L. *et al.* (2007). Effect of deposition and erosion within the main river channel and large lakes on sediment delivery to the estuary of the Yangtze River, *Journal of Geophysical Research*, **112**, F02005, doi:10.1029/ 2006JF000484.

4 Rivers

Michael Church, Tim P. Burt, Victor J. Galay and G. Mathias Kondolf

4.1 Introduction

Water plays a key role in the transfer of mass and energy within the Earth system. Incoming solar radiation drives evaporation of about 434 000 $km^3\ a^{-1}$ from the ocean surface and 71 000 $km^3\ a^{-1}$ from the land surface, while precipitation delivers about 398 000 $km^3\ a^{-1}$ of water to the ocean and 107 000 $km^3\ a^{-1}$ to the land surface. The balance is redressed through the flow of 36 000 $km^3\ a^{-1}$ of water from the land to the oceans via rivers (data in Berner and Berner, 1996). Environmental change affecting any of these water transfers produces changes in runoff and river flows, hence in the rivers themselves.

Changing climate is intensifying the global hydrological cycle, leading to significant changes in precipitation, runoff and evapotranspiration (Huntington, 2006; Bates *et al*., 2008; see also Chapter 1). Intensification of the hydrological cycle is likely to mean an increase in hydrological extremes (IPCC, 2001). Changes in the frequency distribution of precipitation alter water flows and water availability in the surface environment leading, in turn, to a change in river regimes.

These factors are superimposed upon the effects of human actions associated with land use and with the attempt to control water for various uses that have directly changed river channels and the quality of water flowing in them. Land surface condition mediates quantity of water and the amount and calibre of sediment delivered to rivers which, in turn, influences river sedimentation, morphology and stability. Humans also manipulate the terrestrial hydrological cycle deliberately by construction of reservoirs, abstractions of water for human use, and discharges of water into river courses. Moreover, we directly modify watercourses by realigning them, by river 'training' works, by dredging, by fixing banks and building dykes. All these effects exert a dominating influence over the condition of rivers and of riverine and riparian ecosystems.

This chapter explores some of those effects on waterways and their ramifications. The argument will be that humans are modifying the terrestrial hydrological cycle and waterways in profound ways, indeed have been doing so for a long time, in the course of land use and in order to exploit water resources and to secure protection from water hazards. In comparison with the summary effects of these activities, the direct effects of climate change will remain relatively modest. Climate change will have important regional effects on the total water supply which will be critical where water supply is marginally adequate or already inadequate. However, the occurrence and quality of water and the condition of waterways are, in general, overwhelmingly dominated by human actions.

The chapter will pursue this thesis by considering water in uplands and the origin of runoff to rivers, then the controls and condition of river channels, river sediments and sedimentation, and the important topic of water storage in reservoirs. Contemporary issues of river 'restoration' are also considered briefly for this, in truth, represents a further manipulation of rivers.

4.2 Land surface: runoff production

4.2.1 The hillslope hydrological cycle

Until relatively recently, river engineers took the view that the headwaters of a drainage basin were nothing more than passive source areas for runoff. Being mainly concerned with downstream water management, in particular with flood forecasting, they felt able to ignore the exact processes responsible for generating the flood runoff. However, there were always other hydrologists with an interest in the process of runoff production. These interests have converged, especially since the advent of physically based computer simulation models of flood runoff.

Geomorphology and Global Environmental Change, eds. Olav Slaymaker, Thomas Spencer and Christine Embleton-Hamann. Published by Cambridge University Press.

Moreover, it is highly relevant in the context of this chapter that an interest in river flows is invariably linked to concerns about the potential of the land surface to modulate the relation between precipitation and runoff production.

At the same time as L. K. Sherman was developing a numerical flood forecasting method (Sherman, 1932), the unit hydrograph, R. E. Horton, a hydraulic engineer working mainly in New York State, was much concerned with the impact of runoff from agricultural land on flood generation, soil erosion and sediment transport. Horton pioneered research on links between water infiltration into the soil ('infiltration capacity'), overland flow and erosion (Horton 1933, 1945). At much the same time, in the same region but in the rather different hydrological environment of the Appalachian forests, Charles Hursh showed that the source areas of storm runoff were different in forested lands compared with croplands. He identified the need to establish the relation between forest condition and streamflow in places where erosion was of minor importance (Hursh and Brater, 1941). Only then, Hursh argued, would it be possible to fully understand links between agricultural land use, runoff and erosion. Burt (2008) provides a detailed review of the process studies of Horton, Hursh and others in the establishment of a theory of runoff production.

The pathways taken by hillslope runoff draining to the nearest river channel determine many of the characteristics of the landscape, the uses to which land can be put and the strategies required for sustainable land use management (Dunne, 1978). Many factors influence the exact pathways involved at any location: vegetation cover, soil and bedrock properties, land use and land management practices, as well as the characteristics of the local climate, notably the intensity and duration of rainfall. Nevertheless, as far as storm runoff is concerned, the framework laid down by Horton and Hursh continues to provide a broad framework within which runoff production can be analysed, both in terms of the amount and timing of runoff, and the source areas within the river basin.

The partial source area model: infiltration-excess overland flow

Horton considered that the process of infiltration divided rainfall into two parts, which thereafter pursue different courses: one part moves quickly to the stream channel across the land surface as overland flow and generates storm runoff; the other part infiltrates the soil and flows slowly through soil and bedrock to sustain longer-term 'baseflow'.

High-intensity rain may produce overland flow immediately, while lower-intensity rain may produce overland flow only once infiltration capacity has declined as the soil wets up; low-intensity rainfall may well produce no surface runoff at all. Horton argued that surface runoff would be widespread across a river basin, but we have subsequently learned that this is unlikely to be the case except for small areas where infiltration capacity is relatively uniform and low. Infiltration capacity varies widely across space, as well as through time, producing wide variation in surface runoff production. Betson (1964) captured this localised runoff condition in his *partial area* model, which remains the best guide to the location of runoff source areas for infiltration-excess overland flow. In contrast to the *variable source area* model described below, partial source areas of overland flow appear relatively fixed in location during a given storm event. There may be important changes in infiltration capacity over time in some situations (e.g. in a tilled field over a yearly cultivation cycle) but this is not true for the ultimate 'Hortonian' runoff surface, an urban area, where impermeable surfaces present an unchanging pattern of runoff generation over long periods.

The variable source area model: saturation-excess overland flow

The occurrence of subsurface stormflow must be recognised in order to understand saturation-excess overland flow. Subsurface stormflow may happen in one of two phases: rapid flow through large structural pores or macropores (Jarvis, 2007), and slow seepage through small pores within the soil matrix. Hydrologists have tended to emphasise the latter, although the occurrence of macropore flow can be dominant in certain situations such as peat soils, forest soils and drained clay soils (see below). Lateral subsurface flow through the soil matrix occurs in any situation where soil permeability declines with depth. It was originally thought that subsurface flow was too slow to generate storm runoff but, in fact, rapid responses can be produced via 'capillary fringe' effects, whereby only a small amount of water needs to infiltrate the soil to raise the water table significantly. Significant subsurface flow may occur within a few hours of heavy rainfall on steep slopes with permeable soil which allows subsurface drainage to reach the stream channel relatively quickly (Anderson and Burt, 1978). Such delayed peaks in subsurface flow are particularly encouraged by the convergence of flow into hillslope hollows. If soil water accumulates more quickly than it can drain, then the soil profile may become completely saturated, with the result that water exfiltrates the soil to produce saturation-excess overland flow. The source areas for subsurface flow and saturation-excess overland flow are essentially the same, therefore, in locations where soil water tends to accumulate: at the foot of any slope, especially those of concave form; in areas of thin

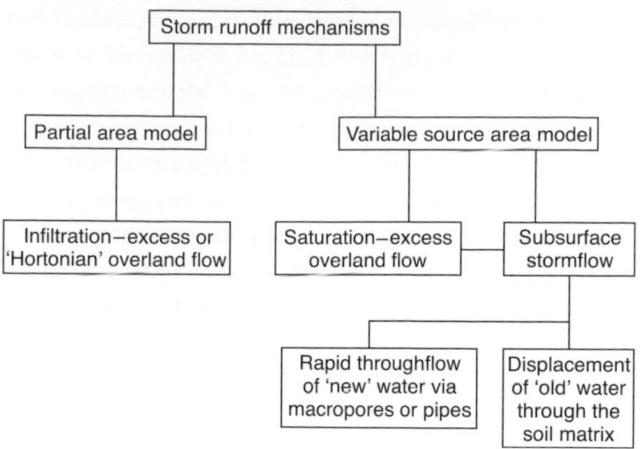

FIGURE 4.1. The partial area and variable source area models for runoff generation.

TABLE 4.1. *Final infiltration rates by Hydrologic Soil Group in relation to soil texture*

US Soil Conservation Service Hydrologic Soil Group	US Department of Agriculture soil textures included in Group	Final infiltration rate (mm hr^{-1})
A	Sand, loamy sand, sandy loam	>7.6
B	Silt loam, loam	3.8–7.6
C	Sandy clay loam	1.3–3.8
D	Clay loam, silty clay loam, sandy clay, silty clay, clay	0–1.3

Source: Based on information contained in Rawls *et al.* (1993).

soils; and, as noted above, in hillslope hollows. Working at the same location as Hursh, J. D. Hewlett (1961) first described the *variable source area* model of storm runoff generation for saturation-excess overland flow. These localised zones of soil saturation expand during a storm, as rainfall adds water to the soil, and seasonally, as soils wet up after the dry season – hence the adjective 'variable'.

The two source area concepts (Fig. 4.1) of storm runoff generation pertain to contrasting process environments: infiltration-excess overland flow (the partial area model) dominates in intensively farmed areas where soil infiltration capacity is likely to be reduced by unfavourable land management practices, and in semi-arid regions. In contrast, saturation-excess overland flow (the variable source area model) tends to dominate in catchments where farming activity is low intensity or non-existent and infiltration capacity remains high.

TABLE 4.2. *The effect of land use on surface runoff at Slapton, Devon, UK*

Land use	Rainfall intensity (mm hr^{-1})	Infiltration rate (mm hr^{-1})
Woodland soil*		180
Permanent pasture*		9 (range: 3 – 36)
Freshly ploughed soil*		50
Temporary grass	12.5	12.3
Barley	12.5	11.0
Rolled, bare soil	12.5	4.0
Lightly grazed permanent pasture	12.5	5.9
Heavily grazed permanent pasture	12.5	0.1

* Results obtained using a rainfall simulator except where indicated by an asterisk.
Source: Data from Heathwaite *et al.* (1990), Burt *et al.* (1983) and Burt and Butcher (1985).

4.2.2 Effects of human activity

Agriculture and runoff

Given the importance of overland flow for erosion, there have been many studies of infiltration in agricultural soils (see review by Rawls *et al.*, 1993). Infiltration capacity depends on soil surface condition and the factors which affect this include soil properties, surface cover and land management practices.

The general relation between soil texture and final minimum infiltration capacity is shown in Table 4.1; fine-textured soils tend to have lower infiltration rates given the general link between texture and pore size distribution.

However, vegetation cover and land use are also crucial influences on infiltration rates, which can consequently vary widely for the same soil in time and space. Table 4.2 shows results from an area of silty clay loam soil; the effect of land use on infiltration capacity is immediately apparent. Under favourable conditions (e.g. woodland soil), silty soils can have a high infiltration capacity, but they are easily compacted; infiltration then falls significantly, and infiltration-excess overland flow occurs even in low intensity rainfall. Very low infiltration capacity can occur if wet soils are compacted by grazing or heavy machinery at inappropriate times. Bare soil is not necessarily the worst-case scenario: freshly ploughed soil has a very high infiltration capacity, for example. However, bare ground is vulnerable to compaction by heavy rain and remains so until a good level of crop cover becomes established (Burt and Slattery, 1996). A 30% cover is often said to be 'safe', although this widely quoted yardstick seems to have no scientific basis.

Understanding spatial and temporal variations of infiltration capacity is thus a complex task. The precise chronology of soil surface changes, even for a single field, varies from season to season, depending on the succession of crops, and management of the crop and inter-crop intervals. More importantly perhaps, inter-annual variation in both erosivity (the tendency to cause erosion) and erodibility (the susceptibility of soil to erosion) depends on climatic conditions. There is a 'window of opportunity' during which a young crop is vulnerable to runoff and erosion as a result of surface sealing by heavy rainfall; whether or not this happens depends entirely on whether heavy rain falls when the crop cover is sparse or non-existent. For example, Boardman *et al.* (1995) describe an erosion event for a maize crop in southern England caused by an unusually heavy thunderstorm for the time of year (late spring). Maize is, in any case, a crop giving rise to conditions of high erodibility, given the wide spacing of rows, but heavy rainfall when the crop had just emerged meant that the soil surface was particularly vulnerable, especially because tillage had formed a fine tilth which was easily sealed by raindrop impact. Even when the soil surface condition is unlikely to generate infiltration-excess overland flow, compacted lines of soil formed by the passage of vehicles ('wheelings' or 'tram lines') may still produce large volumes of surface runoff. Thus, crop management practices (e.g. spraying pesticides or spreading fertiliser) can be damaging even when there is in effect a complete crop cover.

Agricultural land is often drained in order to increase productivity, usually to allow change on clay soils from grassland to arable. Open ditches are dug to extend the channel network while underground drainage in the form of tile drains, pipes traditionally constructed of clay tiles or, nowadays, corrugated plastic, is used to drain excess water from the plant root zone to the nearest channel, usually a drainage ditch. The question of the impacts of agricultural drainage on runoff production and flood peaks has been of interest for a long time. Robinson (1990) summarised the results from many different studies. Because drained soils tend to have a higher capacity to store water, infiltration will increase, reducing surface runoff. Drainage tends to reduce peak flows from clay catchments by decreasing antecedent storage, but may increase peak flows from more permeable soils because of more rapid drainage. Peak discharge decreases because water flows more slowly through the soil to reach the drainage system than it would as surface runoff. However, the overall volume of water lost (surface runoff and tile flow combined) from a drained field may increase (5–15%) compared with fields with surface drainage only; in effect, subsurface discharge increases at the expense of evaporative losses.

Robinson noted that the type of drainage scheme might also be important, with secondary practices such as mole drains or subsoiling giving higher peaks than pipe drains alone, while open ditches produce higher peaks than subsurface drains. Ground condition, due to both agricultural practices and cracking in heavy clay soils, may also be important in controlling the response but, in general, drainage tends to modify the timing of runoff and the peak flow rather than the volume of runoff from a given storm. Soil wetness can also influence the speed of throughflow in cracked clay soils; if the soil is dry, water will infiltrate the soil matrix but if the soil is already saturated, water will flow rapidly through the macropores to the drainage system.

At the catchment scale, it might be thought that land drainage would tend to increase flooding because of its impact on peak flow rates. However, drainage schemes are implemented at the field scale. At larger scales complexities arise because of the effects of routing of the outputs from individual fields, with or without drainage, to a site at risk of flooding. The relative timing of different runoff sources to the stream channel will then be important. It could be that a drainage scheme, by speeding up runoff from an area that before drainage had contributed water directly to the hydrograph peak, would have the effect of reducing the flood peak. Conversely, speeding up runoff from an area that prior to drainage had lagged behind the hydrograph peak could act to increase the flood peak (Beven *et al.*, 2004). Substantial controversy has raged in Europe over whether field drainage has contributed to an increase in the incidence of river floods. The resolution appears to be that, while the duration of intermediate flows may have been extended, drainage has no detectable effect on major flooding, which is controlled by the excessive amounts of precipitation that arrive in a drainage basin in a limited time (see Mudelsee *et al.*, 2003).

Land drainage in peatlands

The hydrological impacts of land drainage in peatlands are much the same as in poorly drained clay soils where intensive agriculture is practised, but both the reasons for drainage and the environmental context are somewhat different in these marginal areas. In the United Kingdom, large tracts of peaty soil in the uplands were drained in the second half of the twentieth century, both to increase productivity (extensive sheep grazing and game) and for coniferous forestry. At the Coalburn catchment in northern England, ribbon plough drainage (creation of large furrows roughly 1 m wide, 75 cm deep and 5 m apart) was carried out in 1972 prior to forest planting. There were significant increases in storm runoff and decreases in the time to peak immediately following drainage, presumably because the generating area for surface runoff was greatly increased. There was a recovery to pre-drainage responses after about 10 years, probably the result of forest canopy closure and a decrease in the efficiency and effectiveness of the surface drains (Robinson *et al.*, 1998).

The impact of open drainage on the hydrological response of peatland catchments was first investigated by Conway and Millar (1960). They concluded that streamflow production in blanket peatlands, catchments that naturally produce large volumes of storm runoff very quickly, was even more rapid where artificial drainage had taken place; there was increased sensitivity to rainfall with peak flows both higher and earlier. Holden *et al.* (2006) showed that, 50 years later, while still 'flashy', the drained catchments produced less overland flow but more throughflow because of long-term changes in peat structure. However, the percentage of rainfall converted to storm runoff was even higher. Whether open ditching in peaty headwater catchments increases the flood risk downstream remains an open question but David and Ledger (1988), studying the effect of plough drainage of deep peat prior to planting with conifers, found that the open drains affected 30% of the area and 50% of the vegetation cover, and acted as major source areas for rapid surface runoff. It is not known whether ditch blocking will allow peat structure to recover eventually. If it does, there might be modest reductions in the speed and volume of stormflow response. A new runoff transition may be imminent, however, inasmuch as areas drained and planted in the second half of the last century are scheduled for harvest early this century.

Urbanisation

In some ways, the effect of urbanisation on runoff generation is simply an extreme form of the changes seen under intensive arable agriculture: the replacement of a permeable, vegetated land cover with a largely impermeable surface with scattered patches of permeable ground. In fact, the percentage of impermeable surface is rarely more than 60%, even in a highly built-up area, but this is still two orders of magnitude more than would be found in a typical rural catchment. Thus, large amounts of surface runoff are generated in urban areas (Endreny, 2005). Because the natural drainage network is greatly extended via roads, underground pipes and culverts, runoff reaches the perennial channel network very quickly. Storm runoff response is greatly enhanced therefore, with higher and earlier flood peaks. Rapid convergence through the drainage system of this increased runoff brings greater risk of flash flooding, especially when downstream drainage capacity is restricted by culverts that are old and unable to cope with extreme events. Moreover, flood protection measures along main river channels can have the unforeseen effect of preventing local drainage into the main channel, generating localised flooding outside the dykes. The quality of urban storm runoff is also very poor and the ecological status of urban watercourses is very likely to be poor too, especially where concrete culverts have replaced natural channels.

Given the problems associated with urban storm runoff, there has been much interest recently in sustainable urban drainage systems (SUDS: www.ciria.org/suds/index.html). Drainage systems are developed in line with the ideals of sustainable development; a balanced approach takes into account quantity, quality and amenity issues. SUDS seek to manage runoff as close to the point of origin as possible through a variety of techniques that attenuate the runoff response via storage ponds, buffer strips and the use of porous surfaces for infiltration. SUDS are more sustainable than traditional drainage systems because they manage peak runoff, reducing the downstream impact of urban runoff, protecting water quality, encouraging infiltration and groundwater recharge (where appropriate), and enhancing the in-stream ecology of urban channels.

Forest management and runoff

Forests cover a substantial portion of Earth's surface and have long been recognised as the principal source of high-quality surface water. They are subject to radical alteration, either in the course of forestry practice or through clearance for other land uses. Forest hydrology is affected by both deforestation and afforestation (not always mirror images in their impact) (National Research Council, 2008). The presence of a tree canopy completely alters the near-surface climate and, because of this, many micrometeorological studies have complemented hydrological research, especially in relation to the study of interception and evaporation (Calder, 1990). Forest harvest imposes a transient but possibly significant effect on runoff which has been extensively studied in deciduous and coniferous northern forests which were the principal sites of industrial forestry well into the twentieth century. Variable findings remain controversial, probably because of the dual role of forests both as interceptors of incoming precipitation, which is then evaporated or transpired back to the atmosphere, and as reservoirs of water in forest soils that contribute to runoff over some time (see Chapter 12).

Forest harvest removes the trees, potentially increasing the total water supply and altering the timing and magnitude of peak flows. Transient increases in water yield of up to 30% have been reported, but the effect rapidly dissipates as new growth occurs and the augmentation of runoff rarely persists beyond a decade (Bruijnzeel, 2005). On a global scale, Bosch and Hewlett (1982) collated the results of 94 experimental studies; they showed that forest removal consistently increased runoff but that the magnitude of the phenomenon varied regionally according to forest type, landscape character and the character of the climate.

Extreme flow effects, much as in agricultural landscapes, have been found to be variable, depending upon antecedent weather, the position of forest harvest in the landscape, and effects on the drainage network of roads and work sites. There is an emerging consensus that effects on extreme flows have more to do with disturbance of the drainage network than with vegetation manipulation.

In recent decades, attention to the effects of forest manipulation has shifted to the tropics, where deforestation proceeds at rates between 0.5% and 1.0% per annum (data from Food and Agriculture Organization, 2005; see also Chapter 8) in contrast to the temperate zone where forest cover is today increasing (see Chapter 2: Fig. 2.6). The implications for tropical hydrology of these rates of change are reviewed in Bonell and Bruijnzeel (2005). For a number of reasons, there is no consensus on the hydrological results at large scale. Firstly, the land conversion may entail the establishment of plantation forestry, or agroforestry, or land clearance for traditional agriculture, each with different hydrological effects. Secondly, it is clear that, at large scale, tropical forests in part create their own climate, so that regional deforestation tends to reduce precipitation with the result that all of the principal elements of the hydrological cycle may decrease in magnitude (see Costa in Bonell and Bruijnzeel, 2005).

4.2.3 Perspective

Changing climate certainly will change the quantity and timing of precipitation around the world. A warmer atmosphere will carry more moisture and a more energetic atmosphere will deliver increased amounts and intensity of storm precipitation in many places. But not in all places – the changing trajectory of weather systems will lead to reduced precipitation in some locations, and changing seasonal occurrence of precipitation will have significant effects on water resources.

Runoff is the residual left from precipitation after evaporation has occurred. In a warmer atmosphere, that quantity will increase too. Hence, some regions will experience increased precipitation yet decreased runoff. The volume and timing of runoff will be further modulated by changing seasonal snow occurrence, and it will eventually be affected by the response of native vegetation to a changed climate. Predicting changes in runoff as the result of expected climate change is therefore a difficult matter (see Bates *et al.*, 2008).

In comparison, it is known that human activities have pervasive impacts on runoff to streams. These effects can be summarised under four major categories:

(a) manipulation of land cover, which affects interception and transpiration losses;
(b) working and trafficking on the land surface, which affects the infiltration capacity of soils;
(c) creation of impermeable surfaces; and
(d) installation of land drainage measures.

These actions have various effects. Mostly, the manipulation of surface cover leads to increases in the volume of runoff; working the land and the creation of impermeable surfaces mostly affects the balance of surface and subsurface drainage, leading to more rapid drainage and a larger volume of runoff, while land drainage measures mostly increase the volume of runoff. The effect on peak flows is complex, since that depends upon the structure of the entire drainage network, as well as on the timing and volume of water delivery from individual land units.

These effects have been developing for thousands of years. For at least a millennium they have had regionally major effects, and within the last two centuries they have become intense everywhere humans have settled. Yet the pace of change has been subtle and, until the mid twentieth century, they remained largely overlooked. Whilst climate change over the next century may have a more immediately noticeable effect on water supply, the cumulative impact on the hydrology by human modification of Earth's surface over the course of human history has undoubtedly far surpassed prospective changes, and will continue.

4.3 River channels: function and management

4.3.1 The form of river channels

River channels are the conduits that drain runoff from the surface of the land. Water flowing over the land surface is capable of eroding soil and even rock material, so that rivers carry and redistribute sedimentary materials as well. In the short term, the sediment transporting activity of rivers creates the channels in which streams and rivers flow and it affects the quality of water in the channels; in the very long term, it creates the landscape.

Channelled water flows originate in upland areas wherever topographic declivities cause soil drainage to converge. Concave hollows on hillslopes are typical sites of channel initiation, but much of the initial concentration of flow occurs in the subsurface in soil macropores (see Section 4.2.1) and comes to the surface at springs and seepage zones – places where subsurface flow convergence creates saturated conditions at the surface. The drainage area required for channel inception varies over Earth's surface according to climate and land surface condition (Dunne, 1978).

The conditions that govern the form and stability of river channels include the volume and timing of the water that flows through the channel, the volume and calibre of the sediment that is carried by the stream, the character of the material that forms the bed and banks of the channel, and the gradient of the valley down which the river flows. The volume and timing of water sets the scale of the channel – determines how large the channel must be in order to convey the larger flows – while the volume and calibre of the sediment set the morphological style of the channel – whether it is a wide and shallow, gravel-bed channel, or a deep and narrow, silt-bound one. Bed and bank materials, which may largely consist of the sediment transported by the stream, are also important determinants of channel form and stability, since only flows that can erode the materials locally forming the channel boundary can cause a change in the shape or position of the channel. Finally, the topographic gradient sets the potential rate at which the energy of the water must be dissipated by the river as it flows downhill. The stability of the river depends upon a balance being achieved between the rate at which the potential energy of the water is expended and the rate at which that energy is consumed in doing the work of moving the water and sediment load over the more or less rough and irregular boundary of the channel.

A simple summary of the scale of river channels is expressed by the so-called equations of downstream hydraulic geometry, which relate channel width and depth to the 'channel forming flow', Q, usually taken to be some relatively large flow such as the bankfull flow. These equations take the form

$$w = aQ^b \tag{4.1}$$

where $b = 0.50$, and

$$d = cQ^f \tag{4.2}$$

where $0.33 \leq f \leq 0.40$; and wherein w is water surface width, $d = A/w$ is the hydraulic mean depth, A is the cross-sectional area of the flow, and Q is the discharge. Since $Q = wdv$,

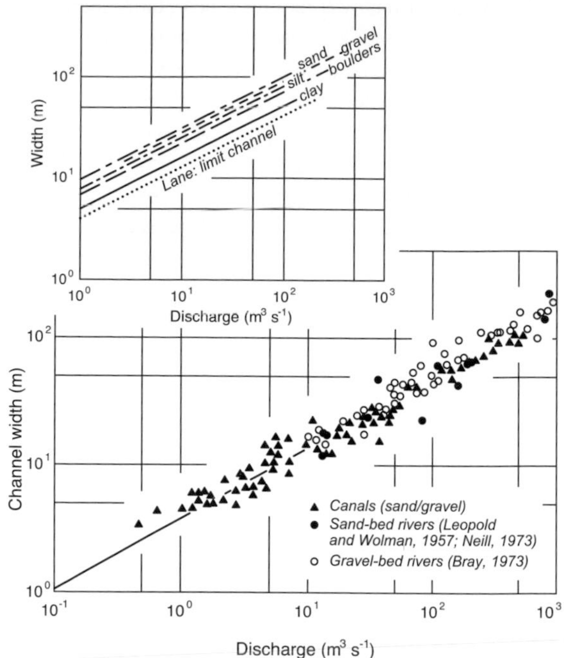

FIGURE 4.2. Scale relation (downstream hydraulic geometry) for alluvial rivers. Inset: variation of hydraulic geometry with materials (modified from Lane, 1957).

wherein v is the mean velocity, the relations above induce a third hydraulic geometry relation,

$$v = kQ^m \tag{4.3}$$

where $m = 1.0 - f - b$; the product $a\ c\ k = 1.0$ and individually vary according to the materials through which the channel flows and the sediments are transported. Figure 4.2 illustrates the channel width equation – the principal scale relation.

The governing conditions vary systematically through a drainage basin in proportion as channels become larger. Headwater streams are small (since they drain a small area) and relatively steep (since water seeks the steepest available line of descent). Because they are steep, they may evacuate most of the sediment supplied to them and flow in channels eroded to bedrock, or through residual materials too large for the stream to move. Farther downstream, channels become larger as tributaries join together in the usual tree-like network; they also become flatter, and tend to deposit a part of their increasing sediment load. Hence, the channels begin to flow in sediments that they may have transported to the site. Such channels are *alluvial* channels. Figure 4.3 illustrates these trends, which are also expressed in the hydraulic geometry, and shows important correlative changes that occur in the drainage system.

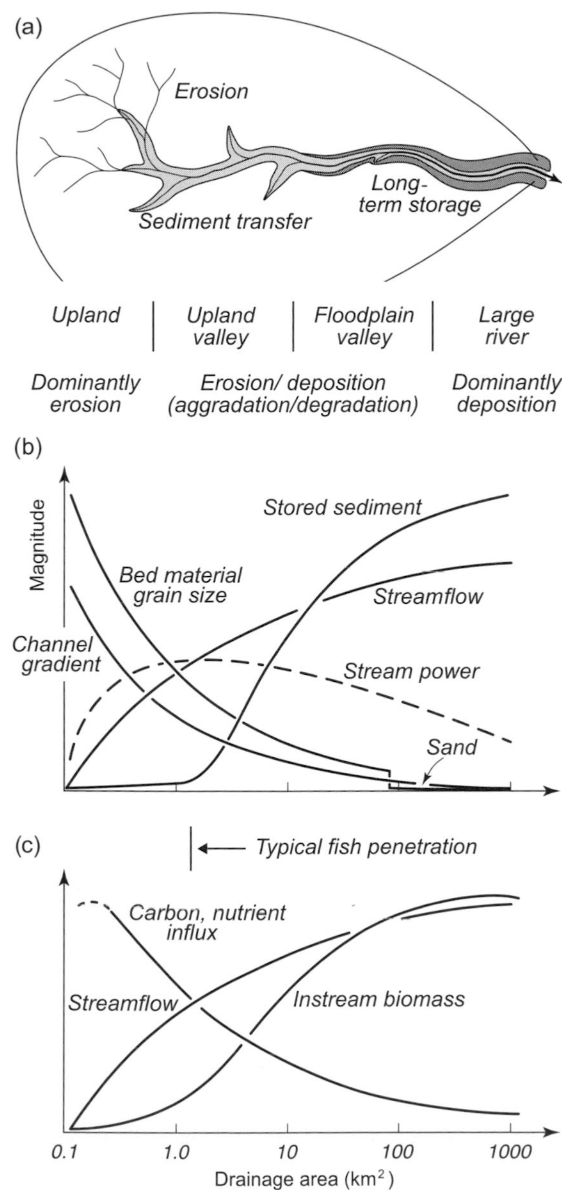

FIGURE 4.3. Variation of flow and river channel properties through the drainage system. (a) Drainage basin map, showing zones of distinctive sediment behaviour (only a fraction of the upland channels is shown); (b) pattern of variation through the drainage basin of principal quantities related to flow, sediment and river morphology; (c) pattern of variation of some ecological quantities that are systematically related to the structure of the drainage system. Area is intended to indicate only the approximate scale. Topography will cause details to vary greatly in individual drainage basins. (Modified from Schumm, 1977.)

This large-scale organisation of the fluvial system permits us to identify an important first order distinction between upland river channels and 'trunk' channels. The former are small, relatively steep, and formed primarily by erosion of the stream into its substrate. Their function is to collect runoff from the land surface and funnel it into the fluvial drainage system. An important variation occurs in flat terrain, where headwater wetlands may drain through channels that are flat, have little erosive capacity and are bounded largely by organic materials. Upland channels occupy about 80% of the total length of the drainage system, simply in view of their high frequency in the landscape. Trunk channels collect water and sediment from upland tributaries and move them toward an end point in a lake or the sea. Sediment is deposited and stored temporarily along the way since most flows are not capable of maintaining the larger part of the sediment load in transport over the reduced valley gradient. Hence, these channels tend to be alluvial ones. A further subdivision can be made between 'transport reaches', those mid-course reaches, often in relatively confined valleys, through which much of the sediment load is moved relatively rapidly, and 'storage reaches', those distal reaches where sediments are deposited and may remain for a long period of time. Such sediment accumulations are the major floodplains and deltas that one finds along the lower course of most major rivers, where much human settlement occurs.

A more detailed view of the function of the fluvial system is gained by considering the pattern of delivery of water through the drainage system and the pattern of sediment movement (see Fig. 4.4). Water arrives relatively rapidly in upland channels (see Section 4.2 above) and drains rapidly. Hence, runoff is highly variable in these channels. Sediment is also delivered very episodically to these channels, either from overbank during periods of overland runoff, or directly from adjacent hillside slopes when slope failures occur. Channels are, accordingly, directly 'coupled' to the adjacent slopes. Flow variation through the year may be of order 100–300 times (becoming infinite in channels that dry up for part of the year), and sediment fluctuations may easily be of order 1000 times. Because they are small, extreme flows and sediment delivery events may create rapid – one might say, 'catastrophic' – changes in such channels. Farther downstream, the integration of drainage from many tributaries, not all of them producing equally extreme runoff simultaneously, modulates the variability of flow, while the deposition of the largest and least mobile sediment grains promotes relatively less extreme variation in onward transport. In the distal parts of large rivers, extreme attenuation may occur, so that flow variations may be reduced to order 3–10 times through the year and sediment fluctuations to order 10–30 times. Alluvial deposits isolate the channels from the adjacent hillsides so sediment delivery to the channels occurs purely by transport along the river from upstream, or by streambank erosion.

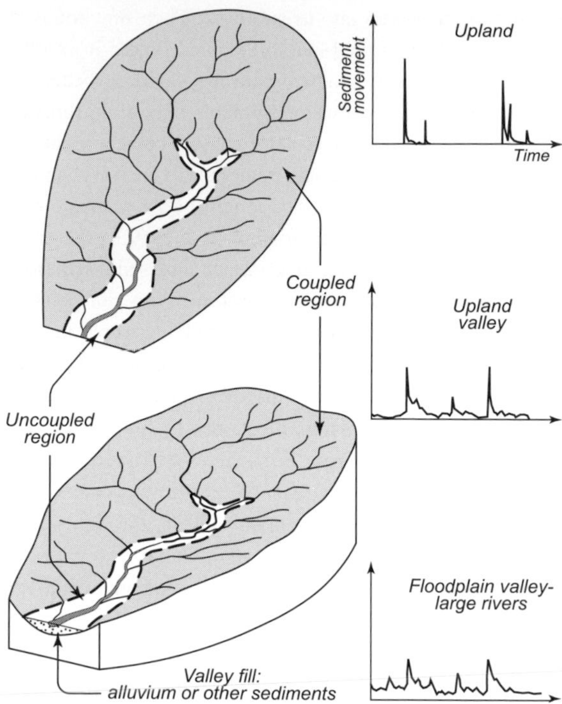

FIGURE 4.4. Schematic view of the drainage network, illustrating the domain of upland channels, where hillslopes are directly coupled to drainage channels, and valley trunk channels, which are buffered from adjacent hillslopes by stream deposits. The graphs show conceptually the variability of flow and sediment transport through the drainage network, sediment transport being more variable even than the flow.

This introductory discussion of function in river systems leads to two conclusions of great importance for understanding the climatic and human impacts on river systems:

- variations occur in the details of river form and function in different parts of the world according to climate, which determines hydrology, and to topography, which determines the available energy gradient for the river;
- drainage basins can be divided into upland and trunk divisions, between which river character changes dramatically.

As an example of the former, we can compare rivers that drain the humid, steep mountains of central Europe with those draining the north European lowlands. Climate – except insofar as it is modulated by altitude – is not too different, but the steepness of the former region creates rivers that transport abundant coarse sediment and typically have wide gravel channels which, in their natural state, are more or less laterally unstable as the coarse sediments are deposited within the channel. In contrast, headwaters of rivers that rise within the North European plain (that is, excepting those with alpine headwaters) mobilise very limited quantities of largely fine organic sediments and flow in relatively deep, narrow channels that may have substantial stability so long as their banks are not severely disturbed.

The second conclusion leads us to recognise that the origins of runoff and sediment on the land surface chiefly control the character of upland river channels, hence that human land use is the major factor that may disturb them whereas, along trunk channels, direct human manipulation of the river (damming it; confining it for flood protection, or to reclaim adjacent land; modifying the channel to improve navigation; diverting significant volumes of water for water resources development purposes) has major impacts on the river.

The history of human impact on river channels in both respects is long and the cumulative impact is large. Relatively few rivers in the world today present anything like their appearance of even 1000 years ago, and humans have been the major driver of change.

4.3.2 Effects of human activities on river morphology

Reinforcement of river banks

One of the earliest modifications of river channels to suit human purposes was the systematic raising and reinforcement of the banks to provide protection against flooding and to protect land by discouraging the lateral movement of the channel. These actions increase the security of people occupying river floodplains which, since the dawn of civilisation, have been the most attractive areas for human settlement because of the richness of the soil for plant growth, the large diversity of plant and animal resources located there, the relative ease of communication (by water or by land), and ready access to the water – arguably the most important of all natural resources.

There is a deep contradiction in these actions, which were expanded with dramatic effect after the advent of powered machinery, and which serve to isolate the river from its floodplain. Most obviously, the isolation of the floodplain from the river eliminates its flood storage function and creates even higher flood peaks within the channelway. More fundamentally, the reason for the resource richness of river floodplains is the connection with the river (Junk *et al.*, 1989). Alluvial soils in floodplains are mixtures of riverborne sediment and organic material that sequester abundant nutrients and to which additions are made each time the river floods. Biological diversity is promoted by the lateral shifts of the river within its floodplain, which erode away mature to decadent floodplain habitats and replace them with fresh deposits and early ecological succession. An active floodplain is a palimpsest of habitats collectively supporting a high diversity of species (Ward *et al.*, 2002). In the long-term view, the floodplain is part of the river: it is the accumulation of sediments on their way through the fluvial system that has entered long-term but, nevertheless, temporary storage. Severing the river from its floodplain ends the regular exchange of water, sediment and nutrients, and severely constrains the exchange of organisms between the two, to the long-term detriment of biodiversity and productivity in the floodplain and the river.

Hardening riverbanks – freezing the position of the river – causes further problems in the channel. If the river transports significant volumes of bed material, that material must be deposited within the constricted channelway, aggrading the bed and creating, again, higher floodwater levels. Furthermore, there will be only limited possibility for aquatic habitat renewal within the laterally constrained channel, leading to a substantially less productive aquatic ecosystem. In effect, the river has been transformed into a ditch.

Modification of channel form

A more ambitious extension of bank modification entails modifying the entire river channel – replacing the original channel by a more or less regular one. There is a wide variety of reasons to do so. This action, accompanied by the construction of strong banks, fixes the position of the channel, thereby protecting occupied riverine land and/or improving access to riverfront lands; navigation is improved; and resources in the form of sand and/or gravel dredged from the river may be provided. The latter activity is widespread in those parts of the world (generally, beyond the margins of Pleistocene glaciation) where high-quality aggregate resources are scarce on the terrestrial surface.

Extensive channel modification commenced in northwestern Europe in the seventeenth century. Purposes included the drainage of wetlands in the attempt to reduce insect-borne diseases, land reclamation and protection, rationalisation of riparian land holdings, and navigation improvements. The history of modification of some of Europe's largest rivers is recounted in Petts *et al.* (1989). Hundreds of kilometres of major rivers, and thousands of kilometres overall, have been forced into rectilinear or gently curved channels with fixed banks. The channels are relatively narrow so that flows are swift and deep to overcome tendencies for sediment to be deposited. Ecologically, they are relatively barren.

One of the most dramatic examples of channel redesign has occurred on the Mississippi River of the USA. Following earlier practice, 14 meanders were cut off between 1929 and 1942 to improve navigation on the lower river and two natural cutoffs occurred (Winkley, 1994). The reach was shortened by 240 km, or about 45%, which more than doubled the gradient in some subreaches. Subsequent bank stabilisation has essentially fixed the reach at about 70% of its former length. As the sandy alluvial channel adjusts to these changes, there has been bed erosion in upstream and mid-reaches and sediment deposition downstream (Biedenharn *et al.*, 2000). The interpretation of these tendencies is made more complex by the changing sediment load delivered to the river as the result of changes in land management in the drainage basin (see Section 4.4.3 for sediment effects).

Today, large-scale channel rectification for flood protection, navigation and water resource control is under way in China, where the importance of flood protection in the summer monsoon areas, the value of waterborne commercial transport and the need for construction materials are all critical spurs for such developments.

Again, the long-term effects may be counterproductive. The replacement of the varied natural topography of the riverbed by an often highly regular cross-section decimates the habitat diversity for aquatic organisms. River channelisation is inevitably followed by a reduction in riverine biodiversity (for perspectives, see Benke, 1990; Sparks, 1995).

More fundamentally, along alluvial rivers (where these actions are most frequent) the change in river channel dimensions that accompanies channelisation or dredging has the effect of replacing the equilibrium cross-section of the channel – that cross-section taken up by the channel as the stable form for the imposed water and sediment loads, and described by the equations of hydraulic geometry – by a channel that is too wide and/or too deep. The river begins to reduce its section by sedimentation so that continuous and possibly costly maintenance activity must be undertaken in order to maintain the engineered channel.

Modifying the flow regime

Damming a river and/or diverting water initiates the most dramatic set of changes of all human actions. Firstly, the hydrological regime may be more or less radically altered, depending on the size and purpose of the project. Early instances, the construction of weirs to trap fish and weirs to direct water to mill-races, effected negligible modification to the river regime. Later, dams constructed for water resource control, for flood control and, most of all, for hydroelectric power generation, have had increasingly radical impacts. The twentieth century has seen a massive project of dam building (see Section 4.5) which has modified a large fraction of all the major rivers of the world.

Today, 60% of the world's large river systems are affected by dams (Nilsson *et al.*, 2005), 36% of these basins covering 52% of their aggregate area being 'strongly affected' according to the criteria of Dynesius and Nilsson (1994), with ongoing developments in many large river basins, worldwide.

Dams inundate more or less extensive land areas, they induce sedimentation and aggradation upstream of the impoundment, and they change the water and sediment regimes downstream, often inducing erosion and degradation of the riverbed (Petts, 1984; Brandt, 2000). Flooding extensive areas produces large transient changes in water quality as various substances are desorbed from the drowned soil. Large power dams are often located along the uppermost trunk channels of a river system – where there is still a significant drop that can be controlled for power generation, but where flow is already quite large. They interrupt passage of aquatic organisms along the river – most notably, anadromous fish – and they stop the transfer of organic nutrients from the extensive network of upland channels where nutrient recruitment is prolific, to the downstream trunk channels, where diverse aquatic ecosystems rely on the nutrient influx. Riverine ecosystems are severely disrupted by dams (Petts, 1984; Ligon *et al.*, 1995).

In 1967, a major dam was closed on the upper Peace River in northwestern Canada establishing what was, at the time, the fifth largest hydropower project in the world. A continuing study of the adjustment of the 1200-km long river channel downstream shows that full adjustment will proceed according to the ability of the river to redistribute sediments along its course. With the highly regulated flow regime, it is expected that this will require of order 1 ka (Church, 1995), although the bulk of the adjustment will occur within the first century. Similar figures have been estimated by Williams and Wolman (1984) for large projects in America.

Major end point deltas may be significantly impacted by upstream dams because of the reduction in sediment supply. The Nile Delta is one such example. Since the closure of the Aswan dams, the shoreline of the delta has been significantly eroded (Stanley, 1988; see Section 4.5.2 for further details). With the closure of the Three Gorges Dam on Changjiang (the Yangtze River of China), a similar drastic reduction in the downstream sediment budget has been detected and similar effects are anticipated in its delta (Yang *et al.*, 2007).

Diversions and canals

Water is sometimes diverted from one drainage basin to another via tunnels, pipes or open channels. The purpose may be delivery of water for resource use in an area of need, or it may be to focus hydroelectric power generation facilities. Canals serve to facilitate freight navigation or to transfer water. Both types of development modify the basic hydrological network.

The direct impact of water diversion is modification of the hydrological regime in both the contributing and receiving waterway. A consequent physical impact is enhanced erosion or sedimentation along one or the other waterway, since sediments are never diverted in proportion to the water diversion. The effects on aquatic ecology that attend the construction of diversions and canals may be serious. The constructed channels represent new routes for the migration of aquatic organisms, which may move either independently, or by attachment to boat traffic along the system. This may lead to species invasions that significantly change the pre-existing ecosystem in the receiving waters.

The heyday of canal building across drainage boundaries occurred in the nineteenth century, before the establishment of high-volume overland transport. Today, most major developments for inland waterborne transport involve modification of single drainage lines, such as the mid-twentieth-century St. Lawrence Seaway into the Great Lakes of eastern North America, or the continued development of the Mississippi–Missouri–Ohio system of the USA. Old-established cross-drainage connections remain a significant ecological concern, however. So, for example, the Chicago drainage canal (properly, the Chicago Sanitary and Ship Canal), which connects the Great Lakes with the Mississippi system, is a major concern for the potential spread into the Great Lakes of a number of major pest organisms that have appeared in the Mississippi system.

An important class of diversions is redirection of water into canals for agricultural irrigation. Perhaps the most famous large-scale diversions were constructed in the Indus basin of then British India (today, divided between Pakistan and India) in the late nineteenth and early twentieth centuries. Similar canals were constructed, as well, in the Nile delta. From these exercises much of the modern theory of river channel behaviour – including the hydraulic geometry – was derived. Water diversion has induced siltation and reduction in size of the river channels, and dramatic modification of water quality (see further discussion in Section 4.5.2).

Diversions for hydroelectric power development and for water resource development continue. The most ambitious development underway in the world today is in China, where the South–North Water Transfer Project is under construction to refurbish the ancient Grand Canal – a 1794-km canal running northeast from Hangzhou (Qiantang River) to Beijing – to supply water, largely from Changjiang (Yangtze River), to the heavily agricultural North China Plain and, especially, to the water-starved capital region

(Stone and Jia, 2006). Most of the rivers between Changjiang and Beijing are, today, virtually 100% diverted for water resource use, with the result that their channels are dry for much of the year. This fate befell even Huanghe (Yellow River) until steps were taken to maintain minimum flows in the lower river.

4.3.3 Perspective

Humans have modified river channels for their particular purposes – protection from floods and command of water resources – for at least 3000 years. Works on a large scale were engineered more than two millennia ago in the Szechwan basin of China. Widespread river channel modifications were undertaken in Europe from the seventeenth century, and large-scale and extensive modification of river channels followed on the development of powered earth-moving and dredging machinery from the mid nineteenth century on. There is a significant reassessment under way today of the wisdom of such large-scale modifications of rivers in the industrial countries (see Section 4.6), but such developments continue unabated in the developing world, where the human condition and the need to control water resources for survival and development generally preclude such rethinking.

The impact of these developments on hydrology and on the rivers is extreme. Channel changes that will accompany the hydrological changes that follow regional or global climate change are predictable using the equations of hydraulic geometry. Thus the relative change in width is related to the change in discharge by

$$\frac{dw}{w} = \frac{b \cdot dQ}{Q} \tag{4.4}$$

where $b = 0.5$ and similarly for d and v; that is, the fractional change in river width will be one-half the fractional change in formative flow – a 20% increase in the magnitude of flood flows will create a 10% increase in the regime width of the river. The exponents b, f (Eq. 4.2) and m (Eq. 4.3) give the fractional changes to be expected in width, depth and mean velocity, respectively.

In comparison, human engineering may alter the size of channels and the flow through them by factors of 2 or more, and may impose artificially simple and fixed geometries on channels. Most of the world's major rivers today are more or less effectively controlled and regulated by human action and, in regions of intense human settlement, nearly all channels have been fixed in position and simplified. The human impact dramatically exceeds the foreseeable consequences of hydroclimatic change and urgently requires comprehensive study to understand the effects at landscape scale on ecosystems, on resource provision, and on human society itself.

4.4 Fluvial sediment transport and sedimentation

4.4.1 Water quality

Water quality refers to the physical and chemical condition of water. In relation to natural waters, it includes the physical properties of the water (temperature, colour, transparency, odour), the content of dissolved minerals (solutes), the quantity and character of particulate matter – both organic and inorganic – carried in the water, and the microbial organisms present in the water. Natural waters always contain some burden of solutes, particulate matter and organisms, and the range may be very wide, but human actions may dramatically augment and change these burdens.

Solutes appear in natural waters as the result of rock weathering, so the chemical character of the water reflects regional geology. Waters draining felsic rock terrains such as granites are acidic in character; ones draining basic rocks such as carbonates are alkaline. Inorganic particulate matter also derives from rock weathering and soil erosion. Mineral grains and rock fragments carried in streams may vary in size from colloidal (<0.5 μm) to boulders of order 1 m in diameter. The latter are moved only on steep gradients in exceptional events. Organic matter comprises plant and animal parts delivered from overbank and transported through the stream system, and organic matter created in the water column and on the streambed – including remains of aquatic organisms, faecal material and metabolites. Living microorganisms that are included in water quality determinations are principally aquatic bacteria.

Humans pervasively affect water quality by land use, which changes water routes and residence times on the land surface and in the soil (see Section 4.2), the exposure to foreign materials, and the direct entrainment of soil materials into runoff. Further, humans inadvertently or deliberately dump materials into water bodies as a means of disposal. The effects exceed those of natural environmental change by orders of magnitude.

The character and stability of rivers – the geomorphology of rivers – is most strongly influenced by the particulate sediment load that they carry, and so that will be the focus of this section.

4.4.2 The fluvial sediment cycle

Mineral particulate matter in rivers is closely tied to the geomorphology of the landscape since this material is the product of erosion of the land surface. Sediments originally mobilised on the land surface may be moved directly into

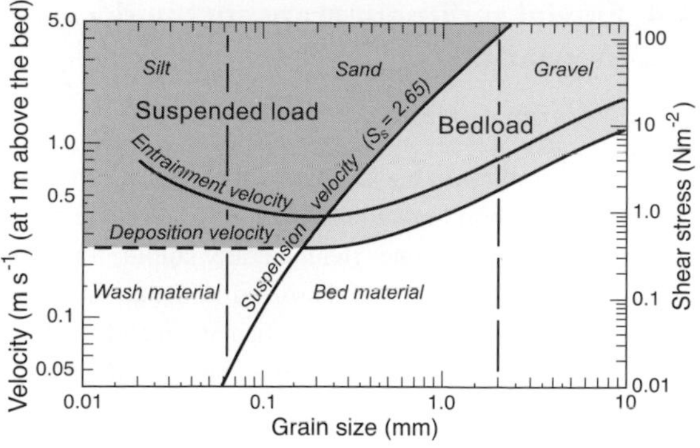

FIGURE 4.5. Diagram to illustrate the water flows necessary to transport particulate sediments of different size and the mode of transport (modified from Sundborg, 1967). S_s is the sediment specific weight.

the river system by surface runoff, or they may go back into storage at field edges or slope base before reaching a stream. Once in a stream, sediments may move a short distance and then be deposited into storage in the channel or overbank on a floodplain surface, or they may move a long way through the system without being redeposited. The distance that material moves is inversely related to grain size – smaller materials, more easily moved by the force of flowing water, are apt to move longer distances in shorter time. Stream systems are, accordingly, sediment sorting machines. Large materials are left behind in upland streams and valleys: fine materials move into floodplains along trunk valleys, to end point deltas, or directly into the receiving water body.

Sorting through the system is reinforced by two processes. Firstly, the manner in which sediment is transported has an important influence on how far and how quickly it moves. Sediments in transport can be divided into two classes. Larger materials move as 'bedload', by sliding or rolling over the bed: the submerged weight of the grains is carried mainly by the bed. Finer grains move in suspension in the water column: their weight is borne by upwardly directed turbulent eddies in the flow, which compensate the settling velocity of the grain. The finer materials move much farther, once entrained. Indeed, very fine material (silt and clay) may move much of the entire distance through the stream system directly to the receiving water body. Accordingly, such material is called 'wash material', in contradistinction to 'bed material', sediment that is found in the bed and lower banks of the channel and determines channel form. Figure 4.5 shows the division of materials according to their mode of transport and persistence in the river channel.

A second process that reinforces sorting is the character of sediment storage places and the amount of time that sediments spend in them. Materials transported as bedload are deposited in bars within the channel, while finer materials are deposited either interstitially within the coarser deposits on river bars, or overbank on floodplain surfaces. The farther the deposition occurs from the central axis of the channel, the longer the material is apt to remain in storage. We are left with two principles for sediment transfer through the river system:

- coarser grains, which move principally as bedload, move relatively short distances in single events and are stored in the channel, where they are apt to be soon re-entrained;
- finer materials, which move intermittently or mainly in suspension, move relatively long distances in single events and are stored toward the channel edges or overbank, where they are apt to remain for a protracted period.

Storage of sediments along the channel can vary from the time between successive sediment entraining flows on bar edges, to periods of years to decades if they are buried deeply within the bars. Overbank storage can vary from periods of years near the channel to millennia near the back of wide floodplains. Figure 4.6 gives a schematic view of the sediment transfer system based on these principles.

An important ancillary process is rock weathering, which is particularly important in the wet–dry, possibly freeze–thaw, environment of river channels and floodplains. Larger materials that enter storage may relatively rapidly be weakened and may break down to finer materials when they are re-entrained into the flow. The net result of these processes is that, although finer sediments are more highly mobile than larger ones, their total time of passage through the fluvial system may not be too strongly different, except that the larger materials will have been eliminated by weathering.

Sediment transfer through the drainage basin nevertheless requires a long time: the virtual velocity of sediments (i.e. the average rate of progress of the sediments, including rest periods) through a drainage basin is on the order of m a^{-1} so that, in large drainage basins, transit time through the basin may be many thousands of years (this is the reason why the adjustment to a dam on a large river may require many centuries to complete). Within this period, important modulations of climate and land surface condition may create significant changes in sediment yield; changes that are buffered by the rate at which sediments move into or out of storage along the river system.

Sediments may be entrained into the river system from two sources: the land surface, or the bed and banks of the river. Sediments go into storage in the channel or in alluvial fan, floodplain or delta deposits of the river itself. During periods when there is a relatively large delivery of sediment

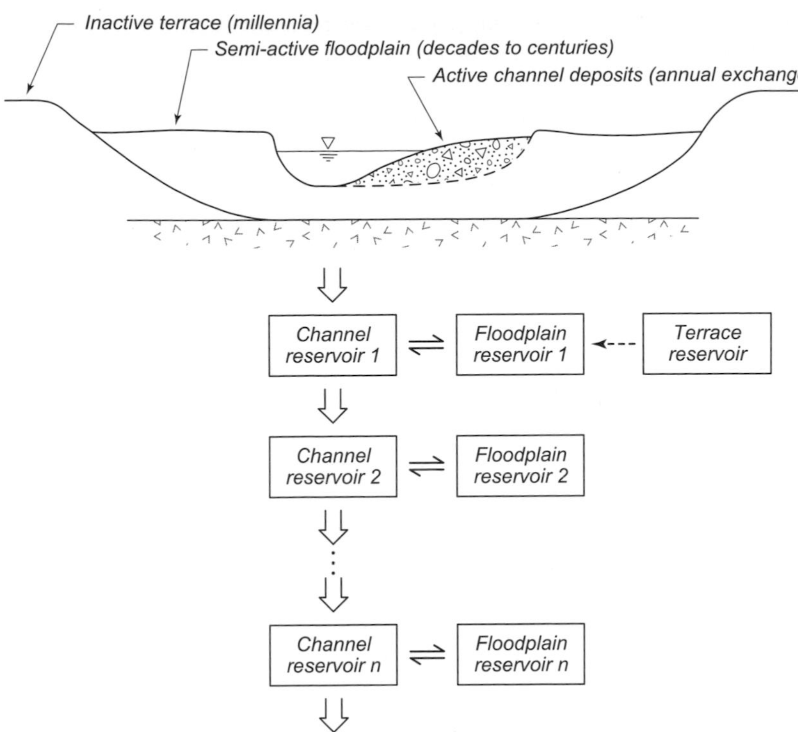

FIGURE 4.6. Cascade diagram of fluvial sediment 'reservoirs'; characteristic storage times given in parentheses.

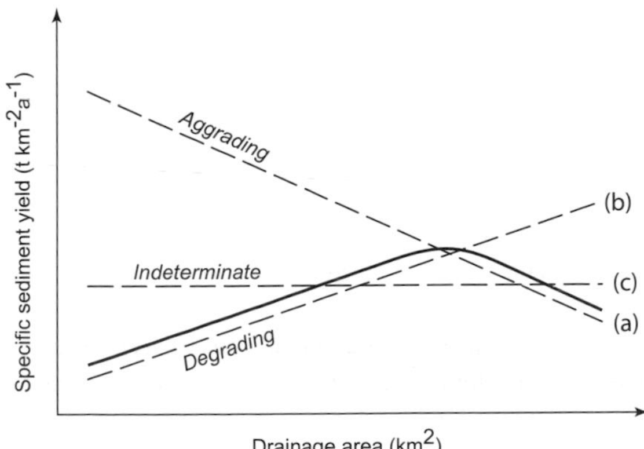

FIGURE 4.7. Trends of sediment yield through the drainage system, illustrating (a) aggrading and (b) degrading patterns. The heavy line illustrates a common pattern whereby degradation of upland channels (net sediment export) is balanced by aggradation along trunk rivers (loss of sediment into floodplain storage). Line (c) indicates a system in sediment yield equilibrium throughout.

from the land surface, there is a net transfer of sediment into storage along the river system – the system is said to be 'aggrading'. During periods when sediment delivery from the land surface is relatively small, the rivers re-entrain material from channel beds and banks and total fluvial storage is reduced – the system is said to be 'degrading'. The tendency of a drainage basin to be aggrading or degrading can be measured by calculating the specific sediment yield – the sediment yield per square kilometre of contributing drainage area. If this declines down the system, then sediment mobilised from the land surface must be lost to storage along the drainage line, and the system is aggrading; in the reverse case, the system is degrading (Church *et al.*, 1999) (Fig. 4.7).

Simple trends of aggradation/degradation are unlikely to hold throughout a large drainage system, for aggradation in one part of the system directly implies degradation wherever the sediment originates. If the sediment originates on the land surface, then the *fluvial* system may indeed exhibit a consistent trend, but if sediment is originating somewhere in the drainage system, then it may be going back into storage farther down the system, hence upstream degradation followed by downstream aggradation is a common landscape signal (as is observed in the Mississippi River example discussed in Section 4.3.2).

The most significant pulse of sediment from the land surface into the fluvial system in recent Earth history was the end-Pleistocene mobilisation of sediment at the time of northern hemisphere deglaciation. This 'paraglacial' sediment pulse was rapidly exhausted in small upland basins, but it dispersed through the larger fluvial system over several millennia (Church, 2002) (see Chapter 2, Section 2.5.9). Lesser pulses of sediment delivery are created by more mundane events, including wildfire, or an individual major landslide on a hillside slope.

Inasmuch as soil is strongly retained by the root systems of vigorous vegetation cover, the condition of the vegetation on the land surface is a prominent modulator of sediment entrainment from the land surface. Since humans both manipulate vegetation in a comprehensive way and directly disturb the soil in the course of forestry, farming and construction activities, we are by far the most important agent influencing sediment mobilisation from the land surface today and, thereby, sediment transfer through the fluvial system.

4.4.3 Effects of human activities on fluvial sedimentation

Agricultural land use

Agriculture in most systems of practice entails the virtually complete replacement of the natural vegetation by managed covers. Where these covers consist of perennial, continuous cover (such as pasture), soil may be well conserved, but most of the world's arable lands are bared for part of the year and crops frequently provide only fractional surface cover. In addition, arable soils are often compacted by heavy machinery, and recently ploughed soils of certain textures may undergo surface sealing as the result of rainsplash (see Section 4.2). The net result may be a significant increase in surface runoff and the effect may be further reinforced by certain ploughing patterns. In these circumstances, soil erosion by runoff can be increased dramatically. The severity of the soil loss may depend on the timing of rainfall in relation to crop planting (see Slattery and Burt, 1996, and Section 4.2.2 for further discussion).

Trimble (1983, 1999) has shown that, in the American Midwest, ploughing of the land between 1850 and 1940 yielded large volumes of sediment to river channels. Upland channels underwent massive aggradation and regional sediment yield followed the curve (a) of Fig. 4.7. Since the late 1930s, the implementation of soil conservation practices has reduced field erosion dramatically. Sediment stored in the upland channels has been re-entrained and is moving into higher order tributaries. The uplands now exhibit a type (b) sediment yield pattern, while regional sediment yield follows the pattern of upland degradation and valley aggradation. Throughout the period, however, less than 10% of all the sediment mobilised was transferred as far as the regional trunk stream, the Mississippi River. Nevertheless, by looking at long-term floodplain aggradation rates inferred from floodplain stratigraphy in the American Midwest, Knox (1989) has determined that average sedimentation rates post-European settlement have been 10–100 times greater than the preceding rates here.

Meade (1982) demonstrated a similar historical pattern of agriculturally dominated sediment yield in the southeastern US Piedmont and coastal plain whereby rivers experienced massive sedimentation between 1750 and *c.* 1900 as forests were cleared, with large sediment yields to the ocean. Hundreds of mill-dams along the smaller streams contributed significantly to sediment trapping, floodplain construction and changes in stream channel character (Walter and Merritts, 2008). During the twentieth century, improved agricultural practice and reafforestation of much of the land reduced sediment influx and the rivers are degrading. There appears, then, to have been a major spike in sediment yield associated with early modern agriculture, a spike that has been largely attenuated with the development of improved land management practices, although sediment losses from agricultural fields have by no means been eliminated.

Nor is modern agriculture exceptional. Examples of agriculturally related soil loss so severe as to cause social collapse are found around the ancient world. Examples come from such different environments as Mesopotamia, prehispanic Central America and Easter Island. Montgomery (2007) has recently given a comprehensive summary of this history.

Forest land use

Forestry has been widely regarded as a significant source of increased sediment yield to fluvial systems (National Research Council, 2008). In fact, in most forest management systems, the effect is highly transient. Three distinct effects are experienced (see Chapter 12 for further discussion). During the preparations for forest harvest, road construction and other site preparations may interrupt natural drainage lines, create exposed road embankments and excavations, and establish important sources of surface runoff and fine-sediment production on unmetalled road surfaces. During harvest, direct disturbance of forest soil by heavy machinery or by log handling may substantially increase soil erodibility. After harvest, the exposure of the forest soil and the decay of root systems may leave soil susceptible to surface erosion or to landslides on hillside slopes. There is a restricted window during which soils may be particularly prone to erosion which ends once a new, young forest is securely established. The time period for this may vary from 5 to 15 years, depending on both forest and site characteristics. During this period, the chance effect of particularly severe weather may be an important factor determining whether or not significant erosion actually occurs. Along river channels, harvesting trees to the streambank may have particularly drastic effects on streambank strength and hence on fluvial erosion. On a regional

TABLE 4.3. *Sediment mobilisation and yield from hillside slopes*

Process	Mobilisation rate		Yield rate to stream channels	
	Forested slopes	Cleared slopes	Forested slopes	Cleared slopes
Normal regime				
Soil creep (including animal effects)	$1\,m^3\,km^{-1}\,a^{-1*}$	2 ×	$1\,m^3\,km^{-1}\,a^{-1*}$	2 ×
Deep-seated creep	$10\,m^3\,km^{-1}\,a^{-1*}$	1 ×	$10\,m^3\,km^{-1}\,a^{-1}$	1 ×
Tree throw	$1\,m^3\,km^{-2}\,a^{-1}$	–	–	–
Surface erosion: forest floor	$<10\,m^3\,km^{-2}\,a^{-1}$		$<1\,m^3\,km^{-2}\,a^{-1}$	
Surface erosion: landslide scars, gully walls	$>10^3\,m^3\,km^{-2}\,a^{-1}$ (slide area only)	1 ×	$>10^3\,m^3\,km^{-2}\,a^{-1}$ (slide area only)	1 ×
Surface erosion: active road surface	–	$10^4\,m^3\,km^{-2}\,a^{-1}$ (road area only)	–	$10^4\,m^3\,km^{-2}\,a^{-1}$ (road area only)
Episodic events				
Debris slides	$10^2\,m^3\,km^{-2}\,a^{-1}$	2–10 ×	to $10\,m^3\,km^{-2}\,a^{-1}$	to 10 ×
Rock failures: falls, slides	No consistent data: not specifically associated with land use			

* These results reported as $m^3\,km^{-1}$ channel bank. All other results reported as $m^3\,km^{-2}$ drainage area.
Source: Generalised results after Church and Ryder (2001), Table 1.

basis, the relative importance of these effects can be gauged from a summary of experience in the Pacific Northwest of North America (Table 4.3).

In well-managed forest harvest operations, erosion associated with initial access and road building is found to be most severe, and mostly associated with the disturbance to natural drainage routes.

Mining and quarrying

Unlike agriculture and forestry, which are distributed across the landscape, mining and quarrying are carried out at individual sites of limited extent. In landscape terms, they are point developments. While huge volumes of material may be moved, very little of it (except product) leaves the site. What does leave the site may, however, be highly troublesome since, by specific purpose, mining is focussed upon materials that are rare in Earth's surface environment and are therefore apt to be toxic to many organisms. Consequently, the complete control of drainage is an important activity at properly designed mining operations. In some parts of the world, however, mine wastes are directly dumped into river systems. Historical mine wastes have often been dumped into rivers, however, and today contaminate floodplain soils (e.g. Lewin and Macklin, 1987).

Waste materials, particularly the waste products of ore concentrating processes, are often slurries that are routed to ponds (called 'tailings' ponds) where the solids settle. The ponds are sealed so that water is recovered for treatment and the solids secured. A significant problem, however, is that mine sites usually are active and supervised for a period of only decades' length, whereas the waste materials remain on the site for ever. After the mine is closed, supervision of wastes ends and there is a significant risk that, eventually, the containment may be breached by natural processes. A number of such failures have been experienced in recent years (see, for example, Grimalt *et al.* (1999) on the 1998 spill in Aznalcóllar, Spain, into the Guadiamar River; Macklin *et al.* (2003) on the 2000 Tisza River inundation in Romania) with disastrous consequences for aquatic and riparian ecosystems.

Placer mining and gravel borrowing

River sediments may deliberately be disturbed or removed for the recovery of precious metals – principally gold – or for industrial use of the river sediments. Placer mining turns over river sediments without, usually, removing the bulk of the material from the river. This activity does, however, mobilise large volumes of fine sediment sequestered amongst coarser materials on the streambed, significantly increasing the fine-sediment load of the river downstream, often with significant impact on fisheries.

Sand and gravel mining from riverbeds has often been presented as renewable resource exploitation. The fact, however, is that almost everywhere rivers have been exploited for aggregate resources the volumes removed have exceeded sediment recruitment by up to an order of magnitude. The result is destabilisation of the channel (Lagasse *et al.*, 1980) and a dramatic change in river morphology, leaving what is essentially a degraded ditch. Accordingly, riverine ecosystems collapse and are replaced by dramatically less productive ones (Kondolf, 1998).

Urban land conversion

On a global basis, urban land conversion is possibly the most severe sedimentary disturbance today. But, like forestry, its initial effects are transient. Urban land conversion and, similarly, the construction of modern communication routes, entails denuding the land surface, exposing the soil for a period of months at least, and the movement and temporary storage of significant amounts of soil. These activities may lead to very large increases in sediment mobilisation and delivery to stream systems (Wolman, 1967; Trimble, 1997).

Developed urban lands are largely covered with unerodible materials, while landscaping of the remaining soil surfaces usually imparts high stability to them. Hence, after urban conversion, sources of sediments become very low. However, a substantial variety of more or less exotic materials is generated in urban areas by industrial, commercial and domestic activities and released deliberately or inadvertently into the surface environment, where it is entrained in urban building and street runoff. Combustion products, petroleum products, paint and metal fragments, fertilisers, pesticides and pharmaceuticals all find their way into the runoff and subsequently into waterways where they may strongly affect water quality.

The contaminant burden is added to by deliberate disposal of industrial and domestic wastes into rivers. This burden includes many synthesised materials that are wholly foreign to the natural environment. Once in the environment, many of these materials are adsorbed onto fine mineral particulates and eventually settle out of the water column with the sediments, to be stored in riverbottom or floodplain sediments. If the contaminant burden of the river is subsequently reduced, a reverse chemical gradient is established from the sediments into the water column and the sequestered material may then be resorbed back into the water. The definitive clearance from a river system of particular contaminants may take a long time. Since rivers were often regarded as natural sewers in the nineteenth and early twentieth centuries, there are many instances of this problem today (see, for example, Wiener *et al.* (1984) on contamination of the upper Mississippi River system).

4.4.4 Perspective

Humans influence riverine water quality and sediment burden in many ways. Comparisons between natural and contemporary sediment loads are difficult to establish because there are no pre-modern (that is, pre-industrial) measurements of riverine sediment loads and, in any case, it is likely that agriculture has had some influence for a very long time, though probably much less than is achieved by modern industrial agriculture in a crowded world. In the absence of measurements Syvitski *et al.* (2005) have attempted to compare contemporary and pre-industrial sediment loads in the world's rivers by modelling pre-industrial sediment yields. They estimate the pre-industrial suspended sediment yield to the oceans of the world's rivers to be 14 000 Mt a^{-1}. In comparison, they estimate the modern yield to be 16 200 Mt a^{-1} – an increase of 16%. However, reservoirs trap approximately 22% of the load, so that the sediment yield to the world ocean is claimed to be reduced to 12 600 Mt a^{-1}.

These figures are based on the sediment yields of relatively large rivers. Given the pervasive influence of human activity in increasing sediment loads, most river systems today exhibit aggradation along their middle and lower courses. Indeed, so pervasive is this phenomenon that, until recently, it was supposed that the regional signal of aggradation (curve (a) in Fig. 4.7) was the standard signal for riverine sediment mechanics everywhere. Wilkinson and McElroy (2007) estimate that as much as 75 000 Mt a^{-1} is being mobilised, mostly from agricultural fields (see Fig. 1.16 in Chapter 1). It is probable, then, that substantially more sediment is being mobilised in headwater regions than is accounted for by statistics based on large river systems. This makes humans incomparably the most important geological agent remaking the surface of Earth today.

4.5 Water control: dams and diversions

4.5.1 Introduction

For centuries the landscape moulded by moving water has been modified dramatically by management and storage of the water (Plate 8). To this day politicians, technocrats and floodplain inhabitants continue to promote large storage dams. Thus, for example, the number of dams along the upper Yangtze River increased from very few in 1950 to approximately 12 000 by the late 1980s and continues to increase (Xu *et al.*, 2006). The purpose for storing water is to hold it during peak flows and then release it during low flows, thus, as the US Bureau of Reclamation (1946) put it, transforming 'a natural menace into a national resource'. But this statement is not entirely appropriate. If the natural water cycle is modified by humans then impacts follow. In addition to storing water, the sediment and nutrients in motion with the river flow are trapped in the reservoir.

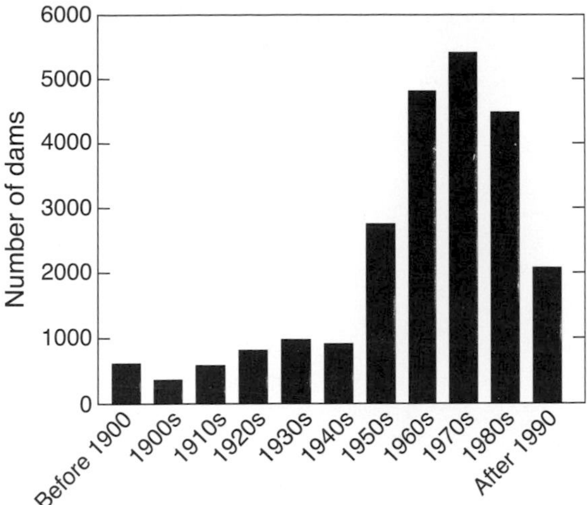

FIGURE 4.8. Numbers of high dams (>15 m high) constructed in the world by decade, excluding China (data of the International Commission on Large Dams).

The downstream river, built from its own sediment, no longer has any building material, but the clear water retains erosive power, resulting in bed lowering (channel incision), bank erosion and changes to riparian vegetation. One of the most dramatic impacts has been the sediment starvation of river deltas – a most striking example being the sinking Mississippi Delta whose flood-protection dykes could not hold back the high waters due to Hurricane Katrina of 2005 (Martin, 2007). Of significance is the combination of deltaic deterioration due to sediment starvation brought on by reservoir storage coupled with global warming causing a rise in sea levels. Such additive impacts have not been accounted for during the planning of large dams; in fact, planners and politicians have scarcely addressed combinations of impacts at all.

The diversion and storage of water developed slowly, beginning *c.* 8 ka BP with the inception of regular crop planting and communal settlements. Within the most recent 200 years, the management of water for multiple uses has become so complex that serious water disputes have resulted between countries as well as communities. The progressive construction of major infrastructure such as dams, canals, flood dykes and pump stations was usually initiated on a small scale, but as the infrastructure expanded, the control of water became a significant component of a region's resource management programme.

The number of high dams (more than 15 m high) has exploded over the last half century (Fig. 4.8). This surge in dam construction has proceeded with inadequate impact analysis and new problems are identified with regularity. The ponding of water behind dams, the spreading of water over fields and the conveyance of floods between dykes has changed fluvial processes and transformed local landscapes. Then, as the landscapes changed, the rainfall and evaporation changed, causing progressive change in local climate. The system is not static. The following discussion presents, through a number of examples, the complexity of landscape changes and the many causative factors. In order to examine the impacts to environments, water resource developments in several large basins (the Colorado, the Indus and the Nile) will be described with illustrations of landscape changes. In addition, future scenarios of the Nile, Mekong and Huanghe (Yellow) rivers will be speculated upon.

4.5.2 Some case studies

Colorado River

The Colorado River basin drains 632 000 km^2. The river originates in Wyoming and Colorado and flows into the Gulf of California in Mexico (Figure 4.9). The Colorado River landscape consists of canyons in the upper basin, floodplains along the lower reaches, and an extensive delta at its lower end. Much of the basin is arid, with less than 80 mm of precipitation per year (Cohen, 2002). The river's reported average annual flow varies with the investigator and source of information – Meko *et al.* (1995) estimated a long-term mean flow of 530 m^3 s^{-1}, based on tree ring records, while Owen-Joyce and Raymond (1996) estimated 590 m^3 s^{-1} based on the past century of instrumental record. Earlier, the US Bureau of Reclamation (1946) had estimated that there was 690 m^3 s^{-1} available for use annually but this value has now been acknowledged to be too high.

Diversion works for irrigation were begun in the late nineteenth century, but it was not until the completion of the Hoover Dam in 1935 that water was stored and flows and sediment downstream were significantly modified. An accidental diversion occurred in 1905 during spring floods which created the inland Salton Sea. Eventually, the Imperial Dam and the All-American Canal were completed to keep the Colorado River flowing toward the irrigated farms of Imperial Valley. The historical sediment load delivered to the delta was about 160 M t a^{-1} (van Andel, 1964) but the present load is almost zero (Carraquiry and Sanchez, 1999). The annual volume of water reaching the Gulf of California has been reduced to about 140 m^3 s^{-1} (Cohen, 2002) since the Glen Canyon Dam was completed in 1963. These dramatic changes to water and sediment have brought about the following significant environmental changes:

FIGURE 4.9. Map of the lower Colorado River basin.

- reversal of the slow accumulation of deltaic sediment as tidal action now removes more sediment from the delta than arrives;
- inversion of the salinity gradient in the upper Gulf of California so that there is now higher salinity in its northern (proximal) part (Lavin *et al*., 1998);
- collapses in several fisheries, namely totoaba, the pacific sharp-nose shark and the shrimp fishery (these events have also been attributed partially to illegal and unreported catches).

It is interesting to note that, despite major modifications to the delta ecosystem, new growth of wetland habitat has emerged after small releases of water from upstream reservoirs (Pitt *et al*., 2000) and in response to the El Niño events of 1982 and 1993 (Lozano, 2006). The fisheries crisis has also been addressed by Lozano, who concluded after simulation studies that a small flow increase – only 1% – may produce an increase of around 10% in the total biomass of the upper Gulf. This indication suggests that further human intervention might trigger significant improvements to the marine ecosystem and to fish species.

The impacts to the middle reaches of the multi-dam Colorado River as reported by the National Research Council (1987) are many. An important summary finding was that, because of the participation of various agencies, with different missions and budgets, the planning process for the Glen Canyon Environmental Study (GCES) did not treat the ecosystem as a whole. The GCES recommendations were aimed at strengthening the database and adopting an ecosystem approach to river management, but they were not adopted by the operating authorities. Subsequently, a controlled flood was created by allowing a release of 1274 $m^3 s^{-1}$ from the Glen Canyon Dam in 1996 in an attempt to restore beaches and rehabilitate habitat. This initiative gathered together an interdisciplinary team of scientists who subsequently published their findings in Webb *et al.* (1999). Conclusions from this research included the following:

- encroachment of riparian vegetation on sand bars (as the result of the cessation of sedimentation) limited use of the bars as campsites;
- after many flow releases sand moved from emergent sand bars to eddy zones, thereby filling aquatic habitats;
- clear, cold water releases changed the trophic structure of the aquatic ecosystem such that the productivity of trout increased; and
- maximising the rainbow trout fishery and habitat for the endangered Humpback chub proved to be mutually exclusive.

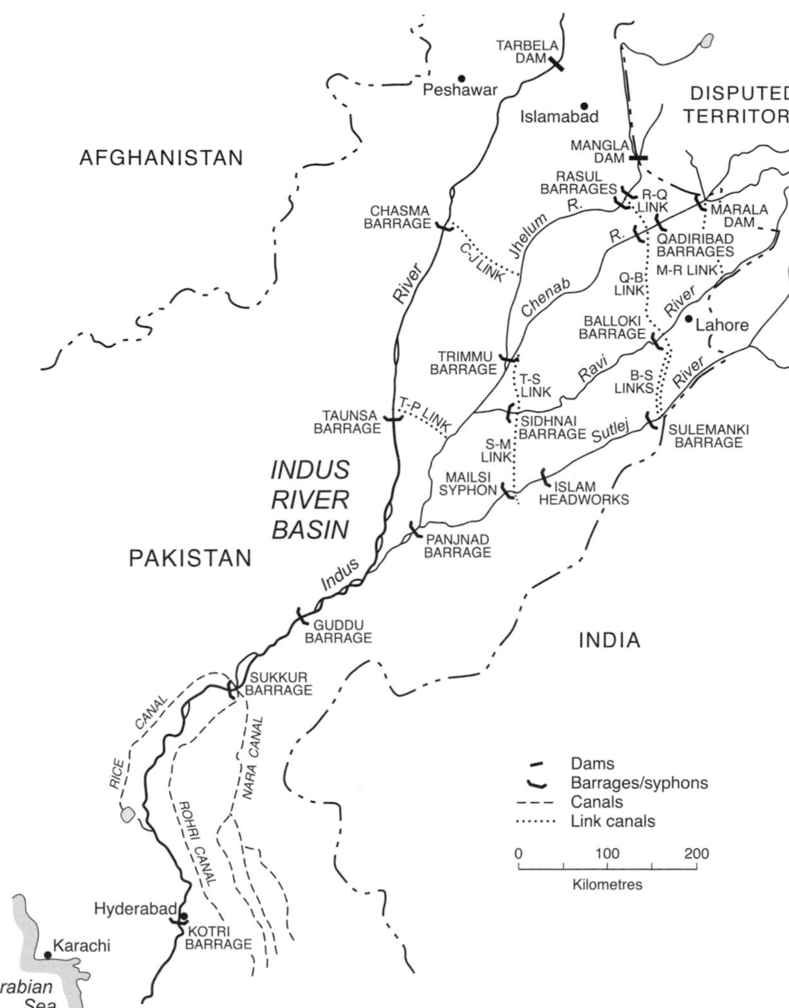

FIGURE 4.10. Map of the Indus River basin, showing the principal tributaries and the principal elements of the Indus water control projects.

The findings on the primary sedimentation process below the Glen Canyon Dam were, however, subsequently contested by Rubin *et al.* (2002) and it appears that research over a longer period is required.

In summary, the main landscape changes have been the transformation of the delta from a deposition feature to an eroding system, the reversal of the salinity gradient in the Gulf of California with higher salinity within its northern zone, and changes in flow and sedimentation through the Grand Canyon that have important ecological effects. The prime causative factor has been the construction of large dams (Hoover and Glen Canyon) resulting in trapping of sediment.

The Indus basin projects

An example of changes to basin-wide environment brought on by the construction of dams coupled with extensive irrigation is found in the Indus River system in Pakistan. The Indus and its main tributaries (Sutlej, Ravi, Chenab, Jhelum and Beas) rise in the Karakorum Range of the Himalayas. The mountains were created by the collision of the Indian Plate with the Asian Plate *c.* 65 Ma BP and subsequent uplift which followed within the last 2 Ma (Pleistocene Epoch). The mountain range forces the monsoon clouds of summer (June to September) to release their moisture in seasonally heavy and persistent downpours and in winter the range prevents cold Tibetan winds from penetrating to the south-facing slopes. The Indus River emerges near Mt Kailash in Tibet, flows northwest and cuts through the Himalayan Range downstream from its confluence with the Gilgit. Thereafter, the river is joined by the main tributaries (Fig. 4.10).

The Indus flows have been modified by dams and diversions constructed under the Indus Water Treaty. Prior to the treaty, the Punjab experienced about 100 years of irrigation development under the British. With partition of the subcontinent into Pakistan and India in 1947, irrigation expanded rapidly under the Indus Water Treaty. Large dams, the Tarbela and the Mangla in Pakistan along with the Bhakra in India, financed through international donors, have also resulted in multi-purpose management of water for irrigation, power, flood control and water supply. The treaty assigned the three western rivers (Chenab, Jhelum and Indus) to Pakistan and the three eastern rivers (Sutlej, Beas and Ravi) to India (Michel, 1967). To replace the waters of the eastern rivers, Pakistan constructed a large system of works, mostly link canals and two high dams, under the Indus Basin Project (Tarar, 1982). The layout of the project is shown in Fig. 4.10. The irrigation system of the Indus Plain now includes two storage reservoirs, 17 barrages, eight link canals and an extensive system of main canals, branches, distributaries and minors.

After the dams, cheap hydropower was used to pump water from the ground to expand the irrigation system. This action resulted in waterlogging and salinisation of soils. Thereafter, drainage systems were added which introduced saline water into the lower rivers. Large barrages diverted the saline water into irrigation systems in the Sind province. Meanwhile, the lower reaches of the Chenab and the Sutlej have shrunk – the widths are much smaller than the widths of these same rivers near the foothills. Nine or ten major link canals transfer flows from the northern rivers to the southern rivers. The expansion of irrigated area has resulted in more salinisation of land and increased evaporation. From anecdotal information, early morning fog is now frequent in northern Pakistan, resulting in cancellation of air flights.

In an attempt to resolve these problems, a multi-agency project (Left Bank Outfall Drain) was initiated in the early 1980s but, after review, the project was rated overall as unsatisfactory (World Bank, 2004). Thereafter, the international donors have proposed another US$785 million to develop a 'National Drainage Program'. Farther downstream, in the Indus delta, the shrunken Indus flows through a single channel with no active distributaries evident. The coastal mangrove zone has shrunk dramatically and both river and coastal fisheries have been impacted. Field inspection indicates new drainage channels to the east of the Indus emptying into the Rann of Kutch. The drainage channels are enlarging their size by headcutting, bringing new water and sediment problems into this tidal zone. Also, destruction of mangroves has impacted local climate and the reduction of silt delivery into the delta has resulted in regular erosion of the delta.

In summary, the introduction of high dams into the upper watershed has provided cheap power for pumping groundwater and, coupled with barrages to divert water, the largest irrigation system in the world. The change to the landscape has been dramatic – vast areas of land have become saline and unproductive while the delta zone is eroding and vegetation is disappearing. The large-scale storage of water has produced problems that, although foreseen prior to the Indus Treaty and acknowledged during the planning and negotiation phase, have been met only slowly with mitigation measures.

The River Nile

For thousands of years, inhabitants of the lower River Nile in Egypt have relied upon the water and its nutrients from the upper basin. The source of water was unknown and Egypt was styled 'a gift of the Nile' (Herodotus). In reality Egypt is a gift of Ethiopia – sediment eroded in Ethiopia is transported down the river and has, through time, filled the floodplain of the Nile Valley with fertile sediment and formed the Nile Delta (Butzer, 1972). The rise of the Egyptian civilisation over a period of about 7 ka was based on annual Nile flows sustaining a prosperous agriculture on the river floodplain and in the delta. Most of the water (about 75%) originates in Ethiopia, and flows down the Blue Nile and the Atbara, while the remainder (25%) originates in the Lake Plateau and flows down the White Nile. During the low flow period in the months March–June, the flow from Ethiopia is low and the White Nile provides about 75% of the total discharge. These flows pass slowly through the Sudd swamps in Sudan and a significant portion of the inflow evaporates.

Proposals for the management of the waters of the River Nile are numerous, with the historical purpose being to feed the floodplain inhabitants and their rulers. An overview of the major river works constructed to 1980 is shown in the profile in Fig. 4.11. Several major river works, mainly dams, were constructed on the Nile with the major impact being downstream from the High Aswan Dam. Prior to the construction of the High Dam many predictions were made and these are contrasted with actual changes in Table 4.4.

After the completion of the High Dam, the water level range in the lower Nile was reduced to about 2 m, with no overflow of the banks. Bank erosion has been significantly reduced since 1950. Studies of the river downstream from the Low Aswan Dam to Cairo indicated that the 1950 eroding bank length was some 500 km, which was 21% of the total bank length. It was reduced to 242 km in 1988 (12% of the reduced, 1988 length). The meandering river path was reduced and out of a total of 150 islands, some 87 joined the main bank, contributing to a substantial reduction in riverine habitat complexity. Also, the bankfull river width was reduced by about 30% from 1950 to 1978 as a result of a flow reduction from 8400 $m^3 s^{-1}$ to about 2600 $m^3 s^{-1}$ in post-High Dam conditions (RNPD, 1991).

TABLE 4.4. *Assessment of environmental changes on the River Nile below the High Aswan Dam*

Prediction	Actual change	Reference
(1) Reduced agriculture production	Increase	Sterling (1972)
(2) Devastation of fish in the Mediterranean	Slow decrease followed by increased numbers by 2004	Sterling (1972) Nixon (2004)
(3) River bed to be lowered by 10 to 20 m and bridges would collapse and banks erode	Bed lowering very small; bed is protected by snail shells and gravel; bank erosion reduced	Simaika (1970) RNPD (1991)
(4) Soil fertility reduced by trapping sediment in reservoir	Process is ongoing, with some fertilisers being applied	
(5) Increase in salinity and waterlogging in the delta	Serious problem partially being alleviated by field drains and pumping stations	Waterbury (1979)
(6) Erosion of coastline	Dramatic erosion after closure of High Dam; coastal defence works being constructed	Mobarek (1972)

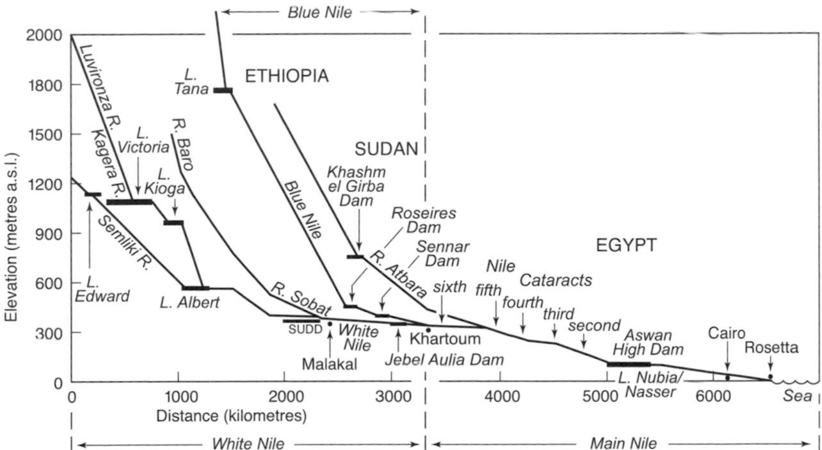

FIGURE 4.11. Long profile of the River Nile showing the principal river control works.

The Nile has become a relatively stable canal. Pre-dam predictions indicated major bed lowering with consequent failure of bridge foundations and landing wharves. However, these predictions did not materialise for a number of reasons:

- the presence of the Low Aswan Dam and six downstream barrages keeps the river bed from being lowered excessively;
- the occurrence of snail shells on the bed directly downstream from several barrages (delta barrages) has armoured the river bed;
- the presence of gravel lenses at locations of wadis entering the main river coarsened and armoured the river bed;
- the significant reduction in flows and velocities resulted in slow removal of bed sediment; and
- the introduction of wind-blown sand into the river caused a rise in the river bed (RNPD, 1991).

Another significant environmental consequence is the erosion of the Nile coastline, primarily as a result of reduced sediment entering this zone. The historical annual sediment load of about 125 Mt a^{-1} has been reduced to about 5 Mt a^{-1}. There is no longer any sediment deposition to perpetuate the growth of the delta. The coastal zone now has to be protected from erosion by large concrete works which are extremely expensive (Mobarak, 1972). The erosion was initially attributed to construction of barrages along the Nile in the nineteenth century but accelerated erosion was noticeable after 1964 with the completion of the High Aswan Dam.

The modification of the sediment budget along the Nile was predictable, but the degree of change to the fluvial landscape (river valley floodplain, delta and coast) were treated as separate topics and predictions were erratic. The changes are closely related to changes in the sediment budget. However, due to the influence of increasing populations reclaiming salt flats in the delta and expanding towns protecting their riverbanks, it is difficult to separate dam-induced impacts from other societal impacts. In summary, changes in sediment budget and river flows led to dramatic changes, some negative and some positive, in landscapes downstream from the High Aswan Dam.

Other significant impacts, such as the decline of fish stocks in the Mediterranean, occurred soon after the construction of the High Dam but, from 1990, the numbers of certain species has increased (Nixon, 2004). The main reason for the larger numbers is that the load of nutrients moving down the Nile – which is predominantly waste from expanding populations – increased, especially downstream from Cairo. As noted by Nixon (2004), the fish are now plentiful but 'may not have the same taste as the pre-dam fish'. Also, waterlogging and salinisation of the soil were predicted but the extent was underestimated.

4.5.3 Twenty-first-century predictions for large water projects

Mekong River

The Mekong River has long been a major water provider for humans, fish and vegetation throughout southernmost China and former Indo-China (Laos, Cambodia, Vietnam, Myanmar and Thailand). The Mekong is the eighth largest river in the world by area with a basin of 795 000 km^2 and an average annual flow of 3800 m^3 s^{-1} at Luang Prabang. The Cambodian floodplains and the Mekong Delta receive most of the water, with some flooding, and 95% of the sediment load. The upper basin in China and Myanmar contributes 18% of the mean annual discharge – this contribution comes during the dry season and is highly important to the lower Mekong system. Since the 1960s, a number of plans for the development of water resources of the Lower Mekong have been presented (White, 1963; Mekong Committee, 1970), but none of the proposed dams had been constructed until China began its aggressive development of eight hydro projects on the upper river in 1993, commencing with a dam at Manwan.

Upon completion, the eight reservoirs will trap about 50% (about 80 Mt a^{-1}) of the total sediment load of the Mekong River. The impacts of the upper river dams can be grouped under hydrological and geomorphological changes and fisheries impacts. The river downstream from the dams will degrade and could become a rock-cut canyon (Gupta and Liew, 2007). Also, local navigation on the river may be hampered. Because dry-season flows will be increased, the lowest lake elevations in Tonle Sap ('Great Lake') will increase in the future. Large areas of seasonally flooded forest will become permanently flooded, resulting in nutrient changes that will affect the fish. With lower peak flows, the maximum flooded areas will decrease and this will impact on the feeding zone for the fish. Hogan *et al.* (2004) have reported on imperilled species, such as the Mekong giant catfish, and the various action plans to save them.

These impacts, however, may be modified if the Upper Mekong Navigation Improvement Project (blasting away of river rapids) is allowed to continue. The Mekong River Commission has reviewed an additional 12 hydropower stations on the Mekong which would certainly block fish migration. Given the pressures for development, there may be inadequate time to study the cumulative environmental impacts of such developments.

The River Nile water budget and impacts due to proposed water projects on the Nile

Historically, one of the main developments proposed for the River Nile is the Jonglei Canal in Sudan – in fact, most of the canal route has been excavated. The canal project was financed by Egypt in order to gain a larger share of the Nile flows by bypassing the Sudd. However, the decision to proceed with the construction was made without consideration of environmental impacts. Studies had been conducted on prospective changes to the Sudd, but the conclusions were unclear. As reported by Howell *et al.* (1988): 'the project unfortunately started in a highly charged political atmosphere. The sudden announcement in 1974 of its impending implementation caused unfavourable reactions among southern Sudanese…'. This is not the only recent project to be constructed under political circumstances: Egypt recently constructed a pump diversion project to irrigate the low-lying area west of Lake Nasser (Kerisel, 1999). The water allocation to irrigate this area has not been negotiated with upstream countries and environmental questions have been raised. Also, hydro projects are currently under construction on the Tekeze River in Ethiopia and on the Nile at Merowe in Sudan. Both of these projects will modify evaporation losses within the basin – evaporation may be less in Ethiopia than from Lake Nasser and with timing of release flows more water could be delivered to the High Aswan Dam turbines. This matter has yet to be negotiated.

Huanghe (Yellow River), China

Huanghe is unique in the world, its annual total sediment load being one of the highest (1900 Mt a^{-1}) with a mean annual flow of 4000 m^3 s^{-1} (Zhao *et al.*, 1989). Frequent floods have resulted in countless deaths and an early response of dyke building to reduce flood impacts. The construction of dykes, however, is not a permanent solution, sediment deposited between the dykes necessitating a programme of progressive dyke raising. The dyked riverbed at some locations is 10 to 15 m above the surrounding floodplain. From 600 BC to 1949, the river changed its course across the North China plain and delta countless times (Xu, 1993).

The first major dam on the system was the high Sanmenxia Dam, built with the assistance of Russian engineers. The reservoir filled with sediment in about 6 years and low-level outlets were subsequently bored through the base of the dam to pass sediment-laden flow. The impact of the dam on the downstream landscape was not drastic – portions of the river degraded, resulting in less frequent flooding while, farther downstream, the sand bed has continued to aggrade (Zhao *et al.*, 1989). Major changes to the floodplain have, however, been ongoing for centuries with the confinement of the river by flood dykes. The main future impact will be due to the recently completed Xiaolangdi Dam which is located about 200 km downstream from Sanmenxia. This reservoir, being quite large, will trap a reasonable portion of the sediment load and could initiate the lowering of the downstream river bed. Also, the delta zone will be relatively stable resulting in the expansion of Port Tianjin to the north and expanded agriculture. The change to the river landscape due to the two dams may, however, be overwhelmed by recent low flows and by the reduction of upstream sediment loads by extensive construction of hillslope terraces and check dams on the minor tributaries of the Loess Plateau (Hassan *et al.*, 2008). It will be difficult to distinguish the impacts of the dams from those of the upstream basin rehabilitation, but neither of these two interferences will 'tame' Huanghe.

Future changes to Huanghe will be dominated by the large-scale inter-basin water transfer from Changjiang (Yangtze River) to Huanghe and farther north to the Beijing area. After a 50-year study the Chinese Ministry of Water Resources announced approval in principle of the General Plan of the South-to-North Water Transfer Project (Shao *et al.*, 2003). The first component will be the East Route, which involves a siphon-type structure under Huanghe. A similar structure will be part of the Middle Route, but a number of potential problems, such as canal leakage, liquefaction of sand and frost heave have been identified. Also, along the East Route, water pollution problems already exist in detention lakes that are to be used as part of the transfer system. The pollution could increase if the diverted water is polluted and soil salinisation could result due to a rise in detention lake levels if the canal is to be used for navigation.

The consequences for Huanghe may be difficult to forecast. The ongoing changes due to terracing in the upper watershed are not yet fully understood. Presently, the lower river runs dry, or nearly so, for several months of the year, so there could also be positive impacts if the south water is diverted into the river.

4.5.4 Perspective

The number and magnitude of river diversions and dams has increased dramatically during the last century, but the environmental impact has yet to be fully documented and understood. Dams and reservoirs have direct climatic impacts themselves inasmuch as evaporation from reservoirs changes the water budget of the river, while regional atmospheric humidity and precipitation may be affected. In a regional climate subject to larger-scale changes, water projects associated with dams and diversions may become uneconomical, and may visit hardship upon those depending on the water supply as the actual amount of disposable water may fall below the designed expectation. This outcome has already been experienced in the Colorado River, where the original projects were based on the most optimistic estimates of the river's flow. Today, even the more recent projects are falling short of the expected water supply.

Withal, there is no sign that the rapid pace of human-induced river changes is slowing – new impacts to the environment probably will supersede earlier impacts. The landscape response will become increasingly complex, such that environmental rehabilitation projects will become the norm and realistic prediction of future impacts will become impossible. Damming a river, of course, creates the most radical changes of all in river hydrology, morphology and ecology. No primary effects on a river of regional climate change come close to such a radical reorganisation of the fluvial system.

4.6 River restoration in the context of global change

4.6.1 Introduction

River channels and floodplains have been directly altered for purposes such as bank stabilisation, navigation, flood control, relocation for highways and other infrastructure, in-channel mining for sand/gravel and placer deposits, and for hydropower and water supply impoundments (see foregoing sections and Table 4.5). In addition, changes in catchment land use change runoff and sediment load to rivers, which in turn can induce changes in river channels. To mitigate these impacts, a variety of restoration actions can be taken, some of which fit the notion of 'restoration' of ecological attributes, while others might be better viewed as

TABLE 4.5. *Some human impacts on river channels and restoration approaches*

Activity	Impacts	Restoration approach
Navigation	Straightening	Re-meander: allow sinuosity to redevelop by removing riprap and other constraints, or reconstruct channel with meanders
	Deepening (dredging)	Stop dredging, allow channel bed to recover through aggradation
	Stabilising banks (rock riprap)	Remove or don't maintain bank protection
	Training (groynes)	Remove or don't maintain training features
	Removing large wood	Allow large wood to remain and move through the system, less desirable: build structures with wood cabled to banks
	Removing other roughness features (e.g. blasting bedrock)	Cannot restore directly (may be able to compensate by creating other rough features)
	Cutting off side channels	Open side channels to flow through removal of plugs/weirs
	Altered flow regime	Re-regulate flows to achieve more natural flow regime
Flood control	Reservoir reduces flood peaks	Operate reservoir to increase flood peaks to restore system dynamics
	Levees prevent overbank flooding, depriving floodplain of flows	Set back, breach or remove levees to restore lateral connectivity with floodplain
	Levees concentrate flow, scouring and simplifying channel	Introduce set-back levees to reduce shear stress within levees, allowing gravel bars, vegetation and other hydraulically rough features to develop
Channel relocation	Relocated channel typically straighter than original channel	Enlarge river corridor and increase sinuosity
Bank stabilisation	Channel migration prevented, undercut bank habitat lost, cohesive vertical bank habitat lost, new point bars deposition prevented	Remove bank protection, allow channel migration to occur
	Reduced gravel supply to river channel by reduced bank erosion	Remove bank protection to restore gravel supply from bank, or artificially add gravel to channel
In-channel mining	Direct loss of complex channel features	Cease in-channel gravel mining; allow channel form to recover if adequate sediment load is available
	Incision propagates upstream of mine site via headcut migration	Incision may be controlled with grade control structures, but tendency to incise persists and scour downstream of weirs may create barriers to fish migration
	Incision propagates downstream due to sediment starvation, and tends to continue until sediment supply can balance sediment deficit (often occurs through bank erosion induced by bank undercutting)	Increase supply of coarse sediment to downstream reach to reverse incision
Reservoirs	Trap coarse sediment, release sediment-starved water downstream, inducing	
	bed coarsening, loss of salmonid spawning gravels	Pass sediment through (small) reservoirs via low-level outlets or around reservoirs via

TABLE 4.5. *(cont.)*

Activity	Impacts	Restoration approach
		bypass channels, or add coarse sediment artificially to channel below dam
	Incision, lowered alluvial water tables	Increase coarse sediment supply downstream of dam (as above) or install structures such as artificial logjams to raise water levels in the channel
	Reduce flood peaks, inducing vegetation encroachment and channel narrowing	Mechanically remove vegetation and berms established along the channel margins, and introduce more dynamic flow regime including flushing flows designed to scour seedlings before they can establish mature trees

countermeasures to control incision or accelerated bank erosion (Table 4.5). The degradation of ecological qualities of rivers as a result of global change has been noted in the preceding sections of this chapter. Here we consider the role of river restoration in the context of global change.

4.6.2 Extent and scope of river restoration

River restoration is increasingly popular in North America and Europe. In the United States, over 37 000 individual projects with combined costs exceeding US$15 billion have been completed since 1990 (Bernhardt *et al.*, 2005), not including large restoration programmes such as the Kissimmee River (Koebel, 1995) or the Colorado River in the Grand Canyon (National Research Council 1999). Many of these projects can be called 'restoration' in name only, having as principal goals bank stabilisation or imposition of a stable channel of socially preferred form. Remarkably few of these projects have been subject to objective post-project appraisal, severely limiting the potential for the field to mature and for practice to improve, and limiting our ability to assess the overall contribution of restoration projects to river ecosystem vitality (Bernhardt *et al.*, 2005).

River 'restoration' in North America has largely become an industry, with a set of standard approaches broadly applied, in many cases without profound understanding of the history and constraints of a particular river (Wohl *et al.*, 2005). In Europe, fewer projects have been completed to date, but under the Water Framework Directive (adopted by the European Union Parliament in 2000), EU member states are required to achieve 'good ecological status' in rivers by 2015, prompting strong interest in restoration programmes. Moreover, in some cases, water managers have taken approaches based on best available science that can be viewed as more innovative than the standard approach of their North American counterparts.

To many land managers in North America, 'stream restoration' implies reconstructing channels into the culturally desired form of single-thread meandering channels. In regions where this approach is currently popular, hundreds of such projects have been completed since 1990. While these projects are usually justified by arguments about width–depth ratios and 'stream types' specified by a particular channel classification system, they are usually designed by practitioners without strong academic backgrounds in fluvial geomorphology, who instead apply the classification scheme, which has been criticised by the academic and research community (Simon *et al.*, 2006). Moreover, the fact that this approach consistently specifies a single-thread meandering channel (no matter what the context) suggests that there is something deeper at work, that these projects are responding to an ingrained cultural preference for such an ideal channel form (Kondolf, 2006).

In Germany, after early attempts to design and build channel forms and habitats, managers now increasingly attempt to initiate channel dynamics as a more sustainable and less expensive restoration strategy. For example, large wood has been introduced into incised channels in Hesse, Germany, and channel widening to restore multi-thread channel systems and their habitats (and increase flood capacity) has become a popular approach in Austria (Piégay *et al.*, 2008). The Drôme River, France, has incised because of reduced sediment supply from the catchment due to rural depopulation, reduced land pressure and afforestation. To counteract negative consequences of sediment starvation and incision, managers are attempting to reactivate sediment sources to restore channel dynamics (Kondolf *et al.*, 2002).

It is useful to consider the full range of human impacts that have degraded river ecosystems, and to compare the trajectories of degradation with trajectories of restoration. We find that, most commonly, the restoration trajectories do not parallel the degradation trajectories because restoration projects tend to tackle only the changes that are easier (logistically, financially and politically) to reverse (Kondolf *et al.*, 2006). For example, Clear Creek (a tributary to the Sacramento River near Redding, California) had two dams blocking salmon migration: Saeltzer was a small dam (*c.* 1912) to divert water for irrigation, while Whiskeytown is a large dam (*c.* 1963) storing water transferred from the Trinity River via penstocks, en route to the Sacramento River as part of the massive Central Valley Project. Both dams blocked migration of salmon, but only Whiskeytown affected high flows that shape the channel. By trapping gravel it created sediment-starved conditions downstream. Restoration of Clear Creek has involved removal of the smaller downstream dam, reconstruction of a reach disturbed by gravel mining to improve fish passage, and injection of gravel into the channel to increase spawning habitat for salmon, relatively easy activities to undertake in the current environment. To date, operation of the larger dam upstream has not been substantially affected, as to do so would affect politically powerful interests and have larger costs (Kondolf *et al.*, 2006). Thus, the trajectory of restoration in Clear Creek does not parallel the trajectory of its degradation, as only some of the changes have been reversed by restoration activities, a common pattern in river restoration projects.

4.6.3 Mitigating the effects of climate change

In North America, the greatest concentration of restoration projects is found in the western Cordillera, driven by programmes to create habitat for anadromous salmon and trout, many runs of which are listed as threatened or endangered. In the southern end of the range of these species, summer water temperatures commonly limit distribution of the fish. Concern about the potential effects of projected temperature increases on habitat suitability for salmon has led to interest among government and non-government agencies (such as the Nature Conservancy) to explore ways to mitigate temperature increases by buffering temperatures at the reach scale. Increased shading and increased hyporheic exchange through increased channel complexity, creating favourable hydraulic gradients to drive surface–groundwater exchange (Poole *et al.*, 2002), are being considered as strategies to maintain some rivers within the temperature tolerance range of salmonids.

4.6.4 River restoration as a developed world activity

Widespread river restoration is a relatively recent phenomenon, mostly restricted to the developed world, reflecting increased demand for environmental quality in affluent, educated societies, and improvements in water quality brought on by environmental legislation such as the Clean Water Act in the United States. Only after water quality improves does it make sense to encourage greater human contact with rivers or attempt to re-establish native (or otherwise socially desirable) ecological communities in rivers. Ironically, much of the improvement in water quality and redevelopment of formerly industrial riverfronts in developed world societies has been made possible by displacement of heavy industries to developing countries where environmental controls are less stringent or unenforced. For example, recent restoration of the riverfront in Pittsburgh, Pennsylvania, was made possible by the closing of steel mills and related industries, most of which moved abroad (Otto *et al.*, 2004). Thus the much-heralded restoration of formerly industrial rivers in North America and Europe is probably at least matched by increased impacts in rivers elsewhere. Rivers in the developing world are affected not only by industry displaced from developed countries, but also by the increasing demands of expanding populations, with concomitant agricultural and industrial development. These trends imply greater loads worldwide of sewage and other pollutants discharged into rivers, as well as extensive land use change and consequent impacts on flow, sediment load and water quality. When we consider the rate of population increase in many developing world cities that lack sewage treatment, and calculate how many sewage treatment plants would be needed to meet current and future needs, it becomes clear that such developed world approaches cannot realistically be employed in this context, but rather that alternative approaches to sanitation need to be developed and employed. Certainly the notion of river restoration as practised in North America today would be irrelevant in this context of heavy practical demands on waterways in the developing world.

4.6.5 Perspective: relative scales of restoration and degradation

Even in a purely developed world context, when we consider the rate of river degradation, river restoration appears very limited in comparison. For example, the CALFED

Bay–Delta Program, encompassing the estuary of the Sacramento River, San Francisco Bay and the Sacramento–San Joaquin Delta (the San Francisco Estuary) and its watershed in northern California, is one of the largest ongoing restoration programmes in the United States, with more than US$500 million invested in restoration projects from 1997 to 2004 (CALFED, 2005). Yet when we look at the results of these and other restoration efforts to date in the context of habitat losses and fish population declines since European settlement in 1850, it is clear that even a restoration effort on this scale will not reverse large-scale historical changes. Many component projects have involved restoration of tidal marsh in the estuary. The Sonoma Baylands, one of the most successful such projects to date, restored 124 ha in 1996. This area is equivalent to less than 0.2% of the estimated 68 000 ha of tidal marsh lost in the San Francisco Estuary since 1850 (Bay Institute, 1998). Collectively, tidal wetland restoration projects in the San Francisco Estuary have restored about 650 ha – about 1% of the tidal marsh habitat lost since 1850 (Bay Institute, 2003). When we consider that there is no guarantee that restoration projects will work as intended, preservation of intact river ecosystems (with still natural processes of flow, sediment, and floodplain connectivity) takes on greater priority.

Most restoration programmes are directed at smaller rivers, within which reasonably affordable engineering effort might have detectably positive results. In the United States, in recent years, this interest has extended to considering the removal of dams of small or moderate scale, most of them constructed in the late nineteenth or early twentieth centuries and now of questionable integrity in any case. Whilst practical restoration efforts might be extended to include most rivers of European scale, this prominently leaves aside many of the largest and most ecologically diverse rivers of the world, on which dam building and navigation improvements have been conducted at too large a scale to consider reversal and, in any case, the 'improvements' are too closely tied to the economy that they serve to consider ending them.

When we take a global view of the improvements in river ecosystems from restoration projects to date, accounting both for improvements in developed nations and continued degradation in the developing world, it is difficult to avoid concluding that river conditions are deteriorating overall in response primarily to increased human pressures on waterways in the developing world, but also because the benefits of the restoration projects undertaken in the developed world are either small relative to historical degradation, or simply cannot be demonstrated due to lack of adequate documentation and monitoring. This is not to say that we should give up on the enterprise of river restoration, but we should not fall into the trap of believing that the world is getting better because we see apparent improvements in some rivers of the developed world. We must ask how we can rigorously assess the actual ecological effectiveness of these projects, how these demonstrated improvements compare with the scale of historical degradation, and how our progress in the developed world scales in light of continued degradation in the developing world. Changing climate adds another layer to this story but, compared with the effects of increased population and increased pressures on river systems, its role is relatively minor.

4.7 Conclusions

There is no question that climate change is affecting the world's rivers. By changing the total volume and seasonal distribution of flow – the consequence of changing patterns of precipitation on to and evapotranspiration from the land surface – riverine water supply is affected in ways that will have an important impact on regional and global economies. The impact in many places will be unfavourable because it will entail a reduction in the availability of already scarce resources, but it may be unfavourable in most places simply because human societies are ill-adapted to deal with change of any kind.

The consequences of climate change for the morphology of the rivers themselves and of the riverine ecosystems that they support will, however, be far, far less dramatic than are the consequences of long-established and ongoing activities of humans. This chapter has sought to establish this fact by describing consequences for rivers of human activities on the land surface which affect the formation of runoff, the quality of water and the supply of sediment to rivers, and the consequences of direct human manipulation of river channels for diverse reasons. We have closed the chapter with some brief remarks on the limited efficacy of human efforts at 'restoration' or, at least, habilitation of rivers.

A great deal remains to be learned about how humans might properly manage rivers (and, on a crowded planet, active management will remain necessary). One thing is clear. Humans interfere with rivers at all stages from runoff formation on the land surface to the discharge of water through river estuaries and deltas to the world ocean. A successful approach to river management must be an approach that integrates all of the processes and effects through the drainage system. This was a prominent theme in the development-oriented society of the mid twentieth century. It needs to return with a much stronger emphasis upon the conservation and appropriate use of the resources that the world's rivers represent (see Newson, 1997). It will require us to learn far more about the hydrology, geomorphology and ecology of rivers than we know at present. The factors contributing to change in local, regional and global environments – primarily directed by human activities – come to a critical focus on water and waterways.

References

Anderson, M. G. and Burt, T. P. (1978). The role of topography in controlling throughflow generation. *Earth Surface Processes*, **29**, 331–334.

Bates, B. *et al.*, eds. (2008). *Climate Change and Water*, Technical Paper of the Intergovernmental Panel on Climate Change. Available at www.ipcc.ch/ipccreports/tp-climate-change-water.htm

Bay Institute (1998). *From the Sierra to the Sea: The Ecological History of the San Francisco Bay–Delta Watershed*. San Rafael: Bay Institute of San Francisco.

Bay Institute (2003). *The Bay Institute Ecological Scorecard: San Francisco Bay Index*. San Rafael: Bay Institute of San Francisco.

Benke, A. C. (1990). A perspective on America's vanishing streams. *Journal of the North American Benthological Society*, **9**, 77–88.

Berner, E. K. and Berner, R. A. (1996). *Global Environment: Water, Air and Geochemical Cycles*. Upper Saddle River: Prentice-Hall.

Bernhardt, E. S. *et al.* (2005). Synthesizing U.S. river restoration efforts. *Science*, **308**, 636–637.

Betson, R. P. (1964). What is watershed runoff? *Journal of Geophysical Research*, **69**, 1541–1542.

Beven, K. J. *et al.* (2004). Review of impacts of rural land use and management on flood generation: Impact Study Report, Appendix A. In *Review of UK Data Sources Relating to the Impacts of Rural Land Use and Management on Flood Generation*, R&D Technical Report No. FD2114/TR. London: Defra.

Biedenharn, D. S., Thorne, C. R. and Watson, C. C. (2000). Recent morphological evolution of the lower Mississippi River. *Geomorphology*, **34**, 227–249.

Boardman, J. *et al.* (1995). Soil erosion and flooding as a result of a summer thunderstorm in Oxfordshire and Berkshire, May 1993. *Applied Geography*, **16**, 21–34.

Bonell, M. and Bruijnzeel, L. A., eds. (2005). *Forests, Water and People in the Humid Tropics*. Cambridge: Cambridge University Press.

Bosch, J. M. and Hewlett, J. D. (1982). A review of catchment experiments to determine the effect of vegetation changes on water yield and evapotranspiration. *Journal of Hydrology*, **55**, 3–23.

Brandt, S. A. (2000). Classification of geomorphological effects downstream of dams. *Catena*, **40**, 375–401.

Bray, D. I. (1973). Regime relations for Alberta gravel-bed rivers. In *Proceedings of the 7th Canadian Hydrology Symposium*. Ottawa: National Research Council of Canada, pp. 440–452.

Bruijnzeel, L. A. S. (2005). Land use and land cover effects on runoff processes: forest harvesting and road construction. In M. G. Anderson, ed., *Encyclopedia of Hydrological Sciences*, vol. 5. Chichester: John Wiley, pp. 1813–1829.

Burt, T. P. (2008). Valley-side slopes and drainage basins: 1. Hydrology and erosion. In T. P. Burt *et al.*, eds., *The History of the Study of Landforms*, vol. 4, *Quaternary and Recent Processes and Forms (1890–1965) and the Mid-Century Revolution*. London: Institute of Geological Sciences, pp. 325–352.

Burt, T. P. and Butcher, D. P. (1985). The role of topography in controlling soil moisture distributions. *Journal of Soil Science*, **36**, 469–486.

Burt, T. P. and Slattery, M. C. (1996). Time-dependent changes in soil properties and surface runoff generation. In M. G. Anderson and S. M. Brooks, eds., *Advances in Hillslope Processes*. Chichester: John Wiley, pp. 79–95.

Burt, T. P. *et al.* (1983). Hydrological processes in the Slapton Wood catchment. *Field Studies*, **5**, 731–752.

Butzer, K. W. (1972). *Early Hydraulic Civilization in Egypt*. Chicago: University of Chicago Press.

Calder, I. R. (1990). *Evaporation in the Uplands*. Chichester: John Wiley.

CALFED Bay–Delta Program (2005). *Ecosystem Restoration Multi-Year Program Plan (Years 6–9) and Annotated Budget (Year 5)*, 16 June, 2005. Available at www.calwater.ca.gov/

Carraquiry, J. D. and Sanchez, A. (1999). Sedimentation in the Colorado River Delta. *Marine Geology*, **158**, 125–145.

Church, M. (1995). Geomorphic response to river flow regulation: case studies and time-scales. *Regulated Rivers: Research and Management*, **11**, 3–22.

Church, M. (2002). Fluvial sediment transfer in cold regions. In K. Hewitt *et al.*, eds., *Landscapes in Transition*. Rotterdam: Kluwer, pp. 93–117.

Church, M. and Ryder, J. (2001). Watershed processes in the Southern Interior: background to land management. In D. A. A. Toews and S. Chatwin, eds., *Watershed Assessment in the Southern Interior of British Columbia*. Nelson: British Columbia Ministry of Forests, pp. 1–16.

Church, M. *et al.* (1999). Fluvial clastic sediment yield in Canada: scaled analysis. *Canadian Journal of Earth Sciences*, **36**, 1267–1280.

Cohen, M. (2002). Managing across boundaries: Colorado River Delta. In P. H. Gleick, ed., *The World's Water, 2002–2003: The Biennial Report on Freshwater Resources*. Washington, DC: Island Press, pp. 133–147.

Conway, V. M and Millar, A. (1960). The hydrology of some small peat-covered catchments in the Northern Pennines. *Journal of the Institute of Water Engineers*, **14**, 415–424.

David, J. S. and Ledger, D. C. (1988). Runoff generation in a plough drained peat bog in southern Scotland. *Journal of Hydrology*, **99**, 187–199.

Douglass, J. E and Hoover, M. D. (1988). History of Coweeta. In W. T. Swank and D. A. Crossley, eds., *Forest Hydrology and Ecology at Coweeta*. New York: Springer-Verlag, pp. 17–31.

Dunne, T. (1978). Field studies of hillslope flow processes. In M. J. Kirkby, ed., *Hillslope Hydrology*. Chichester: John Wiley, pp. 227–293.

Dynesius, M. and Nilsson, C. (1994). Fragmentation and flow regulation of river systems in the northern third of the world. *Science*, **266**, 753–755.

Endreny, T. A. (2005). Land use and land cover effects on runoff processes: urban and suburban development. In M. G. Anderson, ed., *Encyclopedia of Hydrological Sciences*, vol. 5. Chichester: John Wiley, pp. 1775–1803.

Food and Agricultural Organization (2005). *Global Forest Resources Assessment 2005*, FAO Forestry Paper No. 147. Available at www.fao.org/DOCREP/008/a0400e/a0400e00.htm

Grimalt, J. O., Ferrer, M. and Macpherson, E. (1999). The mine tailing accident in Aznalcóllar. *Science of the Total Environment*, **242**, 3–11.

Gupta, A. and Liew, S. C. (2007). The Mekong from satellite imagery: a quick look at a large river. *Geomorphology*, **85**, 259–274.

Hassan, M. *et al.* (2008). Spatial and temporal variation of sediment yield in the landscape: example of Huanghe (Yellow River). *Geophysical Research Letters*, **35**, L06401, doi:10.1029/GL033428.

Heathwaite, A. L., Burt, T. P. and Trudgill, S. T. (1990). Land-use controls on sediment production in a lowland catchment, southwest England. In J. Boardman, I. D. L. Foster and J. Dearing, eds., *Soil Erosion on Agricultural Land*. Chichester: John Wiley, pp. 69–86.

Hewlett, J. D. (1961). Watershed management. In *Report for 1961 Southeastern Forest Experiment Station*. Asheville: US Forest Service, pp. 62–66.

Hogan, Z. S. *et al.* (2004). The imperilled giants of the Mekong. *American Scientist*, **92**, 228–237.

Holden, J. *et al.* (2006). Impact of land drainage on peatland hydrology. *Journal of Environmental Quality*, **35**, 1764–1778.

Horton, R. E. (1933). The role of infiltration in the hydrological cycle. *Transactions of the American Geophysical Union*, **14**, 446–460.

Horton, R. E. (1945). Erosional development of streams and their drainage basins: hydrophysical approach to quantitative morphology. *Bulletin of the Geological Society of America*, **56**, 275–370.

Howell, P., Lock, M. and Cobb, S., eds. (1988). *The Jonglei Canal: Impact and Opportunity*. Cambridge: Cambridge University Press.

Huntington, T. G. (2006). Evidence for intensification of the global water cycle: review and synthesis. *Journal of Hydrology*, **319**, 83–95.

Hursh, C. R. and Brater, E. F. (1941). Separating storm hydrographs into surface- and subsurface-flow. *EOS, Transactions of the American Geophysical Union*, **22**, 863–871.

IPCC (2001). *Climate Change 2001: Impacts, Adaptation and Vulnerability. Contribution of Working Group II to the Third Assessment Report of the Intergovernmental Panel on Climate Change*. McCarthy, J. J. *et al.*, eds. Cambridge: Cambridge University Press.

Jarvis, N. J. (2007). A review of non-equilibrium water flow and solute transport in soil macropores: principles, controlling factors and consequences for water quality. *European Journal of Soil Science*, **58**, 523–546.

Junk, W. J., Bayley, P. B. and Sparks, R. E. (1989). The flood pulse concept in river-floodplain systems. *Canadian Special Publication in Fisheries and Aquatic Sciences*, **106**, 100–127.

Kerisel, J. (1999). *The Nile and Its Masters*. Rotterdam: Balkema.

Knox, J. C. (1989). Long- and short-term episodic storage and removal of sediment in watersheds of southwestern Wisconsin and northwestern Illinois. In R. F. Hadley and E. D. Ongley, eds., *Sediment and the Environment*, Publication No. 184. Wallingford: IAHS Press, pp. 157–164.

Koebel, J. W. (1995). A historical perspective on the Kissimmee River restoration project. *Restoration Ecology*, **3**, 149–159.

Kondolf, G. M. (1998). Environmental effects of aggregate extraction from river channels and floodplains. In P. T. Bobrowsky, ed., *Aggregate Resources: A Global Perspective*. Rotterdam: Balkema, pp. 113–129.

Kondolf, G. M. (2006) River restoration and meanders. *Ecology and Society*, **11**, www.ecologyandsociety.org/vol11/iss2/art42/

Kondolf, G. M., Piégay, H. and Landon, N. (2002). Channel response to increased and decreased bedload supply from land-use change: contrasts between two catchments. *Geomorphology*, **45**, 35–51.

Kondolf, G. M. *et al.* (2006). Process-based ecological river restoration: visualising three-dimensional connectivity and dynamic vectors to recover lost linkages. *Ecology and Society*, **11**, www.ecologyandsociety.org/vol11/iss2/art5/

Lagasse, P. F., Winkley, B. R. and Simons, D. E. (1980). Impact of gravel mining on river system stability. *Journal of Waterway, Port, Coastal, and Ocean Engineering*, **106**, 389–403.

Lane, E. W. (1957). A study of the shape of channels formed by natural streams flowing in erodible material. In *Missouri River Division Sediment Series No. 9*. Omaha: US Army Corps of Engineers, Missouri River Engineering Division.

Lavin, M. F., Godinez, V. M. and Alvarez, L. G. (1998). Inverse-estuarine features of the Upper Gulf of California. *Estuarine, Coastal and Shelf Science*, **47**, 769–795.

Leopold, L. B. and Wolman, M. G. (1957). River channel patterns: braided, meandering and straight. *US Geological Survey Professional Paper* **282**-B, 39–85.

Lewin, J. and Macklin, M. G. (1987). Metal mining and floodplain sedimentation in Britain. In V. Gardiner, ed., *International Geomorphology 1986*. Chichester: John Wiley, pp. 1009–1027.

Ligon, F. K., Dietrich, W. E. and Trush, W. J. (1995). Downstream ecological effects of dams. *BioScience*, **45**, 183–192.

Lozano, H. (2006). Historical ecosystem modelling of the Upper Gulf of California (Mexico). Unpublished Ph.D. thesis, University of British Columbia, Vancouver.

MacDonald, Sir M. and Partners (Ltd.) (1984). *Left Bank Outfall Drain*, Final Report to Water and Power Development Authority. Islamabad: Islamic Republic of Pakistan.

Macklin, M. G. *et al.* (2003). The long term fate and environmental significance of contaminant metals released by the January and March, 2000 mining tailings dam failures in Maramureş Country, upper Tisa basin, Romania. *Applied Geochemistry*, **18**, 241–257.

Martin, T. E. (2007). How lessons from major catastrophes can guide dam safety practices. *Hydro Review*, **26**(3), 24–31.

Meade, R. H. (1982). Sources, sinks and storage of river sediment in the Atlantic drainage of the United States. *Journal of Geology*, **90**, 235–252.

Meko, D., Stockton, C. W. and Boggess, W. R. (1995). The tree-ring record of severe sustained drought. *Water Resources Bulletin*, **31**, 789–801.

Mekong Committee (1970). *Report on Indicative Basin Plan*, Report E/C11/WRD/MKG/L.340. Bangkok: Mekong Committee.

Michel, A. A. (1967). *The Indus Rivers*. New Haven: Yale University Press.

Milly, P. C. D., Dunne, K. A. and Vecchia, A. V. (2008). Global pattern of trends in streamflow and water availability in a changing climate. *Nature*, **438**, 347–350.

Mobarek, I. E. (1972). The Nile Delta coastal protection project. In *Proceedings of the 13th ASCE Coastal Engineering Conference*, Vancouver, Canada, pp. 1409–1426.

Montgomery, D. R. (2007). *Dirt*. Berkeley: University of California Press.

Mudelsee, M. *et al.* (2003). No upward trends in the occurrence of extreme floods in central Europe. *Nature*, **425**, 166–169.

National Research Council (1987). *River and Dam Management*. Washington, DC: National Academies Press.

National Research Council (1999). *Downstream: Adaptive Management of Glen Canyon Dam and the Colorado River Ecosystem*. Washington, DC: National Academies Press.

National Research Council (2008). *Hydrologic Effects of a Changing Forest Landscape*. Washington, DC: National Academies Press.

Neill, C. R. (1973). Hydraulic geometry of sand rivers in Alberta. In *Proceedings of the 7th Canadian Hydrology Symposium*. Ottawa: National Research Council of Canada, pp. 453–461.

Newson, M. (1997). *Land, Water and Development*. London: Routledge.

Nilsson, C. *et al.* (2005). Fragmentation and flow regulation of the world's largest river systems. *Science*, **308**, 405–408.

Nixon, S. W. (2004). The artificial Nile. *American Scientist*, **92**, 158–165.

Otto, B., McCormick, K. and Leccese, M. (2004). *Ecological Riverfront Design: Restoring Rivers, Connecting Communities*, Planning Advisory Service Report No. 518–519. Chicago: American Planning Association.

Owen-Joyce, S. J. and Raymond, L. H. (1996). *An Accounting System for Water and Consumptive Use along the Colorado River, Hoover Dam to Mexico*, US Geological Survey, Water Supply Paper No. 2407. Washington, DC: US Government Printing Office.

Petts, G. E. (1984). *Impounded Rivers: Perspectives for Ecological Management*. Chichester: J. Wiley.

Petts, G. E., Moller, H. and Roux, A. L., eds. (1989). *Historical Change of Large Alluvial Rrivers: Western Europe*. Chichester: John Wiley.

Piégay, H. *et al.* (2008). Integrative river science and rehabilitation: European experiences. In G. Brierley and K. Fryirs, eds., *River Futures: An Integrative Scientific Approach to River Repair*. Washington, DC: Island Press, pp. 201–219.

Pitt, J. *et al.* (2000). Two countries, one river: managing for nature in the Colorado River Delta. *Natural Resources Journal*, **40**, 819–864.

Poole, G. C. *et al.* (2002). Three-dimensional mapping of geomorphic controls on flood-plain hydrology and connectivity from aerial photos. *Geomorphology*, **48**, 329–347.

Rawls, W. J. *et al.* (1993). Infiltration and soil water movement. In D. R. Maidment, ed., *Handbook of Hydrology*. New York: McGraw-Hill, pp. 5.1–5.51.

RNPD (River Nile Protection and Development Project) (1991). *River Nile Sedimentation Modelling Study*, Working Paper No. 200–11. Cairo: Ministry of Public Works and Water Resources.

Robinson, M. (1990). *Impact of Improved Land Drainage on River Flows*, Report No. 113. Wallingford: Institute of Hydrology.

Robinson, M. *et al.* (1998). *From Moorland to Forest: the Coalburn Catchment Experiment*, Report No. 133. Wallingford: Institute of Hydrology.

Rubin, D. M. *et al.* (2002). Recent sediment studies refute Glen Canyon Dam hypothesis. *EOS, Transactions of the American Geophysical Union*, **83**, 273 and 277–278.

Schumm, S. A. (1977). *The Fluvial System*. New York: Wiley-Interscience.

Shao, X., Wang, H. and Wang, Z. (2003). Interbasin transfer projects and their implications: a China case study. *International Journal of River Basin Management*, **1**, 5–14.

Sherman, L. K. (1932). Streamflow from rainfall by the unit hydrograph method. *Engineering News Record*, **108**, 501–505.

Simaika, Y. M. (1970). Degradation of the Nile bed due to the interception of silt in the High Aswan reservoir. *Transactions of the 10th International Congress on Large Dams*, Montreal, **3**(Q.38), R.60, 1161–1181.

Simon, A. *et al.* (2006). Do the Rosgen classification and associated 'natural channel design' methods integrate and quantify fluvial processes and channel response? *Journal of the American Water Resources Association*, **43**, 1117–1131.

Slattery, M. C. and Burt, T. P. (1996). On the complexity of sediment delivery in fluvial systems: results from a small agricultural catchment, North Oxfordshire, UK. In M. G. Anderson and S. M. Brooks, eds., *Advances in Hillslope Processes*. Chichester: John Wiley, pp. 635–656.

Sparks, R. E. (1995). Need for ecosystem management of large rivers and their floodplains. *BioScience*, **45**, 168–182.

Stanley, D. J. (1988). Subsidence in the northeastern Nile delta: rapid rates, possible causes, and consequences. *Science*, **240**, 497–500.

Sterling, C. (1972). Super dams: the perils of progress. *Atlantic Monthly*, **229**(6), 35–41.

Stone, R. and Jia, H. (2006). Going against the flow. *Science*, **313**, 1034–1037.

Sundborg, Å. (1967). Some aspects on fluvial sediments and fluvial morphology: 1. General views and graphic methods. *Geografiska Annaler*, **49**A, 333–343.

Syvitski, J. P. M. *et al.* (2005). Impact of humans on the flux of terrestrial sediment to the global coastal ocean. *Science*, **308**, 376–380.

Tarar, R. N. (1982). Water management in the Indus River System. In *Rivers '76*. New York: American Society of Civil Engineers, pp. 813–831.

Trimble, S. W. (1983). A sediment budget for Coon Creek basin in the Driftless Area, Wisconsin, 1853–1977. *American Journal of Science*, **283**, 454–474.

Trimble, S. W. (1997). Contribution of stream channel erosion to sediment yield from an urbanizing watershed. *Science*, **278**, 1442–1444.

Trimble, S. W. (1999). Decreased rates of alluvial sediment storage in the Coon Creek basin, Wisconsin, 1975–93. *Science*, **285**, 1244–1246.

US Bureau of Reclamation (1946). *The Colorado River*. Washington, DC: US Department of the Interior.

van Andel, T. H. (1964). Recent marine sediments of the Gulf of California. In T. H. van Andel and G. G. Shor, eds., *Marine Geology of the Gulf of California: A Symposium*, Memoir No. 3. Tulsa: American Association of Petroleum Geologists, pp. 216–310.

Walter, R. C. and Merritts, D. J. (2008). Natural streams and the legacy of water-powered mills. *Science*, **319**, 299–304, doi:10.1126/science.1151716.

Ward, J. V. *et al.* (2002). Riverine landscape diversity. *Freshwater Biology*, **47**, 517–539.

Waterbury, J. (1979). *Hydropolitics of the Nile Valley*. Syracuse: Syracuse University Press.

Webb, R. H. *et al.*, eds. (1999). *The Controlled Flood in the Grand Canyon*. Washington, DC: American Geophysical Union.

White, G. F. (1963). The Mekong River Plan. *Scientific American*, **208**/4, 49–59.

Wiener, J. G., Anderson, R. V. and McConville, D. R., eds. (1984). *Contaminants in the Upper Mississippi River*. London: Butterworth.

Wilkinson, B. H. and McElroy, B. J. (2007) The impact of humans on continental erosion and sedimentation. *Geological Society of America Bulletin*, **119**, 140–156, doi:10.1130/B25899.1.

Williams, G. P. and Wolman, M. G. (1984). *Downstream Effects of Dams on Alluvial Rivers*, Professional Paper No. 1286. Washington, DC: US Geological Survey.

Winkley, B. R. (1994). Response of the Lower Mississippi River to flood control and navigation improvements. In S. A. Schumm and B. R. Winkley, eds., *The Variability of Large Alluvial Rivers*. New York: American Society of Civil Engineers, pp. 45–74.

Wohl, E. *et al.* (2005). River restoration. *Water Resources Research*, **41**, W10301, doi:10.1029/2005WR003985.

Wolman, M. G. (1967). A cycle of sedimentation and erosion in urban river channels. *Geografiska Annaler*, **49**A, 385–395.

World Bank (2004). *Bank Management Response to Inspection Panel Review of the Pakistan National Drainage Project*. Washington, DC: World Bank.

World Bank (2006). *Nile Basin Initiative*. Addis Ababa: World Bank.

Xu, J. (1993). A study of long term environmental effects of river regulation on the Yellow River of China in historical perspective. *Geografiska Annaler*, **75**A, 61–72.

Xu, K. *et al.* (2006). Yangtze sediment decline partly from Three Gorges Dam. *Transactions of the American Geophysical Union*, **87**, 185 and 190.

Yang, S. L., Zhang, J. and Xu, X. J. (2007). Influence of the Three Gorges Dam on downstream delivery of sediment and its environmental implications, Yangtze River. *Geophysical Research Letters*, **34**, L10401, doi:10.1029/2007GL029472.

Zhao, Y. *et al.* (1989). Sedimentation in the lower reaches of the Yellow River and its basic laws. In L. M. Brush *et al.*, eds., *Taming the Yellow River*. Dordrecht: Kluwer, pp. 477–516.

5 Estuaries, coastal marshes, tidal flats and coastal dunes

By Denise J. Reed, Robin Davidson-Arnott and Gerardo M. E. Perillo

5.1 Introduction

For millennia people have valued coastal environments for their rich soils, harvestable food resources and access to the oceans. From ancient times to the present, cities and ports have flourished at the coast and this value continues into the twenty-first century. With globalisation and international trade becoming central to many world economies, coastal populations have continued to grow. In 2003, in the USA approximately 153 million people (53% of the population) lived in coastal counties, an increase of 33 million people since 1980 (Crossett *et al.*, 2004). By the year 2008, coastal population in the USA is expected to increase by approximately 7 million. Eight of the world's top ten largest cities are located at the coast. According to the UN *Atlas of the Coast* (www.oceansatlas.org/), 44% of the world's population (more people than inhabited the entire globe in 1950) live within 150 km of the coast and in 2001 over half the world's population lived within 200 km of a coastline. Coastal cities also have higher rates of growth than many other areas. Clearly the massive population now existing along the world coast and the rapid growth it is experiencing induce a major stress on the local and regional geomorphology as well as on the local resource base.

The importance of coastal areas is intimately linked to their geomorphic character and setting. Many early settlements (e.g. the city of Troy mentioned in the Iliad) and later industrial cities (e.g. Shanghai) are located at river mouths where flat land, ample fresh water and transportation linkages to both inland and overseas have facilitated urban growth. Port cities developed on naturally deepwater but somewhat sheltered inlets and estuaries, e.g. Singapore, Seattle and Rotterdam, take advantage of both their coastal setting and their linkage to other continents and oceans. At a finer scale, biogeomorphic aspects of coastal systems (i.e. those components where plants and animals play a role in geomorphic processes) including shallow estuaries, backbarrier lagoons, mangrove swamps and salt marshes provide a natural habitat for fisheries. The essential role of geomorphology in providing the physical template and ongoing processes which support these activities leads to the concept of 'geomorphic services'. The concept of ecosystem services is now widely accepted in ecologically based assessments of how landscapes support human needs (Millennium Ecosystem Assessment, 2005). However, many ecosystem services, e.g. protection from waves, are highly dependent on the geomorphic character of the system. As will be demonstrated here, most societally important aspects of coastal systems are founded on geomorphological processes and structures, and are here termed 'geomorphic services'.

In the twenty-first century coastal systems are subject to the pressures of centuries of population growth and resource exploitation. Increased population density at the coasts often brings pollution and habitat degradation – decreasing the value of many of the resources that initially attract the coastal development. While many geomorphic services are resistant to human activities, e.g. deepwater fjords continue to provide access for shipping, others are threatened by overexploitation and natural system degradation. Mangrove forests that provide essential nursery habitats for fishery species are harvested for firewood and development of aquaculture ponds, producing high-value shrimp, and polluting adjacent estuarine waters with their discharges. Dynesius and Nilsson (1994) note that flow in almost every river in the USA has been altered in some way. Rivers that enter the coastal ocean are dammed for power or irrigation and leveed to prevent flooding of coastal communities (e.g. the Mississippi; see additional discussion in Chapter 4). In these ways, the nature of estuarine and coastal waters, sediment delivery, and natural resources has been fundamentally changed.

Geomorphology and Global Environmental Change, eds. Olav Slaymaker, Thomas Spencer and Christine Embleton-Hamann. Published by Cambridge University Press.

In addition to these human pressures, which once recognised could at least theoretically be mitigated, coastal systems in the twenty-first century face the threats associated with climate change. There is debate over trends in the historical frequency and intensity of tropical cyclones, as the global pattern is complicated by regional variability and decadal scale changes, but some assessments show that at least in the North Atlantic there appears to have been a significant increase in hurricane frequency since 1995 (Webster *et al.*, 2005). This analysis also indicates an increase in the number and proportion of strongest category 4 and 5 hurricanes in the late twentieth century. While many debate the causes of these changes, Timmerman *et al.* (1999) suggest that increasing greenhouse gas concentrations may change the tropical Pacific to a state similar to present-day El Niño conditions. Such conditions would result in the USA experiencing fewer Atlantic hurricanes but more extra-tropical storms on the west coast. More certain perhaps is the trend in sea level where debate centres on the magnitude and rate of the rise around most of the world, the exception being those areas still experiencing falling sea levels due to isostatic rebound. Many coastal island states are clearly vulnerable. Carter and Woodroffe (1994) note that the scares about dramatic sea level rise of the late 1980s dramatically increased awareness and concern about the future of the coast. While the 'lurid and misleading maps' used by many to indicate areas to be flooded by rising seas in the future have been replaced by more considered discussion of the response of coastal dynamics to rising seas (e.g. IPCC, 2007) there is still considerable debate about the amount of sea level rise shorelines will experience in the twenty-first century (see Chapter 1). What is not in doubt is that changes will occur in coastal areas in the future, whether from storm impacts, rising sea levels or changes in the magnitude and frequency of freshwater inflows, and their geomorphological consequences will shape the lives of millions of coastal inhabitants and some of the world's largest cities.

Coastal systems are end members of the sediment cascade (see Chapter 2) which begins with weathering of rocks and transport of sediments down gradient to the ocean. Thus changes in rivers and drainage basins have consequences at the coast. Chapter 4 notes that dams on the River Nile decreased sediment load to the Nile Delta and the Mediterranean coast from 125 million T a^{-1} to 5 million T a^{-1} since historic times. The resulting erosion has been well documented (Stanley and Warne, 1993). Sediments can also be supplied to coastal systems by reworking of seafloor deposits (an important source in areas dominated by deposition during recent glaciations) and by alongshore transport of sediment from areas of coastal erosion. Indeed, many gravel barriers and spits are intimately associated with their fairly local sediment sources (Forbes *et al.*, 1995) and any diminution of that supply or interruption of the transport pathway can have dramatic consequences (Orford and Pethick, 2006). However, it is not simply the amount of sediment reaching any part of the coast that determines its form. Marine processes, most notably waves, continually reshape sediments on open shorelines and Chapter 6 provides a framework for considering the longer-term geomorphic consequences of these short-term events. This chapter focusses on some of the complex interactions amongst riverine and coastal processes and the important role of biota in shaping many coastal landforms.

Coastal marshes, and their adjacent tidal flats, are among the most productive coastal systems in the world (Millennium Ecosystem Assessment, 2005) and by supporting coastal fisheries and migratory bird populations, providing protection from wave attack, and sequestering sediment-associated contaminants they provide important geomorphological and ecological services to local populations. The way in which these services are provided, however, is complex. Barbier *et al.* (2008) examine the role of mangroves, salt marshes, seagrass beds, nearshore reefs and sand dunes in wave attenuation and note that the effect is rarely linear. While other aspects of mangroves valued by local communities, e.g. as a source of wood products, are directly related to the area of mangroves, protection from waves cannot be achieved with small isolated areas of mangroves. Understanding the complex interaction between waves, water levels, mangrove swamp topography and canopy architecture is essential to managing coastal systems to maintain protection for coastal communities. Such examples also illustrate the role of vegetation and its interaction with physical and sedimentary processes in both shaping coastal landforms and contributing to coastal resilience.

The approach taken here is to focus on the processes controlling the fate of sediments. This reflects what some have called the 'reductionist' approach to geomorphology (Kennedy, 2006) of the late twentieth century but one that is essential to understanding the biogeophysical complexity of coastal systems and their response to changing climate and human pressures in the twenty-first century. Coastal geomorphology must provide the linkage between coastal engineering and coastal ecology. The study of coastal geomorphology has its basis in the geologic studies of the eighteenth and nineteenth centuries and the work of Hutton, Playfair, Lyell and Huxley (Woodroffe, 2002). Concepts of time, and change over time, were developed further by Darwin, perhaps most notably in his theory of coral atoll formation. Gilbert's stratigraphic studies of the shoreline of Lake Bonneville were the foundation of the concept of shoreline equilibrium which was also incorporated into

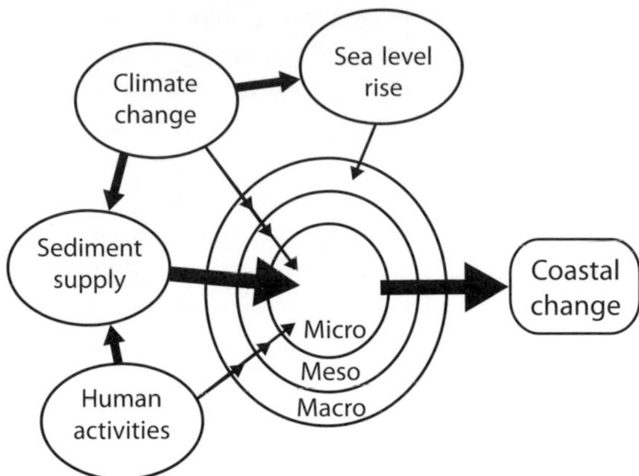

FIGURE 5.1. The influence of important drivers of twenty-first-century coastal change across timescales. Timescales are from Sherman (1995). Macro = greater than decades, meso = months–decades, micro = seconds–months.

TABLE 5.1. *Spatio-temporal scales for coastal environments*

	Megascale	Macroscale	Mesoscale	Microscale
Space	100s km	km	m	cm
Time	century	year/month	days/hours	minutes/ seconds

Source: Perillo (2003).

Davis' (1896) consideration of the development of the spit at Cape Cod in Massachusetts. The Davisian construct of progressive landform evolution was most notably applied at the coast by Johnson (1919) who saw decreasing complexity and diversity in coastal landforms as shorelines 'aged'. These studies of almost a century ago still provide a foundation for the consideration of twenty-first century coastal change. The concept of equilibrium form is fundamental to our understanding of process–form interactions, and adjustments of many coastal landforms to changing sea levels. Woodroffe (2002) and many others have noted how the study of beach dynamics was advanced in the mid twentieth century by the experiences of military actions on beaches during World War II. The process approach was further facilitated by the availability of increasingly detailed measurement devices allowing high-frequency, microscale measurements of water and sediment dynamics. The integration of detailed process understanding into coastal landform development is brought to fruition in the development of the morphodynamic approach (Wright and Thom, 1977) which emphasises constant adjustment between form and process. The longer-term application of these concepts is developed in the 'coastal tract' approach described in detail in Chapter 6.

Throughout the twentieth century the role of spatial and temporal scale in understanding landform dynamics has been central to geomorphology. At the coast, Sherman (1995) has identified three important timescales of adjustment for coastal dunes: decades, months and seconds. To consider how coastal geomorphic systems respond to global change, it is important to identify the temporal scales at which climatic and human influences impact the coast. Figure 5.1 illustrates how some of the important drivers of coastal geomorphic change, i.e. climate, both directly via temperature, precipitation and storm events, and indirectly via sea level rise, sediment supply, and human activities, operate at different temporal scales. While sea level rise can be measured (and to some extent predicted) at an annual timescale, its influence on coastal landforms is really only demonstrable at the scale of decades (at least while rates are below ~1 cm a^{-1}). The role of changes in temperature and precipitation, however, can be expressed in decadal scale or longer changes in runoff, inter-annual droughts and floods, and seasonal–daily changes in moisture availability. Sediment availability is influenced by both changes in climate and human activities and changes at the decadal, inter-annual and shorter timescales.

While many studies focus on temporal scale, few consider that processes in the landscape are also associated with specific spatial scales. Longer-term processes normally affect larger regions (megascale in Table 5.1) whereas faster processes produce smaller and smaller changes in the spatial scale. However, the microscale (Table 5.1) has a key influence on the other scales, because it is frequently the non-linear summation of centimetre and second scale processes that ultimately produce regional-scale modifications in the geomorphology (Perillo, 2003).

Human influences operate at all scales, from dams on rivers blocking sediment or dredging in estuaries which alter tidal circulation, to inter-annual harvesting of wood from mangrove swamps, to holiday weekend trampling of dune vegetation. Importantly, consideration of coastal geomorphic change in the twenty-first century must embrace all three temporal scales (Table 5.1). The assumption here is that trends in drivers at scales of greater than decades have little direct influence on coastal dynamics at the century scale, and are considered as antecedent conditions for the twenty-first century coast.

Given the breadth of previous work, the focus of this volume on twenty-first century global change, and the detailed consideration of broad-scale changes in coastal systems in Chapter 6, the purpose of this chapter is to:

(a) explore two important geomorphic interactions at the coast: interaction between land and sea, and interactions between sediments and biota; and
(b) demonstrate the vulnerabilities both of coastal systems and the geomorphic services they provide in the face of changes likely to occur in the twenty-first century.

This will be achieved through examination of these interactions in three closely related geomorphic environments: estuaries, coastal marshes and tidal flats, and coastal dunes. While these represent a subset of coastal geomorphic systems, they exemplify the nature of the coastal geomorphic interactions (e.g. land–sea, subtidal–intertidal, intertidal–supratidal), embrace processes across a range of spatial and temporal scales and illustrate the complexity of combined human–climate influences.

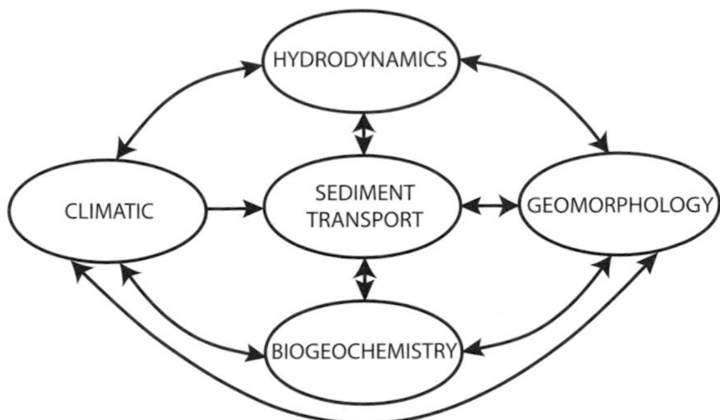

FIGURE 5.2. Integrated relations among the different major processes that act upon an estuary (modified from Perillo *et al.*, 2007).

5.2 Estuaries

Estuaries occur at the interface between terrestrial and marine systems and are thus positioned at a key point in the sediment cascade. They are heavily influenced by land-based processes, especially river flow, and their geomorphology is fundamentally controlled by how riverine inputs (i.e. sediments and water quality constituents) interact with coastal and marine processes, such as tides and waves. This interaction between land and sea frequently results in extensive deposition of both river and coastal derived sediments (see details below). Such deposition has for centuries been seen as a problem for many human activities dependent on estuaries, especially navigation and trade. For example, sedimentation in the mouth of the Mississippi River and its impact on navigation resulted in major engineering of the river as early as the mid nineteenth century.

5.2.1 Driving processes

Perillo *et al.* (2007) highlight the main controls that act upon an estuarine environment (Fig. 5.2). Antecedent geomorphology influences all other factors as it controls the initial condition of formation, has shaped any underlying geological control, and remains in most instances a controlling factor over estuarine dynamics, namely hydrodynamic, sedimentological and biogeochemical processes. Classification of estuaries has been addressed from both hydrodynamic and geomorphologic perspectives. Dyer (1986) distinguishes estuarine types on the basis of the interaction between fresh and saline waters. Salt wedge estuaries are those with little mixing between the two water bodies and a stratified water column while well-mixed estuaries show no stratification and a gradual gradient in salinity towards the ocean. Partially mixed estuaries show higher salinity gradients in their middle reaches. The type of estuary is dependent on the relative strength of tidal currents, sometimes adequately reflected by the local tidal range, and river flow that determines the missing characteristics. Fjords, due to their great depth, are considered separately due to only limited tidal mixing near the surface and an almost motionless bottom layer, perhaps the extreme case of geomorphology controlling estuarine hydrodynamics (Fig. 5.2). However, antecedent geomorphology and its change over time are seldom considered in assessment of estuarine processes. Geomorphology provides the base upon which the dynamic and biogeophysical processes act; similar processes acting upon different geomorphologies will produce totally different results.

Dalrymple *et al.* (1992) use the relative energy of river and wave–tide influences to identify three zones of estuarine facies (Fig. 5.3). However, they then consider only the relative roles of tides and waves in identification of a gradient in estuarine sedimentology and geomorphology between two end member states. Dalrymple *et al.* dismiss the need for a category of estuaries dominated by fluvial processes as they consider that rivers only determine the rate of estuarine infilling rather than the fundamental morphology of estuaries. This contention is disputed by Cooper (1993) who argues on the basis of detailed studies of estuaries in South Africa that river-dominated estuaries develop an equilibrium sediment volume (under constant sea level rise conditions) where sediment accumulation is punctuated by removal during flood events (Fig. 5.4). The processes controlling the fate of sediments within estuaries are important for the development and dynamics of tidal flats. Importantly, as noted in Fig. 5.2, classification schemes based on physical or sedimentological properties often ignore the role of biogeochemical processes. While these may operate at finer temporal and spatial scales than the geologic considerations of

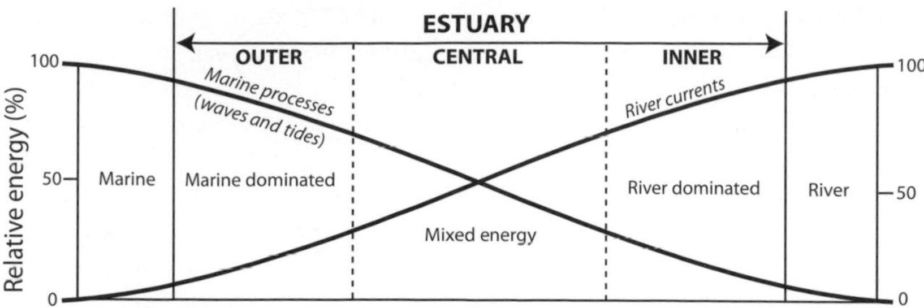

FIGURE 5.3. Schematic distribution of the physical processes operating within estuaries, and the resulting tripartite facies zonation (from Dalrymple et al., 1992).

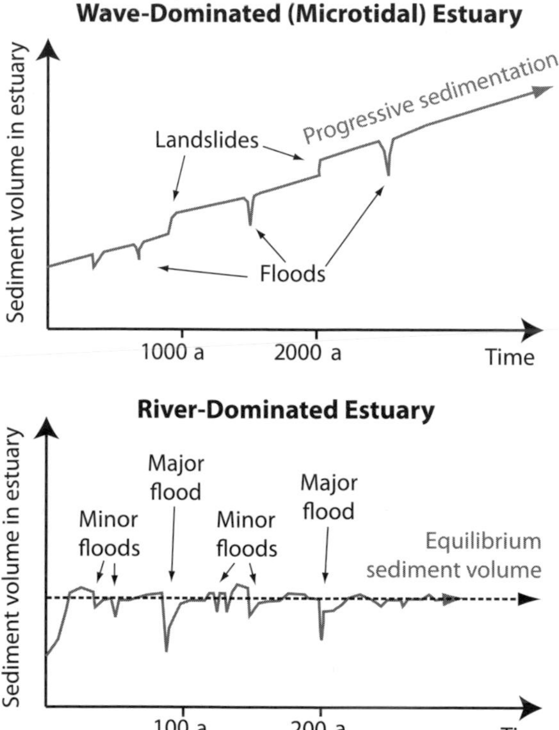

FIGURE 5.4. Variation in sediment volume over time in wave-dominated and river-dominated estuaries under a stable relative sea level (from Cooper 1993).

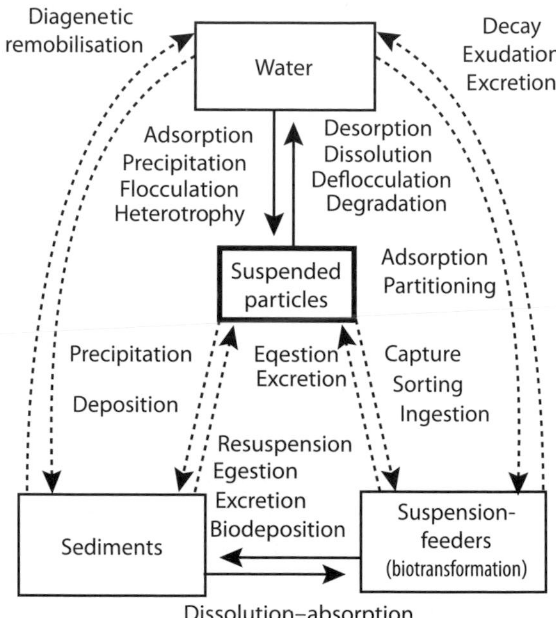

FIGURE 5.5. The role of suspended sediment in estuarine biogeochemical processes (from Turner and Millward, 2002).

some workers, their role in the importance of these environments for local communities is via nutrition, food webs and contaminant uptake. Turner and Millward (2002) argue that suspended sediments represent the focus of estuarine biogeochemical processes (Fig. 5.5). In turn, this points to the dynamic linkage between estuarine hydrodynamics, sediment transport and biogeochemistry illustrated in Fig. 5.2.

5.2.2 Twenty-first century changes

Climate forces (e.g. wind regime, temperature, precipitation) and hydrodynamic forces (i.e. tides and tidal currents, waves, river discharge) are closely intertwined, but act on the geomorphology of estuaries in a non-linear fashion especially with regard to the fate and transport of sediment. Such non-linearity must be considered in attempts to predict future change and develop appropriate management strategies. Estuaries are constantly subjected to changes in their geomorphology, at the scale of millimetres and seconds. Over time, these imperceptible changes cumulate to larger macroscale changes perceived by people.

Many aspects of climate change are likely to influence the functioning of estuaries. These include changes in freshwater runoff, temperature changes, alterations in salinity and sea level rise (see Chapter 1). Differences in the response of estuaries to climate are dependent on regional differences in the response of the upstream terrestrial systems to climate change. Estuaries vary widely in terms of their geomorphology and bathymetry, relative size of the watershed, physical circulation patterns and water residence times, and each of these can affect the response to climate change.

Overall, the greatest direct effect of climate change on estuaries is likely to result from increased temperatures and from changes in the physical mixing characteristics, which are affected by changes in freshwater runoff, sea level rise and tides (Boesch *et al.*, 2000). For temperature, large regional differences in the physical drivers are expected. The relative importance of ocean temperature changes versus the effects of local air temperature (as mediated by continental-scale weather patterns) will vary according to whether estuaries are open or semi-enclosed, the depth of nearshore waters and the magnitude of tidal influence. Estuaries that are shallow and with restricted exchange to the ocean will be more influenced by changes in air temperature and thus, the effect of change in climate will vary seasonally. The effect of temperature increase will likely be greater in winter with some effect on the timing of autumn and spring temperature changes, as heating during the warm summer months tends to be moderated by evaporative cooling, as demonstrated for the Bahía Blanca Estuary, Argentina (Cervellini and Piccolo, 2007).

Temperature changes will combine with changes in salinity to influence estuarine circulation and sediment transport. Future salinities will be influenced by sea level rise and runoff from adjacent terrestrial systems. The effect of sea level rise on salinity in estuaries will depend upon the current pattern of exchange with the ocean, with less of an impact on circulation in estuaries that are already well mixed. In many cases the main effect may be further landward penetration of saline water and the effect on circulation will be dependent on bathymetry and river inflow changes.

Due to complexities of precipitation and land cover changes, runoff predictions are challenging even in small catchments and future freshwater inputs to estuaries are difficult to predict. The IPCC indicates that the runoff in the high latitudes of North America and Eurasia will increase 10–40% (IPCC, 2007). Importantly at the coast however, the timing of river flows is also expected to change. Warming changes the seasonality of river flow especially in areas where winter precipitation is dominated by snowfall. Winter flows will increase and summer flows will decrease potentially amplifying seasonal changes in salinity patterns within estuaries. The role of climate change in altering runoff regime, e.g. timing of discharge peaks, has been demonstrated by Dettinger *et al.* (2004) for several sub-basins of the Sacramento–San Joaquin watershed which supplies freshwater inflow to San Francisco Bay. They note that a 2.5 °C increase in temperature in the twenty-first century would hasten snowmelt and advance the seasonal cycle in discharge by up to a month. Furthermore, these changes in freshwater inflow also influence many aspects of hydrodynamics and ecology in the San Francisco Estuary, including a potential effect on gravitational circulation as changes in freshwater inflow move the salinity gradient across complex bathymetric transitions. It is not yet clear what the implications of such changes in seasonal runoff patterns and estuarine dynamics will be for sedimentation, especially in the extensive tidal flats in the upper parts of the San Francisco Estuary, close to the mixing zone between fresh and saline waters.

In regions with little or no snowfall, changes in runoff are dependent much more on changes in rainfall than on changes in temperature. IPCC (2007) also suggest that in rain-dominated catchments flow seasonality will increase, with higher flows in the peak flow season and either lower flow during the low flow season or extended dry periods. Booij (2005) modelled the impact of climate change on flooding in the River Meuse on a daily basis using spatially and temporally changed climate patterns. He found great differences in the ability to predict average discharges (a small underestimation against historical data) versus a considerable underestimation of extreme discharge behaviour. As noted previously, some models of estuarine morphodynamics (e.g. Cooper, 1993) identify an important role for episodic river floods in modulating sediment accumulation within estuaries so as yet unpredictable changes in extreme events can have important geomorphic implications.

Despite such uncertainties regarding runoff, circulation and their influence on sediment movement, there is some evidence that climate forcing can dominate the movement of sediment throughout the entire sediment cascade – from mountain headwater through estuaries to seabed deposition. Goodbred (2003) examined source to sink variations in the 3000-km long Ganges sediment dispersal system and found a tight coupling between source area, catchment basin and coastal and marine depocentres. The southwest (summer) monsoon appears to exert overwhelming control on regional hydrology. About 80% of the Ganges discharge and 95% of its sediment load are delivered to the coastal margin during only 4 months of the year. The regional scale of the monsoon means that it influences the entire basin rather than individual tributaries or subsystems. The Ganges example illustrates the potential overarching influence of climate (Fig. 5.2). Despite the large spatial scale of the system sediment pulses can be transferred rapidly from source to sink indicating that the effects of twenty-first-century change can be readily manifest at the system scale.

5.2.3 Implications of geomorphic change

As shown in Fig. 5.2 the geomorphic changes in estuaries in the twenty-first century will result from the interaction of climate on hydrodynamics, sediment transport and biogeochemical processes, and feedbacks among these factors. While predictions of climatic change are available, there remains great uncertainty regarding the volume, and especially the timing, of water and sediment discharges from rivers to the coastal zone – primary influences on estuarine circulation and sedimentation. Estuaries have been shown here to be dynamic geomorphic environments, in large part due to daily changes in hydrodynamic forcing and inundation driven by the tide, with important seasonal and interannual (or longer) cycles superimposed. The implication of future change is determined by the net trend in the cumulative result of these processes. This can be illustrated by considering two important uses of these environments: the supply of fresh water for industries and communities, and the productivity of estuaries for shellfish.

The effect of warming and changes in snowpack on freshwater inflows to the Sacramento–San Joaquin Delta, which captures 42% of California's runoff, has been described above. These waters ultimately flow into San Francisco Bay and are managed to prevent saline incursion to the delta. Importantly, the Bay–Delta system also provides drinking water for two-thirds of California's population (22 million people), irrigation supplies for at least US$27 billion in agriculture (45% of the US fruit and vegetable produce), and is a primary water source for California's trillion-dollar economy. About 5 $km^3 a^{-1}$ of fresh water is pumped from the delta by two water projects to supply municipal and agricultural water demands in southern and central California. The risk of salt penetration into the delta and the potential threat to these water supplies as a combination of reduced precipitation, diminished snowpack and sea level rise has been highlighted by Lund *et al.* (2007), and plans to 'replumb' the water supply system of California are being considered. The threat of salinity penetration into drinking water supply caused by sea level rise into the Delaware Estuary has also been examined by Hull and Titus (1986) who identified a potential threat to both water intakes and aquifer recharge. Whether surface water management in either system, through increased water use efficiency, expanded storage or other means, can cope with expected changes has yet to be seen.

The value of estuaries for shellfish harvest is known around the world. Changes in water quality, sediment inputs and salinity in estuaries associated with changes in climate or watershed land use will particularly influence benthic organisms which cannot readily evade unfavourable conditions. Najjar *et al.* (2000) note how changes in water temperature in estuaries can alter the existing range of common species such as *Mya arenaria*. One of the best-documented instances of temperature change impacting commercially important species such as oysters is the way in which the parasite that causes Dermo disease in the eastern oyster *Crassostrea virginica* has extended its range north along the US mid-Atlantic coast since 1990 due to warmer winters. Higher salinities associated with decreased runoff can also impact the occurrence of oyster diseases and predatory snails. Conversely, sudden freshening, such as occurred in Chesapeake Bay in 1972 during Tropical Storm Agnes can also result in oyster mortality.

5.3 Coastal marshes and tidal flats

Coastal marshes are valuable ecosystems. By providing refuge and forage opportunities for wildlife, fishes and invertebrates, marshes are the basis of the economic livelihoods of many communities. Shallow ponds and seed-producing vegetation provide overwintering habitat for millions of migratory waterfowl, and the role of wetlands in absorbing nutrients and reducing loading to the coastal ocean is widely recognised, as is their value for protecting local communities from flooding, either by damping storm surges from the ocean or by providing storage for riverine floodwaters. Climate change and variability compound existing stresses from human activities such as dredging and filling for development; navigation or mineral extraction; altered salinity and water quality resulting from activities in the watershed; and the direct pressures of increasing numbers of people living and recreating in the coastal zone.

Marshes occur in a variety of coastal settings (Woodroffe, 2002). On coasts dominated by marine processes, marshes form where protection from the waves and longshore currents allows infilling by sediments and the development of tidal flats. The transition from tidal flat to coastal marsh occurs when flat elevation reaches a level within the tidal range that supports salt- and flood-tolerant emergent vegetation (typically found in the shelter of spits and barrier islands, and in protected bays and estuaries). While tidal flats occur in a variety of energy regimes, marshes are found on exposed open coasts only in specific settings. Their geomorphic setting can in turn influence their ability to withstand change during the twenty-first century.

5.3.1 Driving processes

The character of coastal marshes is determined by the process regime of the precursor tidal flat as well as the biogeomorphic interactions within the marsh as it develops. As noted by French and Reed (2001, p. 181) 'the interaction of physical and biological factors gives rise to a remarkable diversity in form and function within what are often considered to be rather uniform, featureless, environments'. The nature of this interaction varies globally according to climate, tidal forcing, and soil conditions that influence both physical and biological controls. French and Reed (2001) further identify four main sets of physical factors – fine sediment regime; tidal conditions; coastal configuration; and relative sea level history – which define the geomorphic context for coastal marsh development and survival during the twenty-first century. Mature coastal marsh systems can be considered to include a number of subsystems: vegetated plains, tidal courses, pans and ponds, as well as the adjacent intertidal zone. The dynamics of these subsystems has been examined by a number of workers and their interdependence has been frequently noted (see Perillo, 2009 for a summary). For an examination of coastal marsh futures during the twenty-first century the focus here will be on the transition from tidal flat to vegetated marsh and the character and dynamics of the marsh plain. While the evolution of marsh creeks has recently been the topic of great study (e.g. Fagherazzi and Sun, 2004) and many important ecosystem functions associated with marshes are associated with the creek–plain interface, the occurrence of marshes and their persistence depends upon the creation and sustainability of a vegetated surface.

Tidal flat characteristics control vegetation colonisation and marsh establishment. However, the processes driving tidal flat geomorphology have not been well studied. Benchmark studies of Postma (1961) on sedimentary processes and Evans (1965) in identifying waves and biotic factors as modifiers of the current driven processes provided a foundation for understanding sediment dynamics, while Amos (1995) provided an integrated view of their geomorphology and sedimentology. Dyer (1998) developed a typology for mudflats (i.e. tidal flats dominated by fine sediments) based on a conceptual understanding of the processes influencing mudflat development including tidal range, exposure, waves and sediment type. This typology was largely affirmed by a statistical analysis of mudflat characteristics from northwest Europe (Dyer *et al.*, 2000) which identified tidal range, exposure and mudflat slope as the most important driving factors. Mudflat slope has been shown by Kirby (2000) to vary between concave upwards profiles (with mean mudflat level below the mean tide) when the system is net erosional and convex profiles (with the maximum slope close to the low water mark) when there is net deposition. However, detailed studies of tidal flat profiles (O'Brien *et al.*, 2000) show seasonal variation in the sedimentary status of macrotidal flats in the Severn Estuary, UK and a seasonal cycle between erosional and depositional states; the differences largely attributed to variations in forcing by waves and tides.

Even when elevations are too low in the tidal frame for colonisation by emergent macrophytes tidal flat, physical–biological interactions have an important role in geomorphology. Blanchard *et al.* (2000) in a study of tidal flats in the Humber Estuary, UK found that geomorphological features influenced biological processes in a way which exacerbated the physical processes. In an area of ridge and runnel topography they found that the regularly exposed and apparently stabilised surface of the ridges favoured microphytobenthos growth and carbohydrate production which further increased sediment stability. In the runnels which are dominantly drainage structures with a high sediment water content, microphytobenthos were prevented from building up a carbohydrate pool by the sediment character (they were dissolved by the high water content). This represents a synergistic effect between physical and biological processes on mudflat ridges which acts to stabilise the sediment surface. Such studies illustrate the complex interplay among hydrodynamic, sedimentological and biogeochemical processes shown in Fig. 5.2.

The transition from tidal flat to coastal marsh is marked by colonisation by emergent macrophytes, such as *Salicornia* spp. (e.g. Pethick, 1980) or *Spartina* spp. (Pestrong, 1965). Steers (1960) described the accumulation of fine sediments in sheltered coastal areas as being a necessary precursor for marsh initiation and specifically mentions the likely presence of patches of algae, such as *Enteromorpha* spp., and their role as foci for sediment accumulation. However, these patches of algae can shift with storms allowing redistribution of any accumulated sediments and seasonal variation in tidal flat sediment accumulation, as noted above, can be influenced by abiotic and biotic factors. Nonetheless, it has long been recognised (e.g. Steers 1948) that the critical phase of intertidal flat stabilisation occurs with the spread of emergent vegetation, representing a more effective trap for suspended sediment than surficial microphytobenthos or potentially mobile macroalgae. Moreover, Langlois *et al.* (2001) have found that *Puccinellia maritima* growth, especially stem elongation and the production of adventitious roots, was stimulated by burial with up to 8 mm of sediment, again emphasising the interaction between biological and physical processes.

The colonisation of tidal flats by vegetation is widely thought to be dependent on elevation, and the availability of appropriate soil drainage and or exposure times to allow propagules to survive or seeds to germinate. Studies of colonisation on newly restored marshes show vegetation colonisation can be controlled by local dispersal mechanisms (e.g. Wolters *et al.*, 2005) or elevation (Williams and Orr, 2002). Ursino *et al.* (2004) in their study of subsurface flow through tidal flat substrates note that where pioneer plants, resistant to severely anoxic conditions, begin to grow (usually in well-drained areas adjacent to channels),

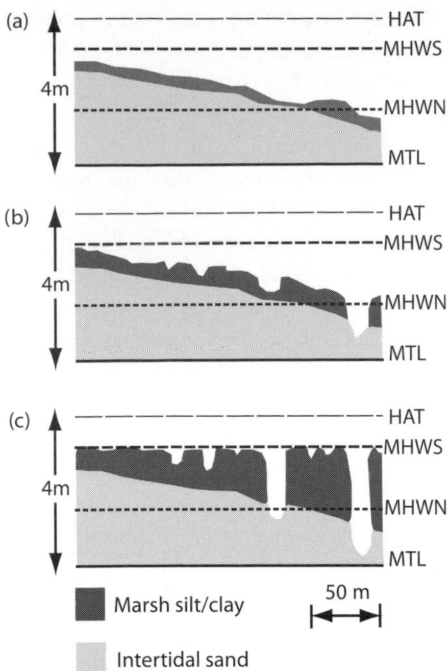

FIGURE 5.6. Stages in the infilling of the intertidal profile (from French, 1993). HAT, Highest Astronomical Tide; MHWN, Mean High Water Neap; MHWS, Mean High Water Spring; MTL, Mean Tide Level.

they increase evapotranspiration rates and thus oxygen availability, creating a more favourable environment for the development of other vegetation. Such a positive feedback when combined with the sediment trapping ability of the plants reinforces this critical biogeomorphic threshold in coastal marsh development.

Where annual *Salicornia* spp. are the colonisers, as is frequently observed in northwest European marshes (e.g. Steers, 1948; Pethick, 1980), enhanced trapping of sediments may be limited, both because of die-back during the winter (frequently a period of increased suspended sediment availability) and because of the susceptibility of the plant morphology to disruption by strong waves and tidal currents (Houwing *et al.*, 1999), which have also been found to disturb seedlings of *Sueda maritima* (Tessier *et al.*, 2001). Consequently, the initial build-up of the emergent marsh surface is slow. French (1993) shows that in the early stages of infilling of the intertidal profile, a thin veneer of fine sediments accumulates over the pre-existing intertidal surface (Fig. 5.6). The approximate time taken for marsh initiation in North Norfolk has been inferred by Pethick (1980) using historical maps. For several marshes on Scolt Head Island, the transition from intertidal flat with no creek or marsh boundary shown on the map, to a marsh shown as a vegetated surface with defined creeks, occurs in less than 100 years.

The rate of transition from intertidal to emergent marsh depends upon the supply of sediment but also the vigour of the vegetative growth. Pestrong (1965) observed *Spartina* spp. colonising tidal flats in San Francisco Bay and described luxuriant growth of dense stands. The efficacy of some *Spartina* species in trapping suspended sediments has been demonstrated in many areas, especially where invasive species have quickly covered large intertidal areas and increased elevations. Reed (1989) has observed initial marsh formation in hollows between recently exposed moraine ridges close to a tidewater glacier in southern Chile. In this environment, where supplies of suspended sediment are low and both slow vegetative growth and physical disruption by rafting icebergs limit vertical development, the transition rate from intertidal to emergent marsh is low compared to estuarine marshes within the same coastal region. On the other side of the South American continent, Perillo *et al.* (1996) observed a very fast growth of a tidal flat/salt marsh system in the Rio Gallegos Estuary based mostly in sediment retention of relatively low suspended sediments but very efficient trapping by the marsh even in high wind and wave conditions. These differences in marsh development merely represent variations in the rate of transition from intertidal to emergent marsh. The essential role of emergent vegetation in trapping and stabilising sediments during the initial development of the marsh surface is common across latitudes and continents.

5.3.2 Twenty-first century changes

While stresses on coastal systems during the twenty-first century will include increased population pressure and further modification of terrestrial inputs as well as climate change, the dominant threat is seen by many as sea level rise (Boesch *et al.*, 2000). The response to sea level rise depends on the balance between submergence, erosive forces and sediment supply, and is mediated by climatic influences on biotic processes (Reed, 1995).

The resistance of tidal flats to erosion is heavily dependent on the biota. Widdows *et al.* (2004) estimate that up to 50% of the sediment accumulation on a tidal flat in the Westerschelde Estuary was due to biostabilisation. The timescale of influence of various biotic factors on tidal flat stability is conceptualised by Widdows and Brinsley (2002) who identify longer-term changes associated with the presence of persistent biota (e.g. mussels) and shorter-term cyclic changes in the balance between microphytobenthos and sediment destabilisers, such as burrowing clams (Fig. 5.7). They also noted that climate is a major driver of inter-annual variation in tidal flat biota. Cold winters caused an increase in abundance on tidal flats of the clam *Macoma balthica*, which is widespread on temperate flats

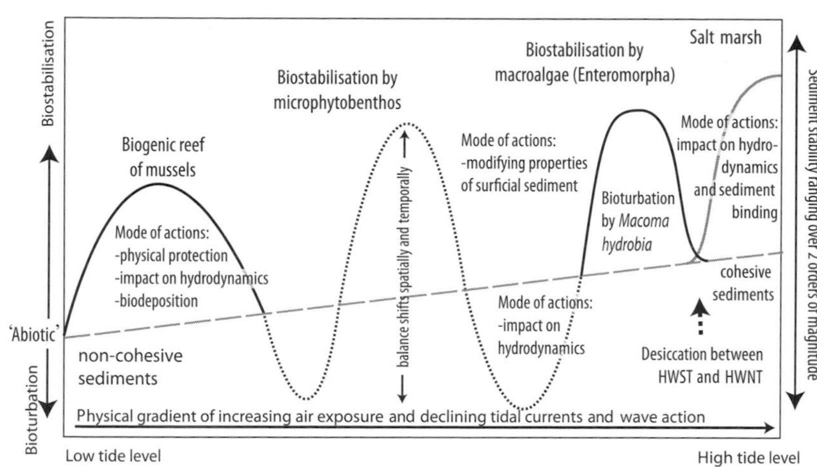

FIGURE 5.7. Conceptualisation of the major biological and physical factors influencing sediment stability in the intertidal zone. Dotted line represents general shoreward increase in sediment stability with increasing exposure and decreasing tide–wave action. Solid line represents the long-term role of biota and the dashed oscillating line shows shorter-term switches from biostabilisers to destabilisers (from Widdows and Brinsley, 2002).

in the northern hemisphere. This increases sediment erodibility on the tidal flats, and in years following mild winters the flats were more stable due to well-developed diatom biofilms and reduced clam abundance. Biofilms are also important in stabilising bedforms (Fig. 5.8).This suggests an increase in sediment stability associated with warmer winters resulting from global climate change. However, the inter-annual variability found by Widdows and Brinsley (2002) implies that periodic cold winters could dramatically change mudflat biota and if accompanied by sufficient wave and tidal energy might lead to 'resetting' of the mudflat profile and a release of sediment for deposition in adjacent marshes and subtidal areas, or to the coastal ocean.

Another aspect of tidal flat and marsh geomorphic processes sensitive to changes in climate is the role of seasonal ice. Pejrup and Anderson (2000) in studies in the Danish Wadden Sea found that sediment transported by incorporation into drifting ice floes, which was subsequently deposited during melt-out, significantly influenced both the sediment balance and the morphology of intertidal flats. This supports the findings of Dionne (1998) in the St Lawrence Estuary where he found evidence of scouring and compression by ice rafts in tidal flat deposits. In contrast, and higher in the tidal profile, Argow and Fitzgerald (2006) found that normal loadings of ice on marsh peats did not appear to cause permanent compaction. Pejrup and Anderson (2000) emphasise that sediment movement by ice rafting acts to redistribute sediments that are already stored within the tidal flat–marsh system. While increased temperatures during the twenty-first century may reduce the effect of winter transfers of sediment from flats to adjacent marshes, how this will combine with changes in summer biota to affect the overall geomorphology of higher-latitude flats has yet to be determined.

FIGURE 5.8. Example of biostabilisation of ripples on a tidal flat of the Bahía Blanca Estuary (Argentina), although the effect of erosive wave activity affects the preservation of the biofilm (photograph by Lea Olsen).

The redistribution of sediments within and between tidal flats and marshes is also strongly influenced by storm surge and waves, the frequency and intensity of which may increase with climate change. The severity of storm surge impacts is highly correlated with the speed and path at which the storm approaches the shore. Hurricane Hugo caused little geomorphic change to the salt marsh at North Inlet, South Carolina because it crossed the shore rapidly with little rainfall. By contrast, Hurricane Andrew caused widespread damage to coastal marshes in Louisiana because it crossed the shore slowly and at an oblique angle, and Hurricanes Katrina and Rita are estimated to have caused the loss of over 500 km^2 of coastal marsh in Louisiana (Barras, 2006). Storm surges can introduce sediment into the marsh and redistribute sediment as portions of the marsh are eroded (Cahoon, 2006) resulting in substantial sediment accretion and geomorphic changes. Tropical and extra-tropical storms are potentially important both in providing a pulse of sediment over a broad geomorphic scale to balance the effects of subsidence, particularly for deltas, and in enhancing coastal marsh survival for areas remote from a sediment source, although Bartholdy (2001) notes that the amount of sediment deposition in coastal marshes depends on the prevailing meteorological and hydrographical conditions as well as on previous events. The source of storm sediment deposited in marshes is rarely clear (e.g. Burkett *et al.*, 2007) and storm erosion or deposition on tidal flats is difficult to observe directly or by analysing satellite images; detailed geomorphologic surveys pre- and post-storm are required to elucidate how these systems are linked during extreme events.

On the vegetated marsh plain, the effects of climate change are felt directly by the vegetation and indirectly through changes in watershed hydrology. Many coastal wetland plants have adapted to hot and dry conditions through alterations in the way in which they fix carbon in the Calvin cycle; commonly these plants are referred to as C4 plants (a minority of plant species), whereas most species are C3 plants. Studies of short-term response of photosynthetic rate to elevated carbon dioxide indicate that the response will be greater in C3 plants than in C4 plants; indicating how the response of salt marshes depends greatly upon the community composition. Curtis *et al.* (1989) showed an increase in primary productivity of *Scripus olneyi* (C3) under increased carbon dioxide. Increases in salinity of marshes resulting from decreased freshwater discharge (and subsequent intrusion by more saline offshore waters) may result in decreased productivity of coastal marshes, but should not necessarily result in marsh loss. It is widely documented that *Spartina alterniflora* tolerates a wide range of salinities (Webb, 1983). Similarly, the variation in salinities found by Visser *et al.* (1998) in marshes dominated by the cordgrass *Spartina patens* was also wide (average salinity 10.4 ± 5.8), and overlapped with the zone dominated by *Spartina alterniflora* (average 17.5 ± 5.9) further demonstrating the wide tolerance range for some of the dominant saline marsh plants. Experimental studies of European salt marsh species have also shown that a growth reduction associated with increased salinity can be reduced to some extent by carbon dioxide enrichment (Rozema *et al.*, 1991) and other workers have noted increased salt tolerance.

The response of coastal marshes to sea level rise must be considered in the context of these changes in sediment dynamics, mediated by both physical forcing and biotic factors, and plant growth. Salt marsh surfaces are frequently considered to be in an equilibrium relationship with local mean sea level (see Allen, 2000 and references therein), and indeed dating of buried salt marsh peats is a frequently used technique for reconstructing past sea level changes. That marsh surface elevation changes to keep pace with sea level rise is well established; unfortunately researchers are unsure whether the elevation change is driven by sea level change or whether the space provided by a rising sea level merely allows the increase in elevation to occur. The latter would imply that elevation change is limited by sea level conditions. The projection of salt marsh sustainability under future climate scenarios is a complex issue and depends on: the relative importance of organic matter to marsh vertical development; the complexities governing organic matter accumulation during rising sea level; the importance of subsurface processes in determining surface elevation change; and the role of storm events and hydrologic changes in controlling sediment deposition, soil conditions and plant growth.

Some insights into the ability of marsh systems to cope with sea level rise may be found where subsidence already produces rapid rates of relative sea level rise for coastal marsh areas. In the Mississippi Deltaic Plain some studies have suggested that marshes cannot keep pace with relative sea level rise rates approaching 1 cm a^{-1} (Baumann *et al.*, 1984). Many parts of this system are being rapidly degraded, for a wide range of reasons (Boesch *et al.*, 1994). On the other hand, tide gauge records show these rates of sea level change have been occurring for decades in some parts of coastal Louisiana, yet vast areas of marsh still exist. These marshes are diminishing in area by marginal erosion rather than submergence effects. This suggests that under optimum conditions of sediment supply, salinity and water quality, salt marshes may be able to survive rates of sea level rise as high as 50 cm in 50 years, rates higher than the expected rise in sea level estimated for the next 100 years. Local subsidence or hydrologic changes, however, could increase the rise in sea level experienced by individual marshes perhaps exceeding the local threshold of marsh vertical development. For Louisiana, Swenson and Swarzenski (1995) noted that for periods of several years at a time in the late 1960s and early 1970s tide gauges have recorded rates of relative sea level

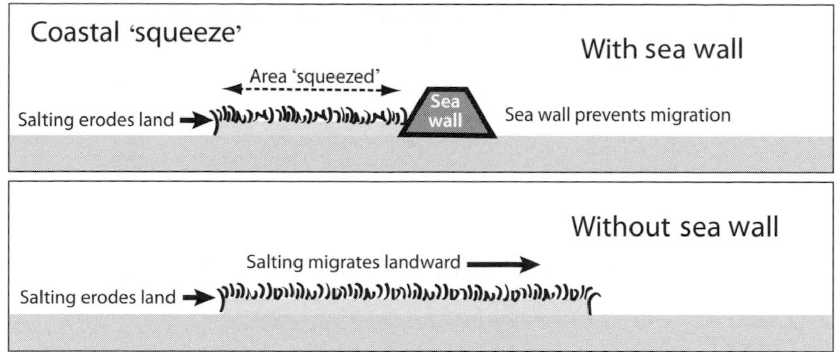

FIGURE 5.9. The concept of 'coastal squeeze' whereby sea walls or other structures limit the ability of marshes to migrate inland, causing them to narrow over time as sea level rises and erosion proceeds at the seaward margin.

rise as high as 6 cm a^{-1}, and Shinkle and Dokka (2004) have used survey data to show similarly high but spatially variable rates of subsidence. Such high rates of relative sea level rise are considerably greater than global long-term trends in sea level. The cause of these short-term and localised perturbations is unclear and may be associated with subsurface compaction and/or fault activation associated with oil and gas withdrawals (Chan and Zobak, 2007). Yet this period does coincide with a period of high land loss in coastal Louisiana confirming the notion of an upper limit of the rate of sea level rise beyond which marsh deterioration occurs. Even if thresholds are exceeded only long enough for marsh vegetation to be stressed and die, the consequences for coastal marshes are serious.

5.3.3 Implications of geomorphic change

The loss of coastal marsh has important implications for ecosystem productivity and coastal protection. Marshes on coasts with high rates of sediment supply and remote from physical disturbances associated with coastal development will probably be sustainable during the twenty-first century. Reed *et al.* (2008) indicate that although many coastal marshes in the US mid-Atlantic region will be sustained under future sea level rise rates 2 mm a^{-1} greater than current rates, many become vulnerable if the increase reaches 7 mm a^{-1}. The fate of many marshes considered by Reed *et al.* (2008) is unclear due to the unpredictable nature of floods and storms which in many cases provide important pulses of sediment to maintain marsh elevation.

Diminished areas of coastal marsh in the twenty-first century, or a dramatic shift in their geomorphic character, would have important consequences for coastal fisheries in many parts of the world. Recent studies have emphasised direct linkages between the geomorphic character of creeks and marsh systems and their use by fisheries species (e.g. Allen *et al.*, 2007), and importance of these linkages in supporting sustainable harvests of these organisms has also been established (e.g. for penaid shrimp and blue crabs in the northern Gulf of Mexico: Zimmerman *et al.*, 2000). Maintaining the biogeomorphic processes that sustain coastal marshes and provide for specific geomorphic habitat characteristics should thus be an important management goal for estuarine-dependent fisheries.

Perhaps of more immediate concern to coastal managers is the potential for loss of coastal marshes adjacent to coastal protection works. The concept of 'coastal squeeze' is now well established (Pethick, 2001), whereby coastal protection works which border marshes limit the onshore migration of marshes as sea level rises. Where the seaward margin of the marsh is erosional, the amount of marsh fronting the protection works decreases over time (Fig. 5.9). Thus the wave attenuation provided by the marshes also decreases over time and the protection works become more subject to wave attack, as well being vulnerable to overtopping as sea level rises. In southeast England coastal squeeze has led to renewed calls for restoration (e.g. Wolters *et al.*, 2005) stimulated to some extent by scientific studies that clearly show the effect of marshes in preventing overtopping (e.g. Moeller *et al.*, 2001). Figure 5.10 illustrates the effect of salt marsh replacement by unvegetated sand flat on the occurrence of overtopping for a sea wall at Stiffkey, North Norfolk, UK. The relative frequency of light overtopping is slightly higher with a 180-m wide salt marsh fronting the sea wall than if the salt marsh was replaced by a sand flat. Importantly however, this study also shows that the occurrence of high overtopping discharges, which can result in damage to the sea wall, increases from 16% to 30% when the salt marsh is replaced by an unvegetated sand flat. This builds on an economic value assessment of coastal marshes by King and Lester (1995). They examined the incremental changes in value associated with various widths of salt marsh, and found higher values for narrow widths of salt marsh (Fig. 5.11), based on the concept that without any marsh fronting the sea wall at all, the wall would need to be replaced. King and

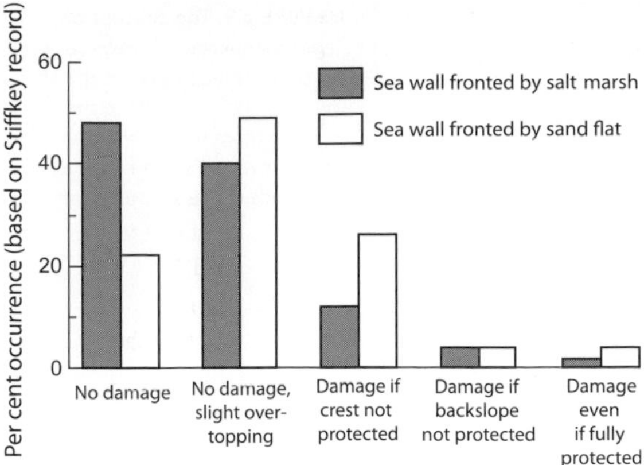

FIGURE 5.10. The likelihood of sea wall overtopping and associated damage levels for sea wall with a crest height of 4 m above Ordnance Datum where the sea wall is fronted (a) by a salt marsh, and (b) by a sand flat (from Moeller et al., 2001).

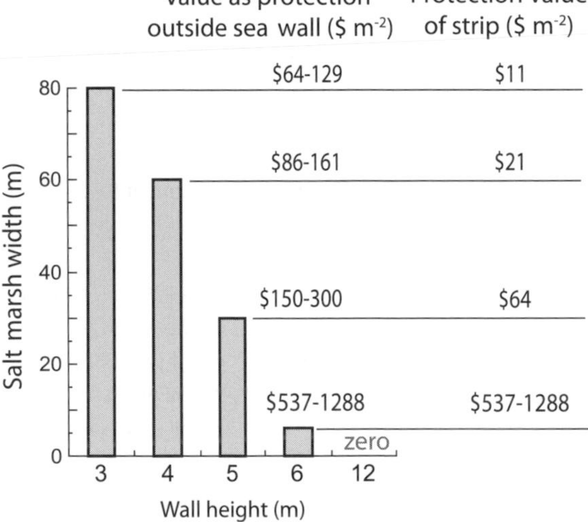

FIGURE 5.11. The height and capital cost of sea walls required to provide the same level of protection with different widths of salt marsh fronting the sea wall (from King and Lester, 1995).

Lester (1995, p. 187) conclude that 'It has been established that land for wildfowling purposes has the highest direct value, however the highest single value of salt marsh is the enormous saving it can afford to capital sea defence costs'. While further work is necessary to quantify the use values of coastal marshes, it is clear that their geomorphic character has many benefits to society, and while the effects of global change will be manifest differently among various geomorphic settings their vulnerability to global change in the twenty-first century should be taken seriously by coastal managers and policy-makers alike.

5.4 Coastal sand dune systems

Coastal sand dunes are sedimentary deposits formed by the transport of sediment inland from the beach by wind action. Because fine sediments are scarce in active sandy beaches, and because they tend to be transported out of the system in suspension, most coastal dune deposits have very small amounts of sediment <0.15 mm in size. Likewise, because of the limited ability of wind to transport particles >1 mm, coarse particles are also scarce and, where present, they are usually emplaced by storm waves rather than wind. A distinction can be made between dunes that are largely fixed in place by vegetation (impeded dunes) and dunes where vegetation is limited or absent with the result that the dune form reflects primarily patterns of wind flow (free dunes). In practice there is a continuum of coastal dune forms from those that are continuously covered by vegetation both temporally and spatially, through forms with varying degrees of stabilisation to completely unstabilised forms in regions where there is too little moisture to support vegetation or where the sediment supply and transport is so large that vegetation cannot colonise the unstable sand surface.

Coastal dunes offer important recreational opportunities around the world. However, the very attraction they provide becomes their greatest threat through overuse, urban expansion, and urban sprawl. Martínez *et al.* (2004) note that dune systems and their functions are often completely replaced by high-rise buildings, residential development, cottages, tourist resorts and recreation parks. The eventual consequence of such activities may be the destruction of the geomorphic system or its replacement by a highly constrained and managed one (Nordstrom, 2000). This section presents a description of coastal dune systems demonstrating their variability in form and process including the influence of human activities and interactions with vegetation, and describes their likely future in the face of twenty-first-century climate change.

5.4.1 Driving processes

Foredune system

Where there is sufficient vegetation to trap sediment blowing inland off the beach it is useful to distinguish between the primary dune system fronting the beach and the dune field inland of this. The primary dune system consists of the foredune ridge, parallel to the beach, and associated embryo dunes seaward of it as well as any unvegetated dune ramp (Fig. 5.12a; Plate 16). The foredune ridge is built up by sand blown off the beach and trapped by the vegetation on the stoss slope. Initially sand is trapped near the base of the dune but this vegetation may become buried

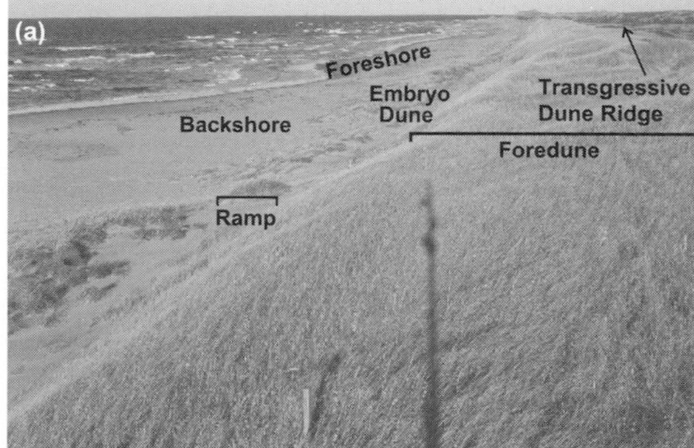

FIGURE 5.12. Photographs of the beach and foredune system at Greenwich Dunes, Prince Edward Island, Canada taken from a fixed camera mounted on a mast on the foredune crest. (a) Photograph taken on 18 October 2007 with vegetation cover at a maximum. (b) Photograph taken on 24 January 2008 showing snow cover on the foredune and backshore, shorefast ice on the foreshore and inner nearshore, and drift ice cover offshore. Presently sand transport from the beach to the dune and wave action are curtailed for 2–4 months a year. The main morphological zones are labelled, including a transgressive dune ridge formed between 1953 and 1997. (See also Plate 16 for colour version.)

by sand, eroded by waves, or may die back during winter months in temperate regions. During high wind events sand is transported up the stoss slope to the crest where it may be deposited near the crest or on the lee slope. Where the vegetation cover near the crest is thin or becomes partially buried sand can be transported onto the lee slope and during high wind events small amounts may be transported several tens of metres beyond the toe of the lee slope.

Embryo dunes (also known as incipient dunes) form seaward of the stoss slope of the foredune where sand accumulates within and behind individual plants on the beach. Vegetation growth onto the beach may result from vegetative propagation of plants from the base of the dune in the form of surface runners or subsurface rhizomes; or from the establishment of new plants from seed or vegetation fragments. During beach progradation the embryo dunes coalesce and grow in height, eventually forming a continuous dune ridge that develops into a new foredune and cuts off the supply of sediment to the older dune landward. On stable and eroding shorelines the embryo dunes are ephemeral features that develop over a period of months to several years and are then removed by wave action during large storms (Fig. 5.12a; Plate 16).

There is usually considerable interaction between the foredune system and the beach system in terms of geomorphic form, vegetation development and especially in terms of the exchange of sediment. This 'beach–dune interaction' involves both the transport of sand by aeolian processes from the beach to the dune (and sometimes from the dune back onto the beach) and removal of sand from the dune by wave action during storms. Aeolian transport into the dunes represents a net loss of sand from the littoral sediment budget and ultimately leads to narrowing of the beach and greater susceptibility to wave action unless there is sufficient replenishment from updrift or offshore. During storms, elevated water levels and high waves lead to erosion of the embryo dunes, scarping of the foredune and, in extreme events, to destruction of the foredune itself. Wave erosion returns sand to the littoral system and thus is an input to the littoral sediment budget. The return of sediment to the beach as well as the temporary protection offered by the dune as a physical barrier to waves act to protect the area landward of the dune from flooding and storm wave action. However, where waters overtop the foredune sand is moved inland producing a depositional overwash fan and providing a base further inland for new foredune establishment. This process is particularly important for preserving a sand supply for foredune formation in the face of sea level transgression.

Transport from the beach to the dunes does depend on the existence of winds with sufficient strength to entrain sand from the beach and to transport it into the vegetated portion of the foredune. The potential aeolian sand transport can be estimated from regional wind statistics using one of the standard sediment transport formulae (Bagnold, 1941; White, 1979) though there is quite a wide range of predictions from these. Regional wind statistics can be used to determine the sand drift potential (Pearce and Walker, 2005) which allows for comparison between different locations. This approach gives the maximum potential transport and must be modified by the cosine of the wind angle to shore perpendicular to convert the transport rate into deposition per metre length alongshore. Even so, measured values of deposition are often much lower than predicted (Davidson-Arnott and Law, 1990, 1996) as a result of factors which act to limit sediment supply from the beach

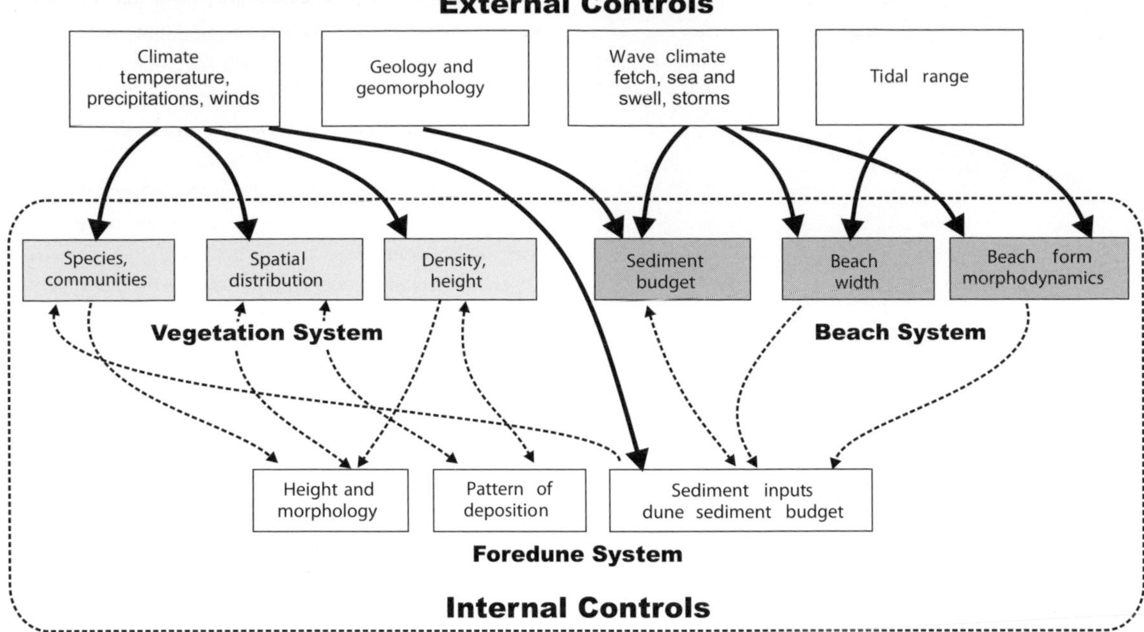

FIGURE 5.13. Internal and external controls on coastal sand dune development.

surface (Nickling and Davidson-Arnott, 1991). These 'supply-limited' factors include beach surface moisture, lag pebble and cobble cover, snow and ice (Fig. 5.12b) and the effects of fetch restrictions (Davidson-Arnott and Law, 1990; Gillette *et al.*, 1996). The beach width provides a minimum fetch for aeolian transport and this in turn reflects the operation of some of the controlling factors noted in Fig. 5.13. In general wide beaches are associated with a positive sediment budget, high wave energy and large storm events, and increasing tidal range. The fetch distance on a beach is a function of the beach width and of the angle of wind approach, with the shortest fetch associated with direct onshore winds and the fetch becoming infinite for alongshore winds. These factors have been included in a framework for modelling sediment transport from the beach outlined by Bauer and Davidson-Arnott (2003).

Beach morphodynamics also influences the potential transport of sand from the beach and, on a decadal scale, differences in supply may be a response to the influence of beach width and beach profile form (Short and Hesp, 1982). At one end of the morphodynamic continuum, dissipative beaches are characterised by high potential for aeolian sediment transport and a high frequency of foredune destabilisation by storm waves which initiates episodes of transgression. At the other end of the continuum, low-energy reflective beaches are characterised by low potential sediment transport and few erosional events leading to stable foredune development and limited inland transport of sand. Intermediate beach types have higher probabilities of foredune destruction and the development of large-scale parabolic dunes (Short and Hesp, 1982).

Littoral sediment supply, beach sediment budget and long-term progradation or retrogradation (shoreline recession) as well as sea level changes also affect both the beach and foredune forms and dune sediment budget. Sea level fall or beach progradation leads to the development of a series of low, parallel foredune ridges, while a stable or rising sea level and/or negative beach sediment budget tend to produce shoreline recession and a single, large foredune (Psuty, 2004). The result is that foredune evolution on a decadal to century scale is complex and is driven by a range of local and regional factors that influence sediment supply, beach form and climatic factors (Hesp, 2002) (Fig. 5.14).

The development of a large foredune system and of extensive dunefields requires the presence of large amounts of sand. In the classic case a progradational dunefield develops under relatively stable sea level and littoral processes with a constant supply of sand being delivered to the beach and a positive beach and dune sediment budget. Such a system is responsible for the formation of the 3-km wide dunefield that has developed over the past 2500 years at Long Point on Lake Erie (e.g. Davidson-Arnott and

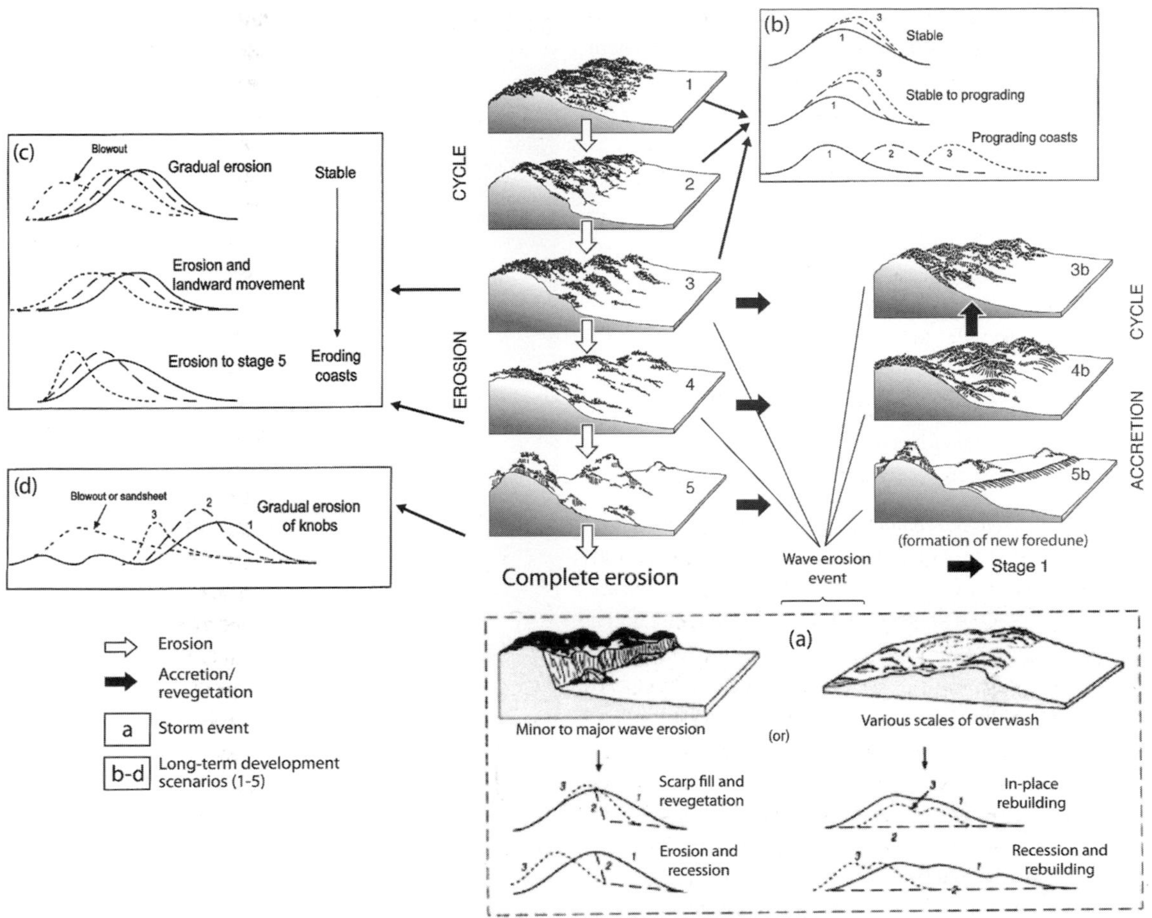

FIGURE 5.14. A model of established foredune morphology, dynamics and evolutionary trends on stable, accreting and eroding coasts. A foredune may develop towards and remain in a particular morpho-ecological stage (types 1 to 5), or it may evolve to another stage over time (e.g. from types 1 to 2 in the erosional cycle, or types 3 to 3b in an accretional/revegetational cycle). Box (a): Wave erosion events may occur at any time leading to minor to major scarping and/or overwash. Foredunes may recover, rebuild landwards or possibly be completely removed. Box (b): In the long term, foredune may be relatively stable gradually building in place (stable coasts), slowly build up and seawards (stable to prograding coasts) or be replaced by new foredune development (prograding coasts). Other long-term evolutionary scenarios are indicated in Boxes (c) and (d) for the increasingly erosional foredune stages (modified from Hesp, 2002, their Figure 3).

Fisher, 1992). The relict foredune ridges that make up the dunefield are time transgressive with sediment being added continuously and lake level and wave climate remaining within fairly narrow bounds. In contrast to this uniformitarian perspective, the foredune system and inland dunefield at Greenwich Dunes, Prince Edward Island (Hesp *et al.*, 2005) have formed since a catastrophic mid-latitude storm event around 1930 which produced a continuous overwash terrace extending 300–1000 m inland along a 5-km long stretch of coast (S. Mathew, personal communication, 2007). The overwash transported large volumes of sediment inland and the presence of extensive areas of bare sediment eventually resulted in the formation of nebkha fields, and active transverse and barchan dune forms 15–20 m high formed by the prevailing westerly winds and by northwesterly and northeasterly winds during the passage of storms. These transgressive dunes (Fig. 5.12a) migrated inland across agricultural fields and forest until the re-establishment of vegetation on the backbeach led to development of a new foredune, cutting off sand supply to the dunefield and permitting the establishment of vegetation on the dunefield. Similar events are produced by tropical storms, notably the hurricanes that frequently produce extensive overwash events on the barrier islands of the US East and Gulf coasts (Stockdon *et al.*, 2007).

Dunefields

Dunefields may be found inland of the primary dune system and result from either:

(a) seaward migration of the shoreline due to a positive littoral sediment budget and/or a fall in sea level, leading to the development of a new primary dune system seaward of the original foredune. As a result the old foredune ridges form one or more parallel relict foredune ridges landward of the existing foredune with their age increasing away from the beach; or
(b) disturbance of the vegetated foredune system and/or of stabilised dune deposits on the foredune plain leading to blowouts, parabolic dunes and the inland migration of an unvegetated or partially vegetated transgressive dunefield.

The growth of a new foredune ridge results in the reduction of sand supply to the old ridge landward, as well as increased protection from salt spray. These reduced stresses lead to the growth of vegetation that is less tolerant of burial and salt spray and ultimately to colonisation by bushes and trees. In some locations a continuous vegetation cover is maintained and the relict foredune ridges may be preserved intact for decades or centuries. Deflation of sand between the dune ridges or an increase in the water table height results in the creation of wet areas or ponds in the swales between the ridges and promotes the growth of vegetation that is tolerant of high water content or waterlogging.

In many areas dunefields develop landward of the foredune as a result of destabilisation of vegetation on the foredune or older dune ridges which in turn leads to sand being blown inland. In dry areas, or areas where colonising plants are unable to withstand large burial depths, the foredune may always be sparsely vegetated. This limits the amount of sand deposition in the foredune zone, and large amounts of sand are transported directly inland. In such areas there is a nearly continuous supply of sand to the inland area and the development of freely migrating or partially stabilised dunes. Where the migrating dune forms are more than a few metres high the sand tends to bury and kill vegetation as it moves.

Destabilisation of vegetated dunes can occur locally as a blowout, defined as a saucer-, cup- or trough-shaped depression (Hesp, 2002) with depth and elongation increasing from the relatively shallow, round saucer shape. Vegetation disturbance leading to blowout development may be initiated by physical processes such as storm wave erosion of the front face of the foredune, topographic acceleration of airflow over the dune crest or around topographic irregularities or local burial of plants by excessive sand deposition. It may also result from drought and fire, or from the effects of trampling and overgrazing by animals; and often from a wide range of human activities (Hesp, 2002). Once formed, the morphology of the blowout affects air flow in the vicinity, leading to topographic steering and local flow accelerations which often enhance the deepening and elongation of the blowout (Gares and Nordstrom, 1995; Hesp, 2002). Blowouts in vegetated dunes may stabilise over time depending on the particular combination of wind and weather conditions, sand supply and vegetation propagation. However, where there is a large supply of sand in the foredune or in older deposits on the dunefield blowouts may evolve into parabolic dunes.

FIGURE 5.15. Development of a parabolic dune in a transgressive dunefield, New South Wales coast, Australia. (a) View landward along the erosional upwind end of the parabolic dune showing the steep sidewalls of the trailing arms and the erosion of a basin down to the water table. The active depositional lobe can be seen in the background. (b) Photograph of the active slip face of the parabolic dune showing burial of vegetation. (See also Plate 17 for colour version.)

Parabolic dunes may be initiated in the foredune system and migrate landward into older dune deposits or they may be initiated in the dunefield itself. Parabolic dunes consist of a high, U-shaped migrating dune with an active slip face at the landward margin. They have an elongate erosional basin on the upwind side (Fig. 5.15a) which supplies sand to the active depositional lobe (Fig. 5.15b; Plate 17), and trailing ridges which are U-shaped or hairpin-shaped and form the sides of the basin (Hesp, 2002). Over time the upwind end of the trough tends to become stabilised as sand movement is reduced by erosion to the water table. Eventually sand supply to the slip face and forward migration of the dune ceases as a result of the spread of vegetation. The height of the migrating dune is a function of the volume of sediment supply/availability as well as of the height of the vegetation on the landward side. Where the vegetation cover consists of grasses and low bushes the slip face is also low and the rate of forward migration can be several tens of metres per year but if the dune is advancing through forest cover then the reduction in wind speed at the forest edge results in the building of the front wall to form a precipitation ridge that results in burial and death of the trees. The rate of landward migration is then much slower but can take place over many decades and reach several kilometres inland (Clemensen *et al.*, 2001; Hesp and Martínez, 2007).

In many cases periodic disturbance on a timescale of several decades to centuries produces a succession of overlapping dune forms with sediments from the older episodes being buried beneath younger deposits (Clemmensen *et al.*, 2001). In New South Wales, Australia, transgressive dunefields show multiple phases of development with periods of stability and foredune growth followed by disturbance and inland migration of sand leading to burial of older transgressions (Lees, 2006). The episodic transgressive dunefield development may result from disturbance caused by large storm(s), drought or human activities and may be moderated by the need to develop a critical mass of sand in the foredune so that the migrating dune front is able to bury vegetation landward of it.

Where large volumes of sand are supplied to the dune (large positive littoral sediment budget and consequently positive dune sediment budget) foredune development may be either discontinuous in space or time depending on the relative ability of the plant communities to withstand burial. Transgressive dunefields develop characterised by sand sheets and large-scale sand seas or dunefields with the downwind borders consisting of precipitation ridges and the development of deflation basins and plains upwind (Hesp and Martínez, 2007). Some of the largest transgressive dunefields have been described from the coast of Brazil (Hesp *et al.*, 2007) and others from the east coast of Australia (Hesp and Thom, 1990), and South Africa (Illenberger and Rust, 1988). The Alexandria dunefield in South Africa has a nearly continuous sand supply and limited vegetation cover close to the coast producing a continuously expanding dunefield and downwind migration of the precipitation ridge forming the landward boundary.

Vegetation–dune interaction

Coastal foredunes develop in response to the presence of vegetation which serves both to promote deposition by reducing wind flow near the bed and to stabilise the resulting dune deposit by limiting subsequent entrainment. The form and evolution of the dune on a scale of years to decades is greatly influenced by the form and growth habits of the colonising vegetation, by its ability to adapt to a number of types of disturbance (e.g. burial by sand, sand blasting, wave erosion) and to factors which inhibit growth (such as salt spray, high surface temperatures, moisture availability and lack of nutrients). In most dunes there is a distinct gradient away from the beach, both in the exposure to disturbance and to limiting factors, and this often leads to the development of a well-defined sequence of vegetation associations along this gradient (Doing, 1985; Hesp, 1991; Dech and Maun, 2007).

The classic model of plant succession on dunes (Cowles, 1899) emphasised the role of a succession of plant species in increasing stabilisation of the surface through time and the improved moisture retention and nutrient availability associated with the replacement of grasses by bushes and then forest. The model has some validity for progradational dune systems but it is now recognised that temporal evolution at any one location is probably more a reflection of spatial (morphological) evolution of the dune system leading to a reduction in disturbance and limiting factors such as exposure to salt spray and moisture stress (Miyanishi and Johnson, 2007). Thus, in the absence of salt spray, cottonwood trees (*Populus deltoides*) are a common coloniser on Great Lakes coastal dunes even in areas with a high sediment supply. Trees that are adapted to salt spray (such as coconut and manchineel) are common close to the shoreline on low-energy tropical beaches where dune accretion rates are small. Where the beach sediment budget is near neutral and the dune budget is positive there may be little change in the gradient of salt spray and sand burial over decades, and consequently little change in the plant communities. On the other hand, disturbances due to erosion by waves, blowout and parabolic dune development and rising water tables can quickly lead to a change in the dominant vegetation. Thus, primary dune colonisers such as *Amophila* or *Ipomoea* can be found in inland locations where there has been local disturbance and reactivation of sand movement, e.g. associated with blowout development. Alternatively foredune growth may lead to an increase in the water table in the dune slack landward of the bottom of the lee slope and thus to replacement of plants associated with dry conditions by plants such as rushes and sedges.

Doing (1985) divided the foredune system into six zones and summarised the zonation of plant communities within these for a range of locations worldwide. The basis for the zonation is the degree of sheltering from burial by sand and to a lesser extent the effect of disturbance by wave action and salt spray. The zones tend to be compressed in areas where sand supply is limited (narrow beaches on rocky coasts; wind climate with a relatively small proportion of winds capable of onshore sand transport) or where the trapping ability of the pioneer species is high. In areas with high sand supply and/or plant communities with relatively poor ability to withstand burial the zones may be quite wide and diffuse as disturbance extends inland for some distance. The morphology and evolution of the embryo dune and foredune are influenced by the growth form of the particular plant species that are characteristic of a zone in a particular climatic region. Thus, for example, in tropical and subtropical areas the dominant vegetation of the embryo dune zones (2 and 3) is *Ipomoea pes-caprae* and *Spinifex*, both of which have surface runners which cover the sand surface and tend to produce a broad, relatively flat embryo dune. In contrast the dominant coloniser of these zones in eastern Canada, the Great Lakes and the northeastern United States is marram grass (*Ammophila breviligulata*) which grows in high, dense clumps and extends seaward through the growth of subsurface rhizomes. Foredune development associated with marram tends to produce much narrower, sharp crested and hummocky ridges compared to those associated with *Ipomoea* and *Spinifex*.

Foredune development and dune stability also depend on the ability of the dune plant communities to trap sand, and in areas of high sand supply to withstand burial to depths up to >0.5 m. *Ammophila breviligulata* is able to survive burial of 1 m or more (Maun and Lapierre, 1984) and propogates rapidly. As a result, foredunes in these areas tend to be relatively stable and major episodes of instability and parabolic dune formation often result from the impact of human activities including agriculture and recreation rather than natural phenomena. Where the pioneering species are less tolerant of burial transgressive dunefields may be more common. The classic Coos Bay transgressive dunefield described by Cooper (1958) developed because of the inability of native species to withstand burial and thus foredunes were nearly absent. The introduction of *Ammophila arenaria* from Europe has led to much greater stabilisation and the subsequent formation of a foredune system. Similarly in New Zealand the native pingao grass is relatively slow growing and unable to withstand high rates of burial, leading to generally higher degree of foredune instability and development of transgressive dunefields. Introduction of *Ammophila arenaria* has changed the characteristic morphology of the foredune system and led to greater stabilisation.

There is considerable interaction between the airflow and vegetation characteristics such as form, height and density. The stems and leaves of vegetation act to retard the flow close to the bed and this effect is increased with increasing vegetation height and density. As might be expected the rate of deposition at the leading edge of the foredune increases with increasing plant height and density and thus tends to build a narrow, high ridge (Hesp, 1983; Arens *et al.*, 2001). Thus, as noted earlier, foredunes formed under a cover of *Ammophila* tend to be higher and narrower than those formed under a cover of *Spinifex* or *Ipomoea*. In temperate regions where there is a winter die-back in vegetation, the effect of this is somewhat reduced because the effective height and density of the vegetation is reduced as sand accumulates and this is reinforced by leaf loss as the plant dies off for the winter, permitting sand to be transported further landward over the late autumn and early winter. The result is that there are marked seasonal variations in the pattern of sand movement and deposition and in the sheltering or stabilising effect of vegetation (Law and Davidson-Arnott, 1990). The actual effectiveness of vegetation as a trap for sand blown inland from the beach is also controlled by its rate of growth. In tropical areas low, rhizomatous plants such as *Ipomoea* may not have a large storage capacity at any instant in time, but because they grow rapidly all year long they may be as effective as tall grasses, such as *Ammophila*, in temperate regions where the vegetation is actively growing for only a few months in the year. Finally, the physical form of plants also affects how they trap blowing sand. Saltating sand grains tend to collide with upright, tall grass stems and then fall to the ground at the plant base (though this effect is reduced by the flexibility of the leaf stems which allows them to bend until they are nearly parallel to the wind producing a streamlined form that reduces the forces on the plant). The relatively broad, flat leaves of *Ipomoea* tend to promote splash rebound of sand grains and transport may occur for a greater distance inland before all of the sand is removed from the airflow.

5.4.2 Twenty-first century changes in dune processes and landscapes

The nature of human impacts on coastal dunes in the twenty-first century will depend in large part on the effectiveness of local management and conservation programmes, with great variability across systems. The impacts of global climate change on coastal foredune systems and on inland dunefields however can be generalised and are most likely to reflect changes in temperature and precipitation, intensity and frequency of storm events and increasing rates of relative sea level rise.

The increase in global temperatures and changes in precipitation can potentially have an impact on the growth and density of plants, both in the foredune system where disturbance is relatively high, and in areas where there is an extensive dunefield landward of this system. Increased temperature, especially if accompanied by reduced precipitation has the potential to reduce the density of plant cover and thus to destabilise the foredune system. This effect is likely to be greatest outside the humid tropics, and may enhance the presence of unvegetated dunes around the margins of the latitudinal desert zones where moderate rainfall is accompanied by a distinct dry season. The impact may be relatively modest for foredune systems where the vegetation is already adapted to high temperatures and moisture deficiencies but it may increase the frequency of disturbance due to fire and drought years, especially in places such as the east coast of Australia and Mediterranean systems such as those in southern Europe and California. Such sensitivity to climate variation has been documented for several coastal dunefields as well as for inland dunefields such as those in the Great Plains of Canada (Hugenholtz and Wolfe, 2005) and western Argentina (Tripaldi and Forman, 2007).

Increasing temperatures may also lead to poleward migration of plant species which in turn may affect the nature of plant communities, most significantly those in the embryo dune and foredune zones where stresses are greatest. Because temperature increases are likely to be small in the tropics, the most significant effect is likely to be at high latitudes and especially in the Arctic (Kaplan and New, 2006) and on the east and west coasts of Canada and Alaska. The most significant impacts are probably an increase in plant density in some areas and thus greater trapping efficiency near the beach, and the possibility that species with a high trapping ability and resistance to burial such as *Ammophila* and *Elymus* will extend their distribution towards higher latitudes. Reduced ice and snow cover will affect beach/dune dynamics, leading to enhanced sand transport into the dune but also increased exposure to storms over a year and thus greater potential for dune erosion (Fig. 5.12b).

Warming in the Arctic is already leading to reduced ice cover and increasing both the duration of open water and the open water fetch (Shimada *et al.*, 2006). This will have the effect of increasing wave action leading to greater coastal erosion (especially in the western Arctic) and this in turn may provide a greater supply of coarse sediments for beaches and increased beach width thus enhancing the potential for sand supply to coastal dunes. Presently in the western Arctic dune development is rare, in part because of the extensive ice cover and in part because of the relatively sparse vegetation cover near the beach – the exception being small dunes developed around the margin of thermokarst lakes which may be exposed at the shoreline (Ruz, 1993). This is in contrast to the east side of Hudson Bay where foredune development occurs in areas where rivers bring fresh sediment supply to the coast and foredune stabilisation is associated with the presence of *Elymus arenaria* (Ruz and Allard, 1994). With increasing temperature conditions may be more favourable for the spread of *Elymus* northward along the shoreline of Hudson Bay and westward thus potentially promoting greater dune development in areas where sand is abundant.

With the exception of the Arctic regions most direct impacts of effects of climate change – temperature and precipitation – on coastal foredune systems are likely to be relatively small and difficult to predict locally because of the dominance of other controlling factors and the relative resiliency of vegetation communities.

One of the scenarios for global warming is the potential for increased frequency and/or intensity of storms. Increased storm frequency might decrease the recovery time between storms, especially the ability of pioneering vegetation to recolonise the base of the stoss slope of the foredune and for the development of new embryo dunes on the backshore. The result would be an increase in the proportion of the foredune that is scarped (both spatially and temporally) and this in turn may increase the potential for sand transport over the foredune crest and onto the lee slope. Increased scarping may be accompanied by increased dune ramp development which can aid the process of landward sand transfer. Increased intensity of tropical storms, an increased frequency of intense storms (e.g. category 4 and 5 hurricanes) and increased intensity of mid-latitude cyclones also potentially impact dune systems. On stable sandy mainland coasts it is likely that the beach profile would adjust to such changes and the impact on dunes may be negligible. However, combined with sea level rise, increased storm intensity, and particularly increased storm surge will likely result in increased barrier overwash and thus an increase in the pace of onshore migration and the proportion of space and time that the barriers are occupied by washover fans and regenerating dunes.

Probably the greatest potential impact of climate change is due to enhanced global sea level rise. The impact of this is global in scale though its significance varies locally – for example in eastern Canada and parts of northwest Europe where this is countered by ongoing isostatic rebound, and locally where a large sediment input to the littoral system produces progradation which acts to counter the effects of transgression caused by sea level rise. On low-lying coasts characterised by sandy beach and dune systems, simple inundation due to sea level rise will produce a large displacement of the shoreline and wave erosion will result in further landward displacement. A key component of coastal response to sea level rise is the fate of the sediments in the dune systems on mainland and barrier coasts. The Bruun model (Bruun, 1962) predicts offshore movement and storage on the outer nearshore profile – the sediment is thus lost from the dune system. The model is only of limited applicability to coastal systems due to the important role of nearshore and coastal morphology in sediment movement, and in the case of dunes recent work shows that much of the sediment eroded during sea level transgression will probably move landward through inlets, overwash and landward migration of the foredune (Dean and Maurmeyer, 1983; Davidson-Arnott, 2005). In effect, the foredune system and dunefield acts as a sink in the same way that tidal inlets, lagoons and estuaries do (Davidson-Arnott, 2005).

The most likely effect of a relatively small increase in sea level (<1 m) over the next century is that we can expect an increase in the frequency and magnitude of storm events leading to erosion and scarping of foredunes on mainland coasts and to overtopping and washover production on barrier coasts. The foredunes in most areas will likely migrate landward in pace with the sea level rise and most will increase in size. On coastlines with strong gradients in littoral transport, the presence of deep embayments or other complicating factors, the rate of transgression and the response of coastal foredunes will be complicated and difficult to predict. The challenge is to permit landward migration to occur and to manage human activities so as to permit this to occur naturally while providing sufficient flexibility to accommodate the complexity of local shoreline response.

5.4.3 Implications of geomorphic change

Understanding the dynamics of coastal dune systems and how they can be resilient to the changes in drivers associated with twenty-first century climate change provides an important message to coastal managers: dunes can be maintained under sea level rise as long as disturbance by people is limited and they are given a fighting chance. As noted by van der Meulen *et al.* (2004) there was a switch in management approaches in the late twentieth century away from nature conservation focussed on species and their habitats, to an increased focus on the physical and biological processes driving the provision of those habitats, and toward a more landscape approach. These authors also note that at the start of the twenty-first century, public expectations of the products or services that dune systems provide for them are increasing.

The Meijendel Dunes, north of The Hague in the Netherlands, provide a good example of the need for dune management that appreciates the variety of geomorphic services they provide. This dune system is close to major population centres and thus a focus for recreation, as well as being a nature reserve, a drinking-water catchment for those same population centres, and part of the coastal defence system for the low-lying western Netherlands. The solution to these potentially conflicting uses of the dune system was addressed in 2000 in a new management plan which focussed on zoning (van der Meulen *et al.*, 2004). Different parts of the 2000-ha dune system are designated for coastal defence, drinking water, nature conservation and recreation. This includes the identification of a 500-ha Natural Core Area where the emphasis is on natural hydrologic and geomorphological processes.

Such planning processes that appreciate the multiple geomorphic services dunes provide are likely all the more necessary in the twenty-first century. Current understanding of both dunes and human use of coastal systems indicates that agriculture will intensify in these sandy, low-lying well-drained areas, port facilities and urbanisation will have continuing demands for coastal lands, infrastructure pressure will increase especially near tourist centres and climate will alter the sediment–vegetative interactions so essential to these systems (Martinez *et al.*, 2004). Dunes are threatened, but in contrast to estuaries, tidal flats and marshes where climate change imposes alterations less amenable to local solutions, their future depends on the ability effectively to incorporate our understanding of geomorphic change into effective coastal management plans.

5.5 Managing coastal geomorphic systems for the twenty-first century

Many coastal systems live with the legacy of centuries of human activity. This influences geomorphic processes at a variety of timescales (Fig. 5.1) from estuarine circulation to sediment deposition and in some systems exerts an overriding control on process–form relations. Importantly, these alterations are in many cases the result of human exploitation of the geomorphic services, from navigation routes to drinking water to fisheries, provided at the coast. A common feature of all three coastal environments examined in this chapter is that they all have the ability to migrate and change with changing sea levels. Estuaries, marshes, mudflats and coastal dunes are each capable of adjusting to sea level rise of the magnitude forecast over the next century by transgressive movement over adjacent terrestrial systems as there are no landward constraints on that movement. This reiterates one of the messages found elsewhere in this volume: that in the twenty-first century the direct consequences of human actions on geomorphic systems, including the coast, are probably more significant than the indirect effects of climate change.

The discussion in this chapter has demonstrated that in the twenty-first century it is essential not only to learn from the past and mitigate to the extent possible past detrimental changes, but also to consider the interactive effects of human activity and global climate change (Fig. 5.1). In many cases, this interaction will be mediated through alteration in sediment supply to the coast, sediment dynamics at the coast, and the role of biota in stabilising or disturbing coastal sediments. For example, this chapter has shown how both estuarine dynamics and the geomorphic services they provide can be influenced by changes in river inflows; how the protection provided by coastal marshes in the future will depend in many areas on sediment supplies; and how dune systems are sensitive to the effectiveness of plants in trapping available sediments. While examining the process interactions by geomorphic system provides insight into their complexity and geographic variation, the challenge for twenty-first century coastal management is the application of our geomorphic knowledge to place-based problems involving many geomorphic subsystems.

On most coasts, the problems faced by estuaries, marshes and dunes interact, and the pressures of both past and human activities cut across these environments. There may also be important trade-offs among the geomorphic needs of these systems as well as between human use and geomorphic sustainability. This had led to a variety of approaches to managing coastal systems that variously incorporate the 'human dimension' into site-specific problem-solving (e.g. integrated coastal zone management (Clark, 1992), community-based coastal management (McClanahan *et al.*, 2005) and ecosystem-based management (McLeod *et al.*, 2005)). Notably, the solution to many coastal problems often requires consideration of factors beyond the coast or the local area, and integration of understanding across geomorphic systems. The complexities of twenty-first century coastal management and the need to provide geomorphic services into the future is illustrated here by the Danube Delta.

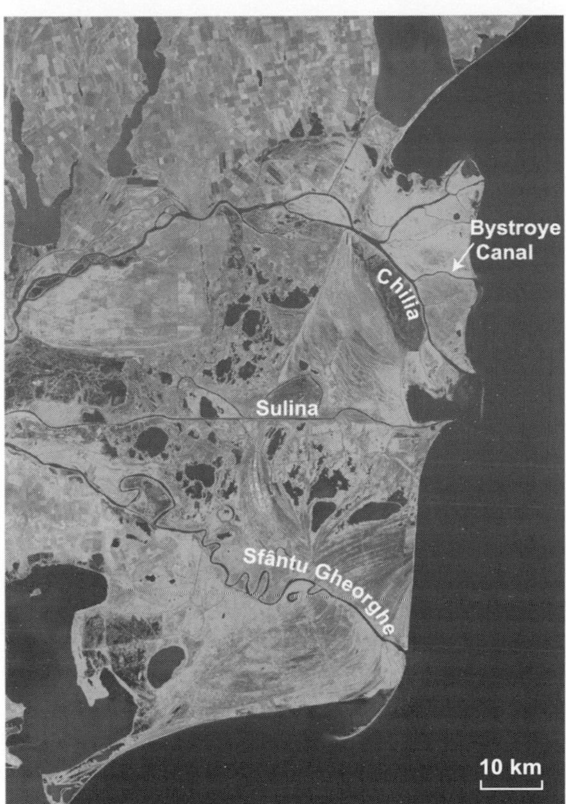

FIGURE 5.16. The Danube Delta. (See also Plate 18 for colour version.)

5.5.1 Danube Delta case study

The Danube Delta on the Black Sea coast of Romania has long been of importance to people. The very long history of trading along the Danube is evident from remains of Greek and Roman settlements (including a lighthouse). It flows through major cities of Eastern Europe like Vienna and Budapest. The Danube is navigable by ocean ships from the Black Sea to Brăila in Romania and by river ships to Kelheim, Bavaria. Since the construction of the German Rhine–Main–Danube Canal in 1992, the river has been part of a trans-European waterway from Rotterdam on the North Sea to Sulina on the Black Sea (3500 km).

The Danube Delta began forming around 11 000 BC in a gulf of the Black Sea, when the sea level was 50 to 60 m lower than today. An offshore bar blocked the gulf into a lagoon, which subsequently led to its filling with sediments. The entire delta region now covers 799 000 ha of which 679 000 ha are in Romania and 120 000 ha in the Ukrainian SSR. The three main distributaries in the delta are the Chilia, the Sulina and the Sfântu Gheorghe (Fig. 5.16; Plate 18). The Chilia is the youngest of the three delta lobes and naturally receives most of the flow (3800 $m^3 s^{-1}$) but has the longest channel. Large-scale modification of the Delta began with the Sulina in 1862 and continued throughout the twentieth century. Overall, the length of the Sulina, already the shortest of the three, was reduced from 92 to 64 km, and its flow more than doubled, to make it suitable for large-vessel navigation. Jetties on the mouth of the Sulina split the coast into two separate geomorphic units, and have resulted in both new spit and lagoon features to the north as well as shoreline erosion to the south (Stanica *et al.*, 2007). The length of the Sfântu Gheorghe was also reduced to 70 km from 108 by cutting off meanders, and its flow slightly increased. The Chilia region of the Delta remains the least disturbed by human activity as most of the older settlements and the ports are further south.

The Delta has also been subject to change upstream. The construction of the Iron Gate I and II dams, which began operation in 1970 and 1984, reduced sediment discharge on the Danube by ~50% at the delta (Panin and Jipa, 2002). The total average Danube sediment discharge is now less than 25–35 million t a^{-1}, of which 4–6 million t a^{-1} are sand. Panin and Jipa (2002) note that after the changes on the River Danube and the decreasing sediment discharge, the Danube Delta became mainly inactive, apart from two restricted sections (Kilia distributary delta and the small St Georghe distributary delta). Giosan *et al.* (1999) have documented the shoreline change resulting from the reduction in sediment delivery to the coast.

By 1990, 25% of the Danube Delta was constrained by levees and embankments, including 400 km^2 used for agriculture. However, the extensive wetlands of the Danube Delta, including the largest reed-bed in the world covering 180 000 ha, remains. The overall basic hydrological and ecological systems of the Delta, although strongly degraded, are considered intact and continue to provide breeding and foraging habitat for pelicans and over 300 other bird species, and supporting over 3400 species of aquatic fauna (over 98% of the European total).

In 1991, the Danube Delta was designated a Wetland of International Importance under the 1971 Convention on Wetlands of International Importance (commonly known as the Ramsar Convention); the Delta was inscribed on the World Heritage List in 1991 and internationally recognised as a Biosphere Reserve under UNESCO's Man and the Biosphere Program in 1992. Within the Delta Biosphere Reserve (covering some 679 222 ha) 18 145 ha are included in a separate biosphere designation (the core zone covers two-thirds of the area, the peripheral areas forming the buffer zone) and 43 790 ha in seven nature reserves (two of which overlap with the Biosphere Reserve).

The Danube Delta shows the legacy of previous human activities on a geomorphic system valued for trade, fresh water and food. The role of upstream changes on the delta illustrates the sensitivity of coastal systems to changes in forcing from land or sea. It is also an area of important biogeomorphic interactions. Due to the limited tidal range of the Black Sea, extensive tidal flats are not an important system component. However, the extensive marshes attest to the key role of river and sediment influence on biota and the significant resources that result. Less well known are the influences of the river and biogeomorphic interactions in controlling dune building on the Sfântu Gheorghe coastline. Here, Vespremeanu-Stroe and Preoteasa (2007) have identified seasonal variations in coastal water levels driven by Danube discharge as a key control on beach width and thus sediment supply to the dunes. Further, the interaction between these river-driven cycles, seasonal wave activity and annual variations in vegetation density lead to distinct intra-annual changes in foredune volume: increasing during April–December and then rapidly decreasing due to storm activity from December to April (Vespremeanu-Stroe and Preoteasa, 2007).

The recognition of the value of this system in the 1990s set a new course for its management in the twenty-first century. There are already signs of change. Between 1994 and 2003 approximately 15% of the area previously enclosed by levees has been restored to natural interaction with the river. However, the value of this system for the economies of the surrounding area still exerts pressure. In 2004, the Ukrainian government began construction of the Bystroye Canal which was planned to pass through the Ukrainian Danube Delta Biosphere Reserve. The idea was to revitalise the shipping industry in the Ukrainian part of the Delta and to provide navigation access in addition to the Sulina Canal in Romania. There were many objections from environmental groups and other governments to the project. Phase I was completed and larger, seagoing ships started using the Bystroye Canal. However, the canal quickly silted up and became unusable. Dredging to reopen the canal began in November 2006 and was completed in April 2007. Since then, ships have been once again using the canal. Phase II of the project, which includes deepening of the canal as well as construction of a dyke into the Black Sea, began in summer 2007.

Global climate change predictions for this area indicate the potential for an increase in coastal storms in the Black Sea and continued vulnerability to sea level rise along the coast, while the Danube drainage basin is likely to see increased frequency and magnitude of winter floods (IPCC, 2007). Sea level changes in the Black Sea have been closely correlated with the North Atlantic Oscillation (Stanev and Paneva, 2001) suggesting complex cycles of forcing in the future. The outcomes of interactions amongst these potential changes in inflow regime, coastal forcing and increases in air temperature are not the only challenge facing the Danube Delta in the twenty-first century. Pringle (2001) notes that long-term management of this system requires both integrated watershed management and international cooperation. Such cooperation must occur not only among the 12 countries in the Danube's watershed, but also among those in the Black Sea watershed that encompasses the Danube drainage. In such cases scientific understanding of the complexities of estuarine, wetland and dune response to twenty-first century climate change across the basin and out to sea can provide a solid foundation for future planning.

5.5.2 Conclusions

Conflict between environment and economy is common in coastal systems. The importance of the geomorphic services coastal systems provide for local, regional and international economies cannot be overstated. Climate change will impact coastal systems not simply through sea level rise, a driver of change frequently considered in assessments of coastal marshes, mangroves and beaches. It will also be manifest through changing patterns of water and sediment delivery to the coast, alterations in water temperature and moisture availability at the coast, and changes in flora and fauna from microphytobenthos to wading birds. As shown by the many examples in this chapter, these changes will occur at a variety of temporal scales from diurnal to multidecadal. And importantly, none of these changes occurs in isolation from the others. Human pressures on coastal systems will not diminish in the future. The Danube Delta illustrates the value coastal systems can have both environmentally and economically through the provision of geomorphic services such as navigation, fisheries and water filtration. The case study also shows that thus far there are few ways to relieve the tension between immediate exploitation and the sustainability of service provision. Multiple jurisdictions, across the land–sea interface, among countries and even among governmental agencies within countries, further confound the problems of coastal management. Ecosystem-based management, and its focus on ecosystem services, has thus far been most successfully applied to marine systems and the development of Marine Protected Areas which balance the needs of the fish, the wider ecosystem and the fishery. As shown in this chapter, at the coast, changing landforms and geomorphic processes also have direct consequences for society.

Improved management of coastal systems in the twenty-first century will require an emphasis on geomorphic services. This will necessarily embrace the land–sea and biogeomorphic interactions emphasised in this chapter, as well as an explicit recognition of the needs of local communities and national and international economies. Estuaries, marshes and dunes must be assessed in terms of their geomorphic dynamics and their value to society, be it for navigation, fisheries or tourism. Geomorphologists must be ready for this challenge and bridge the communication gap between ecologists and engineers. The future of many coastal systems depends on it.

Acknowledgements

The authors thank Heather Viles, Patrick Hesp and an anonymous reviewer for their helpful comments on an earlier version of this chapter. Brendan Yuill is acknowledged for his assistance with figures, permissions and final manuscript preparation. This article is a contribution to the SCOR-LOICZ-IAPSO WG 122 'Mechanism of Sediment Retention in Estuaries.'

References

Allen, D. M. *et al.* (2007). Geomorphological determinants of nekton use of intertidal salt marsh creeks. *Marine Ecology Progress Series*, **329**, 57–71.

Allen, J. R. L. (2000). Morphodynamics of Holocene salt marshes: a review sketch from the Atlantic and Southern North Sea coasts of Europe. *Quaternary Science Reviews*, **19**, 1155–1231.

Amos, C. (1995). Siliciclastic tidal flats. In G. M. E. Perillo, ed., *Geomorphology and Sedimentology of Estuaries*. Amsterdam: Elsevier, pp. 273–306.

Arens, S. M. *et al.* (2001). Influence of reed stem density on foredune development. *Earth Surface Processes and Landforms*, **26**, 1161–1176.

Argow, B. A. and FitzGerald, D. M. (2006). Winter processes on northern salt marshes: evaluating the impact of *in situ* peat compaction due to ice loading, Wells, ME. *Estuarine, Coastal and Shelf Science*, **69**, 360–369.

Bagnold, R. A. (1941). *The Physics of Blown Sand and Desert Dunes*. London: Marrow.

Barbier, E. B. *et al.* (2008). Coastal ecosystem-based management with nonlinear ecological functions and values. *Science*, **319**, 321–323.

Barras, J. A. (2006). *Land Area Change in Coastal Louisiana after the 2005 Hurricanes: A Series of Three Maps*, US Geological Survey Open-File Report No. 06–1274. Washington, DC: US Government Printing Office.

Bartholdy, J. (2001). Storm surge effects on a back-barrier tidal flat of the Danish Wadden Sea. *Geo-Marine Letters*, **20**, 133–141.

Bauer, B. O. and Davidson-Arnott, R. G. D. (2003). A general framework for modelling sediment supply to coastal dunes including wind angle, beach geometry and fetch effects. *Geomorphology*, **49**, 89–108.

Baumann, R. H., Day, J. W. and Miller, C. A. (1984). Mississippi deltaic wetland survival: sedimentation versus coastal submergence. *Science*, **224**, 1093–1095.

Blanchard, G. F. *et al.* (2000). The effect of geomorphological structures on potential biostabilisation by microphytobenthos on intertidal mudflats. *Continental Shelf Research*, **20**, 1243–1256.

Boesch, D. F. *et al.* (1994). Scientific assessment of coastal wetland loss, restoration and management in Louisiana. *Journal of Coastal Research Special Issue*, **20**, 1–89.

Boesch, D. F., Field, J. C. and Scavia, D., eds. (2000). *The Potential Consequences of Climate Variability and Change on Coastal Areas and Marine Resources: Report of the Coastal Areas and Marine Resources Sector Team, US National Assessment of the Potential Consequences of Climate Variability and Change, US Global Change Research Program*, NOAA Coastal Ocean Program Decision Analysis Series No. 21. Silver Spring: National Oceanic and Atmospheric Administration.

Booij, M. J. (2005). Impact of climate change on river flooding assessed with different spatial model resolutions. *Journal of Hydrology*, **303**, 176–198.

Bruun, P. (1962). Sea-level rise as a cause of shore erosion. *Journal of Waterways Harbors Division, American Society of Civil Engineers*, **88**, 117–130.

Burkett, V., Groat, C. and Reed, D. (2007). Hurricanes not the key to a sustainable coast. *Science*, **315**, 1366–1367.

Cahoon, D. R. (2006). A review of major storm impacts on coastal wetland elevation. *Estuaries and Coasts*, **29**, 889–898.

Carter, R. W. G. and Woodroffe, C. D. (1994). *Coastal Evolution: Late Quaternary Shoreline Morphodynamics*. Cambridge: Cambridge University Press.

Cervellini, P. M. and Piccolo, M. C. (2007). Variación anual de la pesca del langostino y camarón en el estuario de Bahía Blanca. *Geoacta*, **32**, 111–118.

Chan, A. W. and Zoback, M. D. (2007). The role of hydrocarbon production on land subsidence and fault reactivation in the Louisiana coastal zone. *Journal of Coastal Research*, **23**, 771–786.

Clark J. R. (1992). *Integrated Management of Coastal Zones*, Report No. 327. Rome: Food and Agriculture Organization of the United Nations.

Clemmensen, L. B. *et al.* (2001). Sedimentology, stratigraphy and landscape evolution of a Holocene coastal dune system, Lodbjerg, NW Jutland, Denmark. *Sedimentology*, **48**, 3–27.

Cooper, J. A. G. (1993). Sedimentation in a river-dominated estuary. *Sedimentology*, **40**, 979–1017.

Cooper, W. S. (1958). Coastal sand dunes of Oregon and Washington. *Geological Society of America Memoir*, **72**.

Cowles, H. C. (1899). *The Ecological Relations of the Vegetation of the Sand Dunes of Lake Michigan*. Chicago: University of Chicago Press.

Crossett, K. M. *et al.* (2004). *Population Trends along the Coastal United States: 1980–2008*, Coastal Trends Report Series. Silver Spring: National Oceanic and Atmospheric Administration.

Curtis, P. S., Drake, B. G. and Whigham, D. F. (1989). Nitrogen and carbon dynamics in C3 and C4 estuarine marsh plants grown under elevated CO_2 in situ. *Oecologia*, **78**, 297–301.

Dalrymple, R. W., Zaitlin, B. A. and Boyd, R. (1992). Estuarine facies models: conceptual basis and stratigraphic implications. *Journal of Sedimentary Petrology*, **62**, 1130–1146.

Davidson-Arnott, R. G. D. (2005). A conceptual model of the effects of sea-level rise on sandy coasts. *Journal of Coastal Research*, **21**, 1166–1172.

Davidson-Arnott, R. G. D. and Fisher, J. D. (1992). Spatial and temporal controls on overwash occurrence on a Great Lakes barrier spit. *Canadian Journal of Earth Sciences*, **29**, 102–117.

Davidson-Arnott, R. G. D. and Law, M. N. (1990). Seasonal patterns and controls on sediment supply to coastal foredunes, Long Point, Lake Erie. In K. F. Nordstrom *et al.*, eds., *Coastal Dunes: Form and Process*. Chichester: John Wiley, pp. 177–200.

Davidson-Arnott, R. G. D. and Law, M. N. (1996). Measurement and prediction of long-term sediment supply to coastal foredunes. *Journal of Coastal Research*, **12**, 654–663.

Davis, W. M. (1896). The outline of Cape Cod. *Proceedings of the American Academy of Arts and Sciences*, **31**, 303–332.

Dean, R. G. and Maurmeyer, E. M. (1983). Models of beach profile response. In P. Komar and J. Moore, eds., *CRC Handbook of Coastal Processes and Erosion*. Boca Raton: CRC Press, pp. 151–165.

Dech, J. P. and Maun, M. A. (2007). Zonation of vegetation along a burial gradient on lee slopes of Lake Huron sand dunes. *Canadian Journal of Botany*, **83**, 227–236.

Dettinger, M. D. *et al.* (2004). Simulated hydrologic responses to climate variations and change in the Merced, Carson, and American River basins, Sierra Nevada, California, 1900–2099. *Climatic Change*, **62**, 283–317.

Dionne, J. C. (1998). Sedimentary structures made by shore ice in muddy tidal-flat deposits, St. Lawrence estuary, Quebec, *Sedimentary Geology*, **116**, 261–274.

Doing, H. (1985). Coastal fore-dune zonation and succession in various parts of the world. *Vegetatio*, **61**, 65–75.

Dyer, K. (1986). *Coastal and Estuarine Sediment Dynamics*. Chichester: John Wiley.

Dyer, K. (1998). The typology of intertidal mudflats. *Geological Society of London Special Publications*, **139**, 11–24.

Dyer, K. R., Christie, M. C. and Wright, E. W. (2000). The classification of intertidal mudflats. *Continental Shelf Research*, **20**, 1039–1060.

Dynesius, M. and Nilsson, C. (1994). Fragmentation and flow regulation of river systems in the northern third of the world. *Science*, **266**, 753–762.

Evans, G. (1965). Intertidal flat sediments and their environments of deposition in the Wash. *Quarterly Journal of the Geological Society*, **121**, 209–240.

Fagherazzi, S. and Sun, T. (2004). A stochastic model for the formation of channel networks in tidal marshes. *Geophysical Research Letters*, **31**, L21503.1–L21503.4.

Forbes, D. L. *et al.* (1995). Morphodynamic evolution, self-organization, and instability of coarse-clastic barriers on paraglacial coasts. *Marine Geology*, **126**, 63–85.

French, J. R. (1993). Numerical simulation of vertical marsh growth and adjustment to accelerated sea-level rise, north Norfolk, UK. *Earth Surface Processes and Landforms*, **18**, 63–81.

French, J. R. and Reed, D. J. (2001). Physical contexts for saltmarsh conservation. In A. Warren and J. R. French, eds., *Habitat Conservation: Managing the Physical Environment*. Chichester: John Wiley, pp. 179–228.

Gares, P. A. and Nordstrom, K. F. (1995). A cyclic model of foredune blowout evolution for a leeward coast: Island Beach, New Jersey. *Annals of the Association of American Geographers*, **85**, 1–20.

Gillette, D. A. *et al.* (1996). Causes of the fetch effect in wind erosion. *Earth Surface Processes and Landforms*, **21**, 641–659.

Giosan, L. *et al.* (1999). Longshore sediment transport pattern along the Romanian Danube delta coast. *Journal of Coastal Research*, **15**, 859–871.

Goodbred, S. L. (2003). Response of the Ganges dispersal system to climate change: a source-to-sink view since the last interstade, *Sedimentary Geology*, **162**, 83–104.

Hesp, P. A. (1983). Morphodynamics of incipient foredunes New South Wales, Australia. In M. E. Brookfield and T. S. Ahlbrandt, eds., *Eolian Sediments and Processes*. Amsterdam: Elsevier, pp. 325–342.

Hesp, P. A. (1991). Ecological processes and plant adaptations on coastal dunes. *Journal of Arid Environments*, **21**, 165–191.

Hesp, P. A. (2002). Foredunes and blowouts: initiation, geomorphology and dynamics. *Geomorphology*, **48**, 245–268.

Hesp, P. A. and Martínez, M. L. (2007). Disturbance processes and dynamics in coastal dunes. In E. A. Johnson and K. Miyanishi, eds., *Plant Disturbance Ecology*. San Diego: Academic Press, pp. 215–247.

Hesp, P. A. and Thom, B. G. (1990). Geomorphology and evolution of active transgressive dunefields. In K. F. Nordstrom, N. P. Psuty and R. W. G. Carter, eds., *Coastal Dunes: Form and Process*. Chichester: John Wiley, pp. 253–288.

Hesp, P. A. *et al.* (2005). Flow dynamics over a foredune at Prince Edward Island, Canada. *Geomorphology*, **65**, 71–84.

Hesp, P. A. *et al.* (2007). Morphology of the Itapeva to Tramandai transgressive dunefield barrier system and mid- to late Holocene sea level change. *Earth Surface Processes and Landforms*, **32**, 407–414.

Houwing, E. J. *et al.* (1999). Biological and abiotic factors influencing the settlement and survival of *Salicornia dolichostachya* in the intertidal pioneer zone. *Mangroves and Salt Marshes*, **3**, 197–206.

Hugenholtz, C. H. and Wolfe, S. A. (2005). Biogeomorphic model of dunefield activation and stabilization on the northern Great Plains. *Geomorphology*, **70**, 53–70.

Hull, C. H. J. and Titus, J. G. (1986). *Greenhouse Effect, Sea Level Rise, and Salinity in the Delaware Estuary*, Report No. EPA-230-05-86-010. Washington, DC: US Environmental Protection Agency and Delaware River Basin Commission.

IPCC (2007). *Climate Change 2007: Impacts, Adaptation and Vulnerability. Contribution of Working Group II to the Fourth Assessment Report of the Intergovernmental Panel on Climate Change*. Parry, M. L. *et al.*, eds. Cambridge: Cambridge University Press. (pp. 173–210, Kundzewicz, Z. W. *et al.*; pp. 315–356, Nicholls, R. J. *et al.*; pp. 541–580, Alcamo, J. *et al.*)

Illenberger, W. K. and Rust, I. C. (1988). A sand budget for the Alexandria coastal dune field, South Africa. *Sedimentology*, **35**, 513–522.

Johnson, D. W. (1919). *Shore Processes and Shoreline Development*. New York: Hafner.

Kaplan, J. O. and New, M. (2006). Arctic climate change with a 2 °C global warming: timing, climate patterns and vegetation change. *Climatic Change*, **79**, 213–241.

Kennedy, B. A. (2006). *Inventing the Earth: Ideas on Landscape Development since 1740*. Chichester: John Wiley.

King, S. E. and Lester, J. N. (1995). The value of salt marsh as a sea defence. *Marine Pollution Bulletin*, **30**, 180–189.

Kirby, R. (2000). Practical implications of tidal flat shape. *Continental Shelf Research*, **20**, 1061–1077.

Langlois, E., Bonis, A. and Bouzille, J. B. (2001). The response of *Puccinellia maritima* to burial: a key to understanding its role in salt-marsh dynamics? *Journal of Vegetation Science*, **12**, 289–297.

Law, M. N. and Davidson-Arnott, R. G. D. (1990). Seasonal controls on aeolian processes on the beach and foredune. *Proceedings of the Canadian Symposium on Coastal Sand Dunes, 12–14 September 1990*, Guelph, Ontario. Ottawa: National Research Council of Canada, pp. 49–68.

Lees, B. (2006). Timing and formation of coastal dunes in northern and eastern Australia. *Journal of Coastal Research*, **22**, 78–89.

Lund, J. *et al.* (2007). *Envisioning Futures for the Sacramento–San Joaquin Delta*. San Francisco: Public Policy Institute of California.

Martinez, M. L., Maun, M. A. and Psuty, N. P. (2004). The fragility and conservation of the world's coastal dunes: geomorphological, ecological and socioeconomic perspectives. In M. L. Martinez and N. P. Psuty, eds., *Coastal Dunes: Ecology and Conservation*. Berlin: Springer-Verlag, pp. 355–370.

Maun, M. A. and Lapierre, J. (1984). The effects of burial by sand on *Ammophila breviligulata*. *Journal of Ecology*, **72**, 827–839.

McClanahan, T. R., Mwaguni, S. and Muthiga, N. A. (2005). Management of the Kenyan coast. *Ocean and Coastal Management*, **48**, 901–931.

McLeod, K. L. *et al.* (2005). *Scientific Consensus Statement on Marine Ecosystem-Based Management*, signed by 221 academic scientists and policy experts with relevant expertise and published by the Communication Partnership for science and the sea. Available at www.compassonline.org/pdf_files/EBM_Consensus_Statement_v12.pdf

Millenium Ecosystem Assessment (2005). *Ecosystems and Human Well-Being: Current State and Trends*. Washington, DC: Island Press.

Miyanishi, K. and Johnson E. A. (2007). Coastal dune succession and the reality of dune processes. In E. A. Johnson and K. Miyanishi, eds., *Plant Disturbance Ecology*. San Diego: Academic Press, pp. 215–247.

Moeller, I. *et al.* (2001). The sea-defence value of salt marshes: field evidence from North Norfolk. *Journal of the Chartered Institution of Water and Environmental Management*, **15**, 109–116.

Najjar, R. G. *et al.* (2000). The potential impacts of climate change on the mid-Atlantic coastal region. *Climate Research*, **14**, 219–233.

Nickling, W. G. and Davidson-Arnott, R. G. D. (1991). Aeolian sediment transport on beaches and coastal sand dunes. In *Proceedings of the Symposium on Coastal Sand Dunes*. Ottawa: National Research Council of Canada, pp. 1–35.

Nordstrom, K. F. (2000). *Beaches and Dunes of Developed Coasts*. Cambridge: Cambridge University Press.

O'Brien, D. J., Whitehouse, R. J. S. and Cramps, A. (2000). The cyclic development of a macrotidal mudflat on varying timescales. *Continental Shelf Research*, **20**, 1593–1619.

Orford, J. D. and Pethick, J. (2006). Challenging assumptions of future coastal habitat development around the UK. *Earth Surface Processes and Landforms*, **31**, 1625–1642.

Panin, N. and Jipa, D. (2002). Danube River sediment input and its interaction with the north-western Black Sea. *Estuarine, Coastal and Shelf Science*, **54**, 551–562.

Pearce, K. I. and Walker, I. J. (2005). Frequency and magnitude biases in the Fryberger model, with implications for characterizing geomorphically effective winds. *Geomorphology*, **68**, 39–55.

Pejrup, M. and Andersen, T. (2000). The influence of ice on sediment transport, deposition and reworking in a temperate mudflat area, the Danish Wadden Sea. *Continental Shelf Research*, **20**, 1621–1634.

Perillo, G. M. E. (2003). *Dinámica del Transporte de Sedimentos*, Publicación Especial No. 2. La Plata: Asociación Argentina de Sedimentología.

Perillo, G. M. E. (2009). Tidal courses: classification, origin and functionality. In G. M. E. Perillo *et al.*, eds., *Coastal Wetlands: An Integrated Ecological Approach*. Amsterdam: Elsevier, pp. 185–209.

Perillo, G. M. E. *et al.* (1996). The formation of tidal creeks in a salt marsh: new evidence from the Loyola Bay salt marsh, Rio Gallegos Estuary, Argentina. *Mangroves and Salt Marshes*, **1**, 37–46.

Perillo, G. M. E. *et al.* (2007). Estuaries and the sediments: how they deal with each other. *INPRINT*, **3**, 3–5.

Pestrong, R. (1965). *The Development of Drainage Patterns on Tidal Marshes*. Palo Alto: Stanford University Publications.

Pethick, J. S. (1980). Salt-marsh initiation during the Holocene transgression: the example of the North Norfolk marshes, England. *Journal of Biogeography*, **7**, 1–9.

Pethick, J. S. (2001). Coastal management and sea-level rise. *Catena*, **42**, 307–322.

Postma, H. (1961). Transport and accumulation of suspended matter in the Dutch Wadden Sea. *Netherlands Journal of Sea Research*, **1**, 148–190.

Pringle, C. M. (2001). Hydrologic connectivity and the management of biological reserves: a global perspective. *Ecological Applications*, **11**, 981–998.

Psuty, N. P. (2004). The coastal foredune: a morphological basis for regional coastal dune development. In M. L. Martinez and N. P. Psuty, eds., *Coastal Dunes: Ecology and Conservation*. Berlin: Springer-Verlag, pp. 11–28.

Reed, D. J. (1989). Environments of tidal marsh deposition in Laguna San Rafael area, southern Chile. *Journal of Coastal Research*, **5**, 845–856.

Reed, D. J. (1995). The response of coastal marshes to sea-level rise: survival or submergence? *Earth Surface Processes and Landforms*, **20**, 39–48.

Reed, D. J. *et al.* (2008). Site-specific scenarios for wetlands accretion as sea level rises in the mid-Atlantic region. In J. G. Titus and E. M. Strange, eds., *Background Documents Supporting Climate Change Science Program Synthesis and Assessment Product 4.1*, EPA 430R07004. Washington, DC: US Environmental Protection Agency, pp. 133–174.

Rozema, J. *et al.* (1991). Effect of elevated atmospheric CO_2 on growth, photosynthesis and water relations of salt marsh grass species. *Aquatic Botany*, **39**, 45–55.

Ruz, M. H. (1993). Coastal dune development in a thermokarst environment: some implications for environmental reconstruction, Tuktoyaktuk Peninsula. *Permafrost and Periglacial Processes*, **4**, 255–264.

Ruz, M. H. and Allard, H. (1994). Foredune development along a subarctic emerging coastline, eastern Hudson Bay, Canada. *Marine Geology*, **117**, 57–74.

Sherman, D. J. (1995). Problems of scale in the modeling and interpretation of coastal dunes. *Marine Geology*, **124**, 339–349.

Shimada, K. *et al.* (2006). Pacific Ocean inflow: influence on catastrophic reduction of sea ice cover in the Arctic Ocean. *Geophysical Research Letters*, **33**, L08605.

Shinkle, K. D. and Dokka, R. K. (2004). *Rates of Vertical Displacement at Benchmarks in the Lower Mississippi Valley and the Northern Gulf Coast*, Technical Report NOS/NGS 50. Silver Spring: National Oceanic and Atmospheric Administration.

Short, A. D. and Hesp, P. A. (1982). Wave, beach and dune interactions in southeastern Australia. *Marine Geology*, **48**, 259–284.

Stanev, E. V. and Peneva, E. L. (2001). Regional sea level response to global climatic change: Black Sea examples. *Global and Planetary Change*, **32**, 33–47.

Stanica, A., Dan, S. and Ungureanu, V. G. (2007). Coastal changes at the mouth of the Danube River as a result of human activities. *Marine Pollution Bulletin*, **55**, 555–563.

Stanley, D. J. and Warne, A. G. (1993). Nile Delta: recent geological evolution and human impact. *Science*, **260**, 628–634.

Steers, J. A. (1948). *The Coastline of England and Wales*. Cambridge: Cambridge University Press.

Steers, J. A. (1960). Physiography and evolution. In J. A. Steers, ed., *Scolt Head Island*. Cambridge: W Heffer and Sons, pp. 12–66.

Stockdon, H. F. *et al.* (2007). A simple model for the spatially variable coastal response to hurricanes. *Marine Geology*, **238**, 1–20.

Swenson, E. M. and Swarzenski, C. M. (1995). Water levels and salinity in the Barataria–Terrebonne estuarine system. In D. J. Reed, ed., *Status and Trends of Hydrologic Modification, Reduction in Sediment Availability, and Habitat Loss/Modification in the Barataria–Terrebonne Estuarine System*, BTNEP Publication No. 20. Thibodaux: Barataria–Terrebonne National Estuary Program.

Tessier, M., Gloaguen, J. C. and Lefeuvre, J. C. (2001). Factors affecting the population dynamics of Suaeda maritima at initial stages of development. *Plant Ecology*, **147**, 193–203.

Timmerman, A. *et al.* (1999). Increased El Niño frequency in a climate model forced by future greenhouse warming. *Nature*, **398**, 694–696.

Tripaldi, A. and Forman, S. L. (2007). Geomorphology and chronology of Late Quaternary dune fields of western Argentina. *Palaeogeography, Palaeoclimatology, Palaeoecology*, **251**, 300–320.

Turner, A. and Millward, G. E. (2002). Suspended particles: their role in estuarine biogeochemical cycles. *Estuarine, Coastal and Shelf Science*, **55**, 857–883.

Ursino, N., Silvestri, S. and Marani, M. (2004). Subsurface flow and vegetation patterns in tidal environments. *Water Resources Research*, **40**, W05115.

van der Meulen, F., Bakker, T. W. M. and Houston, J. A. (2004). The costs of our coasts: examples of dynamic dune management from Western Europe. In M. L. Martinez and N. P. Psuty, eds., *Coastal Dunes: Ecology and Conservation*. Berlin: Springer-Verlag, pp. 259–277.

Vespremeanu-Stroe, A. and Preoteasa, L. (2007). Beach–dune interactions on the dry–temperate Danube delta coast. *Geomorphology*, **86**, 267–282.

Visser, J. M. *et al.* (1998). Marsh vegetation types of the Mississippi River Deltaic Plain. *Estuaries*, **21**, 818–828.

Webb, J. W. (1983). Soil water salinity variations and their effects on *Spartina alterniflora*. *Contributions to Marine Science*, **26**, 1–13.

Webster, P. J. *et al.* (2005). Changes in tropical cyclone number, duration, and intensity in a warming environment. *Science*, **309**, 1844–1846.

White, B. R. (1979). Soil transport by winds on Mars. *Journal of Geophysical Research*, **84**, 4643–4651.

Widdows, J. and Brinsley, M. (2002). Impact of biotic and abiotic processes on sediment dynamics and the consequences to the structure and functioning of the intertidal zone. *Journal of Sea Research*, **48**, 143–156.

Widdows, J. *et al.* (2004). Role of physical and biological processes in sediment dynamics of a tidal flat in Westerschelde Estuary, SW Netherlands. *Marine Ecology Progress Series*, **274**, 41–56.

Williams, P. B. and Orr, M. K. (2002). Physical evolution of restored breached levee salt marshes in the San Francisco Bay estuary. *Restoration Ecology*, **10**, 527–542.

Wolters, M., Gabutt, A. and Bakker, J. P. (2005). Plant colonization after managed realignment: the relative importance of diaspore dispersal. *Journal of Applied Ecology*, **42**, 770–777.

Woodroffe, C. D. (2002). *Coasts: Form, Process and Evolution*. Cambridge: Cambridge University Press.

Wright, L. D. and Thom, B. G. (1977). Coastal depositional landforms: a morphodynamic approach. *Progress in Physical Geography*, **1**, 412–459.

Zimmerman, R. J., Minello, T. J. and Rozas, L. P. (2000). Salt marsh linkages to productivity of penaeid shrimps and blue crabs in the northern Gulf of Mexico. In M. P. Weinstein and D. A. Kreeger, eds., *Concepts and Controversies in Tidal Marsh Ecology*. Dordrecht: Kluwer, pp. 293–314.

6 Beaches, cliffs and deltas

Marcel J. F. Stive, Peter J. Cowell and Robert J. Nicholls

6.1 Introduction

Coastal areas are densely populated and highly productive regions and any changes to them could have profound direct and indirect impacts on coastal societies. The geological and historical record shows that coasts are always evolving. Hence, global environmental change (climate, sea level, water use and land use changes) expected this century (Crossland *et al.*, 2005; IPCC, 2007a) would modify coastal evolutionary behaviour at local, regional and landscape scales with important consequences. These modifications may lead to (1) accelerations in rates of coastal change (e.g. acceleration in erosional (dominant) and accretional (exceptional) trends), (2) reversals in historical trends (e.g. accreting coasts becoming erosive), or (3) initiation of state or mode changes in coastal behaviour (e.g. dune breaching and formation of new tidal basins).

Prediction of accelerations may and should draw on historical data to extrapolate future trends in some settings. Trend reversals and state or mode changes, however, will require sufficient understanding of coastal morphodynamics to allow predictions of future change from the combined application of historical geomorphologic data and morphodynamic models. As well as best estimates, there is a need to evaluate uncertainty in terms of probabilistic risk, and to generate future scenarios to characterise consequences at different levels of risk. The objective of risk-based prediction is to express forecasts in a form that bridges the gap between science and policy. Coastal change can cause major problems for society, especially if reversal of trends and state or mode changes of behaviour occur on coasts mistakenly regarded by the community as 'fixed'. Methodological approaches such as those outlined in this chapter are essential to explore the full range of possible changes and hence to underpin design and choice of appropriate science-based management options.

Socioeconomic changes over the coming century are expected to be so great that direct anthropogenic effects are likely to be drivers of comparable or possibly even greater significance for coastal geomorphology than climate change impacts (Valiela, 2006; IPCC, 2007b, pp. 315–356). These direct societal drivers already exert effects through altered sediment supply and river hydrographs, with a general tendency for reduction (Syvitski *et al.*, 2005), as well as through other more locally impacting engineered interventions, including reclamation, coastal structures and beach nourishment. In this chapter we will discuss case studies that show that in some cases the impact of socioeconomic changes is as large as or even larger than climate change impacts. Moreover, socioeconomic changes on the climate change timescale are likely to alter the landscape completely in terms of exposure of people and economies to hazards associated with geomorphic change. Thus, coastal impacts of climate change relevant to society can only be contemplated meaningfully in the context of likely changes to human settlement patterns and other coastal use projected over the same timescale (Nicholls *et al.*, 2008).

This chapter explores how the understanding of the Holocene morphological evolution of coasts provides a basis for projecting coastal responses to future accelerations in climate change impacts and other environmental modifications, including the combined influence of sea level rise, modified wave climates, and human intervention on the open coast.[1] This understanding is systematised and quantified in a model concept, the *coastal-tract cascade* (Cowell *et al.*, 2003a, 2003b), and used to make projections, including uncertainties, about selected types of coastal system evolution under climate change, accelerated

[1] Enclosed environments, such as estuaries and lagoons, are disregarded here, since Reed *et al.* treat these in Chapter 5. We do include though a delta case (the Ebro Delta), but with a focus on its coastal zone.

Geomorphology and Global Environmental Change, eds. Olav Slaymaker, Thomas Spencer and Christine Embleton-Hamann. Published by Cambridge University Press.

sea level rise and increased human impact. The selection of these illustrative types of coastal systems is driven by two considerations. Firstly, we choose coastal systems that are most likely to be affected by environmental change. Secondly, we select coastal systems for which we can provide quantified evolution. This implies that we will not provide a global vulnerability analysis to environmental change (instead we refer to such efforts, e.g. Hoozemans *et al.*, 1993; Agardy *et al.*, 2005; UNEP, 2005; McFadden *et al.*, 2007a). Rather, we will highlight some selected cases that exemplify coastal evolution in response to environmental change that serve as a showcase with applicability to the many other coastal systems.

As a starting point of this selection we discuss coastal classification (taxonomy) efforts to provide us with an insight into the variety of coastal systems over the globe. An understanding based on classification can impede an understanding in terms of universally applicable evolutionary sedimentation processes. Nevertheless, classification provides a simplified qualitative insight as an introduction to the range of potential responses and sensitivities to climate change impacts. We thus include this qualitative insight as a conceptual bridge to the more systematic process-based concepts in the subsequent section on the coastal-tract cascade.

The bridging role of taxonomy derives also from the reality that classification is inherent in the terminology used to describe coastal features of functional significance with respect to coastal processes. Feature recognition is fundamental to application of the coastal-tract cascade concept, the purpose of which is to identify morphologically related processes that must be included or may be excluded from a site or regionally specific analysis of coastal change and its prediction. Moreover, consistent terminology is also necessary to document and explain observations and predictions through verbal communication. This type of communication is especially important in explaining differences in coastal vulnerability to the broader community, an audience that lacks the expertise necessary to understand morphodynamics. Hence, while we acknowledge and will address the shortcomings of 'static' classification we find it necessary and useful to discuss classification, and subsequently integrate this with an approach that better acknowledges the evolutionary nature of coastal systems.

6.2 Coastal classification

Coastal processes and sensitivity to climate change impacts vary geographically depending on the geological setting, gross physiography and regional climatic regime. An obvious differentiation splits coastal types into those in which morphology comprises unconsolidated sediments as opposed to those comprised of erosion-resistant bedrock. Such resistance in reality can exist by degree, ranging from 'hard' lithologies like granite that are extremely resistant to weathering and erosion, through to 'soft' friable bedrock comprising weakly cemented sediments such as unconsolidated sands, silts and clays. Nevertheless, this fundamental differentiation is warranted to distinguish regions potentially subject to erosion from regions at minimal risk of coastal change, other than through elevation-dependent inundation, when subjected to alteration of environmental conditions (e.g. changes in sea level, wave climate and storm track, frequency and intensity). Another obvious differentiation is between mainland or fringing beaches and barrier coasts due to the respective absence and presence of estuarine–deltaic processes that complicate coastal responses to climate change.

Many different kinds of classification have been applied to coasts in attempts to characterise dominant features in terms of physical or biological properties, modes of evolution, or geographic occurrence. Some of the earlier general classifications were broad in scope but lacked specificity while other specialised systems were narrowly focussed, providing uneven coverage of taxonomic units for coastlines of the world. Finkl (2004) has summarised both the development and range of large-scale classifications of the physical coastal environment. For more than a century, until the 1960s, the almost universal approach of the physical coastal scientist was to attempt a classification, or typology of coastal forms. This approach involved the study of types, the realities that they typify, and the correspondence between types (Pethick, 1984). Two forms of typological study can be identified: those based on a descriptive analysis (e.g. McGill, 1958; Davies, 1973), and those that take a generic form and attempt to indicate the evolution of the feature under study (e.g. Johnson, 1919; Shepard, 1937, 1948, 1973; Valentin, 1952; Inman and Nordstrom, 1971; Bird, 1976). Coastal typology development, in both its forms, was of central importance to the evolution of large-scale analysis of coastal morphodynamics.

During the 1960s interest in classification of coastal environments diminished with the realisation that focus on static categories was retarding development of a truly scientific analysis (McFadden *et al.*, 2007b). In its place, process-driven coastal analysis became important, with the primary emphasis of research targeting detailed site analyses to understand the physical mechanisms of coastal change. This has led to a furthering of knowledge concerning small-scale process and form within the coastal system. However, until recently there were no methodologies that could effectively translate this small-scale process knowledge to larger scales, including those relevant here. This was an important limitation on large-scale modelling, and it was recognised that research is required across the range of scales linking smaller-scale processes to low-order coastal change processes (Stive *et al.*, 2002). This relatively limited understanding of the relations between large-scale processes and the landforms of the coastal system constrained our ability to assess the threat of the range of increasing pressures on coastal environments at these regional to global scales (McFadden *et al.*, 2007b). As a result, the coastal-tract cascade concept was introduced (Cowell *et al.*, 2003a, 2003b) to provide a modern large-scale coastal system alternative to static classification, acknowledging the dynamic and transient nature of coasts that should allow proper use in a global assessment. We will introduce this concept in the next section.

However, as we will argue, sufficient arguments exist to select some elements of the published classifications that reflect the important environmental controls acting under global change, and merge these taxonomic concepts with the coastal-tract cascade concept. Here, we draw on the study of Finkl (2004), who comprehensively reviews and analyses the many classifications proposed in the literature. In terms of process-related classifications Finkl distinguishes two main types, namely (1) coastal geotectonic classification and (2) classification based on relative sea level trends, including sedimentation/erosion. The geotectonic classification acknowledges the role of plate tectonics in coastal typology resulting in the main types of leading and trailing edge coastal systems (Inman and Nordstrom, 1971). Submergence and emergence play a dominant role here.

Shepard (1937, 1948, 1973) abandoned the high-level classificatory submergent–emergent dichotomy of previous workers by (1) placing submergent coasts at a lower level in the classification and (2) suggesting that emergent coasts can be ignored. The basis (highest-level distinction) in this classification is the differentiation between coasts shaped mainly by terrestrial agencies (primary, youthful coasts) and those modified by marine processes (secondary, mature coasts). This classification (Table 6.1 as interpreted by Finkl, 2004) is reasonably comprehensive and has much to recommend it. However, the lack of a category for emergent coasts, as noted by Cotton (1954), is an obvious disadvantage. Some workers also refer to the perceived difficulty in the application of this coastal classification to actual examples. This observation is, however, common to most coastal classification systems (Finkl, 2004), and reflects the motivation behind the coastal-tract cascade rationale advanced later in this paper. Quoting Finkl (2004): '*King (1966), for example, asks the question, how does one determine the precise moment when a coast is sufficiently altered by marine agencies to allow it to be classified in the second major group?*' (p. 172). The classification

TABLE 6.1. *Classification of coasts as primary (shaped by non-marine processes) and secondary (moulded by marine agencies): interpreted by Finkl (2004), after Shepard (1948)*

(I) Coasts shaped primarily by non-marine agencies (primary or youthful shorelines)
- (A) Shaped by terrestrial agencies of erosion (subaerial denudation) and drowned by the recent (Flandrian) marine transgression or down-warping of the land margin
  - (1) Ria coast (drowned mouths of river valleys)
    - (a) Parallel trend between structure and coast (Dalmatian type)
    - (b) Transverse trend between structure and coast (southwest Ireland type)
  - (2) Drowned glacial erosion coast
    - (a) Fjord coast (drowned glacial valleys)
    - (b) Glacial trough coast
- (B) Shaped by terrestrial (subaerial) deposition
  - (1) River deposition
    - (a) Delta coast – convex out
    - (b) Alluvial plain coast – straight
  - (2) Glacial deposition
    - (a) Partially submerged moraine
    - (b) Drumlins – often partly drowned
  - (3) Wind deposition
    - (a) Dune coast
  - (4) Vegetation coast
    - (a) Mangrove coast
- (C) Shaped by volcanic activity
  - (1) Volcanic deposition coast – lava flow – convex out
  - (2) Volcanic explosion coast – concave out
- (D) Shaped by diastrophism
  - (1) Fault coasts or fault scarp coasts
  - (2) Fold coasts, due to monoclinal flexures

(II) Coasts shaped primarily by marine agencies (secondary or mature coastlines)
- (A) Shaped by marine erosion
  - (1) Cliffed coasts made more regular by marine erosion
  - (2) Cliffed coasts made less regular by marine erosion
- (B) Shaped by marine deposition
  - (1) Coasts straightened by marine deposition
  - (2) Coasts prograded by marine deposition
  - (3) Shorelines with barriers and spits – concavities facing ocean
  - (4) Organic marine deposition – coral coasts

also cuts across the commonly used system of cyclic description. Coasts can be described in terms of the state of development by marine processes, from an initial form, which is defined in the classification. This new classification uses what would normally be considered as the youthful stage of development as the second group of the classification, so that a coast which is put into the second group automatically loses any reference to its initial form at the beginning of the cycle, which is running its course at the present time (Bird, 1976). Shepard modified and elaborated his classification in 1973, but retained its basic structure.

The way in which base level varies through time is crucial to the problem of coastal classification, as it is to the whole concept of a marine cycle of erosion or accretion. The fundamental distinction of process here is between advancing and retreating coasts, noting that advance may be due to coastal emergence and/or progradation by deposition, while retreat is due to coastal submergence and/or retrogradation by erosion. The concept of a changing base level consisting of periods of rapid change alternating with periods of sea level stillstand forms the basis of Valentin's (1952) classification (Table 6.2 as interpreted by Finkl, 2004). This system was devised for use on a map scale (1: 50 M) of global coastal configurations. The emphasis in this classification is on temporal change. This system features present coastal change rather than the initial form of the coast before modification by marine processes.

Important aspects of the Valentin classification are the recognition that marine forces are continually active, and influence the coast even during changes in base level that should, on the basis of the older classifications, initiate a new cycle of erosion involving a new coastal type (Bird, 1976). According to Valentin (1952), the possibility is left open that changes of base level are continually operating and that stillstands are the exception rather than the rule. Valentin's classification is expressed graphically by means of a diagram on which each of four axes represents one of four possibilities: coastal erosion and submergence on the negative side and coastal outbuilding and emergence on the positive side. This classification does not specifically consider coastal morphology nor provide groupings of coastal features in a hierarchical system. Nevertheless, it does provide a rudimentary yet useful frame of reference for the conceptualisation of the basic role of coastal advance and retreat (Finkl, 2004).

Finkl (2004) proposes a new comprehensive classification, which incorporates virtually all materials, processes, forms or coastal environmental properties that are considered in earlier classification systems. Although Finkl's classification is probably the most complete classification available, we refrain from using it here because the extent and detail of his classification obscures the main processes that are relevant in the present context; i.e. we consider certain processes to be especially relevant since they are impacted by environmental change, namely relative sea level change, marine processes, terrestrial processes and shoreline position. These elements are emphasised in both Shepard's and Valentin's classifications. We therefore repeat their classifications (in the interpretation of and slightly adapted after Finkl, 2004) as useful for our present discussion.

In the following sections of this chapter, we exemplify our concepts with reference to the Holland coast, Australian strand plains, the Ebro Delta and UK cliffed coasts, again noting that we do not consider enclosed coastal systems. Table 6.3 links our examples to the relevant classifications

TABLE 6.2. *Classification of coasts in relation to shoreline advance (by emergence or sediment accumulation) or retreat (by submergence or erosion): interpreted by Finkl (2004), after Valentin (1952)*

- (I) Coasts that have advanced (moved seaward)
  - (A) Due to emergence (coasts on newly emerged seafloor provided the new coastline has not been eroded back to the original line)
    - (1) Emerged seafloor coasts
  - (B) Outbuilding coasts
    - (1) Organic deposition
      - (a) Phytogenic (formed by vegetation): mangrove coast
      - (b) Zoogenic (formed by fauna): coral coasts
    - (2) Inorganic deposition coasts
      - (a) Marine deposition under weak tides: lagoon-barrier and dune-ridge coasts
      - (b) Marine deposition under strong tides: tideflat and barrier island coasts
    - (3) Fluvial deposition: delta and outwash coasts
- (II) Coasts that have retreated (moved landward)
  - (A) Submergence of glaciated landforms
    - (1) Confined glacial erosion: fjord-skerry coasts
    - (2) Unconfined glacial erosion; fjard-skerry coasts
    - (3) Glacial deposition: morainic coasts
  - (B) Submergence of fluvially eroded landforms
    - (1) On young fold structures: embayed upland coasts
    - (2) On old fold structures: ria coasts
    - (3) On horizontal structures: embayed plateau coasts
  - (C) Coasts of retrogression (coasts eroded back to a continuous line of cliffs, provided reversal processes have not caused deposition in front of the cliff)

TABLE 6.3. *Cross-reference table for selected examples of coastal systems described in the chapter linking evolutionary stages for these examples to Shepard's and Valentin's classification*

Coastal system	Shepard (Table 6.1)	Valentin (Table 6.2)
Holland coast	II-B-3 > II-B-2 > II-B-1	II-C > I-B-2a > I-B-2b > II-C
Australian strand plains	II-B-2	I-B-2a
Ebro Delta	I-B-1 > II-B-1	I-B-3 > II-C
UK 'soft' cliff coasts	II-A-1	II-C

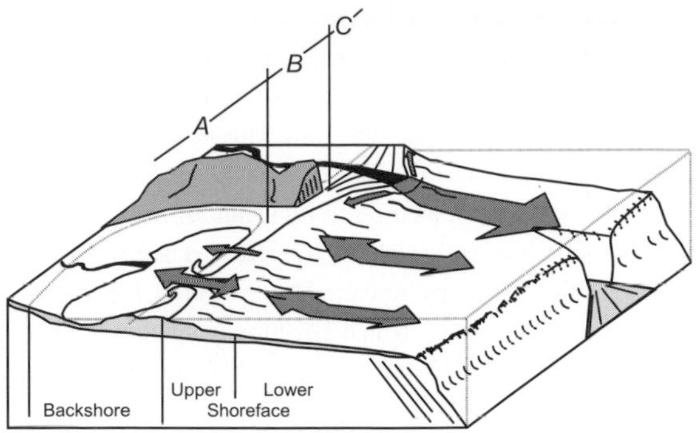

FIGURE 6.1. Physical morphology encompassed by the coastal tract (from Cowell et al., 2003a; see text for explanation).

of Shepard (Table 6.1) and Valentin (Table 6.2) in relation to the evolution of these examples over the Holocene.

The shortcomings of 'static' classification are evident in Table 6.3 where half of the coastal systems cannot be uniquely classified because morphological evolution over the Holocene is an accumulation of evolutionary processes that change over time. This does not mean that static classification is irrelevant; rather it means that static classification is not capable of capturing the evolutionary dimension of coastal systems; it is a momentary classification of a coastal state.

6.3 The coastal-tract cascade

6.3.1 Introduction

Coasts have been responding to climate change and sea level rise continually throughout recent human and geological history. The existing understanding of morphological responses of coasts to climate and socioeconomic drivers draws upon established morphodynamic principles, derived from research into biophysical processes, and knowledge about coastal change over the Holocene, derived from field investigations. The general principles outlined here apply to all coastal categories outlined in the previous section. However, morphological change is continuous, cumulative and subject to the magnitude–frequency characteristics of the drivers both leading to forced and free behaviour. Thus, any systematic understanding of coastal change must be organised in relation to scale. The coastal-tract cascade concept was developed for this purpose (Cowell *et al.*, 2003a).

The coastal-tract concept distinguishes a hierarchy of morphodynamic forms and processes to define coastal systems across the complete range of scales. These scale-related systems are defined through identification of morphologies that interact dynamically by sharing a common pool of sediment on any particular scale. Long-term prediction (decades or longer) of shoreline changes involve movements of the upper shoreface due to its sediment sharing interaction with the lower shoreface and backshore environments (Fig. 6.1). Interaction with the backshore depends on whether this environment comprises a barrier beach/dune and lagoon (A), mainland beach (B) or fluvial delta (C). On this basis, coastal responses can be investigated in terms of morphodynamic constituents whose changes can be further resolved in terms of mean trend and fluctuating components (Stive *et al.*, 1991; Cowell and Thom, 1994). These components, which relate to chronic and acute erosion hazards respectively, have different implications for coastal impacts and the potential management responses.

Chronic erosion hazards are typically associated with imbalances in sediment budget that have ongoing effects, and may become progressively worse over time. Spatial gradients in littoral sediment transport and the effects of sea level rise are the most common cause of chronic coastal instability. Acute erosion hazards arise due to episodic erosional events, such as storms, with subsequent slower recovery of morphology to previous conditions. Such fluctuating changes do not cause coastal management problems unless coastal development has been undertaken imprudently in the active zone, or if mean trend changes displace the region occupied by fluctuating morphology to coincide with pre-existing coastal development. Often the event changes are seen as the problem under these circumstances because property damage usually occurs episodically associated with storms. However, the actual problem under these circumstances is the underlying mean trend change. The only circumstances under which the fluctuating changes become a new hazard in their own right is when the magnitude of maximum fluctuations increases due to changes in environmental conditions, such as an increase in storm intensity and corresponding water level and wave energy.

Thus the question of scale is fundamental to defining the nature and mode of coastal change. For example, the time

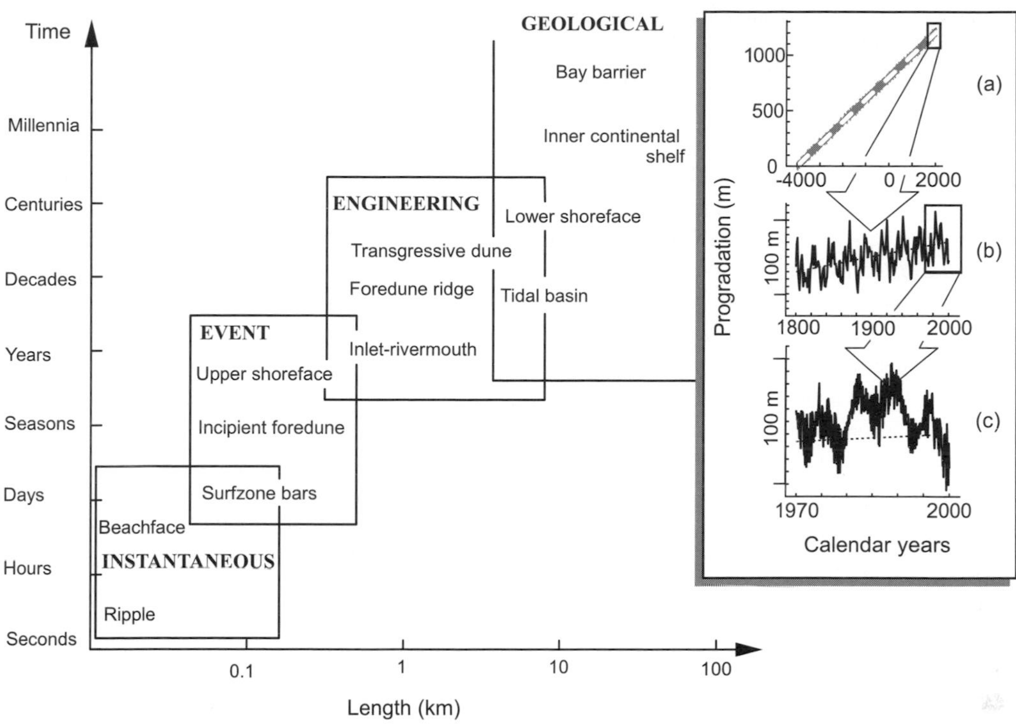

FIGURE 6.2. Scales of coastal change and typical time series associated with the three lower-order (longer) morphological scales (modified from Thom and Cowell, 2005). The underlying progradation can be resolved (a) for the geological timescale using radiometric dating of strand plains and (b) is of key significance to the engineering timescale, but (c) may not be resolvable from measured fluctuations in recurrent surveys spanning several decades. For practical purposes in coastal management, coastal change generally is measured in relation to horizontal movements of the shoreline (e.g. often standardised to the mean high water elevation due to its common legal significance), or alternatively quantified in terms of beach volume.

series in Fig. 6.2 represents the progradation distance during the postglacial sea level high stand (i.e. the last 6 ka) with superimposed shoreline fluctuations due to storm erosion and recovery cycles. The time series is synthetic, rather than entailing measured data, but the parameters are based on generalised data from southeastern Australia (Thom and Hall, 1991; Roy *et al.*, 1994; Cowell *et al.*, 1995).

The mean trend rate of progradation is 0.2 m a^{-1} (or 200 m ka^{-1}), consistent with rates evident in late Holocene barrier progradation, for example, in the Netherlands, Southeastern Australia and the Northwest Pacific USA (Cowell *et al.*, 2003b). The synthetic fluctuations involve two sets of sinusoids with periodicities representing a PDO ($T = 23$ years) and ENSO ($T = 7$ years). The amplitudes of shoreline fluctuations for both cycles were set at 30 m and both were subjected to a chaotic gain factor:

$$G = kG_{\text{last}}(G_{\text{last}} - 1), \qquad 0 \leq G \leq 1 \qquad (6.1)$$

in which the non-linear parameter, k, was set to $k = 4$ with an initial (seed) value of $G = 0.2$ for the PDO and $G = 0.3$ for ENSO. Thus, independent values of G were applied to the PDO and ENSO. The top two graphs plot synthetic annual ('averaged') shoreline position and the bottom graph plots monthly 'averages'.

For these parameters, the maximum theoretical shoreline fluctuation during the 6 ka high stand is 120 m. This value is comparable to maxima in the Narrabeen and Moruya beach survey data sets from southeastern Australia (Cowell *et al.*, 1995). Since these data sets span only 30 years, the modelled maxima are probably on the low side. Centennial and millennial fluctuations associated with shoreline rotation and curvature changes due to shifts in dominance of directional components of wave climates in southeastern Australia are even larger (Goodwin *et al.*, 2006). Furthermore, since the maximum is attained only when successive values occur of $G \cong 0$ and $G \cong 1$ (or $G \cong 1$ and $G \cong 0$) simultaneously for PDO and ENSO, occurrence of maximum fluctuations are rare events.

The time series show mean trend and fluctuating components of large-scale coastal behaviour. The trend line shown in each of the plots (dashed) is for the average progradation rate (0.2 m a^{-1}). Over the full 6 ka of the sea level high stand, the mean trend component is so large that the fluctuations become insignificant (grey smudge in Fig. 6.2a). For the 200-year and 30-year time series, the fluctuating component dominates the signal (Fig. 6.2b, c). For the 30-year record, the mean trend behaviour could not be resolved statistically from the data (Fig. 6.2c). Indeed, for the 30-year data, a linear least-squares best fit (not plotted) gives an average progradation rate of 0.04 m a^{-1} compared to the actual trend of 0.2 m a^{-1}. Depending upon the span of data analysed, it is possible to obtain negative-trend estimates (i.e., shoreline retreat), even though the underlying *geological* trend is progradational.

Thus, as Cowell *et al.* (1995) observe, historical data sets generally do not permit resolution of the underlying tendencies in long-term shoreline change because the fluctuating component of large-scale coastal behaviour overwhelms the signal on the decadal (*engineering*) timescale. Cowell *et al.* (1995) showed that, on the coast of southeastern Australia, mean trend shoreline change becomes comparable to the decadal fluctuations only after the passage of about 180 years: whereas the longest high-resolution data set now spans little more than 30 years (Thom and Hall, 1991; McLean and Shen, 2006). Although longer low-resolution data sets of up to 150 years from historical shoreline analysis exist in a few locations such as the UK and USA (e.g. Crowell *et al.*, 1991; Taylor *et al.*, 2004), these many events and significant fluctuations can be missing from such records. For these reasons, mean trend shoreline changes are more reliably estimated from radiometric dating of progradational sequences (Beets *et al.*, 1992; Roy *et al.*, 1994). For coasts on which progradation is absent, mean trend estimates are less accessible and can only be derived from historic maps or other records, as demonstrated by Dickson *et al.* (2007), or photogrammetric measurement from quality survey aerial photographs.

On any particular scale, net morphological stability reduces to fundamental morphodynamic questions of marine and fluvial sediment availability (see Shepard's classification: Table 6.1), the accommodation space available for deposition, and their spatial organisation. Sea level rise for instance increases accommodation space creating an additional demand for sediments, which are redistributed accordingly. Changes in direction and intensity of wave climates, and magnitude–frequency characteristics of storms, also affect the spatial distribution of accommodation space. These changes would also influence morphological response rates through effects on net sediment fluxes.

Collectively, the balance between the effects of accommodation space, and hence sea level rise, and sediment supply govern modes of coastal change to the extent responsible for the differentiation between coastal types (see Valentin's classification: Table 6.2). The principal types include mainland beach and barrier coasts, where the latter in turn exist as a continuum of estuarine–deltaic morphologies. Position in the continuum depends largely on sediment supply relative to the accommodation space. On this basis, for example, deltaic lowlands in which estuaries have been filled by sediments during the Holocene sea level high stand (Fig. 6.1, Section A) may revert to estuarine environments due to accelerated sea level rise, depending on sediment supply. More generally, the available accommodation space also depends on the energy regime, including the relative influence of tides and waves, and the predominant size fraction of the sediment supply (Swift *et al.*, 1991). However, changing wave climates can also play a marginal role in transforming coasts in terms of where they lie in the typological continuum.

The gross kinematics of the coastal tract are constrained and steered by sediment–mass continuity. The rate of coastal advance or retreat is determined quantitatively by the balance between the change in sediment accommodation space, caused by sea level movements, and sediment availability. If the lower shoreface is shallower than required for equilibrium (negative accommodation), then sand is transferred to the upper shoreface so that the shoreline tends to advance seaward. This tendency also occurs when relative sea level is falling (coastal emergence). Coastal retreat occurs when the lower shoreface is too deep for equilibrium (positive shoreface accommodation). This sediment sharing between the upper and lower shoreface is an internal coupling that governs first-order coastal change. The upper shoreface and backbarrier (lagoon, estuary or mainland) are also coupled in first-order coastal change. Sediment accommodation space is generated in the backbarrier by sea level rise (and reduced by sea level fall), but the amount of space is also moderated by the influx of mud from coastal sources (e.g. by cliffs), or sand and mud from fluvial sources. Remaining space can then be occupied by sand transferred from the upper shoreface causing a retreat of the latter, which is a chronic erosion problem where it occurs (cf. the below example of the US east coast sections influenced by tidal inlets).

6.3.2 The role of sea level rise and sediment availability on large-scale, low-order coastal behaviour during the Holocene

The widely reported results of Bird (1985) that globally 70% of sandy beaches are erosive, triggered the widespread view that present sea level rise, even though moderate on the mid to late Holocene scale, is the most probable cause (see Vellinga and Leatherman, 1989; Leatherman *et al.*, 2000). This causal link has been reinforced by more detailed and confident observations of widespread erosion in both the USA where 90% of beaches are reported to be retreating (National Research Council, 1990; Leatherman 2001), and more recently in the European Union where erosion also appears to be dominant (Eurosion, 2004). Simple if not simplistic views on the process behind this erosion are often solely based on the Bruun effect (see e.g. Zhang *et al.*, 2004), predicting erosion rates of 50 to 100 times the rate of sea level rise (referred to as the Bruun Rule). We will address this effect and its rule below, and note that the textbook view of its application only applies under a restrictive set of conditions and morphologies and that the effect is very small under present near stillstand conditions, which explains the widespread criticism of the Bruun effect found in literature (see for example Cooper and Pilkey, 2004; Pilkey and Cooper, 2004).

As an illustration of this view, we note that there are many coastal systems that have been accretive at some stage during the mid to late Holocene, even though sea level was rising at rates larger than present. A few examples are the Australian coast (Short, 2003), deltaic coasts, by definition, (e.g. Mississippi, Ebro, Po, Yangtze) and composite (i.e. a combination of barrier and deltaic elements) coasts such as the US Northwest Washington coast and the Dutch coast (Cowell *et al.*, 2003a). What then can we conclude from this evidence? Besides the Bruun effect there must be a number of other processes that can exceed the Bruun effect to such an extent that coasts are accretive. Also, many other coasts experience larger erosion than is explained from the Bruun effect. Note for example most of the coastal sections along 2200 km of US east coast that Zhang *et al.* (2004) indicate as inlet influenced show shoreline erosion rates that far exceed the erosion rates along the sections which are judged as being under the Bruun effect. This inconsistency triggers the important question as to what these other processes are, and whether they are impacted by accelerated sea level rise. Stive and Wang (2003) show that the Bruun effect is likely to be overridden by backbarrier basin accommodation processes as soon as the basin sizes exceed 10 km^2. A basin of 100 km^2 would result in a shoreline retreat of the adjacent coast that is 10 times larger than the Bruun Rule states. If this is the case, then the additional impact of global warming on coastal evolution cannot be quantified from the Bruun effect alone. Rather an understanding of all processes involved in the determination of the total sediment budget and the resulting translation of the upper shoreface is required.

In an aggregated way, these other processes may be indicated collectively as sediment availability through cross-shore and alongshore sediment diffusion and advection. These processes are implicitly included in earlier kinematic models for first-order approximation of long-term coastal change (Curray, 1964). Swift (1976) extended Curray's (1964) ideas into a general framework for long-term coastal change entailing *transgression* (landward retreat) and *regression* (seaward advance) of the shoreline due to sea level rise and fall, with corresponding tendencies toward *retrogradation* and *progradation* due to net sediment losses or gains cross-shore and/or alongshore. This corresponds closely to Valentin's (1952) classification, which uses sediment availability and submergence (relative sea level rise) or emergence (relative sea level fall) as discriminating factors (see our later introduction of evolution of the central Netherlands coast as a time trajectory in sediment supply/accommodation phase space; see Section 6.4.2 below). The Valentin, Curray and Swift concepts are dynamically incorporated into the coastal-tract cascade concept and can be quantified, as discussed in Section 6.3.4. First, we will discuss the Bruun effect and Rule in the next section.

6.3.3 The applicability of the Bruun Rule

As discussed above and as will be substantiated and quantified in the next section the Bruun effect is applicable only to the rare, one-dimensional situation in which sea level rise is the only process operational on an alongshore uniform sedimentary coastal cell. This implies that there are no alongshore gradients, no exchanges with the backbarrier (e.g. lagoons and overwash processes) and the shoreface (e.g. canyons, marine feeding processes) and no terrestrial sources. In this particular case Bruun (1962) derived that under the assumption of an invariant upper shoreface the shoreline recession rate, c_p, due to sea level rise is described by:

$$c_p = -\frac{\partial MSL}{\partial t}\left(\frac{L_*}{h_*}\right) \tag{6.2}$$

where h_* and L_* are the height and length of the morphologically active profile.

Mimura and Nobuoka (1995) and Zhang *et al.* (2004) validated the erosion rate, c_p, empirically to be of the order of 50 to 100 times the rate of sea level rise. Assuming an active depth of 10 m, this erosion rate relates to a virtual loss amounting to between 500 and 1000 times the sea level rise rate (in m^3 m^{-1} per unit time). Under present near stillstand conditions, rates of sea level rise of about 0.2 m 100 a^{-1} (0.2 cm a^{-1}) typically occur on otherwise stable coasts (e.g. IPCC, 2007a). Such rates thus imply a tendency for shoreline retreat of 0.1 to 0.2 m a^{-1} due to the Bruun Rule, with an associated virtual loss of sand of 1 to 2 m^3 m^{-1} a^{-1}.

These very small values indicate that the Bruun effect under present near stillstand conditions will generally be

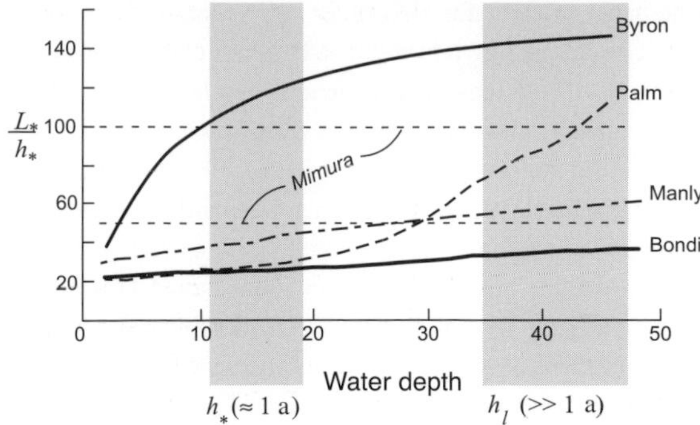

FIGURE 6.3. Variability of shoreface slope in Eq. (6.8) (L_*/h_*) in relation to water depths indicative of limiting depths (shaded bands) of the active upper shoreface and the slowly evolving lower shoreface following Hallermeier (1981) for four well-known beaches in southeastern Australia, illustrating the limited applicability of a 50 $\leq L_*/h_* \leq$ 100 bed-slope factor in sea level induced recession (the *Mimura* range), implied via Eq. (6.2) from results of Mimura and Nobuoka (1995) and Zhang *et al.* (2004).

negligible compared to other effects and it can be debated whether the effect can indeed be inferred from measurements under present conditions. Moreover, the empirical validations referred to above are not readily transferable to other coasts. For example, the shoreface slope term, L_*/h_*, is subject to strong geographic variability. This variability is illustrated in Fig. 6.3 with respect to the active zone, and the less active lower shoreface, with indicative water depths for these zones based on the criteria of Hallermeier (1981). Figure 6.3 also emphasises the strength of intra-regional geographic variability, given that all the beaches except for Byron, lie along the metropolitan coast of Sydney. This variability indicates the limited applicability of the empirical generalisation that c_p is typically 50 to 100 times sea level rise.

Nevertheless, the conclusion must be that the Bruun effect is insignificant under current sea level rise rates and likely to be overridden by more substantial effects, like that of sediment accommodation in lagoons and estuaries.

6.3.4 The quantitative coastal tract

Cowell *et al.* (2003a) show how the Valentin–Curray–Swift concept can be quantified and related back to the Bruun Rule (Bruun, 1962). Cowell *et al.* (2003a) consider the sediment balance of the upper shoreface and adopt the assumption that the upper shoreface to a time-averaged first approximation is form invariant relative to mean sea level over time periods (>>1 a) for which profile closure occurs (Nicholls *et al.*, 1998). This implies that one can derive a kinematic balance equation for the upper shoreface, provided erosion resistance due to lithification effects (e.g. Thieler *et al.*, 1995; Stolper *et al.*, 2005) can be ignored, and all the material is beach grade so that there are no additional losses due to fines going into suspension. The upper shoreface is represented by an arbitrary, but usually concave-up, profile $h(x)$ to a depth h_* (a morphologically active depth) and a length L_*, in which x is the distance from the shore (Dean, 1991). Sediment volume conservation for profile kinematics requires (for a Cartesian coordinate system with seaward and upward directions positive) that

$$\frac{\partial h}{\partial t} + c_p \frac{\partial h}{\partial x} = 0 \tag{6.3}$$

or, via $h = \text{MSL} - z_b$, where MSL is mean sea level and z_b is the bed level,

$$\frac{\partial z_b}{\partial t} + c_p \frac{\partial z_b}{\partial x} = \frac{\partial \text{MSL}}{\partial t} \tag{6.4}$$

where c_p is the rate of horizontal profile displacement. The sediment–transport balance equation for a fixed spatial control volume is

$$\frac{\partial z_b}{\partial t} + \frac{\partial q_x}{\partial x} + \frac{\partial q_y}{\partial y} + s = 0 \tag{6.5}$$

where $q_{x,y}$ are the cross-shore and alongshore sediment transports, and s is a local sink ($s > 0$) or source ($s < 0$). Equations (6.3), (6.4) and (6.5) may be combined to yield

$$c_p = -\frac{\partial \text{MSL}}{\partial t}\left(\frac{\partial h}{\partial x}\right)^{-1} - \frac{\partial q_x}{\partial h} - \frac{\partial q_y}{\partial y}\left(\frac{\partial h}{\partial x}\right)^{-1} - s\left(\frac{\partial h}{\partial x}\right)^{-1} \tag{6.6}$$

or, after cross-shore integration over L_*,

$$c_p h_* = \frac{\partial \text{MSL}}{\partial t} L_* - \left(q_{x,\text{sea}} - q_{x,\text{dune}}\right) - \frac{\partial Q_y}{\partial y} - s \tag{6.7}$$

in which Q_y is the alongshore transport integrated over L_*.

Equation (6.7) is a quantitative expression applicable to an infinitely small alongshore section of the upper shoreface that describes the sediment balance and results in the shoreline rate of change c_p. It is a general expression applicable to any sedimentary coastal system as long as the upper shoreface is form invariant relative to mean sea level. We note that it is similar to the Dean and Maurmeyer (1983) extension of the Bruun Rule. The shoreline change rate is determined quantitatively by the balance between the 'sink' term, for accommodation space generated due to sea level rise (first term on the right-hand side), and sediment availability (being the sum of sinks and sources, the last three terms on the right-hand side of Eq. (6.6)). The relative sea level change term is a virtual sink/source term since there is no absolute loss, although the response is comparable to the impact of a real source/sink regarding horizontal movements of the upper shoreface. The other three terms are real source or sink terms in the budget, respectively cross-shore exchanges with the lower shoreface

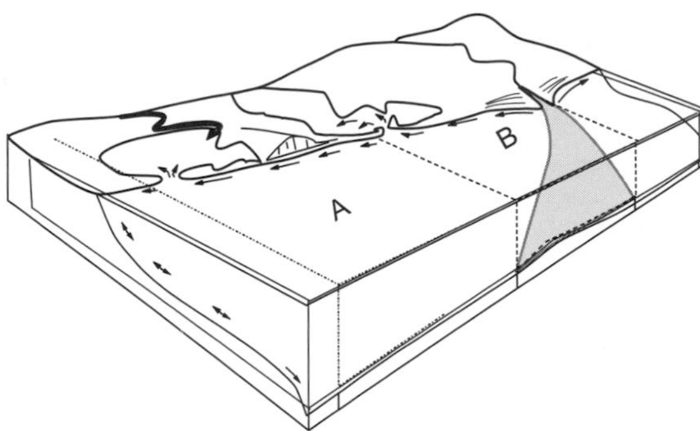

FIGURE 6.4. Two coastal tracts in which magnitudes of sediment fluxes (size of arrows) exemplify cases of alongshore homogeneity in which uniform alongshore flux gradients can be integrated spatially to obtain S in Eq. (6.8). Cell A: alongshore flux gradient is a net sink ($S > 0$) causing retrogradational tendencies (left-hand quadrants in Fig. 6.5). Cell B: alongshore flux gradient is a net source ($S < 0$) causing progradational tendencies (right-hand quadrants in Fig. 6.5).

and the terrestrial beach or dune, alongshore redistribution due to alongshore transport gradients and fluvial, estuarine or lagoonal sources or sinks. Two modifications to Eq. (6.7) may need to be introduced, namely when we are dealing with substantial amounts of fines in the beach barrier, which will enlarge the recession (this is e.g. the case for the Mississippi barrier islands) or when we are dealing with erosion resistant cliffs, which will reduce recession (see section on UK cliffs).

In the context of low-order coastal change we now suggest a fundamental reduction of the quantitative kinematic sediment balance for the active nearshore zone (Eq. 6.7), by aggregating in such a way that alongshore redistribution by alongshore transport gradients and backbarrier exchanges can be ignored, so that we are left with an alongshore uniform response determined by two components only, namely a sea level rise/fall virtual sink/source term and a sediment availability term comprised of two components namely real sources/sinks due to cross-shore exchanges and backbarrier exchanges (cf. Valentin–Curray–Swift concept). That is, we choose our geomorphologic conditions and alongshore boundaries such that details of the sand fluxes are not relevant since we are interested at this stage in large-scale low-order coastal change (Fig. 6.4). The extent of alongshore aggregation depends on the degree to which alongshore homogeneity can be assumed in the time-averaged flow fields and sediment transport gradients over the long term. The homogeneity assumption is fundamental in discrimination of a coastal tract (i.e., the littoral cell used as a sediment control volume).

Under the above assumptions the position of the shoreline in the coastal-tract cascade concept (Fig. 6.4) can be inferred by integrating in the alongshore direction over a uniform length, so that local alongshore and cross-shore redistribution processes are irrelevant, such that Eq. (6.7) reduces to:

$$c_p h_* = -\frac{\partial \mathrm{MSL}}{\partial t} L_* - \left(Q_{x,\mathrm{sea}} - Q_{x,\mathrm{dune}}\right) - S \qquad (6.8)$$

where Q_x is the alongshore integrated, cross-shore sediment transport, and S is the alongshore integrated and averaged local source or sink. Equation (6.8) is the quantitative coastal-tract sediment budget equation for the upper shoreface.

Thus, if the longshore transport gradient is uniform within Cell A in Fig. 6.4, then it and the losses to the lagoon through the inlet and via overwash can be subsumed simply as a sediment sink into S in Eq. (6.8). Similarly, in Cell B in Fig. 6.4, the fluvial supply is subsumed as a source into S. Note that, if uniform alongshore flux gradients exist, Cell A satisfies the alongshore homogeneity assumption required for application of Eq. (6.8) despite segments of the coastal tract comprising barrier and mainland beach morphologies.

The h_* term also includes the effects of dune height. Dunes are often variable in height alongshore on embayed coasts where headlands cause alongshore variation in degree of exposure to ocean storm winds. Despite apparent implications of alongshore variation in dune height in Eq. (6.8), the definitive sediment-sharing properties of the coastal tract mean that c_p is likely to be uniform alongshore. Any tendencies for shoreline rotation induced by slower retreat along segments of the tract where dunes are higher are likely to be suppressed. This effect can be expected due to enhanced alongshore transport divergence from any incipient protrusions that develop in the shoreline wherever c_p is initially retarded.

In particular cases, i.e. when a coastal stretch is bounded and where the alongshore uniformity assumption is not strictly satisfied, we still may apply Eq. (6.8) when we average out alongshore non-uniformity and look at the alongshore averaged budget and associated rate of coastline change c_p. The below cases of Australian mainland beaches and the Ebro Delta are such cases. In the case of erosion-resistant cliffs such as in the UK the recession rate is retarded.

6.4 Applications of the quantitative coastal tract

6.4.1 Introduction

This section discusses four case studies that are exemplary for open coast responses over the Holocene and for which we use the coastal-tract equation (6.8) to quantify the

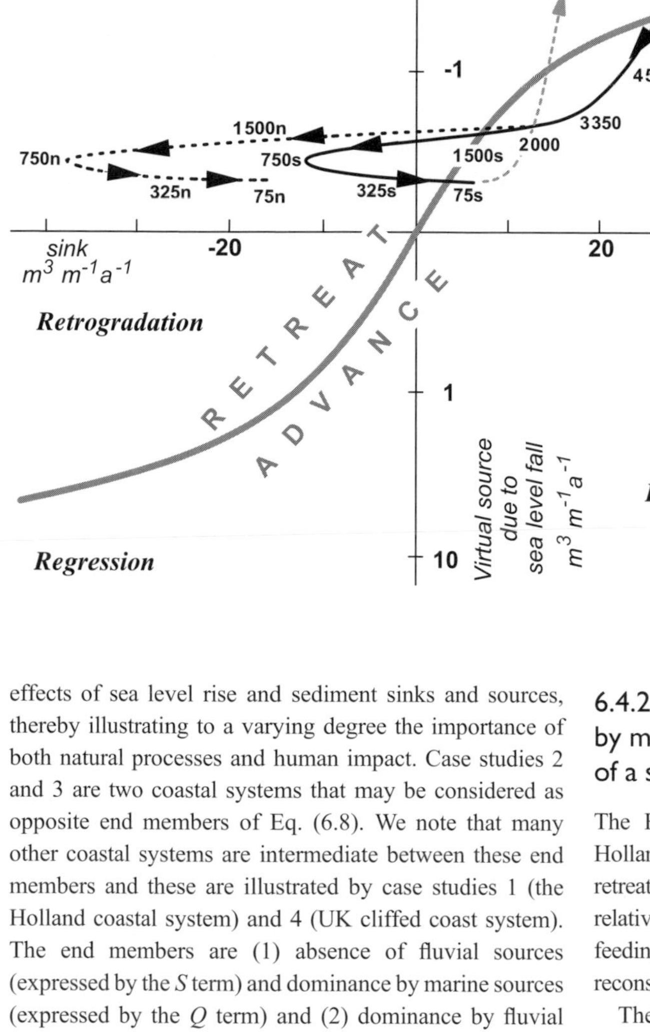

FIGURE 6.5. Evolution of the central Netherlands coast (Hoek van Holland to Den Helder) as a time trajectory in sediment supply/accommodation phase space (abscissa and ordinate respectively, scaled in $m^3\ a^{-1}$ per metre of shoreline). Numbers along the trajectory indicate time (years BP); suffixes n and s denote coast north and south of Haarlem. The bifurcation to the dashed trajectory after 2 ka BP represents the situation in the north (n) where the sediment budget is dominated by tidal inlets of the Friesian Islands.

effects of sea level rise and sediment sinks and sources, thereby illustrating to a varying degree the importance of both natural processes and human impact. Case studies 2 and 3 are two coastal systems that may be considered as opposite end members of Eq. (6.8). We note that many other coastal systems are intermediate between these end members and these are illustrated by case studies 1 (the Holland coastal system) and 4 (UK cliffed coast system). The end members are (1) absence of fluvial sources (expressed by the S term) and dominance by marine sources (expressed by the Q term) and (2) dominance by fluvial sources and weak marine sources. These end members are the two highest levels in the hierarchy of Shepard's classification: namely (1) coasts shaped primarily by marine agencies, secondary or mature coastlines, and (2) coasts shaped primarily by non-marine agencies, primary or youthful coastlines. An interesting illustration – the second system discussed here – of a secondary or mature coastline is the Australian coast, which lacks significant fluvial sources over the Holocene. The category of a primary, youthful coastline is well illustrated by deltas, like those of the Ebro and the Po, that experienced strong fluvial sources at certain stages of their Holocene evolution. The Ebro case is the third system discussed. We close our tract analysis with cliffed coast systems, including a UK example.

6.4.2 The Holland coast primarily shaped by marine agencies (complex example of a secondary or mature coast)

The Holland coastal stretch (stretching from Hoek van Holland to Den Helder) has undergone a succession of retreat – advance – retreat until about 2 ka BP due to changing relative importance of sea level rise driven retreat and marine feeding driven advance. Plate 19 presents some geologic reconstructions of these various moments.

The qualitative Valentin–Curray–Swift model applied to the Holland coastal evolution is quantified by Cowell *et al.* (2003b) as a time trajectory in sediment source/sink phase space: e.g. evolution of the well-documented central Netherlands coast between Hoek van Holland and Den Helder (Plate 19, Fig. 6.5). Figure 6.5 recasts the Curray (1964) phase diagram more systematically in terms of orthogonal drivers (sea level change and sediment supply, which can both be positive or negative) and their conjugate coastal responses: regression–transgression (related to sea level change) and progradation–retrogradation (related to sediment supply). Each conjugate pair of responses is associated respectively with an *advance* of the coastline into the sea, or a *retreat* of the coastline in the landward direction (Fig. 6.5).

The phase trajectory in Fig. 6.5 is based on (a) estimates derived from radiometric data by Beets *et al.* (1992), listed in the Annex of Cowell *et al.* (2003b), for the period 5000–0 BP; and (b) the results of simulated reconstruction for 7200–5000 BP (Cowell *et al.*, 2003b). The line separating advance and retreat of the coast is fitted for the trajectory in the top right quadrant, with its mirror image assumed for the bottom left quadrant in the absence of other data. The trajectory bifurcates after 2000 BP because differences develop in rates of shoreline change averaged alongshore north and south of Haarlem (approximately in the middle of the 120-km long Holland coastal section). The shape of the advance/retreat threshold curve demonstrates that coastal evolution is governed mainly by (a) sediment supply (+/−) under near stillstand sea level conditions (such as those predominating during the late Holocene), and (b) change in accommodation space when sea level changes rapidly (such as during global deglaciation following the last glacial maximum).

The time trajectory for the Netherlands coast shows that a static coastal classification is inapplicable. As the rate of sea level rise declines and sources and sinks change, so does the behaviour mode of the coastal system. With quantitative reference to Eq. (6.8) we note that before 5200 BP primary marine and secondary fluvial sources (the second and third terms) although being quite high (25 to 30 $m^3\ m^{-1}\ a^{-1}$) were overruled by the effect of sea level rise (the first term). When the rate of sea level rise dropped after 5200 BP the balance between the three terms led to advance until about 2000 BP, which turned into retreat afterwards, mainly due to the accommodation space created by the Wadden Sea. A tripling of present sea level rise rates will increase retreat significantly, not only due to the Bruun term but also particularly due to the increase in the Wadden Sea accommodation space.

6.4.3 Australian beaches with significant marine supply and negligible fluvial supply (example of a secondary or mature coast)

Measurements of Holocene barrier volumes on the Australian coast (Short, 2003) (Plate 20) have strengthened the hypothesis (Cowell *et al.*, 2001) that middle shoreface wave-induced sediment transport is generally onshore on concave shaped shorefaces (Fig. 6.6a). Short (2003) made a quantitative inventory of the alongshore-averaged net sedimentation over the late Holocene on parts of the Australian coast that can be considered laterally closed (Fig. 6.6b). This author states that, for the whole Australian coast, mean Holocene onshore sediment transport rates range from 0.1 to 10 $m^3\ m^{-1}\ a^{-1}$. We note that this order of magnitude is confirmed by findings over the Holocene for the Dutch and the Washington State prograded barrier coasts (Cowell *et al.*, 2003b; Hinton and Nicholls, 2007). While fluvial supply is small in general along the southeast and northern coasts, it is totally absent along the southern and southwestern Australian coasts. It is along these parts of the coast where waves are energetic and wave directions are onshore. Waves deliver massive volumes of carbonates from the shelf to the shore, ranging from 3 to 9 $m^3\ m^{-1}\ a^{-1}$ over the Holocene. The net marine sediment supply by waves is associated with wave asymmetry and wave boundary layer induced net flow (Bowen, 1980).

What does the above mean for impacts of increased rates of sea level rise on this type of marine-sand-fed, outbuilding coast? From discussion above in relation to Eq. (6.8), eustatic sea level rise on such coasts under present conditions, at typical rates of 0.2 cm a^{-1}, leads to a virtual loss of 1 to 2 $m^3\ m^{-1}\ a^{-1}$. Such a loss is easily compensated by the marine supply along Australian southern and southwestern strandplain beaches. However, if the rate of eustatic sea level rise triples, then approximately half of these beach systems may become erosive, while the other half may stay accretive.

Geographic variability however is expected due to gross variations in characteristic shoreface depths under present conditions. If active shoreface morphology is assumed to be in equilibrium with time-averaged flow fields under existing conditions, then the shallowest shorefaces (lowest slopes) may exert a larger demand on dune sand due to the virtual losses offshore. On the other hand, if shallowest shorefaces remain too shallow with respect to equilibrium, then continued shoreface sand supply to beaches might be expected even with sea level rise (e.g. Byron in Fig. 6.3). The effect of increased sea levels will however reduce the rate of sand supply, due to a diminished disequilibrium stress. The reciprocal considerations apply to the steepest shorefaces (e.g. Bondi in Fig. 6.3).

6.4.4 The Ebro Delta, a deltaic system with negligible marine supply and significant fluvial supply (example of a primary, youthful coast)

The deltas of the Ebro (Mediterranean coast of Spain) and the Po (Adriatic coast of Italy) can be considered as prototypes of primary coastal systems during their evolution since Roman times until quite recently (Fig. 6.7). At the present stage, due to river regulation and catchment area changes this has ceased to be the case. We will illustrate this by a quantitative evaluation of the sediment balance of the Ebro Delta.

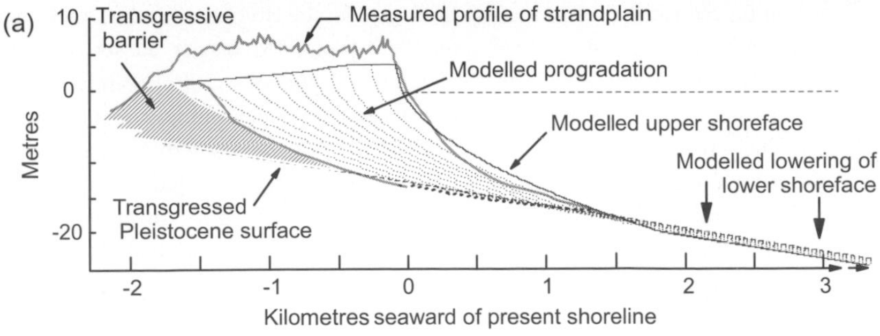

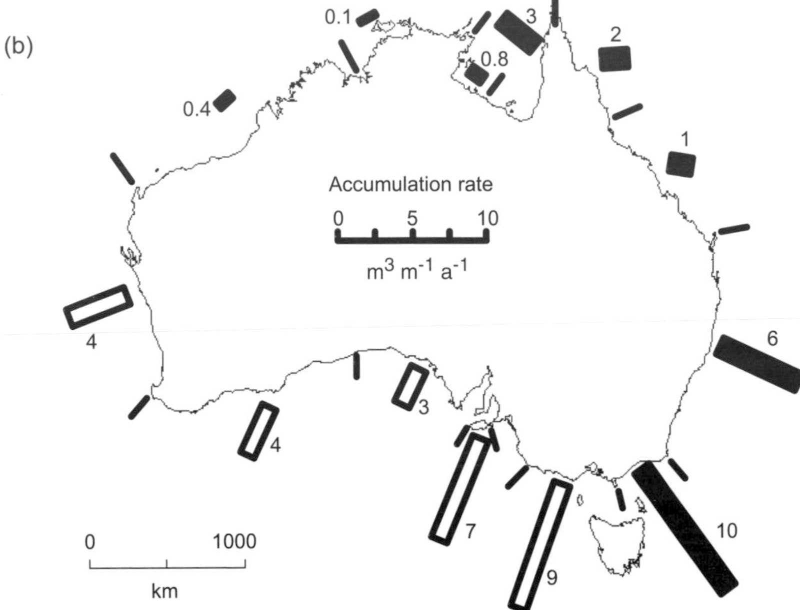

FIGURE 6.6. Sand supply for barrier progradation from the shoreface: (a) modelled and measured supply, at $1.35\ m^3\ a^{-1}\ m^{-1}$ for 6 ka, from Tuncurry (Plate 20), southeastern Australia (modified from Cowell *et al.*, 2003b); (b) volume estimates for segments of the Australian coast in which solid and hollow bars respectively signify siliclastic and carbonate sand supply (modified from Short, 2003).

Iberinsa (1992) portrays the reconstructed evolution of the lower Ebro Delta since Roman times (Fig. 6.7). A continuous strong surface areal growth is observed, roughly a doubling every 1000 years. Jimenez (1996) presents the quantitative surface aerial growth over several periods since 1749, which we use to derive an estimate of the alongshore-averaged accretion (see Table 6.4) in order to compare this with other sources and sinks. We note that a dramatic decrease of sediment supply to this delta has occurred, which is typical for many deltas in which increased catchment management, especially dams, has been undertaken (IPCC, 2007b).

Compared to the net marine sediment supply on Australian beaches it is clear that we are dealing with a youthful river-fed deltaic coast, at least until 1957. Moreover, since we are dealing with a low-wave-energy coast we can expect less marine feeding, which is confirmed by Jimenez (1996) presenting estimates of the marine sources and using different approaches to reach an estimate of $4\ m^3\ m^{-1}\ a^{-1}$, which is estimated to be of the same order as aeolian losses. Similarly, Jimenez (1996) estimates the virtual loss by present relative sea level rise of the subsiding delta (0.2–0.5 cm a^{-1}) as in the range 1 to $5\ m^3\ m^{-1}\ a^{-1}$. We conclude that an increase of relative sea level rise into the range of 0.6 to 1 cm a^{-1} and a further decrease of fluvial supply will change the Ebro Delta from an accretive to an erosive delta, and expect this to be the case for many regulated deltas globally.

6.4.5 UK cliffed coasts (example of a secondary or mature coast)

Cliffed coasts are fairly simple systems compared to other coastal morphology types, as by definition they indicate erosion, a deficiency of beach material and usually a long-term retreating tendency through the mid to late Holocene. There is no direct sedimentary record of their evolution, although they are sources of new sediment to neighbouring coasts where sedimentary records might develop. On

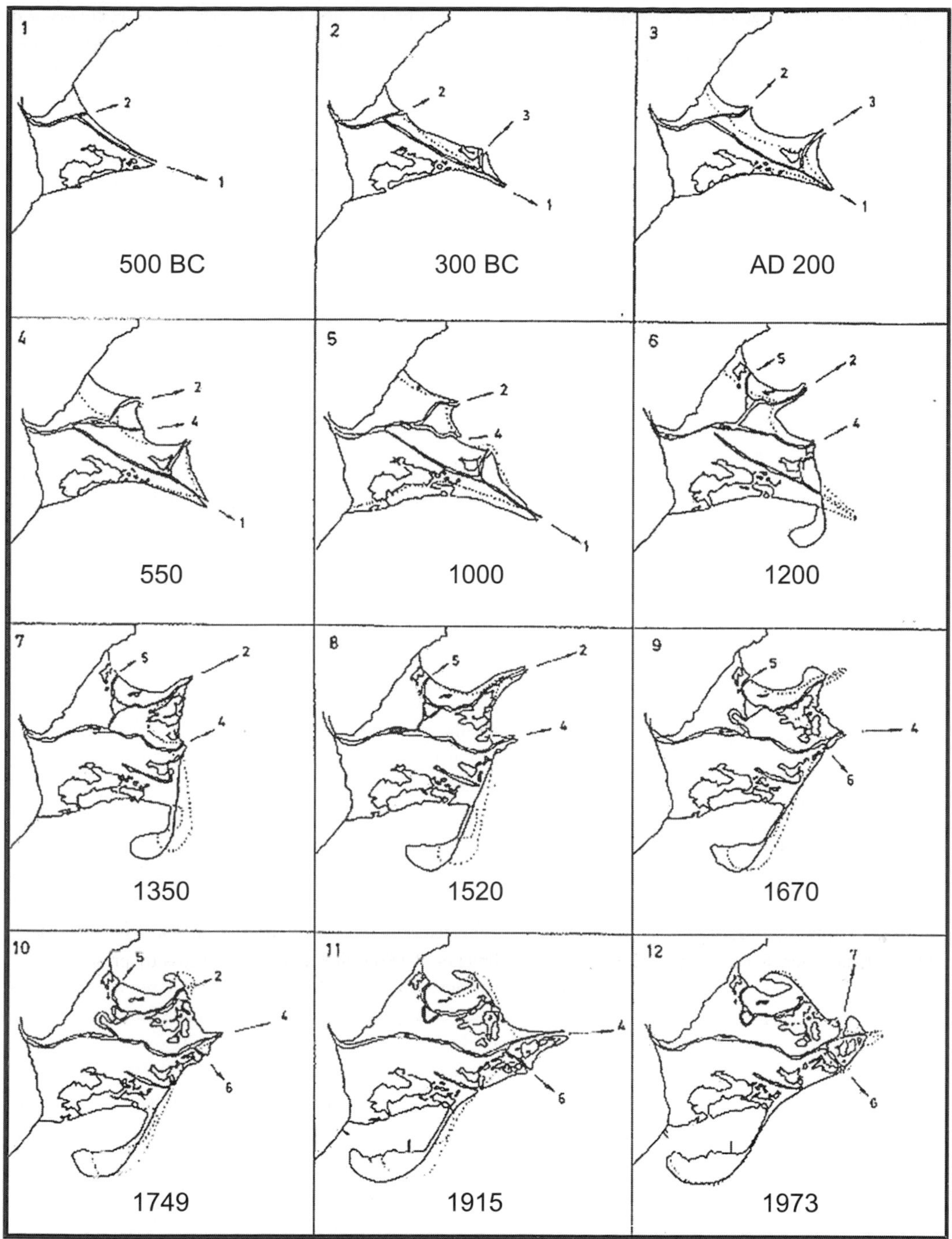

FIGURE 6.7. Reconstructed evolution of the Ebro Delta (modified from Iberinsa, 1992, plate 1).

soft cliffs (Plate 21) this source can be significant with examples found around the world, including the east and south coasts of the UK (e.g. Clayton, 1989; Dickson *et al.*, 2007).

When considering Eqs. (6.7) and (6.8) in the context of a coastal tract comprising continuous cliffs, erosion resistance of the cliff (and shore platform) material is an important additional parameter that needs to be considered.

TABLE 6.4. *Surface area increase of the lower Ebro Delta transformed into alongshore-averaged accretion rate, using an average shoreline length of 40 km and an averaged shoreface depth of 30 m*

Period (AD)	Surface area increase ($m^2\ a^{-1}$)	Alongshore-averaged accretion rate ($m^3\ m^{-1}\ a^{-1}$)
1749–1915	150 000	113
1915–1957	50 000	38
1957–1973	25 000	19
1973–1989	12 000	9

Source: Jimenez (1996).

Hence when considering cliffs where beaches are not important, Eq. (6.8) can be reformulated as

$$c_p = -\left(\frac{\partial \mathrm{MSL}}{\partial t}\right)^a \left(\frac{L_*}{h_*}\right) \tag{6.9}$$

where $a < 1$ to allow for the erosion resistance. The work of Walkden and Dickson (2008) suggests a value of a of about 0.5, but this requires further investigation. In addition, there is a feedback that retreat produces sediment, which protects the cliff from further erosion until this material can be removed. The magnitude of this source term is controlled by retreat rate, cliff height (above h_*) and cliff composition as fines are lost to suspension. Hence, the cliff is a source term and when beaches become significant, this will further reduce the retreat rate compared to that predicted in Eq. (6.9) as captured by the sink term in Eq. (6.7) and demonstrated in studies such as Dickson *et al.* (2007). Many beaches are underlain by cohesive material, and more work is required to fully understand when erosion resistance is an important parameter to consider, or not.

6.4.6 Longshore redistribution and backbarrier exchanges or inlet influences

In the previous section we aggregated gains or losses in beach sand volume through alongshore distribution, estuarine sinks and deltaic sources. The aggregation involved alongshore integration of the sources or sinks to a single term S (Eq. 6.8). If alongshore integration of sediment exchanges is undertaken over a stretch of coast that spans one or more tidal inlets, then S includes the backbarrier sink. This sink is especially important during rising sea levels because accommodation for sedimentation increases in the backbarrier basin, creating a sand demand that is primarily responsible for barrier rollover (Dean and Maurmeier, 1983; Leatherman, 1983; Swift *et al.*, 1991): i.e., the mechanism driving transgressive barriers (Fig. 6.4).

Generally, the main higher-order processes involved in barrier rollover involve redistribution of beach sand alongshore and through tidal inlets into the tidal basin. Other mechanisms include the landward migration of dunes and washover (Swift *et al.*, 1991). Transgressive barriers, including barrier islands, are particularly susceptible to extensive overwash in areas such as the US east coast and Gulf of Mexico as a result of the large waves and extreme storm surge associated with hurricanes (e.g. Stockdon *et al.*, 2007). Washover deposits are possible only if dune heights are less than the maximum elevations attainable by wave runup and surf zone setup during storm surge. Under conditions of rapid and sustained sea level rise, the opportunity for vertical growth of dunes is limited due to incessant overwash, so the prevalence of overwash becomes self-reinforcing. These tendencies are likely to arise more commonly at lower latitudes because dune heights generally are least well developed within tropical regions.

In the mid-latitudes, particularly on parts of the coast exposed to the strongest winds, evidence exists to suggest that massive transgressive dune sheets and parabolic dunes develop (Thom *et al.*, 1981; Lees, 2006; Hesp *et al.*, 2007), preventing any possibility of overwash. Some of the largest transgressive dunefields have been described from the coast of Brazil (Hesp *et al.*, 2007) and others on the east coast of Australia (Thom *et al.*, 1981) and South Africa (Illenberger and Rust, 1988). Under these circumstances, the wind-driven inland migration of the dunes itself becomes a third and significant mechanism driving barrier rollover (Swift *et al.*, 1991). Beach erosion is known to cause foredune instability and initiate dune migration inland (Hesp, 2002). However, the development of large foredune systems and extensive dunefields requires the presence of large amounts of sand.

Encroachment of a transgressive beach into the seaward edge of the dunefield during ongoing sea level rise is likely to maintain foredune instability, promoting aeolian sand feeding to landward. In a review of the dynamics of beach/dune systems, Sherman and Bauer (1993) made use of the conceptual framework of coastal classification proposed by Valentin (1952) in relating progradation and retrogradation to the behaviour of dunes. For example, foredune systems and the inland dunefield at Greenwich Dunes, Prince Edward Island (Hesp *et al.*, 2005) have formed since a major storm around 1930 which overwashed the foredune along 5 km length of shoreline with wave action penetrating up to 300 m inland (Mathew, 2007). The extensive overwash resulted in the formation of transgressive dunes up to 20 m high. Similar events are produced, for example, by hurricane impacts on the barrier islands of the US east and Gulf coasts (Stockdon *et al.*,

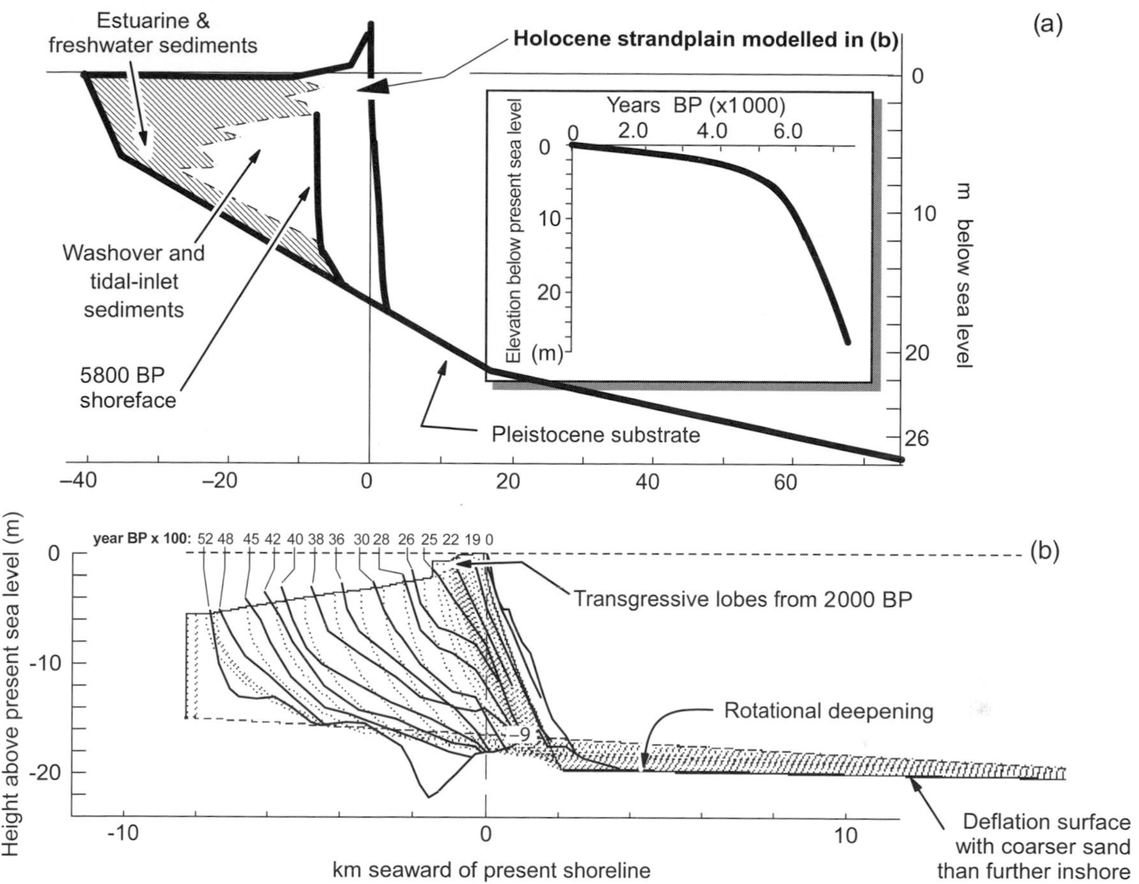

FIGURE 6.8. Seaward translation of the upper shoreface in the central Netherlands (Haarlem) driven by littoral sediment feed and a deepening of the landward 10 km of the lower shoreface, with the deepening decreasing progressively in the offshore direction: (a) generalised stratigraphic reconstruction based on data, with transgressive sand and estuarine mud deposits landward of the 5800 BP shoreface; and (b) simulated shoreface history (dashed lines) overlaid on isochrons (solid lines) determined from ^{14}C-dated core samples (Beets *et al.*, 1992) (from Cowell *et al.*, 2003b).

2007). Overwash produces an extensive area of unvegetated sands, which can quickly be reworked. Moreover, transgressive dune mobility is enhanced especially when combined with the effects of possible increased storm erosion (Christiansen and Davidson-Arnott, 2004). More intense tropical and extra-tropical storms are possible over wide areas due to climate change (IPCC, 2007a), promoting a tendency towards greater dune instability with other compounding factors possible. As an example, in southeastern Australia, projected increases in frequency and intensity of droughts, heat waves and wildfires will likely exacerbate dune instability.

Where transgressive dune activity was strongest on the southeastern Australian coast toward the end of the mid Holocene sea level rise, evidence exists to suggest that dune migration was the dominant mechanism of barrier rollover (Thom *et al.*, 1981). Similarly, the onset of trangressive dune activity around 2000 BP seems to have stalled the progradation of the central Netherlands coast (Beets *et al.*, 1992) which had been under way since about 5800 BP despite ongoing sea level rise throughout the late Holocene (Fig. 6.8b).

The tipping point from transgression to progradation on the central Netherlands coast at 5800 BP (Fig. 6.8a) is thought to have occurred because of a declining rate of sea level rise relative to continued marine sediment supply, including both sand and mud (Beets *et al.*, 1992; Cowell *et al.*, 2003b). That is, the decelerating sea level rise led to a diminishing availability of backbarrier accommodation relative to the sediment supply that eventually caused tidal inlets to become choked with the surfeit of sediments. By 5200 BP, the inlets finally closed, so that the continued supply of sand from the middle shoreface went into driving barrier progradation rather than being transferred to the

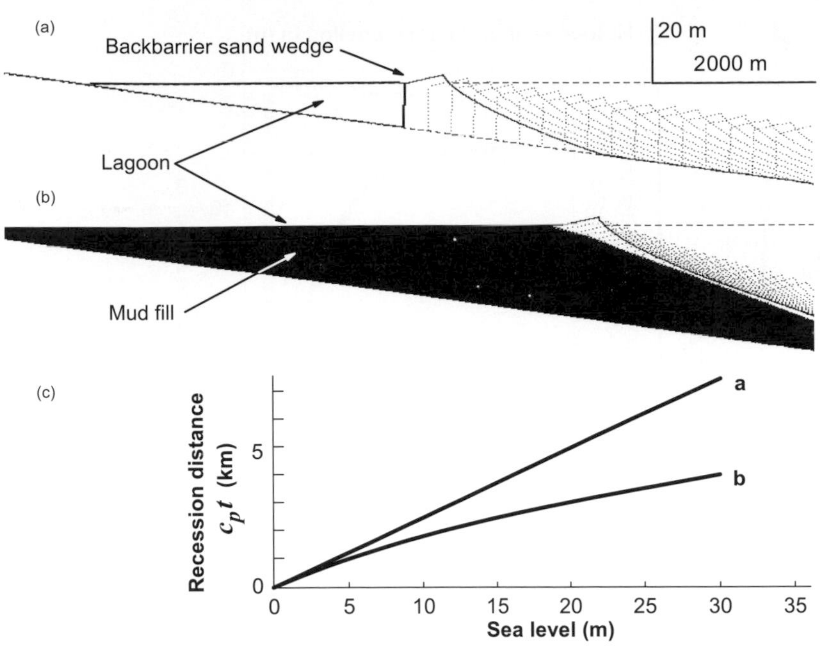

FIGURE 6.9. Moderating effects on rate of barrier transgression due to deposition of fine sediments in the backbarrier under constant rates of sea level rise over a 0.2 substrate: (a) barrier rollover with no estuarine mud deposition, for which the rate of translation is identical to the rate of simple inundation on a sloping plane; (b) effect of mud accumulating vertically at the same rate as sea level rise; (c) comparative rates of horizontal recession of the shoreface (from Cowell et al., 2003b).

backbarrier (Fig. 6.8b): hence the *c.* 5200–1500 BP trajectory in Fig. 6.5.

The progradation simulated in Fig. 6.8b began with a littoral feed of 30 m^3 m^{-1} a^{-1} of coastline from 5200 to 3600 BP, then declining by 5 m^3 m^{-1} a^{-1} until 2000 BP when it was stabilised at -0.5 m^3 m^{-1} a^{-1}. The cross-shore transfer of sand to the upper shoreface involved a deepening of the upper–lower shoreface boundary (L_* in Eq. (6.8)) at a rate of 0.17 m 100 a^{-1}: consistent with, although less than, rates of measured bathymetric change (Stive *et al.*, 1991; Hinton and Nicholls, 2007). The inferred lower shoreface deepening extended 10 km seawards at rates that decreased linearly with distance. The simulated process (Fig. 6.8b) provided 49% of the strandplain volume.

Following closure of the tidal inlets at 5200 BP on the central Netherlands coast, the fine-grained marine sediments previously bypassed to the backbarrier could no longer accumulate in the coastal tract. The shoreface was the only remaining option to accommodate these sediments, but its depositional environment is too energetic to allow mud accumulation. Until inlet closure, however, the availability of fine sediments for deposition in the vast lagoon of the central Netherlands played a crucial role in moderating the rate of barrier transgression due to the rapidly rising sea levels of the early to mid Holocene (Cowell *et al.*, 2003b).

The effectiveness of mud deposition on the moderation of barrier trangression rates is illustrated in Fig. 6.9. Mud accumulation, from fluvial or marine sources, induces an increased gradient over which the barrier must transgress, thus reducing L_* in Eq. (6.6). A side effect is that the mud accumulation occupies accommodation space that tidal inlet, washover or transgressive dune sands would otherwise fill on the seaward margin of the lagoon. The reduced sand accommodation in the lagoon thus leads to a marked reduction in the sand volume of the transgressive barrier and a lower rate of shoreline retreat (Fig. 6.9c). The resulting stratal geometry is typical of transgressive environments like those well-known classic geological studies in the USA on the southeastern coast (Kraft and John, 1979) and Gulf coast (Penland *et al.*, 1985).

6.5 Risk-based prediction and adaptation

Nothing about climate change is certain except that climate has always changed, it always will change, and that change is now likely to accelerate because of global warming (IPCC, 2007a). Predicting the impacts of climate change is even more uncertain than predicting climate change itself. The additional uncertainty for impacts arises because of limits to knowledge on long-term biophysical responses to climate, the role of non-climate drivers of change, and because seldom enough data exist to fully characterise site-specific conditions. These issues pertain to all aspects of climate change impact assessment, and prediction of coastal impacts is no exception. Uncertainty about the present and the future can impede rational decision-making that requires comparative assessment of opportunities and liabilities pertaining to enterprises competing for allocation of scarce resources (Manne and Richels, 1995; Webster, 2002).

Conventional methods exist however to manage uncertainty in providing transparent support for determinations. Although these methods are well established in other fields (e.g. Webster and Sokolov, 2000) such as information science, climate science, financial and project management, as well as some aspects of engineering, including design of coastal structures (Bakker and Vrijling, 1980; Dover and Bea, 1980) they have scarcely been applied to decision-making in coastal management (Vrijling and Meijer, 1992; Titus and Narayanan, 1996; Thorne *et al.*, 2007). The IPCC (2007a) sea level rise scenarios include probability ranges, but as not all the components of global sea level rise are included in these ranges, in terms of application to coastal management, the range of these numbers fails to capture all the uncertainty.

Coastal managers under conditions of uncertainty can derive risk-based forecasts for practical use. Forecasts differ from deterministic predictions in that the latter provide a single estimate, whereas the former indicate the range of feasible impact projections. Moreover, forecasts assign the likelihood of occurrence across the range of impact magnitudes. The advantage of this approach over deterministic methods is that risk can be compared for different projects competing for limited resources despite uncertainty. In addition, support for decisions based on quantified estimation of risk provides greater transparency in administration. Furthermore, single deterministic predictions do not indicate the spread of uncertainty and hence, the consequences of the accepted estimate being wrong.

Deterministic models based, for example, on principles summarised in Eq. (6.7) can be applied through repetitive simulations that vary input parameters at random. These Monte Carlo type simulations are performed using randomly sampled model-input variables. The samples are drawn from probability distributions that characterise uncertainty about the inputs as well as the modelled processes.

The construction of probability distributions to characterise the uncertainty pertaining to input parameters is the key to transforming qualitative uncertainty into quantified risk. This pivotal aspect of the methods can exploit heuristic reasoning and thus draw from qualitative and uncodified expertise. The approach has been demonstrated, for example, for preliminary forecasts of Manly Beach in Sydney (Fig. 6.10) where typical uncertainty exists about climate change impacts (Cowell *et al.*, 2006). Sources of uncertainty include future sea level rise, future changes in the active morphology in response to shifts in wave climate and variable resistance of material comprising strata in a substrate that includes rock reefs and a sea wall.

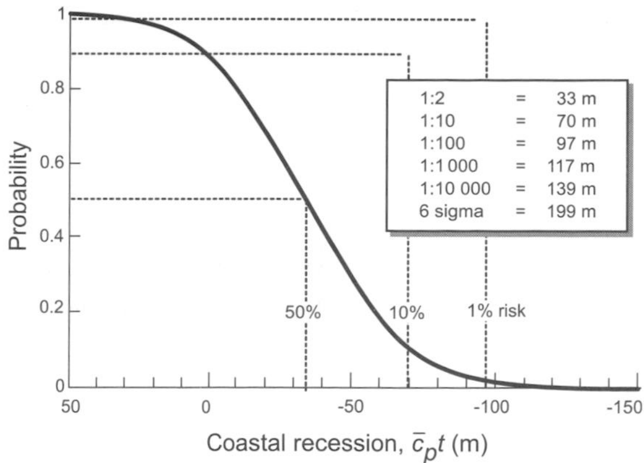

FIGURE 6.10. Risk curve: probability of shoreline recession by year 2100 due to climate change impacts at Manly Beach, Sydney, based on simulations in which model variables were sampled from probability distributions for each of the model parameters during 2000 simulations (based on Cowell *et al.*, 2006).

In terms of probability, therefore, the best estimate is merely the most likely estimate: i.e. the mean or modal (or peak) value in a probability distribution. For Manly Beach simulations, the mean recession distance by 2100 is 33 m, which corresponds to the best estimate and is typical of estimates derived from deterministic application of Eq. (6.7). However, 50% of possible estimates are greater than this 'best estimate'. A more *risk-averse* approach to coastal management would therefore be to adopt estimates with a much lower probability of realisation, such as 1: 10 or 1: 100. For example, at these lower, more acceptable levels of risk, coastal recession estimates for Manly Beach (Fig. 6.10) are more than two to three times greater than the deterministic estimate.

Having presented a methodology that recognises chronic versus acute erosion hazards, including estimates of uncertainty, a firmer scientific basis exists for deciding between available adaptation strategies. While there is a wide range of adaptation options for coastal areas, these can be grouped into a small number of more generic option types that are widely used (Klein *et al.*, 2001; Nicholls and Klein, 2005):

- protect – to reduce the risk of the event by decreasing the probability of its occurrence;
- accommodate – to increase society's ability to cope with the effects of the event; and
- retreat – to reduce the risk of the event by limiting its potential effects.

For chronic erosion hazards, protection will need to be ongoing and protection costs may well grow with time, while accommodation measures would also have to be progressively upgraded. Similarly, a retreat would need to be progressive and ongoing. In contrast, acute erosion hazards do not require a progressive retreat, although set-back approaches may still involve a once-off retreat, and protection and accommodation measures would have lower costs and a greater likelihood of success. Moves to shoreline management planning where strategic choices on 'hold the line' versus 'retreat the line' options can also be informed by the methods presented here. Thus, future geomorphic changes on the coast, forecast through application of coastal-tract cascade concepts, can be readily integrated with strategic adaptation within coastal management. Final decisions will also be strongly influenced by socioeconomic considerations.

6.6 Conclusions

This analysis shows that global environmental change threatens major coastal changes in the coming century and beyond, and the drivers include direct effects such as sea level rise and climate change, as well as more indirect effects such as reduced sediment flux to the coast due to dam construction and river regulation. To understand the likely trends and their magnitudes, a model-based framework, that takes account of the different processes shaping the coast, is essential.

A starting point of such an analysis is coastal classification, and the classic work of Shepard (1948) and Valentin (1952) provides a useful basis for coastal taxonomy to address the issues posed in this chapter. However, while we acknowledge the usefulness of taxonomy for the sake of communication, describing the large variety in appearance of coastal systems, we note a shortcoming of classification. This may become apparent when classification is applied to actual examples. For example, at which precise moment is a coast sufficiently altered by marine agencies to be classified as Shepard's Type II (Table 6.1)? More generally, classification implies integration over a continuum of processes, including subjective judgement to determine the classification boundaries.

As an extension of 'static' classification, we outline the coastal-tract concept, which highlights the dynamic nature of coastal systems. In particular, the coastal tract confers three benefits. Firstly, the coastal tract operates on a landscape scale that acknowledges the processes and dynamic linkages between sediment-sharing coastal elements. Secondly, it has the capacity to describe the continuum of coastal landscapes, providing for specific coasts to evolve smoothly between classes in any arbitrary classification. Thirdly, it introduces the application of sediment balance equations to quantify coastal change.

Using the coastal-tract concept we have evaluated quantitatively four typical systems in the light of environmental change. Alongshore fairly homogeneous coastal systems shaped mainly by marine agencies are represented by Australian strand plain beaches and by UK soft cliffed coasts, the latter exemplified by the Norfolk cliffs. The Australian beach system has been an accretive system due to marine feeding over the Holocene when we average over littoral cells. UK cliffed coasts are shaped by marine agencies and chronically erode, but in doing so, naturally nourish neighbouring beaches. Alongshore inhomogeneous coastal systems are represented by the Ebro Delta and by the Holland coast. The Ebro Delta has been shaped mainly by fluvial agencies, but recently suffers from the impact of river regulation in reducing sediment supply making marine agencies more significant. The Holland coastal system indicates a complex evolution where periods of advance and retreat alternate over the Holocene, depending on the relative importance of the rate of sea level rise and marine feeding, including mud supply to the barrier lagoon and sand supply to the barrier.

Analysis of these systems in the context of climate change and other environmental changes lead to two conclusions. Coasts shaped by marine agencies will generally experience a direct increase in the rate of erosion due to accelerating sea level rise, although on cliffed coasts, neighbouring areas could benefit. Coasts shaped by fluvial agencies will also suffer increasing tendencies for erosion as sea level rise accelerates, but indirect effects due to anthropogenic reduction of sediment supply are a greater threat now and will continue to be so in the next few decades.

The tendency for erosion can be quantified in terms of risk, allowing more informed management decisions about the appropriate response than when considering a single prediction. Application of the coastal-tract cascade concept provides results that can also be linked to strategic adaptation options within coastal management. These options include retreat, accommodate or protect; and 'hold the line' versus 'retreat the line'.

Acknowledgements

Robin Davidson-Arnott is acknowledged for providing text material and discussions on dune issues. Ian Townend and two other anonymous reviewers are thanked for their constructive critique on our draft manuscript.

References

Agardy, T. J. *et al.* (2005). Coastal systems. In *Millenium Ecosystem Assessment: Ecosystems and Human Well-Being*, vol. 1, *Current State and Trends*. Washington, DC: Island Press, pp. 513–549.

Bakker W. T. and Vrijling J. K. (1980). Probabilistic design of sea defences. *Proceedings of the 17th International Conference on Coastal Engineering*, American Society of Civil Engineers, New York, 2040–2059.

Beets, D. J., van der Valk, L. and Stive, M. J. F. (1992). Holocene evolution of the coast of Holland. *Marine Geology*, **103**, 423–443.

Bird, E. C. F. (1976). *Coasts*. Canberra: Australian National University Press.

Bird, E. C. F. (1985). *Coastline Changes*. New York: John Wiley.

Bowen, A. J. (1980). Simple models of nearshore sedimentation, beach profiles and longshore bars. In S. B. McCann, ed., *The Coastline of Canada*. Geological Survey of Canada Paper 80–10, pp. 1–11.

Bruun, P. (1962). Sea-level rise as a cause of shore erosion. *Journal of Waterways Harbors Division*, **88**, 117–130.

Christiansen, M. and Davidson-Arnott, R. G. D. (2004). The effects of dune ramps on sediment supply to coastal foredunes, Skallingen, Denmark. *Geografisk Tidsskrift* (Danish Journal of Geography), **104**, 29–41.

Clayton, K. M. (1989). Sediment input from the Norfolk cliffs, Eastern England: a century of coast protection and its effect. *Journal of Coastal Research*, **5**, 433–442.

Cooper, J. A. G. and Pilkey, O. H. (2004). Sea level rise and shoreline retreat: time to abandon the Bruun Rule. *Global and Planetary Change*, **43**, 157–171.

Cotton, C. A. (1954). Deductive morphology and the genetic classification of coasts. *Scientific Monthly*, **78**, 163–181.

Cowell, P. J. and Thom, B. G. (1994). Morphodynamics of coastal evolution. In R. W. G. Carter and C. D. Woodroffe, eds., *Coastal Evolution*. Cambridge: Cambridge University Press, pp. 33–86.

Cowell, P. J., Roy, P. S. and Jones, R. A. (1995). Simulation of large-scale coastal change using a morphological behaviour model. *Marine Geology*, **126**, 45–61.

Cowell, P. J. *et al.* (2001). Shoreface sand supply to beaches. *Proceedings of the 27th International Conference on Coastal Engineering*, American Society of Civil Engineers, New York, 2495–2508.

Cowell, P. J. *et al.* (2003a). The Coastal-Tract (Part 1): a conceptual approach to aggregated modelling of low-order coastal change. *Journal of Coastal Research*, **19**, 812–827.

Cowell, P. J. *et al.* (2003b). The Coastal-Tract (Part 2): applications of aggregated modeling to low-order coastal change. *Journal of Coastal Research*, **19**, 828–848.

Cowell, P. J. *et al.* (2006). Management of uncertainty in predicting climate-change impacts on beaches. *Journal of Coastal Research*, **22**, 232–245.

Crossland, C. J. *et al.* (2005). *Coastal Fluxes in the Anthropocene*. The Land–Ocean Interactions in the Coastal Zone Project of the International Geosphere–Biosphere Programme Series: Global Change. Stockholm: IGBP.

Crowell, M., Leatherman, S. P. and Buckley, M. K. (1991). Historical shoreline change: error analysis and mapping accuracy. *Journal of Coastal Research*, **7**, 839–852.

Curray, J. R. (1964). Transgressions and regressions. In R. C. Miller, ed., *Papers in Marine Geology*. New York: Macmillan, pp. 175–203.

Davies, J. L. (1973). *Geographical Variation in Coastal Development*. New York: Hafner.

Dean, R. G. (1991). Equilibrium beach profiles: characteristics and applications. *Journal of Coastal Research*, **7**, 53–84.

Dean, R. G. and Maurmeyer, E. M. (1983). Models of beach profile response. In P. Komar and J. Moore, eds., *CRC Handbook of Coastal Processes and Erosion*. Boca Raton: CRC Press, pp. 151–165.

Dickson, M. E., Walkden, M. J. A. and Hall, J. W. (2007). Systemic impacts of climate change on an eroding coastal region over the twenty-first century. *Climatic Change*, **84**, 141–166.

Dover, A. R. and Bea, R. G. (1980). Application of reliability methods to the design of coastal structures. *Proceedings of Coastal Structures '79*, American Society of Civil Engineers, New York, 747–63.

Eurosion (2004). *Living with Coastal Erosion in Europe: Sediment and Space for Sustainability, Part 1, Major Findings and Policy Recommendations of the EUROSION Project: Guidelines for Implementing Local Information Systems Dedicated to Coastal Erosion Management*, Service contract B4-3301/2001/329175/MAR/B3 'Coastal erosion – Evaluation of the need for action'. Brussels: Directorate-General Environment, European Commission.

Finkl, C. W. (2004). Coastal classification: systematic approaches to consider in the development of a comprehensive system. *Journal of Coastal Research*, **20**, 166–213.

Goodwin, I. D, Stables, M. A. and Olley, J. M. (2006). Wave climate, sand budget and shoreline alignment evolution of the Iluka–Woody Bay sand barrier, northern New South Wales, Australia, since 3000 yr BP. *Marine Geology*, **226**, 127–144.

Hallermeier, R. J. (1981). A profile zonation for seasonal sand beaches from wave climate. *Coastal Engineering*, **4**, 253–277.

Hesp, P. A. (2002). Foredunes and blowouts: initiation, geomorphology and dynamics. *Geomorphology*, **48**, 245–268.

Hesp, P. A. *et al.* (2005). Flow dynamics over a foredune at Prince Edward Island, Canada. *Geomorphology*, **65**, 71–84.

Hesp, P. A. *et al.* (2007). Morphology of the Itapeva to Tramandai transgressive dunefield barrier system and mid- to late Holocene sea level change. *Earth Surface Processes and Landforms*, **32**, 407–414.

Hinton, C. L. and Nicholls, R. J. (2007). Shoreface morphodynamics along the Holland Coast. In P. S. Balson and M. B. Collins, eds., *Coastal and Shelf Sediment Transport*, Special Publication No. 274. London: Geological Society of London, pp. 93–101.

Hoozemans, F. M. J., Marchand, M. and Pennekamp, H. A. (1993). *A Global Vulnerability Analysis: Vulnerability Assessment for Population, Coastal Wetlands and Rice Production on a Global Scale*, 2nd edn. Delft: Delft Hydraulics.

Iberinsa (1992). *Estudio de la Regression del Delta del Ebro y Propuesta de Alternatives de Actuación*. Comunidad general de regantes del canal de la derecha del Ebro, Comunidad de regantes-Sindicato agrícola del Ebro.

Illenberger, W. K. and Rust, I. C. (1988). A sand budget for the Alexandria coastal dune field, South Africa. *Sedimentology*, **35**, 513–522.

Inman, D. L. and Nordstrom, C. E. (1971). On the tectonic and morphologic classification of coasts. *Journal of Geology*, **79**, 1–21.

IPCC (2007a). *Climate Change 2007: The Physical Science Basis. Contribution of Working Group I to the Fourth Assessment Report of the Intergovernmental Panel on Climate Change*. Solomon, S. *et al*., eds. Cambridge: Cambridge University Press.

IPCC (2007b). *Climate Change 2007: Impacts, Adaptation and Vulnerability. Contribution of Working Group II to the Fourth Assessment Report of the Intergovernmental Panel on Climate Change*. Parry, M. L. *et al*., eds. Cambridge: Cambridge University Press.

Jimenez, J. A. (1996). Evolucion costera en el Delta del Ebro. Unpublished Ph.D. thesis, Universitat Politècnica de Catalunya.

Johnson, D. W. (1919). *Shore Processes and Shoreline Development*. New York: John Wiley.

Klein, R. J. T. *et al*. (2001). Technological options for adaptation to climate change in coastal zones. *Journal of Coastal Research*, **17**, 531–543.

Kraft, J. C. and John, C. J. (1979). Lateral and vertical facies relations of transgressive barrier. *Bulletin of the American Association of Petroleum Geologists*, **63**, 2145–2163.

Leatherman, S. P. (1983). Barrier dynamics and landward migration with Holocene sea-level rise. *Nature*, **301**, 415–418.

Leatherman, S. P. (2001). Social and economic costs of sea level rise. In B. C. Douglas, M. S. Kearney and S. P. Leatherman, eds., *Sea Level Rise: History and Consequences*. San Diego: Academic Press, pp. 181–223.

Leatherman, S. P., Zhang, K. and Douglas, B. C. (2000). Sea level rise shown to drive coastal erosion, *EOS, Transactions of the American Geophysical Union*, **81**, 55–57.

Lees, B. (2006). Timing and formation of coastal dunes in northern and eastern Australia. *Journal of Coastal Research*, **22**, 78–89.

Manne, A. S. and Richels, R. G. (1995). The greenhouse debate: economic efficiency, burden sharing and hedging strategies. *Energy Journal*, **16**, 1–37.

Mathew, S. (2007). Quantifying coastal evolution using digital photogrammetry. Unpublished Ph.D thesis, University of Guelph.

McFadden, L., Nicholls, R. J. and Penning-Rowsell, E. (2007a). *Managing Coastal Vulnerability*. Oxford: Elsevier.

McFadden, L. *et al*. (2007b). A methodology for modelling coastal space for global assessments. *Journal of Coastal Research*, **23**, 911–920.

McGill, J. T. (1958). Map of coastal landforms. *Geographical Review*, **48**, 402–405.

McLean, R. F. and Shen, J. S. (2006). From foreshore to foredune: foredune development over the last 30 years at Moruya Beach, New South Wales, Australia. *Journal of Coastal Research*, **22**, 28–36.

Mimura, N. and Nobuoka, H. (1995). Verification of Bruun Rule for the estimate of shoreline retreat caused by sea-level rise. In W. R. Dally and R. B. Zeidler, eds., *Coastal Dynamics 95*. New York: American Society of Civil Engineers, pp. 607–616.

National Research Council (1990). *Managing Coastal Erosion*. Washington, DC: National Academies Press.

Nicholls, R. J. and Klein, R. J. T. (2005). Climate change and coastal management on Europe's coast. In J. E. Vermaat *et al*., eds., *Managing European Coasts: Past, Present and Future*, Berlin: Springer-Verlag, pp. 199–255.

Nicholls, R. J., Birkemeier, W. A. and Lee, G.-H. (1998). Evaluation of depth of closure using data from Duck, NC, USA. *Marine Geology*, **148**, 179–201.

Nicholls, R. J. *et al*. (2008). Climate change and coastal vulnerability assessment: scenarios for integrated assessment. *Sustainability Science*, **3**, 89–102.

Penland, S., Suter, J. R. and Boyd, R. (1985). Barrier-island arcs along abandoned Mississippi River deltas. *Marine Geology*, **63**, 197–233.

Pethick, J. (1984). *An Introduction to Coastal Geomorphology*. London: Arnold.

Pilkey, O. H. and Cooper, J. A. G. (2004). Society and sea level rise. *Science*, **303**, 1781–1782.

Roy, P. S. *et al*. (1994). Wave dominated coasts. In R. W. G. Carter and C. D. Woodroffe, eds., *Coastal Evolution: Late Quaternary Shoreline Morphodynamics*. Cambridge: Cambridge University Press, pp. 121–186.

Shepard, F. P. (1937). Revised classification of marine shorelines. *Journal of Geology*, **45**, 602–624.

Shepard, F. P. (1948). *Submarine Geology*. New York: Harper.

Shepard, F. P. (1973). *Submarine Geology*, 2nd edn. New York: Harper & Row.

Sherman, D. J. and Bauer, B. O. (1993). Dynamics of beach–dune systems. *Progress in Physical Geography*, **17**, 413–447.

Short, A. D., 2003. A survey of Australian beaches. Keynote lecture, *Coastal Sediments 2003*, Clearwater, Florida.

Stive, M. J. F. and Wang, Z. B. (2003). Morphodynamic modelling of tidal basins and coastal inlets. In C. Lakhan, ed., *Advances in Coastal Modeling*. Amsterdam: Elsevier, pp. 367–392.

Stive, M. J. F., Roelvink, D. J. A. and de Vriend, H. J. (1991). Large-scale coastal evolution concept. *Proceedings of the 22nd International Conference on Coastal Engineering*, American Society of Civil Engineers, New York, 1962–1974.

Stive, M. J. F. *et al*. (2002). Variability of shore and shoreline evolution, *Coastal Engineering*, **47**, 211–235.

Stockdon, H. F. *et al*. (2007). A simple model for the spatially variable coastal response to hurricanes. *Marine Geology*, **238**, 1–20.

Stolper, D., List, J. H. and Thieler, E. B. (2005). Simulating the evolution of coastal morphology and stratigraphy with a new morphological-behaviour model (GEOMBEST). *Marine Geology*, **218**, 17–36.

Swift, D. J. P. (1976). Continental shelf sedimentation. In D. J. Stanley and D. J. P. Swift, eds., *Marine Sediment Transport and Environmental Management*. New York: John Wiley, pp. 311–350.

Swift, D. J. P., Phillips, S. and Thorne, J. A. (1991). Sedimentation on continental margins: 4. Lithofacies and depositional systems. In D. J. P. Swift *et al.*, eds., *Shelf Sand and Sandstone Bodies: Geometry, Facies and Sequence Stratigraphy*, Special Publication No. 14 of the International Association of Sedimentologists. Oxford: Blackwell Scientific Publications, pp. 89–152.

Syvitski, J. P. M. *et al.* (2005). Impact of humans on the flux of terrestrial sediment to the global coastal ocean. *Science*, **308**, 376–380.

Taylor, J. A., Murdock, A. P. and Pontee, N. I. (2004). A macro-scale analysis of coastal steepening around the coast of England and Wales. *Geographical Journal*, **170**, 179–188.

Thieler, E. R. *et al.* (1995). Geology of the Wrightsville Beach, North Carolina shoreface: implications for the concept of shoreface profile of equilibrium. *Marine Geology*, **126**, 271–287.

Thom, B. G. and Cowell, P. J., 2005. Coastal changes. In M. L. Schwartz, ed., *Encyclopedia of Coastal Science*. Dordrecht: Springer-Verlag, 251–253.

Thom, B. G. and Hall, W. (1991). Behaviour of beach profiles during accretion and erosion dominated periods. *Earth Surface Processes and Landforms*, **16**, 113–127.

Thom, B. G., Bowman, G. M. and Roy, P. S. (1981). Late Quaternary evolution of coastal sand barriers, Port Stephens–Myall Lakes area, central New South Wales, Australia. *Quaternary Research*, **15**, 345–364.

Thorne, C. R., Evans, E. P. and Penning-Rowsell, E. C. (2007). *Future Flood and Coastal Erosion Risks*. London: Thomas Telford.

Titus, J. G. and Narayanan, V. (1996). The risk of sea level rise. *Climatic Change*, **33**, 151–212.

UNEP (2005). *Assessing Coastal Vulnerability: Developing a Global Index for Measuring Risk*, Division of Early Warning and Assessment (DEWA), UNEP/DEWA/RS.05-1. Nairobi: United Nations Environment Programme.

Valentin, H. (1952). Die Küsten der Erde. *Petermanns geographische Mitteilungen, Ergänzungsheft*, **246**, 118.

Valiela, I. (2006). *Global Coastal Change*. Oxford: Blackwell.

Vellinga, P. and Leatherman, S. P. (1989). Sea-level rise, consequences and policies. *Climatic Change*, **15**, 175–189.

Vrijling, J. K. and Meijer G. J. (1992). Probabilistic coastline position computations. *Coastal Engineering*, **17**, 1–23.

Walkden, M. E. and Dickson, M. J. A. (2008). Equilibrium erosion of soft rock shores with a shallow or absent beach under increased sea level rise. *Marine Geology*, **251**, 1–2, doi:10.1016/j.margeo.2008.02.003.

Webster, M. D. (2002). The curious role of 'Learning' in climate policy: should we wait for more data? *Energy Journal*, **23**, 97–119.

Webster, M. D. and Sokolov, A. P. (2000). A methodology for quantifying uncertainty in climate projections. *Climatic Change*, **46**, 417–446.

Zagwijn, W. H. (1986). Nederland in het Holoceen. In *Geologie van Nederland*, vol. 1. Haarlem: Rijksgeologische Dienst, pp. 1–46.

Zhang, K., Douglas, B. C. and Leatherman, S. P. (2004). Global warming and coastal erosion. *Climatic Change*, **64**, 41–58.

7 Coral reefs

Paul Kench, Chris Perry and Thomas Spencer

7.1 Introduction

This chapter examines the implications of global environmental change for the geomorphology of coral reefs and reef-associated sedimentary landforms such as beaches and reef islands. These tropical ecosystems cover an estimated 284 to 300 km^2 of the Earth's surface (Spalding *et al.*, 2001) and are considered among the most valuable on Earth, providing economic goods and services worth in excess of US$375 billion a^{-1} to millions of people (Best and Bornbusch, 2005). They are zones of high biological diversity, habitat for about a quarter of all known marine species, important components of the global carbon cycle, and provide the physical foundation for a number of mid-ocean nation states. Indeed, reefs possess benefits under four categories of ecosystem services (Millennium Ecosystem Assessment, 2005). Such benefits include: *regulation* of incident oceanographic swell conditions to control reef and lagoon circulation, reduce shoreline erosion, protect beaches and coastlines from storm surges, and control beach and island formation; *provision* of food resources (fisheries) and aggregates for building (coral and sand), as well as the provision of land surface area and associated subsurface water resources, especially through reef island construction; *supporting* nutrient cycling and active carbonate production to build reef and reef island structures; and *cultural* benefits that include spiritual identity for indigenous communities and potential for tourism and recreation-based income.

Coral reefs worldwide are considered to be in serious ecological decline as a consequence of anthropogenic impacts, natural stresses and climate change (e.g. Hughes *et al.*, 2003; Buddemeier *et al.*, 2004). Broad-scale assessments (e.g. Wilkinson, 2004) have argued that 20% of the world's coral reefs have been destroyed and that 25% of reefs are under imminent or long-term risk of collapse. In the Indo-Pacific region, it has been calculated that coral cover loss rates were 2% per year (losses of 3168 $km^2\ a^{-1}$) between 1997 and 2003 (Bruno and Selig, 2007). As a consequence of reef ecological decline, it is widely considered that coral reef landforms are also at risk from the same range of stressors as other marine ecosystems, particularly changes in sea surface temperature, ocean chemistry and sea level rise (IPCC, 2007; Smithers *et al.*, 2007). Indeed, such changes are considered to threaten the physical stability of mid-ocean atoll nations rendering small island nations the first environmental refugees of climate change. However, these bleak projections for coral reef landforms are largely based on the assumption that the *short-term* ecological condition of coral reefs is the primary control on their development and change, whilst the *longer-term* geomorphic processes that control reef growth and sediment accumulation are largely ignored.

This chapter explicitly examines the implications of global climate change for the geomorphology of coral reefs and reef sedimentary landforms. It addresses a number of key questions related to the geomorphic future of critical reef landforms:

(a) will coral reefs persist and in what form in the next century?
(b) what will be the geomorphic consequences of bleaching, ocean acidification and ocean temperature change on coral reefs? and
(c) will reef islands become physically unstable?

7.1.1 Geomorphic units and morphodynamic framework for coral reef landforms

To address the above questions, this chapter develops a morphodynamic framework that identifies major process response linkages responsible for the formation of, and changes, to coral reef landforms. Coral reefs possess a

Geomorphology and Global Environmental Change, eds. Olav Slaymaker, Thomas Spencer and Christine Embleton-Hamann. Published by Cambridge University Press.

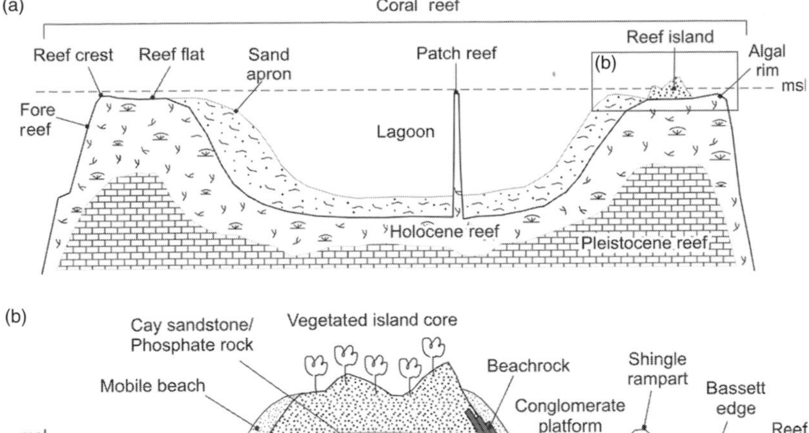

FIGURE 7.1. Nomenclature of coral reef landforms adopted in this chapter. (a) Cross-section of an atoll representing major elements of coral reef structure and large-scale reef geomorphic units typical of many reef settings; and (b) a range of sedimentary landforms deposited on or adjacent to coral reefs.

number of distinct eco-geomorphic units (Fig. 7.1). These units reflect geomorphic development at a range of space and time scales and show varying levels of persistence in the geological record. A primary division is that between *coral reefs* and *reef sedimentary landforms*.

Coral reefs are three-dimensional structures consisting of veneers of living coral and reef-associated organisms that overlie sequences of previously deposited calcium carbonate separated by solutional unconformities (Fig. 7.1a). These structures evolve over geological (millennial) timescales. As geomorphic units, coral reefs range in area from less than 1 km^2, in the case of smaller patch reefs, to more than 100 km^2 in extent. Networks of reefs can form barrier complexes up to 2400 km in length, comprising the largest biological constructions on Earth (e.g. the Great Barrier Reef).

Reef sedimentary landforms are surficial accumulations of unconsolidated sediment deposited by wave and current processes on, or adjacent to, the coral reef structure that include *reef islands* and *beaches* (Fig. 7.1b). On geological timescales, they represent ephemeral stores of detrital material in the carbonate sediment budget.

Both coral reefs and reef sedimentary landforms can be further divided into a suite of distinct geomorphic units that range from the macro- to microscale (Fig. 7.1). In this chapter we focus on a select range of macroscale geomorphic units: the coral reef, reef islands and beaches. It is the potential geomorphic change in these landforms that is fundamental to considerations of future global change as these landforms are critical to societies that both dwell upon or rely intimately on the resources they provide.

The formation and morphological adjustment of coral reefs and reef sedimentary landforms result from the dynamic interaction between biological and physical processes (Fig. 7.2) which operate at a range of time and space scales. In the context of understanding future morphological adjustment of coral reef landforms, three important observations follow. Firstly, alterations to the boundary controls will force change in some reefal landforms. Such changes in boundary conditions can occur not only as a consequence of extrinsic changes in the ocean–atmosphere system but also from intrinsic system state and physical process dynamics. These intrinsic linkages are themselves impacted by anthropogenic activities, such as overexploitation of physical and biological resources (e.g. coral mining) and construction activities. Secondly, the magnitude, mode and time-frames of morphological adjustment (responsiveness) are likely to vary considerably between different geomorphic components of the coral reef landform system (e.g. between the coral reef structure and adjacent reef islands and beaches). Thirdly, morphodynamic feedbacks exist that are both temporally specific but also cascade across timescales to provide a degree of self-organisation to geomorphic development. For example, at centennial to millennial timescales, sea level change modulates reef growth. In turn, this influences wave, current and sedimentation processes that govern the short-term morphological development of reef sedimentary landforms (Fig. 7.2).

The eco-morphodynamic model provides a framework for this chapter. Firstly, recognising the importance of the coral reef 'carbonate factory' in providing the building blocks (skeletal material, sand and gravels) for construction of reef landforms, the chapter examines the primary controls on reef carbonate production and how the carbonate factory may be affected by global change (Section 7.2). Such considerations underpin a detailed examination of the geomorphology and

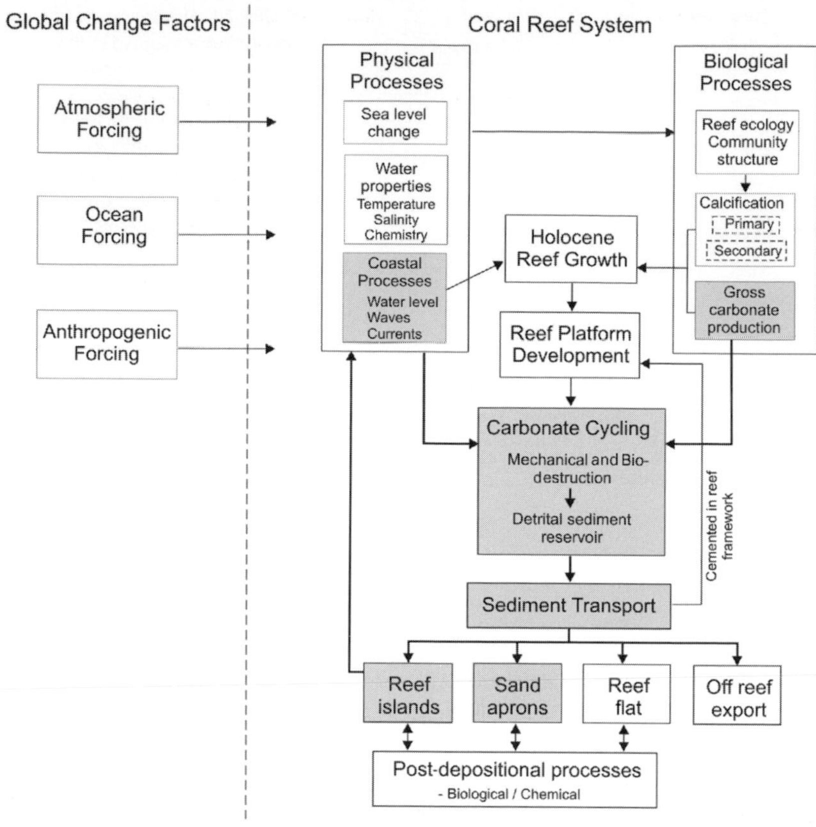

FIGURE 7.2. Structure and function of the eco-morphodynamic model for the coral reef system. The model shows co-adjustment of biological and physical processes, and coral reef morphology and reef sedimentary landforms that operate at a range of timescales. Grey shaded boxes and dark arrows highlight linkages between the contemporary eco-morphodynamic system controlling sedimentary landform development at event to centennial scales, which are embedded in the broader morphodynamic system that controls coral reef development at centennial to millennial scales.

morphodynamics of coral reef structures (Section 7.3) and reef sedimentary landforms (Section 7.4). In particular, these sections: highlight the boundary controls on the geomorphic development of coral reef landforms; evaluate the sensitivity of landforms to perturbations in boundary conditions; and identify the temporal scales of morphodynamic adjustment, based on historical behaviour and modelling efforts. The comparative influence of anthropogenic processes in forcing geomorphic change is explicitly considered in Section 7.5. Collectively, these insights are used to provide a range of assessments for the future geomorphic response of reef systems to global environmental change and anthropogenic stress over the next 100 years (Section 7.6).

7.2 Carbonate production in coral reef environments: the reef carbonate factory

7.2.1 Controls on coral reef carbonate production

Coral reefs and reef sedimentary landforms are unique in that they are composed predominantly of calcium carbonate ($CaCO_3$) that results almost entirely from ecological processes. The primary sources of reef carbonate are the skeletons of corals and other organisms or plants that induce $CaCO_3$ deposition. Furthermore, their provenance is intrabasinal, with sediment production and deposition occurring either on, or in close proximity to, a reef.

The key constructional components within a coral reef are the framework-building corals. Such corals are characterised by the symbiotic relations between the coral animal and single-celled algae, the zooxanthellae, which live within coral tissues. This relationship allows corals to secrete a rigid skeleton via the process of calcification, the biologically mediated mechanism by which calcium (Ca^{2+}) and carbonate (CO_3^{2-}) ions, derived from supersaturated seawaters (Table 7.1), are converted to $CaCO_3$ (Kinzie and Buddemeier, 1996). The development of a reef's structure is therefore fundamentally dependent upon coral growth and the biologically induced precipitation of calcium carbonate. However, additional physical and ecological processes (see below) also play a major role in the cycling of calcium carbonate through the reef system; these processes influence net reef carbonate framework and sediment production rates and are critical for sedimentary landform development.

TABLE 7.1. *Marine environmental parameters influencing the distribution of reef-building (hermatypic) corals and tropical coral reef development. 'Optimal' values for coral growth are shown as well as recorded upper and lower environmental limits. Figures in parentheses are for non-reef-building coral communities*

		Environmental limits	
Environmental parameter	'Optimal' levels	Lower	Upper
Temperature (°C)[a]	21.0–29.5	16.0 (13.9)	34.4 (32.1)
Salinity (PSU)[b]	34.3–35.3	23.3 (20.7)	41.8 (No data)
Nitrate (μmol L^{-1})[c]	< 2.0	0.00	3.34 (up to 5.61)
Phosphate (μmol L^{-1})[c]	< 0.2	0.00	0.40 (up to 0.54)
Aragonite saturation state (Ω-arag)[d]	~ 3.83	3.28 (3.06)	No data
Depth of light penetration (m)	~ 50	<10	~ 90

[a] Weekly data.
[b] Monthly average data.
[c] Overall averages (1900–1999).
[d] Overall averages (1972–1978).
Source: Kleypas *et al.* (1999).

The main environmental controls on tropical $CaCO_3$ production are sea surface temperature, light penetration and the calcium carbonate saturation state of seawater (Table 7.1). Latitudinal and regional variations in these parameters thus exert a major control on the global distribution of coral reefs and on the composition of carbonate sediment-producing biota. Coral reefs typically occur in shallow tropical and subtropical marine settings between 28° N and 28° S, although oceanic upwelling limits reef development along the eastern margins of Africa and South America (Chapter 1, Fig. 1.7). Corals are limited to waters where sea surface temperature rarely drops below 17–18 °C, or exceeds 33–34 °C, for prolonged periods. Coral growth rates and reef-building potential are also progressively restricted, however, as other environmental threshold levels are approached (Kleypas *et al.*, 1999; Perry and Larcombe, 2003). For example, light levels change dramatically with depth and, due to their reliance on photosynthetic energy, reef-building corals are constrained to the photic zone (the depth of water at which surface light level is reduced to 1%). This depth varies greatly depending upon turbidity levels and can range from >90 m to <5 m in highly turbid environments. Corals can also tolerate a range of salinity levels (Table 7.1), but reef growth is constrained close to river mouths where sediment stress may be high and salinity levels low, and in zones of intense evaporation which elevate salinity. The nutrient load of reef waters is a further regulator of coral growth. Corals thrive in low nutrient conditions, but significantly elevated nutrient levels can inhibit coral growth and in extreme cases may lead to replacement of hard coral communities with macroalgae.

Apart from corals, there is a range of flora and fauna that also produce carbonate sediment, including coralline algae, calcareous green algae, molluscs and foraminifera (for a review see Perry, 2007). The distribution of these organisms is controlled by the same set of environmental variables that influence corals. At the global scale, transitions from chlorozoan (tropical) to chloralgal (tropical to subtropical) to rhodalgal (subtropical to warm temperate) assemblages can be delineated, reflecting latitudinal variations in the primary controlling marine environmental parameters (Lees, 1975). From an environmental change perspective, subtle shifts in these parameters can result in marked changes in sediment production. For example, within the Gulf of California, subtle latitudinal shifts in carbonate grain assemblages occur over spatial scales of a few 100 km, with each transition correlating to gradients in mean *chlorophyll a* concentrations (~0.5–1.0 mg Chl a^{-1} m^{-3}) and mean sea surface temperatures (~1.5 °C) (Halfar *et al.*, 2004). In light of the magnitude of predicted sea surface temperature changes, shifts in carbonate sediment production in specific environments can thus reasonably be predicted. This has obvious implications for the composition of, and rates of sediment production within, tropical marine environments and may influence sediment availability for reef and reef sedimentary landform construction.

7.2.2 Cycling of calcium carbonate: influences upon coral reef framework and sediment deposition

Apart from coral growth and carbonate sediment production, critical to future geomorphic change in reef systems is an understanding of the *net* rates and styles of $CaCO_3$ accumulation. Actual $CaCO_3$ accumulation rates on a coral reef and the production of a large proportion of reef-related carbonate sediment are strongly influenced by a range of physical, chemical and biological processes that interact to cycle calcium carbonate. Some of these processes aid reef and reef sedimentary landform development, others result in the removal of coral framework, whilst still others convert this carbonate to sediment (Figs. 7.2 and 7.3). These carbonate 'cycling' processes may thus exert either a 'constructive' or 'destructive' (*sensu* Scoffin, 1992) influence on reef-related carbonate accumulation (for a review see Perry and Hepburn, 2008).

Constructive activities add additional calcium carbonate to the coral framework structure. The key processes are secondary framework production by calcareous encrusters (especially crustose coralline algae, foraminifera and bryozoans) and the precipitation of syn- and early post-depositional cements. Both can contribute significant amounts of carbonate to the reef, thus helping to stabilise the framework constituents (Scoffin, 1992). A range of destructive processes, associated with the effects of either physical or biological erosion, are also important in reef framework development. Bioerosion (biological substrate erosion) is facilitated by a wide range of reef-associated faunas, including species of fish and echinoids, and endolithic forms of sponges, bivalves and worms (Spencer, 1992). These biological agents play two key roles: (i) they directly degrade primary and secondary reef framework, increasing susceptibility to physical and chemical erosion, and (ii) they may produce large amounts of sediment. For example, bioerosion by echinoids and fish may contribute 0.17–9.7 and 0.02–7.62 kg $CaCO_3\ m^{-2}\ a^{-1}$ to the sediment reservoir respectively, and bioerosion by sponges 0.18–3.29 kg $CaCO_3\ m^{-2}\ a^{-1}$ (Scoffin *et al.*, 1980; Hubbard *et al.*, 1990; Bruggeman *et al.*, 1996). Physical disturbance, associated with storms and cyclones, is an important episodic process that influences reef framework development, largely through the generation of coral rubble, the deposition of which is an important reef-building process (Blanchon *et al.*, 1997; Hubbard, 1997).

The relative role of these different processes is clearly crucial to the accretionary potential of coral reefs and to the supply of sediment for reef sedimentary landform construction, a concept that is defined by the carbonate budget approach to conceptualising and quantifying coral reef carbonate production. A carbonate budget is effectively a sum of gross carbonate production from primary (coral) and secondary (encrusters and marine cement) sources (terminology of Scoffin, 1992), as well as sediment produced within or imported into the reef, less that lost through biological or physical erosion, dissolution or sediment

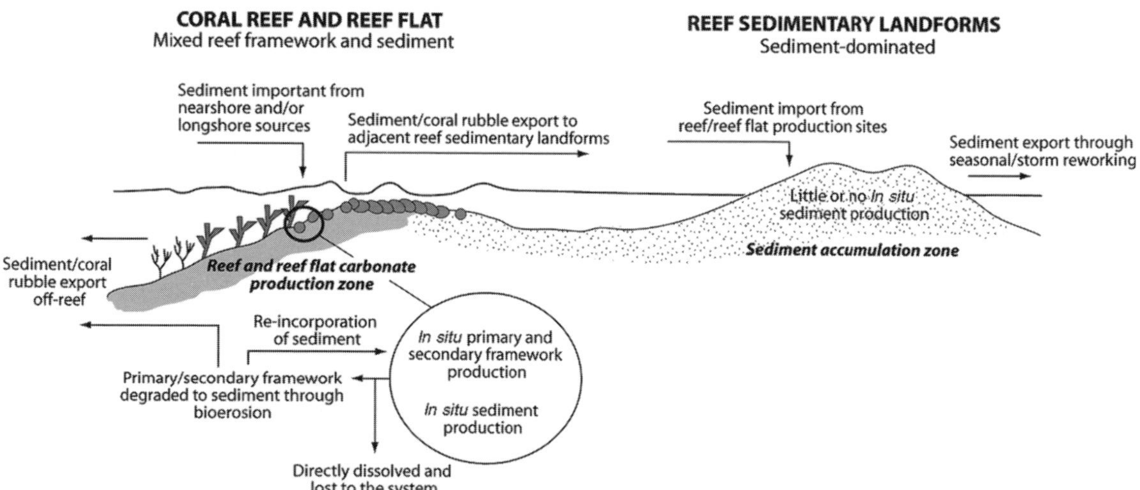

FIGURE 7.3. Key carbonate budgetary components (*in situ* production, sediment import, sediment export) within reefs and reef sedimentary landforms. Note that the budgets associated with reef sedimentary landforms, such as those linked to beaches and reef islands, are essentially dependent upon the external production and import of sediment, balanced against any losses to the system, whilst complex *in situ* carbonate production processes, as well as internal carbonate cycling and sediment import and export processes, characterise reef and reef flat carbonate budgets.

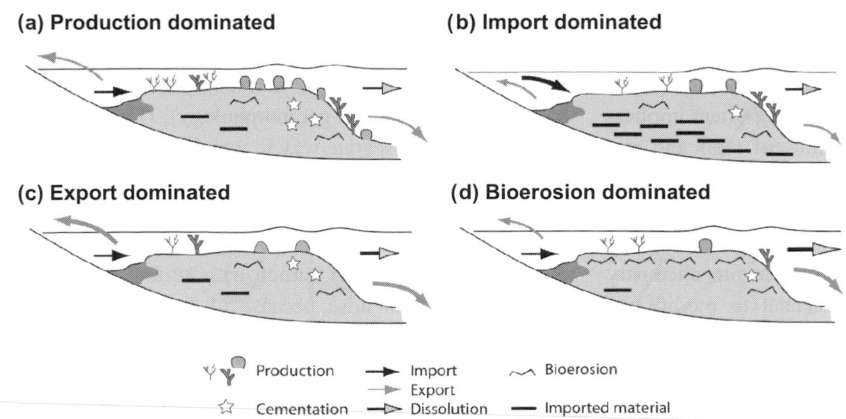

FIGURE 7.4. Conceptual models for the budgetary status of different types of reefs. Each varies in terms of the relative importance of different budgetary components (modified from Kleypas *et al.*, 2001). NB. Of the sediment produced and 'exported' from a reef, a proportion will be available for adjacent reef sedimentary landform construction (beaches, reef islands). This component is more appropriately viewed as a redistribution of sediment into the broader reef depositional system rather than a loss to the system (cf. Figs. 7.2 and 7.3).

export (Fig. 7.3). The balance represents the net accumulation rate of $CaCO_3$. It is relevant to remember that sediment production and its supply to reef sedimentary landforms (where developed) forms an integral part of a reef's total carbonate budget (Fig. 7.2). This approach has great merit for understanding the relative importance of different processes in different reef (and intra-reef) environments and their potential for geomorphic construction.

Kleypas *et al.* (2001) highlight a range of conceptual end members in the spectrum of reef states that are directly linked to variations in rates of carbonate production, sediment import and export, and carbonate cycling (Fig. 7.4). In production-dominated reefs, biological $CaCO_3$ production far exceeds rates of carbonate degradation and thus the carbonate budget is positive, a state that has been common on many reefs through the Holocene (see Section 7.2.3). Import-dominated reefs contain a high proportion of material derived from outside the reefs, typically sedimentary in character and in some cases terrigenous in origin (see Smithers and Larcombe, 2003). These reefs also have positive budgets. Export-dominated reefs may have high primary carbonate production rates but also high rates of carbonate removal, typically due to physical processes. The result may be rather low net accumulation rates. Reefs of this type emphasise the important differences that sometimes exist between the ecological functioning of a reef and its geological performance (Kleypas *et al.*, 2001). Bioerosion-dominated reefs exhibit negative budgets, with carbonate production being exceeded by either direct biological substrate degradation and/or the conversion of framework to sediment that is subsequently exported (see Riegl and Piller, 2000; Benzoni *et al.*, 2003) or, increasingly in the late Holocene, may represent a response to environmental disturbance or reduced rates of primary carbonate production. Although far from exhaustive, these different states (Fig. 7.4) demonstrate how shifts in the relative importance of individual processes, driven by environmental, ecological or anthropogenic change, can fundamentally alter a reef's budgetary status.

Several detailed studies have attempted to quantify the various budgetary elements of reefs. Examples of the resultant net carbonate production estimates include those made on reefs in Barbados (4.48 kg $CaCO_3$ m^{-2} a^{-1}: Scoffin *et al.*, 1980), Jamaica (1.1 kg $CaCO_3$ m^{-2} a^{-1}: Land, 1979), St. Croix (0.91 kg $CaCO_3$ m^{-2} a^{-1}: Hubbard *et al.*, 1990), Hawaii (0.89 kg $CaCO_3$ m^{-2} a^{-1}: Harney and Fletcher 2003) and Indonesia (ranging from 11.68 to −7.6 kg $CaCO_3$ m^{-2} a^{-1}: Edinger *et al.*, 2000). The Barbados and Jamaican examples would, at the time, have been analogous to 'production-dominated reefs', the example from St. Croix to an 'export-dominated' reef. Some of the Indonesian reefs are good examples of 'bioerosion-dominated' reefs. The general paucity of detailed budget studies is however noteworthy. This is particularly the case with respect to more marginal reef-building settings and to the sediment budgets associated with reef islands (but see Hart and Kench, 2007) where the interactions between sediment production, and the transport pathways to islands, are crucial for understanding island construction processes (see Section 7.4) and for predicting geomorphic response of islands to environmental change.

7.2.3 Implications of climatic and environmental change for carbonate budgets and geomorphology

The balance between those processes that produce calcium carbonate and those that remove it/convert it to sediment thus strongly influence rates and styles of reef development

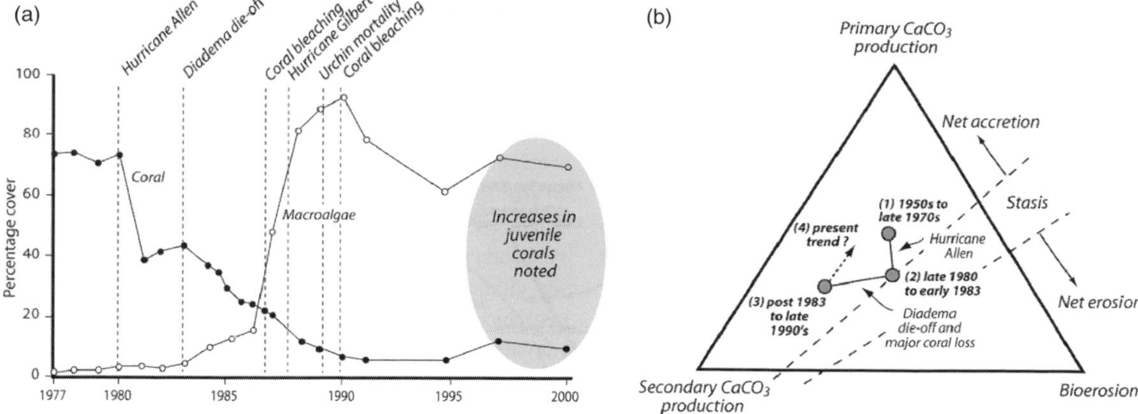

FIGURE 7.6. (a) Temporal trends in reef community structure (per cent cover of corals and macroalgae) and the timing of major disturbance events at Discovery Bay, Jamaica. (b) Temporal trends in the net carbonate production regime at Discovery Bay, based on the framework production status approach illustrated in Fig. 7.5b (from Perry *et al.*, 2008).

function of the coincident reduction in the abundance of bioeroding fish and echinoids that has offset the low coral cover on the reefs, a point that demonstrates the significance of quantifying different budgetary components in the context of reef 'health' assessments. Indeed it seems likely that despite the major ecological changes that have occurred, the reef carbonate budget has remained in an essentially positive state since the mid-1980s (Fig. 7.6b).

These predicted trends for the Discovery Bay reefs are comparable to those illustrated in the ternary diagram above (Fig. 7.5b) where a cessation of disturbance or an adaptation of the coral community (i.e. recruitment of new, better-adapted species) may allow transitions back to conditions of net carbonate production, with either similar (pathway C–B–A) or modified production rates (C–B–A^2). Again, these transitions may occur over a wide range of timescales (10^1 to 10^4 a). Superimposed on these end members of production/degradation are the import and export of sediment. Under both positive and negative net production regimes sediment produced on reefs can be exported and made, in some cases, available for reef sedimentary landform construction (pathway D). In contrast, in some systems relatively low rates of primary and secondary carbonate production may be supplemented by the import of sediment (calcareous or terrigenous), maintaining the system in a positive net accretionary state (pathway E).

Changes in reef biota, driven by either anthropogenic or climate change, or changes in framework production and cycling processes may also fundamentally alter patterns of reef sediment production. For example, studies in Florida have demonstrated clear temporal shifts in sediment production that have been linked to regional water quality changes (Lidz and Hallock, 2000), whilst related studies have documented temporal shifts in the abundance of individual carbonate-producing faunal groups (especially foraminifera). Similarly in Jamaica, clear temporal shifts in sediment constituent abundance have been documented over small (a few km^2) areas of reef and lagoon substrate (Perry, 1996; Perry *et al.*, 2006), with more recent sediment production (since the mid-1980s) dominated by calcareous green and red algae (approximately tripled in abundance) at the expense of corals and molluscs. These changes have occurred in response to reduced coral cover. From a geomorphological perspective, there may be significant implications in environments where just one (or a few) sediment producer groups play a disproportionately important role in reef landform development, e.g. reef islands (Yamano *et al.*, 2000) or carbonate beaches (Harney *et al.*, 1999). How significant changes in sediment production regimes may be for reef sedimentary landform sustainability is not at present clear. If total carbonate production within a reef system falls then negative impacts may reasonably be predicted, but if total production remains relatively unaltered and the response is simply one of a shift to different grain production regimes, the geomorphic implications for beaches and reef islands may be less significant. Such sediment supply dynamics and their influence on landform dynamics are poorly resolved.

7.3 Coral reef landforms: reef and reef flat geomorphology

Most assessments of the future of coral reefs focus on their response (or lack of it) to accelerated sea level rise. However, it may well be the case that the greatest threat to near-future reef performance will come not from sea level rise *per se* but from a wider suite of associated oceanographic changes. These threats include changes in ocean chemistry (Harley *et al.*, 2006) and the direct and indirect effects of changing ocean temperatures and, at more regional and local scales, changes in nutrient levels, salinities, turbidity and atmospheric dust inputs (Buddemeier *et al.*, 2004).

7.3.1 Reef growth – sea level relations at geological timescales

The broad patterning of reefs in the ocean basins and on basin margins, first synthesised by Charles Darwin, can now be explained within the framework of plate tectonic processes and the global arrangement of plate boundaries. Onto this geophysical template must be superimposed the effects of Tertiary and Quaternary fluctuations in sea level which produced alternating periods of subaerial exposure during low sea level stands and reef growth during periods of high sea level. On subsiding basements, growth phases have occupied an accommodation space generated by a combination of slow subsidence and more rapid subaerial erosion during glacial periods. Table 7.2 lists typical Indo-Pacific reef province Holocene reef thicknesses; it is clear that modern reefs form a relatively thin veneer over older structures (Fig. 7.1a). Indeed, at non-subsiding locations, and where there has been tectonic uplift, Last Interglacial limestones may outcrop above present sea level (Woodroffe and McLean, 1998).

As well as these constraints in the vertical, basement characteristics (depth, substrate type and configuration) have also affected the spatial patterning of modern reefs. However, the degree to which these underlying landforms determine the patterning of contemporary reefs is a function of the relations between Holocene sea level rise, the depth of the antecedent platform and rates and styles of coral growth.

Ocean-basin-scale differences in Holocene sea level dynamics have provided important boundary conditions to coral growth and reef development (see Chapter 1). However, within these broad patterns, distinct styles of reef accretion in response to sea level fluctuations have been identified, initially in terms of a tripartite division into 'keep-up', 'catch-up' and 'give-up' reefs (Fig. 7.7). 'Keep-up' reefs track sea level as it rises, maintaining a reef crest at sea level; 'catch-up' reefs initially lag behind sea level rise but, through rapid vertical growth, reach sea level as it slows or stabilises; and 'give-up' reefs find themselves in rapidly increasing water depths and shut down as failed reefs (Neumann and Macintyre, 1985 – although see Blanchon and Blakeway (2003) for a critique of this model). This classification has recently been expanded to include the possibilities of reef backstepping, reef front progradation under intermediate sea level rise scenarios, and reef die-off on emergence (Fig. 7.7; Woodroffe, 2002). Typical patterns of reef stratigraphy and anatomy for these different models have been provided by Kennedy and Woodroffe (2002) and Montaggioni (2005).

Some environmental change scenarios, associated with high rates of sea level rise, envisage coral reef ecosystems being 'drowned out' over centennial timescales. It is clear from the presence of extensive drowned banks, typically at water depths of 30–70 m (e.g. Papua New Guinea: Webster *et al.*, 2004a; Tahiti: Camoin *et al.*, 2007; Great Barrier Reef: Beaman *et al.*, 2008) but down to 100 m or more, that reef systems have previously failed in the Holocene. In the central Indian Ocean, the shallow Great Chagos Bank

TABLE 7.2. *Typical thicknesses (m) of Holocene reef sediments in the Indo-Pacific reef province*

Pacific Ocean oceanic atolls[a]		Indian Ocean oceanic atolls[b]		Great Barrier Reef[c]	
26.4	Funafuti	6–18	Cocos (Keeling) Islands	5–7	Torres Strait
8–17	Tarawa	15–20	Maldives archipelago	4–17	Reefs at 11–16° S
10–14	Enewetak	6–20	Chagos archipelago	15– >28	Reefs at 18–20° S
2–18	Mururoa	10–21.5	Mayotte, Comores archipelago	0–20	Reefs at 21–24° S
15–22	N Cooks	17.7	Réunion Island		
		11–18.8	Mauritius		
		12.9–13.4	Toliar, SW Madagascar		

[a] After Ohde *et al.* (2002).
[b] After Woodroffe (2005) and Camoin *et al.* (2004).
[c] After Hopley *et al.* (2007).

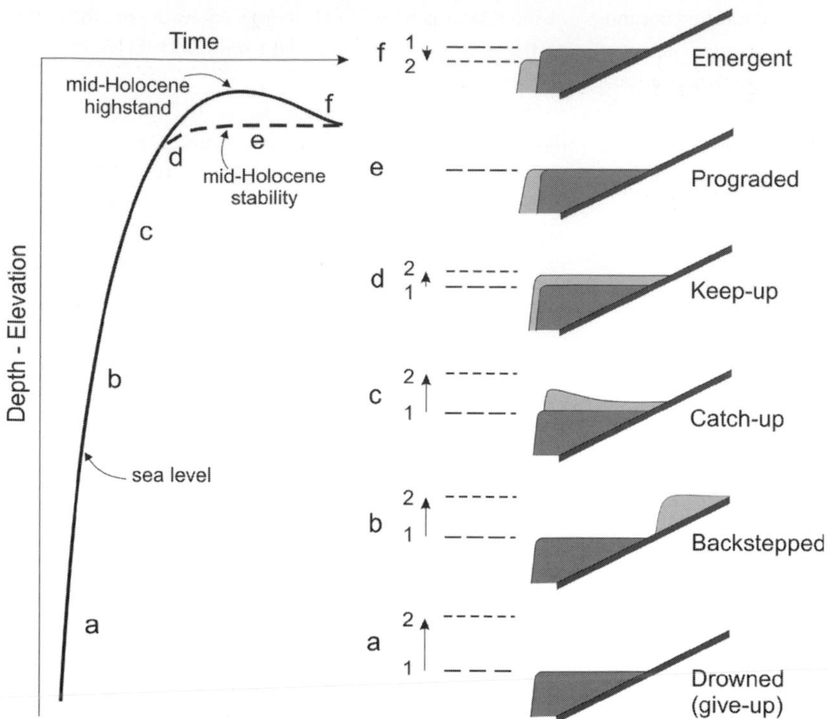

FIGURE 7.7. Relations between sea level and reef growth (from Woodroffe, 2002).

covers an area of 18 000 km^2 but reefs characterise only 5% of its rim; to the southwest, the Nazareth Bank (26 000 km^2) and Saya de Malha (40 000 km^2) of the Mascarene Plateau are largely devoid of modern reefs (Stoddart, 1971). There are also a large number of isolated banks in the central Caribbean Sea and a general paucity of reefs on the Nicaraguan, Honduran and Yucatan shelves (Stoddart, 1977).

Reef growth failure has been variously attributed to sea surface temperature fluctuations and declining water quality (e.g. Dunbar and Dickens, 2003), but it is instructive to assess these features in the context of postglacial sea level rise. Detailed records of the transgression show that the total rise of ~125 m was not smooth but characterised by three periods of extremely fast sea level rise: the termination following the LGM; meltwater pulse 1A (MWP-1A) (14.2–13.8 ka); and meltwater pulse 1B (MWP-1B) (11.5–11.1 ka). These episodes were separated by periods of relative stability related to renewed cooling, and cessation of ice melt, the second of these intervals being associated with the Younger Dryas (Camoin *et al.*, 2004). The best defined, and most widely identified, of these rapid sea level rise episodes is MWP-1A. This appears to have been associated with a sea level rise of at least 15 m over a period of 500 years (a rate of ~40–50 mm a^{-1}). Relict reef-like structures at water depths of 90–100 m on both the Great Barrier Reef and in the Comores, western Indian Ocean have been linked to this episode. Off Hawaii, a reef at –150 m began to fail at ~14.7 ka; by 12 ka, water depths were >60 m over the reef crest, causing the shallow reef-building corals to cease growth and to be replaced by a thin crust of coralline algae (Webster *et al.*, 2004b). In the Atlantic reef province, reef ridges at water depths of 75–90 m and 40–50 m arranged concentrically around the island of Barbados provide a record of the reefs that were left behind as the reef backstepped to successive new positions (Toscano and Macintyre, 2003). Interestingly, in the Indo-Pacific region patterns of reef termination appear less widespread and many reefs, such as those on Tahiti, Vanuatu and on the Huon Peninsula, New Guinea, were able to maintain reef growth, sometimes through all, and at least through the last, of these accelerations in sea level. For reefs where growth was interrupted, water depths of 30–40 m (rather than the 20–25 m of the Caribbean) appear to have been critical for renewed growth (Montaggioni, 2005). In a global environmental change context, sea level rise rates of 30–50 mm a^{-1} are 5 to 8 times greater than the projected globally averaged sea level rise to AD 2090–2099 under the most severe of emissions scenarios (5.9 mm a^{-1}); even allowing for an additional component of rise from enhanced ice flow, it is difficult to see a rate of sea level rise in excess of 7.9 mm a^{-1} (IPCC, 2007; Chapter 1, this volume). The implication, therefore, is that existing coral reefs are unlikely to be 'drowned out' by any near future sea level rise.

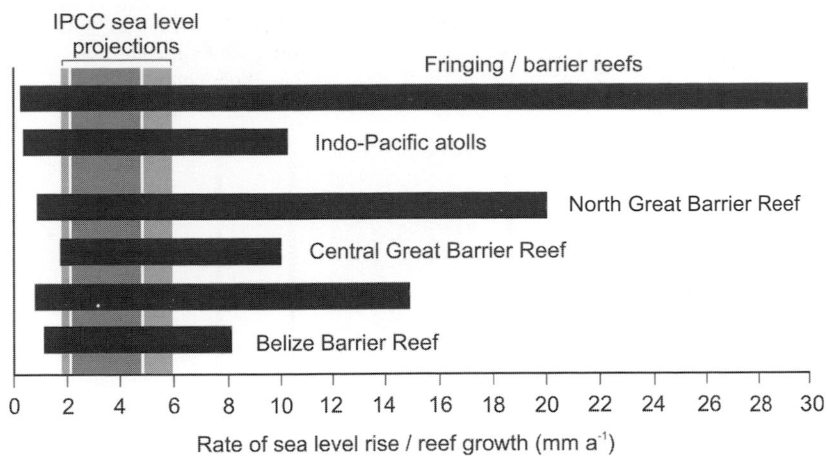

FIGURE 7.8. Range of rates of Holocene vertical accretion (horizontal bars) for reef margins in different reef regions (updated from Spencer, 1995) and rates of near-future sea level rise (vertical bars) based on IPCC (2007) projections (from left to right, scenarios B1 (low estimate), A1B (low estimate), A1B (high estimate), A1FI (high estimate); for further details see Chapter 1)).

This positive perspective is also supported by available long-term net reef accretion data (Fig. 7.8). For example, in the Indo-Pacific reef province, framework-dominated reefs have recorded rates of vertical growth of 1–30 mm a^{-1}, with a modal rate of 6–7 mm a^{-1}. The highest rates relate to high-porosity frameworks of tabular and branching corals whereas the lowest rates come from foliose and encrusting corals and veneers of coralline algae. As coral growth form is related to hydrodynamic energy levels, vertical accretion rates have been greater in sheltered to semi-exposed locations (~9 mm a^{-1}) compared to fully exposed sites (~5 mm a^{-1}). In non-framework, or detrital-dominated, environments, lagoonal carbonate muds have typically accumulated at vertical rates of 1–3 mm a^{-1} and reef flat / backreef sand and coral debris at 4–8 mm a^{-1}. The highest rates of detrital sedimentation, with typical vertical accretion rates of 10 mm a^{-1} but reaching 40 mm a^{-1}, have been the result of the rapid deposition of storm and cyclone deposits (Montaggioni, 2005). The higher rates in such studies, from either 'keep-up' behaviour in the early Holocene when sea levels were rising rapidly (e.g. ~5 mm a^{-1} between 10.5 and 7.7 ka BP) or from subsequent 'catch-up' behaviour once sea level stabilised (far field) or the rate of rise slowed (intermediate field) after 6.5 ka BP, indicate that reefs clearly have the *potential* to keep pace with the magnitudes of sea level change predicted (Spencer, 1995; IPCC, 2007) (Fig. 7.8).

There is, however, an additional issue here and this relates to the localised response modes that reefs will exhibit to the relatively modest near-future increases in sea level outlined above. Critical here is the response (i.e. accretionary) potential of existing reef crests and reef flats. If these environments fail to closely track near-future sea level rises, a factor dependant upon site-specific shallow water carbonate production and accumulation rates, then even water depth increases of ~0.5 m will have important geomorphological implications associated with increased wave, current and tidal energy propagation across reefs. Such response modes of 'catch-up' and 'keep-up', and the speed with which reef systems can react to changes in sea level remain, however, poorly constrained. Smithers *et al.* (2006) have documented the shut-down of fringing and nearshore reef progradation on the Great Barrier Reef between 5.5 and 4.8 ka BP and then again between 3.0 and 2.5 ka BP. It seems likely that the 'turn-off' of reef growth in this region was due to the lack of accommodation space as reefs reached present sea level (which may itself have been falling in this region in the late Holocene). Buddemeier and Hopley (1988) have argued that rates of sea level rise of ~0.5 m by AD 2100 might create new accommodation space and switch reef vertical accretion back on, with calcification rates rising from the current 50 Mt a^{-1} to 70 Mt a^{-1} (Kinsey and Hopley, 1991).

Whether Holocene analogues can be used to infer future performance requires, however, a much deeper assessment of current geomorphological structure and function, and three sets of difficulties arise in any such assessment. Firstly, sea level position in the reef seas has been relatively stable for at least the last 1000 and in some cases the last 6000 years. Many reef structures have slowly adapted to this state of relative stasis. In addition, near-future changes in sea level will be imposed upon structures that are starting from a very different baseline of environmental conditions, certainly in terms of temperature and perhaps in relation to other variables (water quality, nutrient loading, aragonite saturation state) to those experienced in the early to mid-Holocene. Finally, recent global deteriorations in reef community state may have fundamentally altered the resilience of reef systems and their ability to respond quickly and effectively to changing boundary conditions.

7.3.2 Contemporary reef growth and responses to near-future sea level rise

There is a significant difference between the ability of a coral colony to calcify and extend its skeleton and the ability of an entire reef platform to accrete vertically, the latter being the result of aggregated growth of all constructive processes (calcification of all organisms) and destructive processes as outlined in terms of reef carbonate budgets (Fig. 7.2). On any individual reef, there is significant natural variability in reef-related carbonate production rates and thus reef accretion rates and styles. It is, therefore, difficult to apply generic accretion and carbonate production data and this limits attempts to extrapolate models and concepts of reef growth and carbonate production to issues of reef disturbance and change, including responses to accelerated sea level rise. Nevertheless, measurements of changing seawater alkalinity have been used to establish consistent 'standards of metabolic performance' – in terms of photosynthesis, production/respiration ratios and calcification – for a range of coral substrates at a wide range of reef locations (Kinsey, 1983); these rates are consistent with the dating of drill core materials (Davies and Hopley, 1983). Differing rates of vertical accretion can be clearly associated with different reef production zones; when aggregated to the reef scale, the presence of both carbonate source areas and sinks means that overall rates of performance, at ~0.9 mm a^{-1} vertical accretion, are close to an order of magnitude lower than accretion in the most productive within-reef zones (Fig. 7.9). These rates are similar to, or slightly in excess of, globally averaged rates of sea level rise during the period when these measurements were made (0.7 ± 0.7 mm a^{-1}, 1961–2003; IPCC, 2007). A critical question for reef geomorphology, therefore, is whether or not reef growth will be able to accelerate to match the rates of sea level rise predicted for the next 100 years.

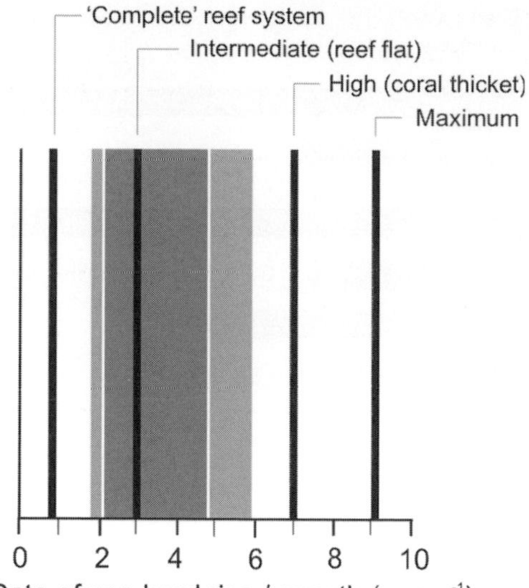

FIGURE 7.9. Comparison of rates of contemporary vertical reef accretion derived from water chemistry measurements (black bars) (from Spencer, 1995) and rates of near-future sea level rise (light and dark grey bars) based on IPCC (2007) projections (from left to right, scenarios B1 (low estimate), A1B (low estimate), A1B (high estimate), A1FI (high estimate); for further details see Chapter 1)).

7.3.3 Coral reefs and increased sea surface temperatures

Although estimates of reefal carbonate production are generally not latitudinally differentiated, within the Hawaiian archipelago vertical accretion rates fall from 11 mm a^{-1} at Hawaii to 0.2 mm a^{-1} at Kure Atoll; beyond this location – termed the 'Darwin Point' by Grigg (1982) – sea level reef growth is not currently maintained. It has been argued, therefore, that ocean warming might extend the region of reef growth into areas that are currently too cool to sustain reef development. Greater poleward extension of Last Interglacial reefs in Florida, and western and eastern Australia, compared to their Holocene counterparts, implies a response to a warmer climate. There is also evidence of more prolific coral growth near latitudinal limits under warmer mid-Holocene conditions in southeast Florida (Toscano and Macintyre, 2003), at Tateyama, near Tokyo, Japan (Veron, 1992) and at Lord Howe Island, southwest Pacific (Woodroffe *et al.*, 2005). However, the steepness of temperature gradients near the current limits and the low availability of suitable substrates for new growth suggest that any range increases are likely to be spatially restricted (Guinotte *et al.*, 2003).

Within the reef seas, increased sea surface temperatures might be expected to increase coral metabolism and increase photosynthetic rates of zooxanthellae, thus aiding calcification. Lough and Barnes (2000), using growth rate data from skeletons of massive *Porites* colonies on the Great Barrier Reef, the Hawaiian archipelago and Thailand, have shown that the relationship between sea surface temperature and growth rate is very consistent: over the temperature range 23–29 °C, for each 1 °C rise in sea surface temperature there is a mean annual calcification increase of 0.33 g cm^{-2} and a mean skeletal extension rate of 3.1 mm a^{-1}. Remarkably, the increase in sea surface temperatures of ~0.25 °C recorded along the Great Barrier Reef in the latter half of the twentieth century are reflected in a statistically significant acceleration in calcification rate in *Porites* colonies (1880–1929. 1.47 ± 0.05 g cm^{-2} a^{-1}

versus 1930–1979: $1.53 \pm 0.07\,g\,cm^{-2}\,a^{-1}$). However, the combination of relatively faster growth rates and slower calcification (discussed below) may result in more fragile skeletons, increased framework degradation rates and thus greater carbonate sediment supply to reef flats and islands (Sheppard *et al.*, 2005). In addition, such findings need to be tempered by (i) the raising of sea surface temperatures to levels at which growth is terminated, temporarily or permanently, by coral bleaching and temperature-related outbreaks of coral diseases; and (ii) the fact that changes in atmospheric carbon dioxide are likely to lower both ocean pH levels and carbonate ion concentrations in surface waters and reduce calcification rates in corals. These controls are now discussed in more detail below.

7.3.4 Ocean temperatures and coral bleaching

Corals respond to thermal stress, and synergistic increases in solar irradiance, by whitening or 'bleaching'. Bleaching is the visible sign of the degeneration and/or loss of zooxanthellae, and/or the loss of cells containing zooxanthellae, from coral tissues as photoprotective mechanisms are lost. The zooxanthellae play a key role in coral metabolism and their reduced function or loss is accompanied by reduced carbon fixation, coral growth and reproductive ability. Bleaching associated with up to several weeks of temperature elevations of +1 to +2 °C above regional seasonal maxima is often species- and/or reef location-specific and repaired after a few months with little coral mortality. However, large temperature excursions of +3 to +4 °C, particularly if they are prolonged can produce 'mass bleaching' of entire reef communities and subsequent coral mortality rates in excess of 90% (e.g. Douglas, 2003; Hoegh-Guldberg *et al.*, 2007).

Major bleaching events took place in 1982–3, 1987–8, 1994–5 and particularly in 1998 when, it has been claimed, 16% of the world's reef-building corals were killed (Walther *et al.*, 2002). There has been a marked upturn in the record of bleaching events from all the major reef provinces in the 1980s which are difficult to explain solely by improved reporting (Glynn, 1993). One argument, therefore, is that the appearance of these impacts represent an early signal of global warming in the oceans, with ENSO triggers to bleaching being superimposed on a secular trend of rising sea surface temperatures of the order of 1–2 °C per century (Williams and Bunkley-Williams, 1990; Hoegh-Guldberg, 1999).

The application of these established sea surface temperature – bleaching relations to the temperature trends seen in large ocean temperature data sets implies that the threshold temperature at which corals bleach will occur more frequently in the near future, potentially to the point on some reefs where bleaching is an annual event. Such scenarios have been used to drive 'time to reef extinction' models (e.g. Hoegh-Guldberg, 1999; Sheppard, 2003). Unfortunately, however, whilst satellite temperature is well correlated with field temperature measurements and bleaching incidence when aggregated over space and time, such correlations often break down at the scale of individual reef systems (McClanahan *et al.*, 2007). This is partly because bleaching is generally correlated with short hot spells rather than mean water temperatures (Berkelmans *et al.*, 2004) and partly because the incidence of bleaching is often 'patchy' in time and space, sometimes down to the scale of the individual coral colony. Bleaching susceptibility can vary dramatically between species (e.g. Marshall and Baird, 2000); between locations, with local hydrodynamics determining upwelling of cool deep waters and wave- and tidal-driven current flows and allowing heating of shallow waters with long residence times (e.g. McClanahan *et al.*, 2005); with variations in water turbidity; and with variations in coral resilience imposed by differential human impacts. Not only do the extinction-type models fail to allow for this small-scale patterning of temperature impacts but they also fail to take account of the potential adaptive responses of corals and/or their algal symbionts to temperature change (both past and predicted). It is well known that corals in different reef areas show differing thermal tolerances which are related to prevailing water temperatures. Developed from these observations is research that suggests that corals may be able to acclimate (an individual, physiological response) or adapt (a genetic response at the population level) to changed thermal regimes, particularly as a result of shifts in coral host–zooxanthellae relations, creating 'new' ecospecies with tougher environmental tolerances and supporting more temperature-tolerant strains of zooxanthellae. In such circumstances, there is likely to be an increase in the thermal threshold to bleaching, a hypothesis supported by field observations which show that past bleaching episodes can indeed provide corals with some measure of resistance to subsequently raised temperatures (e.g. Brown *et al.*, 2002; Baker *et al.*, 2004). It is likely, therefore, that global environmental change will be accompanied by the patchy reorganisation of coral communities and the degradation (but not total loss) of ecosystem function and diversity.

7.3.5 Ocean temperatures, storminess and storm impacts on reefs

There is a strong linkage between patterns of ocean temperature and the frequency and magnitude of tropical storms, cyclones and hurricanes. For reefs that lie within the storm belts (7–25° N and S of the equator and particularly on western ocean margins), coral recruitment and subsequent reef building can be poor and restricted to below wave base (water depths of ~15 m). In such settings, reef accretion is frequently in the form of a patchy veneer less than 1 m thick and vertical growth rates are typically <1 mm a^{-1}, which are below the critical values needed for reef maintenance. If framework does develop its periodic removal constrains reef longevity. Extreme cyclonic rainfall over wide areas can also send freshwater plumes (and hence lower salinities), sediments and nutrients to fringing reefs, and even barrier reefs, leading to a lowering of growth rates or, in extreme cases, coral mortality. On the other hand, storm impacts do promote opportunistic and fast-growing corals over slower-growing forms, 'open up' senescent reefs and provide near-instantaneous height increments for reef surfaces from the accumulation of coral rubble sheets and ridges (see Scoffin, 1993 for review).

It is clear that the resolution of the current debate on the possible changing magnitude, frequency and location of tropical storms with global environmental change (see Chapter 15) has important implications for coral reef systems, which lie both within, and outside, the storm belts. However, the exact nature of this relationship is difficult to define. Firstly, individual storm tracks are generally narrow (<30 km) and thus the chance of a particular location being hit in any one season is low. Secondly, there are sharp thresholds to storm damage. Thus whereas hurricanes with typical wind speeds of 120–150 km h^{-1} result in a patchwork of impacted and non-impacted areas, determined by water depth, reef front aspect and reef topography in relation to storm direction, severe storms, with wind speeds in excess of 200 km h^{-1} may overcome the structural resistance of the reef as a whole, reducing three-dimensional complexity to an unstable rubble plain, unconducive to coral re-establishment, and producing a hiatus to reef recovery lasting for up to 50 years (Stoddart, 1985). Furthermore, storm impacts need to be seen within the context of continued coral growth. Often as much storm damage is caused by the detachment and movement of coral materials. Thus long periods between storm impacts allow considerable carbonate accumulation which, when dislodged, can cause high levels of damage: thus the high levels of damage to both shallow and reef-front reefs by overwash and avalanching respectively at Tikehau Atoll in 1982–3 in an area unaffected by cyclone activity since 1906 (Harmelin-Vivien and Laboute, 1986). Thus if the storm belts widen with future increases in ocean temperatures, reefs currently unaffected by storm impacts may begin to be impacted. Paradoxically, more frequent storms in areas already subject to storms may lead to less morphological impact (Scoffin, 1993). Reefs dominated by more fragile coral growth forms may have high initial sensitivity but as these are replaced by more robust and encrusting forms so impacts may diminish over time.

7.3.6 Ocean acidification and coral reefs

It has been argued that a progressive decrease in seawater pH as a result of enhanced oceanic uptake of carbon dioxide, or 'ocean acidification', will be correlated with both decreased calcium carbonate production in marine organisms and increased calcium carbonate dissolution rates (e.g. Orr *et al.*, 2005). The pH of tropical surface seawater has declined from a value of 8.2 in the pre-industrial period to 8.1 at the present time (representing a ~30% increase in hydrogen ion concentration) and modelling suggests that ocean surface water pH levels may decrease by up to a further 0.5 pH units, and carbonate ion concentration by >30%, over the next 100 years (Royal Society, 2005; International Society for Reef Studies, 2008). Tank and mesocosm experiments suggest that calcification rates in corals will decrease by 30 ± 18% within the next 30–50 years, easily overriding any possible enhancement of calcification by increased sea surface temperature. Comparable impacts are predicted for coralline algae and other calcifying algae. It has been argued that weaker skeletons will make corals more susceptible to storm damage; ultimately, potential changes in the saturation state for aragonite (Table 7.1) may drive values to a point where corals are unable to form skeletons at all (Kleypas and Langdon, 2006). Wider impacts of ocean acidification on coral recruitment and demography are at the present time unknown and the implications of acidification at the scale of the coral reef ecosystem unclear (International Society for Reef Studies, 2008). Interestingly, reduced calcification rates in some massive corals on the inner Great Barrier Reef have been largely seen through decreased linear extension rates rather than changes in skeletal density (Cooper *et al.*, 2008): this suggests that lowered growth rates may reduce the ability of corals to compete for space on the reef and thus for changes in ocean chemistry to be reflected in altered reef community structure. It is reasonable, based on available evidence, to predict that this may have implications for the relative abundance and productivity (calcification) rates of key reef framework constructors. Furthermore, decreased calcification rates in crustose coralline algae may have implications for the cementation and stability of the reef matrix (Diaz-Pulido *et al.*, 2007). The combined effects of reduced calcification and increased dissolution rates are thus likely to be significant for net reef calcium carbonate production. In combination with changed coral cover (linked to a suite of other environmental impacts) the most likely effect of these changes will be to shift some reef carbonate budgets towards states of net erosion. A caveat here is that the magnitude of ocean chemistry changes and the interactions with environmental disturbances are likely to be spatially heterogeneous (Hoegh-Guldberg *et al.*, 2007). Where significant, however, the geomorphic consequences of these changes are likely to be the progressive degradation of reef frameworks and reduced reef topographic complexity (potentially combined with changes in sediment production regimes) – one additional impact of which may be to redefine the near-future reef growth sea level rise relations outlined in Section 7.3.2 above.

7.4 Reef sedimentary landforms

The generation of detrital sediment on reef platforms and its transfer by physical wave and current processes is critical to the construction and modification of reef sedimentary landforms (Figs. 7.1, 7.2, 7.5d). Understanding the co-adjustment of physical processes, sediments and morphology provides a framework to evaluate future change (Fig. 7.2). The focus of this section is on the subaerial landforms created on or adjacent to reefs (islands and coastal plains). A range of these reef sedimentary landforms can be distinguished dependent on the location of sediments relative to the reef platform, reef type and the presence of non-carbonate substrate (Fig. 7.10 a–d).

In fringing and barrier reef settings, carbonate deposition typically occurs toward the leeward edge of reefs at the interface between the reef and terrestrial environment forming coastal plains, beaches, spits and barriers. In these settings, terrigenous sediments delivered to the coast mix with biogenic sediments and contribute to landform development. On fringing reefs, coastal deposits have typically prograded across the adjoining reef surface (Figs. 7.10a, 7.11a; Plate 22). In contrast, in barrier reefs, lagoons separate the reef structure and beaches. Land building in these settings is reliant on the transport of sediment from reefs into lagoons and its subsequent reworking and deposition at the shoreline where mixing with terrigenous sediments also contributes to landform accumulation. On isolated reef platforms (e.g. mid-ocean atolls) sediments accumulate directly on the reef surface or over lagoon sediments, forming reef islands (Figs. 7.10c, d; 7.11c, d). Furthermore, there is a variety of reef island types distinguished by the calibre of sediment from which they are composed (sand cays and sand and gravel *motu*) and the importance of vegetation in promoting the stabilisation of islands (Stoddart and Steers, 1977).

In general, reef islands are typically low-lying, rarely reaching more than 3.0–4.0 m above sea level. However, they do exhibit significant variation in morphology (size, elevation) and sediment composition. This combination of physical factors, low elevation, small areal extent and reliance on locally generated reefal sediment are considered to make reef sedimentary landforms particularly vulnerable to the impacts of climate change and sea level rise. They are inherently unstable landforms that are morphologically sensitive to changes in boundary conditions (sea level, waves, currents and sediment supply). Widespread shoreline erosion is the most commonly cited impact associated with climatic change that threatens the physical stability of reef-associated landforms. In extreme cases total loss of reef islands has been predicted which will undermine the physical foundation of atoll nations such as Tuvalu and the Maldives (Dickinson, 1999; Kahn *et al.*, 2002; Barnett and Adger, 2003). Such assertions are commonly based solely on projections of sea level rise or reef health as the primary controls on landform development and change. Here we evaluate the veracity of these assumptions through consideration of the geomorphic controls on landform development and change.

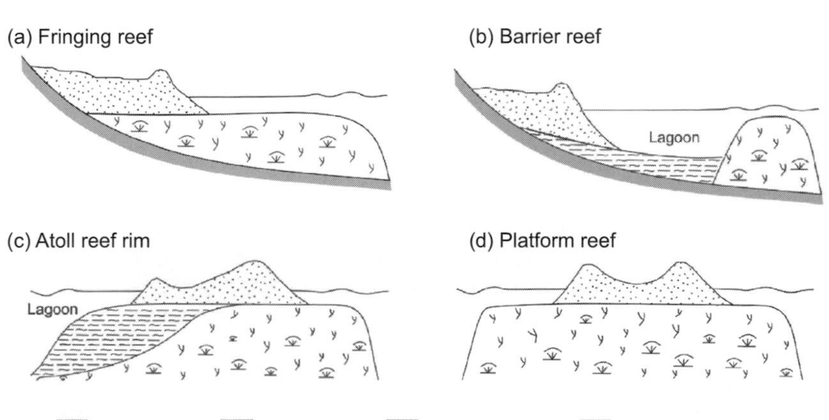

FIGURE 7.10. Reef sedimentary landforms: (a) and (b) coastal plains; (c) and (d) reef islands and their configuration with respect to coral reefs.

Fringing reef, Lizard Island, Great Barrier Reef, Australia.

Barrier reef setting, Mauritius.

Platform reef, Lady Elliot Island, Great Barrier Reef, Australia.

Atoll reef rim and islands, Tarawa Atoll, Kiribati.

Emergent reef flat exposed at low tide, Majuro Atoll, Marshall Islands.

Coral cover on reef platform (2 m depth), Nadi Bay, Fiji.

Multiple overwash layers contributing to island building on a Maldivian reef island since 2003.

Densely urbanised reef platform island, Male, Maldives, Indian Ocean.

FIGURE 7.11. Images of coral reef landforms: (a)–(d) differences in geomorphic state of reefs; (e)–(g) sediment overwash deposition on island margin; (h) example of anthropogenic modification of reefs. (See also Plate 22 for colour version.)

7.4.1 Evolution of reef sedimentary landforms

Underpinning assertions of future morphological change of reef landforms is sea level as the primary control on shoreline stability. However, this assertion oversimplifies a complex relationship between long-term controls on landform development that include:

(a) *sea level change* which, as outlined in Section 7.3, governs gross coral reef development;
(b) *substrate gradient* imposed by structural lithology of continental coastlines and high islands that adjoin fringing and barrier reefs, and *substrate elevation* in the case of reef platform islands;
(c) *accommodation space* which defines the available volume for sediment deposition as controlled by substrate gradient, elevation and sea level (Cowell and Thom, 1994). The upper limit of land building is controlled by storm wave runup processes, which are modulated by relative sea level. For reef islands the lower boundary defining accommodation space is governed by reef margin and reef flat elevation and lagoon depth. However, in fringing and barrier reef coasts accommodation space is also dependent on the gradient of the fixed underlying lithology;
(d) *relative wave energy* which in coral reef settings is modulated by the relationship between reef margin and reef flat elevation, sea level and incident ocean swell; and
(e) *sediment supply* controlled by reef productivity and sediment generation processes (Section 7.2).

Coral reefs are unique in that substrate adjustment (reef growth) occurs in response to sea level oscillations at multidecadal to millennial timescales. Therefore the boundary conditions for land formation also exhibit morphodynamic feedback at timescales that overlap with the processes responsible for island construction and change. This co-adjustment has a number of important implications for the timing and conditions under which sedimentary landforms have been constructed. Understanding these relations can provide useful insights into future landform stability.

Timing of sea level change, reef growth and landform accumulation

Conventional theory suggests that sea level stabilisation, completion of vertical reef growth and landform accumulation occur sequentially (Fig. 7.12a). Evidence for this model is apparent in the Pacific and eastern Indian oceans, where

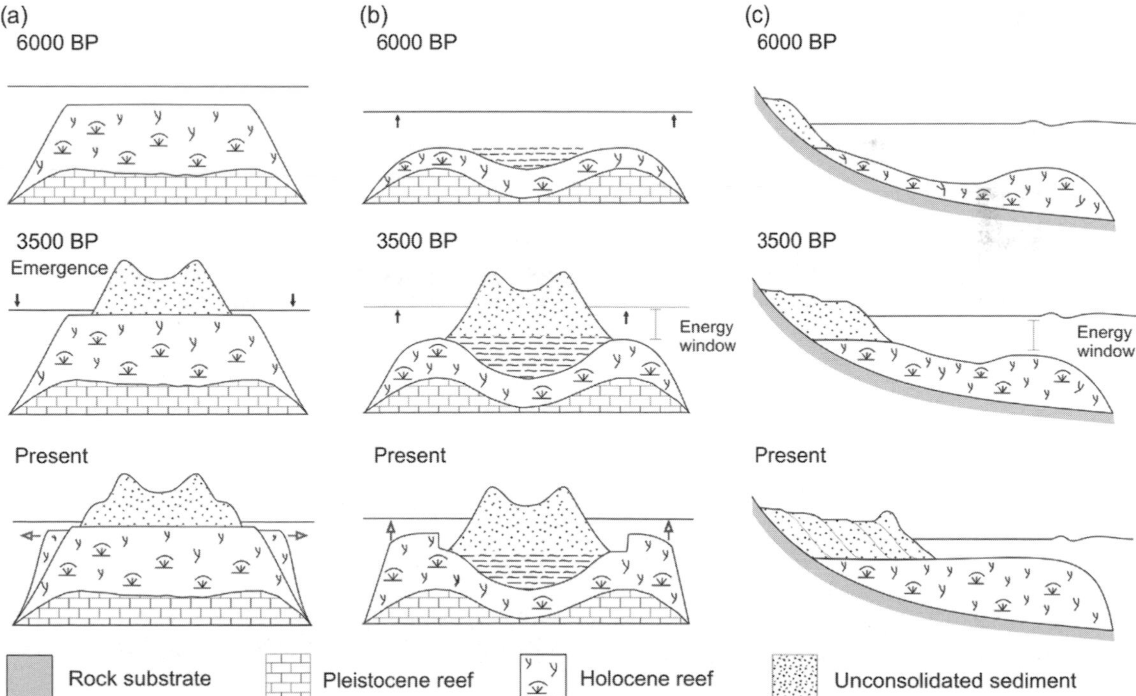

FIGURE 7.12. Contrasting models of the formation of sedimentary landforms on coral reefs. (a) Pacific reef platform model after Woodroffe (2003); (b) central Indian Ocean model after Kench *et al.* (2005); (c) Pacific fringing reef mode. Note differing relations between sea level and reef elevation at time of island formation. Dark and grey open arrows signify direction of sea level change and reef growth respectively.

sea level has been at or slightly higher than present sea level for the past 6000 years (see Chapter 1). In this setting vertical reef growth (in 'keep-up' mode; see Section 7.3) was rapidly constrained and lateral reef accretion became the dominant growth mode resulting in construction of broad reef flats which became emergent as a consequence of late Holocene sea level fall (Woodroffe *et al.*, 1990; McLean and Woodroffe, 1994) (Fig. 7.11d, e). These reef flat surfaces provided the foundation for sediment accumulation and island building in the mid- to late Holocene (Fig. 7.12a). Radiometric dating evidence supports this model with vertical reef platform growth completed prior to land building on reef flats in the Cocos (Keeling) Islands (Woodroffe *et al.*, 1999), Kiribati (Woodroffe and Morrison, 2001) and Tuvalu (McLean and Hosking, 1991), although regionally the timing of island formation differs as a result of contrasting reef growth chronologies. The apparent synchronisation of land formation with late Holocene sea level fall has caused some researchers to suggest that land building was triggered by sea level fall (Schofield, 1977; Dickinson, 1999). The implication from this model is that sea level will force morphological instability as water depths over reefs increase.

However, regional differences in Holocene sea level dynamics (Chapter 1) and reef growth histories have provided contrasting boundary conditions for the onset and accumulation of landforms. Recent studies in the Maldives have shown that reef islands there developed in the mid-Holocene prior to reefs reaching their maximum vertical growth limit. In this model islands formed across submerged reefs (e.g. Fig. 7.11f), in latter stages of 'catch-up' growth mode, and over infilled lagoons (Kench *et al.*, 2005) (Fig. 7.12b). Island formation occurred while water depth across reefs was greater than present and vertical reef growth continued after initial island formation to close down energy processes across reef surfaces. This latter model is likely to have similarities to the Caribbean where landforms have developed under continual rising sea level throughout the late Holocene (Toscano and Macintyre, 2003).

In high island settings landform development has been controlled by the onset of flooding of the non-erodible coastal margin and embayments during the Holocene marine transgression (Fig. 7.12c). Coastal regressive and transgressive sequences have been identified in these settings (Kraft, 1982; Calhoun and Fletcher, 1996) modulated by the pattern of fringing and barrier reef development, mid- to late Holocene sea level and sediment availability.

Collectively these studies from differing reef regions provide critical insights on the role of sea level and reef growth in controlling island formation. Firstly, island formation has occurred under differing sea level change histories including rising sea level. Secondly, sea level fall is not a necessary precondition for island building. Thirdly, reef flat formation at sea level is not a necessary precursor for island formation. Fourthly, island formation can occur in latter stages of reef platform development as it approaches its vertical growth limit.

Holocene high-energy window

In all reef settings the relationship between sea level and reef surface elevation (effective water depth) modulates reef platform processes (waves, currents and sediment transport) that govern sedimentary landform development. Coral reefs act as a filter to oceanic wave energy with the effective water depth determining the amount of residual energy that propagates across reef surfaces and is able to promote sediment transport and deposition (Kench and Brander, 2006a). During catch-up reef growth, water depth across reefs was greater than present allowing higher ocean wave energy to propagate onto reefs, stimulating geomorphic processes (Fig. 7.12b). As reefs caught up with sea level the 'high-energy window' closed. This mid-Holocene high-energy window has been used to account for a range of depositional features in the Great Barrier Reef (Hopley, 1984) and is considered a major control on island formation and the maximum elevation of reef islands in the Maldives (Kench *et al.*, 2005). Of significance for future geomorphic sensitivity of reef sedimentary landforms is whether they were constructed or reached geomorphic equilibrium during past episodes of higher-energy conditions. For those landforms where the high-energy window closed in the late Holocene, the current process regime is less geomorphically active than conditions under which the landforms were constructed. Morphologically these landforms may reflect adjustment to higher-energy processes. Future increase in sea level over reefs will simply reactivate the process regime responsible for land building. In contrast, landforms that post-date reef platform development and/or emergence are likely to have formed and be in morphodynamic equilibrium with the contemporary process regime. Future increases in water level may place such landforms in a higher-energy process regime than previously encountered, promoting greater geomorphic change.

Sediment supply

Land building potential on reefs requires an abundant supply of sediment to fill the accommodation space (Section 7.2). Existing studies indicate that while the onset of land building generally occurred in the mid- to late Holocene in most reef regions (Calhoun and Fletcher, 1996; Woodroffe *et al.*, 1999; Kench *et al.*, 2005), the subsequent accumulation history has shown considerable variation in response to sediment supply. For example, on Warraber Island in the Torres Strait, island development has occurred incrementally over the past 3000–5000 years in response to continued supply of sediment (Woodroffe *et al.*, 2007). At Hanalei Bay, Hawaii, Coulhoun and Fletcher (1996) identify a gradual reduction in the rate of coastal progradation during the Holocene controlled by declining sediment supply. In contrast, Maldivian reef islands appear to have formed (primarily of *Halimeda*) in a 1500-year window in the mid-Holocene and effectively ceased accumulation 3500 years ago (Kench *et al.*, 2005). Episodic land accumulation has also been identified in storm-dominated settings where island building has occurred in discrete depositional phases of storm-derived rubble (Maragos *et al.*, 1973; Bayliss-Smith, 1988; Hayne and Chappell, 2001).

Differences in accumulation history of reef sedimentary landforms may be explained through temporal variations in sediment supply during the Holocene and recognition that the relationship between reef carbonate productivity and sediment generation is non-linear. For example, variations in sediment supply are likely to reflect shifts in the balance between reef growth and reef productivity for both primary and secondary sediment producers (see Section 7.2.2). During rapid catch-up growth mode the reef structure was effective at retaining calcified products in the reef framework. However, as reefs reached wave base and vertical growth was constrained, excess carbonate was shed from the reef system and was either exported from the reef or made available for construction of sedimentary deposits. Consequently, the general trend for studies to identify the onset of land building in the mid-Holocene (e.g. Woodroffe *et al.*, 1999; Harney *et al.*, 2000; Woodroffe, 2003) may coincide with reefs either reaching sea level or reaching wave base and releasing a pulse of excess sediment for landform construction. In many reef settings this may have coincided not only with a transition from vertical to lateral reef growth but also to reef flat emergence (Fig. 7.11e), which may have promoted shifts in reef flat ecology and thus carbonate production. As a consequence, the dominant constituents available for land building may have shifted from sediments derived from frame builders (coral and coralline algae) to those derived from other sediment producers (e.g. foraminifera, calcareous green algae). Yamano *et al.* (2000) identified such ecological change, to a foraminifera-dominant reef platform following sea level stabilisation at Green Island, as a key trigger for the onset of island formation and development. Similar reliance for island building on a narrow range of skeletal constituents has been identified by Woodroffe and Morrison (2001) in Makin Island, Kiribati. The reliance of some landforms on a select number of skeletal constituents also suggests that sediment availability through the late Holocene is also likely to have been influenced by biological perturbations (infestations of bioeroding organisms or mass death of secondary producers; see Section 7.4) that release pulses of sediment to the reef system controlling episodes of land development. Current analogues suggest that landforms that are reliant on a narrow range of constituents, and those landforms which have ceased building and have no apparent significant influx of sediment (i.e. are relict deposits), will be most susceptible to future morphological change in response to negative alterations in sediment supply and altered boundary processes.

Consideration of the importance of sediment supply to reef sedimentary landforms highlights a number of issues fundamental to future morphological change. It is commonly assumed that the reef 'carbonate factory' produces a quasi-continual supply of sediment to build or maintain landforms. However, existing studies suggest that both the supply and composition of sediments available for land building can vary temporally in response to changes in reef growth/reef ecology (Yamano *et al.*, 2000; Woodroffe, 2003) and these in turn influence the accumulation behaviour of coastal deposits. Furthermore, scientific understanding of the processes that 'turn on' and 'turn off' sediment supply as it relates to the construction of landforms is poor and is an urgent priority for research. Preliminary modelling studies suggest that changes in sediment supply, either through alterations in sediment generation or littoral budgets, may be more important than sea level in affecting the stability of reef landforms (Kench and Cowell, 2003; Woodroffe, 2003). Consequently, knowledge of the contemporary rate of sediment generation and likely changes in sediment availability in response to anticipated changes in reef ecology/growth, as a consequence of sea level change, is essential to support assessments of the impact of sediment supply on future geomorphic change.

7.4.2 Morphodynamics of reef sedimentary landforms

Projections of instability and mass inundation of reef islands and coastal margins with sea level rise are commonly founded on inappropriate considerations that such landforms are morphologically static. However, at annual to centennial timescales reef sedimentary landforms are in continual readjustment to changes in climatic and oceanographic boundary conditions and sediment supply. Typical morphological adjustments include shoreline erosion, accretion, sediment washover, shoreline realignment and island migration. The major process mechanism controlling formation and stability of reef landforms is wave action and its interaction with the coral reef surface. This mechanism determines current and sediment transport patterns that ultimately control the upper limit of land building through storm-driven wave runup (Gourlay, 1988; Kench and Brander 2006a). Such processes are modulated by climate and consequently, reef landforms exhibit rapid morphological adjustment to changes in these incident processes at a range of timescales.

At the decadal scale, shifts in prevailing wind fields and their influence on wave propagation (direction and energy) control erosion and accretion patterns on reef islands resulting in island migration (e.g. Verstappen, 1954; Flood, 1986). The Pacific Decadal Oscillation (PDO) and its effect on modulating storm frequency has been found to control multi-decadal fluctuations in longshore sediment transport resulting in erosion and accretion patterns of ± 100 m on the Kihei fringing reef shoreline, Maui (Rooney and Fletcher, 2005). Inter-annual El Niño–Southern Oscillation (ENSO) variations have also been implicated in shoreline erosion and accretion patterns in Kiribati (Solomon and Forbes, 1999).

At seasonal scales, Kench and Brander (2006b) examined the morphological sensitivity of 13 islands in the Maldives to predictable changes in wind and wave conditions controlled by the oscillating monsoons (Kench *et al.*, 2006a). Results identified rapid, seasonal morphological adjustments of island shorelines (up to 53 m of beach change) despite morphological near-equilibrium on an annual basis. However, the magnitude of morphological change, and sensitivity of islands to change, was found to vary between islands as a function of reef platform shape, which controls wave refraction patterns; circular islands were most sensitive to changes in incident wave patterns.

At event scales, storms and hurricanes have both constructional and erosional impacts on reef sedimentary landforms (Fig. 7.13). These contrasting responses reflect differences in storm frequency and texture of island building materials (Bayliss-Smith, 1988). In settings with low storm frequency landforms are typically composed of sand-size sediments, which are susceptible to erosion during extreme events. Stoddart (1963) reported mass destruction of some reef top islands in Belize as a result of Hurricane Hattie. Large differences in island loss were observed depending on the presence or absence of natural littoral vegetation. However, in reef settings with high storm frequency islands are commonly composed of rubble on their exposed margins while leeward and fringing reef fronted coastal plains are composed of sand-size material. In such settings, large volumes of rubble can be generated in single events from coral communities on the outer reef. In a well-documented example Hurricane Bebe (1972) deposited an extensive storm rubble rampart onto the reef flat and islands of Funafuti Atoll, Tuvalu (Maragos *et al.*, 1973). Subsequent storms have reworked this rampart onto island shorelines showing that the hurricane and subsequent storm processes added approximately 10% to island area (Baines and McLean, 1976). Sequential deposition of rubble ridges has also been identified in the late Holocene evolution of gravel islands in the Great Barrier Reef (Chivas *et al.*, 1986; Hayne and Chappell, 2001).

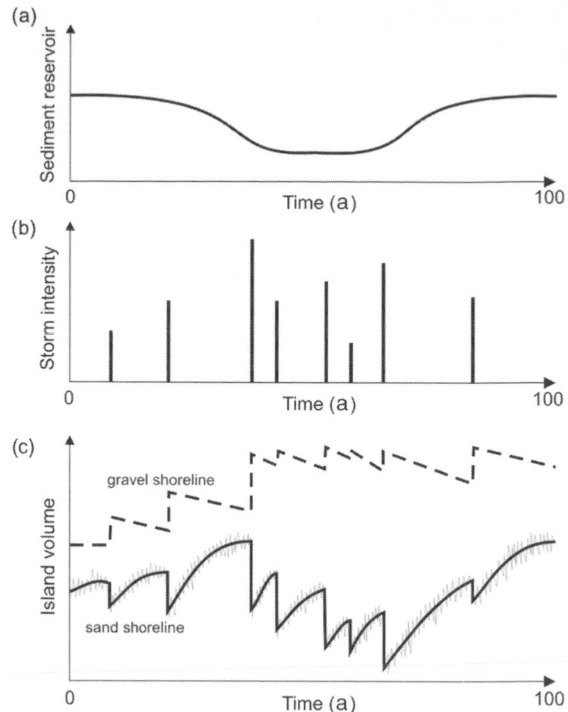

FIGURE 7.13. (a) Medium-term changes in sediment supply and (b) storm frequency and intensity leading to (c) a conceptual model of reef-associated landform response. Note: grey line in (c) depicts short-term shoreline morphodynamics.

Variations in sediment supply either through changes in sediment generation on the reef flat or in net littoral transport gradients (see above) also exert a further control on landform change. While the role of process-driven changes on the sediment budget are somewhat predictable (Rooney and Fletcher, 2005; Kench *et al.*, 2007) (Fig. 7.13), temporal variations in sediment generation of suitable grade for landform development and their influence on reef landform dynamics are poorly understood.

7.4.3 Future trajectories of reef sedimentary landforms: resilience or instability

Attempts to resolve future morphological adjustments of reef sedimentary landforms in response to global climate change have typically relied on interpretation of the

Holocene formation of landforms and analysis of existing geomorphology as analogues to infer future response (e.g. Woodroffe and McLean, 1992, 1993). In these examples, geomorphic attributes of reefs and landforms such as elevation and degree of lithification are used to argue varying levels of resilience or resistance (Woodroffe, 2007). Modifications of sedimentary landforms are also considered in close association with the reef growth response. However, such approaches have not incorporated a full appreciation of the contemporary morphodynamics of landforms nor considered the style and magnitude of changes that may be expected in the future.

In order to explore the mode and magnitude of morphological change expected on low-lying reef landforms Cowell and Kench (2002) adapted the shoreface translation model (STM) specifically to account for non-erodable reef substrates and adoption of perched beach principles, as beaches are commonly truncated by the reef surface (Dean, 1991). This geometric profile model takes explicit account of land building processes (wave runup, elevation and sediment supply) and allows morphological equilibrium to be achieved through cut and fill of sediment according to accommodation space constraints and sediment volume. Model simulations undertaken in Kiribati, Fiji and the Maldives have challenged a number of common perceptions of landform response to sea level change (summarised in Fig. 7.14). Firstly, all simulations identify morphological change, with estimates of shoreline movement ranging from 3 to 15 m for a 0.5 m increase in sea level (Kench and Cowell, 2002). Secondly, simulations indicated that the magnitude and mode of morphological change is highly variable and dependent on initial morphology (elevation, sediment volume) and accommodation space. Higher-elevation coastal deposits and landforms with larger sediment volume are more resilient to morphological change than lower-elevation islands with limited sediment reservoirs. In particular, such simulations challenge the assertion that the presence of lithified sediments (conglomerate platform, cay sandstone, phosphate rock and beachrock) act to increase landform stability. In fact, model results show greater change of shorelines situated on conglomerate platforms (Fig. 7.14b). Where landforms are perched on conglomerate, and are in morphological equilibrium with the process regime, the volume of unconsolidated sediment stored in the landform, available for reworking, is of greater importance for the rate of landform adjustment than the presence of lithified material. The presence of lithified sediments only affords the cemented rocks greater resistance, as observed at numerous field locations where abandoned beachrock outcrops signify older shoreline positions.

Thirdly, results show that conventional interpretations of shoreline erosion require reconsideration. While most simulations identify horizontal displacement of the shoreline, such changes did not necessarily imply erosion. Overwash of entire island surfaces on small islands and inlet bypassing promotes migration of islands on reef platforms while conserving or building the sediment volume (Fig. 7.14a). Field evidence for overwash sedimentation as a process of island geomorphic adjustment has been demonstrated in a number of studies of storm and tsunami impacts on reef islands (Bayliss-Smith, 1988; Kench *et al.*, 2006b). Furthermore, the lagoonward migration of islands has been reported from Belize, where sea level rise has been implicated for the abandonment of beachrock and migration of islands toward and across reef lagoon slopes (Stoddart *et al.*, 1982). One constraint to the extent of island migration is accommodation space. On reef flats this is limited to the platform area, whereas in reef lagoon settings migration is also controlled by the slope of the lagoon sand sheet, with steeper slopes triggering greater rates of geomorphic change. On wider islands and coastal plains, overwash can produce horizontal displacement of the shoreline that reduces island width whilst also conserving the total sediment volume (Fig. 7.14c). Fourthly, overwash deposition provides a mechanism to raise the level of coastal ridges as storm runup processes are elevated by sea level rise (Fig. 7.11g). Fifthly, results indicate that beaches and

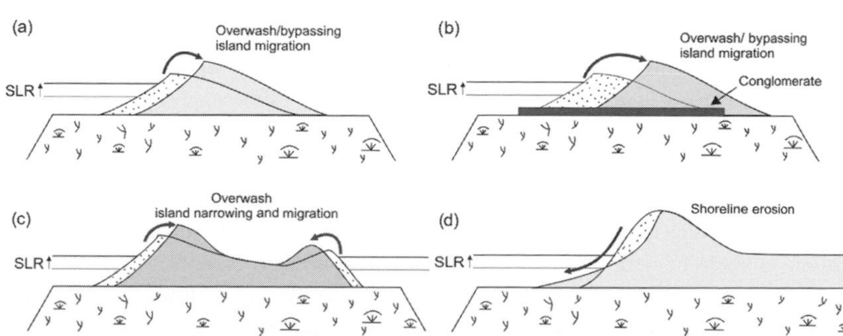

FIGURE 7.14. Summary of modes and relative magnitude of shoreline response to increased sea level. Simulations assume no change in the sediment budget. SLR, sea level rise (from Kench and Cowell, 2002).

the seaward vegetated island margins will be susceptible to reworking. However, central, vegetated island cores are likely to remain largely stable.

Extension of model analysis to the exploration of the morphological sensitivity of reef landforms to changes in sediment supply (Kench and Cowell, 2003) indicates that shifts in the sediment budget may be of greater importance than sea level change in controlling future landform adjustment. Simulations have shown that small negative shifts in the sediment budget accelerate rates of change. The implication of this scenario is that on contemporary reefs with low rates of sediment supply morphological adjustment will have to take place using the finite sediment volume that currently exists within landforms. Alteration in littoral sediment gradients (due to changes in direction of wave approach) can produce substantial alongshore reconfiguration of shorelines (e.g. Rooney and Fletcher, 2005). Negative changes in the littoral budget can produce shoreline adjustments an order of magnitude larger than sea level rise alone. Increases in sediment supply, through a positive change in the littoral budget or increased generation, have been shown to offset horizontal displacement of landforms.

Collectively, therefore, geomorphic investigations and model analysis indicate that reef sedimentary landforms have a greater geomorphic resilience than is commonly recognised. Reef landforms will continue to undergo morphological adjustments (Fig. 7.14) in response to natural variations in climate at seasonal to decadal timescales, global climate change and anthropogenic factors (see Section 7.5). Such adjustments are readily observed under the current process regime and provide useful analogues for the mode and magnitude of change that may occur in the future. The degree of geomorphic resilience of sedimentary landforms is best measured by the ability to reorganise the sediment reservoir and establish a new equilibrium. The pace of such adjustment is regulated by factors that afford event to decadal-scale resistance to geomorphic change (e.g. presence of natural vegetation). Indeed the sensitivity of landforms to change will depend on the interplay of a complex set of factors that include accommodation space, the process regime responsible for the existing equilibrium, existing morphological characteristics and the sediment budget, as well as the magnitude of global environmental change and degree of anthropogenic impacts. These various factors are summarised in Fig. 7.15.

Of additional importance in considering the future morphological behaviour of reef sedimentary landforms is whether environmental thresholds exist beyond which morphodynamic relations are forced along a path of self-organisational behaviour that produces instability of landforms or new morphodynamic equilibrium. An example of such perturbations is the impact of the December 2004 Indian Ocean tsunami on reef islands in the Maldives. Despite the immediate physical effects of the tsunami being minor, Kench *et al.* (2007) show how the tsunami destabilised the seasonal shoreline equilibrium of islands, through perturbations in the sediment budget, triggering medium-term adjustment of reef islands on reef flats. Identifying the level of such thresholds associated with global change is currently speculative. Furthermore, it is probable that such thresholds differ between landforms depending on antecedent morphology, the contemporary sediment budget, anthropogenic impacts and future change in the process regime (Fig. 7.15).

7.5 Anthropogenic effects on sedimentary landforms

Anthropogenic activities have also had a major impact on the ecological and geomorphic functioning of reef systems and have substantially modified the carbonate factory and other boundary controls on landform morphodynamics (Table 7.3). Indeed, the major physical impacts on reef landforms that have been cited (e.g. erosion, island migration) have multiple causes that can be attributed to anthropogenic actions which are difficult to separate from natural causes. Anthropogenic impacts on reefs can be categorised into three types based on the geographic relationship between the human activity and reef system. Some impacts are local and direct (e.g. coral blasting), some are 'proximal' (e.g. resort development, harbour construction) and others are distant but then translated to the reef (e.g. sediment release from land use changes in catchments). These differences influence the time lag of response in the geomorphic system. The importance of understanding the types of anthropogenic impacts on reef systems in the context of global change is threefold. Firstly, to establish key impacts that affect the natural dynamics of reef systems at different timescales, secondly, to understand the antecedent conditions that confront reefs in any future adjustment and thirdly, to allow management strategies to focus at the appropriate space and timescale.

Some anthropogenic stresses, such as over-fishing, have a long history of impact (Jackson *et al.*, 2001) onto which more recent stresses have been superimposed such as direct destruction, littoral habitat modification (e.g. dredging), loss of coastal forests, and land catchment practices (Pandolfi *et al.*, 2003). In Southeast Asia, Burke *et al.* (2002) have calculated that over 20% of reefs are threatened by sedimentation from land-based sources, primarily from poor rainforest logging and agricultural activities, and some of the reefs examined by Edinger *et al.* (2000) in

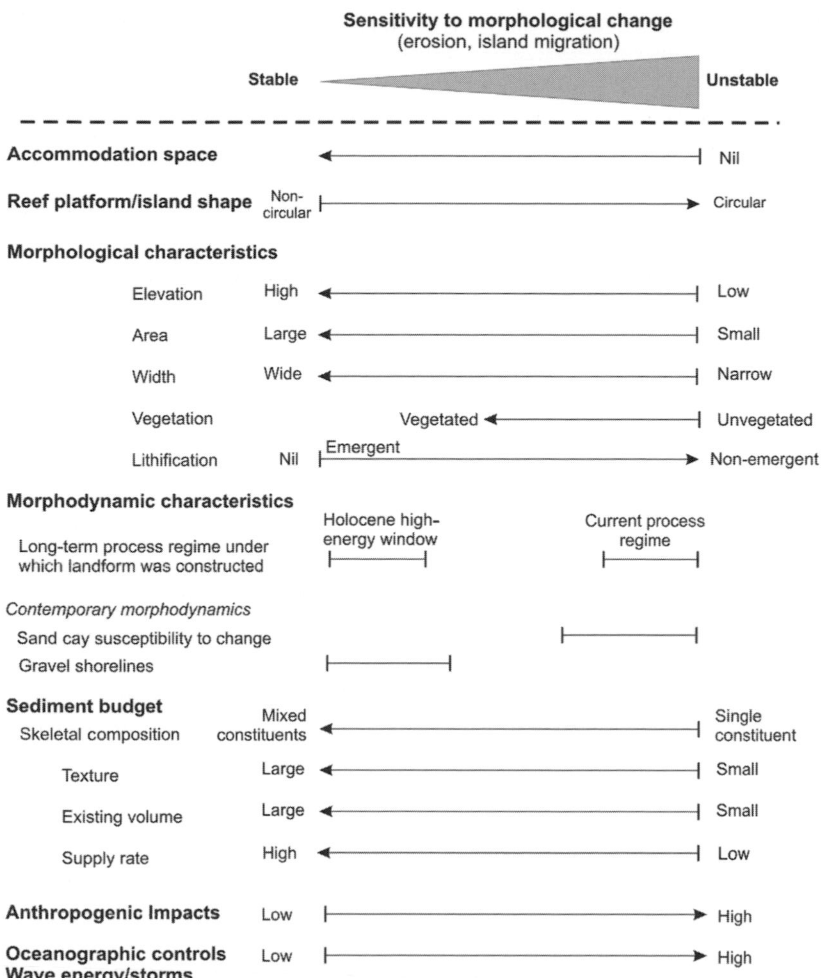

FIGURE 7.15. Factors influencing future sensitivity of reef sedimentary landforms to geomorphic change. Arrows depict increasing values of each characteristic that makes landforms more or less susceptible to change.

Indonesia demonstrate the potential for such water quality changes to drive changes in reef carbonate production. A further layer of stresses has been identified recently, superimposed on the traditional signals of human impact described above. Pathogens that promote coral diseases have triggered complete restructuring of Caribbean shallow-water ecosystems reducing live coral cover from 50% to 10% over the past 90 years (Aronson and Precht, 2001; Gardner *et al.*, 2003).

It is clear that anthropogenically forced changes in reef ecology (e.g. the 'phase-shift') can propagate through to alterations in the reef carbonate factory both in terms of reef framework stability (Eakin, 1996; Edinger *et al.*, 2000; Lewis, 2002) and sediment production (Lidz and Hallock, 2000) (Fig. 7.5). However, evidence for rapid ecological change to drive large-scale reef geomorphic change is relatively rare to date. This may well reflect the multi-decadal to centennial timescales required for subtle alterations in carbonate budgets to be expressed as positive or negative relief changes in reef platforms in many settings. More readily observed anthropogenic impacts (direct and local) that compromise morphodynamic processes within reef systems include direct physical alteration of reefs and shorelines (Table 7.3). The mining of coral from reef surfaces has been well documented in the Maldives with multiple effects on the geomorphic structure and processes within reef systems. Select removal of larger coral heads from Maldivian reef flats has assisted ecological changes (Brown and Dunne, 1988), but has also lowered the elevation of reef surfaces causing an anthropogenically induced higher-energy window to form. Reduced reef elevation in theory allows greater wave energy to propagate across reefs to island shorelines promoting geomorphic change in island shorelines as suggested by Sheppard *et al.* (2005).

While there are multiple anthropogenic impacts on reefs and reef geomorphic systems, of greatest relevance in understanding future landform responses is the degree to which

TABLE 7.3. *Anthropogenic and natural mechanisms affecting coral reef structure and sedimentary landform morphodynamics*

Alterations to controlling factors of landform stability	Natural causes	Anthropogenic causes
Impacts on reef structure		
(i) Changes in reef ecology that shift the carbonate budget of reef platforms	• Global sea level change • Decadal–inter-annual variations in sea level (IPO, ENSO) • Frequency intensity and location of storms • Patterns of reef growth • Disease outbreaks	• Increased nutrient loads • Increased sedimentation • Over-fishing, dynamite fishing • Reef mining • Tourism impacts
(ii) Direct physical impacts on reef structure	• Storms, tsunami	• Reef blasting • Channel construction and dredging • Reef mining
Impacts on reef sedimentary landforms		
Change in hydrodynamic processes		
(iii) Alteration in wave processes (energy and direction): manifest as changes in water depth across reef surface that influence wave energy on reef platform	• Global sea level change • Decadal–inter-annual variations in sea level (IPO, ENSO) • Climate-driven changes in angle of wave propagation • Frequency, intensity and location of storms	• Channel dredging • Reef blasting • Coral mining • Engineering structures at shoreline or on reef (groynes, breakwaters) • Boat wakes
(iv) Change in reef platform current patterns (circulation, flushing)	• Variations in wave processes • Modifications to reef bathymetry	• Causeway construction • Reclamation • Harbour construction
Change in sediment budget		
(i) Alterations in sediment generation	• Changes in reef ecology/ productivity • Ecological perturbations (predator explosions, bleaching)	• Pollution of reef, decreasing productivity • Over-fishing • Insertion of shore perpendicular structures
(ii) Change in littoral reef sediment transport	• Storm magnitude and frequency • Formation of littoral barriers (e.g. beachrock development)	• Scouring of shoreline due to stormwater discharge
(iii) Adjustment in sediment volume at shoreline	• Altered current processes • Redistribution of sediment through oceanographic processes, overwash	• Removal of vegetation • Coral mining • Sand extraction or dredging

anthropogenic activities have compromised the eco-morphodynamic state of reef systems. This provides critical insights into the 'start-up' conditions which confront reefs in responding to future global change and the potential for recovery of morphodynamic processes. At one end of the spectrum there are clearly some reef systems in which the anthropogenic footprint is so great that natural morphodynamic processes have been entirely compromised. For example, the entire perimeter of Male island, Maldives is encircled with hard engineering structures and artificial harbours (Fig. 7.11h). In such settings morphodynamic adjustment is not possible and adaptation strategies will rely solely on continued engineering solutions. At the other extreme are reef systems with little anthropogenic impacts and in which morphodynamic integrity has been preserved (Fig. 7.11). In such systems morphodynamic processes are expected to function and allow morphological changes in reef structure and sedimentary landforms to occur in the near future.

7.6 Synthesis

To date management/adaptation responses to the geomorphic consequences on reef landforms of climate change have been constrained by bleak projections of reef ecosystem collapse and shoreline destruction. However, consideration of the eco-morphodynamics of reef systems outlined in this chapter (Fig. 7.2) provides a new perspective on the future trajectories of reef landforms that differ from conventional prognoses of deterioration in geomorphic condition. Of importance in determining different reef landform trajectories in the context of reef landform management is an understanding of the style and magnitude of morphodynamic responses that can be expected over the coming century; the environmental 'start-up' conditions for individual reefs; and the degree to which morphodynamic processes have been compromised (Sections 7.3–7.5).

Based on detailed assessment of the dynamics of the carbonate factory, reef structure and sedimentary landforms, Table 7.4 presents likely future reef landform responses to global environmental change. A number of scenarios are presented that reflect a range of possible 'start-up' conditions and statements of probability are assessed in light of the best current estimates of future reef ecological condition. The trajectories of landform response highlight a number of important features of future landform dynamics that are pertinent to system management and relevant to questions posed at the start of this chapter.

7.6.1 Morphological trajectories for coral reefs and reef sedimentary landforms

Most coral reefs will persist as geomorphic structures through the next century. However, there is no doubt that reefs will undergo morphological adjustment over the coming century, in response to both climatic and anthropogenic forcing. Analysis of historical geomorphic performance and contemporary morphodynamics provides useful analogues for predicting future reef landform responses (Section 7.3) and indicates a greater degree of geomorphic resilience of most reef systems than implied from interpretations based solely on expected near-future ecological condition.

Here we return to a question posed earlier in this chapter: will reefs accelerate their growth performance toward their geological potential to match the rates of sea level change predicted over the next century? There are some anthropogenically stressed reef systems that are severely compromised in terms of both ecological condition and integrity of morphodynamic processes (Fig. 7.11h). Accelerated reef growth is unlikely in these circumstances, resulting in submergence of the reef platform (opening an energy window) over the next century. In the worst-case scenario such reefs may switch to export dominated budgetary states (Section 7.2) and as a consequence of increased bioerosion (and reduced integrity of coral substrate) lower the existing reef elevation ($<10^0$ m). The net result will be increased submergence of the reef platform by the end of the twenty-first century. In contrast, less impacted reef systems or those where net carbonate production potential is unaltered are likely to exhibit catch-up growth responses. Initial submergence may occur as reefs re-establish vertical growth (compare Fig. 7.11e, f); the degree of submergence being dependant on the time lag before reef accretion either initiates or accelerates. Consequently, emergent and broad Indo-Pacific reef platforms are expected to take considerably longer to accelerate their growth performance compared with those reefs that have only recently achieved their vertical growth limit. The longer-term prognosis for these reefs is closure of the energy window. However, given existing scenarios for the ecological decline of reefs it is difficult to envisage many reef systems maintaining a keep-up growth strategy over the coming century.

The effects of ocean temperature and acidification changes are expected to produce a range of second-order geomorphic responses of coral reefs that include: reduced growth rates; shifts in carbonate budget status; an increase in rocky and rubble surfaces; and reduced structural integrity of reefs (loss of three-dimensional structure). Collectively these scenarios indicate a geographically variable geomorphic response of reefs contingent upon the antecedent geomorphic condition of, and environmental start-up conditions that affect, reefs.

Reef sedimentary landforms will also persist throughout the twenty-first century but are expected to undergo morphological adjustment. Of note, beaches and the marginal seaward zones of islands and coastlines are expected to undergo remobilisation in the next century. This may result in island migration and island narrowing for reef islands and narrowing of the coastal plain in fringing reef settings. Importantly, the vegetated core of islands is not expected to undergo significant modification.

While the styles of reef landform responses are relatively well constrained (Figs. 7.7 and 7.14), the magnitude of change is less well defined. For coral reefs, maximum expected changes in surface elevation due to accretion or destruction are likely to be of the order of ±0.3 m over the next century. However, the sensitivity of sedimentary landforms to sea level change may impose readjustments on the order of $\pm10^1$ m (e.g. Kench and Cowell, 2002), whereas changes in sediment supply, littoral gradients and natural variability may impose larger alterations in shorelines of $\pm10^2$ m (Rooney and Fletcher, 2005). While these magnitudes provide first-order

TABLE 7.4. *Future trajectories of reef ecology and reef landforms adopting IPCC (2007) time steps of sea level and climate change*

Component of reef system	>2050	2100	>2100	Likelihood statement
Reef ecology (IPCC, 2007)	Annual or biannual exceedance of bleaching thresholds (30% decline in coral cover). Beginning of phase switch to algal dominance on Great Barrier Reef	By 2070 reefs reach critical aragonite saturation states resulting in reduced coral cover and greater erosion of reef frameworks. Phase switch completed. Shallow-water corals rare on many tropical coasts	Shallow-water corals increasingly rare on reefs Globally, patchy mosaic of living coral communities	Likely, but depends on coral–zooxanthellae adaptations
Reef structure				
Keep-up	Recolonisation of reef surfaces and transition to production-dominated reef	Reef growth maintains relationship to sea level	Reef growth maintains relationship to sea level	Unlikely under reef ecology scenario
Catch-up	Delayed growth response due to anthropogenic stress and lag to recolonise reef surfaces (Pacific reefs). Increased water depth over reefs of 0.1–0.2 m	Accelerated growth of reef platform to production-dominated status. Closure of water depth window over reef surface	Reef achieves vertical growth limit and maintains relationship to sea level	Likely in reef systems with optimal start-up conditions
Non-responsive	No recolonisation Switch to export-dominated reef system or to condition of stasis Sea level rise controlled increase in water depth across reefs (0.1–0.2 m)	Erosion of reef framework Detection of lowering of reef surface by 2100 (10^0 m) Compounds sea level controlled water depth window across reef (up to 0.6 m depth)	Further deterioration of reef surface and increased submergence of reef surface (up to 1.0 m)	Likely in anthropogenically stressed reef systems
Sedimentary landforms				
Unmodified shorelines (with catch-up reef response)	In response to increased water depth window across reefs: triggers overwash responses, island migration and coastal erosion (0–10 m shoreline displacement)	As reef catches up to sea level: stabilisation of island in new equilibrium position on platforms. Reduction in shoreline erosion on broader islands and fringing reef coasts	Little geomorphic change as consequence of sea level change	Likely; geomorphic services preserved

Unmodified shorelines (with no reef response)	In response to increased water depth window across reefs: triggers overwash responses, island migration and coastal erosion (0–10 m shoreline displacement)	As energy window increases across reefs morphological adjustments (Fig. 7.14) in the range of 5–20 m shoreline displacement	Accelerated erosion and instability of reef islands (>20 m shoreline displacement). Possible loss of smallest islands according to accommodation space constraints	Likely in anthropogenically stressed environments
Unmodified shorelines (with positive sediment supply)	Increased sediment supply buffers islands from instability and erosion Normal morphodynamic processes prevail with overwash sedimentation and possible progradation	Normal morphodynamic processes prevail with overwash sedimentation and possible progradation	Normal morphodynamic processes prevail with overwash sedimentation and possible progradation	Possible where sediment carbonate factory yields maintained or sediment supply is increased
Unmodified shorelines (with negative shift in sediment budget)	Depleted sediment supply combined with sea level change exacerbates island instability Morphodynamic processes promote rapid migration of small islands, hotspots of erosion and deposition on broader islands (5–50 m shoreline displacement)	Continued rapid migration of islands. Some small islands lost as sediment reservoirs fall below critical threshold for accommodation space. Significant redistribution of sediments on broader islands. Shoreline displacement up to 100 m	Continued rapid migration of islands. Small islands lost as sediment reservoirs fall below critical threshold for accommodation space. Chronic erosion hotspots on broader shorelines. Shoreline displacement up to 100 m	Possible where sediment carbonate factory and littoral gradients reduce sediment supply to the coast
Modified shorelines	Natural morphodynamics compromised. Increased erosion and island instability. Increased exposure to flooding	Possible loss of smaller islands. Increased erosion and island instability	Increased erosion and island instability	Likely in highly impacted and engineered shoreline settings

estimates to inform adaptation strategies, refinement of such uncertainties is an urgent research priority.

7.6.2 Varied timescales of geomorphic response and non-linearities

The projections of landform change (Table 7.4) and analysis of reef system eco-morphodynamics highlight a mismatch in timescales of adjustment between reef health and geomorphic response as they relate to reef growth, sediment production and shoreline dynamics. For example, scenarios of decline in reef health over years to decades are typically used to infer deleterious consequences for reef geomorphology. However, the reef structure typically (with acknowledged exceptions) responds at centennial to millennial timescales. Consequently, there is a delayed response for any short-term change in reef health to propagate through the geomorphic system and be expressed as reef adjustment of $\pm10^0$ m (Fig. 7.2). The expected changes in reef ecology imply structural changes in reef platforms that are at the furthest temporal boundary of conventional climate change projections (century scale). Furthermore, the timescale for reef adjustment is not relevant to sedimentary landforms, which exhibit changes one or two orders of magnitude greater than reef structure ($\pm10^1$–10^2 m) at timescales ranging from the event scale to decades.

The relationship between reef ecological condition and sedimentary landform dynamics is also non-linear. For example, negative shifts in reef health through storm impacts, bleaching or anthropogenic activities may paradoxically provide positive geomorphic benefits for sedimentary landforms. Such landforms require an abundant supply of sediment that can be generated either through an increase in carbonate production or an increase in substrate degradation (Section 7.2). Periodic perturbations in reef ecological condition and storm impacts and bleaching episodes may provide the stimulus for pulses of sediment to be released to sedimentary landforms. Furthermore, whereas the ecological response may occur within a season, the conversion of skeletal framework to sediment may occur over years to decades.

7.6.3 Impact on geomorphic services

Predicted adjustments in reef landforms will modify the geomorphic services afforded by reef systems. However, the rate of change in services is closely linked to the timescales of landform adjustment. Consequently, at the multi-decadal to century timescale reef structure will provide comparable geomorphic services in terms of its ability to filter incident oceanic swell and provision of a basement for sediment accumulation. The rate of loss of protective services will accelerate toward the end of the century as bioerosion coupled with sea level rise acts to increase water depth across reefs. The net effect of such changes will be to reopen the energy window across reefs that stimulates geomorphic processes affecting reef ecology and sedimentary shorelines. Consequently, sedimentary landforms may experience a further, delayed, impact of global environmental change as the effect of reef structural change propagates through the geomorphic system.

Under the most extreme scenarios increased destabilisation of sedimentary shorelines and loss of some islands will totally remove provisioning services afforded by physical landforms. However, under more moderate scenarios such services will remain but in altered condition. Thus, changes in the seaward margins of sedimentary landforms, island migration, overwash sedimentation and shoreline erosion of the order of 10^1 m all present unique adaptation challenges to coastal communities but the physical foundation for communities should endure over long timescales.

These suggested modifications in geomorphic services are more subtle than the dramatic loss of ecological services associated with reef ecological decline. This further underscores the point that the ecological condition of coral reefs serves as only a limited guide to the geomorphic performance and future trajectory of reef landforms. Scientific assessment of the eco-morphodynamics of reef systems indicates coral reefs are geomorphically resilient features. While future geomorphic changes are expected to be spatially highly variable and occur across a wide range of timescales, both reef structures and associated sedimentary landforms are expected to persist beyond the twenty-first century.

References

Aronson, R. B. and Precht, W. F. (2001). White-band disease and the changing face of Caribbean coral reefs. *Hydrobiologia*, **460**, 25–38.

Baines, G. B. K. and McLean, R. F. (1976). Sequential studies of hurricane bank evolution at Funafuti atoll. *Marine Geology*, **21**, M1–M8.

Baker, A. C. *et al.* (2004). Corals' adaptive response to climate change. *Nature*, **430**, 741.

Barnett, J. and Adger, W. N. (2003). Climate dangers and atoll countries. *Climatic Change*, **61**, 321–337.

Bayliss-Smith, T. P. (1988). The role of hurricanes in the development of reef islands, Ontong Java Atoll, Solomon Islands. *Geographical Journal*, **154**, 377–391.

Beaman, R. J., Webster, J. M. and Wust, R. A. J. (2008). New evidence for drowned shelf edge reefs in the Great Barrier Reef, Australia. *Marine Geology*, **247**, 17–34.

Benzoni, F., Bianchi, C. N. and Morri, C. (2003). Coral communities of the northwestern Gulf of Aden (Yemen): variation in framework building related to environmental factors and biotic conditions. *Coral Reefs*, **22**, 475–484.

Berkelmans, R. *et al.* (2004). A comparison of the 1998 and 2002 coral bleaching events on the Great Barrier Reef: spatial correlation and predictions. *Coral Reefs*, **23**, 75–83.

Best, B. and Bornbusch, A. (2005). *Global Trade and Consumer Choices: Coral Reefs in Crisis*. Washington, DC: American Association for the Advancement of Science.

Blanchon, P. and Blakeway, D. (2003). Are catch-up reefs an artefact of coring? *Sedimentology*, **50**, 1271–1282.

Blanchon, P., Jones, B. and Kalbfleisch, W. (1997). Anatomy of a fringing reef around Grand Cayman: storm rubble not coral framework. *Journal of Sedimentary Research*, **67**, 1–16.

Brown, B. E. and Dunne, R. P. (1988). The impact of coral mining on coral reefs in the Maldives. *Environmental Conservation*, **15**, 159–165.

Brown, B. E. *et al.* (2002). Experience shapes the susceptibility of a reef coral to bleaching. *Coral Reefs*, **21**, 119–126.

Bruggemann, J. H. *et al.* (1996). Bioerosion and sediment ingestion by the Caribbean parrotfish *Scarus vetula* and *Sparisoma viride*: implications of fish size, feeding mode and habitat use. *Marine Ecology Progress Series*, **134**, 59–71.

Bruno, J. F. and Selig, E. R. (2007). Regional decline of coral cover in the Indo-Pacific: timing, extent, and subregional comparisons. *PLoS ONE*, **2**, e711.

Buddemeier, R. W. and Hopley, D. (1988). Turn-ons and turn-offs: causes and mechanisms of the initiation and termination of coral reef growth. *Proceedings of the 6th International Coral Reef Symposium*, Townsville, **1**, 253–261.

Buddemeier, R. W., Kleypas, J. A. and Aronson, R. B. (2004). *Coral Reefs and Global Climate Change: Potential Contributions of Climate Change to Stresses on Coral Reef Ecosystems*. Arlington: Pew Center on Global Climate Change.

Burke, L., Selig, E. and Spalding, M. (2002). *Reefs at Risk in Southeast Asia*. Washington, DC: World Resources Institute.

Calhoun, R. S. and Fletcher, C. H. III (1996). Late Holocene coastal plain stratigraphy and sea-level history at Hanalei, Kauai, Hawaiian Islands. *Quaternary Research*, **45**, 47–58.

Camoin, G. F., Montaggioni, L. F. and Braithwaite, C. J. R. (2004). Late glacial to post glacial sea levels in the Western Indian Ocean. *Marine Geology*, **206**, 119–146.

Camoin, G. F. *et al.* (2007). *Proceedings of the Integrated Ocean Drilling Program,* vol. 310, *Expedition Reports Tahiti Sea Level*. Washington, DC: Integrated Ocean Drilling Program Management International Inc.

Chivas, A. *et al.* (1986). Radiocarbon evidence for the timing and rate of island development, beach-rock formation and phosphitization at Lady Elliot Island, Queensland, Australia. *Marine Geology*, **69**, 273–287.

Cho, L. L. and Woodley, J. D. (2002). Recovery of reefs at Discovery Bay, Jamaica and the role of *Diadema antillarum*. *Proceedings of the 9th International Coral Reef Symposium*, Bali, **1**, 331–338.

Cooper, T. F. *et al.* (2008). Declining coral calcification in massive *Porites* in two nearshore regions of the northern Great Barrier Reef. *Global Change Biology*, **14**, 529–538.

Cowell, P. J. and Thom, B. G. (1994). Morphodynamics of coastal evolution. In R. W. G. Carter and C. D. Woodroffe, eds., *Coastal Evolution: Late Quaternary Shoreline Morphodynamics*. Cambridge: Cambridge University Press, pp. 33–86.

Cowell, P. J. and Kench, P. S. (2002). The morphological response of atoll islands to sea-level rise: 1. Modifications to the shoreface translation model. *Journal of Coastal Research*, ICS **2000**, 633–644.

Davies, P. J. and Hopley, D. (1983). Growth facies and growth rates of Holocene reefs in the Great Barrier Reef. *BMR Journal of Australian Geology and Geophysics*, **8**, 237–251.

Dean, R. G. (1991). Equilibrium beach profiles: characteristics and applications. *Journal of Coastal Research*, **7**, 53–84.

Diaz-Pulido, G. *et al.* (2007). Vulnerability of macroalgae of the Great Barrier Reef to climate change. In J. E. Johnson and P. Marshall, eds., *Climate Change and the Great Barrier Reef: A Vulnerability Assessment*. Townsville: Great Barrier Reef Marine Park Authority/Australia Greenhouse Office, pp. 153–192.

Dickinson, W. R. (1999). Holocene sea level record on Funafuti and potential impact of global warming on Central Pacific atolls. *Quaternary Research*, **51**, 124–132.

Done, T. J. (1992). Phase shifts in coral reef communities and their ecological significance. *Hydrobiologia*, **247**, 121–132.

Douglas, A. E. (2003). Coral bleaching: how and why? *Marine Pollution Bulletin*, **46**, 385–392.

Dunbar, G. B. and Dickens, G. R. (2003). Massive siliciclastic discharge to slopes of the Great Barrier Reef platform during sea-level transgression: constraint from sediment cores between 15 °S and 16 °S latitude and possible explanations. *Sedimentary Geology*, **162**, 141–158.

Eakin, C. (1996). Where have all the carbonates gone? A model comparison of calcium carbonate budgets before and after the 1982–1983 El Niño at Uva Island in the eastern Pacific. *Coral Reefs*, **15**, 109–119.

Edinger, E. N. *et al.* (2000). Normal coral growth rates on dying reefs: are coral growth rates good indicators of reef health? *Marine Pollution Bulletin*, **40**, 606–617.

Edmunds, P. J. and Bruno, J. F. (1996). The importance of sampling scale in ecology: kilometer-wide variation in coral reef communities. *Marine Ecology Progress Series*, **143**, 165–171.

Flood, P. G. (1986). Sensitivity of coral cays to climatic variations, southern Great Barrier Reef, Australia. *Coral Reefs*, **5**, 13–18.

Gardner, T. A. *et al.* (2003). Long-term region-wide declines in Caribbean corals. *Science*, **301**, 958–960.

Glynn, P. W. (1993). Coral reef bleaching: ecological perspectives. *Coral Reefs*, **12**, 1–17.

Goreau, T. F. (1959). The ecology of Jamaican coral reefs: 1. Species composition and zonation. *Ecology*, **40**, 67–90.

Gourlay, M. R. (1988). Coral cays: products of wave action and geological processes in a biogenic environment. *Proceedings of the 6th International Coral Reef Symposium*, Townsville, **2**, 491–496.

Grigg, R. W. (1982). Darwin Point: a threshold for atoll formation. *Coral Reefs*, **1**, 29–34.

Guinotte, J. M., Buddemeier, R. W. and Kleypas, J. A. (2003). Future coral reef habitat marginality: temporal and spatial effects of climate change in the Pacific basin. *Coral Reefs*, **22**, 551–558.

Guzman, H. M. and Cortés, J. (2007). Reef recovery 20 years after the 1982–1983 El Niño massive mortality. *Marine Biology*, **151**, 401–411.

Halfar, J. *et al.* (2004). Nutrient and temperature controls on modern carbonate production: an example from the Gulf of California, Mexico. *Geology*, **32**, 213–216.

Harley, C. D. G. *et al.* (2006). The impacts of climate change in coastal marine systems. *Ecology Letters*, **9**, 228–241.

Harmelin-Vivien, M. L. and Laboute, P. (1986). Catastrophic impact of hurricanes on atoll outer slopes in the Tuamotu (French Polynesia). *Coral Reefs*, **5**, 55–62.

Harney, J. N. and Fletcher, C. H. III (2003). A budget of carbonate framework and sediment production, Kailua Bay, Oahu, Hawaii. *Journal of Sedimentary Research*, **73**, 856–868.

Harney, J. N. *et al.* (1999). Standing crop and sediment production of reef-dwelling foraminifera on O'ahu, Hawaii. *Pacific Science*, **53**, 61–73.

Harney, J. N. *et al.* (2000). Age and composition of carbonate shoreface sediments, Kailua Bay, Oahu, Hawaii. *Coral Reefs*, **19**, 141–154.

Hart, D. E. and Kench, P. S. (2007). Carbonate production of an emergent reef platform, Warraber Island, Torres Strait, Australia. *Coral Reefs*, **26**, 53–68.

Hayne, M. and Chappell, J. (2001). Cyclone frequency during the last 5000 years at Curacao Island, north Queensland, Australia. *Palaeogeography, Palaeoclimatology, Palaeoecology*, **168**, 207–219.

Hoegh-Guldberg, O. (1999). Climate change, coral bleaching and the future of the world's coral reefs. *Australian Journal of Marine and Freshwater Research*, **50**, 839–866.

Hoegh-Guldberg, O. *et al.* (2007). Vulnerability of reef-building corals on the Great Barrier reef to climate change. In J. E. Johnson and P. Marshall, eds., *Climate Change and the Great Barrier Reef: A Vulnerability Assessment*. Townsville: Great Barrier Reef Marine Park Authority/Australia Greenhouse Office, pp. 271–307.

Hopley, D. (1984). The Holocene 'high energy' window on the central Great Barrier Reef. In B. G. Thom, ed., *Coastal Geomorphology in Australia*. Sydney: Academic Press, pp. 135–150.

Hopley, D., Smithers, S. G. and Parnell, K. E. (2007). *The Geomorphology of the Great Barrier Reef: Development, Diversity and Change*. Cambridge: Cambridge University Press.

Hubbard, D. K. (1997). Reefs as dynamic systems. In C. Birkeland, ed., *Life and Death of Coral Reefs*. New York: Chapman & Hall, pp. 43–67.

Hubbard, D., Miller, A. and Scaturo, D. (1990). Production and cycling of calcium carbonate in a shelf-edge reef system (St. Croix, US Virgin Island): applications to the nature of reef systems in the fossil record. *Journal of Sedimentary Petrology*, **60**, 335–360.

Hughes, T. P. (1994). Catastrophes, phase shifts, and large scale degradation of a Caribbean coral reef. *Science*, **265**, 1547–1551.

Hughes, T. P. *et al.* (2003). Climate change, human impacts, and the resilience of coral reefs. *Science*, **301**, 929–933.

IPCC (2007). *Climate Change 2007: Impacts, Adaptation and Vulnerability. Contribution of Working Group II to the Fourth Assessment Report of the Intergovernmental Panel on Climate Change*. Parry, M. L. *et al.*, eds. Cambridge: Cambridge University Press. (pp. 7–22, Adger, N. *et al.*, eds.; pp. 211–272, Fischlin, A. *et al.*, eds.)

International Society for Reef Studies (2008). *Coral Reefs and Ocean Acidification*, Briefing Paper 5. Available at http://www.fit.edu/isrs/documents/ISRS_BP_ocean_acid_final28jan2008.pdf. Accessed July, 2008.

Jackson, J. B. C. *et al.* (2001). Historical overfishing and the recent collapse of coastal ecosystems. *Science*, **293**, 629–638.

Kahn, T. M. A. *et al.* (2002). Relative sea level changes in the Maldives and vulnerability of land due to abnormal coastal inundation. *Marine Geodesy*, **25**, 133–143.

Kench, P. S. and Brander, R. W. (2006a). Wave processes on coral reef flats: implications for reef geomorphology using Australian case studies. *Journal of Coastal Research*, **2**, 209–223.

Kench, P. S. and Brander, R. W. (2006b). Response of reef island shorelines to seasonal climate oscillations: South Maalhosmadulu atoll, Maldives. *Journal of Geophysical Research*, **111**, FO1001, doi:10.1029/2005JF00323.

Kench, P. S. and Cowell, P. J. (2002). The morphological response of atoll islands to sea-level rise: 2. Application of the modified shoreface translation model (STM). *Journal of Coastal Research*, ICS **2000**, 645–656.

Kench, P. S. and Cowell, P. J. (2003). Variations in sediment production and implications for atoll island stability under rising sea level. *Proceedings of the 9th International Coral Reef Symposium*, Bali, **2**, 1181–1186.

Kench, P. S. McLean, R. F. and Nichol, S. L. (2005). New model of reef-island evolution: Maldives, Indian Ocean. *Geology*, **33**, 145–148.

Kench, P. S. *et al.* (2006a). Wave energy gradients across a Maldivian atoll: implications for island geomorphology. *Geomorphology*, **81**, 1–17.

Kench, P. S. *et al.* (2006b). Geological effects of tsunami on mid-ocean atoll islands: the Maldives before and after the Sumatran tsunami. *Geology*, **34**, 177–180.

Kench, P. S. *et al.* (2007). Tsunami as agents of geomorphic change in mid ocean reef islands. *Geomorphology*, **95**, 361–383.

Kennedy, D. M. and Woodroffe, C. D. (2002). Fringing reef growth and morphology: a review. *Earth Science Reviews*, **57**, 255–277.

Kinsey, D. W. (1983). Standards of performance in primary production and carbon turnover. In D. J. Barnes, ed., *Perspectives on Coral Reefs*. Manuka/Townsville: B. Clouston/AIMS, pp. 209–220.

Kinsey, D. W. and Hopley, D. (1991). The significance of coral reefs as global carbon sinks: response to greenhouse. *Palaeogeography, Palaeoclimatology, Palaeoecology*, **89**, 363–377.

Kinzie, R. A. III and Buddemeier, R. W. (1996). Reefs happen. *Global Change Biology*, **2**, 479–494.

Kleypas, J. and Langdon, C. (2006). Coral reefs and changing seawater chemistry. In J. T. Phinney *et al.*, eds., *Coral Reefs and Climate Change*, Washington, DC: American Geophysical Union, pp. 73–110.

Kleypas, J., McManus, J. and Menez, L. (1999). Environmental limits to coral reef development: where do we draw the line? *American Zoologist*, **39**, 146–159.

Kleypas, J., Buddemeier, R. W. and Gattuso, J. P. (2001). The future of coral reefs in an age of global change. *International Journal of Earth Sciences*, **90**, 426–437.

Kraft, J. C. (1982). Terrigenous and carbonate clastic facies in a transgressive sequence over volcanic terrain. *American Association of Petroleum Geologists Bulletin*, **66**, 589.

Land, L. (1979). The fate of reef-derived sediment on the north Jamaican island slope. *Marine Geology*, **29**, 55–71.

Lees, A. (1975). Possible influence of salinity and temperature on modern shelf carbonate sedimentation. *Marine Geology*, **19**, 159–198.

Lewis, J. (2002). Evidence from aerial photography of structural loss of coral reefs at Barbados, West Indies. *Coral Reefs*, **21**, 49–56.

Liddell, W. D. and Ohlhorst, S. L. (1993). Ten years of disturbance and change on a Jamaican fringing reef. *Proceedings of the 7th International Coral Reef Symposium*, Guam, **1**, 144–150.

Lidz, B. H. and Hallock, P. (2000). Sedimentary petrology of a declining reef ecosystem, Florida reef tract (USA). *Journal of Coastal Research*, **16**, 675–697.

Lough, J. M. and Barnes, D. J. (2000). Environmental controls on growth of the massive coral *Porites*. *Journal of Experimental Marine Biology and Ecology*, **245**, 225–243.

McClanahan, T. R. *et al.* (2005). Effects of geography, taxa, water flow, and temperature variation on coral bleaching intensity in Mauritius. *Marine Ecology Progress Series*, **298**, 131–142.

McClanahan, T. R. *et al.* (2007). Predictability of coral bleaching from synoptic satellite and in situ temperature observations. *Coral Reefs*, **26**, 695–701.

McLean, R. F. and Hosking, P. L. (1991). Geomorphology of reef islands and atoll motu in Tuvalu. *South Pacific Journal of Natural Science*, **11**, 167–189.

McLean, R. F. and Woodroffe, C. D. (1994). Coral atolls. In R. W. G. Carter and C. D. Woodroffe, eds., *Coastal Evolution: Late Quaternary Shoreline Morphodynamics*, Cambridge: Cambridge University Press, pp. 267–302.

Mallela, J. and Perry, C. T. (2007). Calcium carbonate budgets for two coral reefs affected by different terrestrial runoff regimes, Rio Bueno, Jamaica. *Coral Reefs*, **26**, 129–145.

Maragos, J. E., Baines, G. B. K. and Beveridge, P. J. (1973). Tropical cyclone creates a new land formation on Funafuti atoll. *Science*, **181**, 1161–1164.

Marshall, P. A. and Baird, A. H. (2000). Bleaching of corals on the Great Barrier Reef: differing susceptibilities among taxa. *Coral Reefs*, **19**, 155–163.

Millennium Ecosystem Assessment (2005). *Ecosystems and Human Well-Being: Synthesis*. Washington, DC: Island Press.

Montaggioni, L. F. (2005). History of Indo-Pacific coral reef systems since the last glaciation: development patterns and controlling factors. *Earth Science Reviews*, **71**, 1–75.

Neumann, A. C. and Macintyre, I. (1985). Reef response to sea level rise: keep-up, catch-up or give-up. *Proceedings of the 5th International Coral Reef Congress*, Tahiti, 105–110.

Ohde, S. *et al.* (2002). The chronology of Funafuti Atoll: revisiting an old friend. *Philosophical Transactions of the Royal Society of London A*, **458**, 2289–2306.

Orr, J. C. *et al.* (2005). Anthropogenic ocean acidification over the twenty-first century and its impact on calcifying organisms. *Nature*, **437**, 681–686.

Pandolfi, J. M. *et al.* (2003). Global trajectories of the long-term decline of coral reef ecosystems. *Science*, **301**, 955–958.

Perry, C. T. (1996). The rapid response of reef sediments to changes in community structure: implications for time-averaging and sediment accumulation. *Journal of Sedimentary Research*, **66**, 459–467.

Perry, C. T. (2007). Tropical coastal environments: coral reefs and mangroves. In C. T. Perry and K. G. Taylor, eds., *Environmental Sedimentology*. Oxford: Blackwell, pp. 302–350.

Perry, C. T. and Hepburn, L. J. (2008). Syn-depositional alteration of coral reef framework through bioerosion, encrustation and cementation: taphonomic signatures of reef accretion and reef depositional events. *Earth Science Reviews*, **86**, 106–144.

Perry, C. T. and Larcombe, P. (2003). Marginal and non-reef building coral environments. *Coral Reefs*, **22**, 427–432.

Perry, C. T., Taylor, K. G. and Machent, P. (2006). Temporal shifts in reef lagoon sediment composition, Discovery Bay, Jamaica. *Estuarine, Coastal and Shelf Science*, **67**, 133–144.

Perry, C. T., Spencer, T. and Kench, P. S. (2008). Carbonate budgets and reef production states: a geomorphic perspective on the ecological phase-shift concept. *Coral Reefs*, **27**, 853–866.

Riegl, B. and Piller, W. (2000). Reefs and coral carpets in the northern Red Sea as models for organism-environment feedback in coral communities and its reflection in growth fabrics. *Geological Society of London Special Publications*, **178**, 71–88.

Rooney, J. J. B., and Fletcher, C. H. III (2005). Shoreline change and Pacific climate oscillations in Kihei, Maui, Hawaii. *Journal of Coastal Research*, **21**, 535–547.

Royal Society (2005). *Ocean Acidification due to Increasing Atmospheric Carbon Dioxide*, Policy Document No. 12/05. London: The Royal Society.

Schofield, J. C. (1977). Effect of late Holocene sea-level fall on atoll development. *New Zealand Journal of Geology and Geophysics*, **20**, 531–536.

Scoffin, T. (1992). Taphonomy of coral reefs: a review. *Coral Reefs*, **11**, 57–77.

Scoffin, T. (1993). The geological effects of hurricanes on coral reefs and the interpretation of storm deposits. *Coral Reefs*, **12**, 203–21.

Scoffin, T. P. *et al.* (1980). Calcium carbonate budget of a fringing reef on the west coast of Barbados: 1. erosion, sediments and internal structure. *Bulletin of Marine Science*, **30**, 475–508.

Sheppard, C. R. C. (2003). Predicted recurrences of mass coral mortality in the Indian Ocean. *Nature*, **425**, 294–297.

Sheppard, C. *et al.* (2005). Coral mortality increases wave energy reaching shores protected by reef flats: examples from Seychelles. *Estuarine Coastal and Shelf Science*, **64**, 223–234.

Smithers, S. G. and Larcombe, P. (2003). Late Holocene initiation and growth of a nearshore turbid-zone coral reef: Paluma Shoals, central Great Barrier Reef, Australia. *Coral Reefs*, **22**, 499–505.

Smithers, S. G., Hopley, D. and Parnell, K. E. (2006). Fringing and nearshore coral reefs of the Great Barrier Reef: episodic Holocene development and future prospects. *Journal of Coastal Research*, **22**, 175–187.

Smithers, S. G. *et al.* (2007). Vulnerability of geomorphological features in the Great Barrier Reef to climate change. In J. E. Johnson and P. A. Marshall, eds., *Climate Change and the Great Barrier Reef*, Townsville: Great Barrier Reef Marine Park Authority/Australian Greenhouse Office, pp. 667–716.

Solomon, S. M. and Forbes, D. L. (1999). Coastal hazards, and associated management issues on South Pacific Islands. *Ocean and Shoreline Management*, **42**, 523–554.

Spalding, M., Ravilious, C. and Green, E. (2001). *World Atlas of Coral Reefs*. Berkeley: University of California Press.

Spencer, T. (1992). Bioerosion and biogeomorphology In D. M. John, S. J. Hawkins and J. H. Price, eds., *Plant–Animal Interactions in the Marine Benthos*. Oxford: Clarendon Press, pp. 492–509.

Spencer, T. (1995). Potentialities, uncertainties and complexities in the response of coral reefs to future sea-level rise. *Earth Surface Processes and Landforms*, **20**, 49–64.

Stoddart, D. R. (1963). Effects of Hurricane Hattie on the British Honduras reefs and cays, October 30–31, 1961. *Atoll Research Bulletin*, **95**, 1–142.

Stoddart, D. R. (1971). Environment and history in Indian Ocean in reef morphology. In D. R. Stoddart and C. M. Yonge, eds., *Regional Variation in Indian Ocean Coral Reefs*. London: Zoological Society, pp. 3–38.

Stoddart, D. R. (1977). Structure and ecology of Caribbean reefs. *FAO Fisheries Report*, **200**, 427–448.

Stoddart, D. R. (1985). Hurricane effects on coral reefs: conclusions. *Proceedings of the 5th International Coral Reef Congress*, Tahiti, **3**, 349–350.

Stoddart, D. R. and Steers, J. A. (1977). The nature and origin of coral reef islands. In O. A. Jones and R. Endean, eds., *Biology and Geology of Coral Reefs, vol. 4, Geology 2*. New York: Academic Press, pp. 59–105.

Stoddart, D. R., Fosberg, F. R. and Spellman, D. L. (1982). Cays of the Belize barrier reef and lagoon. *Atoll Research Bulletin*, **256**, 1–76.

Toscano, M. A. and Macintyre, I. G. (2003). Corrected western Atlantic sea-level curve for the last 11 000 years based on calibrated ^{14}C dates from *Acropora palmata* and mangrove intertidal peat. *Coral Reefs*, **22**, 257–270.

Veron, J. E. N. (1992). Environmental control of Holocene changes to the world's most northern hermatypic coral outcrop. *Pacific Science*, **46**, 405–425.

Verstappen, H. T. (1954). The influence of climatic change on the formation of coral islands. *American Journal of Science*, **252**, 428–435.

Walther, G. R. *et al.* (2002). Ecological responses to recent climate change. *Nature*, **416**, 389–395.

Webster, J. M. *et al.* (2004a). Drowned carbonate platforms in the Huon Gulf, Papua New Guinea. *Geochemistry, Geophysics, Geosystems*, **5**, 1–31.

Webster, J. M. *et al.* (2004b). Drowning of the –150 m reef off Hawaii: a casualty of global meltwater pulse 1A? *Geology*, **32**, 249–252.

Wilkinson, C., ed. (2004). *Status of Coral Reefs of the World: 2004*. Townsville: Australian Institute of Marine Science.

Williams, E. H. Jr and Bunkley-Williams, L. (1990). The world-wide coral reef bleaching cycle and related sources of coral mortality. *Atoll Research Bulletin*, **335**, 1–71.

Woodley, J. D. (1992). The incidence of hurricanes on the north coast of Jamaica since 1870: are the classic reef descriptions atypical? *Hydrobiologia*, **247**, 133–138.

Woodley, J. D. *et al.* (1981). Hurrican Allen's impact on Jamaican coral reefs. *Science*, **214**, 749–755.

Woodroffe, C. D. (2002). *Coasts: Form, Process and Evolution*. Cambridge: Cambridge University Press.

Woodroffe, C. D. (2003). Reef-island sedimentation on Indo-Pacific atolls and platform reefs. *Proceedings of the 9th International Coral Reef Symposium*, Bali, **2**, 1187–1192.

Woodroffe, C. D. (2005). Late Quaternary sea-level highstands in the central and eastern Indian Ocean: a review. *Global and Planetary Change*, **49**, 121–138.

Woodroffe, C. D. (2007). Critical thresholds and the vulnerability of Australian tropical coastal ecosystems to the impacts of climate change. *Journal of Coastal Research*, **SI 50**, 1–5.

Woodroffe, C. D. and McLean, R. F. (1992). *Kiribati Vulnerability to Accelerated Sea-Level Rise: A Preliminary Study*, ADFA Report to Department of Arts, Sports, Environment and Territories. Canberra: Government of Australia.

Woodroffe, C. D. and McLean, R. F. (1993). *Cocos (Keeling) Islands Vulnerability to Sea-Level Rise*, Report to the Climate Change and Environmental Liaison Branch, Department of Arts, Sports, Environment and Territories. Canberra: Government of Australia.

Woodroffe, C. D. and McLean, R. F. (1998). Pleistocene morphology and Holocene emergence of Christmas (Kiritimati) Islands, Pacific Ocean. *Coral Reefs*, **17**, 235–248.

Woodroffe, C. D. and Morrison, R. J. (2001). Reef-island accretion and soil development, Makin Island, Kiribati, central Pacific. *Catena*, **44**, 245–261.

Woodroffe, C. D. *et al.* (1990). Sea level and coral atolls: late Holocene emergence in the Indian Ocean. *Geology*, **18**, 62–66.

Woodroffe, C. D. *et al.* (1999). Atoll reef-island formation and response to sea-level change: West Island, Cocos (Keeling) Islands. *Marine Geology*, **160**, 85–104.

Woodroffe, C. D. *et al.* (2005). Episodes of reef growth at Lord Howe Island, the southernmost reef in the southwest Pacific. *Global and Planetary Change*, **49**, 222–237.

Woodroffe, C. D. *et al.* (2007). Incremental accretion of a sandy reef island over the past 3000 years indicated by component-specific radiocarbon dating. *Geophysical Research Letters*, **34**, L03602, doi:10.1029/2006GL028875.

Yamano, H., Miyajima, T. and Koike, I. (2000). Importance of foraminifera for the formation and maintenance of a coral sand cay: Green Island, the Great Barrier Reef, Australia. *Coral Reefs*, **19**, 51–58.

8 Tropical rainforests

Rory P. D. Walsh and Will H. Blake

8.1 The tropical rainforest ecological and morphoclimatic zone

This chapter focusses on the likely geomorphological impacts of global warming on the zone with a hot–wet climate covered or formerly covered by tropical rainforest. Defined by Schimper (1903, p. 260) as 'Evergreen, hygrophilous in character, at least 30 m high, but usually much taller, rich in thick-stemmed lianes and in woody as well as herbaceous epiphytes', tropical rainforest represents the vegetational climax of the ever-wet tropical zone (Richards, 1996). It is the biome renowned for its very high species diversity, high biological productivity, continuity of existence extending back to the early Cretaceous (Morley, 2000) and its pivotal role (depending upon how much of it survives) in influencing the future of the world climatic system. Climatic diversity within the zone, a history of climate change and the profound influences which tectonic history and lithology exert within the region mean that old simplistic ideas about a humid tropical morphoclimatic region have been largely discounted (Thomas, 1994; 2006). This chapter, therefore, adopts a modified climatic geomorphological approach, which stresses:

(a) the distinctiveness of some climate-linked features of the tropical rainforest zone;
(b) the influence of diversity in geomorphologically important climatic and bioclimatic variables within the zone;
(c) the influence of climate change and sequences of climate at many different timescales; and
(d) the acute sensitivity of many of its processes, landforms and landscapes to combinations of human activities and climate change.

The importance of changes in extremes as well as means of climate is emphasised. This opening section summarises the main features (including diversity) of geomorphic significance of the climates, vegetation and soils of the zone and the long-term history of the forest biome in relation to climate change. The second section reviews geomorphological processes and landforms within the ever-wet tropics, including the impact of human activities, before later sections consider predicted climate change by the end of the twenty-first century for the zone and the likely geomorphological consequences.

8.1.1 Diversity and distribution of tropical rainforest climates

The distinctive unifying (and defining) features of tropical rainforest climates are continuously high temperatures (mean annual temperature >20 °C; coldest month >18 °C), high annual rainfall (at least 1600 mm and usually >2000 mm) and no or only a short dry season (≤4 months with less than 100 mm rain) (Walsh, 1996a). The zone is also frost-free. These criteria, based upon regional boundaries between rainforest and more seasonal or cooler vegetation types, vary with edaphic conditions and assume that boundaries are adjusted to current climate. The use of a threshold 100-mm monthly mean is supported by measurements of rainforest transpiration (e.g. Jordan and Kline, 1977) as well as recent ecological studies (Maslin *et al.*, 2005).

In reality, the zone contains a range of climates of differing annual rainfall (1600 to over 9000 mm), degree of seasonality and extreme-event magnitude–frequency, with important hydrological and geomorphological consequences. Perhumidity index values, which assess the continuity of wetness of the mean rainfall regime, range from +5 at Wet Seasonal locations on the boundary with seasonal forest and savanna types of the Wet–Dry Tropics to the maximum possible value of +24 at some Superwet locations (where mean rainfalls of all months exceed 200 mm) in Malesia (defined as the biogeographical region straddling the

Geomorphology and Global Environmental Change, eds. Olav Slaymaker, Thomas Spencer and Christine Embleton-Hamann. Published by Cambridge University Press.

Indomalaya and Australasia ecozones), New Guinea and the western Amazon Basin (Walsh, 1996a). Surprisingly little of the rainforest zone is truly aseasonal. Although a few locations have never experienced a month with less than 100 mm rain, most experience frequent short dry periods and some a regular dry season (e.g. Barro Colorado and Manaus). A few locations (for example parts of eastern Borneo and Sulawesi), experience occasional droughts of up to 6 months long during major ENSO events (Walsh, 1996a).

Tropical rainforest climates are mainly found (a) within much of the equatorial zone from 10° N to 10° S and (b) between 10° and 23° latitude on the eastern (windward) sides of continents and islands, where onshore easterly trade winds provide rainfall in the winter as well as summer seasons. In western and interior parts of continents and islands, seasonality increases away from the equator and towards the boundary with the Wet–Dry Tropics. On some eastern sides of continents, however, the boundary with Humid Subtropical forests is determined by declining temperature and frost occurrence (as in southern China) and is more disputed. In the Indo-Australian monsoon area, rain shadow effects of mountain ranges lead to tropical wet–dry season climates in parts of the Philippines and Indonesia even within the equatorial zone.

Although rainfall in the rainforest zone is mostly convectional and linked either to the seasonal migration of the Intertropical Convergence Zone or to disturbances in the trade wind, equatorial and monsoon circulations, the mechanisms involved vary significantly between the major rainforest regions of Amazonia, central and western Africa and Malesia. Local enhancement or dampening of convectional activity is caused, respectively, by low-level convergence or divergence of winds.

8.1.2 Tropical rainforest soils and vegetation

Soil characteristics, which exert strong influences on weathering, slope hydrological and erosional processes and slope stability, vary markedly with lithology, tectonics and topography as well as vegetational and climatic factors within the rainforest tropics. In general, the wet tropics are characterised by ferallitic soils with kaolinite clay minerals in moderately high rainfall areas but allitic soils containing gibbsite and alumina-rich residues in the wettest areas (Thomas, 1994). Baillie (1996) divides the soils of the humid tropics into five main groups: kaolisols; non-kaolisol mature *terra firme* soils; immature *terra firme* soils; poorly drained soils; and montane soils. The first four groups are subdivided on edaphic grounds, yielding 21 sub-groups. Soils tend to be deep and characterised by high infiltration capacities and permeabilities. Horizonisation, however, varies greatly with lithology and relief leading to wide variations in the relative importance of different runoff processes (see Section 8.2). The higher shear strengths imparted by some tropical clay minerals to many tropical soils mean that slopes can be stable at higher angles (often >40°) than in temperate areas (Rouse *et al.*, 1986).

Edaphic variations mean that climatic criteria used to delimit boundaries of the rainforest zone are only rough approximations. Forest formations and their constituent trees can respond to increasing seasonality and associated water shortages in two different ways. Where edaphic conditions preclude access to reliable water in the dry season (as in sloping terrain with relatively shallow soils and underlain by hard lithologies), the response is deciduousness, first to *evergreen seasonal forest* (in which a minority of canopy trees are deciduous in the dry season), then to *semi-evergreen seasonal forest* (the majority of canopy trees being deciduous, but the lower forest remaining evergreen) and then *deciduous seasonal forest* (where both canopy and lower trees are mainly deciduous) (Richards, 1996). The boundary between tropical rainforest and wet–dry forest formations is generally placed between evergreen seasonal and semi-evergreen forest, as the transition is readily visible in the dry season. If edaphic conditions are favourable then rainforest can prevail under a marked dry season climate before there is a change to *dry evergreen* (or 'monsoonal forest') formations with longer dry seasons. This alternative transition to dry evergreen forest is marked by a distinct change in species composition. Favourable edaphic conditions include areas of soft and/or permeable rocks where trees are able to access either shallow groundwater in broad valley bottoms or deeper groundwater via tap roots (Nepstad *et al.*, 1994).

8.1.3 Long-term history of the tropical rainforest biome and its distribution in relation to climate change

Until the 1960s, tropical rainforest areas had been thought largely to have escaped significant Quaternary fluctuations in climate. An abundance of lake level, pollen, sediment, biogeographical, offshore ocean core and geomorphological evidence led to a revision of this view (e.g. Flenley, 1979; Richards, 1996; Mahé *et al.*, 2004). Over large parts of the tropics:

(a) the LGM was marked by significantly reduced precipitation, reductions in temperature of around 2–4 °C and replacement of parts of the rainforest by drier forms of vegetation (Plate 1);
(b) parts of the early Holocene were distinctly wetter and warmer than currently with rainforest covering greater areas than now (Plate 2); and

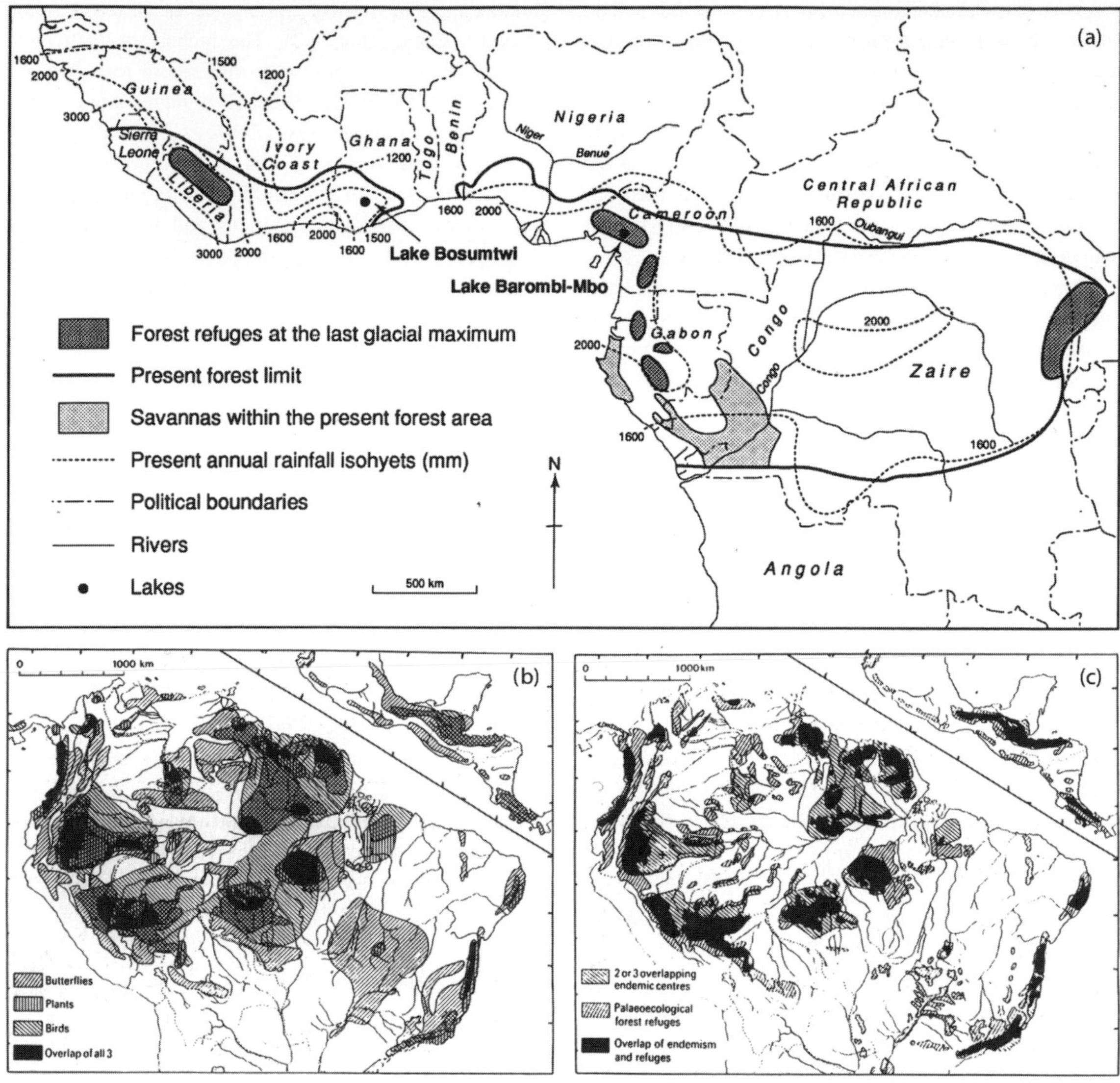

FIGURE 8.1. (a) Refuges for lowland tropical rainforest in tropical Africa during the LGM (after Richards 1996). (b) Pleistocene refuges in tropical America: centres of plant and animal endemism; (c) Overlap of (b) with position of palaeoecological refuges deduced from soils, geomorphology and other evidence (from Prance, 1987).

(c) less wet, somewhat cooler conditions with a contraction of rainforest area commenced by the mid-Holocene (around 5000 BP). Direct and unambiguous evidence from lowland areas of the inner tropics, however, remains poor, mainly because of a paucity of lakes and mires, poor pollen preservation and ambiguities in interpreting tree and grass pollen records. Most lines of evidence have come from extrapolation from tropical highlands, rainforest margins and the adjacent currently Wet–Dry Tropics.

The greatest changes in rainforest extent probably occurred in Africa, reflecting the fact that the climate of much of the area is marginal to rainforest. Figure 8.1 shows how the rainforest is thought to have contracted to just seven enclaves during the LGM (based on lake pollen, biogeographical and geomorphological evidence), with dunes active in part of the Congo Basin. In the early Holocene, in contrast, the rainforest spread beyond its current area with a rise in precipitation to above current levels, before a sharp contraction to its current area around 5000 BP. The arrival of *Homo sapiens sapiens* and fire (favouring savanna types of subclimax vegetation) has obscured the question as to whether rainforest would have later recovered. Changes in the extent and distribution of tropical rainforest in Malesia in the Last Glacial would have reflected not only global climate change, but modifications to regional climates resulting from the conversion of the region from a maritime area of islands and peninsulas into the contiguous continental land mass of 'Sundaland' (Plate 1) as a result of a eustatic lowering of sea level of about 125 m. Some previously (and currently) coastal areas backed by mountains were thereby transformed into rain shadow areas distant from the sea. Pollen evidence (e.g. Morley and Flenley, 1987) suggests that parts of Peninsular Malaysia and southern Borneo were seasonally dry with semideciduous or deciduous forest, whereas northern Borneo and perhaps other smaller areas were refuges where the rainforest survived. In the climatic optimum of the early Holocene, rainforest climates and vegetation are envisaged to have expanded well beyond their current limits into southern China, and Indo-China (Plate 2).

The extent and character of changes in the Amazon Basin continues to be hotly debated. In particular the contraction of rainforests into refugia in drier, more seasonal epochs coinciding with Pleistocene Glacial Maxima (Haffer, 1969; Prance, 1987; Haffer and Prance, 2001) has been questioned (e.g. Colinvaux and de Oliveira, 2000; Maslin *et al.*, 2005). Figure 8.1 summarises evidence for the refugia hypothesis from distributions of endemism of birds, butterflies and plants and the degrees to which they overlap with each other and with the positions of palaeoecological refuges deduced independently from geomorphological, soil and other evidence. Critics (see Maslin *et al.* (2005) for a full discussion) have argued that the palaeoecological evidence for expansion of savanna between the refugia in the core areas of Amazonia is slim with in some cases no evidence of a decline in arboreal pollen. In particular there was no change in the carbon isotopic composition of organic matter received by the Amazon Fan throughout the Last Glacial/Interglacial Cycle, whereas one might have expected a significant change if there had been an increase in C4 (mainly tropical grasses) compared with C3 plants (mainly trees) as the two plant types fractionate carbon differently during photosynthesis (Kastner and Goni, 2003). It has been suggested that lower temperatures (and evapotranspiration rates) in glacial periods may have mitigated the effects of reduced rainfall and lower carbon dioxide and allowed C3 trees to remain competitive with C4 grasses. An alternative explanation for the apparently contradictory findings is that the vegetational change in Amazonia was from evergreen rainforest to either closed canopy semi-evergreen seasonal forest or dry evergreen forest formations rather than to savanna (Pennington *et al.*, 2000). This would account for both continued arboreal pollen as well as a modest increase in grass pollen – and also the marked vegetational changes necessary to account for the refugia evidence.

8.2 Geomorphological characteristics of the rainforest zone: a synthesis

8.2.1 Introduction

The evolution of geomorphological ideas about the tropics has been explored in detail elsewhere (Thomas, 1994, 2006). It is widely agreed that simplistic views that humid tropical climates, even if divided into seasonally wet (see Chapter 9) and ever-wet types, produce distinctive landforms and landscapes are unrealistic (Douglas and Spencer, 1985; Thomas, 2006). This scepticism about a distinctive 'tropical' geomorphology stems from four concerns:

(a) doubts about whether climate, rather than lithology or tectonic history, should be given the premier position in the hierarchy of factors governing geomorphology that climatic geomorphology gives it;
(b) that climatic changes within the humid tropics have been too frequent for many, particularly larger-scale landscape components, to become adjusted and that landforms and landscapes are instead mostly either polygenetic or the products of change itself rather than an individual climate;
(c) that geomorphological diversity within the ever-wet tropics derives not only from differences in non-climatic factors, but from intra-zonal differences in climate and in soil and vegetational characteristics that vary with both climatic and non-climatic factors; and
(d) the continuing inadequacies of process and landscape databases available for testing hypotheses about the zone.

As a result, the utility or validity of any kind of climatic geomorphological approach in the tropics has sometimes been questioned (e.g. Thomas, 1994; 2006). The view adopted here is that a modified climatic geomorphological approach that takes into account the above points is needed when considering the geomorphic impacts of future climate change affecting the rainforest zone. In particular, the approach places emphasis on

(a) the geomorphological significance of some distinctive climatic and bioclimatic features of the tropical rainforest zone;
(b) the significance of diversity in some geomorphologically important climatic and bioclimatic (soil/vegetational) variables within the tropical rainforest zone;
(c) the influence of climate change and sequences of climate at many different timescales; and
(d) the acute sensitivity of many of its processes, landforms and landscapes to combinations of human activities and climate change.

Traditional views about the tropical rainforest zone stressed the importance of chemical over physical denudation (reflected in deep weathered mantles), relatively low erosion rates due to the protective role of the rainforest vegetation, but the potential for greatly enhanced erosion rates with human disturbance. Some of these early ideas have been questioned and revised in recent decades with the spread of process and form studies and the evidence of climate change. Major differences in processes and landforms between territories have also been linked to contrasts in lithology and current and past tectonics. Both current process studies (e.g. Douglas *et al.*, 1999) and longer timescale investigations (e.g. Rodbell *et al.*, 1999) have indicated a key role played by extreme events in the ever-wet tropics in three ways:

(a) geomorphic work is disproportionately concentrated into such events;
(b) key landscape attributes such as drainage density, channel cross-sectional size and slope form appear to be linked to rare rather than frequent events;
(c) extreme events (as opposed to mean events) tend to be the instruments by which climate change largely initiates and achieves landscape change. This emphasis on extreme events forms a key element both of a modern view of humid tropical morphoclimatic environments and assessments of likely geomorphological consequences of predicted climate change in the zone.

Table 8.1 lists some of the key aspects of tropical rainforest climates and bioclimates respectively and the geomorphological and hydrological processes and landscape variables that they influence. It will be referred to in the rest of this section. Features that vary significantly with the climatic or bioclimatic variable in question within the zone are marked (with a superscript *a*). Together they arguably provide elements of a distinctive ever-wet tropical morphoclimatic environment.

Arguably of key geomorphological significance are the high rainfall intensities, high magnitude–frequencies of rainstorms and high record daily rainfall of the rainforest zone compared with most other parts of the world. These are linked to the high water vapour capacity of the hot tropical atmosphere and more active evaporation over warm seas and actively transpiring rainforest. The frequency of large rainstorms and the magnitude of record daily rainfall, however, vary markedly within the rainforest

TABLE 8.1. *Distinctive climatic and bioclimatic features of the geomorphological climatology of tropical rainforest locations and some geomorphological and hydrological variables that they influence*

Features	Geomorphological/hydrological responses
Climatic features	
Perennially hot–wet conditions	High chemical weathering rates; deep weathered mantles; high rates of breakdown of bedload
	Very high transpiration and evapotranspiration
High annual rainfall[a]	Potential for high chemical weathering rates and solute yields[a]
	Enhanced rainsplash[a]
	Moderate/high runoff (streamflow per unit area)[a]
High rainfall intensities	Positive effect on rainsplash and slopewash rates, pipe erosion rates and sediment yields
High rainstorm magnitude–frequency[a]	Enhanced slopewash, pipe erosion and landsliding
	High frequency of competent river flows
	High sediment loads and active fluvial systems
High extreme rainstorm totals[a]	High drainage density[a] and landslide frequency[a] leading to:
	– Finely dissected, high slope-angle landscapes
	– Enhanced sediment yields
Absence of frost	Absence of frost/ice-related processes
Tropical cyclone events[a]	Enhanced drainage density and landslide frequency (in cyclone-prone areas only)
History of climatic change[a]	Large-scale landforms tend to be polygenetic and hence products of sequences of climate (notably periods of landscape disequilibrium and enhanced geomorphic activity during transitions in climate)
	Many landscape elements may be unadjusted to current climate

TABLE 8.1. *(cont.)*

Features	Geomorphological/hydrological responses
Bioclimatic features	
Complex of vegetation–soil factors	High infiltration capacities and water storage capacities of many but not all soils or topsoils, leading to:
	– Muted streamflow responses to rainstorms
High forest productivity and tree-fall frequency	Enhanced debris dam activity, episodic sediment transport, and frequent channel switching (avulsion)
High leaf biomass	Very high transpiration and (with interception) total evapotranspiration
	Low river flow per unit area in marginal rainforest areas, but sharp rate of increase as annual rainfall increases
Leaf characteristics (drip-tips, leathery/waxed surfaces, etc.)	Comparatively low interception (because of reduced leaf storage capacities and enhanced leaf drainage)
Leaf litter cover[a]	Promotion of some Hortonian overland flow
	Protection against excessive slopewash and splash
Shallow root mat[a]	Roofing for piping systems
	Protection against excessive slopewash and splash
	Some Hortonian overland flow via 'thatched roof' effect
	Tight nutrient cycling and reduction of solute losses
	Maximum soil creep below root mat
Absence of tap roots[a]	Increased susceptibility to water shortages
	Reduced anchorage role and enhanced slope instability
High decomposition rates	Enhanced potential for weathering via acids produced
Intense soil faunal activity	Enhanced soil CO_2 levels and weathering potential
	Increased vertical permeability of soils
	Reduction in response times of pipeflow systems
	Possible role in pipe formation and development
Properties of some tropical clay minerals and regoliths	Enhancement of shear strength of soils and slope stability
	Enables slopes to be stable at angles often >40°

[a] Indicates significant spatial variability in the climatic or bioclimatic parameter and/or the geomorphological response *within* the rainforest zone.

zone (Table 8.2), notably in relation to proximity to the sea, topography and aspect (Walsh, 1996a). In addition, some maritime rainforest areas experience tropical cyclones, but frequencies vary and in equatorial and interior continental areas they are absent.

8.2.2 Geomorphological processes in rainforest areas

Data on geomorphological processes in the rainforest zone have expanded greatly in the past two decades. Caution, however, is needed in its interpretation for four reasons:

(a) the short duration of most process assessments (whereas many processes are episodic and dominated by rare extreme events);
(b) the difficulties in placing process information into meaningful spatial and temporal geomorphic contexts (in particular whether the landscape is in equilibrium or adjusting to change);
(c) the unevenness of data availability for different processes; and
(d) the unreliability of data on some processes (notably nutrient fluxes, weathering and solute transport) because of human influences on rainwater quality and atmospheric pollution.

TABLE 8.2. *Variations in the frequency of large daily rainfalls and the magnitude of maximum recorded daily rainfalls within the rainforest zone*

(a) Mean annual frequency of large daily falls

Station	Mean annual rainfall	Annual frequency of days >25.4 mm	Annual frequency of days >76.2 mm
Roseau, Dominica	1922	18	1.4
Danum, Sabah	2833	34	2.7
Mulu, Sarawak	5087	66	8.0
Wet Area, Dominica	5432	70	7.0

(b) Highest recorded daily rainfalls

Station	Record (years)	Annual rainfall (mm)	Highest rainfall (mm)	Type of location
Amazonia				
Uaupes	10	2677	117	Continental interior
Manaus	25	1811	119	Continental interior
Santarem	22	1979	175	Coastal
Peninsular Malaysia				
Penang	48	2736	241	West coast island
Cameron Highlands	26	2644	160	Interior
Kuala Lumpur	19	2441	145	Interior
Kuala Trengganu	15	2921	465	East coast
Borneo				
Kuching	23	4036	414	NW coast, Sarawak
Marudi	43	2716	560	Near N coast, Sarawak
Lio Matu	15	3456	136	Inland, NE Sarawak
Sandakan	94	3101	465	NE coast, Sabah
Danum	23	2833	182	Inland, Sabah
Balikpapan	43	2228	88	E coast, Kalimantan
Pontianak	63	3175	74	W coast, Kalimantan

Source: Modified after Walsh (1996a).

As most geomorphological processes in the rainforest zone are water-related, hydrological processes are first reviewed.

Hydrological processes

Evapotranspiration

A key and distinctive hydrological feature of tropical rainforest areas is very high evapotranspiration (generally in the range 950–1700 mm, but lower on tropical mountains) compared with the rest of the world, reflecting hot–wet conditions prevalent through all or most of the year. Transpiration is considerably greater than interception. Thus in the Manaus area, transpiration (688 mm) was found to be three times greater than interception (228 mm) (Lloyd *et al.*, 1988). Interception losses tend to be relatively low for three reasons:

(a) the smooth, leathery or waxy nature of many rainforest leaves, which reduce the storage capacity and enhance drainage efficiency during rainstorms;
(b) high rainfall intensities, the large droplets of which tend to dislodge water off leaves; and
(c) much of the annual rainfall falls in moderate to large rainstorms, in which much smaller fractions of rainwater are intercepted than in short showers (Jackson, 1975).

Interception tends to be higher in trade wind areas because much of the rain falls in short showers. Transpiration losses tend to fall somewhat in very wet rainforest areas and more markedly in montane forest areas because of reduced solar radiation receipts, fewer sunny days and (in the frequently misty conditions of montane areas) very high relative humidities.

River flow

Mean annual runoff (streamflow per unit area) varies greatly within the rainforest zone. Thus in the catchment data set in Fig. 8.2, annual runoff increases tenfold with a

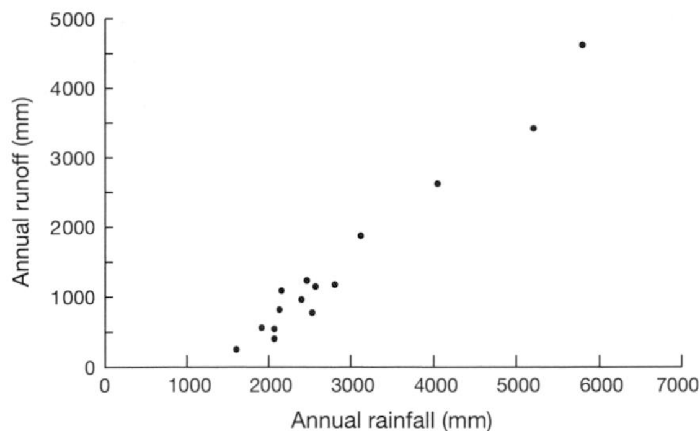

FIGURE 8.2. The relation between annual runoff and annual rainfall for 14 undisturbed tropical rainforest catchments (data from Walsh, 1996b).

2.8-fold increase in mean annual rainfall from 400 mm in the Barro Branco catchment (annual rainfall: 2075 mm) in Amazonia to 4649 mm in the Guma catchment in Sierra Leone (annual rainfall: 5795 mm). Thus changes in annual rainfall will lead to disproportionately large changes in river flow particularly in less wet rainforest areas. The high frequency of large rainstorms by world standards tends to result in a higher frequency of bankfull flows compared with temperate areas. For example, over a 16-month period in 1977–78, bankfull flows in the primary forest Melinau catchment in Sarawak were recorded on six occasions (Walsh, 1982b), considerably more frequently than in the majority of United States and British rivers.

Hillslope runoff processes

Views on slope runoff processes need some reconsideration in the light of recent studies. Earlier schematic models and reviews (e.g. Walsh, 1980; Douglas and Spencer, 1985) linked diversity in storm runoff processes (and in particular the spatial extent and frequency of saturation overland flow) within rainforest areas to contrasts in subsoil permeability and topsoil porosity and depth related to climatic, lithological and topographical factors. Where a soil horizon of low saturated permeability occurs at 20 cm or less beneath the surface, as found in Queensland (Bonell and Gilmour, 1978) and western Amazonia (Elsenbeer and Vertessy, 2000), then widespread saturation overland flow over slopes in intense storm events is the storm runoff process response. By contrast, in environments such as Reserve Ducke in central Amazonia (Nortcliff and Thornes, 1981), the upper permeable layer is far deeper, rendering saturation of the soil difficult except locally in hollows and concavities – the result being the dominance of throughflow and localised saturation overflow and more muted streamflow response to rainstorms. This 'traditional picture, however, now requires some modification given recent studies quantifying the significance of pipeflow' (Sayer *et al.*, 2004, 2006; Sidle *et al.*, 2004) and Hortonian overland flow (Wierda *et al.*, 1989; Clarke and Walsh, 2006) in generating the stormflow of rainforest streams. In both cases, the roles played by distinctive features of the rainforest bioclimatic environment are significant (Table 8.1). At Danum (Sabah), Hortonian overland flow occurs in widespread fashion on slopes, apparently due to the 'thatched roof' effect of a combination of the near-surface fine root mat and leaf litter (Clarke and Walsh, 2006). In the same area quick-responding soil pipe networks are also found at 80–100 cm depth and are supplied by vertical preferential flow (notably around canopy trees) bypassing the soil matrix (Sayer *et al.*, 2006).

Chemical weathering and erosion

The traditional view that chemical denudation is highest in the ever-wet tropics (because of its perennially hot–wet conditions, high soil carbon dioxide levels and production of distinctive acids) has been qualified in recent decades. It is more accurate to state that chemical denudation rates are *potentially* high, but that only in young, usually mountainous landscapes are regoliths sufficiently shallow and unweathered for this potential to be achieved. A decay curve in weathering rate applies as regoliths increase in depth (Stallard and Edmond, 1987) and a strong positive influence of active orogeny on weathering rates has been reported (Stallard, 1995). In stable, old landscapes with deep regoliths, however, there is little weatherable material remaining, little percolating water reaches the deep weathering front and tight nutrient cycling by the rainforest ecoystem in nutrient-poor areas greatly restricts losses to the stream. Thus, in the Amazon Basin, the contrast between the very low solute yields of the eastern Amazon tributaries draining the geologically old Guiana and Mato Grosso massifs and the high solute yields of the western basin tributaries draining the Andes is striking (Gibbs, 1967; Summerfield, 1991).

Two studies, in East Africa (Dunne, 1978) and Dominica (Walsh, 1993), demonstrate the scale with which chemical denudation rates (as indexed by solute yields) vary within the ever-wet tropics with differences in mean annual runoff and rainfall (Fig. 8.3). The study of Dunne affords also a comparison with the wet–dry tropics, where chemical denudation rates are much lower.

Slope erosional processes

It is normally considered that the high *potential* for slopewash (the combination of rainsplash and erosion by overland flow) stemming from high rainfall intensities and high frequency of large rainstorms of the rainforest tropics is

TABLE 8.3. *Annual slopewash (S) (g m^{-2}), overland flow (OF) (mm) and rainfall (mm) in undisturbed forest and on revegetating landslide scars on slopes on Cretaceous sandstones and siltstones and Tertiary quartz diorite in Luquillo Experimental Forest, Puerto Rico, using Gerlach troughs below unbounded plots*

	Cretaceous sandstones/siltstones					Tertiary quartz diorites				
	Forest		1992 Scar			Forest		1989 Scar		
Year	S	OF	S	OF	Rain	S	OF	S	OF	Rain
1992	11.2	7.2	349.2	24.2	3506	80.4	14.0	99.5	9.6	4447
1993	7.3	4.4	18.7	17.6	2966	44.6	12.3	25.3	10.1	3726
1994	8.0	3.7	2.7	13.6	2895	18.2	6.2	7.6	10.3	3380
1995						41.1	24.0	3.5	12.3	3110
Mean	8.8	5.1	123.5	18.5	3122	46.1	14.1	34.6	10.6	3666
Sites	10	10	2	2		5	5	2	2	

Source: Modified from Larsen *et al.* (1999).

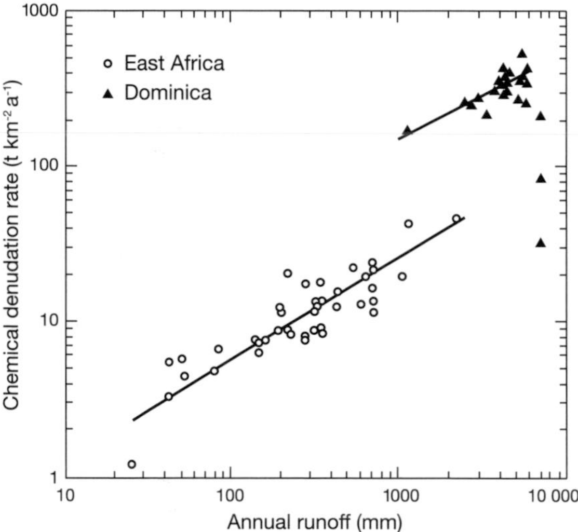

FIGURE 8.3. Rates of chemical denudation and annual runoff for catchments in East Africa on silicate-rich igneous rocks (Dunne, 1978) and in Dominica on andesitic volcanics (Walsh, 1993).

largely offset by the high infiltration capacities, permeabilities and water storage capacities of tropical rainforest soils (reducing proneness to overland flow) and the ground protection afforded by dense forest vegetation, its surface root systems and thin but widespread leaf litter. The array of measurements of slopewash (from unbounded and bounded erosion plots and changes in ground level) suggests that rates can be rather high but that they vary with soil type (via its influences on aggregate stability and proneness to overland flow), slope angle and annual rainfall (Walsh, 1993; Clarke and Walsh, 2006). Douglas (1999, p. 1726) argued that 'there is probably greater variation in rates of runoff and rates of soil erosion under natural conditions than there is at one site before and after disturbance'. Thus Larsen *et al.* (1999) in Puerto Rico demonstrated slopewash rates to be over five times higher on a sandy soil on quartz diorite than on finer-grained soils of higher aggregate stability developed on sandstones and siltstones; also slopewash on a fresh 1992 landslide scar was 30 times higher than under forest in the first year, but in both areas such localised enhancement was reduced by the second year with revegetation (Table 8.3).

Long-term measurements using the erosion bridge technique over periods of 3–12 years at Danum (Sabah) demonstrated a sharp rise in slopewash rates with increasing slope angle from 0.4 mm a^{-1} on 18° slopes to 3.4 mm a^{-1} on slopes averaging 29° (Clarke and Walsh, 2006). The study also demonstrated that most of the erosion was episodic and associated with three extreme rainstorms of 135–177 mm in a day. This suggests that rates derived from shorter-term studies can be unrepresentative; it also points to the importance of considering changes in rainstorm magnitude–frequency in predicting slopewash consequences of climate change.

Although soil piping has been reported from many locations in the rainforest tropics (see Walsh and Howells (1988) for an early review), pipe erosion rates have only recently been systematically assessed, with studies in Sabah (Sayer *et al.*, 2004, 2006) and Peninsular Malaysia (Sidle *et al.*, 2004) demonstrating the process to be an important contributor to stream suspended sediment budgets. In the Sabah study, sediment yields for two monitored pipes monitored for 9–16 months were 24.4 and 47.6 t km^{-2} a^{-1} compared with a stream sediment yield in the same area of 112.7 t km^{-2} a^{-1} and individual pipe tunnels were shown to be actively enlarging and collapsing to form channels. Peak suspended sediment concentrations of pipeflow and streamflow were of a similar order of magnitude. Pipe erosion is thus clearly an important process in those areas where it has been measured, but how widespread it is in the rainforest zone is currently unclear.

Rapid and slow mass movements

Landslides are generally considered to be particularly important on steeper terrain in the wet tropics both as an erosional process (Douglas and Spencer, 1985) and in landscape development (Thomas, 1994). Dykes (1995) estimated that they currently account for 16.5% of erosion in the Ulu Tembulong area of Brunei, though such estimations depend on the longer-term representativeness of the short period of record upon which they are based. Reasons for their importance in the wet tropics include:

(a) the development of deep regoliths through the dominance of chemical weathering over other erosional processes;
(b) proneness to extreme rainstorm totals (and in some areas to tropical cyclones);
(c) lack of root anchorage to bedrock in forest areas without deep tap roots; and
(d) the high drainage densities of the tropics, which tend to maintain finely dissected relief with high slope angles even in areas of relatively modest altitudinal relief.

Paradoxically, the threshold slopes at which failure occurs tend to be higher (over 40°) in the wet tropics than in other soil-covered environments; this is linked to the higher shear strength of many tropical soils associated with the properties of tropical clays (Rouse *et al.*, 1986). Thus, in the shale and sandstone Mulu mountains in Sarawak, Day (1980) mapped recent landslides and found that all occurred on slopes exceeding 40°, with 18 on slopes exceeding 50°. Planar shallow translational slides (in which the soil and regolith are stripped from the bedrock) tend to dominate in mountainous terrain with steep straight slopes, whereas in deeper regolith areas rotational slides are more characteristic. Failure may also be favoured by high vertical permeability of most wet tropical soils coupled with enhanced ponding in extreme events caused by sharp reductions in permeability within some subsoils or at the soil–rock interface (Rouse *et al.*, 1986). Little is known about soil creep in the humid tropics. Lewis (1974), however, found in Puerto Rico that the vertical profile of creep movement was very different than that reported in other environments, with maximum creep movement not at the surface, but below the near-surface root zone, demonstrating the retarding effect of a tropical fine root mat.

Fluvial processes

Rivers play a major role in ever-wet tropical landscapes, though few features of the fluvial landscape can be said to be distinctive from other climatic zones. Most general fluvial form and process theory is readily transferable to rivers in tropical rainforest zones, and is covered in Chapter 4. Two interrelated features of the rainforest fluvial environment are perhaps distinctive:

(a) the episodic nature of sediment delivery during high-energy tropical rainstorms (Douglas *et al.*, 1992); and
(b) the role of rainforest vegetation in channel processes through its influences on debris dams, sediment supply and slope and channel residence times (Spencer *et al.*, 1990; Gomi *et al.*, 2006).

Analyses of collected catchment data (Douglas and Spencer, 1985; Thomas, 1994) have demonstrated great variation in sediment yield within the rainforest zone, but most of this variation is linked to non-climatic variables of relief, tectonics, lithology and land use. There is also a scale effect with sediment yields tending to fall with catchment size (Thomas, 1994) (though this can be due in part to the accompanying decline in mean relief). Although there is a general belief that (with other variables held constant) sediment yields increase with annual runoff (and hence also with annual rainfall) within the zone, the form of such a relation has yet to be convincingly demonstrated. In one of the few systematic studies, Dunne (1979) derived a curve for forested catchments on igneous rocks in East Africa that suggests that the sharp rate of increase in sediment yield with increasing annual runoff applicable in the wet–dry zone becomes more muted within the rainforest zone (Fig. 8.4),

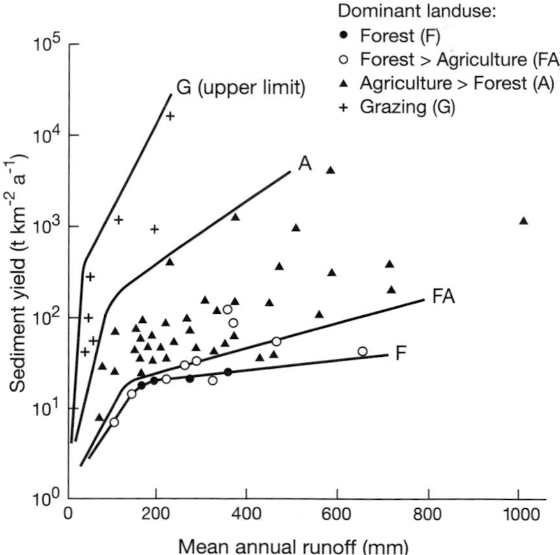

FIGURE 8.4. Relation between sediment yield and annual runoff for forested catchments and three other land use categories in East Africa (from Dunne, 1979).

but the dataset only extends into marginal rainforest areas. Douglas (1999) reported yields varying with tectonic activity and lithology from 10 000 t km^{-2} a^{-1} in New Guinea and Taiwan to 10 t km^{-2} a^{-1} in parts of eastern Amazonia and central Africa. Within Southeast Asia, yields vary from 10 000 t km^{-2} a^{-1} in tectonically active parts of island arcs and around 1000 t km^{-2} a^{-1} on the weak Tertiary mud rocks of eastern Borneo to 50–100 t km^{-2} a^{-1} on the deeply weathered Mesozoic granites of the Thai/Peninsular Malaysia/Bangka intrusive complex (Douglas, 1999). Thus one should consider a spectrum of relations between suspended sediment yield and annual runoff for different regional lithologic/tectonic situations rather than single curves.

Key features in rainforest areas are the episodic nature of suspended sediment and bed material transport and the dominant role of some (but not all) extreme rainfall/runoff events. Thus 8.3% of the sediment transport total of the Segama river (Sabah) over the 8.4-year period February 1988–June 1996 was moved on a single day (19 January 1996) following a storm of 178 mm in 11 hours, and six events accounted for 25.6% of the load over the same period (Douglas *et al.*, 1999). In contrast, in Puerto Rico, the Hurricane Hugo event of 1989, while transporting 32% of the annual load of two monitored catchments of that year, resulted in low peak and weighted-mean storm suspended sediment concentrations compared with other large storms (Gellis, 1993). This may reflect an exhaustion effect affecting the sediment concentrations of the massive overland flow response in the storm, as a concentration of only 33 mg l^{-1} was recorded at peak flow of a small tributary compared with a normal response of 3000 mg l^{-1}. Other important features of the ever-wet tropical environment that contribute to this episodic and variable character are landsliding and debris dam cycles. Spencer *et al.* (1990) describe the cycle by which dams (initiated by fallen trees and in some cases landslide material) can form and build (reducing downstream suspended sediment responses during this period) until the logs decay sufficiently for a future large rainfall/runoff event to induce a dam-burst, and often a chain of connected dam-bursts, and a disproportionately large sediment load response. Where the river possesses a floodplain, the cycle may be interrupted by avulsion once the river has aggraded to the higher base level of the dam, with the river either creating a new, lower channel or reoccupying a previous channel, as happened in the Melinau River in Sarawak in 1977 (Walsh, 1982b).

Traditional views about the unimportance of bedload in the humid tropics because of hypothesised absence of supply from slopes mantled with deep, thoroughly weathered regoliths have been revised given numerous studies highlighting its significance in hilly areas of the tropics. In Peninsular Malaysia, bedload accounts for 20–30% of the total load of most Main Range streams (Lai, 1993). Downstream rates of breakdown of bed material, however, may be more rapid along rainforest rivers than in other areas, as demonstrated by Rose (1984) for the Melinau River in Sarawak. This may lead to a scale effect whereby bedload transport rates in large catchments are relatively small (Douglas, 1999).

The main sources of bedload and suspended sediment are generally considered to be landsliding of deep weathered regolith during high-magnitude rainstorms and channel bank erosion (sometimes involving reworking of fluvial deposits) at times of peak streamflow (Terry, 1999; Douglas and Guyot, 2005), though the latter is constrained to some extent by root stabilisation by riparian vegetation. Recent work, however, has indicated that pipe erosion can also be important (Sayer *et al.*, 2004). Landslide sources may be largely confined to steeper parts of catchments.

8.2.3 Landforms and landscape development including the influence of climate change

This section concentrates on distinctive features of tropical landscapes and their development, as many fluvial landform features found in the ever-wet tropics (from the relatively few studies undertaken) appear to differ little from those found in other humid areas. Perhaps the most important feature of ever-wet tropical areas is the high drainage density compared with most other climatic regions of the world, though values vary greatly both between and within regions with rainstorm magnitude–frequency, soil factors, relief and geology (Walsh, 1996c). Figure 8.5 shows how drainage density values increase at first steeply, then shallowly and then steeply once more with increasing annual rainfall and daily rainfall magnitude and frequency in granitic parts of Sri Lanka (C. Madduma Bandara, unpublished, in Chorley *et al.*, 1984) and the volcanic eastern Caribbean islands of Dominica and Grenada (Walsh, 1985; 1996c). These variations reflect the direct and indirect influence of climatic variables. The direct influence is that of large rainstorm magnitude, as it is only the large runoff events which are associated with such rainstorms that are capable of eroding fingertip sections of the drainage net. The indirect influence of climate is exerted via the influence of annual rainfall, rainfall seasonality and vegetation on soil characteristics (notably topsoil storage capacity and subsoil permeability). Wetter rainforest soils (such as the allophane latosolics of Dominica) tend to have deeper topsoils and sometimes high subsoil permeability, both of which tend to reduce saturation overland flow. Reduced saturation overland flow tends to offset the impact of higher rainstorm

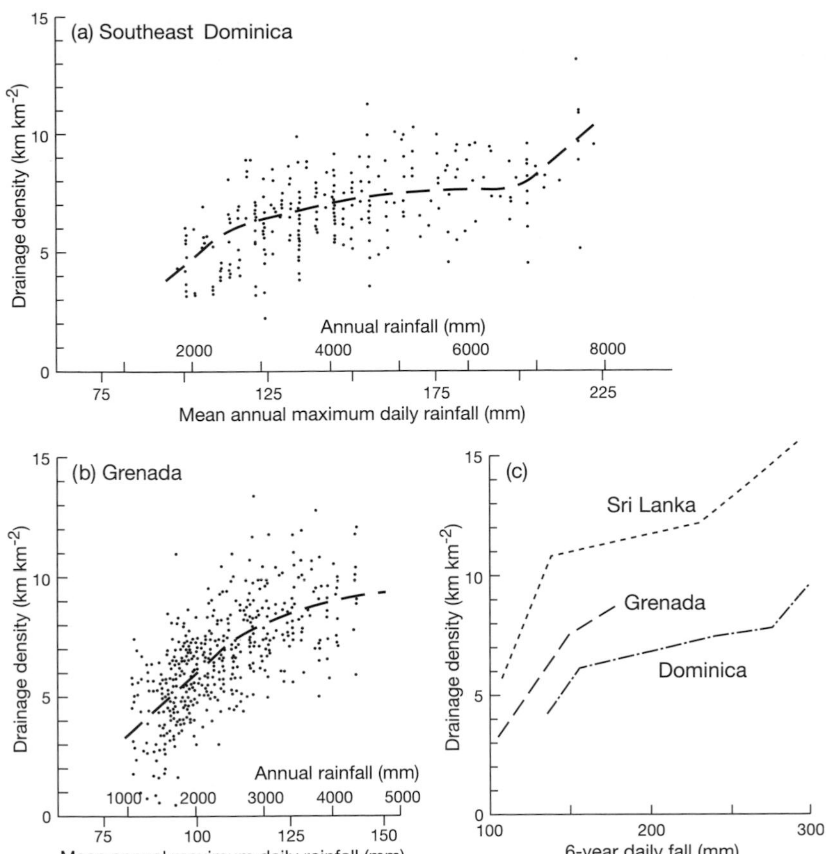

FIGURE 8.5. Relations between drainage density (km km^{-2}) and rainfall parameters (after Walsh, 1996c). (a) and (b) Southeast Dominica and Grenada (volcanic terrain of mid-Pleistocene age and older). (c) Drainage density and 6-year return period daily rainfall for Sri Lanka (C. Madduma Bandara, unpublished, in Chorley *et al.*, 1984), Grenada and Dominica.

magnitude until the effect is exhausted, after which drainage densities again rise sharply in the wettest environments. Individual rainstorm magnitude does not increase with annual rainfall in all areas. Higher drainage textures (source frequencies) in coastal mountain and offshore islands of Peninsular Malaysia compared with the mountainous interior have been linked to higher extreme rainstorm totals of coastal areas. The higher drainage densities to the north (11.1 km km^{-2}) than to the south (7.9 km km^{-2}) of Morne Seychellois on the granitic island of Mahé in the Seychelles are associated with much higher 6-year daily rainfall on the northern compared with southern sides at similar annual rainfalls (Walsh, 1996c). In these studies of drainage density, however, three major questions remain:

(a) what return period of rainstorm/runoff event is responsible for controlling the drainage network (as rainstorm magnitude parameters are often correlated and equally good correlations with drainage density emerge)?
(b) what are the processes responsible? and
(c) how quickly do networks develop or adjust to changes in climate?

Various drainage density models appear to be applicable to different locations in the rainforest tropics either emphasising different slope runoff processes and their spatial patterns, inheritance from collapsed pipe networks, channel development in landslide scars (Chorley *et al.*, 1984) or even the simultaneous operation of uncoupled pipe collapse and erosion by overland flow, as observed influencing different channel heads in the same terrain in Sabah.

The high drainage densities of the rainforest tropics have some implications for slope profiles. A logical implication of fine dissection by the stream network is that slope angles will be steeper (and slope lengths shorter), even in areas of moderate altitudinal relief, than in other humid environments with lower drainage densities. This makes landsliding more likely and rainsplash erosion (heavily dependent on slope angle) more effective. Some authors, notably Garner (1974), have suggested that the bedrock or boulder-mantled slopes of many inselbergs may result from widespread landsliding and dissection during a humid tropical phase. Others, however, propose that ever-wet tropical conditions provide the conditions for deep weathering phases prior to stripping in less wet or semi-arid phases.

Thomas (1994; 2006) considers the whole question of the role of tropical conditions in the long-term development of inselbergs and etched landscapes in detail and concludes that, although the humid tropics constitute the most favourable environment for the formation of deep, clay-rich saprolites, there is no reason to limit etching concepts or theory to the tropics or subtropics. He also stresses that in all environments, large-scale landscape features, including inselbergs, have such a long history of development that they will inevitably have been the product of many climates and their transitions. On the Quaternary timescale, Thomas (2004; 2006) gives a series of tropical examples of the importance of climatic shifts (to both wetter and drier climates) leading to periods of concentrated geomorphic activity and landform genesis during the transitions. It is the *sequences* of different climates and associated disequilibrium situations (rather than the climates *per se*) that are primarily responsible for the landscape features involved.

8.2.4 The effects of human activity

Geomorphic effects of anthropogenic disturbance are potentially particularly high in the humid tropics and, given the scale of current and projected future forest clearance and land use change, will arguably continue to greatly outweigh those of climate change. This reflects three features of rainforest environments:

(a) high potential availability of sediment provided by the deep regoliths, transport-limited slopes and supply-limited fluvial systems of undisturbed humid tropical landscapes;
(b) high erosional energy associated with the high annual rainfall, rainfall intensities and rainstorm magnitude–frequency of the zone; and
(c) high potential for increased erodibility given that runoff and erosional responses in undisturbed forest are muted as a result of high infiltration capacities and ground protection by surface roots and leaf litter.

Disturbance of the forest canopy and soil surface, during forestry or mining operations or shifting agriculture, leads to pronounced modification of hillslope hydrological processes, enhanced sediment delivery to the stream network and downstream changes in the sediment transport and storage dynamics within the fluvial system. Similarly, conversion of forested land to alternative land use such as grazing land or plantations will lead to short-term instability of hillslopes and a prolonged period of fluvial adjustment to new hydrological conditions. Although erosion rates for most types of land use are higher than for undisturbed forest (Tables 8.4 and 8.5), differences can be reduced with the adoption of simple conservational measures, particularly those aimed at avoiding disturbance of steep slopes, maintaining a groundcover and high organic matter, and reducing bare areas to a minimum. Often, however, such measures are not adhered to and Douglas (1994) has demonstrated how most land use types can easily lead to land degradation.

In a recent calculation of terrestrial sediment flux to the global coastal ocean, Syvitski *et al.* (2005) estimated that, in pre-human times, tropical rivers contributed annually ~99 t km^{-2} compared with 192 t km^{-2} from warm temperate rivers (Table 8.6), reflecting catchment stability despite an active hydrological cycle. However, modern annual suspended sediment loads were estimated to have increased to 130 t m^{-2} for tropical rivers and reduced to 170 t km^{-2} for warm temperate rivers largely reflecting the influence of land conversion in the tropics and sediment retention by reservoirs in the temperate zone. Reservoir retention in the tropics is still relatively small (16%) and the 31% increase in tropical suspended loads is of some concern considering continued deforestation and future predictions of rainfall extremes in the context of global warming.

Shifting agriculture and other traditional agricultural systems

Geomorphological impacts of shifting and other traditional agricultural systems vary with the spatio-temporal intensity of the system, forest clearance methods (e.g. mechanical versus by hand and with or without fire) and the crop types and agricultural practices adopted. If fire is not used, tree root systems are left intact and steep slopes are avoided, erosional impacts may be minor. As long as fallow periods are sufficiently long, then net geomorphic impacts are sustainable, as the moderately increased erosion rates incurred during the 2–3 years' duration cultivated phase are outweighed by very low rates in the fallow period because of the high groundcover afforded by the dense low regrowth. Increasingly over the past 40–50 years, however, shifting cultivation has become over-intensive and spread to steep slopes. Douglas (1994) relates how slash and burn has become a cheaper short-term alternative to buying costly fertilizers to maintain nutrient levels and crop yields of permanently cultivated land. Degradation involves increased erosion rates, reduced soil nutrient and carbon levels and a poorer quality of fallow regrowth (including savannisation) and in severe cases can lead to rilling and gullying (on deep regoliths, unconsolidated deposits and erodible lithologies) and landsliding (if steep terrain is involved).

In southwestern Nigeria, Odemerho (1984) used a hydraulic geometry approach to assess downstream effects

TABLE 8.4. *Slope and catchment erosion rates under natural forest and different land use types in the humid tropics: minimum, median and maximum rates of slope erosion*

	Rate of erosion (t km^{-2} a^{-1})		
Land use[a]	Minimum	Median	Maximum
Natural forest	3	30	620
Shifting cultivation: cropping phase	40	280	7000
Shifting cultivation: fallow regrowth phase	5	20	740
Tree crops: clean weeded	120	4800	18300
Tree crops: with a cover crop or mulch	10	80	560
Multistorey tree gardens	1	10	150
Plantations	2	60	620
Forest plantations: litter removed or burned	590	5300	10500
Forest plantations: young with agricultural intercropping	60	520	1740

[a] Rates: (i) do not include differences in landslide susceptibility; (ii) are mostly for established land uses and that varying soil losses during forest clearance and conversion (e.g. bench terracing) may have incurred.
Source: Wiersum (1984).

TABLE 8.5. *Slope and catchment erosion rates under natural forest and different land use types in the humid tropics: catchment suspended sediment yields under natural forest and cultivation*

Location	Forest (F)	Cultivated (C)	C: F Ratio
Adiopodoumé, Ivory Coast	242.5	4250	17.5
Mbeya Range, Tanzania	17.3	73.8	4.3
Northern Range, Trinidad	4.5	40.0	8.9
Cameron Highlands, Malaysia	52.7	257.8	4.9
Gombak, Malaysia	60.0	1157.5	19.3
Cilutung, Java	2250.0	4750	2.1
Barron, Australia	14.3	34.0	2.4
Millstream, Australia	15.5	30.8	2.0

Source: Douglas (1994).

TABLE 8.6. *Estimated sediment loads for tropical compared with other climatic zones in pre-human and modern times*

Climatic zone	Area (Mkm^2)	Discharge (km^3 a^{-1})	Pre-human load (Mt a^{-1})	Modern load (Mt a^{-1})	Difference (Mt a^{-1})	Modern load retained in reservoirs (%)
Tropical	17	7 110	1 690 ± 480	2 220 ± 360	+530	16
Warm temperate	47	21 110	9 070 ± 2600	8 030 ± 1250	–1 040	15
Cold temperate	17	4 760	1 940 ± 250	1 460 ± 160	–480	47
Polar	24	5 560	1 330 ± 170	900 ± 120	–430	6
Global	105	38 540	14 030	12 610	–1 420	20

Source: After Syvitski *et al.* (2005).

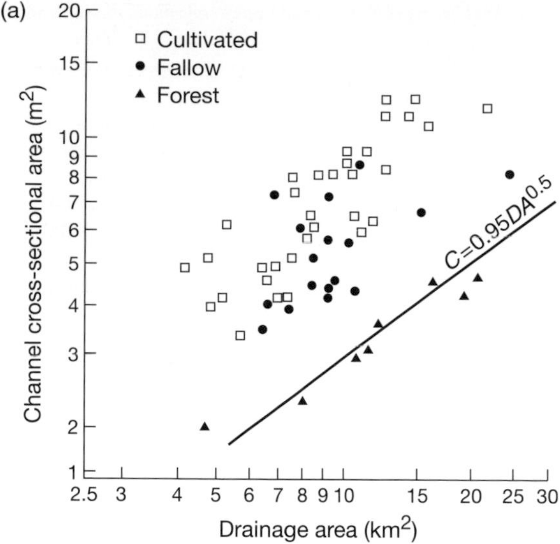

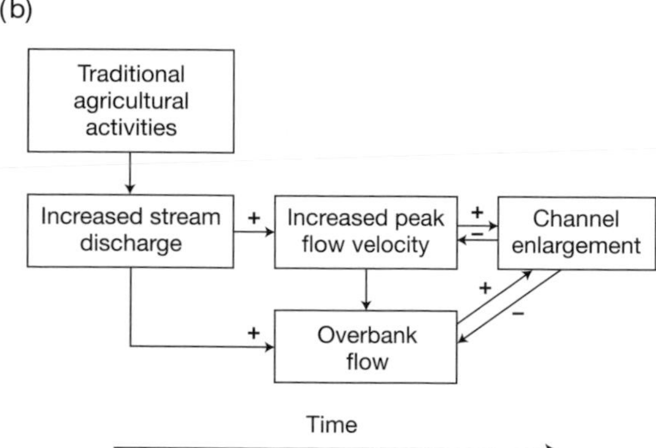

FIGURE 8.6. (a) The increase and then decrease in channel cross-sectional area associated with cultivated and fallow stages in traditional farming practice compared with undisturbed rainforest in southwestern Nigeria; and (b) the sequence of channel adjustments involved (after Odemerho, 1984).

of traditional shifting cultivation on river channels (Fig. 8.6). His data showed how bankfull cross-sectional area increases three- to fourfold during the clearing and cultivation phase, but then progressively declines to around twice the pre-disturbance forest sizes if the land is left fallow and forest is left to recover. Similar temporal hydraulic geometry data are unfortunately lacking in relation to the impacts of logging and replacement land uses in the wet tropics.

Logging

Logging can be selective (whereby some to all canopy trees are felled, but smaller trees are left and the forest is allowed to regenerate) or involve clear-felling (where the land is either left to regenerate, followed by grazing, or converted to plantation or other large-scale agriculture, peasant agriculture, forestry plantations or urban land use). Large-scale logging tends to be carried out by logging companies, but historically and in some areas currently forest clearance has been carried out piecemeal by peasant farmers with or without the use of fire. The short-term impacts of commercial logging on erosion at the slope and small catchment scales, and their variation with logging practice, are well established (Anderson and Spencer, 1991; Douglas, 1999; Sidle *et al.*, 2004; Thang and Chappell, 2004), but much less is known about downstream effects at the large catchment scale or about long-term effects. Erosional recovery is quicker than hydrological recovery. For example, compaction of the soil by vehicles and associated reductions in infiltration capacity can persist many years after vegetation recolonisation has greatly reduced erosion rates (Malmer, 1990).

The long-term erosional impacts of selective logging have been investigated in detail for the Baru catchment in Sabah (Douglas *et al.*, 1999; Chappell *et al.*, 2004; Clarke and Walsh, 2006; Walsh *et al.*, 2006). The previously undisturbed Lowland Dipterocarp Forest catchment (Fig. 8.7) was monitored for 6 months prior to logging, then through different parts of the logging phase in December 1988 – April 1989 and throughout its recovery phase, when slope and gully erosion were also measured; the control West Stream primary forest catchment was also monitored throughout. A mixture of conventional tractor logging and high lead logging techniques was employed. Changes in the sediment yield and sediment sources for the Baru catchment from 1988 to 2005 are summarised in Fig. 8.8 and Table 8.7. The main erosion phase lasted less than 2 years, with peak catchment erosion rate reaching 18 times that of the primary forest catchment. The main sources in this period were bare areas (around 25% of the surface area) created by the logging (notably surfaced and feeder logging roads, skid trails created by bulldozers extracting individual logged trees, and log marshalling areas). Rapid revegetation of much of bare areas and reductions in availability of easily eroded sediment (rather than reductions in overland flow) were the reasons for the decline in sediment supply from many of these sources. A secondary peak in erosion 6 years after logging was linked to two extreme rainfall/runoff events and biological decay of logs in road bridges, culverts and stream debris dams that had been created at the time of logging; sediment sources switched to road-linked landslides (and subsequently landslide scars), debris dam-bursts and re-excavated first- and second-order channels – along with continued headward eroson of knickpoints of gullies of rutted feeder roads that were connected to the main stream network. Recovery after logging is thus more complex than formerly thought and

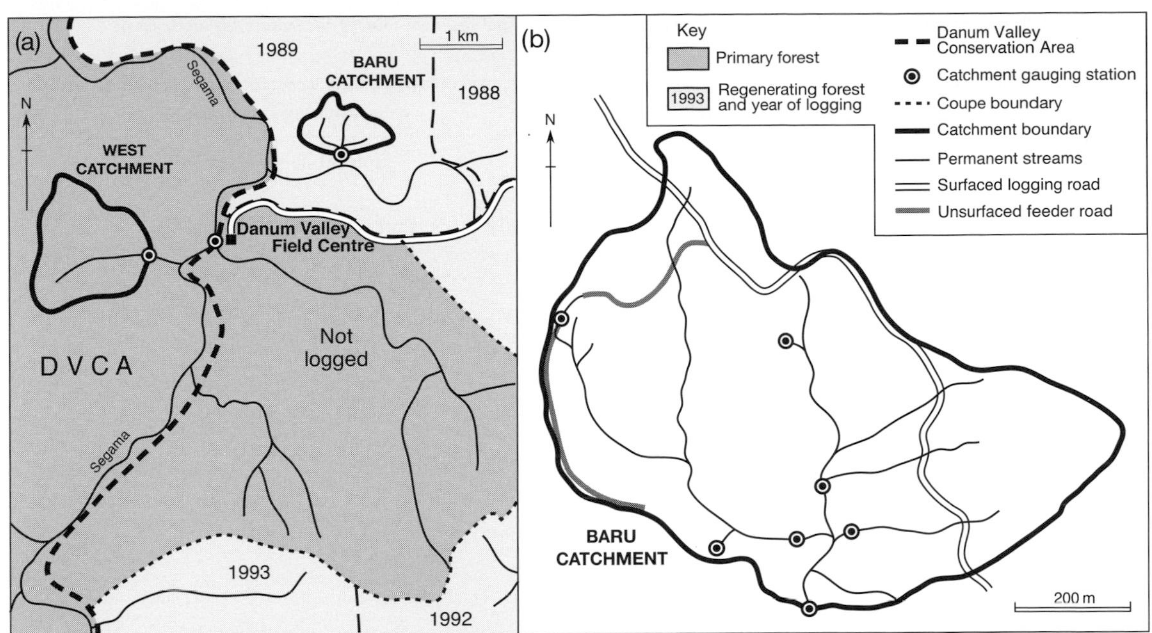

FIGURE 8.7. (a) Long-term study catchments and other erosion monitoring sites at Danum Valley, Sabah (DVCA, Danum Valley Conservation Area); (b) the selectively logged Baru catchment.

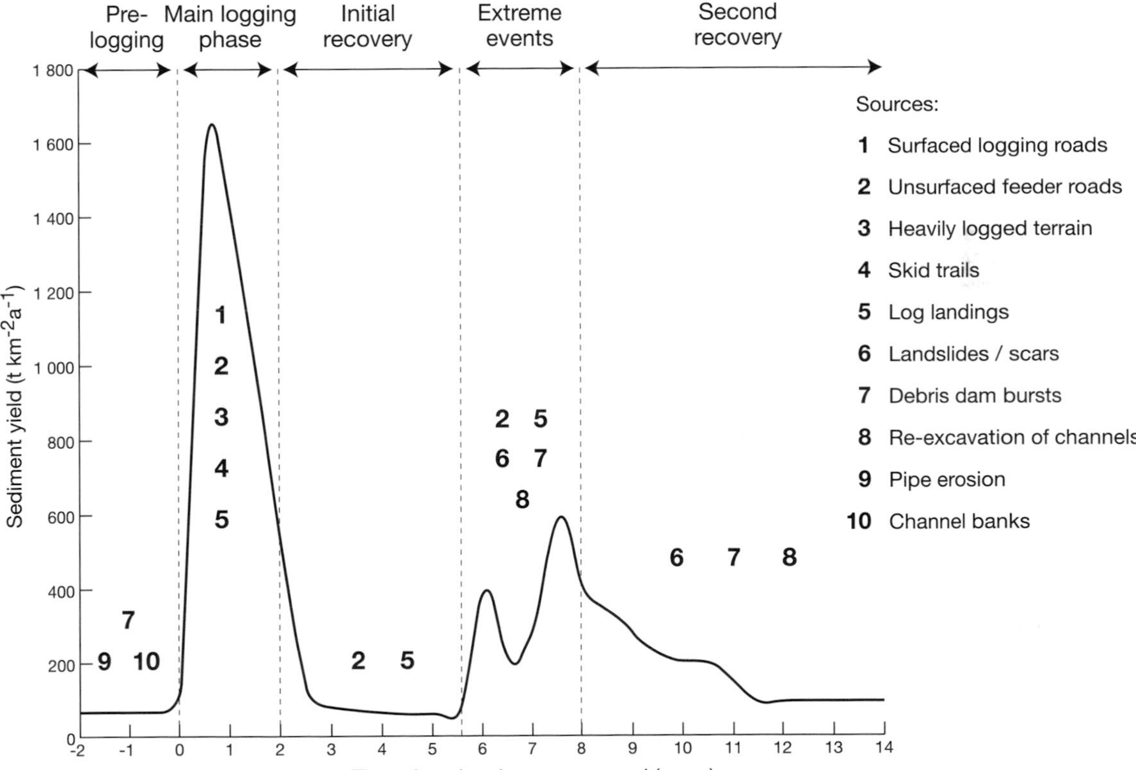

FIGURE 8.8. Changes in sediment yield and principal sediment sources in the selectively logged Baru catchment (Sabah, Borneo) from pre-logging time in 1988 to 14 years post-logging in 2003 (modified after Clarke and Walsh, 2006).

TABLE 8.7. *Trajectory changes following landscape disturbance: changes in principal sediment sources and storage/lag mechanisms in the Baru catchment in the 14 years following selective logging in 1988–89*

Phase	Principal sediment sources	Storage/lag mechanisms
Pre-logging/ primary forest	Slopewash Pipe erosion Landslides (where slope angles are sufficiently steep) Channel bed and banks (including debris dams)	Natural debris dams
During and up to 2 years	Surfaced logging road construction Creation and rutting/gullying of unsurfaced feeder logging roads Skid trails Log landing areas Heavily logged slopes Enhanced debris dam-bursts Only limited mass movement activity	Enhanced debris dam creation Choking of headwater channels Channel aggradation
2–5 years post-logging	Headward retreat of rejuvenated sections of gullied unsurfaced roads Skid trails – rapidly reducing Log landing areas – reducing Heavily logged slopes – reducing Enhanced debris dam-bursts No known landslide activity	Debris dam activity Aggradation and re-excavation of channels
6–12 years post-logging	Road-related landslides (5–7 years post-logging) Gullying of landslide scars and toe deposits Headward retreat of rejuvenated sections of gullied unsurfaced roads Old, and new landslide-related, debris dams in extreme events	Continued debris dam activity enhanced by creation of landslide-related dams Re-excavation of aggraded channels

Source: Walsh *et al*. (2006).

characterised by a decline that is punctuated by secondary rises in erosion triggered by biological decay and extreme events.

In Peninsular Malaysia, Gomi *et al*. (2006) found that bankfull channel width was larger in logged compared with unlogged streams. They also found that sediment and woody debris dam accumulations increased by four to five times with logging and stressed their role in delaying downstream sediment transfer, reducing peak transport rates and thereby buffering against major shifts in channel processes and patterns downstream.

Erosional impacts of logging can be easily reduced with the adoption of an array of simple techniques and procedures (Anderson and Spencer, 1991; Thang and Chappell, 2004). In the Baru catchment example, above, much of the secondary erosional peak was linked to ten landslides associated with a logging road aligned along the contour at a mid-slope position (Fig. 8.7c); positioning roads along ridge tops, as in the rest of the catchment, would have avoided the problem. Reduced Impact Logging (RIL) involves the adoption of a wide range of measures designed to reduce erosional and ecological impacts and enhance forest recovery. These include careful planning of the road and skid trail network, supervision of logging, lifting of bulldozer blades, retention of riparian forest and the avoidance of logging of slopes greater than 25°. The last is perhaps the most important (R. Ghazali, personal communication, 2008). He has demonstrated an enormous difference in landslide occurrence between adjacent areas of similarly steep terrain that had been logged conventionally in 1992–3 (147 landslides in the 1992 Coupe, 72 landslides in the 1993 Coupe) and with RIL techniques (just five landslides).

Regenerating forest is also at a heightened risk of fire because of a higher ground fuel load from logging debris and increased human contact. Fire leads to an extra erosional episode, which Malmer (2004) has demonstrated can rank with the logging phase in terms of enhanced suspended sediment transport. Selective logging with sustainable levels of timber extraction and logging frequency would be a reasonable conservational strategy if RIL protocols were properly adhered to.

The effects of large-scale forest clearance, land-use change and urbanisation

The last few decades have been marked by an acceleration of large-scale conversion of rainforest to either grazing, tree crop plantations, subsistence or cash crop cultivation or forestry plantations. Geomorphic effects including post-conversion erosion rates vary enormously with:

(a) the method of forest clearance – in particular whether fire was used and whether conservation measures were employed to retain the nutrient status of the soils;
(b) the extent to which a new 'engineered' post-clearance landscape was created (as with bench terracing systems of palm oil plantations that have spread across large swathes of Malaysian Borneo and Indonesia over the last two decades);
(c) the land use itself (including the details of land management practices adopted); and
(d) the vulnerability of the landscape (notably relief and the erodibility of the soil, regolith, any unconsolidated deposits and lithology).

Although erosion rates in most land uses are higher than under primary forest, they are only excessive if a protective ground cover is not maintained, land is overgrazed or (as is increasingly the case) steep sloping terrain is involved. Thus median rates of erosion (Table 8.4) on clean-weeded tree crop land (4 800 t km^{-2} a^{-1}) are 60 times higher than where the tree crops have a mulch or cover crop cover (80 t km^{-2} a^{-1}) and 160 times higher than under natural forest (30 t km^{-2} a^{-1}). Median erosion rates are ten times higher (5200 t km^{-2} a^{-1}) for forestry plantations with no litter cover than for young forestry plantations with an inter-cropped cover crop. At the catchment scale, sediment yields in cultivated areas are 2–19 times higher than under forest. Erosion rates under pastoral agriculture can be particularly high and in East Africa (Dunne, 1979) sediment yields are much higher on grazing land than under agriculture and two to three orders of magnitude higher than under forest (Fig. 8.4).

As in other parts of the world, geomorphic and hydrological impacts of urbanisation vary with the character of urbanisation, the preceding non-urban environment and topography. River channel changes tend to be greatest where urbanisation occurs on sloping but previously very permeable terrain. The speed and magnitude of runoff responses to rainfall are the most noticeable increases. Impacts tend to be least in impermeable, low-relief terrain, where runoff responses tend to be less enhanced compared with pre-urban times. The principal urban impacts include: enhanced landsliding (in cities containing steep slopes); channel enlargement and shifting within and downstream of urbanised areas; and enhanced surface erosion, gullying and suspended sediment transport resulting from urban construction. Impacts in humid tropical cities tend to be particularly enhanced because of the high rainfall intensities and storm totals (see Table 8.2) and the often unplanned nature of urbanisation and lack of storm drainage infrastructure. Paradoxically, however, the latter tends to reduce the magnitude of downstream flood peaks and resultant channel enlargement.

8.3 Recent climate change in the rainforest zone

8.3.1 Rainfall and temperature changes

All parts of the ever-wet tropics show upward shifts in mean temperature consistent with global warming (with seven estimates of change for the period 1979–2004 in the range +0.04 to +0.15 °C per decade), but with night minima increasing more than day maxima (IPCC, 2007). Differing trends in rainfall between northern (negative) and southern (positive) Amazonia have been linked to:

(a) a southward shift in the South American Monsoon System and
(b) the development of positive sea surface temperature (SST) anomalies and rainfall in the western subtropical South Atlantic (IPCC, 2007).

Relatively modest reductions in annual rainfall have been reported over the period 1948–2003 over some parts of the rainforest tropics, including parts of Malesia, equatorial parts of West and Central Africa, Central America, Southeast Asia and eastern Australia (IPCC, 2007). In

TABLE 8.8. *Changes over the last 100 years in the frequency of large rainstorms in Sabah*

Station	Mean annual frequency of days with >50 mm			
	1910–40	1947–79	1980–2007	1999–2007
(a) Changes in frequency of daily events >50 mm				
Sandakan	16.2	14.8	15.8	18.6
Tawau	5.8	4.6	6.1	6.9
Kota Kinabalu	13.2	13.3	12.1	14.4
	Mean annual frequency of days with >100 mm			
	1910–40	1947–79	1980–2007	1999–2007
(b) Changes in frequency of daily events >100 mm				
Sandakan	3.3	0.4	3.3	5.4
Tawau	0.6	0.3	0.5	0.8
Kota Kinabalu	1.7	2.3	2.1	2.3
	1986–98	1999–2007	Increase (%)	
(c) Recent increases in annual rainfall and large daily rainfalls at Danum Valley				
Daily events >50 mm	7.8	10.9	+39.7	
Daily events >80 mm	1.7	2.9	+70.6	
Daily events >100 mm	0.85	1.11	+30.6	
Annual rainfall	2663.9	3076.1	+15.5	

West and Central Africa, rainfall and river flow in 1971–89 in the coastal rainforest belt from Sierra Leone to central Cameroon were 5–8% and 12–24% respectively below the 1951–89 mean, but only 2% and 5–6% respectively below the mean for the Congo Basin, southern Cameroon, Gabon and Congo–Brazzaville (Mahé *et al.*, 2004). In trade wind rainforest areas, temporal patterns of annual rainfall (and lengths of dry season) have largely followed the pattern found in the Sahel. Thus the eastern Caribbean experienced very high rainfall in the late nineteenth century, a 25% drop in the early twentieth century, relatively high rainfall in the period 1929–58 and a renewed decline (with longer dry periods) from 1959 (Walsh, 1985). At Barro Colorado (Panama) the recent decline in rainfall has been accompanied by a sharp increase in the intensity of the mid-December to mid-April dry season (Condit, 1998).

Changes in ENSO events and dry period magnitude–frequency relations

ENSO events have changed in magnitude and frequency both over the past and in recent decades. The ocean–atmosphere phenomenon is measured by the Southern Oscillation Index (SOI), which is defined as the anomaly of the monthly sea level pressure difference of Tahiti minus Darwin, normalised by the long-term mean and standard deviation of the Monthly Sea Level Pressure (MSLP) difference. Thus SOI is negative indicating El Niño conditions when monthly pressure at Tahiti is lower than normal and/or monthly pressure at Darwin is higher than normal; SOI is positive (indicating La Niña conditions) when pressure at Tahiti is higher and/or at Darwin is lower than normal. Many tropical rainforest areas are drier than normal during El Niño phases and wetter than normal during La Niña phases. IPCC (2007) reported that ENSO events have become more intense and of longer duration since a global climatic shift in 1976–77, which was marked by a shift to above-normal SSTs in the eastern and central Pacific. The 1982–83 and 1997–98 events were particularly strong. ENSO events were markedly less frequent in the mid twentieth century, but strong events were recorded in the late nineteenth/early twentieth century. Changes in the magnitude–frequency of dry periods in northern Borneo and the surrounding region and at Manaus in Amazonia demonstrated a similar pattern (Walsh and Newbery, 1999).

Rainstorm magnitude–frequency changes

The evidence for change is constrained by the paucity of reliable long-term daily rainfall records for many parts of the tropics, with most studies confined to trends since the 1960s (e.g. Manton *et al.*, 2001; Alexander *et al.*, 2006). In Sabah (northeast Borneo), daily rainfall series compiled from current and archival data sources extend back to around 1910 and allow longer-term comparisons to be made (Table 8.8).

Although frequencies of daily rainfall >50 mm and >100 mm at Sandakan and Tawau in 1980–2007 are higher than in 1947–79, they are little different from frequencies in 1910–40 and there has been little change at Kota Kinabalu. At all three stations, however, there has been an upsurge in frequencies since 1999, a trend also apparent at the shorter-term interior station of Danum Valley. There the frequencies of daily rainfall of >50, >80 and >100 mm have increased in 1999–2007 by 39%, 70% and 30% respectively compared with 1986–98 values, with annual rainfall increasing by 15%.

A return period approach to the above changes using overlapping 20-year data series gives a rather more complex picture, however (Fig. 8.9). The magnitudes of daily rainfalls of differing return periods have changed in different ways. Thus at Sandakan, whereas 1- and 2-year daily rainfalls have increased in recent years (reflecting the sharp increase in >50 mm and >100 mm falls noted in Table 8.8), the sizes of the 5- and 10-year return period daily rainfall have failed to increase after a reduction from peak levels in the 1970s and 1980s. In contrast, in recent years at Tawau, whereas the 1-year daily rainfall has only modestly increased, 5- and 10-year daily rainfalls have increased strikingly. Different patterns of change apply at different locations, reflecting perhaps the importance of aspect and local winds in relation to the monsoonal regional winds in a tropical region. (Sandakan is open to the winter northerly monsoon, whereas Kota Kinabalu and Tawau's large rainstorms occur mainly in the equinox transition months and the summer southwest monsoon.) As different geomorphological features are influenced by events of differing return period, there is potential for differential changes in extremes to have contrary effects on different geomorphological features.

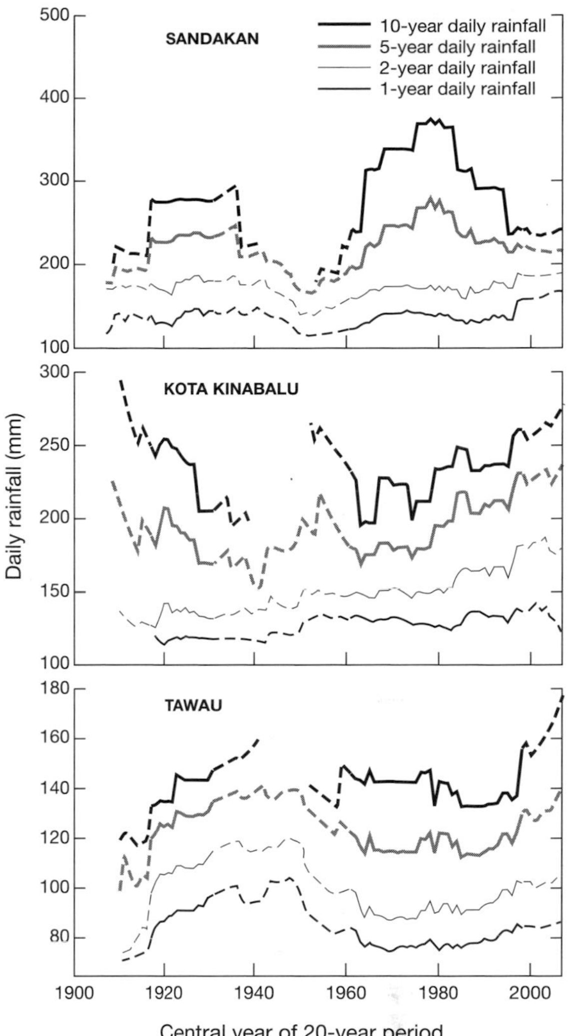

FIGURE 8.9. Changes in the magnitudes of 1-, 2-, 5- and 10-year daily rainfalls for overlapping 20-year periods between 1908 and 2007 for Sandakan, Kota Kinabalu and Tawau in Sabah.

Such changes are not unprecedented elsewhere in the tropics in the recent past. In the eastern Caribbean (Walsh, 1985), at St Thomas Police Station, Barbados, frequencies of rainfall events exceeding 76 mm fell from 2.3 to 0.6 a^{-1} between the periods 1889–1906 and 1907–25 and events exceeding 127 mm from 0.56 to 0.13 a^{-1}. After a partial recovery in 1926–58, frequencies returned to low levels again in 1959–72. At the marginal rainforest station of Roseau (Dominica), mean annual frequencies of 76 mm and 127 mm daily rainfall likewise fell between 1929–58 and 1959–76 from 1.7 and 0.37 to 0.8 and 0.22 respectively.

Tropical cyclone magnitude, track and frequency changes

Tropical cyclones have long been forecast to increase in frequency, and perhaps widen their spatial coverage, because of the increase in SST accompanying global warming. Until very recently, however, there was no evidence of an overall increase in their frequency (Walsh, 2000). In the Caribbean/North Atlantic region, frequencies were significantly lower in 1959–95 than in 1929–58 and 1871–1901 (Walsh, 2000). There was also evidence of significant northward and eastward shifts in *predominant* tracks from the late nineteenth century to the present day, which means that temporal trends in frequency vary greatly between different parts of cyclone regions. Analysis of longer series back to 1650 for the Lesser Antilles demonstrated major peaks in frequency in the late eighteenth and early nineteenth centuries and troughs in the 1850s and 1860s and prior to 1760 (Walsh, 2000).

TABLE 8.9. *The recent upsurge in tropical cyclone magnitude–frequency relations in the North Atlantic*

Years	Named storms	Hurricanes	Major hurricanes (category 4 or 5)
1970–94	8.6	5.0	1.5
1995–2004	13.6	7.8	3.8
2005	28	15	4

Sources: Webster *et al*., (2005); IPCC (2007).

There is currently some evidence of an upsurge in both tropical cyclone frequency and their intensity, which, it has been suggested, may be part of the long-awaited rise linked to global warming (Goldenberg *et al*., 2001; Emanuel, 2005; Webster *et al*., 2005), though it is unclear whether the rise will be sustained. Thus Webster *et al*. (2005) found that the years 1995–2000 saw a doubling of overall activity for the whole North Atlantic basin, a 2.5-fold increase in major hurricanes (greater than or equal to 50 m s^{-1}) (Table 8.9) and a fivefold increase in hurricanes affecting the Caribbean. Webster *et al*. (2005) also noted a large increase in the frequency and percentage of hurricanes reaching categories 4 and 5 globally since 1970. The record number of tropical storms recorded in 2005 was associated with an amplified subtropical ridge at upper levels in the central/east Atlantic, reduced vertical windshear in the central North Atlantic and record-high SSTs in the tropical North Atlantic. The troughs of the mid nineteenth century, early twentieth century and the 1960s to early 1990s and the peaks of the 1870s to 1890s and mid twentieth century can also be linked in part to negative and positive SST anomalies relative to the 1961–90 mean (IPCC, 2007). The year 2004 also saw the first ever recorded hurricane affecting the South Atlantic and the Brazilian coast.

8.4 Approaches and methods for predicting geomorphological change: physical models versus conceptual/empirical approaches

Two fundamental considerations that need to be accepted in considering approaches and methods to predict future geomorphological consequences of climate change in the wet tropics are:

(a) that uncertainties in climatic predictions are particularly high for much of the rainforest zone (see Section 8.5); and
(b) that anthropogenic changes both independent of and consequent to climate change will be of greater importance than (and also obscure) climate change itself in influencing geomorphic change within the zone.

Two groups of approaches could be adopted:

(a) physical modelling approaches; and
(b) a conceptual/empirical approach.

The role of physical-based models in prediction of geomorphological effects of climate change in the zone will be severely restricted for two related reasons:

(a) the deficiencies in current knowledge and understanding of even basic geomorphic processes and landscape systems within the zone (as is apparent from Section 8.2); and
(b) the data demands of such models.

They are most likely to play a role in considering slope and catchment-scale hydrology; slope-scale geomorphological processes, notably slope stability; and the supply of sediment from slopes to stream systems. Although there is a pressing need for models to predict downstream changes and patterns of sedimentation, erosion and sediment delivery within large catchments, the knowledge and expertise base required to accomplish such a task is currently undeveloped and unavailable.

The alternative conceptual/empirical approach that will be adopted here includes the use of:

(a) established empirical relations between geomorphological variables (process rates or landforms) and climatic and climate-linked variables;
(b) theory and empirical relations to predict systems change;
(c) past analogues of response to different types and sequences of climate change; and
(d) knowledge about the influence of past human activities on geomorphic systems to predict impacts of future human activities (including human responses to climate change).

As Section 8.2 has demonstrated, any attempts to predict change in the wet tropics must take into account the contrasts in geomorphology that result from non-climatic factors, notably the tectonic setting, lithology, current relief and the history of human activity in an area. These factors and the history of climate change in a region influence what may be termed the 'inherited geomorphological setting', which determines in turn the distribution and degree of connectivity of landscape units of varying degrees of resilience and vulnerability to disturbance by climate change and further human disturbance. Approaches may differ with the spatial scale of the landscape unit being considered. Smaller scales are potentially easier to tackle, as our

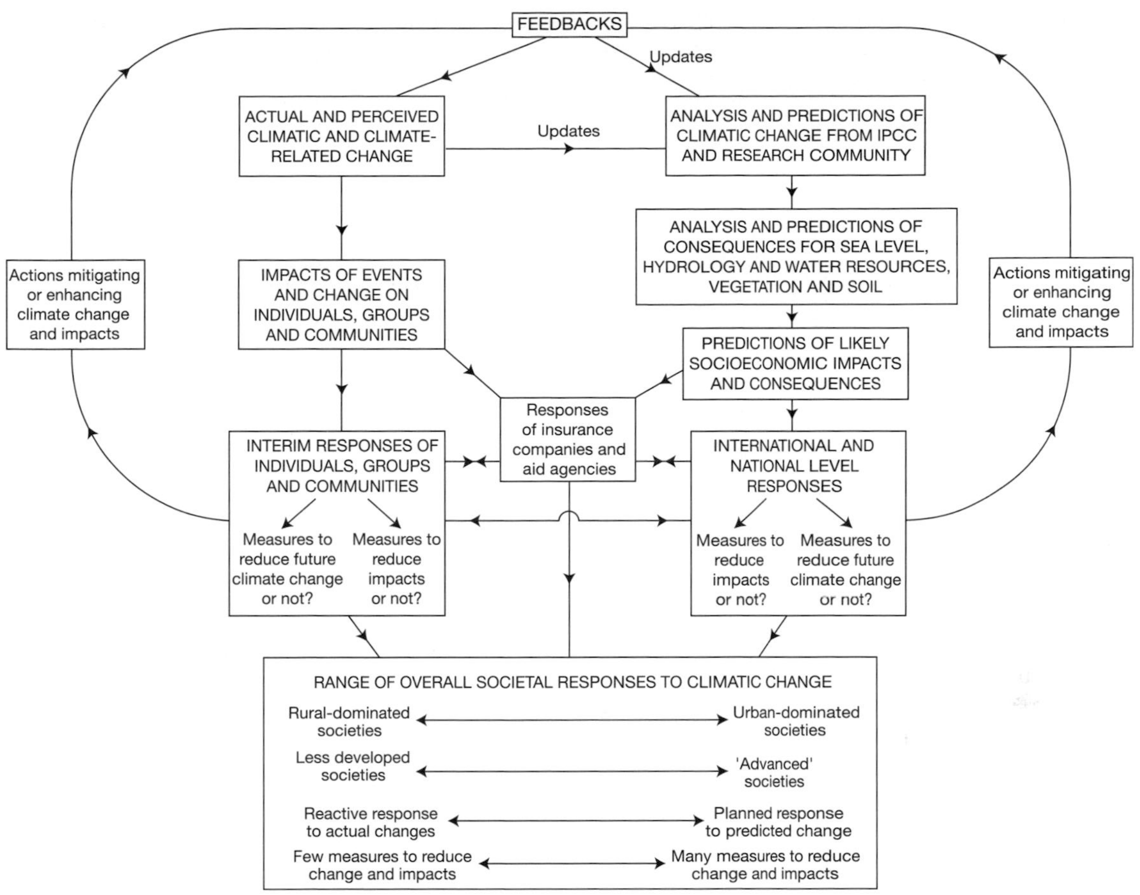

FIGURE 8.10. A conceptual model of the complex response of human communities to current and future climate change (from Walsh, 2007).

current state of knowledge is biased towards the slope and small catchment scales, but it is the larger-scale systems that are of greatest importance to predict, since populations are concentrated in the lower parts of large catchments.

Predictions of future geomorphological change need to incorporate future human activities and their response to climate and socioeconomic change. Analogies derived from past human responses to climate change are of limited use as the nature of human response has changed fundamentally through time. For the first time in human history, human response to climate change now includes the responses (of international organisations, governments, companies, communities or sections of communities, and individuals) to *predicted* change, rather than simply to perceived and actual climate change and events (Fig. 8.10). This is particularly pertinent when the details of IPCC predictions (particularly at the local scale relevant to people and communities) are so uncertain. It is also the case that the balance of response to predicted change and actual perceived change (usually experienced via extreme events) will vary with the societal element involved. Because of the dominance of poor populations in most wet tropical territories, the role that human response to predicted climate change is likely to play will probably be mostly confined to the impacts of policies and strategies adopted at international level by governments.

8.5 Potential ecological, hydrological and geomorphological responses to predicted future climate change in rainforest areas

8.5.1 IPCC predictions for tropical rainforest areas and their uncertainties

IPCC (2007) predictions of changes in temperature and rainfall for the end of the twenty-first century (2080–99) compared with 1980–99 are summarised for the principal

TABLE 8.10. *Summary of IPCC predictions for 2080–99 (compared with 1980–99) for the A1B scenario for the rainforest tropics and adjacent humid subtropics (derived from IPCC, 2007). Annual and seasonal values are median values of the 21 global circulation model outputs used by the IPCC. Seasons are Dec–Feb, Mar–May, Jun–Aug and Sep–Nov. The model range gives minimum and maximum model outputs for predicted changes in annual rainfall. The IPCC regions containing current rainforest areas are in bold*

		Temperature change (°C)		Rainfall change (%)		
Region	Latitude & longitude	Annual	Seasonal range	Annual	(Model range)	Seasonal range
Sahel region	18° N–30° N, 20° E–35° E	+3.6	+3.2 to +4.1	–6	(–44 to +57)	–18 to +6
West Africa	**12° S–22° N, 20° W–18° E**	**+3.3**	**+3.0 to +3.5**	**+2**	**(–9 to +13)**	**–3 to +6**
East Africa	**12° S–18° N, 22° E–52° E**	**+3.3**	**+3.0 to +3.5**	**+7**	**(–3 to +25)**	**+4 to +13**
Southern Africa	35° S–12° S, 10° E–52° E	+3.4	+3.1 to +3.7	–4	(–12 to +6)	0 to –23
SE USA	25° N–50° N, 85° W–50° W	+3.6	+3.3 to +3.8	+7	(–3 to +15)	+1 to +12
Central America	**10° N–30° N, 116° W–83° W**	**+3.2**	**+2.6 to +3.6**	**–9**	**(–48 to +9)**	**–16 to –4**
Caribbean	**10° N–25° N, 85° W–60° W**	**+2.0**	**+2.0 to +2.2**	**–12**	**(–39 to +11)**	**–20 to –6**
Amazonia	**20° S–12° N, 82° W–34° E**	**+3.3**	**+3.0 to +3.5**	**nc**	**(–21 to +4)**	**–3 to +4**
Southern S America	56° S–20° S, 76° W–40° E	+2.5	+2.4 to +2.7	+3	(–12 to +7)	0 to +1
India/Sri Lanka	5° N–50° N, 64° E–100° E	+3.3	+2.7 to +3.6	+11	(–15 to +20)	–5 to +15
Southern China	20° N–50° N, 100° E–145° E	+3.3	+3.0 to +3.6	+9	(+2 to +20)	–9 to +11
SE Asia/ Malaysia	**11° S–20° N, 95° E–115° E**	**+2.5**	**+2.4 to +2.7**	**+7**	**(–2 to +15)**	**+6 to +7**
N/NE Australia	30° S–11° S, 110° E–155° E	+3.0	+3.0 to +3.2	–4	(–25 to +23)	–14 to +1

rainforest regions and adjacent subtropical areas in Table 8.10. Maps of predicted temperature and precipitation changes are given in Plate 4 in Chapter 1. The uncertainties involved in the IPCC modelling predictions for the tropics remain very high. The temperature changes are more confidently predicted than the rainfall changes. There are particular difficulties with the predictions for Africa, where the multi-model dataset (MMD) models have proven unable even to postdict rainfall variations in the twentieth century (IPCC, 2007). Also vegetation feedbacks from dust aerosol production are not included and possible future land surface modification is also not taken into account in the projections (IPCC, 2007, p. 866).

Under the A1B scenario of IPCC (2007), rises in annual temperature of 2.0–3.6 °C are predicted by 2080–99 for rainforest parts of the world (Table 8.10), with the smaller rises predicted for the more maritime areas (Southeast Asia/ Malesia and the Caribbean) and higher rises for continental areas of Amazonia and Africa. Differences between seasons in part reflect rainfall predictions, with lower rises predicted for seasons where rainfall increases are predicted and vice versa. The predicted increases in temperature would mean a poleward spread of rainforest climatic conditions into some subtropical east-margin areas such as southern China, Florida, southeastern Brazil, southeastern Africa and eastern Australia.

Current IPCC predictions are for the Inter-Tropical Convergence Zone (ITCZ) to become more active over the equatorial zone with increased rainfall, but for the subtropical high pressures to intensify and expand poleward. Rainfall predictions are more uncertain than those for temperature and in many regions there is disagreement even as to whether rainfall will increase or decrease. Thus for the latitude/longitude zone containing Amazonia (Table 8.10), the median prediction for annual rainfall is no change, but the predictions of individual models range from –21% to +4%. Increased rainfall, particularly in the northern summer, is predicted for the western Amazon Basin, the equatorial Andes and Pacific coast areas, but reduced rainfall (exceeding 30% in the Mato Grosso region) for eastern Amazonia. The IPCC summarise the situation as: 'It is uncertain how annual and seasonal mean rainfall will change over northern South America, including the Amazon forest' (IPCC, 2007, p. 850). Reductions in annual rainfall and both winter and summer rainfall are predicted by most models for Central America, South America north of the equator and the Caribbean.

A sizeable increase is predicted as likely for the East Africa zone (median +7%; model range −3% to +25%), whereas a more marginal increase (+2%) is less confidently predicted for West Africa (Table 8.10). Rainfall is predicted to decline and become more seasonal in southern Africa (−4% overall). Hence a poleward expansion of the rainforest climatic zone into the seasonal tropics cannot be predicted with any confidence, though an expansion of rainforest climatic conditions within East Africa is likely.

In South and Southeast Asia the projected warming and associated higher saturation vapour pressure are expected to be accompanied by an increase in atmospheric moisture flux and its convergence/divergence intensity. Thus a general increase in rainfall is predicted over the Malesia and Southeast Asia (+7%), Indian subcontinent (+11%) and southern China (+9%) latitude/longitude zones (Table 8.10). Boer and Faqih (2004), however, found very contrasting patterns of change across Indonesia from five general circulation models (GCMs) and concluded that no generalisation could be made on impacts of global warming on rainfall in the region.

Many GCM models find it difficult to incorporate ENSO and its changes into future climatic predictions. Although those that do all predict the continued existence of an ENSO cycle, 'there is no consistent indication of future changes in ENSO amplitude or frequency' (IPCC, 2007, p. 780).

Increases in tropical cyclone and heavy rainstorm magnitude and intensity are predicted by IPCC (2007) with more confidence than are changes in annual rainfall in the wet tropics. The predictions are based on a mixture of theoretical considerations, modelling results and recent trends, with the evidence of recent upswings in tropical cyclone intensity and frequency and extreme rainstorms (see Section 8.3) accounting for some of the increased confidence. The theoretical bases for increased extremes are essentially simple. A hotter atmosphere is capable of holding more water vapour and sustaining higher rainfall intensities for longer durations (IPCC, 2007). Higher SSTs will mean that the SST threshold criterion for hurricane development is exceeded for more of the time and over an expanded area of the world's oceans and will be capable of sustaining more powerful hurricanes. Predictions of the scale and spatial distribution of tropical cyclone frequency changes at the regional level, however, remain vague (IPCC, 2007).

It is generally considered that what happens to the remaining forest and replacement land use in the humid tropics will be an important determinant and modifier of both regional and global future climate change. Key issues involved include:

(a) rates of deforestation and the effectiveness of international initiatives to support rainforest conservation and sustainable management;
(b) evapotranspiration and greenhouse gas cycling dynamics of dominant replacement land covers;
(c) the response of remaining forest to climate change and increased fire risk;
(d) the scale of dam and reservoir construction on the Amazon and other rivers in the rainforest zone and their effects on evapotranspiration and biogeochemical cycling;
(e) anthropogenic impacts on aerosols and hence on the heat budget and cloudiness and rainfall patterns; and
(f) rates of urbanisation.

Some of the above issues are little understood. For example even basic data on flux emissions of some of the major plantation land uses such as palm oil are lacking (D. Fowler, personal communication, 2008), but are essential to provide meaningful inputs to GCMs that are simulating different scenarios for the wet tropical part of the globe. Others are uncertainties that will influence the appropriate scenarios to use for future climate predictions. The Amazon reservoir issue is an interesting one, as potentially it could (together with creating swamp land areas) provide a means of enhancing evapotranspiration (and countering its reduction due to deforestation and enhanced carbon dioxide levels) and maintaining Amazonian rainfall levels. All the above will fundamentally affect global climate and the degree of validity of the current IPCC predictions.

8.5.2 Rainforest responses to climate change

Critical both to geomorphic responses and feedbacks to global warming will be the response of the tropical rainforest biome to predicted climate change. Although past responses to climate change can yield some insight (see Section 8.1), 'The forest is now, however, entering a set of climatic conditions with no past analogue' (Maslin *et al.*, 2005, p. 477).

Key issues include:

(a) the impact of rising atmospheric carbon dioxide levels on tree growth and forest extent;
(b) the impact of higher temperatures on tree growth and forest extent;
(c) the impact on forests of increased dry-period magnitude–frequency;
(d) the impact (especially in marginal areas) of predicted increases and decreases in annual rainfall;
(e) the increased threat of fire to an increasingly fragmented, drought-prone forest;

(f) rates of forest loss to other land uses;
(g) speed and nature of forest responses to altered climates; and
(h) the question of 'tipping points' beyond which rainforest is unable to survive.

Some of these issues are discussed in more detail below.

Rising atmospheric carbon dioxide

Higher atmospheric carbon dioxide should mean that photosynthesis becomes more efficient such that either transpiration will be reduced or carbon assimilation and growth rates increase. Evidence to support the latter comes from 59 long-term forest plots in Amazonia (Baker *et al.*, 2004) where the above-ground biomass in trees above 10 cm diameter has increased by 1.22 ± 0.43 Mg ha^{-1} a^{-1} over the previous 20 years, with the clear implication that the Amazonian forest is acting currently as a net carbon sink. Whether this will be sustainable and result in taller, denser forests, or simply mean shorter life cycles of rainforest trees, is unclear, as the Amazonian plots also recorded increased mortality rates; in addition much of the increased carbon was stored in the canopy trees and lianes rather than in the understorey (Phillips *et al.*, 2004). It may also be the case that increased carbon storage is only possible because the Amazonian forest has ample water supply from perennial rainfall in western Amazonia and from taproot access to deep soil water and groundwater in eastern Amazonia. Such conditions might not apply in more seasonal or drought-prone rainforest locations.

Higher temperatures

Higher temperatures (typically increased by 2.5–3.6 °C by the late twenty-first century) should lead to significant poleward expansions of the frost-free zone and rainforest conditions into currently humid subtropical areas. The 18 °C coolest month isotherm is likely to advance to 23° N in southern China, 31° S in southeastern Brazil, 30° N in the southeastern United States, 30° S in eastern Australia and 32° S in southeastern Africa. Morley (2000) considered how rainforest would theoretically expand under two scenarios of temperature rise (greenhouse and super greenhouse) and the above advances approximate to his greenhouse scenario. In these marginal areas, actual rainforest expansion would be severely limited by forest clearance and fragmentation. The likely impact in existing rainforest areas is more disputed. Higher temperatures may accelerate growth and nutrient cycling, but also lead to increased respiration and transpiration. Key questions include:

(a) whether tropical rainforests are already close to an upper temperature limit or optimum, above which productivity and viability may decline or collapse; and
(b) whether nutrients are available to sustain increased growth, which may only be the case in geologically young areas.

Also higher transpiration may be sustainable in areas with continuous water availability, but not in areas on the semi-evergreen seasonal forest margins, where soil water storages are limited because of shallow soils, steeper relief and non-aquifer underlying rocks.

More frequent and intense dry periods and fire risks with a possibly more intense ENSO cycle

Some (but by no means all) GCM models predict that the ENSO cycle may intensify leading to more frequent and severe droughts, even in areas with increased annual rainfall. Studies of the impacts of the severe 1982–83 and 1997–98 ENSO droughts suggest that dry periods play essential but varying roles in the natural dynamics of rainforests and often indicate a resilience to drought. Also it is not the absolute severity of a dry period that is important, but whether it is unusual for the location in question. Table 8.11 summarises the findings of some of these studies. In the case of East Kalimantan, where ENSO droughts tend to be particularly severe, large-scale canopy tree death and crown die-back leading to numerous canopy gaps occurred in 1982–83 (Leighton and Wirawan, 1986). At Danum Valley, where extreme droughts are much shorter, a dry period of just two successive months with less than 100 mm following lower than average rain the previous winter resulted in over 50% leaf-fall of canopy trees but little canopy tree death, and enhanced growth rates of some understorey species in response to partial die-back of canopy tree crowns. In the 2005 drought in eastern Amazonia, the canopy trees remained green and transpiration actually increased as taproots were able to continue to draw on shallow groundwater. In all three cases, the droughts were regarded as part of the long-term climate and playing essential roles in the long-term dynamics of each forest. The great danger, however, is that fire will more often accompany droughts than formerly as a result of encroachment by other land uses and the greater fuel loading provided under logged forest. The ecological consequences of drought and drought plus fire are very different. As the East Kalimantan example demonstrated, whereas drought alone leads to a successor cohort comprising pre-drought sapling trees, drought plus fire preferentially destroys the understorey trees and gap recolonisation comes from seed germination. Species composition tends to be radically altered. Geomorphological consequences include a subsequent erosional and nutrient flux episode.

TABLE 8.11. *Forest responses to major droughts and dry periods associated with ENSO events in East Kalimantan (Leighton, personal communication, 1984); Sabah (Walsh and Newbery, 1999; Newbery and Lingenfelder, 2004) and southwestern and central Amazonia (Saleska et al., 2007)*

Location	ENSO drought	Immediate impacts and long-term role
East Kalimantan	1982–83	Drought only: – 37–71% canopy tree mortality on slopes/ridges – 11% canopy tree mortality in valley bottoms – Survival of understorey and lianes – Numerous canopy gaps – Cohort of shade-bearer sapling-derived trees in successor forest – Forest with species composition and age–size distribution adapted to episodic drought
	1982–83	Drought and fire: – Similar canopy tree mortality as for drought – Destruction of understorey saplings and lianes – Regrowth from seed (and root-resprouters) – Cohort of light-demanding seed-germinated trees in future forest – Less biodiverse forest with species composition and age–size distribution adapted to fire survival
Danum, Sabah	1997–98	Drought but only two successive months <100 mm (most severe in 25 years) – Considerable leaf-fall from canopy trees – Partial shut-down of the transpiration stream – Partial opening up of the canopy with branch death – Preferential growth of some understorey tree species – Uneven forest age–size distribution due to history of past droughts (some probably more severe)
SW and C Amazonia	2005	Moderate ENSO event – Increased greenness detected by remote sensing during the drought (except in human-affected and edaphically unfavourable areas) – Enhanced transpiration and carbon assimilation facilitated by taproots drawing on deep water – Degree of resilience to drought indicated, but query as to response to larger or more frequent droughts – Fire and deforestation seen as greater threats

Overall prospects for the tropical rainforest if high rates of deforestation continue

Maslin *et al.* (2005) see continued deforestation and its impact on regional rainfall as the greatest threat to the remaining Amazonian rainforest. Some 16% of the Amazon forest was lost in the twentieth century, predominantly to pasture and soya bean. The current rate of loss of 0.38% a^{-1} is predicted to rise with road development programmes and increased demand for biofuels from agricultural crops. The impact on regional rainfall is still hotly debated and uncertain. If the Salati and Nobre (1991) hypothesis that deforestation will lead to reduced transpiration, increased river flow and a drier Amazon atmosphere is accepted, then continued deforestation of the Amazonian forest should lead to a decline in rainfall in the region, thereby also affecting any remaining forest. Simulations of the UK Hadley Centre Model suggest that much of eastern Amazonia, which is already somewhat seasonal and marginal, is under threat of slipping into a permanent El Niño state with sharply reduced rainfall, leading to the large-scale replacement of rainforest by savannas (Cox *et al.*, 2004). Based on this scenario, Maslin *et al.* (2005) have argued that the remaining Amazonian rainforest is likely to contract to a narrower latitudinal band approximately 5° N and S of the equator as in the Last Glacial, though for very different reasons. They therefore see the effects of regional decline in rainfall exceeding the reduction in transpiration resulting from further increases in atmospheric carbon dioxide. It has also been argued that forest clearance may reduce rainfall by increasing the number of smoke and dust nuclei and reducing the chances of each developing into large raindrops as happens over forest. Such predictions are clearly unproven, but they do highlight the urgency of the need for international strategies including financial incentives to conserve the remaining forest and to encourage replacement land uses (such as some multistorey tree crop combinations) with high evapotranspiration.

8.5.3 Hydrological responses to climate change

The high degree of uncertainty of predicted changes in different climatic parameters and their spatial patterns within and adjacent to the rainforest zone, together with uncertainties about vegetational and human response, place severe constraints on the hydrological and geomorphological predictions that can be made. Some hydrological changes are more likely and definite than others. IPCC predicted impacts on soil moisture, runoff (river flow per unit area) and evaporation are given in Plate 5.

Changes in interception losses in rainforest areas will be dependent upon changes in annual rainfall, rainstorm frequency and rainstorm size distribution. Annual interception loss will increase (decrease) in those areas where annual rainfall and rainfall frequency increase (decrease), but increases (decreases) will be reduced if there is a change to a higher (lower) proportion of annual rainfall falling in large storm events. There may also be a marginal increase resulting from increased canopy drying rates as a consequence of increased temperature. Changes in transpiration are more difficult to predict. Rising atmospheric carbon dioxide levels mean that less water loss is needed in transpiration for the same carbon gain. Some have argued, therefore, that transpiration losses may fall (Costa and Foley, 2000; Cox *et al.*, 2004). The Amazonian plot evidence referred to earlier has so far indicated a net increase in forest biomass, albeit mostly within lianes and canopy trees. Much will also depend upon:

(a) any changes in the relative frequency of sunny versus rainy days, as transpiration losses are generally much higher on sunny days; and
(b) the availability of soil water for transpiration, which will vary with soil (notably depth) as well as rainfall factors.

It follows also that evapotranspiration changes may differ significantly from the simplistic IPCC predictions of Plate 5 of increases in areas with predicted increased annual rainfall and reductions in areas (like southeastern Amazonia) where reduced rainfall is predicted.

As Fig. 8.2 demonstrates, annual runoff (and hence also mean river flow) is highly sensitive to changes in annual rainfall, particularly in drier rainforest areas. Thus an increase in annual rainfall from 1500 to 2000 mm would be likely to result in a more than doubling in annual runoff from 200 to 500 mm. Plate 5 predicts increases of up to 150 mm per annum (0.4 mm day^{-1}) in parts of East Africa, Malesia and western equatorial South America, but reductions in parts of Central America and West Africa, where rainfall is predicted to fall or show little change. Changes in high flow and flood magnitude–frequency will be dependent upon changes in rainstorm magnitude–frequency. Although larger extreme rainstorms are generally predicted, the recent changes documented in Fig. 8.9 for stations in Sabah demonstrate that there may be very different trends in rainfalls of differing return period.

8.5.4 Geomorphological changes

Slope process changes

Significant changes in slope process rates are likely even in primary forest areas, but different processes are likely to respond to changes in different rainfall parameters. Weathering and chemical denudation rates will depend largely on changes in annual rainfall and annual runoff and only to a marginal extent on the rise in annual temperature. In principle, regional log-linear relations between chemical denudation rates (as derived from river solute loads) and annual runoff (Fig. 8.3) can be used to predict changes in river solute loads and chemical denudation in the short term. In the longer term, however, application of such a procedure may become inappropriate if the area is simultaneously being affected by a 'stripping' phase of slope instability involving landslides and gullying. Such a phase may enhance chemical denudation rates by increasing contact with less weathered material or even exposed parent rock, though this could be offset by the reduced contact effects of an increased proportion of overland flow and a reduced rock and particle surface area being attacked.

Slopewash rates will be affected by changes in annual rainfall, rainfall intensities and large rainstorm magnitude–frequency. The rainsplash component is primarily linked to the first two factors, whereas overland flow erosion is biased towards extreme events. The predicted increase in extreme rainstorms should lead to higher slopewash rates, but more particularly on soils prone to frequent overland flow. The often deep regoliths in the humid tropics may

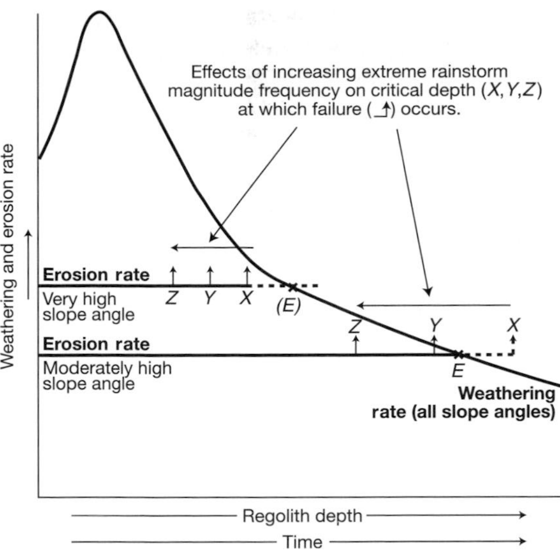

FIGURE 8.11. The critical depth of regolith at which slope failure occurs as a function of increasing extreme rainstorm magnitude and frequency.

sustain accelerating rates of gullying in cases where that occurs – at least until vertical erosion is replaced by declining rates of lateral erosion once dissection reaches bedrock. Under natural vegetation such rapid dissection is perhaps unlikely on a centennial timescale, but it is very possible with disturbance by logging or mining. The high erosion rates and highly dynamic nature of current piping systems (Sayer *et al.*, 2006) are likely to be enhanced if rainstorm magnitude and frequency increases. It is logical that pipe collapse rates may increase, thereby also providing a means of rapid drainage network extension.

In steep terrain, increased landsliding would be a logical outcome of predicted increased magnitude and frequency of rainstorms and (in some areas) increased tropical cyclone frequency and their spread to areas previously unaffected by them (such as eastern Brazil). Some currently stable slopes would become unstable at their current angles and regolith depth – as shown schematically in Fig. 8.11 by a leftward move in the critical regolith depth at which a landslide will be triggered. Saturated zones above impeding layers such as the soil–rock interface should achieve greater vertical extents in larger extreme rainstorms, thus rendering loss of cohesion and landsliding more likely at shallower regolith depths than hitherto. Such a landsliding phase would have important influences on the other parts of geomorphological systems, notably in terms of large, sudden increases in supply of bedload and suspended sediment to river systems, enhancing subsequent slopewash rates through bare areas of landslide scars, and potentially providing a mechanism for drainage network extension via conversion of linear landslide scars into valley-side ephemeral channels.

Fluvial processes and landscape-scale changes

Fluvial processes will be affected by any change in the amount and regime of runoff and sediment supplied to the river system. In areas remaining under rainforest, of key importance will be the predicted rise in the magnitude and frequency of large rainstorms as this influences both flood magnitude–frequency relations and sediment supply to channels from landslides in particular and also tunnelling, gullying and bank erosion. Consequences will include increased channel size (particularly if river systems were previously supply-limited) and higher suspended sediment, bedload and solute yields. Existing relations between suspended sediment yield and discharge variables will be arguably of very limited use, particularly in areas where landslides are significant, as the landscape will be in a state of rhexistasy and transition sediment yields are likely to exceed pre-existing relation predictions.

The most likely landscape variable to be affected by IPCC predictions of climate change is drainage density. Established relations between rainstorm magnitude and frequency and drainage density (Fig. 8.5) could be used to predict drainage density changes, but not the timescale involved, as so little is known about speeds and modes of network adjustment to past climate change or human activities in the wet tropics. The little evidence available suggests that adjustment rates can be rapid, but are likely to vary with the channel head processes that currently operate and the depth and erodibility of regolith or substrate. Rapid expansion could occur via either channel development in landslide scars or pipe extension, enlargement and collapse. Where overland flow erosion is the main driver of channel extension, however, adjustments may be much slower, because of the ground protection afforded by surface and near-surface root systems and the litter cover.

It is very difficult, given our current lack of knowledge of large catchment behaviour in the humid tropics, to predict the scale and character of downstream consequences of a more erosive phase stemming from increased large rainstorm magnitude–frequency. Factors that may be important include:

(a) Whether pre-existing river sediment transport rates are supply- or transport-limited in the upper and lower parts of catchments. Associated with this is whether floodplains in middle/lower catchment areas are currently actively flooded or incised and rarely flooded. Sediment transport in many rivers of the Australian tropics is considered to be supply-limited and hence increased hillslope disturbance in steep headwaters can be readily accommodated in terms of sediment transport through the catchment and channel enlargement without sedimentation and major channel shifting (Amos *et al.*, 2004), but this is not universally true in the humid tropics.

(b) The connectivity between new or enhanced sediment sources and the channel (and subsequently along river channels) is clearly important when considering the influence on downstream fluvial processes. Connectivity between slope and channel tends to be enhanced by the high landscape drainage densities of the humid tropics, which mean that distances from slope to the nearest channel (permanent or ephemeral) are often very short compared with in other humid environments.

(c) Steep, potentially unstable slopes: their presence/absence and spatial distribution within the catchment will determine the degree of susceptibility to a large-scale 'stripping' phase and the acute disequilibrium chain of geomorphic activity that could ensue.

(d) Susceptibility of the catchment to rapid drainage network extension and/or dissection. This will be influenced by distribution of deep regoliths, piping and steep, landslide-prone slopes or (in moderate/low relief and floodplain parts of catchments) reworkable alluvial or colluvial deposits.

8.5.5 Geomorphic change in anthropogenically disturbed areas

Geomorphic change in areas affected by human activities is likely to increase greatly during the twenty-first century regardless of climate change with continued forest conversion and other land use changes. Responses to predicted climate change are likely to be greatly magnified compared to in areas under natural vegetation, as the terrain involved is either already in disequilibrium or the geomorphological thresholds for catastrophic impacts (such as landslides, floods and radical river channel change) are more easily exceeded. In general the following human factors will determine the overall geomorphic impact within catchments:

(a) History of forest disturbance: balance between newly disturbed/converted, established land use and undisturbed forest terrain units.

(b) Degree of contrast of land use(s) with the natural forest environment: notably per cent bare area of land use(s) and whether the replacement landscape has been 'engineered'.

(c) Percentage of a catchment affected by disturbed/replacement land uses.

(d) Spatial distribution of disturbed/replacement land uses within catchments in relation to relief and naturally vulnerable or connected parts of the inherited landscape.

(e) The spatial extent and effectiveness of any soil conservational measures adopted.

(f) The degree of protection given to exceptionally vulnerable landscape areas, notably steep slopes, headwater areas and riparian zones.

Possible impacts in managed forest areas

There are three main ways in which current impacts of logging may be enhanced or modified via predicted IPCC climate change:

(a) increased rainstorm magnitude and frequency implies shorter return periods of extreme events and therefore an increased likelihood that an extreme rainstorm event (and hence enhanced erosion) will occur during periods when logged forest terrain is at its most vulnerable, namely (i) during and for the first 2 years after logging, when the percentage bare area is high and (ii) up to at least 8 years after logging as regards landslide risk (Walsh *et al.*, 2006);

(b) the possibility of an intensified ENSO cycle would bring an increased risk of fire. The net effect would be significantly increased erosional effects of rainforest logging and a less productive forest; and

(c) impacts of climate change will depend fundamentally on the degree of adoption of conservational forest management practices by the countries involved.

This in turn will depend largely upon the outcome of economic initiatives and incentives (paid in part at least by rich, developed countries) for tropical countries and their local communities to retain their forests and manage them in a sustainable way. The knowledge to accomplish this exists both in terms of the logging practices (RIL protocols) and forest certification organisations to approve and monitor the implementation of good practice, but the adoption of conservation practices needs to be seen by people as economically advantageous. The most likely scenario is that of a spectrum of responses across the humid tropics, with the more developed, educated and politically stable countries increasingly implementing more sustainable policies, but poorer and politically unstable or corrupt countries continuing to exert little control over land use and forest conservation.

Likely impacts in replacement land use areas

Erosional impacts of predicted increases in rainstorm magnitude frequency would be particularly severe where agriculture spreads into areas of steep relief, as widespread landsliding would be likely (see Plate 23 for erosional impacts of oil palm cultivation at Danum, Sabah). Effects are likely to be greater for peasant agriculture than for company plantations, because of the latter's greater investment in and maintenance of bench terracing, drainage systems and soil quality. Increases in sediment yield are likely to be very high. Trajectories of slope erosion will vary with sediment availability. In areas of deep regolith and unconsolidated lithology, erosion may remain transport-limited and could increase if gullying develops; in areas with cohesive subsoils or hardpans, periods of high erosion rate may be shortlived. Downstream consequences on fluvial systems of predicted climate change are likely to be greatly enhanced compared with both natural forest and logged forest cover.

Likely impacts in urban areas

Key impacts of predicted enhanced extreme rainstorm (and where applicable) cyclone events are likely to include:

(a) susceptibility to landslides, particularly but not exclusively in poorer settlements in cities;
(b) enhanced bank erosion, flooding and floodplain sedimentation due to the inability of current natural channels within cities to carry enhanced storm flows.

Landsliding is probably the most dangerous geomorphic hazard involved. Engineering adjustments and protective measures can be made to slope drainage to try and prevent critical parts of soils becoming saturated and losing the cohesive fraction of their shear strength; such measures include deliberately sealing soil surfaces so as to encourage maximum overland flow response and reduce the supply of percolating water to critical subsurface horizons. Thus in economically advanced cities like Singapore and Hong Kong, concreting over steep slopes that are criss-crossed by roads can be observed.

8.6 Research gaps and priorities for improvement to geomorphological predictions in the humid tropics

8.6.1 Climatic and geomorphological research priorities

There are seven research priorities as follows.

(a) Improved climatic modelling for tropical areas is an acute need. Greater precision and agreement between models is required on predicted rainfall changes in rainforest regions. Also, although increases in the magnitude of some extreme events (rainstorms and tropical cyclones) are confidently predicted, predictions remain vague and predictions regarding the future of the ENSO cycle are particularly inconclusive.
(b) More research is needed on (i) the climatology of large rainstorms in different parts of the tropics including their varied synoptic causes and (ii) changes through time using old meteorological archives. Studies also need to analyse areal as well as point rainstorms.
(c) There is a need for baseline fluvial studies in the wet tropics on channel patterns, bedload and channel morphology both in undisturbed rainforest and in spatially extensive replacement land use types, such as oilseed and maize in Amazonia and palm oil in Malesia.
(d) A major need is for more large catchment research focussing on (i) responses to past climatic changes, (ii) responses to long-term anthropogenic change, (iii) sediment delivery, floodplain sediment storage and the impact of confluences, and (iv) terrestrial components of biogeochemical fluxes.
(e) There is considerable, but as yet largely untapped, potential for using a sediment fingerprinting approach to investigate both current, and historical changes in, sediment sources, processes and sediment delivery. The prospects for multi-proxy sediment fingerprinting are promising. Thus Fletcher and Muda (1999) used comparisons of downstream changes in the trace metal content of bed sediment of unlogged and logged rivers in Sabah to assess the impact on logging on surface and landslide inputs. More recently Blake *et al.* (2006) demonstrated the potential for using different combinations of fine bed sediment properties to investigate sediment sources at first-order to large-catchment scales in Sabah.
(f) Synthetic analytical studies of existing empirical data. As in many areas of science, there has been an explosion in empirical research, but too little attention is given to synthetic analysis of the findings. This chapter has identified a few aspects of geomorphology where variations in process rates or landscape variables have been analysed and related (albeit not entirely satisfactorily) to variations in governing variables (including climatic factors). Such an approach needs to be extended to other processes and landscape variables.
(g) Modelling studies associated with the above priorities.

8.7 Summary and conclusions

The following nine conclusions provide an appropriate summary of findings:

(a) The tropical rainforest zone contains a range of climates of differing annual rainfall, degree of seasonality and extreme-event magnitude–frequency relations.

(b) Unifying features of the geomorphology of the wet tropics include high potential for chemical denudation, high drainage densities, the episodic nature of sediment transport, transport-limited slopes and supply-limited river sediment systems, and the importance of landsliding.
(c) Geomorphological attributes that vary greatly with climatic variables (or hydrological factors linked to climate) *within* the zone include: solute yields (with annual rainfall and annual runoff); slopewash and suspended sediment yields (with annual rainfall and the frequency of large rainstorms); and drainage density (with large rainstorm magnitude–frequency). To some degree impacts of climate change can be predicted for these variables using these relations.
(d) In contrast, very little is known about the dynamics and controls of channel morphology, channel patterns and sediment delivery within large catchments.
(e) The rainforest zone is highly sensitive to anthropogenic disturbance, which is likely to continue to outweigh the impacts of climate change.
(f) The IPCC (2007) predictions of annual rainfall and ENSO cycle intensity by the end of the twenty-first century in the rainforest zone are very uncertain. Increases in rainstorm extremes and tropical cyclone frequency are confidently but imprecisely predicted and are of concern given evidence of their increase in recent years – and their potentially major geomorphological impacts.
(g) Given the gaps and uncertainties in knowledge and understanding of landscape in the humid tropics – and the uncertainties about predicted climate change in the wet tropics – predictions of landscape change can only be speculative.
(h) Increased slope instability, leading to a landslide stripping phase and downstream fluvial disequilibrium, is considered to be the main possible geomorphological consequence for undisturbed forest areas of the predicted significant increase in rainstorm magnitude–frequency. Consequences would be greatly magnified in anthropogenically disturbed areas.
(i) A key identified research need is for large catchment research focussing on (i) responses to past climatic changes, (ii) responses to long-term anthropogenic change, (iii) sediment delivery, floodplain sediment storage and the impact of confluences and (iv) terrestrial biogeochemical fluxes.

References

Alexander, L. V. *et al.* (2006). Global observed changes in daily climate extremes of temperature and precipitation. *Journal of Geophysical Research*, **111**, D05109, doi:10.1029/2005JD006290.

Amos, K. J. *et al.* (2004). Supply limited sediment transport in a high-discharge event of the tropical Burdekin River, North Queensland, Australia. *Sedimentology*, **51**, 145–162.

Anderson, J. M. and Spencer, T. (1991). *Carbon, Nutrient and Water Balances of Tropical Rainforest Ecosystems Subject to Disturbance: Management Implications and Research Proposals*, MAB Digest No.7. Paris: UNESCO.

Baillie, I. (1996). Soils of the humid tropics. In P. W. Richards, ed., *The Tropical Rain Forest: An Ecological Study*. Cambridge: Cambridge University Press, pp.256–286.

Baker, T. R. *et al.* (2004). Increasing biomass in Amazonian forest plots. *Philosophical Transactions of the Royal Society of London B*, **359**, 353–365.

Blake, W. H. *et al.* (2006). Quantifying fine-sediment sources in primary and selectively logged rainforest catchments using geochemical tracers. *Water, Air and Soil Pollution: Focus*, **6**, 615–623.

Boer, R. and Faqih, A. (2004). *Current and Future Rainfall Variability in Indonesia*, AIACC Technical Report No. 021. Available at http://sedac.ciesin.colombia.edu/aiacc/progress/AS21_Jan04.pdf.

Bonell, M. and Gilmour, D. A. (1978). The development of overland flow in a tropical rainforest catchment. *Journal of Hydrology*, **39**, 365–382.

Chappell, N. A. *et al.* (2004). Sources of sediment within a tropical catchment recovering from selective logging. *Hydrological Processes*, **18**, 685–701.

Chorley, R. J., Schumm, S. A. and Sugden, D. E. (1984). *Geomorphology*. London: Methuen.

Clarke, M. A. and Walsh, R. P. D. (2006). Long-term erosion and surface roughness change of rainforest terrain following selective logging, Danum Valley, Sabah, Malaysia. *Catena*, **68**, 109–123.

Colinvaux, P. A. and de Oliveira, P. E. (2000). Amazon plant diversity and climate through the Cenozoic. *Palaeogeography, Palaeoclimatology, Palaeoecology*, **166**, 51–63.

Condit, R. (1998) Ecological implications of changes in drought patterns: shifts in forest composition in Panama. *Climatic Change*, **39**, 413–427.

Costa, M. H. and Foley, J. A. (2000). Combined effects of deforestation and doubled atmospheric CO_2 concentration on the climate of Amazonia. *Journal of Climate*, **13**, 35–58.

Cox, P. M. *et al.* (2004). Amazonian forest dieback under climate-carbon cycle projections for the twenty-first century. *Theoretical and Applied Climatology*, **78**, 137–156.

Day, M. J. (1980). Landslides in the Gunong Mulu National Park. *Geographical Journal*, **146**, 7–13.

Douglas, I. (1994). Land degradation in the humid tropics. In N. Roberts, ed., *The Changing Global Environment*. Oxford: Blackwell, pp. 332–350.

Douglas, I. (1999). Hydrological investigations of forest disturbance and land cover impacts in South-East Asia: a review. *Philosophical Transactions of the Royal Society of London B*, **354**, 1725–1738.

Douglas, I. and Guyot, J. L. (2009). Erosion and sediment yield in the humid tropics. In M. Bonell and L. A. Bruijnzeel, eds., *Forests, Water and People in the Humid Tropics*. Cambridge: Cambridge University Press, pp. 407–421.

Douglas, I. and Spencer, T. (1985). Present-day processes as a key to the effects of environmental change. In I. Douglas and T. Spencer, eds., *Environmental Change and Tropical Geomorphology*. London: George Allen and Unwin, pp. 39–73.

Douglas, I. *et al.* (1999). The role of extreme events in the impacts of selective tropical forestry on erosion during harvesting and recovery phases at Danum Valley, Sabah. *Philosophical Transactions of the Royal Society of London B*, **354**, 1749–1761.

Dunne, T. (1978). Rates of chemical denudation of silicate rocks in tropical catchments. *Nature*, **274**, 244–246.

Dunne, T. (1979). Sediment yield and land use in tropical catchments. *Journal of Hydrology*, **42**, 281–300.

Dykes, A. P. (1995). Regional denudation by landslides in the tropical rainforest of Temburong District, Brunei. *International Association of Geomorphologists Southeast Asia Conference*, Singapore, 18–23 June 1995, p. 39.

Elsenbeer, H. and Vertessy, R. A. (2000). Stormflow generation and flowpath characteristics in an Amazonian rainforest catchment. *Hydrological Processes*, **14**, 2367–2381.

Emanuel, K. A. (2005). Increasing destructiveness of tropical cyclones over the past 30 years. *Nature*, **436**, 686–688.

Flenley, J. (1979). *The Equatorial Rainforest: A Geological History*. London: Butterworth.

Fletcher, W. K. and Muda, J. (1999). Influence of selective logging and sedimentological process on geochemistry of tropical rainforest streams. *Journal of Geochemical Exploration*, **67**, 211–222.

Garner, H. F. (1974). *The Origin of Landscapes*. New York: Oxford University Press.

Gellis, A. (1993). The effects of Hurricane Hugo on suspended-sediment loads, Lago Loiza basin, Puerto Rico. *Earth Surface Processes and Landforms*, **18**, 505–517.

Gibbs, R. J. (1967). The geochemistry of the Amazon river system: 1. The factors that control the salinity and concentration of the suspended solids. *Bulletin of the Geological Society of America*, **78**, 1203–1232.

Goldenberg, S. B. *et al.* (2001). The recent increase in Atlantic hurricane activity: causes and implications. *Science*, **293**, 474–479.

Gomi, T. *et al.* (2006). Sediment and wood accumulations in humid tropical headwater streams: effects of logging and riparian buffers. *Forest Ecology and Management*, **224**, 166–175.

Haffer, J. (1969). Speciation in Amazonian forest birds. *Science*, **165**, 131–137.

Haffer, J. and Prance, G. T. (2001). Climate forcing of evolution in Amazonia during the Cenozoic: on the refuge theory of biotic differentiation. *Amazoniana*, **16**, 579–607.

IPCC (2007). *Climate Change 2007: The Physical Science Basis. Contribution of Working Group I to the Fourth Assessment Report of the Intergovernmental Panel on Climate Change.* Solomon S. *et al.*, eds. Cambridge: Cambridge University Press. (pp. 237–336, Trenberth, K. E. *et al.*, eds.; pp. 747–843, Christensen, J. H. *et al.*, eds.)

Jackson, I. J. (1975). Relations between rainfall parameters and interception by tropical forest. *Journal of Hydrology*, **24**, 215–238.

Jordan C. F. and Kline J. R. (1977). Transpiration of trees in a tropical rainforest. *Journal of Applied Ecology*, **14**, 853–860.

Kastner, T. P. and Goñi, M. A. (2003). Constancy in the vegetation of the Amazon basin during the late Pleistocene. *Geology*, **31**, 291–294.

Lai, F. S. (1993). Sediment yield from logged, steep upland catchments in Peninsular Malaysia. *International Association of Hydrological Sciences Publication*, **216**, 219–229.

Larsen, M. C., Torres-Sánchez, A. J. and Concepción, I. M. (1999). Slopewash, surface runoff and fine-litter transport in forest and landslide scars in humid-tropical steeplands, Luquillo Experimental Forest, Puerto Rico. *Earth Surface Processes and Landforms*, **24**, 481–502.

Leighton, M. and Wirawan, N. (1986). Catastrophic drought and fire in Borneo tropical rain forest associated with the 1982–83 El Niño Southern Oscillation event. In G. T. Prance, ed., *Tropical Rain Forests and the World Atmosphere*. Boulder: Westview, pp. 75–102.

Lewis, L. A. (1974). Slow movement of earth under tropical rainforest conditions. *Geology*, **2**, 9–10.

Lloyd, C. R. *et al.* (1988). The measurement and modelling of rainfall interception by Amazonian rain forest. *Agricultural and Forest Meteorology*, **42**, 63–73.

Mahé, G., Servat, E. and Maley, J. (2004). Climatic variability in the tropics. In M. Bonell and L. A. Bruijnzeel, eds., *Forests, Water and People in the Humid Tropics*. Cambridge: Cambridge University Press, pp. 267–286.

Malmer, A. (1990) Stream suspended sediment load after clear-felling and different forestry treatments in tropical rain forest, Sabah, Malaysia. *International Association of Scientific Hydrology Publication*, **192**, 62–71.

Malmer, A. (2004). Streamwater quality as affected by wild fires in natural and manmade vegetation in Malaysian Borneo. *Hydrological Processes*, **18**, 853–864.

Manton, M. J. *et al.* (2001). Trends in extreme daily rainfall and temperature in Southeast Asia and the South Pacific: 1961–1998. *International Journal of Climatology*, **21**, 269–284.

Maslin, M. *et al.* (2005). New views on an old forest: assessing the longevity, resilience and future of the Amazon rainforest. *Transactions of the Institute of British Geographers (NS)*, **30**, 477–499.

Meehl, G. A. *et al.* (1976). The role of climate in the denudation system: a case study from West Malaysia. In E. Derbyshire, ed., *Geomorphology and Climate*. Chichester: John Wiley, pp. 317–343.

Morley, R. J. (2000). *Origin and Evolution of Tropical Rain Forests*. Chichester: John Wiley.

Morley, R. J. and Flenley J. R. (1987). Late Cenozoic vegetational and environmental changes in the Malay archipelago.

In T. C. Whitmore, ed., *Biogeographical Evolution of the Malay Archipelago*. Oxford: Clarendon Press, pp. 50–59.

Nepstad, D. C. *et al.* (1994). The role of deep roots in the hydrological and carbon cycles of Amazonian forests and pastures. *Nature*, **372**, 666–669.

Newbery, D. M. and Lingenfelder, M. (2004). Resistance of a lowland rainforest to increasing drought intensity in Sabah, Borneo. *Journal of Tropical Ecology*, **20**, 613–624.

Nortcliff, S. and Thornes, J. B. (1981). Seasonal variations in thc hydrology of a small forested catchment near Manaus, Amazonas, and their implications for its management. In R. Lal and E. W. Russell, eds., *Tropical Agricultural Hydrology*. Chichester: John Wiley, pp. 37–57.

Odemerho, F. O. (1984). The effects of shifting cultivation on stream channel size and hydraulic geometry in small headwater basins of south-western Nigeria. *Geografiska Annaler*, **66**A, 327–340.

Pennington, R. T., Prado, D. A. and Pendry, C. (2000). Neotropical seasonally dry forests and Pleistocene vegetation changes. *Journal of Biogeography*, **27**, 261–273.

Phillips, O. L. *et al.* (2004). Pattern and process in Amazon tree turnover 1976–2001. *Philosophical Transactions of the Royal Society of London B*, **359**, 381–407.

Prance, G. T. (1987). Vegetation. In T. C. Whitmore and G. T. Prance, eds., *Biogeography and Quaternary History in Tropical America*. Oxford: Oxford Scientific Publications, pp. 28–44.

Richards, P. W. (1996). *The Tropical Rain Forest: An Ecological Study*, 2nd edn. Cambridge: Cambridge University Press.

Rodbell, D. T. *et al.* (1999). An ~15 000-year record of El Niño-driven alluviation in southwestern Ecuador. *Science*, **283**, 516–520.

Rose, J. (1984). Contemporary river landforms and sediments in an area of equatorial rain forest, Gunung Mulu National Park, Sarawak. *Transactions of the Institute of British Geographers (NS)*, **9**, 345–363.

Rouse, W. C., Reading, A. J. and Walsh, R. P. D. (1986). Volcanic soil properties in Dominica, West Indies. *Engineering Geology*, **23**, 1–28.

Salati, E. and Nobre, C. A. (1991). Possible climatic impacts of tropical deforestation. *Climatic Change*, **19**, 177–196.

Saleska, S. R. *et al.* (2007). Amazon forests green-up during 2005 drought. *Science*, **318**, 612.

Sayer, A. M. *et al.* (2004). The role of pipe erosion and slopewash in sediment redistribution in small rainforest catchments, Sabah, Malaysia. *International Association of Hydrological Sciences Special Publication*, **288**, 29–36.

Sayer, A. M., Walsh, R. P. D. and Bidin, K. (2006). Pipeflow suspended sediment dynamics and their contribution to stream sediment budgets in small rainforest catchments, Sabah, Malaysia. *Forest Ecology and Management*, **224**, 119–130.

Schimper A. F. W. (1903). *Plant Geography upon a Physical Basis*, trans. W. R. Fisher, P. Groom and I. B. Balfour, eds. Oxford: Clarendon Press.

Sidle, R. C. *et al.* (2004). Sediment pathways in a tropical forest: effects of logging roads and skid trails. *Hydrological Processes*, **18**, 703–720.

Spencer, T. *et al.* (1990). Vegetation and fluvial geomorphic processes in south-east Asian tropical rainforests. In J. B. Thornes, ed., *Vegetation and Erosion: Processes and Environments*. Chichester: John Wiley, pp. 451–469.

Stallard, R. F. (1995). Tectonic, environmental and human aspects of weathering and erosion: a global view using a steady-state perspective. *Annual Review of Earth and Planetary Science*, **23**, 11–39.

Stallard, R. F. and Edmond, J. M. (1987). Geochemistry of the Amazon: 3. Weathering chemistry and limits to dissolved inputs. *Journal of Geophysical Research*, **92**, C8, 8293–8302.

Summerfield, M. A. (1991). *Global Geomorphology*. Harlow: Longman.

Syvitski, J. P. M. *et al.* (2005). Impact of humans on the flux of terrestrial sediment to the global coastal ocean. *Science*, **308**, 376–380.

Terry, J. P. (1999). Kadavu island, Fiji: fluvial studies of a volcanic island in the humid tropical South Pacific. *Singapore Journal of Tropical Geography*, **20**, 86–98.

Thang, H. C. and Chappell, N. A. (2004). Minimising the hydrological impact of forest harvesting in Malaysia's rain forests. In M. Bonell and L. A Bruijnzeel, eds., *Forests, Water and People in the Humid Tropics*. Cambridge: Cambridge University Press, pp. 852–865.

Thomas, M. F. (1994). *Geomorphology in the Tropics*. Chichester: John Wiley.

Thomas, M. F. (2004). Landscape sensitivity to rapid environmental change: a Quaternary perspective with examples from tropical areas. *Catena*, **55**, 107–124.

Thomas, M. F. (2006). Lessons from the tropics for a global geomorphology. *Singapore Journal of Tropical Geography*, **27**, 111–127.

Tricart, J. (1985). Evidence of Upper Pleistocene dry climates in northern South America. In I. Douglas and T. Spencer, eds., *Environmental Change and Tropical Geomorphology*. London: George Allen and Unwin, pp. 197–217.

Walsh, R. P. D. (1980). Runoff processes and models in the humid tropics. *Zeitschrift für Geomorphologie (N. F. Supplement)*, **36**, 176–202.

Walsh, R. P. D. (1982). Hydrology and water chemistry. In A. C. Jermy and K. P. Kavanagh, eds., *Gunung Mulu National Park, Sarawak: An Account of its Environment and Biota Being the Results of The Royal Geographical Society/Sarawak Expedition and Survey 1977–78*. Sarawak Museum Journal **XXX**, 51, S.I.2, 121–181.

Walsh, R. P. D. (1985). The influence of climate, lithology and time on drainage density and relief development in the tropical volcanic terrain of the Windward Islands. In I. Douglas and T. Spencer, eds., *Environmental Change and Tropical Geomorphology*. London: George Allen and Unwin, pp. 93–122.

Walsh, R. P. D. (1993). Problems of the climatic geomorphological approach in the humid tropics with reference to drainage density, chemical denudation and slopewash. *Würzburger geographische Arbeiten* **87**, 221–239.

Walsh, R. P. D. (1996a). Climate. In P. W. Richards, ed., *The Tropical Rain Forest: An Ecological Study*. Cambridge: Cambridge University Press, pp. 159–205.

Walsh, R. P. D. (1996b). Microclimates and hydrology. In P. W. Richards, ed., *The Tropical Rain Forest: An Ecological Study*. Cambridge: Cambridge University Press, pp. 206–236.

Walsh R. P. D. (1996c). Drainage density and network evolution in the humid tropics: evidence from the Seychelles and the Windward Islands. *Zeitschrift für Geomorphologie, Supplementband*, **103**, 1–23.

Walsh, R. P. D. (2000). Extreme weather events. In M. Pacione, ed., *Applied Geography: Principles and Practice*. London: Routledge, pp. 51–65.

Walsh, R. P. D. (2007). Adapting to climate variability and change. In I. Douglas, R. Huggett and C. Perkins, eds., *Companion Encyclopaedia of Geography*, vol. 2. Abingdon: Routledge, pp. 663–682.

Walsh, R. P. D. and Howells, K. A. (1988). Soil pipes and their role in runoff generation and chemical denudation in a humid tropical catchment in Dominica. *Earth Surface Processes and Landforms*, **13**, 9–17.

Walsh, R. P. D. and Newbery, D. M. (1999). The ecoclimatology of Danum, Sabah, in the context of the world's rainforest regions, with particular reference to dry periods and their impact. *Philosophical Transactions of the Royal Society of London B*, **354**, 1869–1883.

Walsh, R. P. D. *et al.* (2006). Changes in the spatial distribution of erosion within a selectively logged rainforest catchment in Borneo, 1988–2003. In P. N. Owens and A. J. Collins, eds., *Soil Erosion and Sediment Redistribution in River Catchments: Measurement, Modelling and Management*. Wallingford: CAB International, pp. 239–253.

Webster, P. J. *et al.* (2005). Changes in tropical cyclone number, duration and intensity in a warming environment. *Science*, **309**, 1844–1846.

Wierda, A., Veen, A. W. L. and Hutjes, R. W. A. (1989). Infiltration at the Tai Rain Forest (Ivory Coast): measurements and modelling. *Hydrological Processes*, **3**, 371–382.

Wiersum, K. F. (1984). Surface erosion under various tropical agroforestry systems. In *Proceedings of the Symposium on the Effects of Forest Land Use on Erosion and Slope Stability*, Honolulu, pp. 231–239.

9 Tropical savannas

Michael E. Meadows and David S. G. Thomas

9.1 Introduction

Since our human lineage evolved in the open woodlands and grasslands of Africa, savanna landscape images may lie in our genetic memory and, indeed, may be subliminally familiar to us all. Certainly they are among the more widespread of the global zonal systems that are known as major biomes and their extent and diversity lends importance to the need to understand their possible responses to environmental change. Vast areas of the globe are characterised by seasonal tropical climates that support a characteristic range of landforms and geomorphological processes. 'Savanna' (alternatively savannah) is a word of Amerindian origin initially used in the sixteenth century to describe grasslands in the West Indian islands (Shorrocks, 2007). The term is now most frequently associated with extensive plains in the continental interiors of Africa, South America, India and Australia where there are abundant mixed tree and grass vegetation formations. Although in essence the word has an ecological or biogeographical meaning, the strong interrelationships between climate, vegetation, soils, geomorphology and even geology (Fig. 9.1) entail that it is applied to the description of landscapes in general and therefore has geomorphological meaning. The word is interpreted differently in different parts of the world and, because savannas are extraordinary repositories of both biological and geomorphological diversity – due in no small part to their widespread geographical nature – there is no simple or universally agreed definition. Nevertheless, common ecological elements of savannas would likely include a continuous C4 grass layer with varying proportions of C3 trees and shrubs, while geomorphologically they are characteristically associated with plains and inselbergs on deeply weathered surfaces. Given the seasonal nature of the rainfall regime in these regions, fire is a consistent component of the natural environment and of considerable ecological and, through a complex of interactions, even geomorphological importance. The proportion of woody elements in savannas varies greatly and the definition certainly includes pure grasslands, while in central Africa the *miombo* is sometimes described as forest. Given their juxtaposition between the tropical rainforests and the mid-latitude deserts, tropical savannas are very sensitive to climate change and have responded so repeatedly in the geological past. This chapter, therefore, argues that current rates and patterns of global change represent a particularly significant threat to the maintenance of their geomorphological and biological integrity – the more so because savannas are the dominant biome in so many developing nations that may lack the infrastructure and economic flexibility to adapt to or mitigate the change.

9.1.1 Distribution, nature and complexity

Although the principles of climatic geomorphology are not without problems, the global distribution of tropical savanna landscapes (Plate 1) is integral to the spatial extent of climates with strongly seasonal summer rainfall regimes at low latitudes. As Thomas (1994) points out, the argument for a distinctive tropical geomorphology, and in this case for a *savanna* geomorphology, is based on the nature of the balance between weathering, involving the accumulation of residual materials, and the means and rate by which these products are removed by erosion. Since many such processes are chemical or biogeochemical, it is hardly surprising that there is an identifiable element of climatic control. Thus, the contemporary distribution of tropical savanna *ecosystems* is mirrored in the distribution of tropical climates with a pronounced dry season or seasons, although the obvious influence of former environmental conditions does mean that the *landscape features* associated with those ecosystems may be inherited from the past, or be polygenetic

Geomorphology and Global Environmental Change, eds. Olav Slaymaker, Thomas Spencer and Christine Embleton-Hamann. Published by Cambridge University Press.

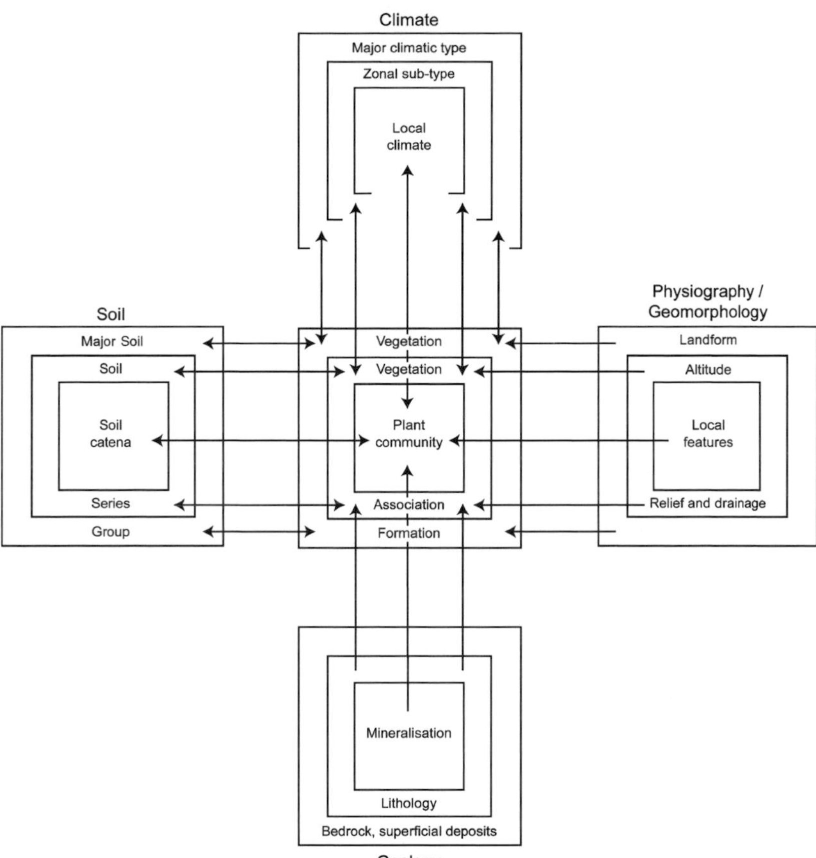

FIGURE 9.1. Interrelations between climate, vegetation, soils, geomorphology and geology in the savanna (modified from Cole, 1986).

in nature. Indeed, there is 'no neat zone of tropical landforms' (Thomas, 2006, p. 113) and there is limited justification for the concept of morphogenetic regions that are static in time and space.

The modern distribution of savanna ecosystems is, nevertheless, to some extent reflected in the mid-Holocene biome map of the introductory chapter (Plate 2), although this is something of an artificial view because the nature and extent of human activity, especially during the last 100 years or so, has markedly disturbed and even destroyed savanna vegetation in many of the regions indicated, while at the same time leading to its expansion into regions that were tropical rainforest during the altithermal. Such a scene is in any case only a coarse-focus 'snapshot' in time and must be seen against the longer-term history of the Earth's dynamic environments. The distribution in the present is also reflective of climates that are drier now in many regions than during the mid-Holocene. Broadly speaking, savannas are today found across a fifth of the Earth's terrestrial land surface and represent some 33 M km^2 (Woodward *et al.*, 2004) including large areas surrounding the Amazon Basin in South America, across substantial parts of Africa, especially through central north Africa and southern Africa, virtually throughout the Indian subcontinent and in significant parts of Southeast Asia and Australia.

Climatic variables clearly underlie this distribution which is characterised by a seasonal rainfall regime in which there is available moisture during the summer when conditions are most favourable to productivity and significant moisture deficits during the dry cooler months. The nature and length of the dry season distinguishes savannas from neighbouring biomes. At the wetter end of the moisture gradient, where the rainfall is aseasonal, or at least more evenly distributed, savanna merges into tropical rainforest, while at the more arid end there is a transition to desert scrub. At the margin of the equatorial forests, mean annual precipitation may approach 2000 mm, while annual values of only around 250 mm characterise the boundary with the deserts. This is a substantial range and, together with variations in geology, soil, topography and in the nature and intensity of land use, accounts for the very substantial range of vegetation types associated with this biome (Cole, 1986). Indeed, as Collinson (1988, p. 232) notes, 'no better example of the complex, intricate mosaic patterning of communities can be found than those which exist in the savannas'.

The fundamental description of different forms of savanna is essentially an ecological one made on the basis of the nature of the relationship between climate, more specifically precipitation quantity, and vegetation and generally revolves around the relative balance of woody and grass cover. There is conventionally a classification of savannas into 'wetter' and 'drier' forms on the basis of the mean annual rainfall and the length of the dry season or seasons and their ecophysiological manifestation. Wetter savannas are dominated by panicoid and andropogonoid grasses and generally occur where rainfall exceeds 1500 mm or drainage is impeded and waterlogging and flooding occur annually for extended periods. At the lower end of the moisture spectrum, dry savannas are characterised by aristidoid and eragrostoid grasses that appear especially on sandy and skeletal soils and at most sites receiving less than 500 mm annual rainfall (Johnson and Tothill, 1985). However, while this coarse classification into 'wetter' and 'drier' forms is a useful broad model of such manifestation, it markedly oversimplifies the relations involved. Werger (1983) presents a widely accepted classification of savanna vegetation with four divisions in which grass cover is uniformly present and the trees and/or shrubs are deciduous:

- Grassland (trees less than 1% surface cover)
- Savanna (trees 1–10% surface cover) possibly with thickets or bush clumps
- Dense savanna (trees or shrubs 10–50% cover)
- Savanna woodland (trees dominant with 50–90% cover).

Cole's (1986) version is broadly similar, although reflective of greater complexity and demonstrates the importance of soils, hydrology and long-term landscape evolution in addition to contemporary climate and ecological factors. She identifies an intermediate formation (presumably between savanna and dense savanna) known as 'savanna parkland' and furthermore demarcates a category in which only low trees and shrubs form the canopy above the grass layer. According to such a model, it is the interaction between mean annual rainfall, soil texture, moisture availability and geomorphological situation that determines the particular vegetation formation that prevails in any location (Fig. 9.2); thus at 750 mm mean annual precipitation, clay soils support savanna grassland, whereas at the same rainfall on sandy soils they may support thicket. At the drier end of the spectrum, low tree and shrub savanna and grass cover becomes increasingly ephemeral as mean annual rainfall decreases below 300 mm. Woody vegetation only really becomes dominant where precipitation exceeds 500 mm annually, although clearly other factors such as fire, grazing

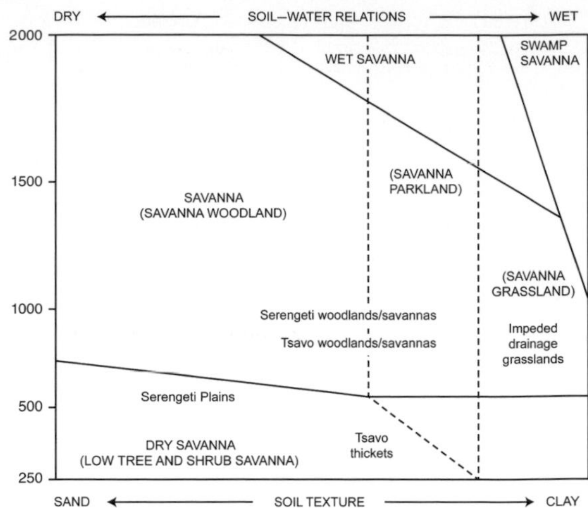

FIGURE 9.2. Relations between savanna types and savanna grasses and rainfall, soil water, soil texture and geomorphological position (modified from Johnson and Tothill, 1985; Adams, 1996).

pressure and soil characteristics are also involved. Above a mean annual precipitation of *c*. 650 mm, tree cover in African savannas is constrained by fire and grazing impacts, whereas below this threshold, woody biomass appears to be linearly correlated with rainfall (Sankaran *et al*., 2005). There have been a number of attempts to model these obvious complexities and these are reviewed by Adams (1996) who demonstrates that the well-recognised positive correlation between rainfall and biomass is a sometimes inappropriate simplification. Geomorphologically, however, it is arguably sufficient for us to employ the generic 'wetter' and 'drier' classification, since conditions associated with pronounced aridity are distinctively different from those at the moist end of the environmental gradient which are more akin to those that promote humid tropical type weathering and associated geomorphic processes. Accordingly, pans and aeolian-associated erosion and depositional landforms may dominate the geomorphology of the drier savannas, whereas deep weathering phenomena and plains and inselberg landscapes characterise the wet savannas.

9.1.2 Key environmental drivers

Climate, geomorphology, soils and landscape evolution
As highlighted above, climate, principally in the form of precipitation amount and seasonality, is a key determinant of ecological and geomorphological pattern in savannas. The length of the dry season, or seasons, is a highly variable parameter, depending on geographical position in relation to major moisture-bearing air masses and also on altitude. Large-scale climatic patterns have, of course, varied over

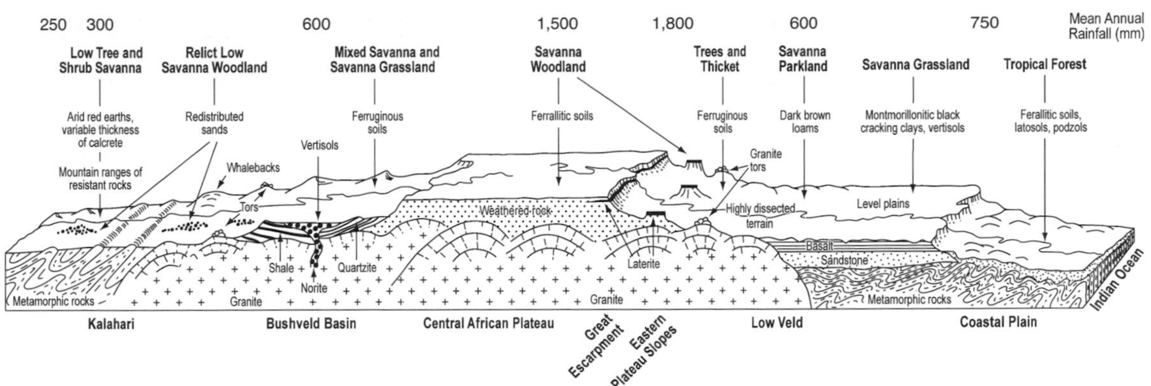

FIGURE 9.3. Relations between vegetation, climate, soils, geomorphology and geology of southern Africa (modified from Cole, 1986).

geological time, indeed, increasing aridity during the last 25 Ma in Australia coupled with repeated ice age oscillations during the Quaternary are probably at the root of savanna evolution on that continent (Bowman and Yeates, 2006). It should be acknowledged that the geological (and therefore climatic) histories of savannas on different continents constituting the present-day fragments of Gondwanaland may be quite different. For example, South America and Africa may have experienced relatively high temperatures and humidity through the Cretaceous and upper Cenozoic while Australia and India, both tracking northwards relatively rapidly under the influence of moving tectonic plates, were experiencing progressively warmer temperatures at their lower latitude positions. Aridification from the Miocene onwards is widely reported from all these areas, for example in Australia (Martin, 2006) and may have played a role in the emplacement and activation of the huge sand seas of southern Africa and Australia. Nevertheless, there are also smaller-scale dynamics of climate, such as the El Niño–Southern Oscillation (ENSO), that are variable over shorter timescales and impact on a range of physical environmental factors in the savannas (Plisnier *et al.*, 2000). Temperature in these tropical latitudes is of secondary importance, although at altitude, for example across much of high eastern and southern Africa and in the elevated plateau regions of Bolivia and Colombia, frost may be considered ecophysiologically important, especially after fire. Soil conditions, in part related to moisture relations, are also drivers of vegetation and this is illustrated classically in the bush clumps associated with soils of higher nutrient content on termite mounds in otherwise grassy savannas in parts of central Africa (Werger, 1983). Halomorphic soils and soils rich in metals usually support distinctive communities, a phenomenon that Cole (1982) has put to economic use in terms of mineral prospecting through remote sensing. On the *campos cerrados* of Brazil, nutrient-impoverished soils derived from deeply weathered granite support evergreen trees that with their deep roots are able to tap into a perennial water supply; meanwhile on richer basalt-derived soils under similar rainfall conditions, shallower-rooted tree species are unable to access the groundwater and are deciduous (Motta *et al.*, 2002). Drainage is also a factor, since where water tables rise seasonally close to the surface many tree species are excluded and this produces the distinctive pattern of *miombo* woodland (dominated by *Brachystegia* spp.) interspersed with linear grasslands in the lower lying *dambos* (see Section 9.2.5) found across much of south central Africa (Goudie, 1996). Cole's (1986) greatest contribution to an understanding of savanna dynamics was surely the recognition that longer-term geomorphological history is a significant landscape-scale driver of regional vegetation patterns; the *miombo* woodlands are most characteristically associated with the ferrallitic and oligotrophic soils developed on the weathered Tertiary planation surfaces of south central Africa. In contrast, post-Tertiary erosion has removed the lateritic cover and the exposed bedrock favours renewed soil formation promoting the development of savanna grassland and parkland (Fig. 9.3). Similar processes have been described in South America and Australia (see Cole, 1986).

Fire, grazing and human activity

The complex interrelationships between the three important ecosystem factors of fire, grazing and human activity are geomorphologically significant in savannas because, through their impact on vegetation type and cover, they strongly influence the balance between groundwater dynamics, weathering, soil formation and soil erosion. Fire has certainly played a fundamental role in the origin and evolution of the tropical savannas; indeed C4-grass-dominated savanna grasslands emerged rapidly onto the ecological stage around 8 Ma BP during the late Miocene in response to feedback loops involving changes in the fire regime, atmospheric carbon dioxide concentrations and even cloud microphysics (Beerling and Osborne, 2006). Lower atmospheric carbon dioxide levels during the Miocene promoted the evolution of the C4 photosynthetic pathway, so prominent in tropical grasses, and in combination with the co-evolution of large grazing herbivores, the associated physiological advantage led to the rapid expansion of fire-adapted savannas. The increased fire frequency in turn influences the hydrological cycle, since black smoke aerosols absorb radiation resulting in a warmer troposphere which slows evaporation and cloud formation and decreases precipitation. Through multiple positive feedback loops, these systems became effectively self-sustaining in preventing the encroachment of C3 forest tree elements which sequester carbon less efficiently at lower partial pressures of carbon dioxide; tree seedlings grow more slowly and thereby fail to reach the minimum height required to render them sufficiently resistant to fire (Bond and Midgley, 2000). Moreover, under cooler, drier climates, such as those of the Last Glacial Maximum (LGM) in many savanna regions (e.g. southern Africa: see Scott, 2002), a reduction in forest tree cover reduces evapotranspiration and may promote the further aridification of climate. Such pathways 'feed into and accelerate the closed positive feedback involving fire, tree mortality, forest cover and C4 grass cover' (Beerling and Osborne, 2006, p. 2027). The emergence of novel communities of large ungulate herbivores such as horses, antelope and elephants during the later Miocene introduced additional feedbacks and, although the balance between obligate grazers (e.g. springbok) and browsers (e.g. giraffe), and even mixed feeders (e.g. elephant), is known to influence the grass and tree composition, the effects are complex and have not yet been quantitatively modelled. In Australia, the role of fires initiated by Aboriginal hunter–gatherers has long been recognised (Bowman, 1998) as a factor in the evolution and maintenance of the balance between trees and grasses, although the details have been much debated and the relative roles of long-term aridification of climate during the Tertiary and Quaternary and the role of aboriginal burning is not fully resolved (Bird *et al.*, 2004). Hope *et al.* (2004) posit that the occupation of Australia prior to 45 ka BP may indeed have occurred but that their impact on vegetation was limited until a subsequent period of more extreme climate variation. Whatever, grazing and fire (itself, in effect, a form of herbivore: see Bond and Keeley, 2005) and the interrelationships between these two factors combined with the manner in which humans interact with them are fundamental controls of vegetation and other physical environmental attributes in savanna regions.

Termites are important herbivores in the wet and dry savannas and, indeed, are significant geomorphological agents through the construction of termitaria. Termites and ants consume enormous quantities of biomass and as part of their physical activity fine particulate material is migrated to the surface. Goudie (1996) quotes Drummond in 1888 suggesting that 'The soil of the tropics is in a state of perpetual motion. ' *Macrotermes* species construct large mounds that are in themselves substantial landforms and can reach 10 m in height (Pullan, 1979). Termitaria are both ecologically significant due to their higher moisture and nutrient content and also geomorphologically very prominent. For example, in the Brazilian *cerrado* they are important and persistent elements of the landscape known as *campos de murundus* (de Oliveira Filho, 1990). Larger termite mounds in Africa are found at densities of a few per hectare (Meyer *et al.*, 1999) but are nonetheless very obvious surface features. Smaller termitaria constructed by other species may achieve densities of more than 1000 per hectare in some savanna regions and there is no doubting their role in nutrient cycling and bioturbation; Holt *et al.* (1980) estimate that termites can turn over as much as 400 kg ha^{-1} of fine material. Furthermore, the subterranean network of galleries associated with termite mounds may act as macropores facilitating greater water infiltration and transfer (Gillieson, 2006). Such activity may indeed complicate attempts to interpret the chronology of soil and sediment accumulations in the savannas (Bateman *et al.*, 2007).

Human activities, through primary impacts on land use and land cover which have associated impacts on the grazing and fire regime, are also clearly important ecosystem determinants in the savannas. Biomass burning is a major form of land use management in the tropics generally and changes in population density coupled with developments in the nature and extent of land utilisation have had impacts on savanna landscapes at a range of scales (Eva and Lambin, 2000). People manipulate the grazing and fire regime and, indeed, have done so for a very long time in some savanna regions (e.g. in Gabon: see Abbadie *et al.*, 2006). The balance between tree and grass cover is, to a large extent, controlled by the interactive effects of fire and grazing (van Langevelde *et al.*, 2003) and, in turn, these drivers are influenced by land use. The effect of anthropogenic forcing on these landscapes is widely observed – the savannas are of course home to millions of farmers and pastoralists, although it is possible to misinterpret the effects of human agency in these regions. For example, islands of rainforest trees in the savannas of Guinea were widely held to be degraded relicts of formerly more extensive forests but Fairhead and Leach (1996) present compelling evidence to suggest that the pattern emerged over a considerably longer period of land use management and that the landscape has in effect been 'misread' by

Eurocentric foresters, scientists and policy-makers. Giertz *et al.* (2005) document significant hydrological and soil physical effects following the clearance and subsequent cultivation in a savanna woodland catchment in Benin and, through secondary impacts on soil fauna and other parameters, the change in land use results in substantially higher surface runoff and soil erosion in affected areas (see Chapter 2, Section 2.2.1). Meanwhile, in central Kenya, Lamb *et al.* (2003) find relatively little evidence for significant alteration of savanna vegetation due to human influence and point, instead, to the dominance of precipitation changes. Without doubt, however, human activity, in some cases over extended periods of time, directly through land use change and indirectly through climate change, needs to be considered in unravelling the complexities of any future geomorphic response.

9.1.3 Uncertainties

Aside from the uncertainties regarding the nature and rate of climate change in the savannas that are related to the (un) reliability of the models employed by the IPCC (IPCC, 2007), the complexity of these systems poses a substantial challenge to developing a robust understanding of the geomorphological response of savannas. There have been few attempts to model the impact of changing climate on geomorphic attributes per se, although there is more abundant literature considering possible effects on, for example, vegetation and fire regime of these regions which can, in turn, affect landforms and associated processes. The complex, even cryptic nature of these interactions, however, means that the uncertainties in prediction are very pronounced indeed and render any definitive geomorphological statements as to the future of the savannas frustratingly generalised or even speculative.

Future modelled climate scenarios of the savannas are themselves prone to considerable uncertainty, not least because of the extended geographical coverage of the biome and the complex nature of contemporary environmental conditions. Estimates of the absolute magnitude – and even the *direction* – of change for many climatic parameters differ from region to region. Thus, although all climate models predict mean annual temperature increase across the savannas, the absolute magnitude of that increase varies from between 2 and 5 °C (IPCC, 2007) over the twenty-first century. Precipitation is an especially difficult parameter to predict; some regions may be expected to receive increased rainfall while still others are likely to experience drier climates. For example, much of inter-tropical Africa, including large areas that are today classified as savannas, appears to be very likely to endure significant rainfall *increase* under IPCC regional climate change scenarios, whereas less rainfall is predicted for much of the Indian subcontinent and relevant parts of Southeast Asia. Climate models generally also suggest future increased *variability* in climate with higher frequencies of geomorphically significant events, such as droughts and floods. Accordingly, the geomorphic response resists any simple level of analysis. The matter is discussed further in relation to thresholds below but, in brief, it means that – even if we reduce 'climate change' to a few simple variables involving mean annual precipitation and temperature (clearly an oversimplification) – the nature and direction of the geomorphic response is unlikely to be the same, or even similar, across the range of savanna environments. Climate, of course, does not simply become 'wetter' or 'drier' and many other geomorphologically key variables, such as the frequency and magnitude of extreme events and variations in seasonality, further complicate the situation.

The complexity of the interrelationships and feedbacks between climate, fire, herbivory and the ratio of grass to tree cover outlined above accentuates the uncertainties of the magnitude and direction of climate change predicted for the future. There are several published attempts to model some of these interactions in savannas, for example at Nylsvley in South Africa (Scholes and Walker, 1993) and Lamto in West Africa (Abbadie *et al.*, 2006). At the continental scale, Sankaran *et al.* (2005) have modelled the determinants of woody cover in African savannas and successfully link resource availability and disturbance regimes. Nevertheless, many of the linkages involved in such studies remain inadequately understood and constrain the appropriately accurate prediction of how the various elements involved may respond to climate change into the future. Beerling and Osborne (2006) illustrate this in their efforts to integrate biomass burning, grazing and climate features in the savannas (see Fig. 9.4) and conclude that a number of positive feedbacks, the magnitude of which is very difficult to quantify with any degree of precision in a dynamic climate situation, are apparent. Moreover, the complex array of ecological factors is not considered in the context of landforms, geomorphic processes, soils or weathering phenomena. Thus, multiple uncertainties concerning, for example, the linkages between fire, climate and vegetation are further accentuated when we attempt to establish their geomorphic relevance. To illustrate this conundrum consider a scenario whereby mean annual precipitation decreases across the savannas of, say, central India. Were this to be the only parameter that changes, a reduction in woody biomass cover could be anticipated. Under such circumstances this would result in increased fire frequency followed by a further reduction in woody vegetation cover (Hoffman and Moreira, 2002). But what would be the land use management

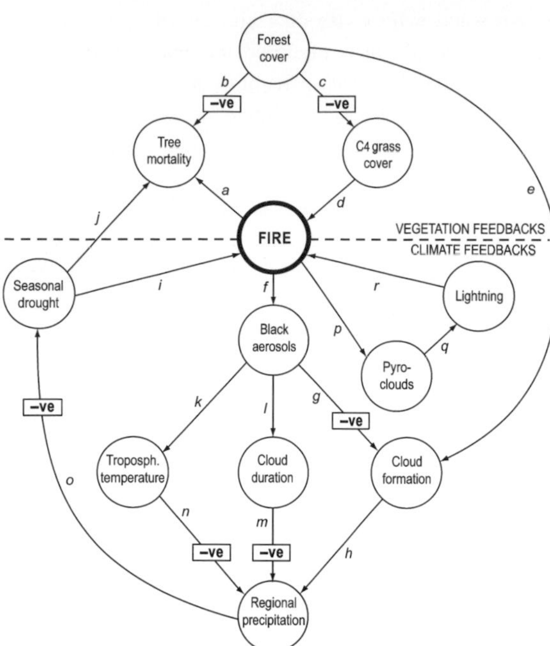

FIGURE 9.4. A systems analysis of fire–climate feedbacks. Arrows originate with causes and end at effects. Plain arrows indicate direct responses, and arrows marked with −ve symbol show inverse responses. Closed loops with an even number of −ve labelled arrows or solely plain arrows are positive feedbacks, and those with an odd number of arrows with −ve labels are negative feedbacks (modified from Beerling and Osborne, 2006).

response to this change? Would the grazing capacity of the now more extensive grasslands be greater, resulting in lower groundcover thus reducing fuel load and, accordingly, fire susceptibility? On the other hand, would the increased partial pressures of atmospheric carbon dioxide promote the growth of trees and shrubs, accelerate so-called bush encroachment and thereby increase flammable biomass? And, if so, what would the land use management impact be? Clearly there are key uncertainties at almost every link in the chain of possible events. Any attempt to place this into a geomorphic response model therefore further reduces the degree of confidence with which forecasts can be made in this biome. For example, in the same hypothetical case, would the reduced rainfall, coupled with increased temperatures, increase or decrease chemical weathering rates? Some of these issues are explored in a case study of African savannas below.

A further substantial cause of uncertainty is the possible land use/cover change and land use management response. There are some published attempts to explore the land use and land cover implications of global change in the savanna or adjacent biomes and it is worthwhile briefly reflecting on these in order to assess the extent of their influence. There is an additional important reason to consider land use change as, through effects on surface albedo for example, there are significant feedback effects of this process and the future climate itself may change further as a result. Indeed, DeFries *et al.* (2002) note that future land use changes are likely to be especially marked in the tropics, including the savannas; for example, rangeland and cropland in Africa alone are predicted to increase by more than 5.6 M km^2 by the year 2050. The modelled effects of these land use changes for the savannas suggest reductions in carbon flux, increases in albedo and augmentation of predicted temperature increases by up to 1.5 °C (DeFries *et al.*, 2002).

Future land use changes are, however, especially difficult to predict (Lambin *et al.*, 2001) and there is a paucity of quantitative, spatially explicit data on how land use and land cover has evolved under human utilisation in the recent past that severely hampers our ability to model how it may change over the remainder of the century. The problem seems particularly pertinent in the savannas, where the relations between vegetation, fire, grazing and climate are strongly integrated and where human utilisation can impact markedly on each of these interconnected variables. Lambin and Geist (2001) document the various ways in which our understanding of the complexities of changing land cover and land use has evolved in the recent past, in particular the fact that the focus has now moved away from temperate regions and tropical lowland forests to highlight the scale of transformation in other parts of the tropics, including the rangelands and savannas. Any reliable prediction of the nature and rate of land use changes in the savannas in the future does, however, require a good understanding of the underlying and proximal causes, which are themselves numerous and interrelated (Table 9.1). For example, while the immediate cause of land use change in the savannas of, say, Tanzania, may be agricultural extension, there are underlying structural processes at play that may involve demography, technology, economics at global, regional and local scale, political and institutional factors and cultural or sociopolitical factors. Human mobility, an increasing trend across the developing world incorporating much of the savannas, is an especially dynamic force that is recognised as the most important of the three basic demographic variables of fertility, mortality and migration (Lambin and Geist, 2001). Throwing the vagaries of climate change into this mix and trying to assess what the geomorphic impact might be raises even more intangibles. For example, Brook and Bowman (2006) provide evidence in parts of Australia that savannas may have reverted to closed woodland during the twentieth century as a consequence of increased partial pressures of atmospheric carbon dioxide.

TABLE 9.1. *Changes in the understanding of land cover and land use changes with reference to the savannas*

Previous understanding	Contemporary understanding
Mainly associated with temperate regions or tropical forests	Impacts all cover types, including rangelands, savanna woodlands, etc.
Landscape assumed to have been pristine prior to change	Landscapes altered by people for millennia, indeed may be a product largely of human impact
Changes permanent	Changes complex and potentially reversible; land cover is a constantly dynamic entity
Changes spatially heterogeneous	High degree of spatial heterogeneity; landscape fragmentation especially important
Major driver is population growth	Drivers are complex and influenced by population response to economic policy and resultant opportunity as well as biophysical and socioeconomic triggers
Impacts mostly local	Local effects are amplified (or attenuated) by globalisation: there is a strong local–global interplay
Major mechanism is agricultural expansion	Land use intensification and diversification includes agricultural, pastoral and urban land uses
Impacts largely restricted to carbon cycle	Impact incorporates all aspects of biophysical environment (including geomorphology, biodiversity, hydrological cycle, etc.)
Impacts a function of magnitude of biophysical change	Impacts expressed biophysically but manifest themselves according to vulnerability of affected populations
Changes are widespread, but diffuse	Impacts are widespread but spatially concentrated in 'hotspots' of change

Source: Modified from Lambin and Geist (2001).

9.2 Key landforms and processes

9.2.1 Geomorphological inheritance

As is undoubtedly the case for other landscapes globally, the evolution of landforms over time influences their contemporary morphology in important ways and this means that climatic geomorphological features may be inherited from former climatic regimes. An obvious example would be the distinctively shaped valleys carved by glaciers that have long since retreated but leave behind both the morphology and the sedimentary deposits emanating from the glacial period. No less significant are the now vegetated and, largely, stabilised sand dunes that characterise much of the semi-arid savannas of Africa and Australia and even wet savannas in Angola and northern Nambia, which suggest considerably less vegetative cover under drier and perhaps windier conditions of the past. Thus the landform shapes might indicate a particular climatic condition that no longer actually prevails at the site of the landforms themselves. This is the issue of geomorphological inheritance and is prominent in savanna landscapes, particularly those that are marginal to the deserts, where the inheritance of arid zone features is possible, or those that are marginal to the tropical forests, where past conditions of less pronounced seasonality and greater moisture availability may be expressed in, for example, weathering profiles. The concept is developed in Thomas (2004) in respect of tropical areas and using examples from the Quaternary as a model as to how landscapes respond to 'rapid' change. Some important questions are raised as a result; for example, are the deeply weathered regoliths and duricrusts that underlie many savanna regions products of contemporary processes or of past conditions or, perhaps, a combination of both?

Kniveton and McLaren (2000) ponder the cumulative effects of changing climate on geomorphological and sedimentological processes and note that spatial and temporal scales influence the answer. The problem has also been addressed by Meadows (2001) who assesses the relative roles of Quaternary environmental change and human activity in determining landform processes and concludes that, for semi-arid southern Africa at least – and this includes substantial areas of savanna – shorter-term changes are reflected mainly in smaller-scale landforms and geomorphic processes. The linear dunefields of the Kalahari may represent an exception. The landforms here comprise macroscale features that appear to have been markedly impacted, at least in terms of surface activity, by short-term climate changes of the late Quaternary and also potentially strongly altered by human activity both within the last few decades and into the future. In a sense, there are some entire landscapes that are more obviously a product of Quaternary, most importantly late Quaternary, events; these include the aeolian features of southern Africa's savannas and semi-arid regions (Meadows, 2001). Inheritance is therefore an important legacy in regard to the relationship between geomorphology and ecology and in terms of landscape sensitivity to extreme events such as droughts and floods; of course it is also a key element of the explanation of the present distribution of features in the savannas. As Thomas (2006, p. 122) notes, 'over long time-scales, landscapes become more complex palimpsests of repeated imprints of climate, extreme events and tectonics' (see Chapter 1, Section 1.11.1).

9.2.2 Deep weathering, groundwater and inselbergs

Although the landforms of the savannas are extremely varied, they are characterised by a number of commonly recurring geomorphic features that, while not restricted to the seasonal tropics, are especially prominent in such regions. Paramount among these are deeply weathered regoliths that are especially widespread on the geologically old, tectonically stable regions that are based on Precambrian cratons and to be found at the core of the African, Indian, South American and Australian continents (Reading *et al.*, 1995). The seasonality of rainfall which induces fluctuations in the groundwater table coupled with relatively high mean annual temperatures is a combination of factors that could be expected to increase the rate of chemical reactions. In wetter parts of the savannas seasonal water excess may be sufficient to remove solutes and allow for ongoing chemical alteration, while in drier parts such processes become inhibited by increased concentrations of the reaction products that then accumulate as surface or near-surface crusts. In the oldest parts of the tropics, especially where there is subdued relief, chemical alteration may have been operative over geologically extended periods of time and, assuming that conditions favour the continual flushing of solutes, extremely mature weathering profiles may prevail across large areas (Thomas, 1966, 1974). These landscapes are also associated, although again not exclusively, with distinctive rock outcrops, typically interpreted as residual hills and widely known as inselbergs, products of long-term landform evolution, most likely a response to longer-term climate changes of the Tertiary and Quaternary and in particular cooler and drier climates that may have stripped off some of the accumulating mantle (Thomas, 1994). Chemical weathering, especially of granites, may penetrate to extraordinary depths, indeed perhaps as much as 1000 m (Thomas, 1974), although values of 30 m to 60 m would be regarded as

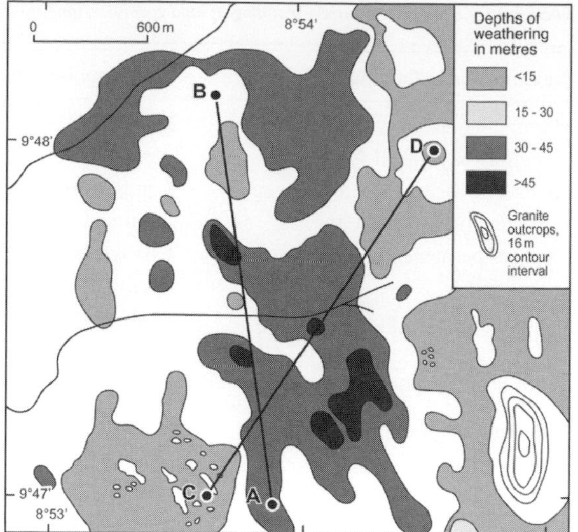

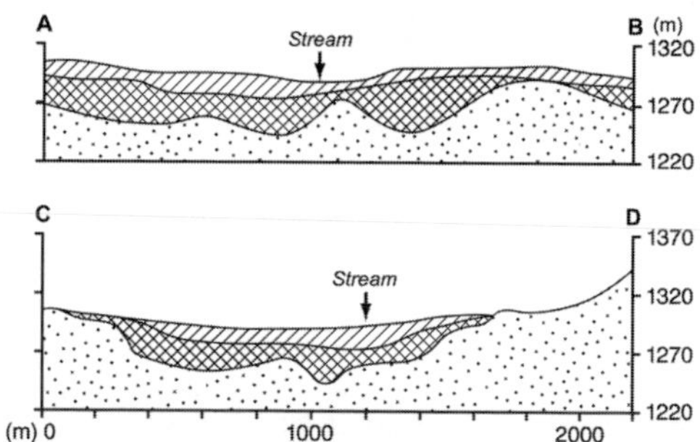

FIGURE 9.5. Deep weathering patterns in fine-grained biotite granite, near Jos, northern Nigeria. On the cross-sections (A–B, C–D) saprolite is cross-hatched; alluvial–colluvial fill is hatched (modified from Thomas 1966, 1994).

more typical (Goudie, 1996). The nature and depth of the weathering front is spatially highly variable as illustrated by Thomas (1966) for the Jos Plateau in Nigeria and shown here as Fig. 9.5; such variation relates to differences in bedrock chemistry and structure, long-term geological evolution, groundwater chemistry and spatial dynamics along with a host of other physical environmental parameters.

The relative roles of geological time and actual rates of chemical weathering have been debated by geomorphologists working in the tropics but it is now clear that the concept of a direct correspondence between contemporary tropical climatic conditions and the aggressive alteration of substrates to regolith is a myth. The availability of unaltered rock for weathering, for example through tectonic uplift and/or erosion, may be as important, or more important, than the existence of climatic or hydrological circumstances that might intuitively be thought to accelerate such alteration. Persano *et al.* (2002) suggest that the deep weathering profiles in Australia (a continent that became subject to tropical, in the sense of latitudinal position, climates during the Cenozoic) are probably linked to periods of rapid denudation following the break-up of Gondwanaland in the Cretaceous. Naturally, this complication severely hampers any attempt to understand what may happen to weathering rates and weathering profiles under future short-term climate scenarios. For example, if climate change were to interact with savanna landscapes so as to increase the rate of erosion – certainly conceivable in regions where rainfall is predicted to be augmented and the frequency of extreme events to increase – this might accentuate chemical weathering. In turn, through the drawdown of atmospheric carbon dioxide in the process – given that chemical weathering is effectively a carbon sink – this acts as a negative feedback loop. Moreover, increased mean annual temperatures and moisture supply, coupled with the increased amplitude of seasonal groundwater movements, would also favour chemical alteration. Of course, the opposite may well apply in savanna landscapes where climates become drier. Given the above discussion, it may be reasonable to conclude that the impacts of climate change on deeply weathered profiles and residual hills are likely to be localised. In respect of inselberg development in Zimbabwe, Römer's (2007) simulation model exercise revealed that, despite integrating several phases of significant environmental (including climate) change, during their development, these changes were unable to 'divert the general trend of the process–response system away from its steady course' (p. 349). Nevertheless, changing frequency and magnitude of seasonal oscillations in groundwater per se, could have very significant implications for catchment hydrology and, through impacts on land use, produce effects on geomorphic processes such as soil erosion.

9.2.3 Slopes, colluvium and stone lines

Although many savanna landscapes are characterised by low relief, slopes and slope processes are important geomorphic elements. Rapid mass movement processes resulting in slope failure are restricted to areas of higher relief but other more subtle mechanisms of sediment movement on slopes are ubiquitous. This has led to the widespread occurrence of colluvium (up to 20% of Africa south of the Zambezi is characterised by such material; see Watson *et al.*, 1984) which may or may not exhibit complex stratigraphy and palaeosol development indicative of landscape instability (see, for example, Botha and Federoff, 1995). These deposits seem to provide contrasting signals according to locality, since some appear to be highly responsive to environmental change (for example the work of Erikson *et al.* (2000) in central Tanzania), while still others have perhaps been developing more or less undisturbed since before the LGM (Thomas, 2004).

Stone lines are also prominent geomorphic elements in these landscapes. Concentrations of, typically, quartz or quartzitic clasts, these layers may be up to 1 m thick and occur at varying depths below the contemporary land surface. Most frequently assumed to be lag deposits, they are so widespread in savanna landscapes that it is meaningless to attempt to attribute any single mode of formation. In fact, their occurrence is due to a range of processes, perhaps working in combination, including bioturbation (in particular by termites), long-term land surface lowering and soil creep (Thomas, 1994). In Uganda, Brown *et al.* (2004) find little evidence to support a biological mode of formation and instead argue for sedimentation of finer material indicative of punctuated landscape dynamics rather than stability. Of course, this does not rule out bioturbation as a key formation mechanism elsewhere, and Breuning-Madsen *et al.* (2007) demonstrate the transport, by termites, of gravel-free soil to the surface in Ghanaian savanna woodland resulting in the accumulation of pisolitic concretions at depth.

It therefore goes without saying that any influence of global climate change on the distribution and activities of termite species (see Bignell and Eggleton, 1995) could result in substantial landform impact in the savannas. Should, for example, temperature and moisture conditions induce population decline and the abandonment of termitaria in some areas, the subsequent exposure and erosion of the mounds, which can certainly degrade rapidly under certain conditions (Williams, 1968; Pullen, 1979) may generate additional sediment yield in the short term. Secondary impacts of climate change involving alteration of land use and land cover will also affect termite activity; for example Okwakol (2000) illustrates how woodland clearance for agriculture dramatically reduces termite diversity and biomass. Given the distribution and complex stratigraphy of much of the colluvial cover on slopes, it is probable that this too would be highly susceptible to accelerated erosion under climates envisaged in IPCC scenarios. Over longer periods, anticipated vegetation change, accompanied by secondary impacts on fire regime, herbivory and land use, is difficult to assess but would undoubtedly express itself in terms of slope processes in the savannas.

9.2.4 Duricrusts

Horizons containing a high proportion of compounds that tend to form compact and erosion-resistant duricrusts are strongly associated with long-term landform evolution and the development of deeply weathered regoliths in savanna regions. Depending on the chemistry of the parent material, duricrusts may be enriched in iron oxides (referred to as laterite if the accumulation is autochthonous or ferricrete if allochthonous: see Widdowson, 2007), aluminium oxides (alucrete or bauxite), silica (silcrete), calcium carbonate (calcrete) or calcium sulphate (gypcrete). Their formation within the weathering profile is arranged according to the mobility of the predominant cations, such that there is spatial patterning of the typical indurated profiles along a moisture gradient. More humid regions appear to be associated with duricrusts rich in aluminium sesquioxides, whereas calcrete and gypcrete form at the arid end of the gradient with laterite, ferricrete and silcrete prominent in intermediate rainfall regions (Thomas, 1994), although hybrids dominated by more than one cation also occur, e.g. silcrete–calcrete intergrade duricrusts in the Kalahari (Nash and Shaw, 1998; Kampunzu *et al.*, 2007). Ferricrete or lateritic weathering profiles, with thicknesses of the hardpan layer varying from 1 to 10 m at up to 2 m below the surface, are commonly underlain by mottled clay horizons in turn underlain by what is described as the pallid zone and these appear to be especially widespread in the savannas. This bears testimony to the role of rainfall and the seasonal or longer-term fluctuations of groundwater in their evolution, although they may also form pedogenetically. Iron-oxide-rich duricrusts take various morphologies, evolving from relatively weakly structured cementations to harder nodules and on to conglomerates as mobilised iron proceeds to replace kaolinitic clay minerals and quartz in the profile. The precise nature and position of these erosion-resistant, indurated, dark red or purple crust layers in the landscape may vary due to uplift and incision and they may also be subject to dissolution and physical disintegration as conditions change (McFarlane, 1983). Duricrust development is subject to a complex interplay between temporal, climatic, biotic, hydrological, geochemical and geomorphic factors.

The prominence of primary (i.e. rainfall and temperature) and secondary (i.e. groundwater hydrology) climatic factors in the formation of duricrusts suggests that changing climates could impact on them markedly. Increasing or decreasing iron mobility may result in either the enhancement or attenuation of ferricrete horizons but the geomorphic significance, at least in the short term, seems marginal. The amplitude and timescale of predicted climate changes in the twenty-first century is unlikely to manifest any substantial geochemical or structural adjustments to duricrusts and their associated weathering profiles over the short term because their development proceeds over geological rather than ecological timescales. On the other hand, loss of superficial topsoil due to accelerated erosion as a consequence of reduced vegetation cover could result in the exposure of duricrust horizons at or nearer to the land surface and this process could occur relatively rapidly. There is recent geological precedence for such processes, as witnessed by the lateritic cap rocks in some desert margin situations that surely formed during periods of greater moisture availability (Thomas, 1994). The uncovering of indurated horizons in this way would certainly have a major potential impact on land use, since tilling in such material is considerably more difficult than in workable superficial soils and, in addition, drainage would be markedly affected. The emergence of infertile stone pavements at the surface of savanna landscapes is widely recorded, as, for example, in the lateritic *bowals* of West Africa (Goudie, 1996) and similar situations would arise in the case of stone lines.

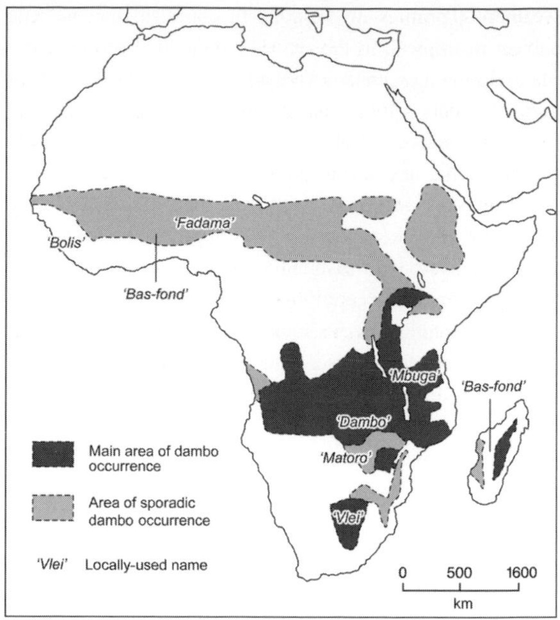

FIGURE 9.6. The distribution of dambos and dambo-like forms in Africa (from Goudie, 1996).

9.2.5 *Dambos*

Shallow, seasonally flooded treeless depressions in catchment headwaters that lack a permanent channel (Boast, 1990) are distinctive and prominent features of savanna landscapes across substantial regions of the tropical and subtropical latitudes. As Goudie (1996) points out, these features are particularly, but not exclusively, abundant in Africa (Fig. 9.6), where they are variously known as *dambos* (Central Africa), *mbugas* (East Africa), *fadamas* or *bolis* (West Africa), *matoros* and *vleis* (southern Africa) and *bas-fonds* (Madagascar). The evolution and dynamics of dambos – the most widely applied geomorphological term – has attracted considerable attention, not least because of their economic importance as they play an important role as a water resource and in dry-season cropping (McFarlane and Whitlow, 1990). Although dambos are characteristically distributed in areas with seasonal climates and where rainfall amounts range between 600 and 1500 mm a^{-1}, geomorphological factors are among their primary determinants since they are strongly associated with areas of low relief that are usually assumed to be ancient erosion surfaces (Goudie, 1996). Accordingly, slopes are generally very shallow, rarely exceeding 0.6° in long profile and 5° in cross profile (von der Heyden, 2004). Although characteristically there are no channels in their headwaters, channels may form and disappear again downstream, although their presence generally increases in downflow direction. As with other geomorphic features of the savannas, the evolution and development is complex and much debated, in particular whether or not such features act as sponges and attenuate the amplitude of the downstream hydrograph (von der Heyden and New, 2003).

Processes of dambo formation are 'intriguing and still imperfectly understood' (Goudie, 1996, p. 153), although there are two main schools of thought. One interpretation is that they are channel-free extensions of drainage basins formed by headwater erosion and infill by colluvial inwash supplemented by periodic alluvial action. The characteristic distribution of sediment types across dambos is supportive of this view, since there is often a marked marginal 'wash zone' of sandy material grading into finer-grained, usually vertisolic clay material, downslope (Mäckel, 1974, 1985). However, this explanation fails to take into account the possible role of intense chemical weathering or 'etching' and its ability to mould the landscape surface. Quite possibly, dambos develop as a response to both surface runoff and subterranean chemical corrosion but cut-and-fill processes have been widely reported and suggest the importance of climate change in their evolution. Meadows (1985) demonstrated how the sequence of sediments in dambos of the Nyika Plateau of Malawi responded to changing climate and vegetation during the late Pleistocene and Holocene and this clearly begs the question as to how such features might respond to changes in precipitation regime in the future. It seems possible, for example, that on the one hand a reduction in rainfall and increased frequency and magnitude of storm events could augment sandy sediment supply to the dambo beyond the wash zone (i.e. accentuate the 'fill' phase). On the other hand, the same scenario could result in channel formation and in the process initiate a phase of downcutting that might significantly alter downstream runoff and sediment yield. Not only is there uncertainty, then, in the climate scenarios *per se*, but in the case of dambos the trajectory of geomorphic response appears to be especially cryptic. Given their agricultural significance to the vulnerable rural populations of large areas of savanna, modelling of dambo response to possible climate futures is arguably one of the key scientific challenges in the biome.

9.2.6 Pans

Pans are, with some important exceptions, small-scale enclosed basins generally associated with drainage networks and most prominently distributed at the arid end of the savanna landscape moisture spectrum, although clearly their evolution and origin over long periods may again mean that their locations are not in equilibrium with contemporary environmental conditions. Often strongly associated with adjacent crescentic duneforms, these depressions are variously floored with sediments rich in clay or silt (Shaw and Thomas, 1998). While climatic factors are critical, their occurrence is also controlled by the availability of susceptible surface conditions (Goudie and Wells, 1995). Suitable surfaces in the context of savannas would include palaeolacustrine basins, palaeodrainage lines and interdunal areas, while the absence of a well-connected drainage system and lack of any major contemporary aeolian accumulations are additionally favourable. That such a combination of circumstances is reasonably common in dry savannas is indicated by the widespread distribution of such features globally (Fig. 9.7) and under certain circumstances small pans may achieve densities of 200 km^{-2} (e.g. Zimbabwe: see Goudie and Wells, 1995).

Morphologically they are distinctive in shape, frequently, as Goudie and Wells (1995) put it, 'likened to a clam, a kidney or a pork chop' (p. 1), usually elongated in the plain that lies perpendicular to the prevailing or formative winds in a region, although they may display a diversity of form depending on, among other things, climate setting (Bowler, 1986) or structural control (Shaw and Thomas, 1998). Pans are frequently accompanied by lee-side crescentic mounds, or lunette dunes (Bowler, 1973; Lawson and Thomas, 2002; Telfer and Thomas, 2006), sediment supply to which is clearly a function of pan floor characteristics, although the precise nature and chronology of sediment delivery is controlled by factors that extend beyond the local to include regional groundwater fluctuations. Pan formation and development has been the focus of much geomorphological attention, with several mechanisms put forward to explain their distribution and diversity of forms and relationship to surrounding features (Bowler, 1986). Given the (by no means exclusive) juxtaposition of

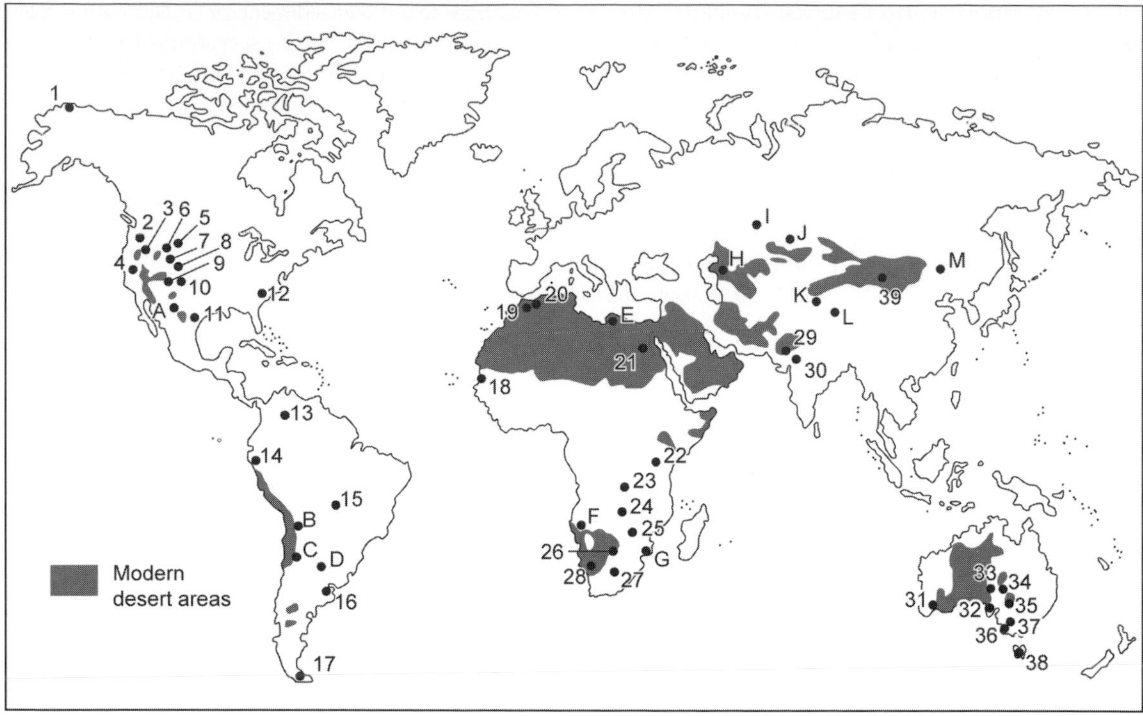

FIGURE 9.7. The global distribution of pans as determined from the literature (numbers) and from remote sensing (letters) (modified from Goudie and Wells, 1995).

many pans in, or at least marginal to, subtropical deserts, wind deflation is an obvious candidate factor. The fact that grazing animals and their associated predator species are known to concentrate on the pan floors for the purposes of salt-licking or, in periods of flooding, as water sources, serves to strengthen the argument for a deflation mode of formation, since trampling of the surface aids erosion and transport of the sediment. However, it has long been recognised that the occurrence of pans is a response to a complex array of factors and that groundwater movement and chemical, more specifically salt, weathering are also very strongly involved.

The observation that some pans globally appear to be in a state of disequilibrium with respect to contemporary environments makes it difficult to assess the potential impact of global climate change into the future. Nevertheless, Bowler (1986) developed a model to explain how, for pans, the interaction of groundwater and surface water controls adjust various formative processes along a hydrological gradient. Warming, accompanied by aridification, would certainly result in changes in these interactions at individual pans including less frequent pan floor inundation and possibly exposure of sediments to aeolian transport. A similar effect could be anticipated in the adjacent lunettes which might exhibit loss of vegetation cover and potential reactivation. Higher temperatures accompanied by greater seasonal moisture could serve to intensify chemical weathering, increase the duration and depth of inundation and likely lower sediment transport from the pan floor. Certainly at the arid end of the savanna moisture gradient, increased evaporation under higher temperatures might alone be sufficient to induce deflation, particularly if there were greater grazing animal pressures consequent upon reduced carrying capacity due to lower grass biomass and productivity. The integrated nature of pan development factors (Fig. 9.8) does indicate caution as to any simplistic assumptions about the nature of pan and lunette responses to the various scenarios of global climate change. Notwithstanding this caveat, some of the larger pans globally, for example those of Etosha in Namibia and Makgadikgadi in Botswana, are already substantial sources of global dust (Washington *et al.*, 2003; Bryant *et al.*, 2007) and this would surely intensify under elevated temperatures and reduced moisture, more especially if accompanied by increased windiness.

9.2.7 Soil erosion by water

Erosion of soils by surface wash, subsurface flow and gullying is expressed prominently in the savannas. Whatever mechanism, or more likely combination of mechanisms, is

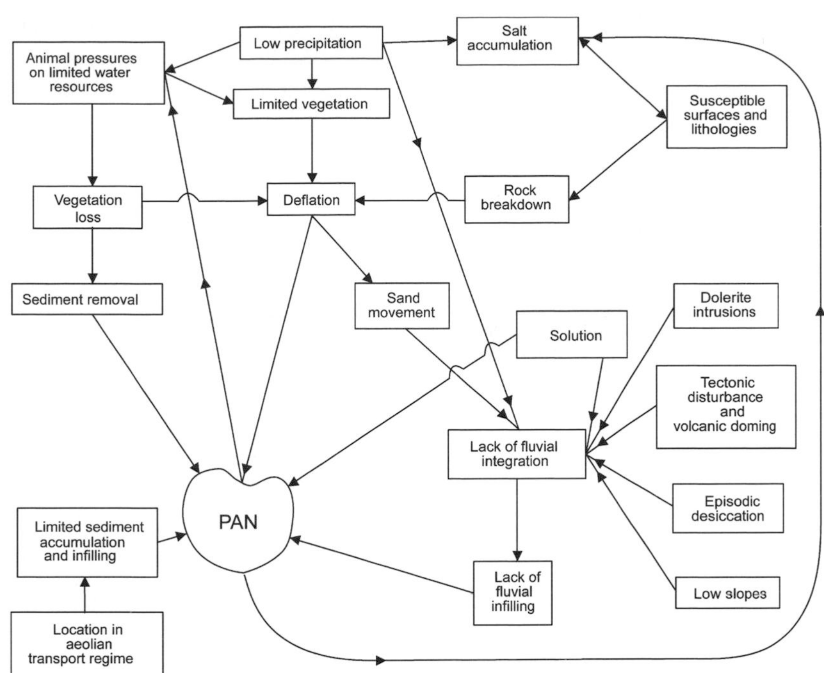

FIGURE 9.8. An integrated model of pan development (modified from Goudie and Wells, 1995).

involved, the deep weathering profiles and colluvial deposits that blanket many savanna areas are certainly vulnerable to accelerated removal. The relative importance of surface to subsurface flow, and of both in relation to gully erosion is variable, even debatable (Thomas, 1994). Nevertheless, soil piping appears to be especially common in the seasonal tropics (Thomas, 1994) due to the wet-season downward flux of soil particles and associated dissolved material being counterbalanced by desiccation and clay flocculation in the dry season. The nature and direction of the relationship between soil pipes and gullies (known as *dongas* in southern Africa) is not fully resolved, but there have been numerous studies of both that testify to the susceptibility of savannas to accelerated soil loss and sediment transport by such means. Gullies vary in distribution and morphology but are striking landscape features in many savanna landscapes that are draped in colluvium. Physical and chemical characteristics of this colluvial material influence gully development, although the extent to which it does has been the subject of vigorous debate (see the contrasting views of Yaalon (1987) and Watson *et al.* (1987) on the matter). Ironically, perhaps, even soil conservation structures may result in gully formation, for example in Lesotho (Showers, 2005). Notwithstanding this, Quaternary changes in the seasonality and intensity of rainfall have at times led to the acceleration and at still other times stabilisation of gullies and this points to their overall sensitivity to fluctuating climate (Shakesby and Whitlow, 1991) which may well be aggravated by human activity.

Because accelerated soil erosion has such important on- and off-site effects on agricultural and grazing productivity, and also because human activity appears to play an integral, if uncertainly quantified, role in its initiation and development, there have been intensive attempts to understand and model the processes. Possible thresholds in respect of physical environmental factors controlling gully initiation and activity have been the subject of considerable scrutiny (Poesen *et al.*, 2003; Valentin *et al.*, 2005a, 2005b). In Swaziland, Morgan and Mngomezulu (2003) identify the key factor as surface runoff velocity and that this in turn corresponds to particular land cover parameters, although different land systems appear to induce different gully types and there is a need to exercise caution in developing predictive models. As Valentin *et al.* (2005a, p. 129) point out: ‘Long term erosion monitoring is thus essential to observe possible transient and non-equilibrium responses to climatic and land use changes’, but such data are available for only a few catchments and not at all within the tropics. Sidorchuk *et al.* (2003) successfully model gully erosion and landscape response in Swaziland and this highlights the need for a greater degree of quantification and accuracy in predicting where such accelerated erosion problems occur.

Undoubtedly, anthropogenic disturbance, including inappropriate cultivation and grazing regimes, irrigation, log haulage, road construction and urbanisation, can markedly alter savanna slope erosion systems in exceeding the threshold for gully initiation or expansion. Changing rainfall frequencies and intensities, especially extreme storm events, have the potential to dramatically extend gully systems even if this is difficult to model and predict (Larson *et al.*, 1997). Global climate change over the coming century in such regions, whether involving more or less rainfall, may well serve to accelerate both surface runoff and subsurface transfer of water and sediment. However, the degree to which this will lead to greater gully frequency or, in cases where they already exist, gully expansion, is not yet quantifiable. Intuitively, accelerated erosion of susceptible savanna landscapes seems a very likely outcome with associated negative impacts on the vulnerable peasant farmers that occupy them.

9.2.8 Wind erosion and deposition

The effect of aeolian activity, both erosional and depositional, is dealt with in greater detail in Chapter 10. Nevertheless, the juxtaposition of the savannas, at the arid end of the moisture gradient, with the subtropical deserts is conspicuous and warrants mention here. Savanna landscapes in some regions, particularly in southern Africa, Africa immediately south of the Sahara, Australia and in parts of the Indian subcontinent, are developed on unconsolidated sandy sediments that have been sculpted into various dune bedforms which are frequently vegetated in the present day. The widespread occurrence of vegetated aeolian bedforms beyond the present-day desert margin, for example longitudinal or seif dunefields south of the Sahara margin in the Sahel, implies contemporary inactivity and, therefore, stability. This is, perhaps, the classic case of an inherited landscape, since the large-scale geomorphic features have, in the main, been developed under conditions different from those of the present day. Indeed, there appear to have been numerous phases of dune activity and inactivity in the late Quaternary (Stokes *et al.*, 1997b), suggesting that unravelling the complex record would be more than a useful clue to understanding the possible landscape response under climate change scenarios.

Arguably, the most spectacular example is in the Kalahari which has been studied intensively in relation to dune dynamics in response to climate changes during the late Quaternary (Thomas and Shaw, 2002). The sensitivity of this subcontinental-scale dunefield is amply demonstrated by the increasingly detailed chronology of activation based on luminescence dating of dune stratigraphy, although it is clear that dune mobilisation is a complex process that is not simply a response to greater aridity. For example, at Tsodilo Hills in Botswana (Thomas *et al.*, 2003), phases of late Quaternary dune activity (implying aridity) sometimes overlap with elevated lake levels (implying increased precipitation to evaporation ratios) and it appears that models need to take account of variations in sediment supply and windiness as well as aridity in this context. An additional problem is that reduced partial pressures of atmospheric carbon dioxide around the Last Glacial may have reduced the vigour of C3 shrubs and trees and led to grassland expansion, thereby also influencing dune dynamics (see Bond *et al.*, 2003; Harrison and Prentice, 2003). The implications of future climate change are very significant, since it is landscape-scale geomorphic dynamics that are impacted here rather than merely individual landforms. The matter is explored further as a case study below.

9.3 Landscape sensitivity, thresholds and 'hotspots'

9.3.1 Landscape sensitivity

The foregoing section indicates that some elements of the contemporary savanna landscapes may be especially sensitive to change. This is perhaps most appropriately illustrated by the dynamics of dune bedforms during the late Pleistocene, not only in the Kalahari but also in semi-arid Australia (Hesse and Simpson, 2006; Fitzsimmons *et al.*, 2007), Argentina (Tripaldi and Forman, 2006) and India (Juyal *et al.*, 2003). The importance of these studies is that they represent truly landscape-scale geomorphic responses over sub-millennial temporal scales. In other words, the mobilisation of entire land surfaces is possible under circumstances of rapid climate change provided that surface conditions are conducive.

As Brunsden (2001, p. 99) notes: 'The landscape sensitivity concept concerns the likelihood that a given change in the controls of a system or the forces applied to the system will produce a sensible, recognisable, and persistent response. The idea is an essential element of the fundamental proposition of landscape stability.' Landform sensitivity in the savannas, however, needs to be considered in light of two important caveats. Firstly, geomorphic response is really a secondary effect of vegetation change rather than being directly due to climate change *per se*. Indeed, vegetation change may itself be thought of as an integration of fire, grazing and rainfall; the reaction of savanna landforms and processes is, therefore, highly complex. Even an apparently simple parameter, such as the ratio of tree to grass cover, is controlled by an assortment of factors so that predicting change following a shift in climate is demanding and forecasting the landscape outcome verges on the intractable. Interestingly, while biomes are shown to be responsive to simulated changes in climate, Hély *et al.* (2006) conclude that, for the African tropics at least,

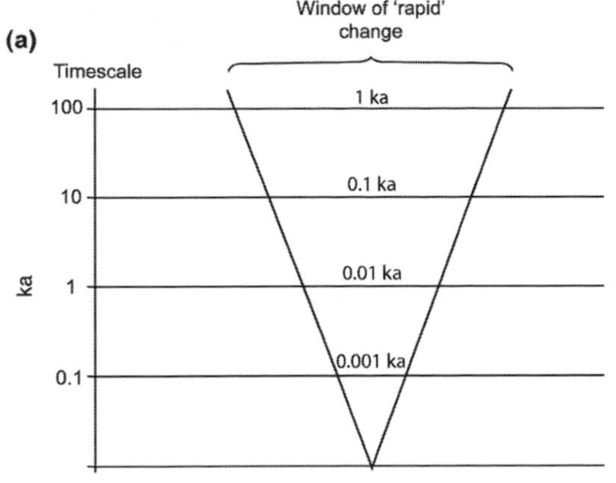

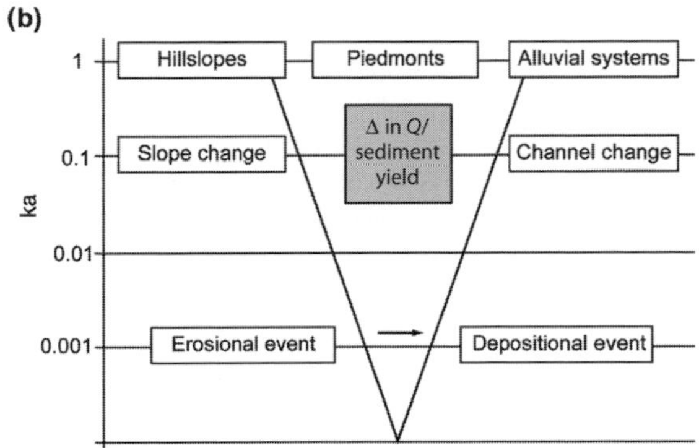

FIGURE 9.9. (a) The perception of 'rapid' change in relation to the timescale of enquiry; (b) the response of geomorphic systems to rapid change over different time periods (modified from Thomas, 2004).

tropical rainforest appears to be more sensitive to IPCC scenario changes than savannas. Secondly, there is the thorny issue of scale, both in the spatial and temporal sense (see Chapter 1, Section 1.3.2). This issue has been the subject of analysis by Thomas (2001, 2004). In essence, the line of reasoning goes that some geomorphologically significant timescales of change are lengthy, perhaps of the order of tens of thousands of years, for example chemical weathering or soil formation, whereas others, such as slope failure, are virtually instantaneous. Moreover, landforms are also scaled geographically, since some are spatially extensive (e.g. high-order catchments) while still others are very small (think of ripples on the surface of a dune). Naturally, these two scales, spatial and temporal, overlap so that large-scale landforms tend to respond only to long-term environmental changes. The visual model designed by Thomas (2004) to illustrate this is shown in Fig. 9.9 that attempts to demonstrate what geomorphologists mean when they use the term 'rapid environmental change' and what components of a landscape are most likely to alter quickly under such circumstances. As Thomas (2004, p. 112) notes: 'to alter the morphology and patterns of sedimentation along a piedmont zone, for example, simultaneous response is needed in all catchments and hillslope categories, over a period long enough to redistribute the loci of sediment stores in the landscape'. Put straightforwardly, larger-scale landforms tend to manifest themselves at millennial rather than decadal scale. Nevertheless, the stabilised dunes of, for example, the Kalahari are an agglomeration of relatively small to medium spatial scale landforms that can and do respond to small-scale fluctuations in climate and vegetation cover that, by virtue of the extensive geographical coverage of appropriate surface characteristics, are expressed at the larger, landscape scale. Clearly, it may not take much of a change in, say, precipitation or windiness, to exceed the threshold for sand movement and reactivate the widespread dunes. However, for the savannas this may be a relatively unusual situation, since here most of the characteristic landforms and geomorphic processes lie in closer accord with the generalised model. Thus, inselbergs, deep weathering phenomena, duricrusts and other prominent features of the savanna landscape are large scale and, accordingly, relatively resilient features that are not automatically sensitive to rapid environmental change unless these prove to be of a magnitude that exceeds that experienced over the longer term. Soil erosion phenomena, including gullies and pipes, on the other hand, are more inclined to adjust promptly to shifts in boundary conditions. Moreover, climate changes predicted over the next century may not manifest themselves in all landforms and processes immediately but still result in adjustments that accumulate incrementally over time.

There are other issues in relation to geomorphological processes associated with the concept of threshold. Langbein and Schumm (1958) developed a model of sediment yield in relation to climate and demonstrated how the balance between erosivity and erodibility changes with mean annual precipitation. Thus, a shift in climate from 'wetter' to 'drier' may increase or decrease sediment yield (a measure of catchment-scale erosion) depending on where the catchment lies in relation to the approximately 350 to 400 mm mean annual precipitation threshold (see Chapter 1, Section 1.11.2). Moglen *et al.* (1998) suggest a similarly complex set of responses for drainage density around a similar threshold but that environments at the arid end of the spectrum appear to be more sensitive to changes in rainfall whatever the sign. Future climate change may well be rapid and trigger non-linear responses in geomorphic processes, as indeed is noted for fluvial systems over the Holocene and historical past by Knox (1984, 2000) and for rivers of the arid savannas in particular by Zawada and Hattingh (1994).

9.3.2 The Quaternary archive: some examples from the savannas of Africa and beyond

A clearer idea of the sensitivity of savanna landscapes can be gleaned from a consideration of how they responded to the environmental changes of the Quaternary. Certainly the Quaternary is a useful archive of ecosystem response, although the tropics have in general been rather poorly studied in this regard relative to temperate latitudes. Indeed, until recently the view was widely held that climate changes of only low amplitude had characterised tropical regions and that conditions there had remained reasonably stable over millions of years (Thomas, 2006). Even so, the paucity of, for example, fossil pollen evidence in the savannas necessitates the interpolation of their late Pleistocene palaeoecology from the rather few reliable records that do exist, many of which are fragmented in nature and spatially remote. Notwithstanding this lack of spatial and temporal contiguity in the palaeoenvironmental reconstructions for many savanna regions, there is now convincing evidence that the Quaternary represented a period of dramatic shifts in the distribution of tropical biomes over millennial timescales (see Chapter 8, Section 8.1.3).

For the African savanna regions, reliable late Quaternary pollen sequences have been retrieved from wetlands and other archives in Zambia (Livingstone, 1971) and South Africa (Wonderkrater, Equus Cave, the eastern Free State and Pretoria Salt Pan: Scott, 1982, 1987, 1989 and 1999 respectively) and also from Madagascar (Gasse and van Campo, 1998). These are supplemented by long lacustrine sequences from both Lake Tanganyika (Vincens, 1991) and Lake Malawi (de Busk, 1998). Significantly different vegetation characterised many areas in response to lower temperature conditions around the time of the LGM, but the moisture signal appears to be complex in both geographical and temporal dimensions. Evidence of greater aridity around this time is indeed apparent at several localities. Pollen evidence from two cores from the southern part of Lake Tanganyika (Vincens, 1991) reveals changes in vegetation patterns related to climatic fluctuations from around 30 ka BP onwards. Before 18 ka BP, open and poorly diversified savanna woodland prevailed at low and mid altitudes, with local patches of montane forest including *Podocarpus* spp. indicating cooler and drier climatic conditions than today, with frost during the night at low altitude. The records for Lake Malawi, Lake Massoko and Lake Rukwa all point to lower levels, and correspondingly drier climates, around the LGM (Barker and Gasse, 2003). A similar picture emerges from the savannas of southern Africa, for example around the Pretoria Salt Pan (now known as the Tswaing Crater), where Scott (1999) documents cooler and drier conditions associated with the lowering of vegetation belts by up to 1000 m as well as changes in the relative abundance of C3 and C4 grasses (Scott, 2002) possibly in response to lower carbon dioxide concentrations. In Madagascar, Gasse and van Campo (1998) outline a cool and dry LGM phase, although this appears to have terminated abruptly at around 17 ka BP with significantly increased moisture availability, a change that may have preceded the equivalent in the northern hemisphere by as much as 2 ka. In summary, Late Pleistocene temperature signals in the savannas of Africa are relatively well resolved but the precipitation and evaporation trends have proved much more difficult to secure. The problem of reconstructing rainfall amounts and seasonality is no doubt exacerbated by the effect of reduced atmospheric carbon dioxide concentrations, especially in this biome, since woodland and grassland dynamics appear to be particularly sensitive to such changes (Bond *et al.*, 2003).

The details can be fleshed out to some extent using marine pollen records that have been obtained off the southwest African coast (Shi and Dupont, 1997; Shi *et al.*, 2000). Marine core GeoB 1023–5 is from off the Cunene River mouth at around latitude 18° S but yields information on the dynamics of Namib desert vegetation from around 25 ka BP onwards (Shi *et al.*, 2000) and indicates the prominence of desert and scrub vegetation under colder and more arid conditions, although the presence of certain elements of the fynbos flora that are today associated with winter rainfall conditions that prevail further south is intriguing. A longer temporal perspective, up to 300 ka, is provided by other cores in the region (Shi and Dupont, 1997; Shi *et al.*, 2001) and suggests several periods in the late Pleistocene during which desert and semi-desert scrub vegetation changes were driven by pronounced SST cooling and associated intensification of the southeast trade winds.

Elsewhere in the savannas there are further indications of the sensitivity of vegetation to climate change during the late Quaternary. Brazilian grasslands, for example, were considerably more widespread during cooler and drier glacial conditions and largely replaced the modern semi-deciduous and cerrado woodlands as indicated for southern and southeastern Brazil by the pollen analyses of Behling (2002). Indeed, the modern vegetation appears to have been established only relatively late in the Holocene in response to increased precipitation suggesting that a return to more arid conditions would again favour the spread of grasslands. Moisture balance is a major vegetation determinant of, for example, the Colombian savannas and has been shown to vary significantly, even at sub-millennial timescales, during the Holocene producing corresponding major shifts in the distribution of biomes (Marchant *et al.*, 2006). The rainforest–savanna boundary appears to be particularly responsive to changes in the frequency of El Niño and La Niña phenomena (Wille *et al.*, 2003) in these regions so that any future change in the frequency or magnitude of these ocean–atmospheric conditions is likely to induce vegetation dynamics (see Chapter 8, Section 8.5.1). In Australia, Quaternary climates have fluctuated in relation to precessional and other orbital factors but do appear to have become progressively arid through time and culminated in a cooler and drier LGM across large areas of the continent (Moss and Kershaw, 2000; Hesse *et al.*, 2004; Haberle, 2005).

Clearly then, as these African, Australian and South American examples illustrate, climate change during the Quaternary produced a marked vegetation response. But was this *geomorphologically* significant? At the arid end of the savanna moisture gradient, the answer appears to be strongly in the affirmative as the ever-emerging record of dune mobility from the Kalahari and other regions testifies (see Kalahari case study below). However, for other elements of the landscape, the same may not hold true. Sørenson *et al.* (2001), in central Tanzania, found that colluvial sediments appeared to be relatively unresponsive to the amplitude and rate of environmental change at least during the Holocene. Thomas and Murray (2001) meanwhile suggest there is indeed evidence of episodes of sediment accumulation during the last glacial cycle but that these have as yet proved difficult to fix chronologically. In contrast, there are clear time-constrained geomorphic manifestations of climate change that produced major vegetation shifts in Australia (Moss and Kershaw, 2000) which are paralleled by changes in the nature and rate of colluvial sedimentation in northeast Queensland (Thomas, 2006). These studies suggest that a reduction in rainfall, such as occurred around the time of the LGM, appears to induce alluvial fan development and increase sediment yield.

9.3.3 Thresholds

At the arid zone margin, the 150-mm isohyet has been frequently cited as the threshold below which dune activity occurs and this assumption has long been used to reconstruct late Quaternary palaeoenvironments in regions where there are widespread fossil dunefields. The significance of this 150-mm rainfall threshold was initially employed as a means to reveal the nature and extent of precipitation changes during the later Quaternary at the Sahara desert–savanna margin in central Africa (Grove and Warren, 1968) and in southern Africa (Grove, 1969) presumably on the basis that the contemporary limit of active aeolian bedforms appears approximately to coincide with the 150-mm isohyet. However, given likely changes in other climatic parameters, such as temperature and the wind field, and non-climatic factors such as sediment supply, such a threshold can really only be considered a very approximate guideline. Furthermore, aeolian activity is also a function of wind erosivity which is difficult to ascertain from palaeo records. At the opposite end of the savanna moisture spectrum, the transition from semi-deciduous woodland to tropical moist forest is usually assumed to occur around the 1500-mm isohyets, but again the complexities of the relations between rainfall, vegetation, land use and fire are such that any simplistic assumptions as to the nature of changes in the position of this isohyet should also be eschewed. Accordingly, therefore, there can be no guarantee that a future shift in the position of the 150-mm isohyet would per se have any direct causal impact on aeolian activity and the same could be said for a change in the rainfall circumstances at the rainforest boundary. Vegetation cover is clearly a key variable and Wiggs *et al.* (1994) indicate that a threshold cover value of 14% is sufficient to inhibit sand mobility. Therefore, rather than considering an absolute mean annual precipitation value as a threshold limit, an integrated understanding of the relations between climate, vegetation, land use, fire, herbivory and sediment dynamics is necessary before any particular geomorphic response to climate change can be invoked. Fortunately, these parameters have been adequately modelled at an appropriate scale for the Kalahari region of southern Africa and this provides the basis for the case study of climate change impacts on geomorphology that follows.

9.4 A case study in geomorphic impacts of climate change: the Kalahari of southern Africa

One of the most striking examples of how changing climate impacts on landforms in the savannas is provided in the geomorphological studies of southern Africa's Kalahari. This extensive dryland environment is underlain largely by unconsolidated sand sediments that have been sculpted, to a greater or lesser degree, into a range of dune bedforms by aeolian activity over extended periods of time. The dunes, along with other landscape elements, provide an archive of the climatic and other environmental changes that are recorded, both in the landforms per se and in their constituent sediments, and have been the object of close scientific scrutiny. The detailed record of change over time, especially during the later Quaternary, has facilitated the construction of models (Knight *et al.*, 2004; Thomas *et al.*, 2005) that can be applied to predicted climate scenarios with a view to revealing the nature of future surface conditions.

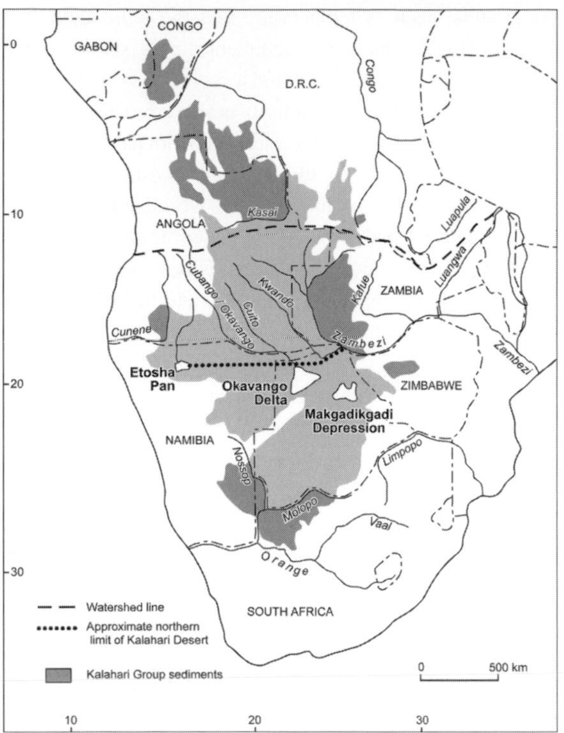

FIGURE 9.10. The Kalahari regions. The shaded area is the so-called Mega-Kalahari; the contemporary Kalahari 'desert' corresponds to that part between the Orange River in the south and the Etosha–Okavango–Zambezi wetland zone in the north (modified from Thomas and Shaw, 1991).

9.4.1 The Kalahari environment

Thomas and Shaw (1991) comment on the 'unusual and elusive' nature of the Kalahari, which is romanticised as a place uninhabited except by Bushmen living a Stone Age existence surrounded by a vast landscape of gently undulating, largely vegetated, dunefields on which roam the remnants of enormous herds of grazing mammals and their associated carnivorous companions. Frequently described as a 'desert', the Kalahari is really an enormous area of savanna that covers a substantial part of southern Africa (Fig. 9.10). It is delimited (at least in the present day for, as described below, aeolian activity has extended way beyond such limits in the geological past) in the north by the Etosha–Okavango–Zambezi swamp zone and in the south by the Orange River (Thomas and Shaw, 1991). The eastern margin coincides with the Kalahari–Limpopo watershed while the uplands ascending to the Great Escarpment approximate the western boundary. The physiographic and sedimentological uniformity characterising this zone forms but part of a much more extensive region better termed the 'Mega-Kalahari' (Thomas and Shaw, 1991) consisting of a downwarped basin in which terrestrial sediments have accumulated since the Jurassic and over an area of some 2.5 M km^2 spanning 30° of latitude. Far from being an archetypal 'desert', since this is no sea of barren shifting sand, there is nevertheless a notable lack of permanent, or even seasonal, water courses (the Okavango, Zambezi and Chobe rivers being exceptions, although they merely traverse the Kalahari and their water sources are located entirely beyond the region itself). Contemporary climate is characterised by warm to hot summers and winters with warm days and cool or cold nights, while the mean annual precipitation, strongly summer seasonal, ranges from 150 mm in the southwest to more than 600 mm in northeast Botswana (Thomas and Shaw, 1991). The current wind regime is one of relatively low energy, as it is in most of the Australian arid zone, so that the dunefield is relatively stable (Knight *et al.*, 2004). The evaporative response to these circumstances exceeds 2000 mm annually meaning that most parts of the Kalahari are in a permanent state of water deficit. Surface characteristics are dominated by the quartzitic sandy parent material (Kalahari sand: see Wang *et al.*, 2007) which has the potential to retain moisture and support permanent vegetation. Typically, vegetation responds to the northeast–southwest moisture gradient, dominated by grasses and shrubs at the arid end of this spectrum with increasing size of trees at the wetter end, although bush encroachment in drier areas results in an overall negative relationship between rainfall and woody cover (Ringrose *et al.*, 2003).

Knowledge of the geomorphology of what early travellers interpreted as a largely featureless and homogeneous landscape has dramatically been improved through aerial photography and satellite imagery (Thomas and Shaw, 1991). Major drainage features include perennial rivers in the north, the best known of which is the Cubango/Cuito/Okavango system which culminates in an enormous inland delta via the fault-defined panhandle. Other major depressions are occupied by pans including those of the Makgadikgadi (Ringrose *et al.*, 2005), Ngami (Burrough *et al.*, 2007) and Etosha (Brook *et al.*, 2007), all three of which display impressive evidence of significant and rapid hydrological change during the recent geological past. Perhaps the most characteristic geomorphological features of the Kalahari are, however, the extensive aeolian landforms which have provided scientists with the fascinating challenge of understanding the combination of ancient and modern environmental conditions which led to their development. As Thomas and Shaw (1991, p. 141) point out, the key problems relate to whether 'these dunes are currently geomorphological active or relict features, and, in the case of the latter, the considerable problem of dating the time of development'. Six major dune types are found in the region, namely parabolic dunes, blowouts,

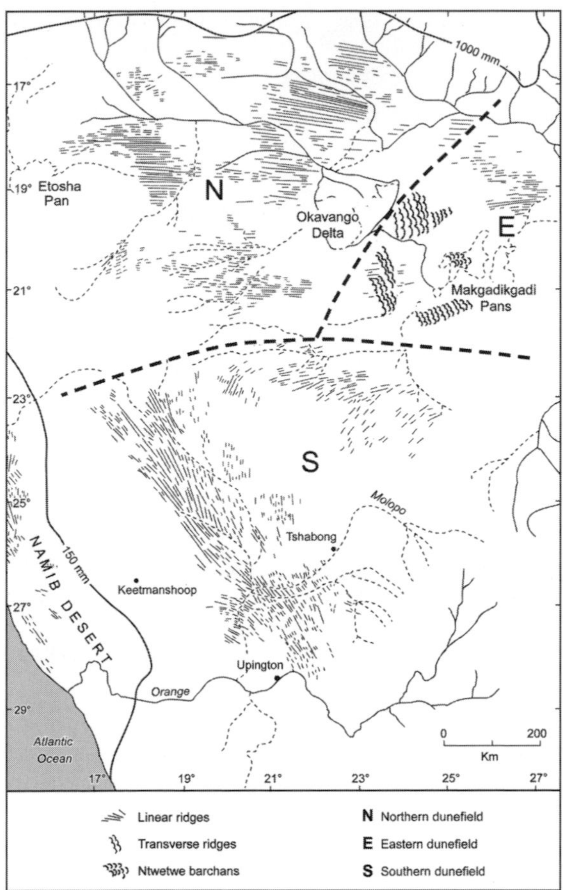

FIGURE 9.11. The three major dunefields of the Kalahari dominated by linear dune forms (modified from Thomas and Shaw, 1991).

barchans, transverse ridges, linear ridges and seif dunes, although the most widespread are the linear or longitudinal dune forms (Fig. 9.11, Plate 24) which cover some 85% of the Kalahari and represent 99% of all dune forms present (Fryberger and Goudie, 1981). These dunes, with crest heights of up to 20 m and running semi-continuously for up to 100 km, are mostly vegetated (more especially in the subhumid northern and eastern Kalahari) with woodland dominating the dune crests and grassland the interdunal depressions. As discussed in the following section, however, this situation may have changed frequently and dramatically at various stages during the later part of the Quaternary and may well be set to change again in the near future.

9.4.2 Landform sensitivity in the Kalahari: a Quaternary perspective

Vegetated linear dunes of the Kalahari, and elsewhere, have long attracted the attention of geomorphologists as, more

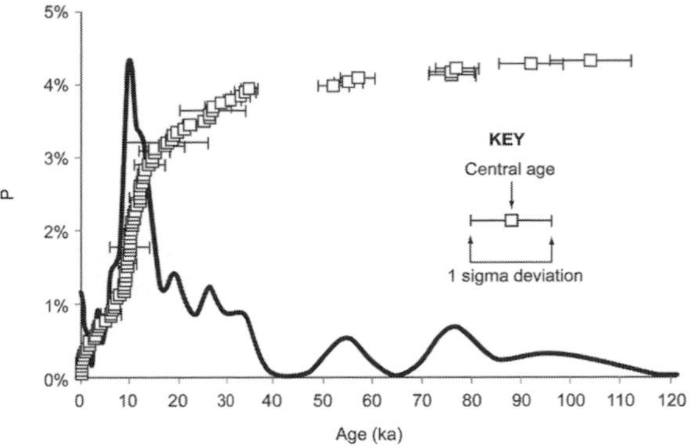

FIGURE 9.12. Distributions of luminescence ages from the linear dunes of the southwestern Kalahari (modified from Telfer and Thomas, 2007).

particularly since the onset of optically stimulated luminescence (OSL) dating to establish a reasonable chronology of accumulation and activity, they offer an opportunity to reconstruct palaeoenvironmental conditions over extended periods (Stokes *et al.*, 1997a). While previously interpreted from temporal gaps in proxies of humid chronologies fixed through radiocarbon dating, the timing of evolution and development of the extensive linear dunes is now feasible through optical dating. By implication, it has become possible to reconstruct associated palaeoclimates in respect of, for example, wind regime and precipitation and there are now more robust late Quaternary chronologies indicating dune dynamics in, for example, the northern Kalahari (Thomas *et al.*, 2000) and the southwestern Kalahari (Telfer and Thomas, 2007). The most recent collation of geochronostratigraphic evidence for the southwestern part of the region is shown in Fig. 9.12 (Telfer and Thomas, 2007) and reveals considerable complexity in the record of aeolian activity that needs to be interpreted with caution due to factors such as bias induced by limitations in sampling depth. As Telfer and Thomas (2007) note, there remains uncertainty regarding the spatial and temporal patterns of chronostratigraphy. Indeed, the mode of formation of linear dunes in general is debated and it is difficult to distinguish between patterns of sediment accumulation that represent intermittent and intense deposition from those representing continuous low-intensity deposition. These authors suggest that both modes of accumulation are evident in the data from linear dunes near Witpan in South Africa. Regarding the southwestern Kalahari in general, Stokes *et al.* (1997a, 1997b) document six phases of dune activity with the most recent period focussed on the period 20–10 ka BP. Some elements of this temporal pattern contrast with the situation in other parts of the Kalahari which may be attributed to the fact that different parts of the region (study sites are up to

1200 km apart) are subject to different environmental stimuli (Thomas and Shaw, 2002). Figure 9.12 suggests that linear dunes have been active to some extent throughout most of the last glacial cycle, with a prominent peak in aeolian activity at around 10 ka. The data set offers support for the idea that the period around the LGM (23–19 ka BP) was one of reduced aeolian accumulation, this being in agreement with other proxies suggesting a relatively humid phase around this time (e.g. Stuut *et al.*'s (2002) offshore sediment record). Some 1.5 m of sand accumulates in some locations leading Telfer and Thomas (2007) to question whether or not these dunes are really to be considered inactive in the present day and, may in fact, be close to a state of reactivation. It may be concluded that the linear dune forms of this part of southern Africa are highly responsive to changes in aridity, wind energy, wind direction and sediment supply. The following section explores the implications of such a conclusion for the future of the Kalahari under scenarios of global climate change.

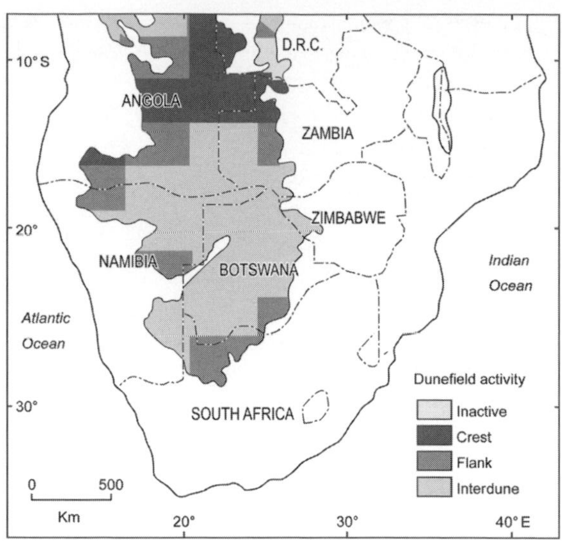

FIGURE 9.13. Predicted 3-month block Kalahari dunefield activity after 2070 based on Hadcm3 runs using various emission scenarios. Note the significant increase in dune mobility especially during May, June and July (southern hemisphere winter) (modified from Thomas *et al.*, 2005).

9.4.3 The Kalahari in the future

As is evident from the above discussion, there has been considerable attention paid to the dynamics of the Kalahari dune systems in the late Quaternary. Thomas *et al.* (2005) note that the interplay between dune *surface erodibility*, which is basically a function of vegetation cover and moisture availability, and *atmospheric erosivity*, characterised by wind energy, is the prime relationship in determining whether a dune is likely to be active or inactive. This relationship is, of course, highly susceptible to global climate change that potentially impacts any or all of these key parameters. In an attempt to explore the possible impact of future climate change, Knight *et al.* (2004) and Thomas *et al.* (2005) employ a set of general circulation model (GCM) simulations to predict the activity of the Kalahari dunefield.

In the present day, the Kalahari linear dunes, particularly in the more arid southwestern part of the region, are prone to mobility during droughts, as was the case in the period 1960–90 when annual precipitation totals were only in the range of 50% of the long-term mean (Bullard *et al.*, 1997). If, as seems to be the case for much of the region under IPCC (2007) climate change scenarios, droughts become more frequent as temperatures – and correspondingly, evaporation – are higher and soil moisture values lower, then mobility seems set to become more frequent and more intense. Thomas *et al.* (2005) attempt to assess the extent to which such changes may impact on dunefield activity through an analysis of the outputs of several GCMs and their effects on a standardised measure of mobility. Mobility indices typically indicate susceptibility to sediment movement based on parameters associated with wind energy, for example the length of time wind exceeds a particular velocity threshold coupled with indicators of moisture and potential evaporation. Although there are a number of significant challenges to estimating the effects of climate change on dunefield mobility, not least issues around the nature and spatial scale of GCM outputs, Knight *et al.* (2004) develop novel solutions to many of these constraints and outline the feasibility of applying climate model outputs to an estimate of regional changes in sand mobility. Notwithstanding the fact that the method does not successfully model some important elements of dune dynamics, such as the implications of elevated carbon dioxide concentrations on vegetation productivity, Knight *et al.* (2004) propose the means whereby changes in monthly mobility indices can be computed from the outputs of four GCMs under various emission scenarios.

The Thomas *et al.* (2005) method involved establishing potential future Kalahari dunefield activity by integrating monthly determinations of surface erodibility and wind erosivity into a measure of mobility. The results (Fig. 9.13) are striking as 'All modelled outputs project marked increases in dune activity during the twenty first century in all (Kalahari) dunefields, including, after 2040 in the (currently more stable) northern dunefield … and in the eastern dunefield' (Thomas *et al.*, 2005, p. 1220). Predicted values of the mobility index after 2070 exceed a critical threshold in many, especially winter, months in a number of scenarios and the 'environmental and social consequences of these changes will be drastic' (p. 1221).

Across vast swathes of Angola, Botswana, South Africa, Zambia and Zimbabwe, the large population of, mainly, poor subsistence farmers and pastoralists reliant on the Kalahari rangelands for their livelihoods face a mounting challenge if the basic tenets of these predictive models hold. This case study clearly demonstrates that, under particular atmosphere and surface conditions, geomorphological dynamics under rapidly changing climate can have extremely severe consequences for huge numbers of vulnerable individuals. Globally, although dunefields such as that of the Kalahari occur in both developed and developing nations, it is those 'with the least capacity to adapt that could suffer the most financial, social and developmental consequences' (Knight *et al.*, 2004, p. 198).

The predictive model presented by Thomas *et al.* (2005) is not without its assumptions and simplifications and is especially uncertain in respect of the direct and indirect effects of the climate change scenarios on vegetation cover which is 'the weakest part of the equation when trying to calculate monthly dune mobility values' (Knight *et al.*, 2004, p. 210). The model also fails to take account of any possible land use change or management response to future changes in climate and associated environmental characteristics. Although land cover and land use are remarkably difficult parameters to predict, being dependent not only on physical environmental drivers but also on socioeconomic and political factors, their importance in gauging the geomorphological impact of climate change is such that attempts to model these into the future are clearly also required, as indeed is argued in the concluding discussion.

9.5 Concluding remarks

Based on the preceding discussion, it is apparent that some elements of the savanna landscape are likely to be highly responsive to future climate change on the timescales envisaged. Nevertheless, there are many uncertainties and constraints to appropriately and accurately predicting the geomorphic (and other) impacts of such change because the system is itself highly complex and will respond in a multifaceted and possibly non-linear fashion. Not only are there uncertainties as to precisely how, and how quickly, the geomorphologically significant components of climate will change (IPCC, 2007) but the integrated nature of the processes involved renders the prediction of geomorphic responses extremely challenging. As noted above, there are numerous determinants of landscape in the savannas and their interrelationships are imperfectly understood. Moreover, the likely human response to future dynamics is cryptic, for example, how will land use and land use management factors be impacted and, in turn, how will this influence land cover, biomass, grazing and fire regime among many other variables?

Notwithstanding these constraints, it is possible to assess how the various landscape elements and processes may respond over the coming century, especially if some assumptions about the rate and nature of climate change are made. Examining the IPCC (2007) regional scenarios for savannas, it is apparent that temperature increase is extremely likely and that this will be reasonably uniform across tropical areas at between 3 to 5 °C, whereas the precipitation predictions are much less consistent between savanna regions. It is therefore helpful to consider the geomorphic response in relation to climates that may become warmer and wetter (up to 25% augmentation of mean annual precipitation) compared to a situation in which climates become warmer and drier (up to 25% attenuation of mean annual precipitation). In line with IPCC (2007) models, these two hypothetical climate scenarios for the savannas would both be characterised by greater climate variability and an increased frequency of extreme events and this also needs to be taken into account in assessing any possible geomorphic impacts. Table 9.2 represents an attempt to summarise the projected geomorphological response to the two kinds of scenario and indicates, for each savanna landscape element or geomorphic process, both the direction of response and the likelihood of that response over the time period envisaged, namely the rest of the twenty-first century. Because of the many interacting uncertainties noted above, and because of the complex network of interrelationships between geomorphology, climate, atmospheric chemistry, vegetation structure, plant productivity, biomass, fire, grazing and land use management factors, Table 9.2 needs to be interpreted cautiously as a crude (at best) approximation as to how the savanna landscape may react to climate and associated environmental changes. Moreover, it should also be recognised that increasing or decreasing precipitation accompanied by temperature increase along the lines suggested by the latest IPCC scenarios, may not necessarily have the same geomorphic result across the range of savanna environments. For example, given the existence of thresholds of sediment yield as theorised by Langbein and Schumm (1958), whether erosion is accelerated or impeded by climate change in the savannas could depend on whether or not – and in which direction – the threshold of 350 to 400 mm mean annual precipitation is breached.

Notwithstanding the limitations, there are some key pointers to how climate change may impact savanna geomorphology

TABLE 9.2. *Susceptibility of various landscape elements and processes to climate change in the savannas*

Landscape element or process	Warmer and wetter climate	Warmer and drier climate	Likelihood of change over next 100 years
Deep weathering profiles	Weathering accelerated	Weathering retarded	Low
Groundwater movement	Water table rise	Water table recedes	Moderate (greater seasonality increases amplitude of groundwater movement with possible secondary effects on weathering)
Inselbergs	Uncertain	Uncertain	Low
Colluvium	Uncertain: sediment accumulation on slopes reduced due to increased runoff or increases due to higher vegetation cover	Uncertain: lower vegetation cover favours sediment removal; less runoff increases slope sediment storage	High
Stone lines and duricrusts	Uncertain: stone lines or hardpans at shallower depth if greater runoff accelerates sediment removal	Uncertain: stone lines or hardpans may be buried by sediment if reduced runoff induces slope sediment storage	Moderate
Dambos	Higher water table, sediment accumulation	Lower water table, gully erosion	High
Pans	More seasonal flooding, sediment accumulation	Desiccation due to groundwater lowering, wind deflation of surface sediments	Moderate
Gullies	Depends on vegetation response, accretion of sediments possible, i.e. gully infilling	Depends on vegetation response, reactivated incision	High
Dunes	Greater vegetation cover, less wind erosion and greater dune stability	Lowered vegetation cover, greater wind erosion and dune mobility	High

in the face of projected temperature and precipitation trends. Clearly, some large-scale features that have evolved over long periods of geological time, such as inselbergs and deep weathering phenomena are largely resilient to short-term climate changes. Still other elements, for example the widespread colluvial deposits that flank many shallower savanna slopes, may well be impacted, but the direction and nature of the change is difficult to predict even if the future precipitation conditions are assumed. However, there are certain features and processes that appear to be especially sensitive and may, in a sense, be regarded as geomorphic 'hotspots'. Dambos (or their equivalent), widely distributed on older, deeply weathered, land surfaces, appear to be responsive to changing climate (and land use) parameters and they are potentially subject to incision producing concomitant changes in seasonal water table fluctuations. Given their widespread importance for seasonal grazing and cultivation, for example in Africa (Roberts, 1988), there could be significant deleterious socio-economic effects, particularly within the subsistence economy of affected areas.

Without doubt the most obvious and pervasive geomorphic response to projected climate changes over the remainder of the twenty-first century is in the widespread reactivation of currently metastable longitudinal dune systems. As illustrated graphically in the Kalahari, modelled by Thomas *et al.* (2005) and highlighted in the case study above, almost any combination of temperature increase, precipitation reduction, soil moisture loss and enhanced wind field energy can be expected to result in loss of vegetation cover and remobilisation of sandy surface deposits across vast areas of southern Africa. Equivalent dunefields in the southern margins of the Sahara, in India, Australia and South America may well react similarly. The resultant scale of impact on the agricultural and grazing economies of an enormous, vulnerable and already stressed rural poor population is incalculable, but certainly liable to be severe.

Indeed, the concluding message of this chapter is that, wherever there are, as indicated, pronounced geomorphic impacts of climate change in the savannas, the effects are accentuated because of the fact that most of the people living within these landscapes are directly dependent for their livelihoods, through agriculture and/or grazing, on the functioning of the ecosystem. Because the vast majority of these people are peasant farmers, they are extremely susceptible to landscape dynamics, despite the adoption of flexible livelihood strategies. The consequences for such people, historically and currently marginalised by the combined pressures of political and economic forces that have their roots in colonialism and, more recently, globalisation, may well be dire. Geomorphologists, therefore, have both a scientific and moral imperative to seek a deeper understanding of the nature of savanna landscape sensitivity to global climate changes.

References

Abbadie, L. *et al.* (2006). *Lamto: Structure, Functioning and Dynamics of a Savanna Ecosystem*. Berlin: Springer-Verlag.

Adams, M. E. (1996). Savanna environments. In A. Goudie, W. Adams and A. Orme, eds., *The Physical Geography of Africa*. Oxford: Oxford University Press, pp. 196–210.

Barker, P. and Gasse, F. (2003). New evidence for a reduced water balance in East Africa during the Last Glacial Maximum: implications for model-data comparison. *Quaternary Science Reviews*, **19**, 198–211.

Bateman, M. *et al.* (2007): Detecting post-depositional sediment disturbance in sandy deposits using optical luminescence. *Quaternary Geochronology*, **2**, 57–64.

Beerling, D. J. and Osbourne, C. P. (2006). The origin of the savanna biome. *Global Change Biology*, **12**, 2023–2031.

Behling, H. (2002). South and southeastern Brazil grasslands during late Quaternary times: a synthesis. *Palaeogeography, Palaeoclimatology, Palaeoecology*, **177**, 19–27.

Bignell, D. E. and Eggleton, P. (1995). Termites in ecosystems. In T. Abe, D. E. Bignell and M. Higashi, eds., *Termites: Evolution, Sociality, Symbioses, Ecology*. Dordrecht: Kluwer, pp. 363–387.

Bird, M. I., Hope, G. and Taylor, D. (2004). Populating PEP II: the dispersal of humans and agriculture through Austral-Asia and Oceania. *Quaternary International*, **118–119**, 145–163.

Boast, R. (1990). Dambos: a review. *Progress in Physical Geography*, **14**, 153–177.

Bond, W. J. and Keeley, J. E. (2005). Fire as a global herbivore: the ecology and evolution of flammable ecosystems. *Trends in Ecology and Evolution*, **20**, 387–394.

Bond, W. J. and Midgley, G. F. (2000). A proposed CO_2 controlled mechanism of woody plant invasion in grasslands and savannas. *Global Change Biology*, **6**, 865–869.

Bond, W. J., Midgley, G. F. and Woodward, F. I. (2003). The importance of low atmospheric CO_2 and fire in promoting the spread of grasslands and savannas. *Global Change Biology*, **9**, 973–982.

Botha, G. A. and Federoff, N. (1995). Palaeosols in late Quaternary colluviums, northern Kwa Zulu-Natal, South Africa. *Journal of African Earth Sciences*, **21**, 291–311.

Bowler, J. M. (1973). Clay dunes, their occurrence, formation and environmental significance. *Earth-Science Reviews*, **12**, 279–310.

Bowler, J. M. (1986). Spatial variability and hydrologic evolution of Australian lake basins: analogue for Pleistocene hydrologic change and evaporite formation. *Palaeogeography, Palaeoclimatology, Palaeoecology*, **54**, 21–41.

Bowman, D. M. J. S. (1998). Tansley Review No. 101: The impact of Aboriginal landscape burning on the Australian biota. *New Phytologist*, **140**, 385–410.

Bowman, D. M. J. S. and Yeates, D. (2006). A remarkable moment in Australian biogeography. *New Phytologist*, **170**, 208–212.

Breuning-Madsen, H. *et al.* (2007). Characteristics and genesis of pisolitic soil layers in a tropical moist semi-deciduous forest of Ghana. *Geoderma*, **144**, 130–138.

Brook, B. W. and Bowman, D. M. J. S. (2006). Postcards from the past: charting the landscape-scale conversion of tropical Australian savanna to closed forest during the twentieth century. *Landscape Ecology*, **21**, 1253–1266.

Brook, G. A. *et al.* (2007). Timing of lake level changes in Etosha Pan, Namibia, since the middle Holocene from OSL ages of relict shorelines in the Okondeka region. *Quaternary International*, **175**, 29–40.

Brown, D. J., McSweeny, K. and Helmke, P. A. (2004). Statistical, geochemical and morphological analyses of stone line formation in Uganda. *Geomorphology*, **62**, 217–237.

Brunsden, D. (2001). A critical assessment of the sensitivity concept in geomorphology. *Catena*, **42**, 99–123.

Bryant, R. G. *et al.* (2007). Dust emission response to climate in southern Africa. *Journal of Geophysical Research*, **112**, D09207, doi:10.1029/2005JD007025.

Bullard, J. E. *et al.* (1997). Dunefield activity and interactions with climatic variability in the southwest Kalahari Desert. *Earth Surface Processes and Landforms*, **22**, 165–174.

Burrough, S. L. *et al.* (2007). Multiphase Quaternary highstands at Lake Ngami, Kalahari, northern Botswana. *Palaeogeography, Palaeoclimatology, Palaeoecology*, **253**, 280–299.

Cole, M. M. (1982). Integrated use of remote sensing and geobotany in mineral exploration. *Transactions of the Geological Society of South Africa*, **85**, 13–28.

Cole, M. M. (1986). *Savannas: Biogeography and Geobotany*. London: Academic Press.

Collinson, A. S. (1988). *Introduction to World Vegetation*. London: Unwin Hyman.

de Busk, G. H. (1998). A 35 000 year pollen record from Lake Malawi and implications for the biogeography of Afromontane forests. *Journal of Biogeography*, **25**, 479–500.

DeFries, R. S., Bounoua, L. and Collatz, C. J. (2002). Human modification of the landscape and surface climate in the next 50 years. *Global Change Biology*, **8**, 438–458.

de Oliveira Filho, T. A. (1990). Floodplain *murundus* of central Brazil: evidence for the termite-origin hypothesis. *Journal of Tropical Ecology*, **8**, 1–19.

Erikson, M. G., Olley, J. M. and Payton, R. W. (2000). Soil erosion history in central Tanzania based on OSL dating of colluvial and alluvial hillslope deposits. *Geomorphology*, **36**, 107–128.

Eva, H. and Lambin, E. F. (2000). Fires and land cover change in the tropics: a remote sensing analysis at the landscape scale. *Journal of Biogeography*, **27**, 765–776.

Fairhead, J. and Leach, M. (1996). *Misreading the African Landscape: Society and Ecology in a Forest–Savanna Mosaic*. Cambridge: Cambridge University Press.

Fitzsimmons, K. E. *et al.* (2007). The timing of linear dune activity in the Strzelecki and Tirari Deserts, Australia. *Quaternary Science Reviews*, **26**, 2598–2616.

Fryberger, S. G. and Goudie, A. S. (1981). Arid geomorphology. *Progress in Physical Geography*, **5**, 420–428.

Gasse, F. and van Campo, E. (1998). A 40 000 year pollen and diatom record from Lake Tritrivakely, Madagascar, in the southern tropics. *Quaternary Research*, **49**, 299–311.

Giertz, S., Junge, B. and Diekkrüger, B. (2005). Assessing the effects of land use change on soil physical properties and hydrological processes in the sub-humid tropical environment of West Africa. *Physics and Chemistry of the Earth*, **20**, 485–496.

Gillieson, D. (2006). A commentary on Michael F. Thomas's 'Lessons from the tropics for a global geomorphology'. *Singapore Journal of Tropical Geography*, **27**, 131–133.

Goudie, A. S. (1996). The geomorphology of the seasonal tropics. In A. Goudie, W. Adams and A. Orme, eds., *The Physical Geography of Africa*. Oxford: Oxford University Press, pp. 148–160.

Goudie, A. S. and Wells, G. L. (1995). The nature, distribution and formation of pans in arid zones. *Earth Science Reviews*, **38**, 1–69.

Grove, A. T. (1969). Landforms and climatic change in the Kalahari and Ngamiland. *Geographical Journal*, **135**, 191–212.

Grove, A. T. and Warren, A. (1968). Quaternary landforms and climate on the south side of the Sahara. *Geographical Journal*, **134**, 194–208.

Haberle, S. G. (2005). A 23 000 year pollen record from Lake Euramoo, wet tropics of NE Queensland, Australia. *Quaternary Research*, **64**, 343–536.

Harrison, S. P. and Prentice, C. I. (2003). Climate and CO_2 controls on global vegetation: distribution at the last glacial maximum: analysis based on palaeovegetation data, biome modelling and palaeoclimate simulations. *Global Change Biology*, **9**, 983–1004.

Hély, C. *et al.* (2006). Sensitivity of African biomes to changes in the precipitation regime. *Global Ecology and Biogeography*, **15**, 258–270.

Hesse, P., Magee, J. and van der Kaars, S. (2004). Late Quaternary climates of the Australian arid zone: a review. *Quaternary International*, **118–119**, 87–102.

Hesse, P. P. and Simpson, R. L. (2006). Variable vegetation cover and episodic sand movement on longitudinal desert sand dunes. *Geomorphology*, **81**, 276–291.

Hoffman, W. A. and Moreira, A. G. (2002). The role of fire in population dynamics of woody plants. In P. S. Oliveira and R. J. Marquis, eds., *The Cerrados of Brazil: Ecology and Natural History of a Neotropical Savanna*. New York: Columbia University Press, pp. 159–177.

Holt, J. A., Coventry, R. J. and Sinclair, D. F. (1980). Some aspects of the biology and pedological significance of mound-building termites in a red and yellow earth landscape near Charters Towers, North Queensland. *Australian Journal of Soil Research*, **18**, 97–109.

Hope, G. A. *et al.* (2004). History of vegetation and habitat change in the Austral-Asian region. *Quaternary International*, **118–119**, 103–126.

IPCC (2007). *Climate Change 2007: The Physical Science Basis. Contribution of Working Group I to the Fourth Assessment Report of the Intergovernmental Panel on Climate Change*. Solomon, S. *et al.*, eds. Cambridge: Cambridge University Press.

Johnson, R. W. and Tothill, J. C. (1985). Definitions and broad geographic outline of savanna lands. In J. C. Tothill and J. G. Mott, eds., *Ecology and Management of the World's Savannas*. Canberra: Australian Academy of Science, pp. 1–13.

Juyal, N. *et al.* (2003). Luminescence chronology of aeolian deposition during the late Quaternary of the southern margin of Thar Desert, India. *Quaternary International*, **104**, 87–98.

Kampunzu, A. B. *et al.* (2007). Origins and palaeo-enviroments of Kalahari duricrusts in the Moshaweng dry valleys (Botswana) as detected by major and trace element composition. *Journal of African Earth Sciences*, **48**, 199–221.

Knight, M., Thomas, D. S. G. and Wiggs, G. F. S. (2004). Challenges of calculating dunefield mobility over the twenty-first century. *Geomorphology*, **59**, 197–213.

Kniveton, D. R. and McLaren, S. J. (2000). Geomorphological and climatological perspectives on land surface–climate change. In S. J. McLaren and D. R. Kniveton, eds., *Linking Climate Change to Land Surface Change*. Dordrecht: Kluwer, pp. 247–260.

Knox, J. C. (1984). Fluvial response to small scale climate changes. In J. E. Costa and P. J. Fleisher, eds., *Developments and Applications of Geomorphology*. Berlin: Springer-Verlag, pp. 318–342.

Knox, J. C. (2000). Sensitivity of modern and Holocene floods to climate change. *Quaternary Science Reviews*, **19**, 439–457.

Lamb, H. F., Darbyshire, I. and Verschuren, D. (2003). Vegetation response to rainfall variation and human impact in central Kenya during the past 1100 years. *The Holocene*, **13**, 285–292.

Lambin, E. F. and Geist, H. J. (2001). Global land-use and land-cover change: what have we learned so far? *Land Use/Cover Change Newsletter*, **4**, 27–30.

Lambin, E. F. *et al.* (2001). The causes of land-use and land-cover change: moving beyond the myths. *Global Environmental Change*, **11**, 261–269.

Langbein, W. B. and Schumm, S. A. (1958). Yield of sediment in relation to mean annual precipitation. *EOS, Transactions of the American Geophysical Union*, **39**, 1076–1084.

Larson, W. E., Lindstrom, M. J. and Schumacher, T. E. (1997). The role of severe storms in soil erosion: a problem needing consideration. *Journal of Soil and Water Conservation*, **52**, 90–95.

Lawson, M. P. and Thomas, D. S. G. (2002). Late Quaternary lunette dune sedimentation in the southwestern Kalahari, South Africa: luminescence based chronologies of aeolian activity. *Quaternary Science Reviews*, **21**, 825–836.

Livingstone, D. A. (1971). A 22 000 year pollen record from the plateau of Zambia. *Limnology and Oceanography*, **16**, 349–356.

Lomolino, M. V., Riddle, B. R. and Brown, J. H. (2005). *Biogeography*, 3rd edn. Sunderland: Sinauer Associates.

Mäckel, R. (1974). Dambos: a study of morphodynamic activity of plateau regions of Zambia. *Catena*, **1**, 327–365.

Mäckel, R. (1985). Dambos and related landforms in Africa: an example for the ecological approach to tropical geomorphology. *Zeitschrift für Geomorphologie (Supplement)*, **52**, 1–24.

Marchant, R. *et al.* (2006). Colombian dry–moist forest transitions in the Llanos Orientales: a comparison of model and pollen-based biome reconstructions. *Palaeogeography, Palaeoclimatology, Palaeoecology*, **234**, 28–44.

Martin, H. A. (2006). Cenozoic climatic change and the development of the arid vegetation in Australia. *Journal of Arid Environments*, **66**, 533–563.

McFarlane, M. J. (1983). Laterites. In A. S. Goudie and K. Pye, eds., *Chemical Sediments and Geomorphology: Precipitates and Residua in the Near-Surface Environment*. London: Academic Press, pp. 7–58.

McFarlane, M. J. and Whitlow, R. (1990). Key factors affecting the initiation and progress of gullying in dambos in parts of Zimbabwe and Malawi. *Land Degradation and Rehabilitation*, **2**, 215–235.

Meadows, M. E. (1985). Dambos and environmental change in Malawi, Central Africa. *Zeitschrift für Geomorphologie (Supplement)*, **52**, 147–169.

Meadows, M. E. (2001). The role of Quaternary environmental change in the evolution of landscapes: case studies from southern Africa. *Catena*, **42**, 39–57.

Meyer, V. W. *et al.* (1999). Distribution and density of termite mounds in the northern Kruger National Park, with specific reference to those constructed by *Macrotermes* Holmgren (Isoptera: Termitidae). *African Entomology*, **7**, 123–130.

Moglen, G. E., Eltahir, E. A. B. and Bras, R. L. (1998). On the sensitivity of drainage density to climate change. *Water Resources Research*, **34**, 855–862.

Morgan, R. P. C. and Mngomezulu, D. (2003). Threshold conditions for initiation of valley side gullies in the Middle Veld of Swaziland. *Catena*, **50**, 401–414.

Moss, P. T. and Kershaw, A. P. (2000). The last glacial cycle from the humid tropics of northeastern Australia: comparison of a terrestrial and a marine record. *Palaeogeography, Palaeoclimatology, Palaeoecology*, **155**, 155–176.

Motta, P. E. F., Curi, N. and Franzmeier, D. P. (2002). Relation of soils and geomorphic surfaces in the Brazilian Cerrado. In P. S. Oliviera and R. J. Marquis, eds., *The Cerrados of Brazil: Ecology and Natural History of a Neotropical Savanna*. New York: Columbia University Press, pp. 13–32.

Nash, D. J. and Shaw, P. A. (1998). Silica and carbonate relations in silcrete–calcrete intergrade duricrusts from the Kalahari of Botswana and Namibia. *Journal of African Earth Sciences*, **27**, 11–25.

Okwakol, M. J. N. (2000). Changes in termite (Isoptera) communities due to the clearance and cultivation of tropical forest in Uganda. *African Journal of Ecology*, **38**, 1–7.

Persano, C. *et al.* (2002). Apatite (U/Th)/He age constraints on the development of the Great Escarpment on the southeastern Australian passive margin. *Earth and Planetary Science Letters*, **200**, 79–90.

Plisnier, P. D., Serneels, S. and Lambin, E. F. (2000). Impact of ENSO on East African ecosystems: a multivariate analysis based on climate and remote sensing data. *Global Ecology and Biogeography*, **9**, 481–497.

Poesen, J. *et al.* (2003). Gully erosion and environmental change: importance and research needs. *Catena*, **50**, 91–133.

Pullan, R. A. (1979). Termite hills in Africa: their characteristics and evolution. *Catena*, **6**, 267–291.

Reading, A. J., Thompson, R. D. and Millington, A. C. (1995). *Humid Tropical Environments*. Oxford: Blackwell Scientific Publications.

Ringrose, S. *et al.* (2003). Vegetation cover trends along the Botswana Kalahari transect. *Journal of Arid Environments*, **54**, 297–317.

Ringrose, S. *et al.* (2005). Sedimentological and geochemical evidence for palaeoenvironmental change in the Makgadikgadi subbasin, in relation to the MOZ rift depression, Botswana. *Palaeogeography, Palaeoclimatology, Palaeoecology*, **217**, 265–287.

Roberts, N. (1988). Dambos and development: management of a fragile ecological resource. *Journal of Biogeography*, **15**, 141–148.

Römer, W. (2007). Differential weathering and erosion in an inselbergs landscape in southern Zimbabwe: a morphometric study and some notes on factors influencing the long-term development of inselbergs. *Geomorphology*, **86**, 349–368.

Sankaran, M. *et al.* (2005). Determinants of woody cover in African savannas. *Nature*, **438**, 846–849.

Scholes, R. J. and Walker, B. H. (1993). *An African Savanna: Synthesis of the Nylsvley Study*. Cambridge: Cambridge University Press.

Scott, L. (1982). A late Quaternary pollen record from the Transvaal Bushveld, South Africa. *Quaternary Research*, **17**, 339–370.

Scott, L. (1987). Pollen analysis of hyaena coprolites and sediments from Equus Cave, Taung, southern Kalahari (South Africa). *Quaternary Research*, **28**, 144–156.

Scott, L. (1989). Late Quaternary vegetation history and climatic change in the eastern Orange Free State, South Africa. *South African Journal of Botany*, **55**, 107–116.

Scott, L. (1999). Vegetation history and climate in the savanna biome since 190 000 ka: a comparison of pollen data from the Tswaing Crater (the Pretoria Saltpan) and Wonderkrater. *Quaternary International*, **57–58**, 215–223.

Scott, L. (2002). Grassland development under glacial and interglacial conditions in southern Africa: review of pollen, phytolith and isotope evidence. *Palaeogeography, Palaeoclimatology, Palaeoecology*, **177**, 47–57.

Shakesby, R. A. and Whitlow, R. (1991). Perspectives on prehistoric and recent gullying in Central Zimbabwe. *GeoJournal*, **23**, 49–58.

Shaw, P. A. and Thomas, D. S. G. (1998). Pans, playas and salt lakes. In D. S. G. Thomas, ed., *Arid Zone Geomorphology*. Chichester: John Wiley, pp. 293–318.

Shi, N. and Dupont, L. M. (1997). Vegetation and climatic history of southwest Africa: a marine palynological record of the last 300 000 years. *Vegetation History and Archaeobotany*, **6**, 117–131.

Shi, N. *et al.* (2000). Correlation between vegetation in southwestern Africa and oceanic upwelling in the past 21 000 years. *Quaternary Research*, **54**, 72–80.

Shi, N. *et al.* (2001). Southeast trade wind variations during the past 135 kyr: evidence from pollen spectra in eastern South Atlantic sediments. *Earth and Planetary Science Letters*, **187**, 311–321.

Shorrocks, B. (2007). *The Biology of African Savannas*. Oxford: Oxford University Press.

Showers, K. B. (2005). *Imperial Gullies: Soil Erosion and Conservation in Lesotho*. Athens: Ohio University Press.

Sidorchuk, A. *et al.* (2003). Gully erosion modelling and landscape response in the Mbuluzi River catchment of Swaziland. *Catena*, **50**, 507–525.

Sørenson, R. *et al.* (2001). Stratigraphy and formation of late Pleistocene colluvial apron in Morongoro District, central Tanzania. *Palaeoecology of Africa*, **27**, 95–116.

Stokes, S., Thomas, D. S. G. and Shaw, P. A. (1997a). New chronological evidence for the nature and timing of linear dune field development in the southwest Kalahari Desert. *Geomorphology*, **20**, 81–93.

Stokes, S., Thomas, D. S. G. and Washington, R. (1997b). Multiple episodes of aridity in southern Africa since the last interglacial period. *Nature*, **388**, 154–158.

Stuut, J.-B. W. *et al.* (2002). A 300 kyr record of aridity and wind strength in southwestern Africa: inferences from grain-size distributions of sediments on Walvis Ridge, SE Atlantic. *Marine Geology*, **180**, 221–233.

Telfer, M. W. and Thomas, D. S. G. (2006). Complex Holocene lunette dune development, South Africa: implications for palaeoclimate and models of pan development in arid regions. *Geology*, **34**, 853–856.

Telfer, M. W. and Thomas, D. S. G. (2007). Late Quaternary linear dune accumulation and chronostratigraphy of the southwestern Kalahari: implications for aeolian palaeoclimate reconstructions and predictions of future dynamics. *Quaternary Science Reviews*, **26**, 2617–2630.

Thomas, D. S. G. and Shaw, P. A. (1991). *The Kalahari Environment*. Cambridge: Cambridge University Press.

Thomas, D. S. G. and Shaw, P. A. (2002). Late Quaternary environmental change in central southern Africa: new data, synthesis, issues and prospects. *Quaternary Science Reviews*, **21**, 783–797.

Thomas, D. S. G. *et al.* (2000). Dune activity as a record of late Quaternary aridity in the northern Kalahari: evidence from northern Namibia interpreted in the context of regional arid and humid chronologies. *Palaeogeography, Palaeoclimatology, Palaeoecology*, **156**, 253–259.

Thomas, D. S. G. *et al.* (2003). Late Pleistocene wetting and drying in the NW Kalahari: an integrated study from the Tsodilo Hills, Botswana. *Quaternary International*, **104**, 53–67.

Thomas, D. S. G., Knight, M. and Wiggs, G. S. F. (2005). Remobilisation of southern African desert dune systems by twenty-first century global warming. *Nature*, **435**, 1218–1221.

Thomas, M. F. (1966). Some geomorphological implications of deep weathering patterns in crystalline rocks in Nigeria. *Transactions of the Institute of British Geographers*, **40**, 173–193.

Thomas, M. F. (1974). *Tropical Geomorphology*. London: Macmillan.

Thomas, M. F. (1994). *Geomorphology in the Tropics: A Study of Weathering and Denudation in Low Latitudes*. Chichester: John Wiley.

Thomas, M. F. (2001). Landscape sensitivity in time and space: an introduction. *Catena*, **42**, 83–98.

Thomas, M. F. (2004). Landscape sensitivity to rapid environmental change: a Quaternary perspective with examples from tropical areas. *Catena*, **55**, 107–124.

Thomas, M. F. (2006). Lessons from the tropics for a global geomorphology. *Singapore Journal of Tropical Geography*, **27**, 111–127.

Thomas, M. F. and Murray, A. S. (2001). On the age and significance of Quaternary colluviums in central Zambia. *Palaeoecology of Africa*, **27**, 117–133.

Tripaldi, A. and Forman, S. L. (2006). Geomorphology and chronology of late Quaternary dune fields of western Argentina. *Palaeogeography, Palaeoclimatology, Palaeoecology*, **251**, 300–320.

Valentin, C., Poesen, J. and Yong L. (2005a). Preface. *Catena*, **63**, 129–131.

Valentin, C., Poesen, J. and Yong L. (2005b). Gully erosion: impacts, factors and control. *Catena*, **63**, 132–153.

van Langevelde, F. *et al.* (2003). Effects of fire and herbivory on the stability of savanna ecosystems. *Ecology*, **84**, 337–350.

Vincens, A. (1991). Late Quaternary vegetation history of the South Tanganyika basin: climatic implications in South Central Africa. *Palaeogeography, Palaeoclimatology, Palaeoecology*, **86**, 207–226.

von der Heyden, C. J. (2004). The hydrology and hydrogeology of dambos: a review. *Progress in Physical Geography*, **28**, 544–564.

von der Heyden, C. J. and New, M. (2003). The role of a dambo in the hydrology of a small Zambian catchment and the river network downstream. *Hydrology and Earth Systems Sciences*, **7**, 339–357.

Wang, L. *et al.* (2007). Biogeochemistry of Kalahari sands. *Journal of Arid Environments*, **71**, 259–279.

Washington, R. *et al.* (2003). Dust storm areas determined by the Total Ozone Monitoring Spectrometer and surface observations. *Annals of the Association of American Geographers*, **93**, 297–313.

Watson, A., Price-Williams, D. and Goudie, A. S. (1984). The palaeoenvironmental interpretation of colluvial sediments and palaeosols of the late Pleistocene hypothermal in southern Africa. *Palaeogeography, Palaeoclimatology, Palaeoecology*, **45**, 225–249.

Watson, A., Price-Williams, D. and Goudie, A. S. (1987). Reply to 'Is gullying associated with highly sodic colluvium?' Further comment to the environmental interpretation of southern African dongas. *Palaeogeography, Palaeoclimatology, Palaeoecology*, **58**, 123–128.

Werger, M. J. A. (1983). Tropical grasslands, savannas, woodlands: natural and man-made. In W. Holzner, M. J. A. Werger and I. Ikusima, eds., *Man's Impact on Vegetation*. Dordrecht: Dr W Junk, pp. 107–137.

Widdowson, M. (2007). Laterite and ferricrete. In D. J. Nash and S. J. McLaren, eds., *Geochemical Sediments and Landscapes*. Oxford: Blackwell, pp. 46–94.

Wiggs, G. F. S. *et al.* (1994). Dune mobility and vegetation cover in the southwest Kalahari Desert. *Earth Surface Processes and Landforms*, **20**, 515–529.

Wille, M. *et al.* (2003). Sub-millennium scale migrations of the rainforest savanna boundary in Colombia: ^{14}C wiggle-matching and pollen analysis of core Las Margaritas. *Palaeogeography, Palaeoclimatology, Palaeoecology*, **193**, 201–223.

Williams, M. J. A. (1968). Termites and soil development near Brock's Creek, Northern Territory. *Australian Journal of Science*, **31**, 153–154.

Woodward, F. I, Lomas, M. R. and Kelly, C. K. (2004). Global climate and the distribution of plant biomes. *Philosophical Transactions of the Royal Society of London B*, **359**, 1465–1476.

Yaalon, D. (1987). Is gullying associated with highly sodic colluvium? Further comment to the environmental interpretation of southern African dongas. *Palaeogeography, Palaeoclimatology, Palaeoecology*, **58**, 121–128.

Zawada, P. K. and Hattingh, J. (1994). Preliminary findings of a national palaeoflood hydrological investigation for South African rivers: implications for flood prediction and Holocene climate change. *South African Journal of Science*, **90**, 567–568.

10 Deserts

Nicholas Lancaster

10.1 Introduction

Deserts, defined by lack of water and low density of vegetation, cover some 26.2 million km^2, or about 20% of the Earth's land surface (Ezcurra, 2006) (Fig. 10.1). Despite their geographic extent, there are few and very limited discussions of the effects of future climate change on desert regions as part of the IPCC process (e.g. IPCC, 2007a).

Most deserts are fragile environments (Plate 25), easily affected by natural and human disturbance, and are being impacted by a rapidly growing and increasingly urban population, which is dependent on scarce surface and groundwater. The importance of water to human and natural systems in deserts makes them very sensitive to those changes in climate that affect the amount, type, timing and effectiveness of precipitation. The rich record of past climate changes in desert regions indicates the magnitude of their amplitude and duration as well as their effect on landscapes and ecosystems and allows the assessment of the nature and effects of future climate changes.

Hydroclimatological observations show that changes in seasonal and annual temperature, precipitation, snowmelt and runoff, groundwater recharge and evapotranspiration are occurring today in most deserts (e.g. Dai *et al.*, 2004; IPCC, 2007a) and models predict that they are likely to continue in the future. Many areas are already experiencing significant increases in temperature and a reduction in rainfall over the past two decades, manifested in extended droughts, as in the Colorado River basin since 2000; Australia from 2001–07 (with especially severe drought in 2002–03); Southern Africa from 2001 to 2004; Iraq in 2008; and in Afghanistan from 1998 to 2005. These changes will have significant geomorphic impacts, including changes to the fluvial regime, dust storm frequency and the mobility of sand dunes (Goudie, 2003).

The results of global climate models differ in their predictions of the direction and magnitude of change in arid regions, in large part because prediction of precipitation in global climate models is difficult, but a consistent pattern is emerging (Ezcurra, 2006; IPCC, 2007a) (Fig. 10.2). In some areas, such as China, southeastern Arabia and India, increased monsoon precipitation is predicted, but its effects may be offset by higher evaporation as a result of increased temperatures. In the Sahara, there is support in many climate model predictions for a moistening of the southern and southeastern areas (including the Sahel), but strong drying for the northern and western areas. Some models, however, suggest a strong drying throughout the region (IPCC, 2007a). The differences between model predictions show the complexity of forcing factors for this region, as well as the possible influence of feedbacks between land surface conditions and the atmosphere, which may affect rainfall total, effectiveness and spatial distribution (e.g. Nicholson, 2000; Lau *et al.*, 2006). Most of the interior of southern Africa is also predicted to become drier. In the southwestern USA, higher temperatures are predicted to increase the severity of droughts (Easterling *et al.*, 2007). The region may already be in transition to a new more arid state as a result of anthropogenically influenced climate change (Seager *et al.*, 2007). Desert areas which receive winter rainfall are likely to be especially vulnerable to warming (IPCC, 2007b). The effects of increased levels of carbon dioxide on plant productivity in arid regions are uncertain, but may favour invasive exotic species, with possible effects on fire regimes (Smith *et al.*, 2000). Model results that incorporate carbon dioxide fertilisation of vegetation indicate a reduction in desert areas in the next century, introducing an additional level of uncertainty (Mahowald, 2007).

Geomorphology and Global Environmental Change, eds. Olav Slaymaker, Thomas Spencer and Christine Embleton-Hamann. Published by Cambridge University Press.

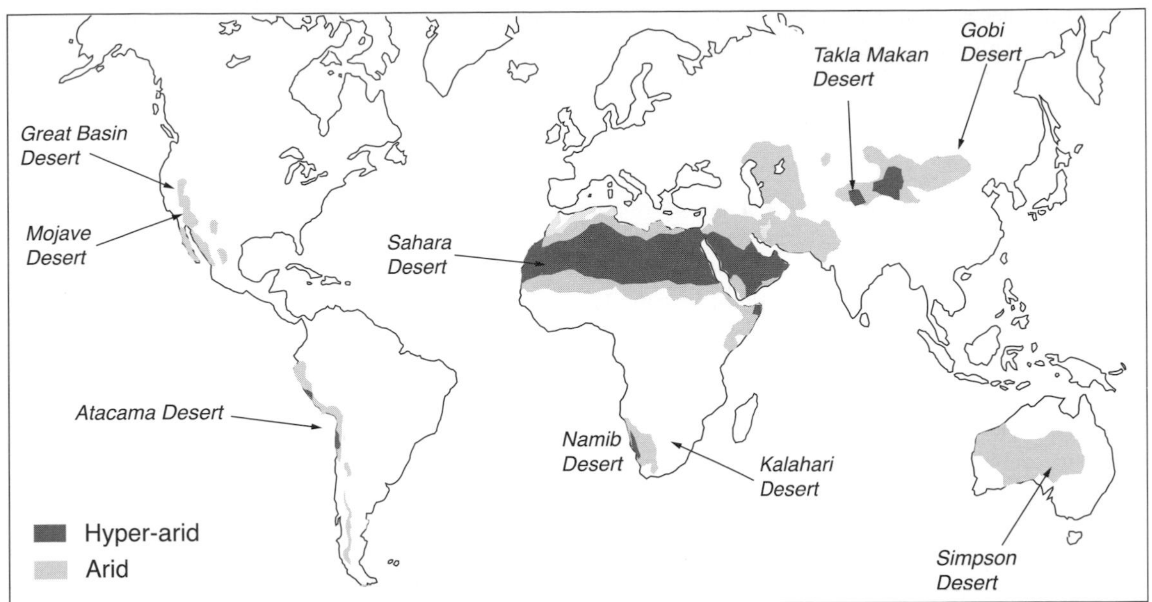

FIGURE 10.1. Extent of arid and hyper-arid areas today. Map compiled from various sources, including Ezcurra (2006).

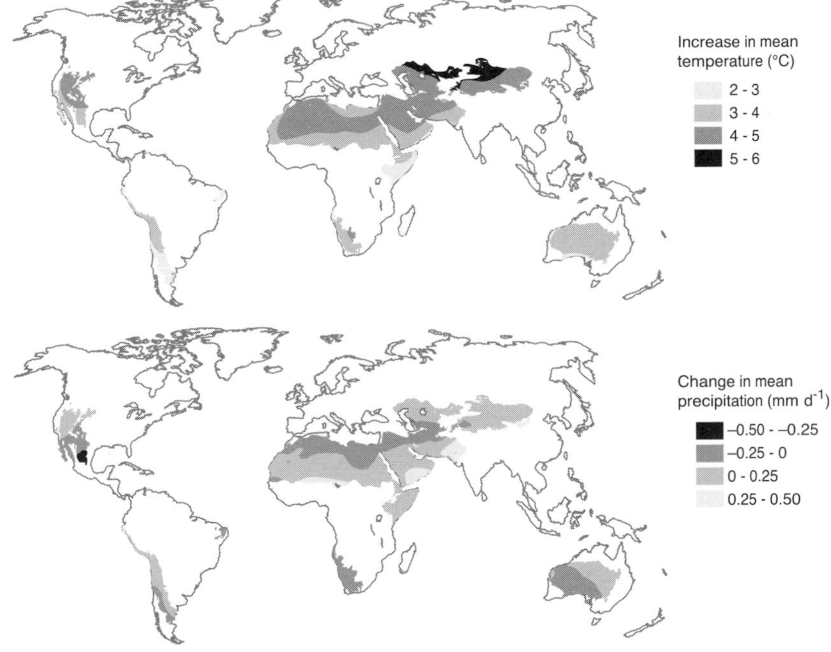

FIGURE 10.2. Climate change scenarios for desert areas. SRES scenarios show the period 2071 to 2100 relative to the period 1961 to 1990, and were performed by atmosphere–ocean general circulation models (AOGCMs). Scenarios A2 and B2 are shown as no AOGCM runs were available for the other SRES scenarios from UNEP/GRID-Arendal (2007).

10.1.1 Causes of deserts

The aridity index, the ratio between mean annual precipitation (P) and mean annual potential evapotranspiration (PET), is commonly used to define areas of hyper-arid climates (P/PET <0.05) and arid climates (P/PET 0.05–0.20). Deserts can also be defined as areas that are characterised by xerophilous vegetation, or by areas of very low vegetation cover identified from satellite image data such as the normalised difference vegetation index, (NDVI). These three criteria are coincident in many areas (Ezcurra, 2006).

Desert climates are characterised by low humidity (except in cool foggy coastal deserts like the Namib and Atacama), a high daily range of temperatures, and precipitation that is highly variable in time and space. The most extensive deserts lie astride the tropics. Descending, dry stable air in the subtropical anticyclonic belts maintain the arid conditions throughout the year. The effects of stable air masses are reinforced by large land masses. Long distances to continental interiors restrict the influence of moist oceanic air in summer, as in the central Asian and African deserts. In winter, large continental areas develop strong high-pressure cells, reducing the influence of frontal systems on the poleward margins of the desert. Mountain barriers block rain-bearing winds and create rain shadows in their lee, especially in the Great Basin Desert of North America and in central Asia, where the Himalaya and other mountain ranges prevent the penetration of the southwest monsoon to the Gobi and Takla Makan deserts. Deserts located on the west coasts of South America and southern Africa (Atacama, Namib) owe their hyper-arid climates to the influence of cold oceanic currents offshore. These reinforce the subsidence-induced stability of the atmosphere by cooling surface air and creating a strong temperature inversion.

10.1.2 Desert geomorphic processes and landforms

Although no landforms or geomorphic processes are unique to deserts, certain characteristics of desert environments have a significant effect on the operation of the major processes of weathering, erosion, transport and deposition (Cooke *et al.*, 1993). A sparse vegetation cover with a high percentage of bare ground results in rapid runoff of water when intense rainfall does occur and allows the wind to erode and transport sand and dust (silt- and clay-sized sediment). Sand accumulates in areas of lower wind velocity and transport capacity to form dunefields and sand seas. Dune form is governed by the availability of sand and the variation in wind direction from season to season. Dust storms transport fine-grained material away from desert regions to be deposited in the oceans and desert margins.

The excess of evaporation over precipitation gives rise to physical or mechanical rather than chemical weathering of rocks, and to upward movement of soil moisture and near-surface groundwater. As a result, water-soluble salts (principally sodium chloride, calcium carbonate and calcium sulphate) accumulate in desert soils forming saline, calcic and gypsic horizons in the subsoil. Insolation weathering and salt weathering dominate processes of rock breakdown. On a regional scale, lack of water gives rise to internal drainage to playas and salt lakes.

TABLE 10.1. *Proportions of landform types in selected deserts (percentage of area covered)*

Landform type	Southwestern USA	Sahara
Desert mountains	38.1	43.0
Playas	1.1	1.0
Desert flats	20.5	10.0
Bedrock fields	0.7	10.0
Regions bordering throughflowing rivers	1.2	1.0
Ephemeral streams	3.6	1.0
Alluvial fans	31.4	1.0
Sand dunes	0.6	28.0
Badlands	2.6	2.0
Volcanic fields	0.2	3.0

Source: After Goudie (2002).

The character of desert landforms is also affected strongly by the regional geologic and tectonic environment. Two end member models can be recognised (Table 10.1). The tectonically stable Old World shield deserts, such as those in the Arabian Peninsula, Australia and southern Africa, are characterised by low relief and extensive rocky plains and isolated hills (inselbergs); in these areas up to 30% of the land surface may be covered by sand dunes or sand seas. Variants of this landscape depend on whether the bedrock is sedimentary, as in the northern and eastern Sahara, or crystalline, as in much of Australia and Namibia. At the other end of the spectrum are the high-relief deserts of the tectonically active areas of the Atacama, the Basin and Range Province of western North America and parts of Central Asia. These deserts have few and generally small areas of sand dunes, but extensive areas of mountains, alluvial fans and basins with internal drainage.

10.2 Drivers of change and variability in desert geomorphic systems

Many desert landforms are sensitive indicators of environmental change on a variety of timescales. Until recently, attention had focussed on the use of landforms as indicators of Quaternary climate change on timescales of millennia. More recent work shows that many desert landforms can provide information on changes that occur on a scale of years to centuries. Whereas pediments and desert pavements may change little over decades (unless affected by humans), desert hillslopes, fluvial systems, playa lakes and aeolian sediment transport and depositional systems appear

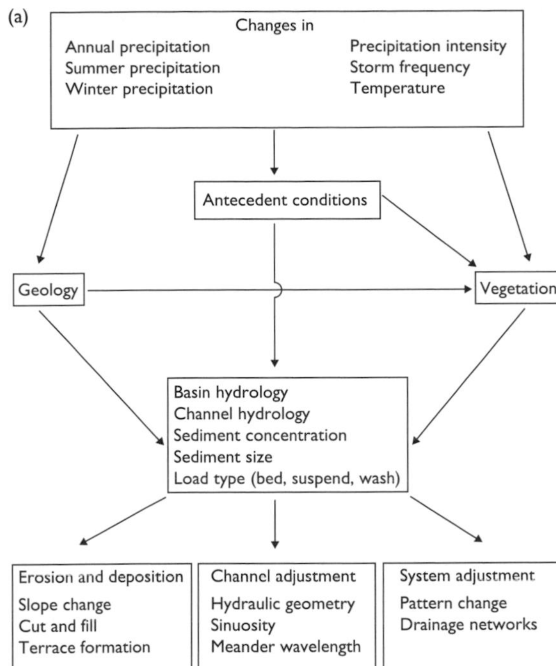

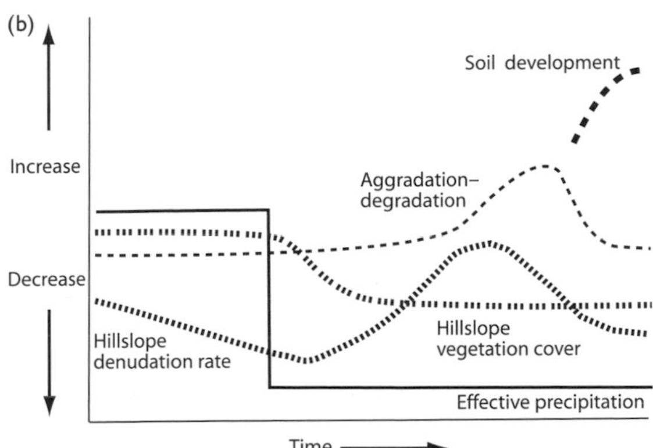

FIGURE 10.3. Conceptual model for response of arid fluvial systems to climate change: (a) fluvial system response (from Lancaster, 1996); (b) leads and lags in response to Holocene climate change (from Bull, 1991).

to react sensitively to environmental change on annual to decadal timescales.

The identification of geomorphic responses to climatic change is based on two approaches: (a) studies of the response of geomorphic systems to modern, short-term climatic changes (such as El Niño events and extended regional droughts). Good observational data enables threshold conditions to be identified and well-constrained process–response models to be developed (see Figs. 10.3 and 10.4); and (b) studies of Quaternary sediments and landforms. Such investigations provide information on the response of landforms to fluctuations that precede the observational record. These may include both ancient, longer-term climatic fluctuations (centuries to millennia) as well as greater extremes in more recent times (such as the Little Ice Age).

A characteristic feature of many natural systems in desert regions is their temporal and spatial variability. As a result, equilibrium concepts of form–process relations developed in humid regions are difficult or impossible to apply to many desert landforms, especially fluvial systems, in which extreme events play a major role (Graf, 2002). Similar constraints exist for desert ecosystems (Whitford, 2002). As a result, non-linear responses are common and thresholds play an important role in landform and ecosystem dynamics (Bull, 1991).

10.2.1 Natural drivers of change

Desert rainfall is highly variable in time and space. Ecological and geomorphological processes are driven by a succession of short pulses of rainfall and abundant water availability that punctuate long periods of drought. The characteristics of these pulses vary, and may include intense summer rains of limited spatial extent, or more spatially extensive but lower-intensity winter rains. Extreme runoff events are temporally and spatially variable in desert river systems, but play a major role in their dynamics (e.g. Graf, 2002). Pulses of rainfall may result in the germination and growth of annual or ephemeral plants, which have a significant effect on surface runoff (Kochel *et al.*, 1997) and aeolian sand mobility (Lancaster, 1997). Runoff may flood playas and contribute sediment that is mobilised by wind in subsequent dry periods (Mahowald *et al.*, 2003; Reheis, 2006; Bryant *et al.*, 2007), so that dust emissions appear to correlate with ENSO cycles in the southwestern USA (Okin and Reheis, 2002). Increased rainfall and runoff increase recharge along ephemeral streams, so that in the Sahel region rates of recharge increased from 30 mm a^{-1} in the drought period of the 1970s and 1980s to 150 mm a^{-1} in wetter periods (Scanlon *et al.*, 2006). On longer timescales, decadal-scale changes in rainfall intensity have been related to arroyo cutting in New Mexico and elsewhere in the southwestern USA (Balling and Wells, 1990; Miller and Kochel, 1999), and floodplain destruction and rebuilding in Chile (Manners *et al.*, 2007). Longer duration periods of increased rainfall have also been recognised as significant factors in Holocene episodes of flooding of rivers in the western USA (Enzel *et al.*, 1992; Ely *et al.*, 1993), which are in turn linked to periods of aeolian deposition as sediments deposited in terminal lake basins are reworked by wind (Clarke and Rendell, 1998). The causes of periods of increased rainfall

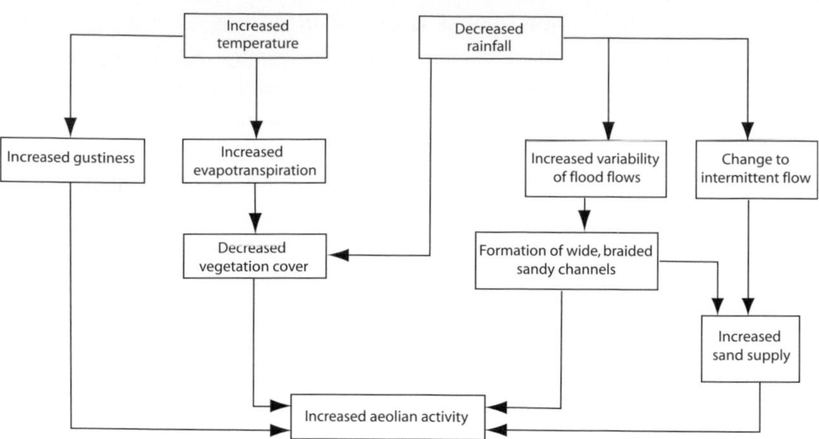

FIGURE 10.4. Conceptual model for response of aeolian systems to climate change (from Muhs and Holliday, 1995).

are often difficult to discern, but the link to ENSO cycles in central Australia (Kotwicki and Isdale, 1991), southern Africa, the southwestern USA and the Atacama Desert seems clear, as is the relation of Sahel rainfall to SSTs (Giannini *et al.*, 2003). ENSO cycles also have a significant effect on groundwater recharge via increased precipitation and flow in ephemeral streams. For example, recharge rates in the southwestern USA were as much as three times higher in periods of more frequent El Niños (e.g. 1977–98) relative to periods dominated by La Niña conditions (e.g. 1941–57) (Scanlon *et al.*, 2006). Enhanced recharge related to ENSO cycles is also documented from Argentina.

The modern climates and landforms of desert regions should be viewed in the context of Pleistocene–Holocene climate changes, which have left a significant legacy of landforms (such as vegetation-stabilised dune systems, palaeolakes, palaeosols and alluvial fans). The area covered by desert biomes has changed significantly from a maximum coincident with the LGM (Plate 1) to a minimum during the early to mid-Holocene (Plate 2) although significant regional variations in the history and pattern of aridity occur, as discussed below.

The millennial-scale response of desert regions to climate change is complex, involving changes in the area of deserts, as well as the intensity of aridity. The examples given below show the diversity of conditions and responses to climate change that have occurred over the past 25 ka. During the LGM (*c.* 20 ka BP), desert regions in the Sahara, Arabia and Australia experienced dry windy climates, with increased aeolian activity leading to higher levels of dust input to ocean sediments (deMenocal *et al.*, 2000; Hesse *et al.*, 2004) and the formation or reworking of sand dunes (Fig. 10.5), as far south as the modern Sahel zone, now characterised by dry savanna (see Chapter 9). Data from dune systems in Mauritania indicate that this period was one of greater trade wind intensity and persistence (Lancaster *et al.*, 2002). By contrast, the northern margin of the Sahara was wetter than today during the same period, as was the southwestern USA, where open woodlands with pinyon, juniper and oak flourished in all but the driest areas, and large lakes formed in basins in the Mojave and Chihuahuan deserts. In the Mojave Desert, increased rainfall and a well-developed vegetation cover led to the development of soils on hillslopes (Harvey and Wells,2003).

The early to mid-Holocene shows similar contrasts in climatic conditions. In North America, the desert regions desiccated, groundwater systems changed from recharge to discharge, and drying lake basins fed aeolian accumulation, especially in the Mojave Desert, while a transition to drier climates and reduced vegetation cover destabilised hillslopes, leading to aggradation of alluvial fans (Wells and Harvey, 2001). In the Sahara and Arabia, the greater insolation resulted in enhanced monsoon activity, leading to the spread of savanna vegetation throughout the Sahara and probably Arabia, accompanied by the stabilisation of dune systems, increased fluvial activity, high lake levels in areas such as Bodélé depression (Lake Mega-Chad), as well as greatly increased human and animal populations (Kuper and Kröpelin, 2006). The African humid period in the Sahara was however not uniformly wetter: a significant dry spell occurred around 8.4–8 ka BP (Gasse and Van Campo, 1994; Gasse, 2000), nor was the increase in moisture spatially coherent (Kuper and Kröpelin, 2006). Some authors have suggested that the early to mid-Holocene period may be an appropriate model for the response of the Sahara region to future climate change and global warming (Petit-Maire, 1990), but forcing of increased rainfall by an insolation maximum in subtropical latitudes is probably not a good analogue for climate change in the

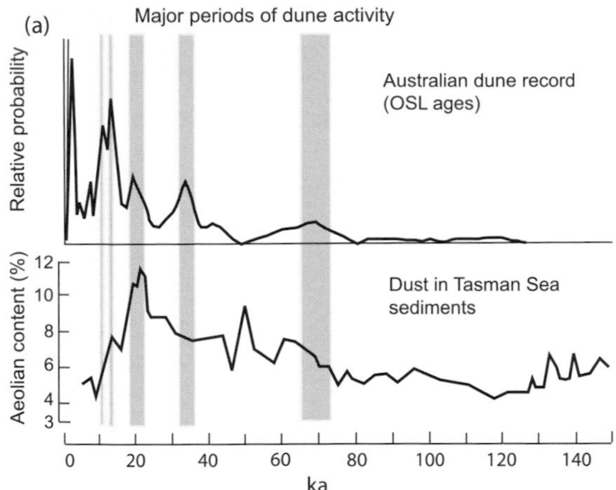

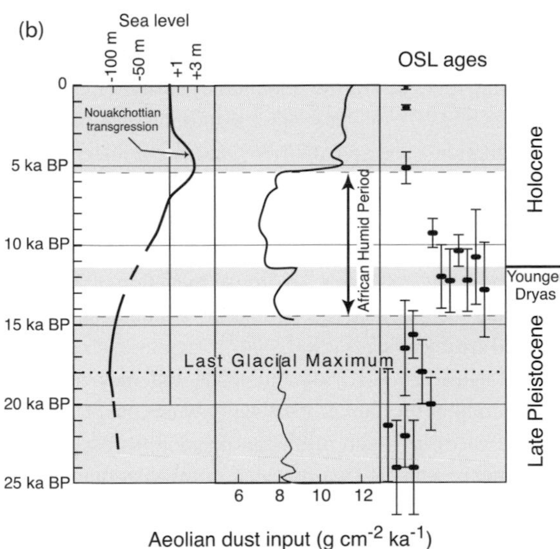

FIGURE 10.5. Relations between periods of dune activity and dust deposition in adjacent oceans. (a) Australia (modified from Fitzsimmons, 2007); (b) Mauritania (modified from deMenocal *et al.*, 2000; OSL ages from Lancaster *et al.*, 2002). This illustrates the connections between different geomorphic systems.

coming decades and centuries. The 'greening' of the Sahara during the African Humid Period may also have been enhanced by feedbacks between the land surface and atmosphere that reduced albedo and increased precipitation effectiveness (Kutzbach *et al.*, 1996), but recent palaeoenvironmental data (Kröpelin *et al.*, 2008) suggest that biogeophysical feedback effects may be relatively weak, in line with modelling results (Liu *et al.*, 2007).

10.2.2 Human influences

Humans have had an influence on the environment and geomorphic processes of many desert regions for millennia, especially in the Middle East, India and China (Roberts, 1998; Sun *et al.*, 2006), but also in the Sahara and its margins (Brooks *et al.*, 2005). Distinguishing between the effects of human activities and climate variability and change on geomorphic systems is often difficult, as indicated by Wang *et al.* (2008) in their assessment of desertification in China. The human impact on desert regions as a whole has until recently been limited by the low density of population. In recent decades, however, the rapidly increasing population of many arid regions, especially in urban areas (Plate 25) has resulted in much higher levels of human impact, such that these were regarded as dominating over climate change by IPCC (2001b). Some examples follow.

Intensive use of groundwater, often at rates exceeding recharge, or groundwater mining, is widespread in desert regions (e.g. Khater, 2003). Rates of withdrawal exceed recharge by more than 500% in Libya and by 374% in Saudi Arabia (Khater, 2003). Excessive groundwater withdrawals result in lowering of groundwater levels (Khater, 2003), leading to increased aeolian activity (as in the Mojave River, California: Laity, 2003), desiccation of groundwater-fed lakes (as in the Fezzan area of Libya: Drake and Mattingly, 2006), springs and wetlands (and the destruction of their phreatophyte vegetation communities) (e.g. Sada and Vinyard, 2002) and to the subsidence of the surface, principally in areas underlain by unconsolidated alluvial or lacustrine sediments. A notable example of subsidence in a desert region is North Las Vegas, Nevada (see Plate 51), where persistent overdraft of groundwater resources since 1950 has lowered groundwater levels by as much as 90 m, leading to subsidence of more than 2 m since 1935 over an area of 1000 km^2 (Bell, 1981). Indirect effects of groundwater withdrawal include the abandonment of irrigated croplands as water becomes too expensive to pump because the depth to water table is excessive. Many of these abandoned fields become sources for aeolian dust (Hyers and Marcus, 1981). Conversely, removal of natural vegetation may increase recharge rates and raise groundwater levels, as in the Murray–Darling basin of Australia (Allison *et al.*, 1990). In Niger and Senegal, land use changes involving replacement of natural vegetation by rain-fed agriculture have an effect far greater than climate variability, resulting in an order of magnitude increase in recharge rates (Scanlon *et al.*, 2006).

The soil and vegetation resources of arid regions are naturally fragile and are of relatively low productivity. When the effects of natural climatic variability (e.g. prolonged drought) are combined with increased human pressures, the results can be loss of productivity and ecosystem degradation (Safriel and Adeel, 2005). Such degradation, if irreversible, results in permanent loss of primary productivity and is then termed

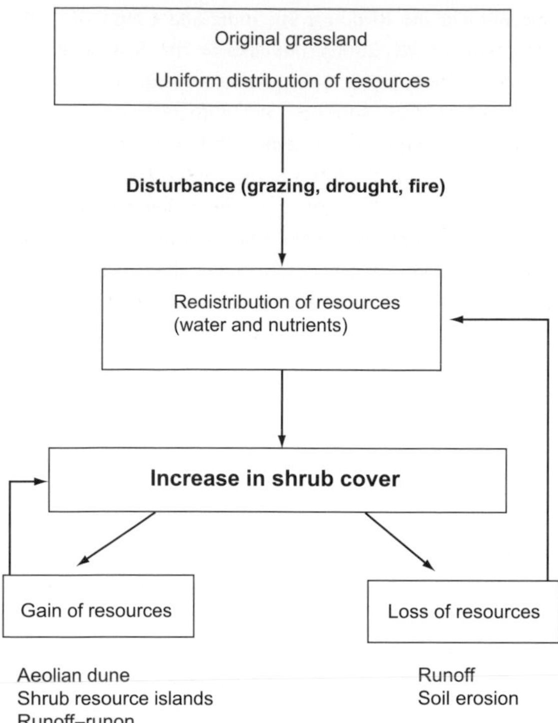

FIGURE 10.6. Conceptual model for progressive degradation of arid ecosystems (modified from Whitford, 2002).

desertification (Thomas and Middleton, 1994). Loss of vegetation cover due to drought and/or overgrazing may lead to reactivation of vegetation-stabilised dunes, as on the southern margin of the Sahara following the prolonged drought of the 1970s (Niang *et al.*, 2008) and to an increased frequency and magnitude of dust events (Prospero and Lamb, 2003). A less extreme result of human impacts is the replacement of grasses and other palatable forage species by woody shrubs. In the Chihuahuan Desert of the southwestern USA and adjacent areas of Mexico, grasses have been gradually replaced by mesquite shrubs over the past 50–100 years, with important consequences for ecosystems and geomorphic processes (Okin *et al.*, 2001). Vegetation change of this magnitude results in irreversible changes to soil structure and hydrology, with a series of positive feedback processes that result in the concentration of nutrients and water in a series of 'islands of productivity' separated by bare ground (Schlesinger *et al.*, 1990) (Fig. 10.6). The intervening bare areas are subject to increased erosion by water and wind (Abrahams *et al.*, 1995; Okin *et al.*, 2001). In the Jornada area of New Mexico, the mesquite shrubs anchor coppice dunes separated by 'streets' of high sand transport rates (Okin and Gillette, 2001; Okin, 2008). Grazing also damages or destroys the biological soil crusts that are important in stabilising soils, reducing wind and water erosion, and fixing nitrogen in many desert regions. In southeastern Utah, even after 30 years without grazing, soils in areas that have been historically grazed have much lower contents of organic matter and primary nutrients, compared to areas that have never been grazed, suggesting that livestock grazing promotes increased wind erosion which reduces soil productivity in native grasslands (Neff *et al.*, 2005).

Surface disturbance by offroad vehicles is becoming increasingly widespread in arid regions, leading to increased erosion by wind and water as a result of compaction, disruption of surface crusts and desert pavements (Webb and Wilshire, 1983). Offroad vehicle use by civilian and military vehicles has resulted in increased aeolian activity and sand enchroachment in Kuwait (Khalaf and Al-Ajmi, 1993), the Mojave Desert (Laity, 2003) and many other areas. The level of disturbance may be linked to increasing affluence of the populations of arid regions and their use of 4WD vehicles rather than traditional means of transportation – the so-called Toyota-isation effect coined by Goudie (2008).

Regulation by dams and water diversion for agriculture and urban use has affected the streamflow characteristics of many desert river systems, leading to reduction of instream flows and the rate recharge of riparian aquifers, as well as reduced sediment loads (Graf, 2002). Trapping of sediment behind dams reduces sediment loads and may result in degradation or incision of the river downstream. There are notable cases in which discharge of additional sediment-free water (e.g. treated waste water, stormwater runoff) into formerly ephemeral channels has resulted in massive channel change downstream. A good example is the Las Vegas Wash, where discharge of treated sewage effluent from the rapidly growing urban area together with storm runoff from El Niño-generated storms resulted in extensive erosion of pre-existing fluvial sediment, incision and destabilisation of the channel in the past 25 years (Buckingham *et al.*, 2004).

Humans have deliberately or accidentally introduced many species of plant into desert regions, a large number of which have become invasive, and have resulted in significant changes in geomorphic processes and responses. In the southwestern USA, tamarisk (*Tamarix chinensis*) was introduced into the region in the nineteenth century as an ornamental shrub and for erosion control. It has since spread rapidly in sandy riparian environments, with minimal competition from native species, and dramatically altered the channels and floodplains of many streams by increasing bank resistance and trapping sediment (Graf, 2002). In the Colorado River system, the rate of invasion is as much as 20 km a^{-1}, resulting in the restriction of channel width by 13% to 55% largely through the development of enlarged stabilised islands and channel bars (Graf, 1978).

10.3 Fluvial geomorphic systems in deserts

10.3.1 Desert rivers

Rivers in desert regions are usually ephemeral, flowing only after significant rainfall. The exceptions are through-flowing rivers that originate outside the arid zone (for example the Nile, Colorado, Indus; see Chapter 4), and some spring-fed streams. Understanding of the dynamics of desert rivers and their response to external drivers, including climate change, is based on observations of the effects of major flood events, as well as studies of past events preserved in the Quaternary stratigraphic record. It is clear that, compared to rivers in more humid regions, large-magnitude, infrequent flood events play a much more important role in effecting change in desert rivers (Knighton and Nanson, 1997; Tooth, 2000; Nanson *et al.*, 2002).

The nature of the flood events in ephemeral desert rivers is closely linked to the nature of the precipitation events that generate them, and is affected by the sparse vegetation cover and thin soils that characterise many deserts, as well as by transmission losses into the bed and banks. The result is a great diversity of hydrologic conditions in dryland rivers (Nanson *et al.*, 2002), summarised in Table 10.2. Many convectional storms produce channel flow with a sharply peaked hydrograph (a flash flood) that lasts for minutes to a few hours. Because the storms are generally limited in their spatial extent and last for a relatively short period of time most affect runoff in basins with an area of 100 km^2 or less (Graf, 2002). Periods of rainfall that are generated by the passage of frontal systems or tropical storms generate single or multiple-peak floods of much greater magnitude and duration, which may affect large areas and a greater length of the river system and accomplish significant geomorphic work. Good examples are the floods in Tunisia (1969); the Wadi Watir (1972) and the Wadi El Arish (both in Sinai) (1975); the Salt River, Arizona (1980); the Santa Cruz River, Arizona (1978 and 1983) (Webb and Betancourt, 1992); and Coopers Creek and the Finke River, Australia in many years (Nanson *et al.*, 2002). A major characteristic of channel changes associated with large floods is channel widening, often with major impacts on the floodplain. In the intervening periods, the floodplain is gradually rebuilt by smaller-magnitude overbank flows, creation of channel bars and trapping of vegetation by riparian vegetation; and the period of recovery usually exceeds the recurrence intervals of the floods. As a result most desert river channels can be considered to be non-equilibrium forms (Graf, 2002).

Climate change affects desert fluvial systems directly through changes in the amount and intensity of precipitation and indirectly through effects on vegetation cover and therefore surface runoff and sediment yield. Water balance studies (e.g. de Wit and Stankiewicz, 2006) indicate a significant loss of the already minimal surface flow in arid regions of Africa. Given the importance of exorheic rivers to human settlement in arid regions, most research on the effects on future climate change on desert fluvial systems has targeted rivers such as the Colorado in the southwestern USA and the Murray–Darling in Australia. The influence of temperature changes on the water balance of such river systems has been highlighted by studies of the Murray–Darling basin in Australia, where a 1 °C rise in temperature results in a 15% decrease in the annual inflow (Cai and Cowan, 2008); and the Colorado River (McCabe and Wolock, 2007), where a temperature increase of 0.86 °C applied to twentieth-century conditions (and with no increase in precipitation) has the effect of reducing flow by 8%. This is equivalent to reducing the flow from that estimated for the wettest century in the tree ring reconstruction of flow from 1490 to 1988 to that estimated for the driest century in this record. A 2 °C temperature increase (with no precipitation change) applied to the twentieth century would result in a 17% decrease in flow, which is unprecedented in the tree ring record.

A commonly observed geomorphic response of fluvial systems to external changes in water and sediment discharge is aggradation or degradation (incision) of stream channels (Bull, 1991) as a result of changes in the balance between stream power and resistance (Bull, 1997) (Fig. 10.3). Aggradation occurs when the stream power available is insufficient to transport the sediment supplied to the stream or discharge decreases (lower effective stream power); incision occurs when stream power exceeds that needed to transport sediment as in cases where sediment supply decreases (land use changes or trapping of sediment behind dams) or discharge increases (land use or climate change). Alluvial fans are particularly sensitive to changes in sediment yield and water discharge (Wells, 1987; Blair and McPherson, 1994). The studies of Blair and McPherson, Harvey, Wells and many others have shown that even minor climatic changes can affect fan sediments (such as the proportions of debris flow to sheet wash facies) and fan morphology (e.g. fan head trenching as against fan aggradation). For example, alluvial fans in the Anza Borrego region of southern California responded in a complex manner to an increase

TABLE 10.2. *Diversity in hydrologic conditions and channel characteristics in desert rivers*

	Channel characteristics		
	Ephemeral	Intermittent	Perennial
	Internally fed	***Combination of***	***Internally fed***
Input	Convective storms ——→	Convective storms + periodic incursions of moist air ——→	Springs
			Externally fed
			Mountain rain and snowmelt
	Highly localised	————→	Larger supply area
	Very variable		Greater reliability
Throughput	Horton overland flow dominant	————→	Mix of surface and subsurface flow
	Rapid initiation of surface runoff	————→	Longer response time
	Transmission losses largely by seepage	————→	Transmission losses via evaporation (and abstraction)
Output	Flash floods ——→	Single and multi-peaked floods ——→	Seasonal floods
	Sharply peaked hydrographs	————→	Broader-based hydrographs
	High variability	————→	More (seasonally) dependable
Channels	Poorly connected drainage network ——→	Better integration of network (often density high) ——→	Mostly dominated by a single large river
	High magnitude floods as major control	————→	More frequent discharges of great significance in channel adjustment
	Transient behaviour dominant	————→	Greater tendency for channels to equilibrate

Source: Modified from Nanson *et al*. (2002).

in the mean annual rainfall from 13 cm to over 30 cm during the period 1978–83 (Kochel *et al.*, 1997). Whereas most of these alluvial fans that were dominated by debris flow deposits showed little activity during the wet period, those fans that were dominated by fluvial sedimentation showed evidence of recent aggradation and/or incision. Two periods of aggradation were recognised: (a) 1938–47 and (b) within the past 10–12 years. Both aggradational phases correspond to wet intervals in the local rainfall record, as well as to regional episodes of flooding in distal playas along the Mojave River system (Enzel *et al.*, 1989, 1992). On a longer (millennial) timescale, early Holocene periods of fan aggradation in the Mojave Desert have been attributed to increased sediment yield as a result of decreased vegetation cover and hillslope erosion (Harvey and Wells, 2003), but more recent work suggests that such changes may have had a limited effect, given the high infiltration capacity of most soils in the region. An increase in extreme events such as tropical storms would have been necessary to mobilise sediment from mountain areas to distal alluvial fans (McDonald *et al.*, 2003).

Many areas of the southwestern USA are characterised by narrow, steep-walled incised channels, known there as arroyos (Plate 26). Similar features occur in other arid and semi-arid regions (as in southern Africa, southern Australia). Many of the arroyos in the southwestern USA formed rapidly in the late nineteenth to early twentieth century and resulted in the severe erosion of valley floors, leading to the loss of crop and grazing land, changes in sedimentation patterns, destruction of irrigation systems and declining water tables (Cooke and Reeves, 1976; Bull, 1997). The 'arroyo problem' has been studied intensively and many possible causes of channel incision have been advanced. These include:

(a) increases or decreases in rainfall totals and/or the seasonal distribution, duration, and intensity of precipitation;
(b) land use changes;
(c) catastrophic events; and
(d) exceedance of internal geomorphic thresholds.

Although the impacts of Anglo-American settlement on streams in the region have been documented by many workers (see Cooke and Reeves, 1976), studies of Holocene alluvial stratigraphy have shown that as many as five, more or less synchronous and regionally extensive cut-and-fill cycles occurred in the region during the Holocene (e.g. Miller and Kochel, 1999; Waters and Haynes, 2001). Studies of modern alluvial stratigraphy and climatic records (Hereford, 1984; Balling and Wells, 1990; Webb, 2007) indicate that there were marked increases in rainfall totals and intensity during the early twentieth century following a long period of drought. The result was a widespread episode of channel widening and incision. Rainfall and mean annual discharge decreased again in the 1940s and 1950s leading to floodplain stabilisation, re-establishment of riparian vegetation and aggradation. A further period of incision and channel widening occurred from the 1970s to the 1990s. Currently, channels are refilling in dry conditions (Webb, 2007). Relations between regional rainfall patterns and flood frequency and magnitude (Hereford and Webb, 1989; Ely *et al.*, 1993) suggest that these patterns may be related to shifts in atmospheric circulation patterns associated with the ENSO. However, the response of arroyo systems to climatic change may vary temporally as a result of antecedent moisture conditions and spatially with sediment yield, and runoff processes that are influenced by basin size and local topography.

10.3.2 Terminal lake basins

Many desert drainage systems are internal or closed and end in terminal basins, some containing saline lakes (for example the Aral Sea; Pyramid and Walker lakes, Nevada; the Dead Sea; Lake Chad; see Chapter 3); others are seasonally or ephemerally flooded (Lake Eyre; chotts of Tunisia and Algeria). In part because of the high evaporation rates in desert regions, such lakes are very sensitive to changes in inflow as a result of climatic change and variability, or diversion of water for agriculture or urban use, resulting in changes in lake level and salinity. On millennial timescales, increased inflow and reduced evaporation resulted in extensive lakes in the now dry basins of western North America (e.g. lakes Bonneville, Lahontan and Mojave: Thompson *et al.*, 1993) and the Middle East (e.g. Lake Lisan: Bartov *et al.*, 2002) during the latest Pleistocene. In the southern Sahara, large lakes existed in the now dry Bodélé Depression of Chad and the Taoudeni basin of Mali during the early to mid-Holocene African Humid Period (Gasse, 2000; Drake and Bristow, 2006). Many of these basins are now major sources of aeolian dust (Washington *et al.*, 2006). Similarly, the desiccation of lakes in the Mojave Desert resulted in extensive deflation and aeolian activity during the early Holocene (Lancaster and Tchakerian, 2003).

In historic and modern times, diversion of water for agricultural use has resulted in catastrophic decline of the level and area of many terminal lake basins including the Dead Sea, or partial or complete drying as of the Hamoun Lakes of Seistan (Iran and Afghanistan) (Whitney, 2006). Owens Lake, in eastern California, changed from a 12-m

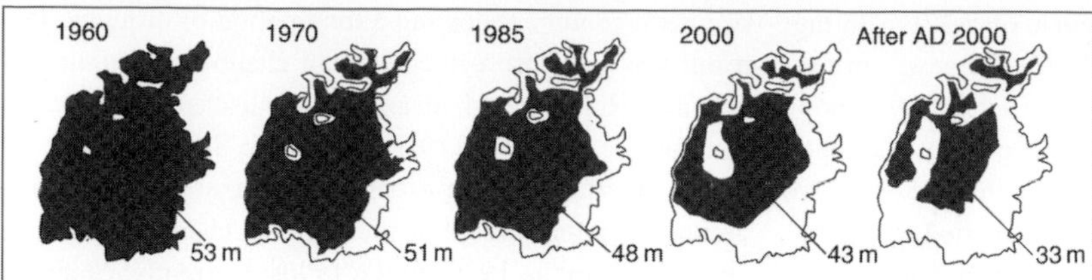

FIGURE 10.7. Changes in the extent of the Aral Sea (from Goudie, 2002).

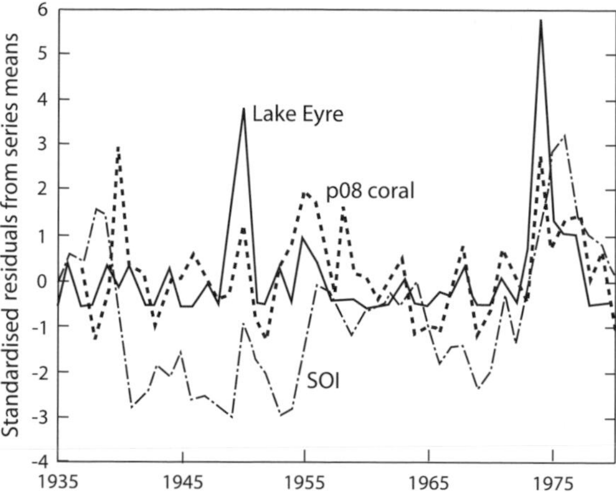

FIGURE 10.8. Relations between inflow to Lake Eyre and ENSO intensity (from Kotwicki and Isdale, 1991).

deep freshwater lake in the early 1900s to a dry saline playa by the 1920s, as a result of diversion of the flow of the Owens River to supply Los Angeles. The lake became the major source of aeolian dust in North America (Gill, 1996), and prompted extensive dust mitigation measures in recent years. The Aral Sea has declined in surface area by about 40% since 1960 as a result of diversion of its two major influent rivers (the Amu Darya and the Syr Darya) for irrigation (Micklin, 1988). The resulting drop in lake level of more than 14 m (Fig. 10.7) has exposed the saline lake bed, which is now a source for large dust storms, which threaten human health over wide areas (Wiggs *et al.*, 2003). Lowering of lake levels and local base level may result in incision of influent streams as in the Dead Sea area (Ben Moshe *et al.*, 2008) and at Walker Lake and Pyramid Lake, Nevada (Adams, 2007).

In a few cases, increased rainfall in the watershed and higher inflow has resulted in high levels of lakes, such as Great Salt Lake in the 1980s. In most cases, episodes in which playas are flooded for long periods of time are related to the occurrence of extreme storms in the headwaters of the influent streams. In the Mojave Desert of California, such storms can be related to anomalous circulation patterns in the North Pacific Ocean, and result in high-magnitude, long-duration flows in the Mojave River that terminate in the Silver Lake basin (Enzel *et al.*, 1989; Enzel and Wells, 1997). Such flood events are documented for 1938, 1969, 1978, 1980, 1983, 1991 and 2005 and resulted in lakes that persisted for months. Extended periods of perennial lakes also occurred in the Holocene around 390 ± 90 and 3620 ± 70 a BP, most probably under similar circulation patterns to those observed in modern episodes of heavy regional winter rainfall (Enzel and Wells, 1997).

Lake Eyre in Australia is also very sensitive to rainfall changes in its contributing river basins in eastern Australia, where enhanced monsoon rains associated with ENSO fluctuations generate runoff that may result in flooding of the playa (Kotwicki and Isdale, 1991) (Fig. 10.8). Notable flood events occurred in 1949–52 and 2000.

10.4 Aeolian systems

Aeolian geomorphic systems play a major role in the landscape dynamics of most desert regions. Aeolian processes, involving erosion, transportation and deposition of sediment are facilitated by sparse or non-existent vegetation cover (either seasonally or all the time), a supply of fine sediment (clay, silt and sand size), and strong winds. Aeolian processes are responsible for the emission and/or mobilisation of dust and formation of areas of sand dunes. Most of their sediment is derived from deposits created by other geomorphic agents (such as rivers) (Bullard and McTainsh, 2003).

The state of an aeolian geomorphic system (Fig.10.9) is controlled by the supply of sediment of a size suitable for transport by the wind (defined as the emplacement of sediment that serves as a source for the aeolian transport system), the mobility of this sediment (controlled by wind conditions) and the availability of sediment for transport (determined by vegetation cover and soil moisture) (Kocurek and Lancaster, 1999). Changes in these external drivers can be the result of climate change and variability or human impacts. Climate change and variability affect the mobility of sediment through variations in wind strength (such as more or less strong storm events); vegetation cover and soil moisture are directly influenced by the amount of

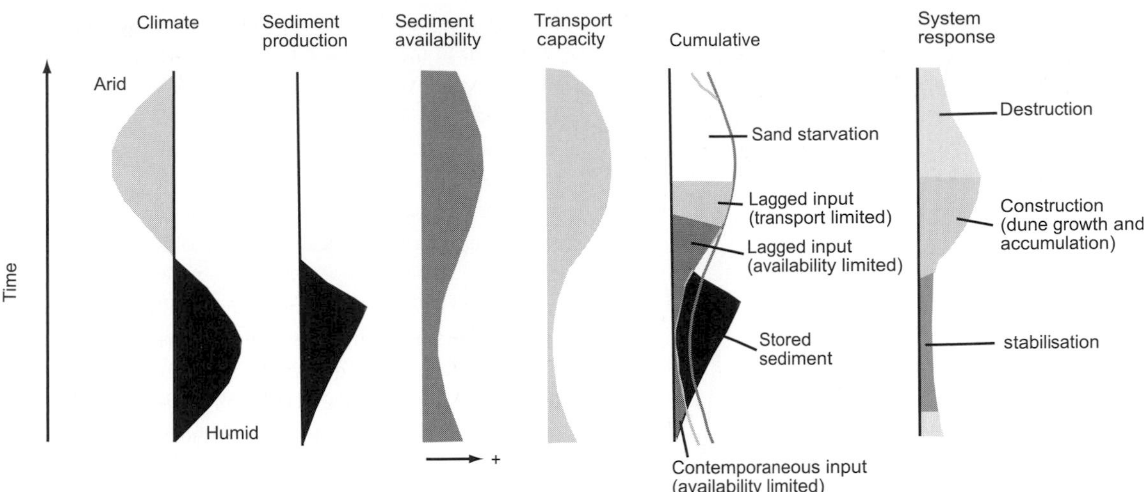

FIGURE 10.9. Changes in state of aeolian sand systems with climate change (modified from Kocurek and Lancaster, 1999).

precipitation; and the supply of sediment may be affected by changes in river discharge. Human activities may change vegetation cover by grazing pressure; surfaces may be disturbed by offroad vehicles or trampling by animals or people; and humans can also directly or indirectly affect sediment.

10.4.1 Dust storms

Dust emitted and transported from deserts is a major linkage between desert and other environments. In addition to important effects on ocean and terrestrial productivity, air quality and atmospheric radiative properties, the most important effect on geomorphic systems is through the addition of dust to desert margin soils (Tsoar and Pye, 1987). Recent interest in air quality and the impact of aerosols on atmospheric radiative properties has promoted intensive study of dust storms and their occurrence in space and time using meteorological records and satellite images (Goudie and Middleton, 2006). Satellite data can be used to identify major dust source areas and to track the dispersion of particulate matter (Plates 27 and 28). Major dust source areas include the Bodélé Depression in Tchad, the desiccated surface of the Aral Sea, southeast Iran and parts of the deserts of China. These areas are characterised by seasonally strong winds and areas of fine-grained sediment (including distal fluvial deposits, playas and lake basins or previously deposited aeolian materials) (Washington *et al.*, 2006). The inter-annual frequency and magnitude of dust storms is strongly linked to climatic variability: in many areas dust emissions are inversely correlated to rainfall, although in a complex non-linear manner (Bach *et al.*, 1996; Prospero and Lamb, 2003; Wang *et al.*, 2004; Reheis, 2006) (Fig. 10.9). Studies in the deserts of the southwestern USA show that although the generation and accumulation of dust is affected by the amount and seasonal distribution of rainfall, different source types (alluvium, dry playas and wet playas) respond in different ways. A major factor determining dust generation is the condition of surface sediments, especially their moisture content, which is often related to the groundwater level (Reynolds *et al.*, 2007). For example, the flux of silt and clay and soluble salt increased following the El Niño events of 1987–88 and 1997–98 at sites close to playas with a shallow depth to groundwater. In this case, evaporative concentration of salts disrupted surface crusts and this increased the susceptibility of surface sediment to wind erosion. The silt and clay flux increased during drought periods at sites downwind of alluvial sources and playas with deeper groundwater as a result of reduced vegetation cover on alluvial sediments, and local runoff events that delivered fresh sediment to playa margins and the distal portions of alluvial fans (Reheis, 2006). Reheis (2006) also noted geographical differences in the response of dust sources to precipitation variability, with a greater range of dust fluxes noted in southern (mostly Mojave Desert) sites.

Human impacts on land cover and the disturbance of desert surfaces through grazing and vehicles may enhance dust production. Many field studies have shown that the threshold wind velocity required for dust entrainment and emission is strongly affected by disturbance (e.g. Gillette *et al.*, 1980; Belnap *et al.*, 2007; Macpherson *et al.*, 2008). The majority of mineral dust is derived from natural surfaces; the contribution of human activities (such as agriculture) to dust loading of the

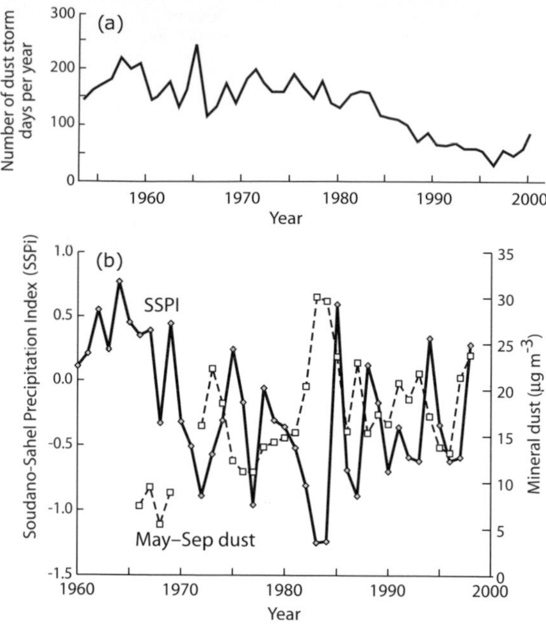

FIGURE 10.10. (a) Time series of dust storm events in China (from Wang *et al.*, 2004); (b) relations between Sahara and Sahel rainfall and dust flux to the Caribbean (from data in Prospero and Lamb, 2003).

atmosphere is uncertain: global estimates vary between 50% (Tegen *et al.*, 1996) to less than 10% (Tegen *et al.*, 2004). In China, Wang *et al.* (2004) attribute dust storms to human activity, especially land degradation; but note an overall decrease in dust storm occurrence since the 1950s and 1960s (Fig. 10.10), possibly due to a decrease in wind energy (Wang *et al.*, 2008). Regionally, variability in meteorological conditions (such as the passage of frontal systems) and climate (such as changes in precipitation) together with varying degrees of human impact appear to have a dominating effect on dust emissions (Wang *et al.*, 2004; Engelstaedter and Washington, 2007). Future changes in dust emissions are therefore likely to be influenced by climate change, most notably as the effect of vegetation cover change, with human activities affecting conditions locally (Mahowald, 2007). The direction of change in dust emissions as predicted by different models is, however, uncertain. In one model, Saharan dust emissions are predicted to increase by 11% as a result of higher wind speed, compared to a decrease of 4% in other simulations, as a result of increased monsoonal rainfall and vegetation cover (Tegen *et al.*, 2004).

10.4.2 Dune systems

Most sand dunes occur in contiguous areas of aeolian deposits called sand seas (with an area of >100 km^2). Smaller areas of dunes are called dunefields. Major sand seas occur in the Old World deserts of the Sahara, Arabia, central Asia, Australia and southern Africa, where they cover between 20% and 45% of the land area. In North and South America there are no large sand seas, and dunes cover less than 1% of the arid zone.

Uncertainty about the relations between dune activity and climate has arisen because of the problems of definition of dunes as 'active', 'inactive', 'mobile' or 'stable' (Lancaster, 1994). 'Active' and 'inactive' dunes represent end members of a continuum of dune dynamics that ranges from mobile (rapidly migrating) active barchan dunes to linear dunes that are stabilised by vegetation or colluvial mantles (Thomas, 1992). In addition, dunes may be seasonally active, or active only after periods of prolonged regional drought that may occur on timescales of decades to centuries (Thomas and Shaw, 1991).

Dunes may be classified as active, dormant or relict forms based upon geomorphic and sedimentary criteria (Table 10.3, Plate 29). Active dunes are those on which contemporary surface sand transport and deposition occurs. Depending on their morphological type, active dunes may be migrating in the net transport direction (crescentic or transverse dunes), extending (linear dunes), or vertically accreting (star dunes) (Thomas and Shaw, 1991; Thomas, 1992). The degree of aeolian activity may vary seasonally, annually or decadally in response to changes in sand supply, wind velocity, vegetation cover and moisture content. Dormant dunes are those on which surface sand transport and deposition are currently absent or at a low level, yet are capable of reverting to an active condition as a result of minor climatic changes (e.g. prolonged regional drought) or disturbance (grazing, fire). Relict dunes are those that are clearly a product of past climatic regimes or depositional environments and have been stabilised for a period of at least 10^3 years. Relict dunes may revert to an active state only as a result of major climatic changes.

Where sediment supply is high and sediment is freely available (a transport-limited system), variations in wind energy and therefore the transport capacity of the wind play a major role in determining episodes of aeolian activity, as in southwestern Africa and Mauritania during the Pleistocene (Lancaster *et al.*, 2002; Chase and Thomas 2007) and during some periods of the development of the Wahiba Sands (Preusser *et al.*, 2002). Variations in wind energy have also been correlated with changes in decadal-scale variations in dune mobility in China (Wang *et al.*, 2008). In some areas, low wind energy limits dune activity. Studies in Australia (Hesse *et al.*, 2004), the Kalahari (Lancaster, 1988) and the Negev (Tsoar and Illenberger, 1998; Yizhaq *et al.*, 2007) indicate that many

TABLE 10.3. *Criteria for identifying active, dormant and relict dunes*

Active dunes	Evidence of contemporary sand transport:
	Wind-rippled surfaces
	Seasonal/inter-annual changes in form
	Dune orientation consistent with modern winds
	Migration/extension of dunes
	Avalanching on lee face
	Vegetation cover less than 14%
Dormant dunes	Sand transport absent or at a very low level:
	Wind ripples rare or absent
	Degraded avalanche faces (lee face < angle of repose)
	No form changes
	Perennial vegetation cover (>14%)
	Biogenic crusts present
	Dune orientation may be consistent with modern winds
Relict dunes	No sand movement:
	No form changes
	Soil development present
	Vegetation cover >50% and includes trees and large shrubs
	Surface partially indurated
	Dune orientation inconsistent with modern winds
	Possible development of lag surfaces and colluvial cover

vegetation-stabilised dunes would be mobile if wind speeds were increased by as much as 30%.

In areas far from external sediment sources, and/or in areas where the sediment supply is low, the effects of changes in sediment supply to the system are relatively small and the major control on dune dynamics is sediment availability. Changes in sediment availability as a result of variations in rainfall and vegetation cover have played a dominant role in episodic development of dunes in the southwestern Kalahari (Telfer and Thomas, 2007) and Australia (Fitzsimmons *et al.*, 2007) during the late Quaternary.

In many areas, Holocene dune activity has been largely determined by changes in sediment availability as a result of climate change and variability or human influence (e.g. Forman *et al.*, 2006: Sun *et al.*, 2006). At Great Sand Dunes, Colorado, periods of aeolian deposition can be correlated with decadal-scale drought episodes recognised in tree ring records (Forman *et al.*, 2006), suggesting an overall control of dune activity by rainfall and vegetation cover in this area. In the Thar Desert, dune mobilty has increased substantially (by a factor of at least 6) in the past century, probably because of increased grazing in the area (Kar *et al.*, 1998). Even in the hyper-arid Namib Desert, construction of large linear dunes near the northern margin of the sand sea was episodic, with three phases of dune building in the Holocene (prior to 5.24 ka BP, 2.41–0.14 ka BP, and today: Bristow *et al.*, 2007). The hiatus in dune construction from 5.24 to 2.41 ka BP is attributed to an increase in vegetation cover which temporarily stabilised the dune during a mid-Holocene period of increased rainfall.

On decadal timescales, Smith (1980) described year-to-year changes in the size of barchan dunes at the Algodones dunefield, California that could be related to variations in the frequency of sand-moving winds as well as to the growth of annual or ephemeral plants in inter-dune areas following rains. A combination of low wind velocities and growth of vegetation in 1978–79 caused dunes to shrink as sand was trapped in inter-dune areas. In the Coachella Valley, dune migration rates vary inversely with regional rainfall (Lancaster, 1997). At Great Sand Dunes, Colorado, migration rates of parabolic and barchan dunes were up to six times higher in drought periods compared to intervening wetter periods. More than 50% of the movement of parabolic dunes occurred during periods of drought that occurred in 1936–41, 1953–66 and 1998–99 (Marîn *et al.*, 2005). They identified a threshold condition with a Palmer drought severity index (PDSI) value <–2, equivalent to a reduction in summer and autumn precipitation of 25%, for increased dune activity.

Studies of dune dynamics in relation to vegetation cover on an annual timescale show that the crests of partially

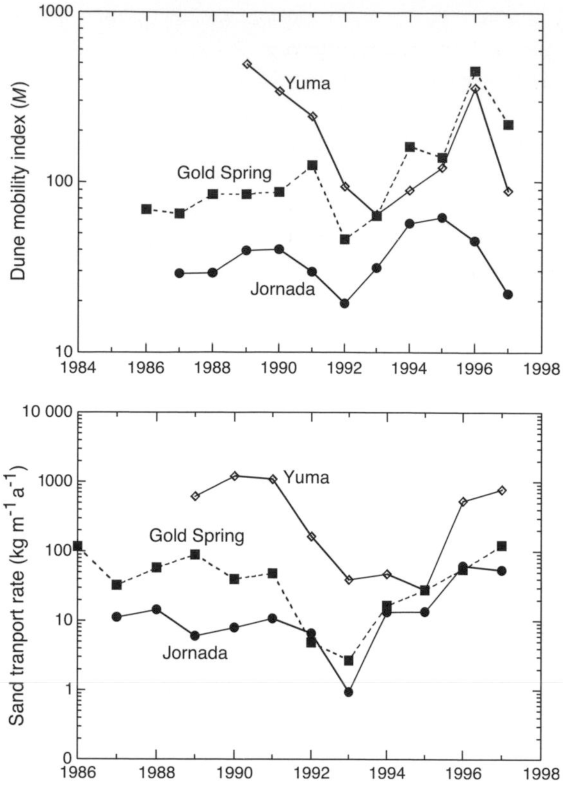

FIGURE 10.11. Empirical test of dune mobility index against measured sand transport rates in southwestern USA.

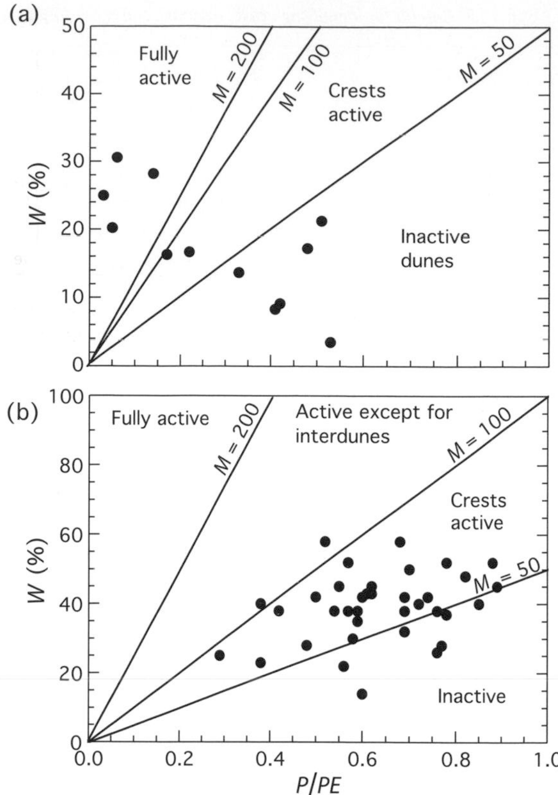

FIGURE 10.12. Values of the dune mobility index for: (a) southern Africa; and (b) Great Plains (data from Muhs and Maat, 1993). Note that dunes can change activity level as a result of changes in effective precipitation and/or wind strength.

vegetated linear dunes can be very active, with relatively large amounts of erosion and deposition, especially in periods of drought or after disturbance, e.g. by fire (Wiggs *et al.*, 1994, 1995; Bullard *et al.*, 1997; Thomas and Leason, 2005; Hesse and Simpson, 2006). As pointed out by Bullard and others, inter-annual and decadal-scale changes in rainfall, temperature and wind strength give rise to significant temporal changes in the amount of vegetation cover and dune surface activity. Empirical studies (Wiggs *et al.*, 1995; Lancaster and Baas, 1998) suggest that a vegetation cover of 14% is sufficient to restrict sand transport on dune surfaces, although studies in Australia suggest that the threshold is much higher at 35% cover (Ash and Wasson, 1983). Thomas and Leason (2005) used Landsat image data to show that the proportion of the area of the southwestern Kalahari dunefield with less than 14% vegetation cover ranged between 10% and 16% for dry years but was only 3–6% for wet years. This suggests that areas of active sand are much more extensive after periods of extended drought.

Given the importance of sediment availability, models for the response of dune systems and sand seas to climate change at various timescales have concentrated on the relations between vegetation cover, wind strength and effective precipitation. These relations are captured in part by the widely used dune mobility index (Lancaster, 1988; Lancaster and Helm, 2000; Hugenholtz and Wolfe, 2005). This index provides a measure of potential sand mobility (*M*) as a function of the ratio between the annual percentage of the time the wind is above sand transport threshold (*W*) and the effective annual rainfall (*P/PE*), where *PE* is the potential evapotranspiration:

$$M = W/(P/PE). \quad (10.1)$$

The index has been tested empirically in a number of dune areas (e.g. Muhs and Maat, 1993; Lancaster and Helm, 2000), and also tested against data on climatic variables and measured sand transport rates (Fig. 10.11). Although not a good predictor of the rate of sand transport from year to year, probably because of the lagged response of vegetation cover to rainfall changes, the index does represent the overall state of the system well, and can therefore be used to assess decadal and longer-term responses to climate change (Fig. 10.12). Modifications to the index that address lags

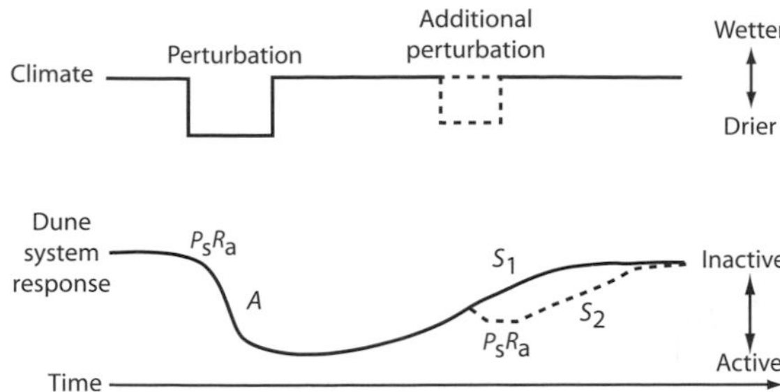

FIGURE 10.13. Conceptual model for the response of dune systems to climate change (modified from Hugenholtz and Wolfe, 2005).

and seasonal effects (Thomas *et al.*, 2005) represent a valuable advance, but are untested outside the region to which they were first applied.

Recently developed models (e.g. Hugenholtz and Wolfe, 2005; Yizhaq *et al.*, 2007) however stress the non-linear response of dunes to climate and vegetation change on annual to decadal timescales (Fig. 10.13). In general, the process of stabilisation of dunes by vegetation growth occurs more slowly than activation of dunes as a result of vegetation reduction. The result is a bi-stable state of dunes – active or stabilised by vegetation – that is dependent on the overall wind regime characteristics. In places with very low wind energy the dune has only one stable state, which is vegetated and fixed, suggesting that human intervention is likely to be the dominant factor in activating such dunes; at intermediate levels of wind energy, both active and stabilised dunes can co-exist; whereas in areas of very high wind energy, active dunes are the stable state (Yizhaq *et al.*, 2007).

10.5 Discussion

The examples of the effects of modern and Holocene climate change and variability, as well as the human impacts on desert geomorphic systems described above, indicate the diversity of responses to external drivers. Such diversity is likely to characterise responses to future climate change and human pressures on desert geomorphic systems. Some possible future trends and scenarios can, however, be identified, based on understanding of documented responses in the historical and recent geological record.

Increased monsoon strength and resulting increases in precipitation in northern and eastern Australia will probably lead to increased magnitude and frequency of flood events in the river systems that flow toward Lake Eyre and Simpson Desert. With increased evaporation as a result of higher temperatures, flood waters will evaporate more rapidly, leaving large areas of fine-grained sediment exposed to deflation, which will probably be more intense as a result of the higher wind speeds that are suggested for this region in IPCC scenarios. Increased aridity of the Simpson Desert is also suggested, leading to more active dunes in a future that may also be windier. The net result will be an increase in dust storm magnitude and frequency in Australia. In the Sahara, increased aridity will affect the northern and western parts of the region. This may reduce the magnitude and frequency of flood events in rivers flowing from the Atlas Mountains. In the southwestern Sahara, drying will probably enhance dune activity on the Sahel margin, but in the southeastern Sahara, increased monsoon rainfall may decrease dune activity. Because of the strong links between summer dust events and convective activity associated with the monsoon (Engelstaedter and Washington, 2007) any increase in monsoon activity will probably increase such dust events in the Sahara. In southern Africa, decreased rainfall and higher temperatures will probably lead to increased dune mobility (Thomas *et al.*, 2005), as discussed in more detail in Chapter 9, but the degree of additional mobility will be limited by wind strength. This will also be the case in other deserts, like the Negev and the southwestern USA, where wind strength is the primary limitation on dune mobility. In the southwestern USA, increased aridity is predicted by regional climate models, leading to significant decreases in streamflow and the flooding of playas, which will probably result in fewer dust events as a result of reduced sediment supply.

These scenarios serve to indicate the possible nature of geomorphic responses to future climate change. They highlight the complexity and non-linear nature of responses, as well as the connections between different geomorphic systems in desert regions. Human impacts are expected to locally or regional enhance the effect of future climate change, by increasing disturbance and pressure on land resources.

Understanding the nature of geomorphic responses to future climate change in desert systems is important to mitigation of and adaptation to climate change by the inhabitants of these regions. Planning for future conditions requires that regional and local effects of climate change can be modelled at a spatial resolution that is appropriate for decision-making. Statistical and dynamical downscaling of global climate model output and development of robust regional climate models is a rapidly emerging field. Linking climate models at different spatial scales to hydrological, landscape and ecological models is a major challenge, but one in which the understanding of geomorphic processes and knowledge of past responses can play a major role, using the approach advocated by Anderson *et al.* (2006) in which dynamic, process-based models are used to provide scenarios for future states of biophysical systems on timescales of years to centuries. These models recognise that the present state of landscapes and ecosystems is a legacy of the history of the system and that modelling the present and future requires knowledge of the past. Data from the past are also essential for validating models and providing understanding of the relative roles of external and internal stressors. This approach provides a conceptual and theoretical basis for integrating models and data across a range of timescales and has been successfully adopted in studies of the effects of climate change on a number of European ecosystems. It has not been applied to arid systems to our knowledge. Application of such predictive models will be an important step towards integrating geomorphic knowledge with decision-making.

References

Abrahams, A. D., Parsons, A. J. and Wainright, J. (1995). Effects of vegetation change on interrill runoff and erosion, Walnut Gulch, Arizona. *Geomorphology*, **13**, 37–48.

Adams, K. D. (2007). Late Holocene lake-level fluctuations and sedimentary environments at Walker Lake, Nevada, USA. *Geological Society of America Bulletin*, **119**, 126–139.

Allison, G. B. *et al.* (1990). Land clearance and river salinisation in the western Murray Basin, Australia. *Journal of Hydrology*, **119**, 1–20.

Anderson, N. J. *et al.* (2006). Linking paleoenvironmental data and models to understand the past and to predict the future. *Trends in Ecology and Evolution*, **21**, 696–704.

Ash, J. E. and Wasson, R. J. (1983). Vegetation and sand mobility in the Australian desert dunefield. *Zeitschrift für Geomorphologie (Supplement)*, **45**, 7–25.

Bach, A. J., Brazel, A. J. and Lancaster, N. (1996). Temporal and spatial aspects of blowing dust in the Mojave and Colorado Deserts of Southern California. *Physical Geography*, **17**, 329–353.

Balling, R. C. J. and Wells, S. G. (1990). Historical rainfall patterns and arroyo activity within the Zuni River drainage basin, New Mexico. *Annals of the Association of American Geographers*, **80**, 603–617.

Bartov, Y. *et al.* (2002). Lake levels and sequence stratigraphy of Lake Lisan, the late Pleistocene precursor of the Dead Sea. *Quaternary Research*, **57**, 9–21.

Bell, J. W. (1981). Subsidence in Las Vegas Valley. *Nevada Bureau of Mines and Geology Bulletin*, **95**.

Belnap, J. *et al.* (2007). Wind erodibility of soils at Fort Irwin, California (Mojave Desert), USA, before and after trampling disturbance: implications for land management. *Earth Surface Processes and Landforms*, **32**, 75–84.

Ben Moshe, L. *et al.* (2008). Incision of alluvial channels in response to a continuous base level fall: field characterisation, modelling, and validation along the Dead Sea. *Geomorphology*, **93**, 524–536.

Blair, T. C. and McPherson, J. G. (1994). Alluvial fan processes and forms. In A. D. Abrahams and A. J. Parsons, eds., *Geomorphology of Desert Environments*. London: Chapman & Hall, pp. 354–402.

Bristow, C. S., Duller, G. A. T. and Lancaster, N. (2007). Age and dynamics of linear dunes in the Namib Desert. *Geology*, **35**, 555–558.

Brooks, N. *et al.* (2005). The climate–environment–society nexus in the Sahara from prehistoric times to the present day. *Journal of North African Studies*, **10**, 253–292.

Bryant, R. G. *et al.* (2007). Dust emission response to climate in southern Africa. *Journal of Geophysical Research, Atmospheres*, **112**, D09207.

Buckingham, S. *et al.* (2004). Accelerated modern erosion in Las Vegas Wash, Clark County, Nevada. *Geological Society of America Abstracts with Programs*, **36**, 514.

Bull, W. B. (1991). *Geomorphic Responses to Climatic Change*. New York: Oxford University Press.

Bull, W. B. (1997). Discontinuous ephemeral streams. *Geomorphology*, **19**, 227–276.

Bullard, J. E. and McTainsh, G. H. (2003). Aeolian–fluvial interactions in dryland environments: scales, concepts and Australia case study. *Progress in Physical Geography*, **27**, 471–501.

Bullard, J. E. *et al.* (1997). Dunefield activity and interactions with climatic variability in the southwest Kalahari Desert. *Earth Surface Processes and Landforms*, **22**, 165–174.

Cai, W. and Cowan, T. (2008). Evidence of impacts from rising temperature on inflows to the Murray–Darling Basin. *Geophysical Research Letters*, **37**, L07701.

Chase, B. M. and Thomas, D. S. G. (2007). Multiphase late Quaternary aeolian sediment accumulation in western South Africa: timing and relationship to paleoclimatic changes inferred from the marine record. *Quaternary International*, **166**, 29–41.

Clarke, M. E. and Rendell, H. M. (1998). Climatic change impacts on sand supply and the formation of desert sand dunes in the south-west USA. *Journal of Arid Environments*, **39**, 517–532.

Cooke, R. U. and Reeves, R. W. (1976). *Arroyos and Environmental Change in the American South West*. Oxford: Clarendon Press.

Cooke, R. U., Goudie, A. S. and Warren, A. (1993). *Desert Geomorphology*. London: UCL Press.

Dai, A., Trenberth, K. E. and Qian, T. (2004). A global dataset of Palmer drought severity index for 1870–2002: relationship with soil moisture and effects of surface warming. *Journal of Hydrometeorology*, **5**, 1117–1130.

deMenocal, P. *et al.* (2000). Abrupt onset and termination of the African Humid Period: rapid climate responses to gradual insolation forcing. *Quaternary Science Reviews*, **19**, 347–361.

de Wit, M. and Stankiewicz, J. (2006). Changes in surface water supply across Africa with predicted climate change. *Science*, **311**, 1917–1921.

Drake, N. and Bristow, C. (2006). Shorelines in the Sahara: geomorphological evidence for an enhanced monsoon from paleolake Megachad. *The Holocene*, **16**, 901–911.

Drake, N. and Mattingly, D. J. (2006). Ancient lakes of the Sahara. *American Scientist*, **94**, 58 65.

Easterling, D. R. *et al.* (2007). Effects of temperature and precipitation trends on U.S. drought. *Geophysical Research Letters*, **34**, L20709, doi:10.1029/2007GL031541.

Ely, L. L. *et al.* (1993). A 5000-yr record of extreme floods and climatic change in the southwestern United States. *Science*, **262**, 410–412.

Engelstaedter, S. and Washington, R. (2007). Atmospheric controls on the annual cycle of North African dust. *Journal of Geophysical Research, Atmospheres*, **112**, 14.

Enzel, Y. and Wells, S. G. (1997). Extracting Holocene paleohydrology and paleoclimatology information from modern extreme flood events: an example from southern California. *Geomorphology*, **19**, 203–226.

Enzel, Y. *et al.* (1989). Atmospheric circulation during Holocene lake stands in the Mojave Desert: evidence of regional climatic change. *Nature*, **341**, 44–48.

Enzel, Y. *et al.* (1992). Short-duration Holocene lakes in the Mojave River drainage basin, Southern California. *Quaternary Research*, **38**, 60–73.

Ezcurra, E., ed. (2006). *Global Deserts Outlook*. Nairobi: United Nations Environment Programme.

Fabre, J. and Petit-Maire, N. (1988). Holocene climatic evolution at 22–23° N from two palaeolakes in the Taoudeni area (northern Mali). *Palaeogeography, Palaeoclimatology, Palaeoecology*, **65**, 133–148.

Fitzsimmons, K. E. *et al.* (2007). The timing of linear dune activity in the Strzelecki and Tirari deserts. *Quaternary Science Reviews*, **26**, 2598–2616.

Forman, S. L. *et al.* (2006). Episodic Late Holocene dune movements on the sand sheet area, Great Sand Dunes National Park and Preserve, San Luis Valley, Colorado, USA. *Quaternary Research*, **66**, 119–132.

Gasse, F. (2000). Hydrological changes in the African Tropics since the Last Glacial Maximum. *Quaternary Science Reviews*, **19**, 189–211.

Gasse, F. and Van Campo, E. (1994). Abrupt post-glacial events in West Africa and North Africa monsoon domains. *Earth and Planetary Science Letters*, **126**, 435–456.

Giannini, A., Saravanan, R. and Chang, P. (2003). Oceanic forcing of Sahel rainfall on interannual to interdecadal time scales. *Science*, **302**, 1027–1030.

Gill, T. E. (1996). Eolian sediments generated by anthropogenic disturbance of playas: human impacts on the geomorphic system and geomorphic effects on the human system. *Geomorphology*, **17**, 207–228.

Gillette, D. A. *et al.* (1980). Threshold velocities for the input of soil particles into the air by desert soils. *Journal of Geophysical Research*, **85**, 5621–5630.

Goudie, A. S. (2002). *Great Warm Deserts of the World: Landscapes and Evolution*. Oxford: Oxford University Press.

Goudie, A. S. (2003). The impacts of global warming on the geomorphology of arid lands. In A. S. Alsharan *et al.*, eds., *Desertification in the Third Millennium*. Lisse: A. A. Balkema, pp. 13–20.

Goudie, A. S. (2008). The Toyota-isation effect. Available at http://home.planet.nl/~jan87536/toyotaisation.htm (accessed July 2008).

Goudie, A. S. and Middleton, N. J. (2006). *Desert Dust in the Global System*. Berlin: Springer-Verlag.

Graf, W. L. (1978). Fluvial adjustments to the spread of tamarisk in the Colorado Plateau region. *Bulletin of the Geological Society of America*, **89**, 1491–1501.

Graf, W. L. (2002). *Fluvial Processes in Dryland Rivers*. Caldwell: Blackburn Press.

Harvey, A. M. and Wells, S. G. (2003). Late Quaternary variations in alluvial fan sedimentologic and geomorphic processes, Soda Lake Basin, eastern Mojave Desert, California. In Y. Enzel, S. G. Wells and N. Lancaster, eds., *Paleoenvironments and Paleohydrology of the Mojave and Southern Great Basin Deserts*. Boulder: Geological Society of America, pp. 207–230.

Hereford, R. (1984). Climate and ephemeral stream processes: twentieth-century geomorphology and alluvial stratigraphy of the Little Colorado River, Arizona. *Bulletin of the Geological Society of America*, **95**, 654–668.

Hereford, R. and Webb, R. H. (1989). Timing and possible causes of late Holocene erosion and aggradation, southwestern Colorado Plateau, USA. *EOS, Transactions of the American Geophysical Union*, **70**, 1124.

Hesse, P. P. and Simpson, R. L. (2006). Variable vegetation cover and episodic sand movement on longitudinal desert dunes. *Geomorphology*, **81**, 276–291.

Hesse, P. P., Magee, J. W. and van der Kaars, S. (2004). Late Quaternary climates of the Australian arid zone: a review. *Quaternary International*, **118–119**, 23–53.

Hugenholtz, C. H. and Wolfe, S. A. (2005). Biogeomorphic model of dunefield activation and stabilization on the northern Great Plains. *Geomorphology*, **70**, 53–70.

Hyers, A. D. and Marcus, M. G. (1981). Land use and desert dust hazards in central Arizona. In T. L. Péwé, eds., *Desert Dust: Origin, Characteristics, and Effect on Man*. Boulder: Geological Society of America, pp. 267–280.

IPCC (2001). *Climate Change 2001: Impacts, Adaptation and Vulnerability. Contribution of Working Group II to the Third Assessment Report of the Intergovernmental Panel on Climate Change*. McCarthy, J. J. *et al.*, eds. Cambridge: Cambridge University Press. (pp. 235–342, Gitay, H. *et al.*, eds.)

IPCC (2007a). *Climate Change 2007: The Physical Science Basis. Contribution of Working Group I to the Fourth Assessment Report of the Intergovernmental Panel on Climate Change*. S. Solomon *et al.*, eds. Cambridge: Cambridge University Press.

IPCC (2007b). *Climate Change 2007: Impacts, Adaptation and Vulnerability. Contribution of Working Group II to the Fourth Assessment Report of the Intergovernmental Panel on Climate Change*. Parry, M. L. *et al.*, eds. Cambridge: Cambridge University Press.

Kar, A. *et al.* (1998). Late Holocene growth and mobilty of a transverse dune in the Thar Desert. *Journal of Arid Environments*, **38**, 175–185.

Khalaf, F. I. and Al-Ajmi D. (1993). Aeolian processes and sand encroachment problems in Kuwait. *Geomorphology*, **6**, 111–134.

Khater, A. R. (2003). Intensive groundwater use in the Middle East and North Africa. In R. Llamas and E. Custodio, eds., *Intensive Use of Groundwater Challenges and Opportunities*. Lisse: A. A. Balkema, pp. 355–386.

Knighton, A. D. and Nanson, G. C. (1997). Distinctiveness, diversity, and uniqueness in arid zone river systems. In D. S. G. Thomas, ed., *Arid Zone Geomorphology: Process, Form, and Change in Drylands*. Chichester: John Wiley, pp. 185–203.

Kochel, R. C., Miller, J. R. and Ritter, D. F. (1997). Geomorphic response to minor cyclic climatic changes, San Diego County, California. *Geomorphology*, **19**, 277–302.

Kocurek, G. and Lancaster, N. (1999). Aeolian sediment states: theory and Mojave Desert Kelso Dunefield example. *Sedimentology*, **46**, 505–516.

Kotwicki, V. and Isdale, P. (1991). Hydrology of Lake Eyre, Australia: El Niño link. *Palaeogeography, Palaeoclimatology, Palaeoecology*, **84**, 87–98.

Kröpelin, S. *et al.* (2008). Climate-driven ecosystem succession in the Sahara: the past 6000 years. *Science*, **320**, 765–768.

Kuper, R. and Kröpelin, S. (2006). Climate-controlled Holocene occupation in the Sahara: motor of Africa's evolution. *Science*, **313**, 803–807.

Kutzbach, J. *et al.* (1996). Vegetation and soil feedbacks on the response of the African monsoon to orbitial forcing in the early to middle Holocene. *Nature*, **384**, 623–626.

Laity, J. (2003). Aeolian destabilisation along the Mojave River, Mojave Desert, California: linkages among fluvial, groundwater, and aeolian systems. *Physical Geography*, **24**, 196–221.

Lancaster, N. (1988). Development of linear dunes in the southwestern Kalahari, southern Africa. *Journal of Arid Environments*, **14**, 233–244.

Lancaster, N. (1994). Controls on aeolian activity: new perspectives from the Kelso Dunes, Mojave Desert, California. *Journal of Arid Environments*, **27**, 113–124.

Lancaster, N. (1996). Geoindicators from desert landforms. In A. Berger and W. J. Iams, eds., *Geoindicators: Assessing Rapid Environmental Change in Earth Systems*. Rotterdam: Balkema, pp. 265–282.

Lancaster, N. (1997). Response of eolian geomorphic systems to minor climatic change: examples from the southern California deserts. *Geomorphology*, **19**, 333–347.

Lancaster, N. and Baas, A. (1998). Influence of vegetation cover on sand transport by wind: field studies at Owens Lake, California. *Earth Surface Processes and Landforms*, **23**, 69–82.

Lancaster, N. and Helm, P. (2000). A test of a climatic index of dune mobility using measurements from the southwestern United States. *Earth Surface Processes and Landforms*, **25**, 197–208.

Lancaster, N. and Tchakerian, V. P. (2003). Late Quaternary eolian dynamics, Mojave Desert, California. In Y. Enzel, S. G. Wells and N. Lancaster, eds., *Paleoenvironments and Paleohydrology of the Mojave and Southern Great Basin Deserts*. Boulder: Geological Society of America, pp. 231–249.

Lancaster, N. *et al.* (2002). Late Pleistocene and Holocene dune activity and wind regimes in the western Sahara of Mauritania. *Geology*, **30**, 991–994.

Lau, K. M. *et al.* (2006). A multimodel study of the twentieth-century simulations of Sahel drought from the 1970s to 1990s. *Journal of Geophysical Reseach*, **111**, D07111.

Liu, Z. *et al.* (2007). Simulating the transient evolution and abrupt change of Northern Africa atmosphere–ocean–terrestrial ecosystem in the Holocene. *Quaternary Science Reviews*, **26**, 1818–1837.

Macpherson, T. *et al.* (2008). Dust emissions from undisturbed and disturbed supply-limited desert surfaces. *Journal of Geophysical Research, Earth Surface*, **113**, F02S04.

Mahowald, N. M. (2007). Anthropocene changes in desert area: sensitivity to climate model predictions. *Geophysical Research Letters*, **34**, L18817.

Mahowald, N. M. *et al.* (2003). Ephemeral lakes and desert dust sources. *Geophysical Research Letters*, **30**, 1074.

Manners, R. B., Magilligan, F. J. and Goldstein, P. S. (2007). Floodplain development, El Niño, and cultural consequences in a hyperarid Andean environment. *Annals of the Association of American Geographers*, **97**, 229–249.

Marîn, L. *et al.* (2005). Twentieth-century dune migration at the Great Sand Dunes National Park and Preserve, Colorado, relation to drought variability. *Geomorphology*, **70**, 163–183.

McCabe, G. J. and Wolock, D. M. (2007). Warming may create substantial water supply shortages in the Colorado River basin. *Geophysical Research Letters*, **34**, L22708.

McDonald, E. V., McFadden, L. D. and Wells, S. G. (2003). Regional response of alluvial fans to the Pleistocene–Holocene climatic transition, Mojave Desert, California. In N. Y. Enzel, *et al.* eds., *Paleoenvironments and Paleohydrology of the Mojave and Southern Great Basin Deserts*. Boulder: Geological Society of America, pp. 189–206.

Micklin, P. P. (1988). Desiccation of the Aral Sea: a water management disaster in the Soviet Union. *Science*, **241**, 1170–1174.

Miller, J. R. and Kochel, R. C. (1999). *Review of Holocene Hillslope, Piedmont, and Arroyo System Evolution in the Southwestern United States: Implications to Climate-Induced Landscape Modifications in Southern California, Southern California Climate Symposium: Trends and Extremes of the Past 2000 Years*. Los Angeles: Natural History Museum of Los Angeles County, pp. 139–192.

Muhs, D. R. and Holliday, V. T. (1995). Active dune sand on the Great Plains in the nineteenth century: evidence from accounts of early explorers. *Quaternary Research*, **43**, 118–124.

Muhs, D. R. and Maat, P. B. (1993). The potential response of eolian sands to Greenhouse Warming and precipitation reduction on the Great Plains of the United States. *Journal of Arid Environments*, **25**, 351–361.

Nanson, G. C., Tooth, S. and Knighton, A. D. (2002). A global perspective on dryland rivers: perceptions, misconceptions, and distinctions. In L. J. Bull and M. J. Kirkby, eds., *Dryland Rivers*. Chichester: John Wiley, pp. 17–54.

Neff, J. C. *et al.* (2005). Multi-decadal impacts of grazing on soil physical and biogeochemical properties in southeast Utah. *Ecological Applications*, **15**, 87–95.

Niang, A. J., Ozer, A. and Ozer, P. (2008). Fifty years of landscape evolution in southwestern Mauritania by means of aerial photographs. *Journal of Arid Environments*, **72**, 97–107.

Nicholson, S. E. (2000). Land surface processes and the Sahel climate. *Reviews of Geophysics*, **39**, 117–140.

Okin, G. S. (2008). A new model of wind erosion in the presence of vegetation. *Journal of Geophysical Research, Earth Surface*, **113**, F02S10.

Okin, G. S. and Gillette, D. A. (2001). Distribution of vegetation in wind-dominated landscapes: implications for wind erosion modeling and landscape processes. *Journal of Geophysical Research*, **108**, 8673–8683.

Okin, G. S. and Reheis, M. C. (2002). An ENSO predictor of dust emission in the southwestern United States. *Geophysical Research Letters*, **29**, doi:10.1029/2001GL014494.

Okin, G. S., Murray, B. and Schlesinger, W. H. (2001). Degradation of sandy arid shrubland environments: observations, process modeling, and management implications. *Journal of Arid Environments*, **47**, 123–144.

Petit-Maire, N. (1990). Natural aridification or man-made desertification? A question for the future. In R. Paepe, ed., *Greenhouse Effect, Sea Level and Drought*. Dordrecht: Kluwer, pp. 281–285.

Preusser, F., Radies, D. and Matter, A. (2002). A 160 000-year record of dune development and atmospheric circulation in Southern Arabia. *Science*, **296**, 2018–2020.

Prospero, J. M. and Lamb, P. J. (2003). African droughts and dust transport to the Caribbean: climate change implications. *Science*, **302**, 1024–1027.

Reheis, M. C. (2006). A 16-year record of eolian dust in Southern Nevada and California, USA: controls on dust generation and accumulation. *Journal of Arid Environments*, **67**, 488–520.

Reynolds, R. L. *et al.* (2007). Dust emission from wet and dry playas in the Mojave Desert, USA. *Earth Surface Processes and Landforms*, **31**, 1811–1827.

Roberts, N. (1998). *The Holocene: An Environmental History.* Oxford: Blackwell.

Sada, D. W. and Vinyard, G. L. (2002). Anthropogenic changes in historical biogeography of Great Basin aquatic biota. In R. Hershler, D. B. Madsen and D. Currey, eds., *Great Basin Aquatic Systems History.* Washington, DC: Smithsonian Institution, pp. 277–293.

Safriel, U. and Adeel, Z. (2005). Dryland systems. In R. Hassan, R. Scholes and N. Ash, eds., *Ecosystems and Human Wellbeing: Current Status and Trends.* Washington, DC: Island Press, pp. 625–662.

Scanlon, B. R. *et al.* (2006). Global synthesis of groundwater recharge in semiarid and arid regions. *Hydrological Processes*, **20**, 3335–3370.

Schlesinger, W. H. *et al.* (1990). Biological feedbacks in global desertification. *Science*, **247**, 1043–1048.

Seager, R. *et al.* (2007). Model predictions of an imminent transition to a more arid climate in southwestern North America. *Science*, **316**, 1181–1184.

Smith, R. S. U. (1980). *Maintenance of Barchan Size in the Southern Algodones Dune Chain, Imperial County, California*, Reports of the Planetary Geology Program, NASA Technical Memorandum No. 81776. Pasadena: NASA.

Smith, S. D. *et al.* (2000). Elevated CO_2 increases productivity and invasive species success in an arid ecosystem. *Nature*, **408**, 79–82.

Sun, J. *et al.* (2006). Holocene environmental changes in the central Inner Mongolia, based on single-aliquot-quartz optical dating and multi-proxy study of dune sands. *Palaeogeography, Palaeoclimatology, Palaeoecology*, **233**, 51–62.

Tegen, I., Lacis, A. A. and Fung, I. (1996). The influence on climate forcing of mineral aerosols from disturbed soils. *Nature*, **380**, 419–422.

Tegen, I. *et al.* (2004). Relative importance of climate and land use in determining presetn and future global soil dust emission. *Geophysical Research Letters*, **31**, L05105.

Telfer, M. W. and Thomas, D. S. G. (2007). Late Quaternary linear dune accumulation and chronostratigraphy of the southwestern Kalahari: implications for aeolian palaeoclimatic reconstructions and predictiosn of future dynamics. *Quaternary Science Reviews*, **26**, 2617–2630.

Thomas, D. S. G. (1992). Desert dune activity: concepts and significance. *Journal of Arid Environments*, **22**, 31–38.

Thomas, D. S. G. and Leason, H. C. (2005). Dunefield activity response to climate variability in the southwest Kalahari. *Geomorphology*, **64**, 117–132.

Thomas, D. S. G. and Middleton, N. J. (1994). *Desertification: Exploding the Myth*. Chichester: John Wiley.

Thomas, D. S. G. and Shaw, P. A., (1991). 'Relict' desert dune systems: interpretations and problems. *Journal of Arid Environments*, **20**, 1–14.

Thomas, D. S. G., Knight, M. and Wiggs, G. F. S. (2005). Remobilisation of southern African desert dune systems by twenty-first century global warming. *Nature*, **435**, 1218–1221.

Thompson, R. S. *et al.* (1993). Climatic changes in the western United States since 18,000 B.P. In H. E. J. Wright *et al.*, eds., *Global Climates since the Last Glacial Maximum*. Minneapolis: University of Minnesota Press, pp. 468–513.

Tooth, S. (2000). Process, form, and change in dryland rivers. *Earth Science Reviews*, **51**, 67–107.

Tsoar, H. and Illenberger, W. (1998). Re-evaluation of sand dunes' mobility indices. *Journal of Arid Lands Studies*, **7S**, 265–268.

Tsoar, H. and Pye, K. (1987). Dust transport and the question of desert loess formation. *Sedimentology*, **34**, 139–154.

UNEP/GRID-Arenal (2007). *Climate Change Scenarios for Desert Areas*. Available at http://maps.grida.no/go/graphic/climate_change_scenarios_for_desert_areas (accessed 10 June 2008).

UN Population Division of the Department of Economic and Social Affairs (2006). *World Population Prospects: The 2006 Revision*. Available at http://esa.un.org/unup/

UN Population Division of the Department of Economic Social Affairs (2007). *World Urbanization Prospects: The 2007 Revision*. Available at http://esa.un.org/unup/

Wang, X. *et al.* (2004). Modern dust storms in China: an overview. *Journal of Arid Environments*, **58**, 559–574.

Wang, X. *et al.* (2008). Desertification in China: an assessment. *Earth Science Reviews*, **88**, 188–206.

Washington, R. *et al.* (2006). Links between topography, wind, deflation, lakes and dust: the case of the Bodélé Depression, Chad. *Geophysical Research Letters*, **33**, L09401.

Waters, M. R. and Haynes, C. V. (2001). Late Quaternary arroyo formation and climate change in the American Southwest. *Geology*, **29**, 399–402.

Webb, R. H. and Betancourt, J. L. (1992). *Climatic Variability and Flood Frequency of the Santa Cruz River, Pima County, Arizona*, US Geological Survey Water Supply Paper No. 2379. Washington, DC: US Government Printing Office.

Webb, R. H. and Wilshire, H. G., eds. (1983). *Environmental Effects of Offroad Vehicles: Impacts and Management in Arid Regions*. New York: Springer-Verlag.

Webb, R. L. (2007). Arroyo formation and filling in the sourhwestern United States: a conceptual model. *Geological Society of America Abstracts with Programs*, **39**, 99.

Wells, S. G. (1987). The complex response of alluvial fans to late Quaternary climatic change, eastern Mojave Desert, California. *Bulletin of the Geological Society of America*, **99**, 821–834.

Whitford, W. G. (2002). *Ecology of Desert Systems*. San Diego: Academic Press.

Whitney, J. W. (2006). *Geology, Water, and Wind in the Lower Helmand Basin, Southern Afghanistan*, US Geological Survey Scientific Investigations Report No. 2006–5182. Washington, DC: US Government Printing Office.

Wiggs, G. F. S. *et al.* (1994). Effect of vegetation removal on airflow patterns and dune dynamics in the southwestern Kalahari Desert. *Land Degradation and Rehabilitation*, **5**, 13–24.

Wiggs, G. F. S. *et al.* (1995). Dune mobility and vegetation cover in the southwest Kalahari Desert. *Earth Surface Processes and Landforms*, **20**, 515–530.

Wiggs, G. F. S. *et al.* (2003). The dynamics and characteristics of aeolian dust in dryland central Asia: possible impacts on human exposure and respiratory health in the Aral Sea basin. *Geographical Journal*, **169**, 142–157.

Yizhaq, H., Askenazy, Y. and Tsoar, H. (2007). Why do active and stabilised dunes coexist under the same climatic conditions. *Physical Review Letters*, **98**, 188001.1–188001.4.

11 Mediterranean landscapes

Maria Sala

11.1 Introduction

Environments characterised by a Mediterranean climate are located on the western rims of continents in the latitudinal bands 30–45° N and S. The basic influences are the tropical (summer) and polar (winter) air masses with their spring and autumn north–south and south–north movement. The most defining climate characteristic is the summer drought; maximum rains occurring during the cold seasons, which can be autumn, winter or spring. Continental winds associated with Mediterranean climates (e.g. *cers*, *bora*, *tramontane*, *mistral*) are dry and often strong. The largest area with these climatic characteristics is on the western side of the Eurasian continent, around the Mediterranean Sea, after which this special type of climate is named. Other areas with this climate are: South Africa, California, Chile and Western Australia (Fig. 11.1). But these different bioclimatic zones have been subjected to completely different anthropogenic processes, resulting in different landscapes. While the Mediterranean basin was the birthplace of western civilisation and has supported human activities over several millennia, the other Mediterranean areas did not become densely populated until the last three or four centuries, and even now they have a lower population density than the Mediterranean basin. Only in their more populated areas do they present similarities with the socioecological landscapes of the 'old' Mediterranean countries. This chapter examines how the main drivers of global environmental change – climate, hydrology, relief, sea level and human activity – have had, and are having, an impact in Mediterranean environments and how these impacts may change in the near future with global climate change.

One of the most marked characteristics of Mediterranean landscapes is the complex mosaic of environments, characterised by important intra-Mediterranean climate variations ranging from the semi-arid to the subhumid, due to their locational proximity to arid and humid lands. This can be seen in Fig. 11.2, where diagrams have been constructed showing (a) on the abscissa the months of the year (from January in the northern hemisphere, July in the southern hemisphere) and (b) on the ordinate: to the right monthly precipitation, P (in millimetres) and to the left average temperature T (in °C) to a scale double that of precipitation. The thermic curve (the line joining the values for mean monthly temperature) and the ombrographic curve (the line joining the values for monthly rainfall) are plotted; when the ombrographic curve sinks below the thermic curve, P is greater than $2T$. The space enclosed by the two curves then indicates the duration and severity of the dry season (Gaussen, 1955; UNESCO–FAO, 1963). Four regimes can be identified: xeromediterranean (Fig. 11.2a) which is warm and dry, with more than 6 months of dryness; thermomediterranean (Fig. 11.2b) with a long dry season and a notable rainy period; mesomediterranean (Fig. 11.2c) with a short dry season and abundant rains; and submediterranean (Fig. 11.2d) which has no dry season but low rainfall in the summer months. There is a strong awareness of the fact that global warming may lead to serious desertification problems in the drier Mediterranean environments and that land use and socioeconomic changes may accelerate this process.

11.2 Geology, topography and soils

Most Mediterranean environments, except for some parts of Western Australia and South Africa, are mountainous and tectonically active. Regularly folded zones are found in the Pre-Alpine and Jurassic zones of the Pyrenees, Alps, southern Apennines, Atlas and Balkans in the Mediterranean Sea area (Fig. 11.3), and also in Chile and California; in the latter, two distinctly mobile and seismically active structural units are separated by the San Andreas Fault system.

Geomorphology and Global Environmental Change, eds. Olav Slaymaker, Thomas Spencer and Christine Embleton-Hamann. Published by Cambridge University Press.

FIGURE 11.1. Global distribution of the Mediterranean biome.

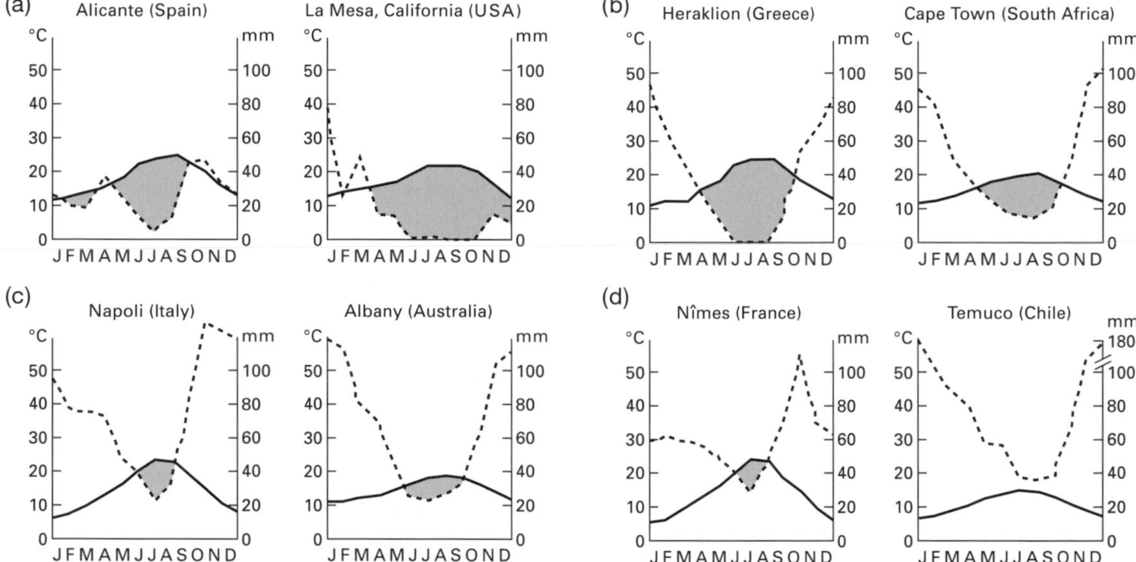

FIGURE 11.2. Examples of the range of Mediterranean climates (see text for explanation of figure construction). (a) Xeromediterranean: Alicante, Spain (left) and La Mesa, California, USA (right); (b) thermomediterranean: Heraklion, Crete, Greece (left) and Cape Town, South Africa (right); (c) mesomediterranean: Napoli, Italy (left) and Albany, Australia (right); (d) submediterranean: Nîmes, France (left) and Temuco, Chile (right).

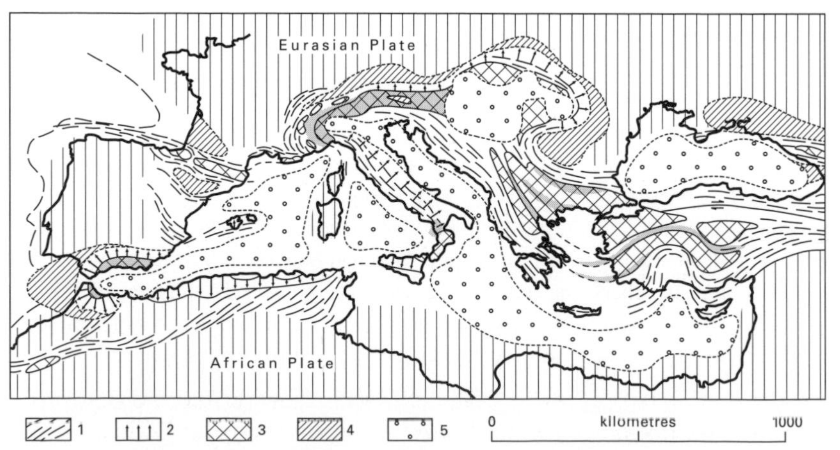

FIGURE 11.3. Geology of the Mediterranean Sea. (1) Orientation of the main fold systems; (2) gravity thrust folds; (3) ancient massifs (Hercynian and Precambrian) within the alpine chain, with the zone of Alpine metamorphism; (4) Molassic basins; (5) post-tectonic continental basins and oceanic floor of the Mediterranean and Black Sea (modified from Mattauer, 1973).

The mountain ranges of the southern Alps, the Apennines and the northern ranges of Sicily are experiencing tectonic uplift at rates of up to 1.4 mm a^{-1} over the last million years, whereas maximum erosion rates, for the same period and at the scale of the mountain range, have been estimated at up to 0.4–0.6 mm a^{-1}. Folded and faulted zones occur in these mountain belts, giving terrain unevenness and steep slope gradients that favour erosion. In addition, there are also wide tablelands or *mesetas*, as in the Iberian Peninsula, North Africa and Mediterranean Australia. There are both extinct and active (e.g. the Aeolian Islands, Mount Etna and Vesuvius regions of southern Italy, Phlegrean area) volcanic landscapes and extensive areas of karst landscapes developed on carbonate geologies.

In Mediterranean environments soils are usually thin, poorly developed and varied in relation to topographic, climatic and lithologic conditions. Topography is responsible for the thinness and poor degree of development of most soils, and lateral leaching processes are sometimes more important than vertical ones. Climate also influences the slow rate of pedological processes because, especially as a consequence of the summer drought, there is not enough water for solution and leaching. Depending on the local conditions and parent materials, soils range from aridisols, sometimes associated with calcareous crusts, to more fertile alfisols, with base saturation and nutrient reserves, located in flat or hilly areas in the main drainage basins. In regions of limestone rocks, the main soil type is erodible *terra rossa*. The considerable redistribution of parent materials by slopewash processes in many parts of the Mediterranean Sea regions during the Holocene has produced landscapes often devoid of soils on the rocky interfluves, interspersed with thick colluvial accumulations in the valleys and hollows. This leads to a complicated soil pattern (Faulkner and Hill, 1977; Ibañez *et al*., 1996).

11.3 Climate, hydrology, vegetation and geomorphological processes

11.3.1 Climate

As well as latitudinal differences (in Chile, for example, there is an annual rainfall increment of 132 mm for every 100 km between 32° and 37° S), regional Mediterranean climates are strongly influenced by relief. In the Iberian Peninsula, relief conditions, namely the coastal ranges around most of the perimeter and the Pyrenees barrier from northwest Europe, determine climatic differences between interior and coastal regions. Rainfall diminishes from north to south, with the highest totals on the windward slopes of the mountain systems. The drier zones, with less than 300 mm a^{-1}, are located in the southeast (Murcia, Almeria) and in the Ebro Basin (Los Monegros). In southeast Spain, Cabo de Gata receives only 130 mm, the Peninsula's lowest measured rainfall. In the Croatian Adriatic Coast, the Dinaric mountain belt forms a climatic partition between the coastal karst region and the inland areas. In Greece, the Pindos mountain chain running in a north-northwest–south-southeast direction separates the country into two parts with different climatic characteristics, especially in rainfall. The continental zone of northern Greece has a climate changing gradually from the pure Mediterranean to the colder climates of central Europe. The coastal regions and the Ionian and the Aegean Islands have a typical Mediterranean climate. The amount of rainfall ranges from 780 to 1280 mm a^{-1} in the western part of Greece, reducing by about half in the eastern part, where the range is from 380 to 640 mm a^{-1}. The dry season is very pronounced in the Eastern Mediterranean and lasts for six months, increasing towards the south and east (Inbar, 1998). In North Africa, the influence of the west–east orientated Atlas Mountains determines the existence of two major zones, with a desert in the south and a föhn effect that can reach the Mediterranean Sea. The climate becomes progressively drier from north to south, with an increase of irregularity and marked spatial variations linked to the orographic effect. In South Africa, as a consequence of the relief, there is enormous variation in annual amounts of rainfall, with some mountain peaks receiving 2500 mm a^{-1}, while areas of the coastal foreland record less than 400 mm. In Chile, the Coastal Range separates the western, humid side, with few thermal fluctuations, from the eastern side, which is drier and has continental thermal influences. Rainfall fluctuates between 200 and 900 mm a^{-1} (Castro and Calderon, 1998).

Another key factor in Mediterranean climates is the torrential nature of rainfall. In the coastal areas of the Iberian Peninsula, daily maximum rainfall can be as high as the annual mean rainfall. Reported extreme values for a 24-hr period are 600 mm in Albuñol (Granada) and Zurgen (Almeria) in October 1973, and 426 mm in Cofrentes (Valencia) in October 1982. But the most extraordinary record is 817 mm in Oliva (Valencia) on 3 October 1987, the highest value for 1-day rainfall of the century. Calabria, southern Italy is characterised by exceptionally intense rainfall events and a single day's rainfall may yield more than 50% of the annual average rainfall total. Hourly rainfall intensities of 138 mm hr^{-1} were recorded in this region in October 1953. In South Australia, rainfall intensities greater than 100 mm hr^{-1} have been recorded in Adelaide, and in California, 228 mm in 4 hours near Truro. In winter, the high pressure cell contracts, mid-latitude cyclonic systems predominate, and storms from the westerly quadrant bring rain and, at higher elevations, snow. Precipitation during storm events is frequently intense and persistent, leading to rapid runoff from mountain slopes and flash flooding in nearby lowlands.

11.3.2 Hydrology and fluvial geomorphology

The rivers which drain into the Mediterranean Sea, except for the Ebro, Rhône and Po, which receive waters from areas outside the Mediterranean, have small drainage areas with short and high-energy watercourses and low water yields. In California (Orme and Orme, 1998), the Coast Ranges are drained by many poorly integrated streams whose swift, erratic courses reflect recent uplift, often flowing along synclinal or faulted troughs. In the north, where the wet season is longer, major rivers are perennial. Further south, as the dry season lengthens and precipitation decreases, summer flows diminish and many rivers become intermittent. Thus drainage networks in Mediterranean regions can be discontinuous, with several months with low or non-existent flows, particularly where dry seasons are long and evaporation rates high. Streams characterised by an extremely high seasonal variability of discharge are often known by regional terms; these include *fiumara* (in Calabria), *jumara* (Sicily) and *rambla* (Spain) (Plates 30 and 31). The nature of Mediterranean rainfall, topographic steepness and thin soils combine to produce irregular discharges, with devastating floods.

In the Iberian Peninsula (Sala and Coelho, 1998), for example, differences between mean, maximum (observed on scale) and peak (measured by a limnigraph) discharges are very marked because floods and droughts are the norm and peaks may last for only a few minutes. The smaller the basin the larger the relative flood peak and the shorter the lag time between rainfall and runoff. The most spectacular floods occur on the south Mediterranean coast, in the streams draining the Betic Cordillera. Data from the October 1973 floods indicate discharge intensities greater than 5 $m^3\ s^{-1}\ km^{-2}$ in *rambla* type streams, with lag times of 2–6 hours, a few minutes peak and a total flood duration of only 3–6 hours (Table 11.1).

Further to the north, the Segura, Jucar and Turia rivers also experience floods, in these cases with big economic losses in relation to the intensive agricultural activity on their alluvial plains. The effects of droughts are also more marked in small basins due to insufficient groundwater reserves. Human impacts in the coastal *ramblas* also have to be taken into account because much urbanisation has taken place, increasing the area of impermeable surfaces and favouring surface runoff.

Floods constitute an important form of land degradation in the south of France and Corsica (Ballais, 1998), a region that experiences very sudden, violent floods. Serious floods

TABLE 11.1. *Hydrological data from selected Mediterranean Spanish coastal rivers, from north to south and Balearic Islands*

					Runoff				
Region	River	Gauging station	Area (km^2)	An (hm^3)	Q_n	Q_c	Q_{ci}	Q_c/Q_n	Q_{ci}/Q_n
Catalonia	Muga	C. Empuries	761	68	2.2	281	450	130	209
Catalonia	Onyar	Girona	295	58	1.8	230	600	125	326
Catalonia	Llobregat	Martorell	4561	680	21.4	2420	2785	114	133
Catalonia	Caldes	La Florida	110	8	0.3	48	131	189	516
Catalonia	Francoli	Montblanc	338	23	0.7	120	192	165	263
Catalonia	Foix	Embalse	279	9	0.3	118	244	413	855
Valencia	Turia	La Presa	6294	500	15.6	2674	3700	171	237
Andalucia	Guadalentín	Puentes	1389	26	0.8	342	2100	417	2625
Andalucia	Algeciras	Librilla	52		0.1	46	310	1150	7150
Andalucia	Almanzora	outlet	1100		0.3	38	3100	127	10333
Andalucia	Nacimiento	outlet	616		0.3	55	220	183	733
Andalucia	Albuñol	outlet	113		0.3		1518		5060
Andalucia	Guadalfeo	outlet	1292		0.3		1142		3805
Balearic	Torrent Gros	Palma	215		0.1	31	103	238	792
Balearic	Major	Soller	50		0.5	16	68	32	136

An, total annual production; *Q*n, average annual runoff ($m^3\ s^{-1}$); *Q*c, maximum annual runoff ($m^3\ s^{-1}$); *Q*ci, maximum instantaneous runoff of the year ($m^3\ s^{-1}$).
Source: Summarised from Sala and Coelho (1998).

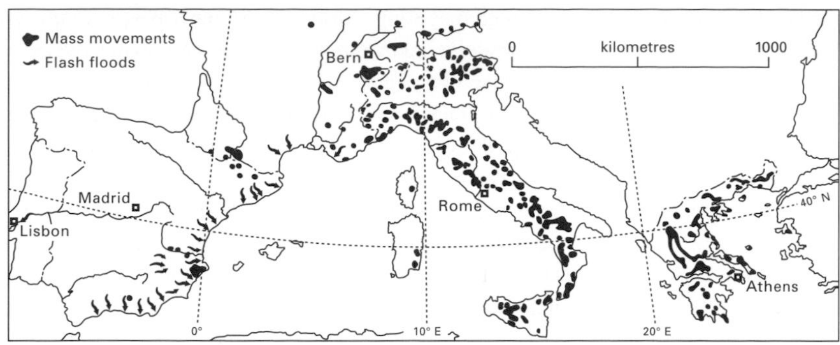

FIGURE 11.4. Geomorphic processes along the north Mediterranean coasts (modified from De Ploey, 1989).

occurred in the south of France in 1940 and 1986, causing the loss of many lives and serious economic damage. In Italy (Sorriso-Valvo, 1998), *fiumara* floods occur nearly every year. Cities located along large rivers, including Rome, Florence, Pisa and Genoa, are prone to flooding. The most vivid floods in living memory were those in 1951 and 1953 in Calabria and in 1966 in Florence. Another consequence of floods is active fans. These cover some 5% of Calabria and it is estimated that 1 to 50 km^2 is affected by debris invasion every year in this region.

In North Africa (Laouina, 1998), rainstorms can result in concentrated, sudden floods (for example, an instantaneous flow of more than 10 000 $m^3\ s^{-1}$ in the Sebou and more than 8000 $m^3\ s^{-1}$ in the Moulouya). The Medjerda can often discharge 20% of its annual flow in 1 day. In small streams, the values are still more impressive. Within the Tellian or Atlas mountains, violent storms and persistent rains may give rise to very violent floods. The Beni Chougran, in Algeria, may receive half its annual rainfall in 3 days, and very rapid floods occur shortly afterwards. Discharge/load ratios vary considerably during floods, according to the bedload materials and the resistance of the banks and slopes that are close to the channel. The river continually shifts its bed while eroding the banks, increasing its load. In wide sections, it spreads its coarse load on the convex bank. In the gorges, it drags its load, behaving like a mudflow, for example, after crossing soft rock outcrops.

Areas of carbonate rocks are characterised by a lack of surface water but often support spring-fed, perennial rivers. Some groundwater reserves in karstic areas are substantial: thus the Fontaine de Vaucluse, the most powerful spring in France, has a discharge of 29 $m^3\ s^{-1}$ and a variable reserve of 80–100 M m^3.

11.3.3 Slope processes and soil erosion

Processes of sheetwash, gullying and mass movements are common geomorphic agents in Mediterranean environments and many studies on these topics can be found dedicated to particular environments, including forests, vegetated slopes and agricultural areas. Most of the research is in local areas of limited extent; these areas do not always provide a reliable representation of regional-scale patterns although they do reflect the variety and fragmentation of Mediterranean terrains and the additional diversity introduced by human impacts. For a general view, De Ploey's (1989) *Soil Erosion Map of Western Europe* reflects the great importance of present-day geomorphic processes along north Mediterranean coasts (Fig. 11.4), showing the great extent of erosion by water all along the northern Mediterranean coast, the general, although less extensive, badland areas and the importance of mass movements in the mountainous regions of Italy and Greece.

For soil erosion, differences in rock type, slope angle, aspect and vegetation type (Sala, 1988; Sala and Calvo, 1990) determine local variations in runoff and erosion processes. Nevertheless at a regional scale some general trends can be described. Soil losses from Spanish drainage basins, calculated by the water authorities in relation to the silting of dams and applying the Universal Soil Loss Equation (USLE), show very high degradation rates, with more than 4000 t $km^{-2}\ a^{-1}$ accumulating behind 19 dams in the coastal rivers (MOPU, 1989). According to the hydrology branch of the Forest Service, a quarter of the territory, mostly located along the coast, suffers from serious erosion, 38% has moderate or low erosion rates and only one-third is not affected by erosion. As for agriculture, the highest erosion rates occur in association with tree crops, followed by areas under annual herbaceous dry farming that are susceptible to autumn torrential rains when the soil surface is bare. In Italy, potential erodibility has been assessed using a modified USLE (Buondonno *et al.*, 1993). The results show erodibilities ranging from less than 26 to more than 312 t ha a^{-1}. In a badland zone of south Calabria, the erosion rate has been assessed at *c.* 40 mm a^{-1} (Sorriso-Valvo *et al.*, 1992). In Croatia, 26% of agricultural and forest Mediterranean lands have been eroded and in the Peljesac Peninsula 5–11 mm of topsoil under Aleppo pine was eroded between September 1979 and January 1980, equivalent to 100 t ha^{-1} (Bilandzija *et al.*, 1998). In Greece, according to the results of the CORINE project (CORINE, 1992), 43% of the land is classified as having a high potential erosion risk, mainly in the south and west, while about 20% of the country, the belt through Macedonia, the Peloponnese and Thessaly, has a low erosion risk (Plate 32).

In North Africa (Laouina, 1998), erosion is most severe in the Riff-Tellian region, for reasons of structural geology, lithology and climate. Rates of soil loss are variable, ranging from insignificant in many cases to a maximum of 54 t ha^{-1} a^{-1}, comparable to sediment loss from gullies and badlands. In the southeast Cape region (Meadows, 1998) the problem of soil erosion by water appears to be less marked than in other Mediterranean regions, although the database is small. Erosion is significant in California (Orme and Orme, 1998) where the clearance of native plant cover has reduced interception, transpiration and infiltration while increasing rainsplash and overland flow. These effects favour sheet and rill erosion and thus the transfer of soil from hillslopes to valley floors and from there downstream to the coast. Erosion is also a serious environmental problem in Chile. In Mediterranean Australia (Conacher and Conacher, 1998b), it has been estimated that 4% of the agricultural land in the Mediterranean zone requires treatment to control water erosion (Matheson, 1986). There is considerable variability in erosion between different time periods, largely coinciding with droughts but also reflecting periods of agricultural intensification.

Mass movements are an important process in mountain environments, as, for instance, along the Apennines in Italy. In North Africa considerable areas are affected by mass movements in the Riff chain, and in California, tree removal and the consequent loss of root strength has commonly led to increased mass movement (see also Chapter 12), further exacerbated by the introduction of pastureland, domestic crops and orchards. Intense gullying occurs in clay and marl terrains subject to the typical Mediterranean rainstorms, producing badland areas where erosion rates are very high during storms.

11.3.4 Wind erosion

Wind erosion represents a serious hazard in Greece, especially in the Aegean Islands (Kosmas, 1998). Strong north winds in combination with weak vegetative protection create favourable conditions for wind erosion, particularly during the summer and/or autumn periods when soils are dry. Erosion of the topsoil in unprotected areas may account for several hundreds of tonnes per hectare each year. In the Cape region of South Africa (Meadows, 1998), widespread drift sands in the western lowlands of the southwestern Cape have been interpreted as indicative of a major wind erosion problem, assumed to be driven by a combination of severe fires, overgrazing and an energetic wind regime. In Southern Australia (Conacher and Conacher, 1998b), wind erosion is particularly variable from year to year and season to season but it is estimated that erosion of topsoil from unprotected, cultivated land may amount to several hundreds of tonnes per hectare each year. In one storm event in 1988, 234 000 ha of soils on the Eyre Peninsula were affected by wind erosion (Community Education and Policy Development Group, 1993).

11.3.5 Land subsidence

In Italy, another distinctive form of land degradation is subsidence, generally due to the extraction of water from unconsolidated sediments underlying alluvial plains and gas withdrawal from deep sediments. In the Romagna coastal plain, for example, subsidence of 1.3 m in 40 years has affected the industrial zone, city centre and countryside of Ravenna (Guerricchio *et al.*, 1976). Subsidence induced by human activity also affects 16 000 km^2 of California, mostly in the Central Valley where maximum subsidence approaches 10 m, and is most extensive in the western and southern San Joaquin Valley and in the Santa Clara Valley south of San Francisco, where it largely results from excessive pumping of groundwater. Other subsidence problems are linked to the consolidation of moisture-deficient deposits, the oxidation of organic soils and the withdrawal of fluids from oil and natural gas fields.

11.3.6 Vegetation characteristics and their geomorphic implications

The geological and hydro-climatic factors described above result in a fragile landscape, with little capability for regeneration due to low productivity, vulnerability to urban and infrastructural interventions and with a low capability for sealing subsequent wounds related to the movement of earth and soil. In agricultural lands erosion is a serious problem, especially during the period of late summer and autumn rainstorms when there is little vegetation cover.

The governing climatic controls on present-day vegetation are mean annual temperature and its range, precipitation total and its distribution, and the ratio between precipitation and potential evapotranspiration (Faulkner and Hill, 1977). The key factor in vegetation development is water availability during the summer drought. In landscape terms, this climatic regime translates into:

- moderate total biomass per unit area with low, open forests with a dominance of chaparral, maquis and matorral formations;
- the perennial and evergreen character of vegetation;
- adaptations to drought by means of short-leaved, sclerophyllous species in order to avoid evapotranspiration;
- severe vegetative consequences of the summer stress, with frequent forest fires; and
- serious erosion, especially during late summer and autumn rainstorms, accentuated by the low vegetation cover.

In relation to the intra-Mediterranean climate variation, several vegetation landscapes can be found, ranging from dense forest, chaparral, maquis, garrigue, matorral and open shrubs. Differences are not only related to climate but also to elevation and exposure, with drier south-facing slopes in contrast to humid north-facing ones (and vice versa in the southern hemisphere), and dry headlands supporting grassy steppe. It is interesting to note the contrasting role of granite and limestone landscapes in relation to their water retention. With similar amounts of rainfall, granite areas are much more vegetated due to the presence of a weathering mantle while in limestone areas karst processes direct water deep under the soil and the vegetation has more of a dry character.

11.3.7 Forest fires and slope hydrology

Fire is another integral element in the natural evolution of Mediterranean ecosystems, producing soils and vegetation adapted to it (Naveh, 1975; Le Houerou, 1987; Inbar *et al.*, 1998). Humans have traditionally used the positive effects of fire for renewal of pastures, pest and species control, and to encourage fertilisation. From a climatic point of view, the key factors are the high summer temperatures and frequent lightning storms. The flammability of the vegetation adds to this natural risk. In relation to erosion potential, rainfall intensities are at a maximum after the dry summer, the period immediately after burnings. Finally, poor soil development and mountainous terrain both increase erosion risk. The destruction of vegetation leads to a decrease in biomass and biodiversity. The absence of obstacles to slope runoff and of protection from raindrop impact, together with the development of hydrophobicity in soils, lead to an increase in surface runoff and erosion (Plate 33). Although absolute values of these phenomena are often difficult to obtain at a slope scale, comparative relative data can give an idea of the order of magnitude of the problem.

Studies undertaken in the Catalan Coastal Ranges, a subhumid environment, provide a good example of the intensified runoff and erosion that result from frequent forest fires. In the cases studied (Soler and Sala, 1992), runoff in the burnt areas was, on average, 14 to 8 times higher than in the forested areas, and 5.8 times higher than in clear-felled areas. Average infiltration rates of 50 mm hr^{-1} in the forested areas and 25 mm hr^{-1} in the burnt plot were measured. As for sediment, the burnt areas produced 16 times more sediment than the forested areas and the clear-felled areas 8.4 times more. Remarkably, high erosion rates were registered in granite areas in comparison to schist areas due to the fact that the granite has been weathered to a sand and silt regolith. Soil texture, porosity, aggregate stability and infiltration capacity are also important factors in the degree of runoff and erosion. A clear relationship exists between fire intensity and the degree of alteration of soils and their erosion (Ubeda and Sala, 1998) (Table 11.2), except in the case of very low intensity fires where burnt leaves act as a mulch, protecting the soil from raindrop impact, runoff and erosion. It is very important to note that the degree of runoff and erosion after fire diminishes drastically in 3 to 6 months (Fig. 11.5). This is because plant regeneration starts soon, usually within the first month after the burning, and vegetation cover reaches about 35% in 6 months (Ubeda *et al.*, 2006). Finally, suspended sediment load measured before and after a fire at the catchment scale (Rovira *et al.*, 1998) increased by an average of two orders of magnitude during a flood immediately after the fire. Analysis of the temporal variations in runoff and erosion associated with this event shows that the runoff coefficient was at its highest during the first 3 months after the fire and that after this peak the values remained higher than in the undisturbed areas, a difference that was maintained for 6 to 9 months. It was not until after 18 months that the rates were reduced to the level of the undisturbed area. Soil erosion also diminished very strongly with time, following the same general pattern as runoff. In addition, spatial variations were also important, due to the fact that water and sediment descended the slope in pulses. However, while runoff infiltrated the soil and thus joined the water cycle after each event, sediment was stored a few metres downslope of the source area and remained there, being available for further downslope transport during the next rainfall/runoff event. Plot differences are thus often more important than longitudinal variations (Sala, 1988).

11.4 Long-term environmental change in Mediterranean landscapes

For the pre-instrumented period, climate changes can be inferred from a range of data sources, including slope and fluvial geomorphology, pollen analysis and historical documents. This information on past climates is very important, in order to be aware (1) of the climatic changes that occurred in both the Quaternary period and in historical times and (2) that this phenomenon is not exclusive to the present day.

TABLE 11.2. *Relations between intensity of burning, runoff and erosion*

	Forest (data from 16/6/93 to 9/12/94)			Burnt forest (data from 7/7/94 to 22/3/95)					
	Closed vegetation	Open vegetation	Date	Low intensity	Date	Medium intensity	Date	High intensity	Date
Total rainfall (mm)	588	588		497		520		512	
Total runoff ($l\ m^{-2}$)	13.10	14.19		1.37		37.21		24.75	
Total erosion ($g\ m^2$)	0.012	0.157		0.014		0.635		3.056	
Max. rainfall intensity ($mm\ h^{-1}$)	30	30	20/10/94	30	20/10/94	30	20/10/94	30	20/10/94
Max. runoff ($l\ m^{-2}$)	3.10	3.20	20/10/94	0.71	22/7/94	6.00	20/10/94	8.00	20/10/94
Max. runoff coefficient (%)	6.35	10.00	7/3/94	8.93	22/7/94	37.43	1/9/94	33.65	14/9/94
Max. erosion ($g\ m^{-2}$)	3.00	131.25	20/10/94	6.33	22/7/94	355.00	20/10/94	3250.00	20/10/94
Max. erosion/runoff ($g\ l^{-1}$)	10.00	41.02	20/10/94	93.33	25/8/94	136.25	22/7/94	464.29	20/10/94
Max. erosion/rainfall ($g\ l^{-1}$)	0.19	0.68	20/10/94	0.79	22/7/94	22.71	22/7/94	16.75	20/10/94
Slope (%)	9	9		9		11		11	
Area (ha)				1.613		5.666		5.380	
Total erosion ($T\ ha^{-1}\ yr^{-1}$)	0.08	1.00		0.20		8.46		40.74	

Source: From Ubeda and Sala (1998).

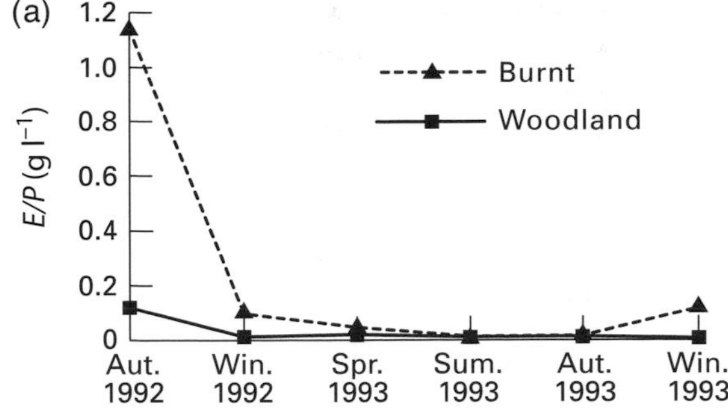

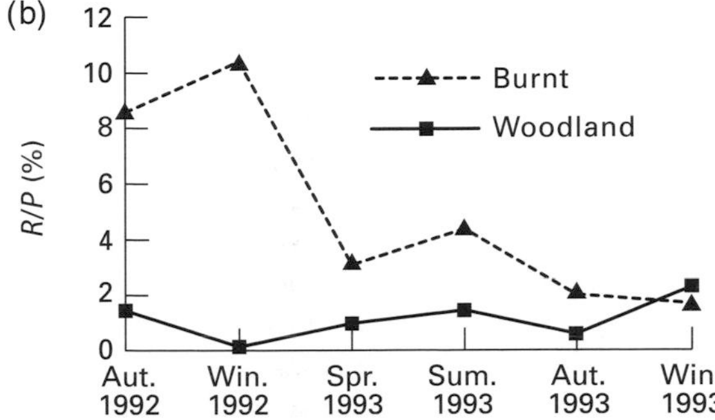

FIGURE 11.5. Erosion rate changes over time in burnt and woodland area, Collserola, Barcelona, Spain. (a) Changing rates of erosion in a burnt area and under woodland in relation to precipitation (*E*/*P*); (b) changing rates of runoff in a burnt area and under woodland in relation to precipitation (*R*/*P*) (modified from Ubeda and Sala, 1998).

11.4.1 Slope processes and landforms

From a geomorphologic point of view, several landforms and associated sediments prove the existence of changes in climate over Quaternary timescales in the western Mediterranean Sea basin.

For slopes, periglacial landforms and sediments have been identified in many mountain environments in France, Spain, North Africa and Italy. Widespread mantles of rock debris and their mobilisation downslope, by both slow and rapid mass movements, have been interpreted as remnants of the gelifraction processes related to colder climates during the Quaternary periods, alternating with phases in which the geomorphologic processes were more similar to those of the present day. At these latitudes, the processes showed the characteristics of an 'attenuated periglacial' (Tricart, 1966; Raynal, 1973). These processes have been studied in the Catalan Coastal Ranges by Llobet (1975), especially in the Montseny Massif, and their lower limit established at 600–500 m. Grèze type deposits have been interpreted as characteristic of this attenuated periglacial period (Raynal, 1973; Panizza, 1978), due to their origin in the micro fragmentation of rocks and the transport of weathered materials by snowmelt-generated slopewash.

A common altitudinal sequence of periglacial forms and deposits is, from upslope to the bottom of the slope, (Raynal, 1977; Sala, 1978) as follows:

- angular rock outcrops surrounded by rock debris due to gelifraction;
- rectilinear slopes due to laminar solifluction;
- niches produced by bowl-slides;
- convexities for accumulation of materials;
- slopes blanketed by general solifluction deposits;
- solifluction lobes; and
- discontinuous grèze litée deposits.

In terms of age, in several places two periods of formation can be observed from the superimposition of two sequences, the lower one with an abundant reddish matrix and the upper one with a scarce yellowish matrix, which may correspond to the penultimate and last glacial periods. In addition to the widespread periglacial features, remnants of glacial forms (cirques and moraines) are present in the Southern Apennines in the Mt. Pollino range of North Calabria.

11.4.2 Fluvial processes and landforms

For fluvial landforms, several terrace levels have been described in Mediterranean Sea river systems which can be related to Quaternary climatic changes. The levels indicate accumulation phases alternating with incision periods. In general, four to six terrace levels can be found in most river systems, with ages ranging from Moulouyen (Upper Villefranchian) to Soltanien (Wurm) (Raynal and Tricart, 1963; Sala, 1978). Accumulations just above present watercourses are composed of much finer materials, arranged in unconsolidated, sorted layers. Raynal and Tricart (1963) attributed these deposits, with their predominance of silts and clays, to a humid recurrence after the Flandrian transgression. They called this phase the 'Rharbien'. Vita-Finzi (1969) also defined this alluvium as finer and better sorted than the deposits associated with older levels and the present channel bed. The presence of well-bedded sediments of a dark grey colouring indicates fluvial channels with less erratic behaviour than either the preceding period or the present day and more humid climatic conditions.

11.4.3 Vegetation change

The character of vegetation is a reflection of climatic conditions. In Mediterranean environments, climate fluctuations during the Pleistocene have determined the existence of species that are relict from past climatic conditions when temperature was lower and rainfall more abundant and more evenly distributed throughout the year. In mountain environments, pollen analysis of bog areas also indicates climatic fluctuations. The PAGES biome maps with distributions around the LGM (Plate 1) and the HOP (Plate 2) show the extreme range of distributions over the past 20 000 years. In addition, human exploitation, mainly of trees, has determined the state of present-day vegetation (Plate 3). Human occupation and use of the landscape has favoured certain kinds of species but has especially determined the extinction of a great number.

11.4.4 Sea level changes

Sea level within the Mediterranean Sea region is predominantly affected by complex local or regional tectonic factors and regional eustatic changes (Mörner, 2005). Eustatic sea level in the eastern Mediterranean is, and has been, affected by changes in precipitation and evaporation. Lambeck and Purcell (2005) explain that sea level change in the Mediterranean during glacial cycles was determined by temporally variable eustatic change and by spatially variable glacio-hydro-isostatic responses of the Earth and ocean to the growth and decay of ice sheets. Superimposed upon these dynamics have been relative changes in sea level from vertical tectonic movements of the land. Lambeck and Purcell (2005) demonstrate that observations from certain regions in the Mediterranean are particularly important in allowing for the separation of these controls. This is supported by a trial analysis of a small observation data set from sites that exhibit some of the desirable features of an ideal data set. Basin-wide predictions of sea level change, palaeo water depth and shoreline locations based on these analyses are presented for selected epochs.

Sneh and Klein (1984) deduced changes in Holocene sea level in the eastern Mediterranean from geological, geomorphological and archaeological data from Dor, Israel. They provided a regional eustatic sea level curve, showing that sea level was *c.* 2 m below its present level at 4 ka BP, rose to 1 m below the present level at 3 ka BP and was 1 m higher than the present level at 1.5 ka BP. It then dropped to 1 m below the present level at *c.* 0.8 ka BP.

11.4.5. Historical climatological and hydrological change

Climate changes can also be inferred from historical documents, a source of information that, as pointed out by Barriendos and Llasat (2003), is assuming importance in the context of the uncertainties that surround the likely natural and man-induced climatic changes that may emerge in the course of the twenty-first century. The IPCC Fourth Assessment (IPCC, 2007) emphasises the lack of reliable local or regional detail in climate projections for precipitation change and for changes in extreme events.

The study of climatic anomalies on the basis of various types of instrumental information and proxy data allows unusual events to be identified. Barriendos and Llasat (2003) explain a hydrometeorological anomaly that occurred between 1760 and 1800 (the Maldá Anomaly) which was characterised by a sequence of both anomalous droughts and floods, and compare it with the features of the second part of the twentieth century. Climatic indices obtained mainly from documentary sources, in addition to instrumental observations made in earlier times and published materials, have confirmed the presence of considerable variation in atmospheric circulation systems, especially between 1780 and 1795. In the western Mediterranean, these variations gave rise to a simultaneous increase in the frequency of droughts and heavy rainfall.

The occurrence of floods is often connected with damage to property and sometimes also with the loss of human lives (Barriendos and Martin Vide, 1998). For these reasons, information about floods is usually included in documentary evidence, including chronicles, diaries and records of economic activity. Despite some limitations, series of reconstructed historical floods from these forms of evidence can be used for the study of fluctuations in flood frequency in the pre-instrumented period. Using this methodology, Brazdil *et al.* (1999) studied flood events of selected rivers in northern Europe and in the Mediterranean (central Italy, south of France, Catalonia and Andalusia). They found a prevalence of floods during the second half of the sixteenth century in central European and Andalusian rivers, while Italian and Catalonian rivers showed a higher occurrence during the first half of the century. However, it should be noted that the evidence for changes in the flooding seasons in both halves of the century was not unambiguous. An analysis on a broader European scale showed floods to be a random natural phenomenon, with a limited areal extent defined by the spatial influence of forcing meteorological factors, such as continuous heavy rains or the sudden melting of a thick snow cover.

In the particular case of floods in the western Mediterranean, the atmospheric processes which are apparently present with unusual frequency during the identifiable climatic oscillations may be the result of the presence in those latitudes of the jet stream, together with its associated band of depressions, resulting from the latitudinal expansion of the circumpolar vortex. The passage of these depressions is reinforced as they reach the Mediterranean by the sharp contrast between cold air masses from the north and warmer and more humid masses already present over the Mediterranean area, with consequent atmospheric instability. But the direct cause of such a noticeable increase in the catastrophic flood frequency over limited periods of time must involve complex processes, as yet unclear.

Ecclesiastical sources offer a variety of interesting possibilities for environmental reconstruction, with documents generated and preserved over the centuries. These include administrative papers, land use records, fiscal documents and books of minutes from parishes, monasteries, cathedrals and dioceses. In particular, the documentary archives of the Spanish and Italian cathedrals contain direct information on the climatic history of the Mediterranean region from which many types of information can be drawn about climatological, hydrological and biological phenomena. One of these sources is the *pro pluvia* rogations ceremonies that each town would organise in an attempt to diminish the effects of drought by means of a *pluvia congruente* (good rain). These ceremonies were common in Mediterranean countries and in Latin America and their analysis makes it possible to reconstruct reliable time series with daily, monthly, seasonal or annual resolution over periods of several centuries, depending upon the intensity with which religious ceremonies were enacted (Barriendos, 1997). The dynamic that is revealed by this methodology is one of relatively brief pulsations (20–40 years) of climate change. For example, the first pulsation in Catalonia coincides with one of the clearest manifestations of the Little Ice Age in Europe (1580–1620) and can be clearly correlated with a notable decrease in the number of *pro pluvia* rogations. The same phenomenon can be seen at the end of the eighteenth century (1760–1800) and at the end of the Little Ice Age (1830–1860). Plurisecular (i.e. repeated during centuries) climatic episodes are not continuous and homogeneous in nature, but brief and spasmodic (Pfister *et al.*, 1999).

11.5 Traditional human impacts in Mediterranean landscapes and nineteenth- and twentieth-century change

The most important impacts of traditional land uses in the Mediterranean have been related to primary activities: forestry, agriculture and mining. In the following discussion a great deal of information on different Mediterranean countries is taken from Conacher and Sala (1998) and reviews the research of the following authors: Sala and Coelho (Iberian Peninsula); Ballais (southern France); Sorriso-Valvo (Italy); Bilandzija *et al.* (Croatia); Kosmas (Greece); Inbar (eastern Mediterranean); Laouina (North Africa); Meadows (South Africa); A. and A. Orme (California); Castro and Calderon (Chile); and A. and J. Conacher (southern Australia).

11.5.1 Forests

Man's impact on Mediterranean forests has been intense, both for the use of wood and with clearcuttings for construction and to enlarge the agricultural areas. *The Mountains of the Mediterranean World* (McNeill, 1992) documents great changes in the nineteenth and twentieth centuries, creating the often barren and depopulated landscapes of the present day. Nevertheless, at present, in the more developed Mediterranean countries, abandonment of forestry due to its low profitability and the consequent depopulation of mountains are producing an important regeneration of forest and an increase in their extent.

In the Iberian Peninsula, some researchers (Ceballos, 1966) state that in pre-Neolithic times 96% of the land area was covered by forest (although there is not total agreement on this statistic because climate may never have allowed such forest coverage to take place: Campos Palacín, 1992). Today forests have been reduced to 13% of the land area. And the same can be said of all the countries around the Mediterranean Sea. The most significant loss of forests took place during the Roman period, with the resulting erosion and sedimentation leading to the seaward extension of many floodplains. Deforestation was also used during times of war, especially in Spain during the reconquest from the Arab invasion in order to create strategically important open ground between the warring factions.

In North Africa, the Tell is described by Roman authors as being abundantly wooded. Arab invasions and the consequent withdrawal of local populations from the plains to the mountains have often been considered as responsible for the extensive recession of the forest cover. But some travellers in the late nineteenth century described green, wooded and sparsely inhabited plains which served mainly as rangelands, with the mountains being intensely occupied and developed, and the first European settlers to arrive in Algeria described the presence of extensive forest tracts. Excessive exploitation of wood during the two World Wars was, however, an important phase of forest decline. Some damage dates back to 1956, just after Independence, when the rural populations of the Riff appropriated lands before the government's delimitation of state forests was applied.

In the southwestern Cape region of South Africa, there is evidence that fires impacted on *fynbos* communities and the activities of later Stone Age hunter–gatherers are widely accepted as having an impact on surrounding vegetation communities. Fire has also long been both a form of, and a primary cause of, land degradation in California. Charcoal found among fluvial, aeolian and marine deposits of Pleistocene age clearly demonstrates the recurrence of fire as a factor in the pre-human Californian landscape. Field evidence indicates that Native American peoples used fire to drive game and encourage new plant growth, although fire may also have escaped accidentally from cooking hearths. Thus along the coast, where indigenous peoples were most concentrated, it cannot be assumed that early European colonists arrived to find uninterrupted expanses of native chaparral. After World War II, timber harvesting spread beyond readily accessible areas to become more general throughout the forested regions. Fire was also used during the Spanish and Mexican periods. Thus the modern Baja California landscape probably resembles more closely that of the late-eighteenth- and nineteenth-century landscape of southern California. In Australia, aboriginal peoples used fire extensively and it seems very probable that the vegetation which was seen by the first explorers was not the same as that which would have been present in the absence of Aborigines.

In Chile, at the beginning of the Spanish colonisation, chronicles of the time described the Chilean landscape as provided with exuberant vegetation. Afterwards the destruction of natural vegetation began, in order to obtain wood for construction, land for agriculture, fuel for homes and in the development of mining operations.

However, many Mediterranean forests today fall under the threshold for economically viable production. Such assessments are the result of the lack of a forestry policy for the European Mediterranean countries and to the inadequacy of the economic model applied to these forests, which is based on the forests of more humid environments. In addition, a tension exists between the aims of landowners and social and recreational demands. On the other hand, forest management has a decisive influence on the water balance, with effects on hydrologic stress, water retention and the availability of water in watersheds. Abandonment of rural areas has led to an increase of tree growth in former agriculture fields and, because trees consume more subsurface water, water harvesting from rivers is decreasing, as is well illustrated by changes in the Ebro Basin (Gallart and Llorens, 2004). On the other hand, in order to fight erosion it often seems that one key solution is afforestation, when it has been demonstrated that trees are not erosion protectors and take much water from the soil (Yair, 1983). Where the tourist industry has become the main resource, building and recreation have taken the place of forestry and agriculture. So in spite of the abandonment of the farms associated with traditional land uses, population pressure is becoming more intense in the countryside, especially in forested areas, the favourite recreation resort for urban populations. Many pine woodlands are the sites of second-residence houses, and new roads have been needed to access these areas.

As a result of these changes in usage, the risk of forest fires has increased and reports from local authorities point clearly to roads and urban areas within forests as the areas where the majority of these fires are initiated. Concern for the danger of forest fires in the Mediterranean is becoming increasingly widespread (De Ploey, 1994). In 1973 Le Houerou estimated the area annually destroyed by fire in the Mediterranean to be around 200 000 ha, a trebling of the affected area in 30 years. Data on forest fires in Greece (Alexandris, 1989) indicate that wildfires have become an important phenomenon in the last 15 years and are one of the country's biggest environmental problems, with an annual destruction of 36 000 ha of woods and thicket. There are similar problems in Algeria and Italy. Fires are particularly frequent in pine-dominated forests. Corsica and Sardinia are strongly affected by fires, especially in *macchia* terrains. Forest fires have increased in recent decades in Israel, with the increase of reforestation, abandonment of marginal mountain areas and the increase in natural reserves; on average, 5% of the forests in Israel are burnt each year (Inbar *et al.*, 1998). In Morocco, fire destroys 3000 ha of forest and alfa (*Stipa* sp.) every year. In many parts of Australia, bushfires, whether accidentally or deliberately lit, are a major concern, and the Mediterranean zones are particularly vulnerable with their hot, dry summers and combustible vegetation types. Large bushfires have raged every 3–5 years in southern Australia over the past 50 years (Department of Primary Industries and Energy, 1990).

11.5.2 Pasture

Livestock farming has been one of the main activities in all the Mediterranean environments of the world, often producing problems of land degradation due to overgrazing and the misuse of fire to stimulate fresh plant growth.

In Spain, extended pastoralism was established during the Middle Ages in the Christian states in a transhumance regime, thus involving both lowland and mountain environments. In more modern times, pastoralism was reduced and often located in mountain environments. It is interesting to note that pastoralism in Spain does not only imply cattle, sheep and goats but also pigs. Overgrazing by ruminants and pigs became apparent in the 1960s due to the overloading of carrying capacity on *montado* and *dehesa*. In Italy, sheep were once the most important livestock, especially in the hills and mountains, and this is still the case in Sardinia and parts of Sicily. In North Africa, there was an increase in livestock numbers in the nineteenth century in Algeria and again following independence.

In the southwestern Cape region, South Africa, herding was introduced approximately 2000 years ago. By the beginning of the European contact period in the mid seventeenth century, there may have been as many as 500 000 cattle and 1 million sheep being pastured, mainly on the Cape forelands. In California, with the American settlers, there was an expansion of cattle ranching along the coast and by 1860 there were approximately 1 million cattle, with a further increase during the Gold Rush due to the local demand. Mexican independence from Spain in 1821 led to mission lands being granted to ranchers, including a growing number of Americans, and to the rapid expansion of cattle ranching.

In Chile, overgrazing of the natural meadows/prairies is considered to have initiated the degradation of natural resources and although in the beginning sheep were dominant, later on the dominant ruminants were goats.

11.5.3 Agriculture

The Mediterranean areas in Israel, Lebanon, Syria and Turkey (the 'Fertile Crescent', with Mesopotamia) were the first sites of widespread domestication of animals and plants. Changes in environmental conditions are reflected in the changing agricultural crops of ancient Mesopotamia, where the increase of soil salinity affected the growth of wheat and led to its replacement by barley. The greatest development of the agro-pastoral economy in the eastern Mediterranean started around 5000 years ago and lasted until the end of the Roman period; in this period, land clearance affected the mountainous areas. With the Muslim conquest of the region, the pastoral nomadism of the Arab tribes replaced the developed hill lands and irrigation ditches, producing increased erosion. Traditional agriculture took place under low technical capabilities which meant that it affected the land and soils extensively, if only superficially.

In Spain, with the privatisation of church property, land was subject to speculation and overexploited for short-term benefits and population pressures caused the establishment of agriculture in marginal lands (mostly on mountain slopes which were later abandoned), being the source of erosion. In Italy, some historians maintain that land degradation was one of the main reasons for the decay of the Roman Empire. It is now acknowledged that the causes of these environmental crises were only partly climatic in nature, being principally of anthropogenic origin. It seems that since the Greco-Roman period, human activity has determined the alternation of erosion and accumulation along the rivers and coasts and, consequently, the slopes, that produced major ecological changes.

In California, the arrival of Spanish colonists after 1769 introduced fundamental changes, with Franciscan missions founded on lowland chaparral and oak savanna which were then converted to grassland. In Chile, in the period from 1870 to 1960, soil erosion from Santiago to Lanquihue reached very high values. International demand for cereals determined that many areas of Chile were cultivated and wheat was exported. Production in virgin forest soils was very high at the beginning of this period but after a while monoculture gave way to accelerated degradation and erosion and as a consequence productivity was lowered. The advance of population towards the south brought wealth to the country but at the same time a great destruction of forests and soils; in order to clear vegetation quickly and economically generalised burnings were used.

Finally, over the last 100 years, population increase and mechanisation and irrigation schemes have produced major ecological changes, including the effects of the opening the Suez Canal, the construction of the High Aswan Dam on the River Nile, and large water diversions and damming projects in the Euphrates and Jordan basins.

11.5.4 The agro-sylvo-pastoral system

A particular agro-sylvo-pastoral system exists in the Mediterranean called *dehesa* in Spain (Plate 34) and *montado* in Portugal. It is largely widespread in the southern and western part of the Iberian Peninsula. These tree-covered rangelands have evolved through centuries of multiple land use. Comparable systems exist also in Italy and Greece, the southern Mediterranean and in several islands. In California it is represented by oak woodland ranches occupying more than 2 million hectares (Huntsinger *et al.*, 2007). Similar landscapes are also found in Chile and Australia.

Although for certain researchers coming from more humid environments it may seem to be a not very efficient system and an important cause of land degradation, Mediterranean historians and local researchers have realised that it is in fact a system very well adapted to Mediterranean climatic and soil conditions. In its best-developed form, the system includes cork oak exploitation in an open forest under which a rotation of cereals or forage cultivation and grazing takes place. In places the trees are holm oaks. This association produces a landscape that resembles a park rather than a forest. Developed through millennia of experience, this form of management permits continuous, but sustainable, exploitation of fragile resources (Le Houerou, 1989). The equilibrium reached by this integrated system was partly lost during the 1960s due to more intensive exploitation, aiming at a self-sufficient production of cereals and importing agricultural techniques from northern Europe (Schnabel and Ferreira, 2004).

The importance of these areas rests on both environmental as well as socioeconomic values. They support outstanding biodiversity, form unique landscapes, are the source of high-quality food derived from animal production, sustain rural population, and constitute important areas for rural leisure and tourism. In Europe these areas have undergone rapid changes during the second half of the twentieth century, shifting from traditional farming systems with very low external energy inputs to more simplified systems causing decreasing diversity of land use and inadequate management techniques. Land degradation is recognised as a significant problem in many of these rangelands, including the lack of tree regeneration, which threatens the future of the woodlands as well as soil erosion, soil degradation and increased runoff production. A common feature is the coexistence of extensification and intensification, causing different problems of degradation. For example, subsidies in the form of headage payments in the European Union led to an unchecked increase in animal numbers, thus increasing the risk of soil and pasture degradation. On the other hand, abandonment of livestock breeding produces vegetation changes, leading to shrub encroachment and an increased risk of wildfires.

11.5.5 Mining

Although mining and quarrying have been important activities in Mediterranean Sea countries since Roman times, it is in the newly colonised areas that their impact has been more noticeable. Nevertheless in Spain they are still important on the margins of the Hercynian block and may have an important impact on the landscape. In Sicily, Calabria, Sardinia and the island of Elba, Italy, mining tunnels and waste disposal have caused a relatively heavy impact on the environment since Ancient Greek times.

Much land degradation in California has been linked historically with mining during Spanish and Mexican times, but it was the Gold Rush that initiated the first major impact, especially during the period of hydraulic mining of river gravels on the western slopes of the Sierra Nevada. More recently, mining for petroleum and natural gas has introduced a new series of problems, notably surface degradation and subsidence.

In Chile's Mediterranean area there are important copper mines in the Andes Mountains, which were exploited by the Spanish colonisers; this is still an important economic resource at the present time. In Australia, mining has been an important element of economic development, but not precisely in the Mediterranean zone. Nevertheless, the world's largest tonnages of bauxite are mined in the Darling Ranges, southeast of Perth, and several major industries based on mineral resources are located in the Mediterranean region.

11.5.6 Urbanisation

Present-day Mediterranean landscapes are constructed ones. Since the nineteenth century a complex network of road, railway, energy and urban structures has been superimposed onto the secular Mediterranean landscape. More recently, from the middle of the twentieth century, in some areas the industrial city has expanded to cover the entire territory. These built landscapes have lost their pre-industrial form, one of compact cities standing in the middle of extended rural areas, or even more or less virgin terrain. Today satellite images show diffuse metropolitan areas, discontinuous urban areas and urban corridors that extend without precise limits.

Unlike territories poleward of 45° N that have experienced similar processes, the capacity of Mediterranean landscapes to absorb these changes – its resilience – has been lower. The typical Mediterranean relief makes it difficult to install infrastructure and so populations and infrastructure become concentrated in scarce plainlands, located along fluvial systems and coasts. These generate corridors of activity and the concentration of infrastructure. Cities, roads, railways and electricity lines are superimposed on fluvial and agricultural landscapes. Easily workable land is at a premium.

11.5.7 Tourism

Tourism is a fundamental economic activity in the Mediterranean Sea countries, especially since the growth of package tourism. From the United Nations World Tourist Organization (2006) we know that the Mediterranean Sea is the main tourist destination on the planet. At present it receives an average of 290 million visitors per year, concentrating 30% of world tourism and 25% of global tourist income. And there is an exponential increase in numbers, expected to reach 440–665 million tourists by AD 2025. More than 75% of tourists go to Spain, France, Italy, Greece and Turkey, with Spain being the favourite destination. With the expansion of the European Union (EU), it is expected that tourism will increase in Malta, Cyprus, Croatia, Tunisia and Morocco.

11.6 Contemporary and expected near-future land use changes

Land use and related land degradation problems are strongly related to socioeconomic and political settings, which, by contrast to the natural environment, are different across the Mediterranean environments of the world. Within the Mediterranean Sea countries, two main divisions can be made: countries belonging to the EU versus countries outside this organisation (in this second group, one exception is Israel because its economic development is similar to that of the EU countries).

In the Mediterranean EU countries, the main land use changes are related to the shift of population from rural to

TABLE 11.3. *Population characteristics and urbanisation in Mediterranean Sea countries*

Mediterranean Sea countries	Total population (× 10^6)	Littoral population (× 10^6)	Demographic growth (%)	Natality (0/00, 2001)	Mortality (0/00, 2001)	Urbanisation rate (2003)	Urban growth 2000–05
Spain	41.1	23.0	0.2	10.0	9	77	0.3
France	60.4	15.0	0.5	13.1	8.9	76	0.7
Italy	57.3	33.0	−0.1	9.4	9.7	67	0
Slovenia	2.0	2.0	negative	9.0	9	51	negative
Croatia	4.4	4.4	negative	10.0	12	59	0.5
Albania	3.2	3.2	0.7	17	5	44	2.1
Greece	11.0	11.0	0.1	11.7	10.5	61	0.6
Macedonia	2.1	2.1	0.5	13.6	10.8	60	0.6
Bosnia	4.2	4.2	1.1	12	8	44	2.2
Turkey	72.3		1.4	22	7	66	2.2
Syria	18.2	12.8	2.4	31	6	50	2.5
Lebanon	3.7	3.7	1.6	21	7	88	1.9
Israel	6.6	6.6	2.0	21.7	6	92	2.5
Palestine	3.7	3.7	3.6			71	4.1
Tunis	9.9	8.0	1.1	17	6	86	2.3
Algeria	32.3	28.0	1.7	5	5	59	2.6
Morocco	31.1	10.0	1.6	25	6	58	2.8
Malta	0.4	0.4	1.0	14.2	7.1	91.7	0.7
Cyprus	0.8	0.8	1.0	16.9	7.5	69.2	1.0

Source: From Toumi (2006) with permission.

urban areas and the increasing tendency of urban people to take holidays in mountain and coastal environments (Table 11.3). This means not only the abandonment of many mountain lands that become recolonised by wild vegetation but also a progressive loss of knowledge as to how the natural environment functions. In addition, in the areas where agriculture is still active it has generally been transformed into intensive industrial agriculture. In relation to forestry, at present, there are a variety of tendencies, ranging from industrial forestry (eucalyptus and pine plantations) to urban/recreational use (second residences, roads) to abandonment of traditional forestry (cork oak) and agro-silvo-forestry (*dehesa*). While traditional forestry was sustainable, the present uses produce either increased erosion or induce forest fires.

Important land use changes have taken place during the last 30 years in relation to economic development, mainly related to tourism and to joining the EU. The economy has shifted from an agro-sylvo-pastoral base to a more tertiary oriented one, although industry and some agricultural products are also important. Mediterranean products such as oranges, wine and olive oil, together with early-season vegetables, are the main agricultural exports. On the contrary, wheat and sheep from the wide central basins are now in recession. The main development trends such as tourism, intensive agriculture, industry and services are mostly located along the Mediterranean coast.

In the southern and eastern Mediterranean Sea countries, present-day changes are based on the movement of population from the mountains to the plain and in the increase of extensive agricultural fields in the lowlands, giving way to gullying, flooding and severe soil loss. In the mountains, extensive pasture held under a public property regime is contributing to the degradation of forests and probably to an increased incidence of forest fires. Clear-felling for cultivation purposes on steep slopes is still an active process. In relation to forestry, traditional cork oak exploitation is still active but, as in the northern EU countries, industrial forestry is increasing. In these countries economies are still oriented towards satisfying basic needs and not to tertiary activities as in the EU countries. Another important trend is the massive emigration to the northern Mediterranean Sea countries.

In the New World Mediterranean countries, the situation shows a mixture of the characteristics described above. Population growth is similar to that seen in the south

Mediterranean Sea countries but in this case it is not paralleled by emigration. In relation to primary activities, timber and mineral overexploitation have to be mentioned. Vegetation decline and species invasion are two of the main problems. Expanded agricultural practices have produced erosion, flooding, sedimentation and, in many cases loss of water quality and secondary salinisation (Plate 35).

Land degradation problems are extensive in Mediterranean environments, mostly related to the seasonality and intensity of the rainfall regime. Soil erosion and deterioration, flooding and forest fires are the most serious problems (Sala and Rubio, 1994; Brandt and Thornes, 1996; Rubio and Calvo, 1996; Geeson *et al.*, 2002). Others are related to human activity, like urbanisation, industrialisation and tourism. Conacher (1998) stresses that whilst there are similarities amongst the Mediterranean regions, particularly in relation to soil degradation, vegetation clearance and forest fires, there are also some marked contrasts. Often differences reflect historical factors, especially the duration of intensive land management in the Mediterranean Sea countries, or Old World, in relation to the New World settlements in South Africa, California, Chile and Australia. Others reflect the nature of the lithology, because some of the most severely eroding areas are underlain by soft marls and mudstones.

Conacher (1998) also stresses that water quality and water shortages are major problems. Sedimentation, salinisation and accumulation of nutrients and toxic wastes in streams, rivers, dams, estuaries and groundwater mean that increasing proportions of previously freshwater resources are no longer suitable for irrigation or for human or animal consumption. For those reasons it is considered that one of the major by-products of land degradation has been rural depopulation, although other, important, social and economic factors are also responsible. Rural depopulation and the associated decline of country towns and villages in turn have further implications for land degradation: lands that are no longer managed carefully, or that have even been abandoned, are perhaps a luxury which a world characterised by rapid population growth and increasing demands for food cannot afford. On the other hand, intensification of agriculture in the more restricted but more fertile areas carries with it a range of actual and potential environmental costs.

11.7 Global environmental change in Mediterranean environments and its interaction with land use change

Given the climatic and land use changes that are taking place in Mediterranean environments, the future can to a certain extent be foreseen. The following sections summarise what is known about future scenarios of climate change; changing water resources; sea level change and flood hazard; geomorphic change; forest fire incidence; and land use change in several Mediterranean areas of the world.

11.7.1 Temperature and precipitation change

In terms of temperature, models predicting future climate change suggest that the southern and eastern shores of the Mediterranean in particular could undergo deleterious changes by the middle of the next century.

An indication of the scale of possible climate changes is given by one scenario based on the output from four climate models (IPCC, 2007). This suggests that temperatures could rise by over 4 °C by AD 2100 over many Mediterranean inland areas and by over half of this amount over the Mediterranean Sea. Over the same period, annual precipitation is projected to decline by 10–40% over much of Africa and southeastern Spain, with smaller changes elsewhere. By way of example, the possible implications of these changes for biomes in South Africa have been studied by Meadows (2006); a reduction of the areas suitable for South African biomes in 2050 appears to be only one-half of their current spatial distribution.

For the Mediterranean Sea regions, the high variability in local climates that masks general trends in climate change has to be taken into account (IPCC, 2007). The analysis of historic surface air temperature averaged over the entire basin indicates an evolution similar to that recorded on both the global and the hemispheric scale: a cooling during the period 1955–75 and a strong warming during the 1980s and the first half of the 1990s. However, the east–west Mediterranean difference in air and sea surface temperature trends is distinctive. Most of the studies concerning air temperature agree that there was a positive trend in western Mediterranean versus a negative trend in eastern Mediterranean temperatures for the periods 1950–90 and 1975–90. The cold period of the year appears to have contributed most strongly to the observed cooling of the eastern Mediterranean. The year 1999 was exceptionally warm in the eastern Mediterranean compared to the 1961–90 average and this was due to high summer and autumn temperatures. The mean temperature in the central west Mediterranean for the twentieth century shows an increase of about 0.008 °C a^{-1}. Sea surface temperature records show a rapid cooling in the 1970s while warming was resumed in the late 1970s.

Since 1900, precipitation has decreased by over 5% over much of the land bordering the Mediterranean Sea, with the exception of the stretch from Tunisia through to Libya where it has increased slightly. A general drying is evident over most of southeastern Mediterranean and Greece up to the early 1990s. A precipitation decrease has been observed as well during the last 50 years in the central west Mediterranean. Although the prediction of changes in precipitation is much less certain than that for temperature, most projections point to more precipitation in winter and less in summer over the region as a whole. Even areas receiving more precipitation may become drier than today, due to increased evaporation and changes in the seasonal distribution of rainfall and its intensity. As a consequence, the frequency and severity of droughts could increase across the region. Changes in large-scale atmospheric circulation – as represented by indicators such as the El Niño–Southern Oscillation index (ENSO) and the North Atlantic Oscillation index (NAO) – may further affect the occurrence of extreme events.

Hotter and drier conditions will extend the area prone to desertification northward and the rate of desertification will be further exacerbated by increases in erosion, salinisation and fire hazard. In addition, inappropriate land use practices will further increase these land degradation processes. As a result, desertification could become irreversible, with very high economic and human costs.

Nevertheless, experimental and model research undertaken by Osborne and Woodward (2002) on the potential effects of rising atmospheric carbon dioxide on vegetation suggest that the rate of water loss from soil under stands of Mediterranean sclerophyllous shrubs may decrease. This is because reduced rates of transpiration will lead to higher water availability for vegetation during the summer drought period (the efficiency of water use in primary production increases with rising atmospheric carbon dioxide because carbon dioxide has the potential to partially alleviate the adverse effects of drought on primary production). Finally, Osborne and Woodward (2002) stress that these responses will vary on a seasonal and inter-annual basis in relation to soil water availability. This finding could be of crucial importance in the future with increasing aridity in the Mediterranean Basin, especially in areas where precipitation is predicted to decline.

11.7.2 Changes in water resources and flood hazard

It is likely that the first impacts of climate change will be felt in the Mediterranean water resource system. Reductions in water availability will hit southern Mediterranean Sea countries the hardest. Even relatively water-rich countries, such as Spain, Greece and Italy, could suffer ever more frequent regional water shortages. Some water supplies could become unusable due to the penetration of salt water into rivers and coastal aquifers as sea level rises. Water pollution – already a major health hazard in the region – will become still worse if pollutants become more concentrated with reductions in river flow.

Mediterranean environments are naturally at risk from flooding events. Seasonality of rainfall determines a concentration of the rainy period into a few months and days of the year and high rainfall intensities are a common characteristic of the climatic regime. The result is high peak flows in the fluvial system. These conditions are natural and cannot be changed. The only possibility is to adapt human activities to them.

There are two environmental constraints to the flooding problem, one related to climate and the other to geomorphology: (1) high rainfall intensities and seasonality; and (2) steep relief along the coast determining short, high-energy streams with, depending on setting, the possibility of short runoff lag times in relation to rainfall. In addition, there are two anthropogenic conditioning actions: (1) increased urbanisation and road construction leading to an increase of impervious surfaces and so to an increase of surface runoff; and (2) frequent location of urban and tourism constructions in floodplains and alluvial fans. The result is an increase of flooding hazard and human vulnerability to flood impacts. It is clear that constructions should be carefully assessed in relation to flood hazard and measures taken to minimise potential near-future impacts resulting from increases in runoff and exposure to more frequent extreme events.

Studies of fluvial dynamics over recent decades have shown that one of the most effective agents of change in river behaviour is the role of human activities, through land use change (Batalla and Sala, 1996; Sala, 2003; and see Chapter 4). The main changes in river watersheds are produced by the removal of vegetation, followed by the regulation (straightening) of channels and the covering of land surfaces by buildings and roads. Although the response of drainage systems to these changes is complex, it is obvious that an increase in the runoff rate and the peak flows will occur due to the greater area of impermeable surfaces and the reduction in lag time between rainfall and runoff. Present-day economic development based on tourism parallels urban growth and growth of the road network in forested areas; both lead to runoff increases downstream due to the decrease in permeable surfaces. In other cases, landscape attractiveness may induce the location of tourist facilities in flood-prone areas.

The effects of increased urbanisation in Spain have been studied by Sala and Inbar (1992), using yearly discharge data for several streams draining the Catalan Coast, both before and after an increase in urbanisation within their watersheds (Fig. 11.6). While the cumulative curves of

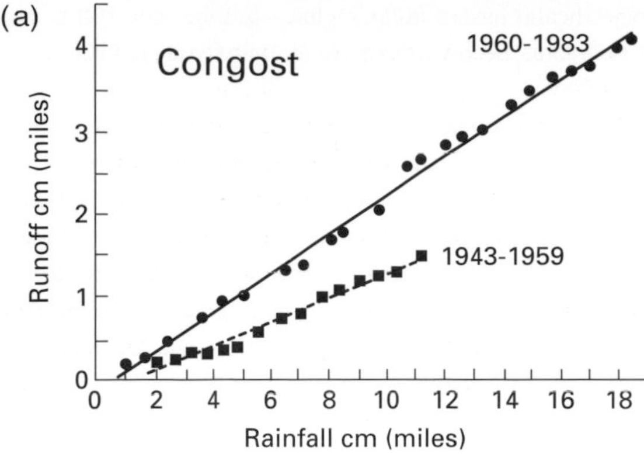

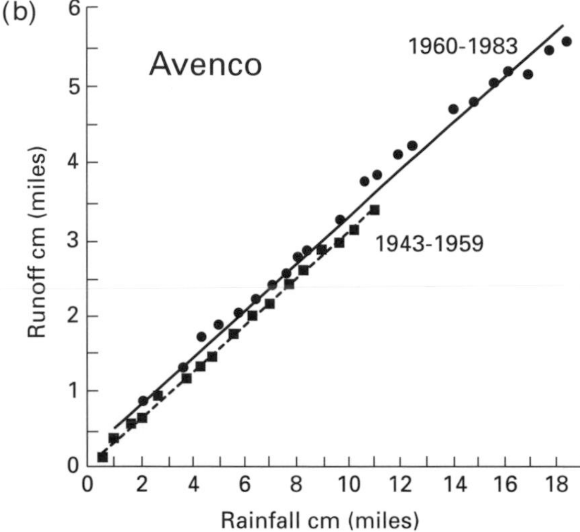

FIGURE 11.6. Rainfall–runoff relation and urbanisation. (a) Changed rainfall–runoff relation, Congost catchment, northeast Spain with increased urbanisation after 1960; (b) no change in rainfall–runoff relation in mountainous Avenco catchment (modified from Sala and Inbar, 1992).

annual precipitation have shown steady increases over the years, thus indicating no change in rainfall during the study period, the cumulative curve of annual runoff shows one or two inflection points that can be related to periods of industrial and urban development. For instance, the degree of urbanisation in 1980–81 for various basins increased by two to five times that of 1951–55 and, as a consequence, floods have become a common feature, especially in the coastal *ramblas*. This flooding can be related to both an increase in peak discharges and a shortening of lag times. Also in relation to urbanisation, road construction, especially on mountain slopes (Batalla and Sala, 1996) affects fluvial behaviour by increasing sediment yields and channel changes. At the present time there is a risk of severe and often catastrophic flooding. In the Catalan coastal catchments alone (Sala, 2003), 296 zones have been estimated to be at risk from flooding in basins where maximum daily rainfall values associated with significant runoff events are in the order of 200 mm over a 24-hr period, with a recurrence interval of 7 to 10 years.

In California, removal of native vegetation has also disrupted the hydrological system, leading to increased runoff which is reflected in increased flooding, erosion, sediment delivery and habitat changes downstream. Logging, mining and agriculture have each generated such responses, most dramatically in the vast sediment yields that were flushed from placer gold workings in the Sierra Nevada foothills during the late nineteenth century.

11.7.3 Sea level changes

Cazenave *et al.* (2001), using altimetry data from the Topex/Poseidon satellite available since early 1993, show that the eastern Mediterranean sea level has been continuously rising over the period 1993–99, at a rate up to 20 mm a^{-1} southeast of Crete. Sea level rise is also observed in the Algerian–Provençal basin as well as in the Tyrrhenian and Adriatic seas. The north Ionian Sea, on the other hand, shows an opposite trend, with a sea level fall during the past 7 years. Sea surface temperature trends are strongly correlated to sea level trends, indicating that at least part of the observed sea level change has a thermal origin. The recent Mediterranean sea level rise may be related to the warming trends reported from hydrographic cruises in the intermediate and deep waters of the eastern basin since the early 1990s, and in the western basin since the 1960s.

The IPCC (2007) states that marked changes in thermohaline properties have been observed throughout the Mediterranean and that the Western Mediterranean Deep Water (WMDW) warming is in agreement with recent atmospheric temperature changes over the Mediterranean. The salt content of the Mediterranean has also been steadily increasing during the last 50 years, and has been attributed to both decreasing precipitation over the region since the 1940s and to anthropogenic reduction in the freshwater inflow. These changes in water properties and circulation are linked to the long-term variability of surface fluxes and have affected the outflow of water into the North Atlantic Ocean at the Straits of Gibraltar. Over the period 1987–91, the Eastern Mediterranean Deep Water (EMDW) became warmer and more saline as a result of the switch of its source water from the Adriatic Sea to the Aegean Sea, most likely related to changes in the heat and freshwater flux anomalies in the Aegean Sea. This switch of source waters has continued and increased in its impact, with increased density of the westward outflow through the Strait of Messina. While there are strong natural variations in the thermohaline properties of the Mediterranean Sea, overall there has been a discernible trend of increased salinity and warmer temperatures in key water masses over the last 50 years and this signal is observable in the North Atlantic Ocean.

Around much of the Mediterranean Basin sea levels could rise by close to 1 m by AD 2100 (Tsimplis, 2005). Sea level changes are important as population density and human activities in the coastal zones have increased and will be affected by changing coastal conditions. As a consequence, some low-lying coastal areas may be lost through flooding or erosion, while rivers and coastal aquifers will show increases in salinity. The worst affected areas will most probably be the Ebro, Rhone, Po and Nile deltas (see also Chapter 4), together with Venice and Thessalonica where local subsidence means that sea levels could rise by at least one-and-a-half times as much as elsewhere.

As the Mediterranean Sea coasts are the world's most popular tourist destination, with 120 million visitors every year, sea level rise may lead to important changes in low-lying coastal settlements, some of which may be lost. This impact is already being seen in certain areas due to the reduction of fluvial sediment reaching the sea following the construction of dams upstream of many Mediterranean rivers, and landslide and erosion control works (Sorriso-Valvo and Terranova, 2006).

11.8 Concluding remarks

Important changes are taking place in Mediterranean environments, particularly in Mediterranean Sea countries, related both to climate change and its implications, but also, and perhaps more importantly, to socioeconomic trends and their land use change component.

11.8.1 General implications of global climate change

Because Mediterranean environments are located close to arid lands, the effects of global climate changes on them can be expected to tend towards the processes associated with desertification. In the Mediterranean Sea countries, vulnerability to climate change is expected to be high and any sustained reduction of winter rainfall, together with increasing temperature and evaporation during the summer months, could lead to near-desert conditions (Perry, 1977). One key issue (Karas, 2003) is that future climate change could critically undermine efforts for sustainable development in the Mediterranean region. In particular, climate change may add to existing problems of desertification, water scarcity and loss of food production whilst also introducing new threats to human health, ecosystems and national economies. The most serious impacts are likely to be felt in North African and eastern Mediterranean countries.

From the perspective of geomorphology, Lavee *et al.* (1998) state that three important aspects of climate should be considered if conditions become more arid: (1) any decrease that might occur in the annual rainfall total; (2) the duration of rainfall events; and (3) any increase in the interval between rainfall events. Less rainfall, of shorter duration and at more widely spaced intervals, together with increasing temperature, will lead to less available water, less biomass and lower soil organic matter contents and hence to a decrease in soil aggregate size and stability. As a consequence, soil permeability will decrease, soils will develop surface crusts and infiltration rates will decrease dramatically. Such changes in vegetation cover and soil structure will lead to an increase in overland flow and in the erosion of fertile topsoil layers.

11.8.2 Implications of socioeconomic and land use changes

Probably the most important driving force of change in Mediterranean environments is human action in relation to socioeconomic activities. At present important changes are taking place that will have different impacts in relation to physical, cultural, social and political backgrounds. According to Thornes (2002), the social and political framework in the Mediterranean Sea countries has changed dramatically in a way that makes a change in the research approach crucial. He also states that land abandonment occurs as a result of both external driving forces and internal changes in soil properties but that this abandonment does not necessarily mean land is no longer used. Rather, it means a change in land use from the traditional or recent pattern to another use.

Mass urbanisation, especially along coasts, will certainly have an effect on the water cycle due to the increasing extent of impervious surfaces, thus decreasing groundwater recharge and increasing surface runoff and flooding. Water demand will increase in regions already poor in such resources, although with variations in the availability of fresh water and the pattern of water requirement, and with increases in water pollution in coastal areas (Smith, 1977). This demand for water has already caused salinisation, due to the lowering of aquifers.

In addition to tourist impacts, changes in the management of rural environments have to be considered. In Mediterranean Sea countries belonging to the EU, the challenge in forestry is the dichotomy between sustainability and profitability, and the lack of European Commission

(EC) policy specifically adapted to Mediterranean forests. Only the cork forestry obtains subsidies in order to increase its economic income, to prevent forest fires and to preserve one of the most typical Mediterranean landscapes. But for other forest types, the low price of wood and, in certain areas, the low added value of many forest products is a serious drawback. Abandonment leads to an increased growth of shrubs and thus an increased risk of forest fires. In some cases profitability has been achieved by the establishment of plantations of eucalyptus and pines, but more and more the forest is being used for recreational purposes by urban populations. All in all, an increase of forest fires may be expected, leading to erosion and lowland floods.

In non-EU countries, forestry is at the level of furnishing basic needs and development. But in most cases there is a lack of effective government control on deforestation, overgrazing and clear-felling for cultivation purposes, even on steep slopes. In these situations, erosion and soil and vegetation degradation occurs. In North Africa, the traditional uses are cork forestry, with free harvesting of remaining branches. If we consider management related to property regime, in contrast to pasture managed in a private property regime, the interest is in maintaining costs at a minimum and revenues at a maximum, so it is more likely that there will not be grazing expansion above a sustainability limit. However, in open access situations, as in the North African countries, herders have unlimited access to the same pasture and each one expands their herding beyond the sustainable limit. In these cases long-term profits will be zero.

As for agriculture, Thornes (2002) mentions a shift in the Common Agricultural Policy (CAP) as a result of changing public awareness of the failures of the agricultural price support system and, specifically, the negotiation in 1992 of the General Agreement on Tariffs and Trade (GATT). It seems that in the EU there is determination to reform the CAP into a more broadly based rural policy, integrating environmental issues.

11.8.3 Where to go

It seems idealistic to claim that it is possible to change socioeconomic trends but some proposals can be made in relation to the mitigation, or constraint, of some of the degradation effects of present land uses. In relation to tourism, there should be controls on the extensive urbanisation of land, especially in coastal areas (Leontidou *et al.*, 1996) and on illegal constructions.

As regards forest fires (Plate 33), some key management approaches were identified during the Conference on Soil Erosion as a Consequence of Forest Fires (Sala and Rubio, 1994). In relation to prevention, and because of summer drought, constructions in the mountains should follow the patterns of traditional practices by leaving wide bare areas surrounding the houses. Concerning burnt vegetation, it seems much better not to remove the remains as clearing activities may be more detrimental to the soil than the burning itself. In relation to erosion, quantitative studies show that the loss of soil is high immediately after burning but that it reduces greatly after 1 year.

In flooding events, the applied structural solutions are always expensive engineering, including the straightening of river courses and the building of concrete levees and breakwaters. But these solutions lead to increased runoff volumes and velocities. Solutions related to the 're-naturalisation' of rivers seem to have been taken into account but have ultimately still been applied with an engineering emphasis (i.e. artificial vegetation and channel banks). Unfortunately, rivers are not seen as natural elements subject to physical laws, which in many aspects have to be respected, but as a totally controllable resource. Floodplains should be preserved, not urbanised.

Several solutions to maintain or improve water supply have been proposed by Smith (1977): increased and more efficient storage of seasonal surpluses; more efficient water use, especially for irrigation and including the use of treated water; more effective distribution of existing resources; and programmes of water leak reduction. More drastic strategies proposed are: importing water; wider use of desalination; rationing of water supplies; and economic restructuring to shift resources away from high-demand sectors.

In forestry, multifunctionality, together with sustainability, seems to be desirable. For reasonable planning, it is necessary to know about ownership (exploitation objectives); society (local views and aims); climate (rainfall, temperature and insolation); and geomorphology (relief, aspect, rocks and soils). Possible actions range from regeneration of the Mediterranean forest to reforestation with adequate species and fire-prevention policies. Maintaining traditional agro-sylvo-pastoral systems (*dehesa*, *montado*) is important (see Section 11.5.4). When forests are used for recreational purposes, ecotaxes should be implemented in order to allow management and residential areas should be concentrated in the lowlands.

For rural environments, some of the recommendations issued during the International Conference on The Future of The Mediterranean Rural Environment held in Menemen, Turkey (2000) pointed to major areas of concern. These include: the continuing degradation of the soil resource; major within-region conflicts in the use of land and no apparent strategy for decision-making on the most appropriate use of a particular piece of land; the lack of clear decision-making strategies for the development of the urban–rural interface; the lack of adequate land information systems and of data sets in some regions; and the need for a major involvement and discussion with stakeholders on key decisions affecting their land. It is interesting to note that different groups look for different outputs from the rural environment. The two most contrasting groups are developers and conservationists. For developers, building and irrigation are the key issues. For conservationists, everything should be left under 'natural' conditions, that is with the minimum of human intervention (no treatment of the forest, slow development of trees and forests fully open to the public for recreational purposes) which leads to non-economic forests and thus, abandonment and fires (due to increased fuel load in forests, abandoned fields and absence of fire breaks). These two positions are often related to the land being in private hands or belonging to the government.

Although the key point is sustainable and unsustainable productive agriculture and forestry, and biodiversity conservation, it is also true that some sort of agreement has to be found between developers and conservationists, and between private and government landowners. Planning is necessary, taking into account social and economic interests. Ecologically, the best planning leads to a mosaic, with the coexistence of forestry and agriculture as in traditional land uses, with intensive agriculture and leisure spaces confined to specific locations. For Thornes (2002), one of the major difficulties facing planning operations is the fact that the old Mediterranean Sea landscape is one of the most complicated in the world due to the essential diversity of the landscape arising from both its physical characteristics and its culture. The palimpsest character of this mosaic, which arises from its history, is such that the search for universal truths about causes and remedies for desertification and the appropriate actions to be taken are as diverse as the mosaic of the landscape itself.

11.8.4 Mediterranean perceptions

Some cultural views may help to explain the different perceptions of Mediterranean environments (and see Racionero (1996) for the main dichotomies between Europeans and Mediterraneans). In the context of most studies, and possibly in this chapter, without denying at all the risks of global change, both natural and human-induced, in a fragile environment like the Mediterranean, there is a subtle view that implies the inhabitants of the Mediterranean Sea area cannot properly manage their own environment. North Europeans tend to consider the Mediterranean environment from their humid and stable, often human-constructed, landscapes, and overemphasise land degradation and desertification problems, from the absence of forest vegetation and hydrological regime. Proudfoot and Smith (1977) consider it ironic that an area that has been settled for so long should be in many respects so hostile. One might add: hostile for whom? For the ones that do not understand its complexity? Finally, they write (p. 303) the following:

> Inevitably, our portrayal of these differences has been an exercise in hermeneutics, reflecting our own cultural conditioning and standpoint. Thus Westernising, Eurocentric perspectives tend to stress those tensions and problems that impact most strongly on European identities, economic interests and political security; hence the European Union's growing concern with illegal South–North immigration from the Magreb and Mashreq to Spain, France and Italy, and its attempted involvement in the long-running political instability in former Yugoslavia.

and (p. 305):

> whereas two hundred years ago, European economic power legitimised the assertion of European cultural supremacy, today recipient states like Morocco, Tunisia or Algeria confidently assert their own cultural and political identities, even though these may be challenged from within. It is in this conundrum – the dissonance between cultural identity, economic modernisation, political motivation and social aspiration – that the future of the Mediterranean lies.

References

Alexandris, S. (1989). The impact of fires on the forestry and natural environment. In *The Protection of the Environment and Agricultural Production*. Thessaloniki: Geotechnical Society of Greece, pp. 353–364.

Ballais, J.-L. (1998). The South of France and Corsica. In A. J. Conacher and M. Sala, eds., *Land Degradation in Mediterranean Environments of the World: Nature and Extent, Causes and Solutions*. Chichester: John Wiley, pp. 29–39.

Barriendos, M. (1997). Climatic variations in the Iberian Peninsula during the late Maunder Minimum (AD 1675–1715): an analysis of data from rogation ceremonies. *The Holocene*, **7**, 105–711.

Barriendos, M. and Llasat, M. C. (2003). The case of the 'Maldá' anomaly in the western Mediterranean Basin (AD 1760–1800): an example of a strong climatic variability. *Climatic Change*, **61**, 191–216.

Barriendos, M. and Martín Vide, J. (1998). Secular climatic oscillations as indicated by catastrophic floods in the Spanish Mediterranean coastal area (14th–19th centuries). *Climatic Change*, **38**, 473–491.

Batalla, R. J. and Sala, M. (1996). Impact of land use practices on the sediment yield of a partially disturbed Mediterranean catchment. *Zeitschrift für Geomorphologie*, **107**, 79–93.

Bilandzija, J., Frankovic, M. and Kaucic, D. (1998). The Croatian Adriatic coast. In A. J. Conacher and M. Sala, eds., *Land*

Degradation in Mediterranean Environments of the World: Nature and Extent, Causes and Solutions. Chichester: John Wiley, pp. 57–65.

Brandt, C. and Thornes, J. (1996). Introduction. In C. Brandt and J. Thornes, eds., *Mediterranean Desertification and Land Use*. Chichester: John Wiley, pp. 1–3.

Brazdil, R. *et al.* (1999). Flood events of selected rivers of Europe in the sixteenth century. *Climatic Change*, **43**, 239–285.

Buondonno, C. *et al.* (1993). Carte dell'erosione potenziale massima della Communhità Montana 'Fortore Beneventano'. *Annali Facoltè Scienze Agrarie Università di Napoli, Serie IV*, **27**, 20–33.

Campos Palacin, P. (1992). Spain. In T. Jones and S. Wibe, eds., *Forests: Market and Intervention's Failures – Five Case Studies*. London: Earthscan, pp. 165–200.

Castro, C. and Calderon, M. (1998). Chile. In A. J. Conacher and M. Sala, eds., *Land Degradation in Mediterranean Environments of the World: Nature and Extent, Causes and Solutions*. Chichester: John Wiley, pp. 123–138.

Cazenave, A. *et al.* (2001). Recent sea level change in the Mediterranean Sea revealed by Topex/Poseidon satellite altimetry. *Geophysical Research Letters*, **28**, 1607–1610.

Ceballos, L. (1966). *Mapa forestal de España*. Madrid: Ministerio de Agricultura.

Community Education and Policy Development Group (1993). *The State of the Environment Report for South Australia 1993*. Adelaide: Department of Environment and Land Management.

Conacher, A. J. (1998). Problems of land degradation. In A. J. Conacher and M. Sala, eds., *Land Degradation in Mediterranean Environments of the World: Nature and Extent, Causes and Solutions*. Chichester: John Wiley, pp. 171–175.

Conacher, A. J. and Conacher, J. L. (1998a). Introduction. In A. J. Conacher and M. Sala, eds., *Land Degradation in Mediterranean Environments of the World: Nature and Extent, Causes and Solutions*. Chichester: John Wiley, pp. xxiv–xxviii.

Conacher, A. J. and Conacher, J. L. (1998b). Southern Australia. In A. J. Conacher and M. Sala, eds., *Land Degradation in Mediterranean Environments of the World: Nature and Extent, Causes and Solutions*. Chichester: John Wiley, pp. 155–168.

Conacher, A. J. and Sala, M., eds. (1998). *Land Degradation in Mediterranean Environments of the World: Nature and Extent, Causes and Solutions*. Chichester: John Wiley.

CORINE (1992). *Soil Erosion Risk and Land Resources in the Southern Regions of the European Community*, Commission of the European Communities EUR 13233. Brussels: Office for the Publications of the European Communities.

De Ploey, J. (1989). *Soil Erosion Map of Western Europe*. Reiskirchen: Catena.

De Ploey, J. (1994). Introduction. In M. Sala and J. L. Rubio, eds., *Soil Erosion and Degradation as a Consequence of Forest Fires*. Logroño: Geoforma Ediciones, pp. 13–15.

Department of Primary Industries and Energy (1990). *Public Land Fire Management*. Canberra: Australian Government Publishing Service.

Faulkner, H. and Hill, A. (1977). Forest, soils and the threat of desertification. In R. King, B. Proudfoot and B. Smith, eds., *The Mediterranean: Environment and Society*. London: Arnold, pp. 252–272.

Gallart, F. and Llorens, P. (2004). Observations on land cover changes and water resources in the headwaters of the Ebro catchment, Iberian Peninsula. *Physics and Chemistry of the Earth*, **29**, 769–773.

Gaussen, H. (1955). *Déterminations des climats par la méthode des courbes ombrothermiques*. Paris: CNRS.

Geeson, N. A., Brandt, C. J. and Thornes, J. B. (2002). Preface. In N. A. Geeson, C. J. Brandt and J. Thornes, eds., *Mediterranean Desertification: A Mosaic of Processes and Responses*. Chichester: John Wiley, pp. xv–xvii.

Guerricchio, A., Melidoro, G. and Tazioli, S. (1976). Lineamenti idrogeologici e subsidenza dei terreni olocenici della Piana di sibari. *Numero Speciale Atti 68 convegno Società Geologica Italiana*, Praia a Mare, pp. 77–80.

Huntsinger, L., Bartolome, J. W. and D'Antonio, C. M. (2007). Grazing management of California grasslands. In J. Corbin, M. Stromberg and C. M. D'Antonio, eds., *Ecology and Management of California Grasslands*. Berkeley: University of California Press.

Ibañez, J. J. *et al.* (1996). Mediterranean soils and landscapes: an overview. In J. L. Rubio and A. Calvo, eds., *Soil Degradation and Desertification in Mediterranean Environments*. Logroño: Geoforma Ediciones, pp. 7–36.

Inbar, M. (1998). The Eastern Mediterranean. In A. J. Conacher and M. Sala, eds., *Land Degradation in Mediterranean Environments of the World: Nature and Extent, Causes and Solutions*. Chichester: John Wiley, pp. 79–89.

Inbar, M., Tamir, M. and Wittenberg, L. (1998). Runoff and erosion processes after a forest fire in Mount Carmel, a Mediterranean area. *Geomorphology*, **24**, 17–33.

International Conference on The Future of the Mediterranean Rural Environment: Prospects for Sustainable Land Use and Management, Menemen, Turkey (2000). Organized by the General Directorate of Rural Services, Turkey, and Cranfield University, Silsoe, UK. Conclusions and Recommendations.

IPCC (2007). *Climate Change 2007: The Physical Science Basis. Contribution of Working Group I to the Fourth Assessment Report of the Intergovernmental Panel on Climate Change*. Solomon, S. *et al.*, eds. Cambridge: Cambridge University Press.

Karas, J. (2003). *Climate Change and the Mediterranean Region*. Madrid: Tecnociencia.

Kosmas, C. (1998). Greece. In A. J. Conacher and M. Sala, eds., *Land Degradation in Mediterranean Environments of the World: Nature and Extent, Causes and Solutions*. Chichester: John Wiley, pp. 67–77.

Lambeck, K. and Purcell, A. (2005). Sea-level change in the Mediterranean Sea since the LGM: model predictions for tectonically stable areas. *Quaternary Science Reviews*, **24**, 1969–1988.

Laouina, A. (1998). North Africa. In A. J. Conacher and M. Sala, eds., *Land Degradation in Mediterranean Environments of the World: Nature and Extent, Causes and Solutions*. Chichester: John Wiley, pp. 91–108.

Lavee, H., Imeson, A. C. and Sarah, P. (1998). The impact of climate change on geomorphology and desertification along a Mediterranean-arid transect. *Land Degradation and Development*, **9**, 407–422.

Le Houerou, H. N. (1987). Vegetation wild fires in the Mediterranean basin: evolution and trends. *Ecología Mediterranea*, **13**, 13–24.

Le Houerou, H. N. (1989). Agrosilvicultura y silvopastoralismo para combater la degradacuón del suelo en la cuenca mediterránea. In *Degradación de zonas áridas del entorno mediterráneo*. Madrid: MOPU, Monografía Dirección General del Medio Ambiente, pp. 105–116.

Leontidou, L. *et al.* (1996). Urban expansion and littoralisation. In P. Mairota, J. Thornes and N. Geeson, eds., *Atlas of Mediterranean Environments in Europe: The Desertification Context*. Chichester: John Wiley, pp. 92–97.

Llobet, S. (1975). Materiales y depósitos periglaciares en el macizo del Montseny: antecedentes y resultados. *Revista de Geografia*, **9**, 35–58.

Matheson, W. E. (1986). Soils. In C. Nance and D. L. Speight, eds., *A Land Transformed: Environmental Change in South Australia*. Melbourne: Longman Cheshire, pp. 126–147.

Mattauer, M. (1973). *Les Déformations des matériaux de l'écorce terrestre*. Paris: Hermann.

McNeill, J. R. (1992). *The Mountains of the Mediterranean World: An Environmental History*. Cambridge: Cambridge University Press.

Meadows, M. (1998): The South West Cape. In A. J. Conacher and M. Sala, eds., *Land Degradation in Mediterranean Environments of the World: Nature and Extent, Causes and Solutions*. Chichester: John Wiley, pp. 139–154.

Meadows, M. (2006). Global change and southern Africa. *Geographical Research*, **44**, 135–145.

MOPU (1989). *Medio ambiente en España*. Madrid: Monografías de la Secretaría General del Medio Ambiente.

Mőrner, N. A. (2005). Sea level changes and crustal movements with special aspects on the eastern Mediterranean. *Zeitschrift für Geomorphologie (Supplement)*, **137**, 91–102.

Naveh, J. (1975). The evolutionary significance of fire in the Mediterranean region. *Vegetatio*, **9**, 199–206.

Orme, A. and Orme, A. J. (1998). Greater California. In A. J. Conacher and M. Sala, eds., *Land Degradation in Mediterranean Environments of the World: Nature and Extent, Causes and Solutions*. Chichester: John Wiley, pp. 109–138.

Osborne, C. P. and Woodward, F. I. (2002). Potential effects of rising CO_2 and climatic change on Mediterranean vegetation. In N. A. Geeson, C. J. Brandt and J. B. Thornes, eds., *Mediterranean Desertification: A Mosaic of Processes and Responses*. Chichester: John Wiley, pp. 33–46.

Panizza, M. (1978). Héritages périglaciaires würmiens dans l'Apennin Émilien. *Colloque sur le périglaciarie d'altitude du momain méditerranéen et abords*, Strasbourg, pp. 205–208.

Perry, A. L. (1977). Mediterranean climate. In R. King, B. Proudfoot and B. Smith, eds., *The Mediterranean: Environment and Society*. London: Arnold, pp. 30–44.

Pfister, C. *et al.* (1999). Documentary evidence on climate in sixteenth-century Europe. *Climatic Change*, **43**, 55–110.

Proudfoot, L. and Smith, B. (1977). Conclusions: From the past to the future of the Mediterranean. In R. King, B. Proudfoot and B. Smith, eds., *The Mediterranean: Environment and Society*. London: Arnold, pp. 300–305.

Racionero, L. (1996). *El Mediterráneo y los bárbaros del Norte*. Barcelona: Plaza y Janés.

Raynal, R. (1973). Quelques vues d'ensemble à propos du périglaciaire pleistocène des régions riveraines de la méditerranée occidentale. *Biuletyn Peryglacjalny*, **22**, 249–255.

Raynal, R. (1977). Etagement comparé en altitude des processus périglaciaires actuels dans les hauts massifs du Maroc et d'Iran. *Abhandlungen der Akademie der Wissenschaften in Gottingen, Mathematisch-Physikalische Klasse*, **N31**, 276–289.

Raynal, R. and Tricart, J. (1963). Comparaisons des grandes étapes morphogénetiques du Quaternaire dans le Midi méditerranéen français et au Maroc. *Bulletin de la Société Géologique de France* (7), **5**, 587–596.

Rovira, A., Garcia, C. and Batalla, R. J. (1998). Land use and forest fire hazards in the suspended sediment load of a Mediterranean basin. *Proceedings of the International Seminar on Land Degradation and Desertification*. IGU Regional Conference, Lisbon, 37–50.

Rubio, J. L. and Calvo, C. (1996). Mechanisms and processes of soil erosion by water in Mediterranean Spain. In J. L. Rubio and A. Calvo, eds., *Soil Degradation and Desertification in Mediterranean Environments*. Logroño: Geoforma Ediciones, pp. 37–48.

Sala, M. (1978). Summary: the Tordera drainage basin – a geomophologic study. In M. Sala, ed., *La cuenca de la Tordera: Estudio geomorfológico*. Lleida: Milenio, pp. 151–157.

Sala, M. (1988). Slope runoff and sediment production in two Mediterranean mountain environments. *Catena*, **12**, 13–29.

Sala, M. (2003). Floods triggered by natural conditions and by human activities in a Mediterranean coastal environment. *Geografiska Annaler*, **85**A, 301–312.

Sala, M. and Calvo, A. (1990). Response of four different Mediterranean vegetation types to runoff and erosion. In J. Thornes, ed., *Vegetation and Erosion*. Chichester: John Wiley, pp. 347–362.

Sala, M. and Coelho, C. (1998). The Iberian Peninsula and Balearic Islands. In A. J. Conacher and M. Sala, eds., *Land Degradation in Mediterranean Environments of the World: Nature and Extent, Causes and Solutions*. Chichester: John Wiley, pp. 3–28.

Sala, M. and Inbar, M. (1992). Some hydrologic effects of urbanisation in Catalan rivers. *Catena*, **19**, 363–378.

Sala, M. and Rubio, J. L. (1994). Preface. In M. Sala and J. L. Rubio, eds., *Soil Erosion and Degradation as a Consequence of Forest Fires*. Logroño: Geoforma Ediciones, pp. 9–12.

Schnabel, S. and Ferreira, A. (2004). Prolog. In S. Schnabel and A. Gonçalves, eds., *Sustainability of Agrosilvopastoral Systems – Dehesas, Montados*. Reiskirchen: Catena.

Smith, B. (1977). Water: a critical resource. In R. King, B. Proudfoot and B. Smith, eds., *The Mediterranean: Environment and Society*. London: Arnold, pp. 227–251.

Sneh, Y. and M. Klein (1984). Holocene sea level changes at the coast of Dor, Southeast Mediterranean. *Science*, **226**, 831–832.

Soler, M. and Sala, M. (1992). Effects of fire and of clearing in a Mediterranean *Quercus ilex* woodland: an experimental approach. *Catena*, **19**, 321–332.

Sorriso-Valvo, M. (1998). Italy. In A. J. Conacher and M. Sala, eds., *Land Degradation in Mediterranean Environments of the World: Nature and Extent, Causes and Solutions*. Chichester: John Wiley, pp. 41–56.

Sorriso-Valvo, M. and Terranova, O. (2006). The Calabrian fiumara streams. *Zeitschrift für Geomorphologie, Supplementband*, **143**, 109–125.

Sorriso-Valvo, M., Antronico, L. and Borelli, A. (1992). Recent evolution of badland-type erosion in southern Calabria (Italy). *Geoöko-plus*, **3**, 69–82.

Thornes, J. B. (2002). Evolving context of Mediterranean desertification. In N. A. Geeson, C. J. Brandt and J. B. Thornes, eds., *Mediterranean Desertification: A Mosaic of Processes and Responses*. Chichester: John Wiley, pp. 5–10.

Toumi, A. (2006). Les Méditerranéens. Unpublished manuscript, International Geographical Union Mediterranean Renaissance Programme (MRP), Cairo.

Tricart, J. (1966). Quelques aspects des phénomènes périglaciaires quaternaires dans la Peninsule Ibérique. *Biuletyn Peryglacjalny*, **16**, 313–327.

Tsimplis, M. N. (2005). Global sea level rise: a useful sea level predictor in the Mediterranean Sea? *Zeitschrift für Geomorphologie, Supplementband*, **137**, 103–110.

Ubeda, X. and Sala, M. (1998). Variations in runoff and erosion in three areas with different fire intensities. *Geoökodynamik*, **19**, 179–188.

Úbeda, X., Outeiro, L. R. and Sala, M. (2006). Vegetation regrowth after a differential intensity forest fire in a Mediterranean environment, northeast Spain. *Land Degradation and Development*, **17**, 429–440.

UNESCO–FAO (1963). *Bioclimatic Map of the Mediterranean Zone*. Paris: UNESCO.

United Nations World Tourist Organization (2006). *Facts and Figures*. Madrid: UNWTO/OMT.

Vita-Finzi, C. (1969). *The Mediterranean Valleys: Geological Changes in Historical Time*. Cambridge: Cambridge University Press.

Yair, A. (1983). Hillslope hydrology, water harvesting and areal distribution of ancient agricultural fields in the Northern Negev. *Journal of Arid Environments*, **6**, 283–301.

12 Temperate forests and rangelands

Roy C. Sidle and Tim P. Burt

12.1 Introduction

12.1.1 Hydrological and geomorphic processes in temperate forests and rangelands

Long-term experiments in forest hydrology, often using the paired catchment approach, have shown the striking dependence of streamflow volume on type of vegetation cover. The main change following forest clearance is an increase in stormflow volume on the falling limb of the storm hydrograph; as volumes of storm runoff from many headwater catchments converge lower down the stream network, extensive and serious flooding of lowlands can occur. These problems are exacerbated by the land degradation that follows clearcutting of forest and it is this land degradation and its catastrophic consequences through mass movements which form the core of this chapter.

Landslides are one of the major geomorphic processes affected by land use and climate change in mid-latitude temperate forests and rangelands. Inherent in these discussions are potential changes in hydrological processes that either drive or influence landslides, as well as issues such as land degradation. Landslides are typically episodic and not only impact stream channels and decrease site productivity, but also represent formidable hazards to humans and property. Surficial or slow-moving landslides contribute sediment to streams, decrease site productivity and may damage property, but do not typically endanger people (Sidle and Ochiai, 2006).

Mid-latitude forest and rangeland ecosystems are constantly changing as the result of human activities, movement of population sectors, climate change, and most recently a surge in information technology. This last development is important because it affects public perception of these geomorphic changes and hazards. Given the current easy access to such information, it is imperative that scientists, land managers and planners provide accurate, rational and timely information and assessments on environmental change issues to the public (Sorensen, 2000). Neither apocalyptic nor 'business as usual' scenarios of climate change and forest land use are in the best public interest. Due to close linkages amongst geomorphic processes, climate change and land use, simplifications are often conveyed by the media that mask the complexity of ecosystem behaviour and portray inappropriate cause and effect relations; some of these inappropriate generalisations have made their way into the scientific literature in the past few decades (e.g. Friedman and Friedman, 1988; Portela and Aguirre, 2000; Ericksen *et al.*, 2002). A systematic collection of scientific data together with identification of information gaps and strategic research is needed to address the combined influences of climate change and changing forest and rangeland use on geomorphic processes and hazards. In addition to mid-latitude terrain currently occupied by forest and range vegetation, previously forested land that has been converted to predominantly agricultural land is included. It is in this latter case where many of the most pressing and challenging environmental change issues reside.

12.1.2 Background

Temperate forests and rangelands in mid-latitude regions comprise 16.75% of the global land area between approximately 30° and 55° latitude (depending on elevation) in the northern and southern hemispheres (FAO, 2001) (Plate 36). Temperate mountain ecosystems in the Food and Agricultural Organization's (FAO) classification include some subalpine forest components, but temperate deserts are not included in their calculated percentage. Temperate forests are characterised by well-defined seasons with precipitation occurring year-round. Within temperate regions, maritime environments promote more stable temperatures throughout the year with higher rainfall in winter months,

Geomorphology and Global Environmental Change, eds. Olav Slaymaker, Thomas Spencer and Christine Embleton-Hamann. Published by Cambridge University Press.

while continental environments experience warmer summers and colder winters, often with accumulations of snow. Geomorphic processes and potential hazards are active as much of this terrain is steep and/or experiences intense or prolonged periods of rainfall or snowmelt. Mid-latitude rangelands are often used for cattle grazing and vary from relatively flat to steep terrain. Many are in semi-arid to marginally temperate environments with cold winters, such as the interior mountain region of the USA and the Eurasian steppe, the expanse of grazing land that extends from Manchuria to the eastern boundaries of Hungary.

Landslides, including the mass wasting processes of soil creep, debris flows and dry ravel, are strongly affected by environmental change. Surface erosion, discussed briefly in the section here on anthropogenic effects, is an important and arguably more widespread geomorphic process that has been extensively addressed in recent books (e.g. Sidle, 2002; Boardman and Poesen, 2006; Montgomery, 2007; Blanco-Canqui and Lal, 2008) and comprehensive scientific reviews (e.g. Kirkby and Cox, 1995; Lee *et al.*, 1996; Poesen *et al.*, 2003; Soil and Water Conservation Society, 2003; Nearing *et al.*, 2005). Surface erosion is widely associated with agricultural practices and other intensive land uses and is also affected by climate change. Fluvial geomorphic processes are directly linked to both landslides and surface erosion by nature of sediment supply and transport (e.g. Benda *et al.*, 2003; Imaizumi *et al.*, 2006), and stream and riverbank erosion is dominantly a landslide (mass wasting) process. Glacial and permafrost processes are most affected by environmental change at high latitude/altitude (Haeberli and Burn, 2002). Weathering processes, which can be substantially altered by environmental change, exert a direct effect on landslide initiation (e.g. Chigira, 2002). Thus, many of the other geomorphic processes affected by environmental change have a direct influence on landslides in temperate forests and rangelands.

12.1.3 Some applied forest hydrology

Most hydrological studies of vegetated catchments have been concerned with trees; studies of short vegetation have been much less common. Forest hydrology has covered both deforestation and afforestation (not always mirror images in their impact) and, whereas in the USA the main concern has been the link between deforestation, flooding and erosion, in the UK controversy has raged in relation to afforestation and its impact on water supply. The presence of a tree canopy completely alters the boundary layer meteorology and because of this many micrometeorological studies have complemented hydrological research, especially in relation to the study of interception and evaporation. A full review, albeit from a UK perspective, is provided in Calder (1990).

Frank Law presented the results of his water balance studies in the Lancashire hills of northern England in 1956 (Law, 1956). He argued that the loss of water by evaporation far exceeds the value gained from afforestation; not surprisingly, his results were hotly disputed. To investigate Law's claims further, a paired catchment experiment was established on the Plynlimon uplands of central Wales, where the headwaters of the River Wye (open moorland) and the River Severn (largely coniferous forest) could be compared. After adjusting for the unforested area of the Severn catchment, it was shown that evaporative losses from a catchment covered with mature coniferous forest would be twice that from (low-vegetation) moorland (Table 12.1). Thus, the evaporation loss from the low-vegetation moorland vegetation is 15% of input, whereas the loss from the high-canopy forest is 35%. The difference arises from the high aerodynamic roughness of the forest canopy, which ensures good mixing of air above the canopy. In this location, the canopy is wet sufficiently often that the low resistance of the atmosphere to vapour flux becomes significant compared with the smoother moorland surface where resistance is consequently higher.

Two major experiments at the Coweeta Experimental Forest in North Carolina, USA (Douglass and Hoover, 1988) have examined the effects on water yield of converting deciduous hardwood forest to different vegetation cover. One involved the replacement of hardwoods with white pine, and the other a conversion to grass, followed by natural succession. For 6 years after the planting of white pine, water yield remained about the same but then fell steadily so that after 15 years, water yields were about 200 mm (20%) less than would be expected for a hardwood cover. Transpiration and interception losses in the dormant season accounted equally for the increase in evaporative losses. The grass experiment involved a more complicated sequence of manipulations (Burt and Swank, 1992). When the grass was fertilised and exhibited lush growth, there was little difference from the hardwoods, implying that both interception and transpiration were similar for low and high vegetation under these circumstances. However, as the grass growth declined in successive years, the differences began to increase. When the grass was killed off, the differences were largest, although still less than in some of the clear-felling experiments, presumably because there was still some interception loss from the dead vegetation. Using flow duration curves, Burt and Swank (1992) showed that the increased flow was particularly associated with increased baseflow. Thereafter, the water yields returned to near-normal levels as natural regrowth

TABLE 12.1. *Likely responses of shallow–rapid and slow deep-seated landslides to various climate change scenarios in mid-latitude forests and rangelands*

Landslide type	Annual rainfall	Extreme precipitation	Winter snowpack	Likely temporal change in landslide susceptibility[a]
Shallow, rapid	Decrease	Stable	N/A	←
Shallow, rapid	Decrease	Increase	N/A	**→**
Shallow, rapid	Stable	Stable	N/A	←
Shallow, rapid	Stable	Increase	N/A	**→**
Shallow, rapid	Increase	Stable	N/A	No change
Shallow, rapid	Increase	Increase	N/A	←
Shallow, rapid	Decrease	Stable	Decrease	→
Shallow, rapid	Decrease	Increase	Decrease	←
Shallow, rapid	Stable	Stable	Decrease	**→**
Shallow, rapid	Stable	Increase	Decrease	**←**
Shallow, rapid	Increase	Stable	Decrease	No change
Shallow, rapid	Increase	Increase	Decrease	**→**
Slow, deep-seated	Decrease	Stable	N/A	**←**
Slow, deep-seated	Decrease	Increase	N/A	←
Slow, deep-seated	Stable	Stable	N/A	←
Slow, deep-seated	Stable	Increase	N/A	→
Slow, deep-seated	Increase	Stable	N/A	→
Slow, deep-seated	Increase	Increase	N/A	**→**
Slow, deep-seated	Decrease	Stable	Decrease	**←**
Slow, deep-seated	Decrease	Increase	Decrease	←
Slow, deep-seated	Stable	Stable	Decrease	←
Slow, deep-seated	Stable	Increase	Decrease	No change
Slow, deep-seated	Increase	Stable	Decrease	No change
Slow, deep-seated	Increase	Increase	Decrease	→

[a] Direction and boldness of arrows reflect trajectory and extent of responses.

occurred. Taken together, the Coweeta experiments show the striking dependence of streamflow volume on type of vegetation cover (Swank *et al.*, 1988).

Changes in the size and shape of storm hydrographs have also been studied at Coweeta (Hewlett and Helvey, 1970). The main change following forest clearance was an increase in stormflow volume; storm peak discharge and time to peak were not significantly different. Following Hursh and Brater (1941), Hewlett and Helvey (1970) argued that forest clearance would produce wetter soils, generating larger volumes of subsurface stormflow. However, variable source areas would not expand significantly, given the steep slopes, so that saturation-excess overland flow would not increase to any significant extent. Thus, the main increase in stormflow volume occurred on the falling limb of the storm hydrograph. As far as downstream flooding is concerned, it is the increase in stormflow volume rather than peak discharge from these deforested headwater catchments that is important. Increased volumes of storm runoff from many headwater catchments converge lower down the stream network to cause flooding in the surrounding lowlands. It was these flooding problems, together with land degradation following clearcutting, which prompted the hydrological experiments at Coweeta, rather than issues of water supply, as in the UK. It is the degradation following clearcutting that forms the central topic of this chapter.

12.2 Global distribution of mid-latitude temperate forests and rangelands

12.2.1 Spatial and topographic distribution of temperate forests and steppes

Based on a Forest Resources Assessment Report by FAO (FAO, 2001), temperate ecosystems (including oceanic and continental forests, steppes and mountain systems) comprise similar areas in North America (6.16 Mkm^2), Asia

(5.97 Mkm^2) and Europe (5.75 Mkm^2), and lesser areas in the Asian part of the former Soviet Union (2.73 Mkm^2) and South America (0.834 Mkm^2) (Plate 36).

Within North America the mid-latitude temperate ecosystems include temperate continental (2.02 Mkm^2), mountain (*c.* 1.85 Mkm^2) and oceanic (*c.* 0.17 Mkm^2) forests together with temperate steppes (2.12 Mkm^2) (FAO, 2001). The precise breakdown between mountain and oceanic forests is controversial, with the majority of the oceanic forest located at relatively low elevations of the Coast Mountains, Vancouver Island and the Queen Charlotte Islands, Canada and the Olympic Mountains, USA. Temperate continental forests cover large areas around the Great Lakes and northeastern USA south to Appalachia. Further inland, less extensive mixed stands of deciduous and conifer species exist. Temperate mountain forests are largely coniferous and occupy considerable portions of the western coastal ranges and Rocky Mountains of North America. Temperate steppes largely occur in the Intermountain Region, a relatively dry area between the Rocky Mountains and the mountains near the west coast. Higher, wetter elevations in the interior mountain region support groves of aspen and poplar, but this region is largely vegetated by shrubs and grasses.

South America contains considerably less temperate ecosystems than North America; the most abundant temperate ecosystems in South America are steppes (0.50 Mkm^2) and oceanic forests (0.26 Mkm^2), the former occupying the Patagonian trans-Andean zone (east of the Andes) and the latter largely located to the west of the Andes in Chile (FAO, 2001). Temperate mountain forests constitute less than 10% of the temperate ecosystems (0.077 Mkm^2) in South America and occur mainly in the central Patagonian Andes, at up to 52° S.

Europe contains a similar area of temperate ecosystems compared to North America (Fig. 12.1), dominated by continental forests (2.91 Mkm^2) in parts of southern Scandinavia, the Balkan Peninsula, the foothills of the Crimean and Caucasus Mountains, and much of Eastern Europe (FAO, 2001). Lesser amounts of temperate oceanic forest (1.29 Mkm^2) exist along the coast of Portugal and Spain, the British Isles, southernmost Sweden, Denmark and central Europe (including much of France). Temperate mountain forests occupy about 0.61 Mkm^2 of scattered areas throughout Europe, with mixed beech forests dominating lower elevations and conifers dominating higher elevations (FAO, 2001). Steppes are largely vegetated with grasses and dwarf shrubs and are widespread (0.96 Mkm^2) south of the diagonal between Ufa, Russia (about 54° 45′ N) and Bucharest, Romania (about 44° 25′ N).

Temperate ecosystems in Asia also occupy similar land areas compared to North America and Europe (Plate 36),

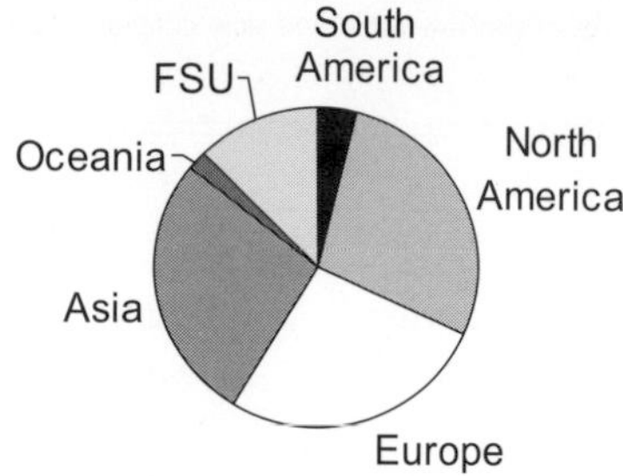

FIGURE 12.1. Global distribution of mid-latitude, temperate forest and steppe biomes; FSU refers to the former Soviet Union.

but the distribution is heavily skewed to mountain forests (3.60 Mkm^2) including those in the vast mountain systems of central Asia and Mongolia, the Tibetan Plateau and the mountainous regions of Japan and Korea. While many of the former forests in China have been exploited, natural forests exist in higher and more inaccessible locations. Because of the recent tectonic history of this region, together with land cover changes and episodic high precipitation, these mountain forests are among the most susceptible to landslides worldwide (Sidle *et al.*, 2004). Mid-latitude continental forests and grass and shrub steppes cover similar land areas in Asia (1.25 and 1.12 Mkm^2, respectively) (FAO, 2001). Mixed continental forests occupy large areas of northeast China, and mixed hardwood forests, with cypress and cedar plantations, occupy much of Japan. Grass and shrub steppes are found in the eastern part of Inner Mongolia and in central and eastern Mongolia.

The total area of temperate ecosystems in Oceania is restricted to 0.41 Mkm^2. Temperate oceanic forest complexes cover the southeastern coast of Australia, Tasmania and the lowlands of South Island, New Zealand (FAO, 2001). Temperate mountain forests are distributed in the Tasmanian Highlands, Southeastern Highlands, Alps and New England Tablelands of Australia, as well as in the Southern Alps on South Island, New Zealand. Because of the geology, tectonic conditions and relatively high rainfall in these regions (especially South Island, New Zealand), these mountain forests are nevertheless highly affected by landslides (Sidle *et al.*, 1985).

In the Asian part of the former Soviet Union, temperate ecosystems occupy less than half the comparable land areas in Europe, Asia and North America (Plate 36). Grassland steppes with inclusion of shrubs and trees comprise the largest temperate ecosystem (1.20 Mkm^2); these are oriented along a belt running from west to east from the Pricaspian lowlands, through the hills of the southern Urals and the northern part of the Turan lowland, to the uplands of the Kazakh steppe and Altai foothills (FAO, 2001). Continental and mountain forests occupy nearly identical land areas (0.76 and 0.77 Mkm^2 respectively). Continental birch and aspen forests comprise a belt that begins east of the Urals in Asian Russia and becomes progressively narrower to the east. Mountain forests occur in the Pamir-Alai and Tien-Shan mountains and the southern Sikhote-alin Mountains of the Russian Far East (Cushman and Wallin, 2000; FAO, 2001).

12.3 Potential climate change scenarios and geomorphic consequences

12.3.1 Likely scenarios in various regions

The projected temperature increases for the twenty-first century are greater than the expected global average over land masses and in the northern hemisphere, with increases being progressively greater when moving from coastlines into continental interiors (IPCC, 2007a). Predicted warming will be least in the southern hemisphere. Depending on carbon dioxide emission projections, warming trends of *c.* ± 0.2 °C per decade are anticipated for the rest of this century, of which about half can be attributed to committed climate change due to the lag effect from past emissions (IPCC, 2007a). While these temperature scenarios are useful for estimating patterns of soil water changes, they need to be coupled with accurate spatial and temporal predictions of distributed rainfall to be meaningful for landslide processes.

Climate models predict future increases in total precipitation in the northern hemisphere over most of the northern portions of North America, Europe and Asia; decreases in precipitation are predicted for Mediterranean Africa and southern Europe (IPCC, 2007a). Precipitation predictions are uncertain for much of the continental USA and northern Mexico. The only mid-latitude regions of the southern hemisphere that are projected to experience increased precipitation during the twenty-first century are South Island, New Zealand and Tierra del Fuego (IPCC, 2007a). Because of the increased winter temperatures, snow cover will decrease, especially in the northern portion of the mid-latitude northern hemisphere; warming will result in increased winter rainfall in some areas at the expense of reduced snowfall and consequent reductions in snowmelt and earlier snowmelt (Leung *et al.*, 2004).

Overall, the frequency and intensity of extreme weather events such as storms and hurricanes are expected to change with scenarios of climatic warming. Recent modelling results from mid-latitude regions of the northern hemisphere indicate general increases in heavy rainfall in northern Eurasia (Khon *et al.*, 2007), France (Déqué, 2007), Great Britain (Fowler *et al.*, 2005), Mediterranean region (Sánchez *et al.*, 2004) and the Cascades and the Sierra Nevada of western USA (Leung *et al.*, 2004). Such increases in intensity and variability of storms (albeit with fewer events in total) are also predicted for mid-latitudinal regions of the southern hemisphere (Pezza *et al.*, 2007).

12.3.2 The relations and uncertainties of specific climate changes to mass wasting

The most important climate parameter influenced by global change that affects rapid landslides and debris flows is short-term rainfall (Sidle and Ochiai, 2006). Unfortunately this is the most uncertain and most difficult variable to estimate at meaningful spatial and temporal resolutions (IPCC, 2007a). Predicted surface temperature increases indirectly affect landslide initiation and movement by altering soil moisture, an important precursor to both shallow and deep-seated landslides (Sidle and Dhakal, 2002). Decreases in soil water (attributed to warming) prior to large storms or even earthquakes may lessen the probability of slope failure. Nevertheless, it is necessary to know the timing and spatial distribution of such decreases in soil moisture to evaluate their effects on slope stability. Current climate change models cannot address such details. While shallow, rapid landslides respond to shorter-term precipitation inputs, regions experiencing increases in annual or seasonal precipitation may well see increased periods of deep-seated mass wasting processes (e.g. earthflows, slumps, soil creep) (Table 12.2); however, these could be offset by increases in evapotranspiration, particularly during the growing season, in a warming climate. Changes in evapotranspiration demands (including rooting depth) of vegetation cover modified by extended warming will also need to be assessed in relation to deep-seated landslides. Smaller snowpacks in winter will generally result in fewer snowmelt-triggered landslides, although the timing and rapidity of the melt is critical in this respect (Sidle and Ochiai, 2006) (Table 12.2). Some of these complex interrelations among climate, vegetation and geomorphic response, as well as anthropogenic forcing, are explored in the next sections.

12.4 Types, trajectories and vulnerabilities associated with anticipated mass wasting responses to climate change

Landslides could respond rather quickly to emerging climate conditions, even though our knowledge of how recent and past climate change has affected geomorphic processes is rather limited (Boer *et al.*, 1990; Schlyter *et al.*, 1993; Evans and Clague, 1994; Dehn and Buma, 1999; Buma,

TABLE 12.2. *Summary of landslide erosion data from temperate forests in unstable terrain in the Pacific Northwest, North America*

Area	Landslide erosion rate (t $ha^{-1}a^{-1}$)		Reference
	Uncut forests	Clearcut forests	
Oregon Cascades, H. J. Andrews Experimental Forest	1.1	3.2	Swanson and Dyrness, 1975
Oregon Cascades, Maple Creek	0.6	1.5	P. H. Morrison, personal communication 1975
Oregon Cascades, Blue River	0.5	4.2	D. A. Marion, personal communication 1981
Oregon Coast Ranges, Mapleton area	0.19	0.70	Ketcheson and Froehlich, 1978
Oregon Coast Ranges, most unstable soil, Mapleton area	0.28	1.13	Swanson *et al.*, 1977
Oregon Coast Ranges, all unstable soils, Mapleton area	0.32	0.62	Swanson *et al.*, 1977
Klamath Mountains, southwest Oregon	0.56	3.7	Amaranthus *et al.*, 1985
Olympic Peninsula, Washington	0.7	0.0	Fiksdal, 1974
Chichagof and Prince of Wales Islands, southeast Alaska	1.1	2.0	Swanston and Marion, 1991
Coastal southwest British Columbia	0.14	0.32	C. L. O'Loughlin, personal communication 1972
Capilano River basin, southwest British Columbia	1.3	5.9	Brardinoni *et al.*, 2003
Idaho Batholith, Zena Creek	0.07	–	Megahan and Kidd, 1972

2000). Long-term changes in average climate conditions (temperature and precipitation) as well as possible shifts in the frequency of extreme events are expected (IPCC, 2007a). In general, climatic models indicate that both evaporation and precipitation will increase in most regions. While some regions may become wetter, in others the net effect of the temperature-modified hydrological cycle will be a loss of soil moisture. The interplay between changes in antecedent soil moisture and precipitation inputs strongly affects landslide initiation (Sidle and Ochiai, 2006). The effect of climate change on other environmental factors, such as vegetation and soil, may introduce more complex interactions and scenarios related to landslide occurrence.

Increases in annual precipitation have been observed in eastern parts of North and South America, northern Europe and north and central Asia during the past century, while drying patterns have been noted in the Mediterranean, southern Africa and portions of southern Asia. However, given the spatial and temporal variability of precipitation, long-term temporal patterns cannot be established at the present time (IPCC, 2007a). Such general changes in precipitation patterns have limited utility for predicting rapid, dangerous landslides and debris flows because these events are typically triggered by shorter-term precipitation inputs. Such rapid slope and channel failures pose considerable threats to people and property (Sidle and Ochiai, 2006). Slow, deep-seated landslides, which typically are activated by longer-term rainfall or snowmelt, generally do not put lives at risk but may impart significant damage to buildings and infrastructure, as well as delivering chronic sediment loads to streams (Wasson and Hall, 1982; Campy *et al.*, 1998) (Fig. 12.2). Regions where drying trends are predicted will probably experience fewer deep-seated mass movements with time (Table 12.2), but steep hillsides (greater than 40°) may experience higher rates of dry ravel (the movement of individual particles resulting from wetting and drying), a process also exacerbated by declining vegetation cover (Sidle and Ochiai, 2006). Additionally, any increases in freeze–thaw and wetting–drying cycles will promote dry ravel (Fig. 12.3).

Already there is scattered evidence for an increased incidence of extreme precipitation events (Mason *et al.*, 1999; Zhai *et al.*, 1999; Pfister *et al.*, 2000; Fowler *et al.*, 2005; IPCC, 2007a; Khon *et al.*, 2007). However, the linkage of such increases to climate change is clouded by the lack of widespread long-term precipitation records, variable data quality, decadal climatic variability and spatially offsetting effects during El Niño–Southern Oscillation

FIGURE 12.2. Slow, deep-seated mass wasting processes in northwestern California, USA contribute chronic sediment to streams. Timing and rate of movement respond to longer-term precipitation inputs. (a) Soil creep; (b) slump.

FIGURE 12.3. Active dry ravel caused largely by freeze–thaw processes on forested hillslopes in (a) Shizuoka and (b) Gunma, Japan.

FIGURE 12.4. The deep-seated Gros Ventre Slide in Wyoming, USA, initiated in June 1925, experiences seasonal movement, particularly during periods of snowmelt.

(ENSO) events (Easterling *et al.*, 1999; Fu and Wen, 1999; Landsea *et al.*, 1999; Trenberth and Owen, 1999; Pezza *et al.*, 2007). Current predictions of increased severe storms are not spatially explicit enough to be useful in the delineation of regions of increasing risk to shallow or deeper rapid landslides (i.e. landslides that respond to individual storms: Sidle and Ochiai, 2006). The predicted increases in intensity of storms in many mid-latitude regions (e.g. Loaiciga *et al.*, 1996; Leung *et al.*, 2004; Pezza *et al.*, 2007) would lead to an increase in rapid landslide potential (Table 12.2) – but the seasonal period of rainfall-induced landslide susceptibility may decrease in mid-latitude regions (Sidle and Dhakal, 2002).

Winter snowpacks are projected to decline in most regions south of 60° N (Rowntree, 1993; IPCC, 2007a). In western North America, decreases in snowpacks are expected to be greatest along the coastal ranges (60–70%), compared to a *c.* 20% decline in snowpack during this century in the northern Rocky Mountains (Leung *et al.*, 2004). As a consequence, landslides and debris flows triggered by snowmelt may decline and the seasonal movement of deep-seated landslides activated by snowmelt may decrease (Fig. 12.4). Portions of Scandinavia and southern Alaska could also have reduced winter snowpacks (Rowntree, 1993). Mid-latitude mountains of South America and Africa will not experience as much warming and therefore snowpacks and resultant landslides will not be as drastically affected (Nogués-Bravo *et al.*, 2007). Declines in the coldest decile of night-time temperatures and number of frost days in mid-latitude regions from 1951 to 2003 support these predictions for continuing reductions in snowpacks (IPCC, 2007a). Coincident with warming and decreased snowpack scenarios are projected increases in cold season rainfall. Leung *et al.* (2004) estimated a 15–20% increase in cold season extreme daily precipitation in the Cascades and the Sierra Mountains of the western USA; in the Columbia River basin such changes were manifested as more frequent rain-on-snow events. The frequency of shallow, rapid landslides could increase in the future if rain-on-snow events increase in mid-latitude mountains. However, landslides caused by both snowmelt and rain-on-snow depend strongly on the timing of melt related to the thermal conditioning of the snowpack and whether a threshold melt rate will be produced that initiates slope failure. Such detailed precipitation–radiation conditions cannot be obtained from current climate change models.

12.5 Anthropogenic effects on geomorphic processes

12.5.1 Overview of land uses and disturbances

Both widespread land use and more concentrated management activities clearly affect the magnitude, frequency and type of geomorphic processes that occur in mid-latitude temperate forests. Landslides and, to some extent, extreme surface erosion (e.g. gullies) are the primary geomorphic processes that are influenced by land use in temperate forest areas. The major land uses affecting these processes include timber harvesting, roads and trails, forest conversion (to agriculture or pasture), recreation, grazing, fire, urban/residential development and mining. The resultant anthropogenic effects alter the thresholds for both initiation and acceleration of certain landslide types, as well as extreme erosion. Shallow, rapid landslides and debris flows are affected by management practices that decrease rooting strength in the soil mantle, including timber harvesting, forest conversion to agriculture or pasture, and fire. These management practices may also increase soil moisture, but such effects are not as significant as reductions in rooting strength (Sidle and Ochiai, 2006). Both shallow and deep-seated landslides in temperate forests are affected by roads and trails, surface mining and residential development as the result of undercutting steep slopes, overloading slopes with unstable fill material, and concentrating drainage water onto steep slopes (Sidle and Ochiai, 2006). The period of movement of deep-seated landslides may be extended due to reductions in evapotranspiration (and therefore wetter soils) associated with conversion of forest vegetation and, in the shorter term, following fire and forest harvesting (with subsequent regeneration). Such changes may cascade down through terrestrial/aquatic ecosystems, generating cumulative effects related to eroded materials and transported sediment (Sidle and Hornbeck, 1991; Dunne, 1998). Here, three important examples of anthropogenic effects are presented; two represent widespread land use effects that influence mass wasting and severe surface erosion (forest conversion to pasture and timber harvesting) whilst the third describes the severe consequences of a concentrated human disturbance (mountain roads). The extent to which various aspects of these management activities and land uses affect slope stability and severe surface erosion are discussed.

12.5.2 Example of forest conversion to pasture: New Zealand

Widespread clearance of woody vegetation with subsequent conversion to pasture or other weak-rooted species has exacerbated landslide and gully erosion during the past few centuries in many mid-latitude regions of the world (e.g. Fairbairn, 1967; Rice *et al.*, 1969; Kuruppuarachchi and Wyrwoll, 1992). An economically significant and highly studied example of such land cover change and degradation is the widespread forest conversion that has occurred in New Zealand. Prior to human occupance, forests covered most of the land area below the alpine treeline; estimates of forest cover for the entire country range between 75% and 82% (Leathwick *et al.*, 2004). The arrival of Polynesians (Maori) around AD 800 initiated widespread clearing of indigenous forests by fire to encourage the growth of bracken fern (*Pteridium aquilinum*) for subsistence, to facilitate hunting of the flightless ratite moa birds and to expedite travel within the country (Blaschke *et al.*, 1992; Ewers *et al.*, 2006). By the time European settlers arrived in the early nineteenth century, about half of the lowland forest cover had already been converted or destroyed (McGlone, 1989; Ewers *et al.*, 2006). Based on pollen and diatom analysis of Holocene lake bed sediments together with tephra chronology and historical evidence from a landslide-prone region of North Island, the total erosion resulting from

FIGURE 12.5. Extensive shallow landslides in young radiata pine plantations in North Westland, New Zealand following a large storm in the late 1970s. Forest cover was previously podocarp beech.

this cover conversion is estimated to have increased five- to sixfold compared to rates when forests dominated the region (Page and Trustrum, 1997).

The colonisation by Europeans increased the rate of forest clearing, primarily for conversion to pasture land. This rate of forest conversion peaked in the early twentieth century. Overall, Page and Trustrum (1997) estimated that European forest conversion to pasture land increased total erosion by 8- to 17-fold compared to indigenous forests, at least twice the impact of the Maori forest conversion. Despite the overwhelming evidence of increased landslide activity in formerly forest land converted to pasture, studies on the east coast of North Island infer that dense surface mats of pasture roots may afford some protection against shallow incipient earthflows (Preston and Crozier, 1999). Extensive plantations of radiata pine (*Pinus radiata*) were established in the 1960s, often on steep slopes for erosion control. While these plantation forests reduced landslide erosion compared to pasture lands (e.g. Eyles, 1971; Fransen and Brownlie, 1995), the weaker root systems of radiata pine made these forests more susceptible to landsliding during major storms compared to native podocarp beech forests (O'Loughlin and Ziemer, 1982; Sidle *et al.*, 1985; Marden and Rowan, 1993). In North Westland, South Island, 97% of the landslides that occurred during the 7-year period from 1974 to 1981 (primarily during two large storms) were in recently clearcut podocarp beech forests that were replanted to radiata pine (O'Loughlin *et al.*, 1982) (Fig. 12.5). Earlier investigations in this area indicate that landslide rates in podocarp beech forests did not exceed 1.2 t ha^{-1} a^{-1}, while rates in recently converted radiata pine forests have ranged from 13 to 48 t ha^{-1} a^{-1} (O'Loughlin and Pearce, 1976).

A detailed examination of the potential delivery of sediment to channels from shallow landslides and gullies in the Weraamaia catchment, Raukumara Peninsula, North Island, revealed that large gullies remain coupled to channels for up to 100 years while shallow landslides become decoupled within 10 years (Kasai *et al.*, 2005). As a result, reforestation of these degraded areas with radiata pine in the 1960s reduced sediment delivery from gullies to a greater extent than delivery from shallow landslides on hillslopes. During Cyclone Bola in 1988 (the 1-in-100 year event), only subsurface drainage under pasture cover formed large gully complexes, whereas many shallow landslides occurred on both forested and grassland hillslopes (Kasai *et al.*, 2005). In this catchment, much steeper slope gradients were required for gully initiation under forest cover compared to pasture (Parkner *et al.*, 2006). In a nearby area (Mangatu Forest), gully erosion affected *c.* 4% of the landscape; following reforestation with Douglas fir (*Pseudotsuga menziesii*), over a period of 24 years, the area affected by gullies was reduced to *c.* 1.5%. However, most of the largest gullies were not fully stabilised after reforestation (Marden *et al.*, 2005), lending support to the concept that gullies are important long-term, chronic sources of sediment to streams (Kasai *et al.*, 2005). Future increases in total precipitation and possible storm intensities predicted for the South Island may further exacerbate landslide and gully erosion on converted lands.

The extent of damage to the landscape incurred by forest conversion in New Zealand has been the focus of several investigations, with an initial comprehensive economic appraisal conducted by the New Zealand Ministry of Works and Development in the mid-1960s in a large area (6500 km^2) of eastern Raukumara Peninsula. Here the conversion of native forests to pasture caused severe landslide and gully erosion after the 1930s, resulting in agricultural declines, aggradation of sediment in streams and rivers, and subsequent flooding and damage to many bridges and roads (New Zealand Ministry of Works and Development, 1970; Kelliher *et al.*, 1995). Options for various conservation practices were assessed along with the continuation of pastoral farming; all had negative present values and rates of monetary return, regardless of the discount rate employed. Pasture production on young (*c.* 2-year-old) landslide scars was only 20% of the production on uneroded sites, while pasture productivity on 15–40-year-old scars gradually increased to 75% of undisturbed rates (Lambert, 1980; Trustrum *et al.*, 1983). During the dry season, the 40-year-old landslide scars yielded only half the pasture production of uneroded sites. Studies in similar terrain found that 7–8-year-old disturbed sites produced only half the forage compared to uneroded sites (Garrett, 1980). These data indicate that the average pasture production during the 15 years after landslide erosion is less than half of the production on uneroded sites and, for several decades thereafter, will only be about 75% of the production on uneroded areas. This estimate corresponds to modelling predictions of a 70% decline in pasture productivity of landslide-prone terrain which stabilises after about 100 years (Luckman *et al.*, 1999). Trustrum and DeRose (1988) likewise noted that productivity of converted pasture land would reach an approximate steady state (albeit at lower than pre-erosion levels) due to thinner soils that would develop under pasture cover which are inherently less susceptible to landslide erosion.

Data from the Wairarapa district during the wet winter of 1977 showed that landslides eroded about 4% of the 1400-km^2 area. This immediate loss of pasture equated to *c.* NZ $600 000 (1980 prices) in the first year (Hawley, 1980; Trustrum and Stephens, 1981). The total cost of pasture lost from this area due to landslides during the subsequent 40 years was estimated at more than NZ$9 M. In the Wairarapa study sites, long-term stripping of the soil mantle by landslides eroded 41–56% of the total area. Such extensive mass erosion could reduce New Zealand's export revenue as much as 5%, or several hundred million dollars (Sidle *et al.*, 1985). The severity of this erosion problem in New Zealand where considerable erosion control investments have been made provides a warning for developing nations intending to convert hillslope forests to agriculture.

Future landslide and severe gully erosion in these forests and pasture lands of New Zealand may be affected by climate change, but estimated changes are highly uncertain. The findings of the Fourth Assessment Report of the IPCC show some inconsistencies related to landslide hazards (IPCC, 2007b). While the Report notes that the 'frequency of heavy rainfall is likely to increase, especially in western areas' and that 'rain events are likely to become more intense', the only evidence presented for tropical cyclones shows that neither the frequency nor the intensity of the large storms that instigate landslides and gully erosion has increased during the period from 1970 to 2006 (Burgess, 2005; Diamond, 2006). Given the projected increases in atmospheric carbon dioxide concentrations, mean temperatures (0.2 to 4.0 °C), and evaporation demands during the century following 1990 (Hennessy *et al.*, in IPCC, 2007b), it is likely that climate change could actually reduce shallow landslides due to increased vegetation growth (in areas that are not water-limited) and the drier soil moisture conditions that will precede storms. This concept contrasts with the general landslide scenarios presented by the IPCC (2007b). Increased fire hazard caused by higher temperatures would increase shallow landslide and gully erosion potential, but to an unknown extent. The future increases in total annual rainfall predicted for western New Zealand (IPCC, 2007b) will probably increase the period of activity of deep-seated landslides (e.g. earthflows), while the lower total rainfall expected in eastern North Island and northern South Island will decrease deep-seated landslide movement rates there. Deep-seated landslide movement rates in the eastern portion of South Island will probably be unchanged in the near future. These adverse effects of climate change on deep-seated landslides will probably be only marginally impacted by changes in vegetation cover, since failure planes are well below the rooting depth of most vegetation; the effects of vegetation will only be related to the benefits of evapotranspiration (Sidle and Ochiai, 2006). The reduction in the number of frost days noted since 1950 (Salinger and Griffiths, 2001) could reduce the rates of dry ravel on exposed slopes, although such changes are expected to be minimal at higher elevations where dry ravel is most prevalent.

12.5.3 Example of forest harvesting effects: Pacific Northwest, North America

Forest harvesting increases the likelihood of landslide initiation by: (1) temporarily increasing water inputs and soil moisture because of decreased evapotranspiration and changes in the volume and rate of snowmelt; and (2) deterioration in the root strength of harvested trees (Sidle and Ochiai, 2006). The first factor is not particularly important for most landslides that occur in temperate forests during an extended rainy season because soils are generally very wet and evapotranspiration is minimal at this time. However, decreases in evapotranspiration after timber harvest could extend the 'window of susceptibility' for landslide activity in temperate forests (Swanson and Swanston, 1977; Sidle *et al.*, 1985). Root strength deterioration following tree removal appears to affect shallow landslides to a much greater extent than changes in soil moisture. Field investigations in temperate forests indicate that root strength reaches a minimum about 3–15 years after timber harvesting depending on the rates of root decay and regeneration of various species (O'Loughlin and Ziemer, 1982; Sidle, 1991, 1992) (Fig. 12.6). Independent tests to assess the effects of timber harvesting on root strength, including mechanical straining of different sizes of roots and shearing of *in situ* soils and soil columns, have confirmed empirical field observations of higher landslide frequencies after timber harvesting (Burroughs and Thomas, 1977; O'Loughlin and Watson,

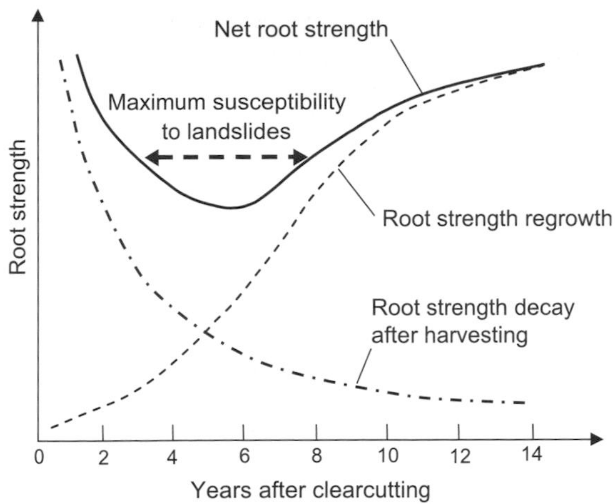

FIGURE 12.6. Changes in net rooting strength after clearcutting (modified from Sidle and Ochiai, 2006).

FIGURE 12.7. Evidence of landslides and debris flows that initiated during a large storm in October 1961 within an area that had been clearcut 6 years previously. These highlighted sites are now colonised by alder (photograph taken in 1998).

1979; Abe and Ziemer, 1991). Additionally, recent catchment-scale models have confirmed that the period of maximum landslide susceptibility is between 3 and 15 years after clearcutting (Sidle and Wu, 1999; Dhakal and Sidle, 2003). In the forests of continental interiors where substantial snowpacks accumulate, large canopy openings created by harvesting allow more snow to accumulate on the ground and generally promote a more rapid melt rate, leading to a higher probability of both shallow landslides and possibly an earlier and longer period of activity of deep-seated landslides (Sidle and Ochiai, 2006).

The most comprehensive field investigations of the effects of clearcut harvesting on landslide erosion rates have been conducted in the Pacific Northwest of North America (Sidle *et al.*, 1985; Sidle and Ochiai, 2006). One of the first studies to directly link timber harvesting with accelerated landsliding was in a large clearcut on Prince of Wales Island, Alaska (Bishop and Stevens, 1964). Here landslides persisted for about 9 years after clearcutting, with more than half occurring during a large storm 6 years after harvesting (Fig. 12.7). The area affected by landsliding during this period was five times the estimated area disturbed by landslides during a 100-year period before logging. The best harvesting–landslide data available are based on aerial photo interpretation combined with field surveys that cover the period of root strength decay and recovery (i.e. 15–35 years) and a wide range of potential triggering storms. Such studies, summarised in Table 12.2, show that landslide erosion rates from unharvested forests range from about 0.1 to 1.3 t ha^{-1} a^{-1}. Corresponding landslide erosion in recent clearcuts range from 0 to 5.9 t ha^{-1} a^{-1}. Overall these investigations showed that landslide erosion rates from clearcuts in Pacific Northwest temperate forests were 1.8 to 8.7 times higher than in undisturbed neighbouring forests, an average increase of about fourfold; the only exception was a rather short-term investigation in the Olympic Peninsula of Washington (Table 12.2). Virtually all of these studies were concerned with shallow, rapid landslides.

Many other studies in temperate forests of the Pacific Northwest have assessed shallow landslide frequency following large storms. Such event-based data must be interpreted cautiously because without knowledge of past storm and landslide histories it is problematic to assess the effect of forest harvesting on landslides in such a narrow time window. During a large storm in November 1975 in the Oregon Coast Ranges, 77% of the landslides occurred in clearcuts and landslide frequency was 23 times higher in clearcuts than in undisturbed forests (Gresswell *et al.*, 1979). Following two major storms in 1996 in western Oregon landslide density was higher in three out of four recently clearcut forests (0–9-year-old stands: 5.0–8.1 landslides km^{-2}) compared to mature forests (100+-year-old trees: 2.1–5.2 landslides km^{-2}) (Robison *et al.*, 1999). In the Queen Charlotte Islands, British Columbia, landslide frequency was 14 times higher in clearcuts compared to natural forests during a relatively large 5–10-year return period storm (Schwab, 1983).

Other studies have assessed the frequencies of landslides for different forest practices and different site conditions in the Pacific Northwest; erosion rates are, however, not available. In the Klamath Mountains of northwestern California, forest harvesting (including roads) in vulnerable inner valley gorges accelerated landslide frequency by 11- to 26-fold compared to other lands managed for timber (Wolfe and Williams, 1986). A survey in the Queen Charlotte Islands, British Columbia, revealed that average landslide frequency was 10 landslides km^{-2} in all clearcuts, with frequencies of 22.4 and 36.7 landslides km^{-2} in clearcuts in the two most sensitive terrain categories (Schwab, 1988). Montgomery *et al.* (2000) measured 5.8 landslides km^{-2} a^{-1} in a small clearcut near Coos Bay, Oregon, compared to an estimated 0.01–0.03 landslides km^{-2} a^{-1} in pre-logging history in that area. Clearcuts in the unstable Clayoquot Sound region on Vancouver Island, British Columbia, produced an estimated 0.053 landslides km^{-2} a^{-1} compared to 0.0055 landslides km^{-2} a^{-1} in unlogged forests (Jakob, 2000). In three other catchments on Vancouver Island, Guthrie (2002) found that clearcuts produced 0.026 landslides km^{-2} a^{-1} compared to 0.003 landslides km^{-2} a^{-1} in natural forests. Few field studies have assessed the effects of partial cutting and stand tending on landslide occurrence. A survey in the northern California Coast Range indicated that landslide density was about 5- to 9-fold higher in clearcuts compared to thinned and unthinned second-growth forests (T. P. Rollerson, personal communication, 2003), probably due to reinforcement by residual and regenerating roots. Landslide frequency in Idaho increased only slightly as overstorey crown cover decreased from 100% to 11%; however, for crown covers greater than 11% major increases in landslide frequency were found (Megahan *et al.*, 1978).

Almost all of the effects of forest harvesting discussed in this example are related to shallow, rapid landslides where the deterioration of woody roots greatly affects landslide erosion. Despite much discussion on the possible effects of forest harvesting on increased water content in shallow soils and resultant landslide initiation, the only comprehensive study in this region found that maximum piezometric response was little affected by clearcut harvesting during large winter storms (Dhakal and Sidle, 2004a). The effects of forest harvesting on deep-seated landslides is less important because root systems generally do not penetrate through the entire regolith. However, little research has been conducted to document the effects of timber harvesting on deep-seated landslides in the Pacific Northwest. As natural rates of deep-seated landslides respond to seasonal increases in soil moisture (Swanson and Swanston, 1977; Iverson and Major, 1987; Swanston *et al.*, 1995), higher water contents after timber harvesting should accelerate these processes. In the near-coastal areas of the Pacific Northwest, where winters are dominated by rain, increases in soil moisture caused by vegetation removal occur primarily in the autumn and spring and thus may extend the natural winter period of deep-seated landslides. For continental interior sites, where substantial snowpacks accumulate, forest harvesting may cause short-term increases in slump–earthflow activity until a sufficient forest cover regenerates to augment snow interception. A short-term study at a forested site in southwestern Oregon showed that soil creep approximately doubled during the second winter after logging (Swanston, 1981).

The effect of future climate change in temperate forests of the Pacific Northwest on landslide erosion is complicated by spatial uncertainties of regional climate models in this complex terrain (IPCC, 2007a). Higher-elevation sites, like the Cascades, will probably receive less total winter precipitation and therefore support reduced snowpacks. As such, both snowmelt-triggered shallow landslides and deep-seated landslides that are activated by melt should decrease in the future. The general warming trend and relatively unchanged precipitation scenarios projected for autumn months in this region (IPCC, 2007a) will also tend to decrease shallow, rainfall-triggered landslides and reduce the period of activation of deep-seated landslides in coastal mountains. This is because antecedent soil moisture will be lower at the onset of the winter storm season. Using a regional downscaling approach, Leung *et al.* (2004) predict that warming will result in increased episodes of winter rainfall at the expense of reduced snowfall in the Pacific Northwest and that there will be more frequent rain-on-snow events in the Columbia River basin. Increases in extreme rain-on-snow events will increase landslides by this triggering mechanism, with the extent being dependent upon snowpack conditions (which arguably will contain less water in a warmer climate). Thus, the unilateral negative implications related to climate change–landslide interactions in this region contained within the Fourth Assessment Report of the ICCP (2007b) appear unjustified. Where, and if, rain-on-snow landslides do increase, they will be exacerbated by timber harvesting.

12.5.4 Example of mountain roads: northern Yunnan, China

Mountain roads destabilise hillsides by: (1) altering natural hydrologic pathways and concentrating water onto unstable portions of the hillslope; (2) undercutting unstable slopes, thus removing support; and (3) overloading and oversteepening fillslopes, including the road prism (Sidle and Ochiai, 2006). The relative importance of these destabilising factors depends upon the design and construction standards of the road and its associated drainage system, as well as the natural instability of the terrain through which the road is excavated. Typically, roads constructed in mid-slope locations are the most unstable because of the combination of steep slopes, large quantities of intercepted water and unstable fill-and-cut materials that need to be disposed or incorporated into the fillslope (Megahan *et al.*, 1978; Wemple *et al.*, 2001; Sidle and Ochiai, 2006). These roads also contribute sediment via surface erosion from exposed cutslopes and fillslopes, as well as the running surface itself (if unpaved) (e.g. Megahan and Ketcheson, 1996). The effects of roads on these geomorphic processes is particularly severe in the mountainous regions of developing countries where often little attention is paid to proper road location, construction techniques, erosion control measures and maintenance (Bansal and Mathur, 1976; Haigh, 1984; Arnez-Vadillo and Larrea, 1994; Sidle *et al.*, 2006).

In the past decade there has been a significant expansion of both major and small (unpaved) roads in the mountainous terrain of northwestern Yunnan Province, China. While this area is marginally below the temperate zone (latitude *c.* 27–29°), elevation and vegetation cover characteristics typify lower, mid-latitude forests, rangelands and converted hillslopes in this part of Asia. Roads in northwestern Yunnan are being built at a rapid pace with little attention paid to location, construction practices and erosion control, due to the recent surge in economic development and tourism in the region (Krongkaew, 2004; Nyaupane *et al.*, 2006). While such papers extol the importance of these road systems for economic development of the region, virtually no mention is made of the environmental consequences related to potential increases in surface erosion and landslides. The new roads link towns, remote villages, agricultural regions, hydropower plants and mines to cities. Most of the small mountain roads are unpaved and constructed with virtually no engineering standards, in many cases simply blasting into unstable bedrock on the hillsides. Within about a 30-km distance in northeastern Yunnan lie the headwaters of three great river systems: the Salween, Mekong and Jingsha rivers, the former two flowing through other Southeast Asian nations, the latter the upstream reach of the Yangtze River. While land cover in this area has been modified, the roads appear to contribute much more sediment to the headwaters of these major rivers than other land uses. Most of this region was originally covered by forest and shrub forests, but increasing forest clearing and conversion to agriculture, grazing and shifting cultivation has occurred in the past 100 years, as well as forest clearing to produce steel instituted during the Great Leap Forward in 1958 (Xu and Wilkes, 2002). Although some reforestation of agricultural lands is now being encouraged by the provincial government, much steep land near and adjacent to rivers is still under intense cultivation and the remaining forests are not being actively managed, following the 1998 logging ban. Nevertheless, these converted lands appear to have only a small effect on sediment delivery to headwaters compared with roads and trails in the region (Plate 37).

As an example of the level of erosion and sediment produced from these new and expanding road systems in the region, Sidle *et al.* (unpublished data) surveyed a 23.5-km section of the Weixi–Shangri-La road that was constructed in 2002 through steep mountainous terrain along the headwaters of the Mekong River. While another access road existed that generally followed the ridge line, the new road is shorter by ~13 km and more passable during sporadic winter snows. The new road was blasted into the weathered ignimbrite bedrock along the steep slopes, exposing cutslopes up to 80 m high and depositing the waste rock and soil onto the oversteepened fillslopes. Because the steep and uniform hillsides below the road are directly connected to the incised tributary of the upper Mekong River, an estimated 80–95% of the sediment contributions within this reach can be attributed to the road (Sidle, 2007) (Fig. 12.8). Estimated mass erosion during the 4-year period after road construction was a staggering 9608 t ha^{-1} a^{-1}, with a rate of 33 451 t ha^{-1} a^{-1} for the most severely eroded 6-km section of the road. These road-related landslide rates (for the road right-of-way) are the highest ever reported (Sidle and Ochiai, 2006). Average surface erosion rates for the 4-year period were also very high (765 t ha^{-1} a^{-1}), but more than 12-fold lower than landslide rates.

These levels of erosion, and their respective contributions to the Mekong River headwaters, strongly bring into

FIGURE 12.8. Epic levels of landslide erosion along the newly constructed Weixi–Shangri-La road in Yunnan, China.

question development practices in this region. The landslide erosion rates alone measured along the Weixi–Shangri-La road were on average 178 times higher than average rates of landslide erosion (~55 t ha^{-1} a^{-1}) along forest roads in unstable terrain of western North America (Sidle *et al.*, 1985; Sidle and Ochiai, 2006). Similar erosion and sedimentation scenarios are occurring in association with the expanding road networks in the headwaters of the Salween and Yangtze rivers (Plate 38). Such unsustainable practices exert both local impacts, as well as downstream, transnational consequences on the poorer nations of Myanmar, Laos, Thailand, Cambodia and Vietnam. As noted by Sidle *et al.* (2006), not only roads but also trails contribute to high erosion rates. Such trails on steep hillsides are common throughout northwestern Yunnan. When roads and trails are cut into steep slopes that are directly connected to streams, they can be expected to contribute the bulk of the landslide and surface erosion sediment that reaches channels.

Given the extreme nature of the alteration of topography and hydrology by poorly located and constructed mountain roads and trails in northwestern Yunnan, the additive consequences of climate change will likely be minor. Nevertheless, the increased frequency in occurrence of intense rainfall events that has been noted in this region (Cruz *et al.*, 2007), together with the predictions (albeit uncertain and complicated by topography) for increased precipitation during all but the winter seasons (IPCC, 2007a), may further increase road-related landslides and extreme surface erosion from roads and trails in this region in the future.

12.5.5 Comparison of climate-induced and anthropogenic-induced geomorphic change

Overall it is difficult to predict the effects of potential long-term climate changes on landslide and extreme surface erosion activity because these events largely depend on the timing of large-magnitude storms. As noted earlier, long-term changes in average climate conditions (temperature and precipitation) as well as possible shifts in the frequency of extreme events are expected as a result of climate change, but predicting timing and magnitude of extreme events is difficult. While some general scenarios can be estimated (Table 12.1), the extent to which such changes are realised will determine whether they play important roles in influencing landslides in various regions of the world (Evans and Clague, 1994; Wyss and Yim, 1996; Buma, 2000). Also, based on past experience, it is difficult to relate past climate change scenarios to landslide activity (e.g. Innes, 1997). The effect of climate change on other environmental factors, such as vegetation and soil, may introduce more complex interactions and scenarios related to landslide occurrence. On the other hand, the effects of land cover change, road construction, timber harvesting and other land management practices on both surface and mass erosion are much better understood, and such effects have been shown to be real and very damaging to the environment (e.g. Singh, 1998; Slaymaker, 2000; Sidle and Ochiai, 2006; Sidle *et al.*, 2006). Thus, a higher priority must be placed on understanding land use–landslide/surface erosion interactions and applying this knowledge to the management of temperate forests in mountainous terrain (Slaymaker, 2001; Sidle and Ochiai, 2006).

12.6 Techniques for assessing effects of anthropogenic and climate-induced mass wasting

12.6.1 Empirical approaches and models

Multifactor empirical approaches to landslide prediction typically estimate the relative landslide hazard using the relations between past landslide patterns with various site characteristics. Thus, the weighting of site characteristics that affect slope stability is very important. Factors typically considered include topography, geology, vegetation cover or land use, hydrology and soil properties. Trigger mechanisms, such as rainfall and seismic patterns, are usually not included because such hazard assessments focus on conditions predisposing hillslopes to failure. Landslide assessments that use qualitative factor weightings based on professional judgement can be very effective if high-quality and systematically collected distributed data are available, together with adequate professional expertise to interpret these data (e.g. Newman *et al.*, 1978; Nilsen *et al.*, 1979); however, if such data and expertise are not available, the derived factor weighting estimates may vary considerably and lack objectivity. Geographical information systems (GIS) allow for accurate and unbiased development of weighting factors typically used in such analyses (e.g. Carrara *et al.*, 1991; Soeters and van Westen, 1996; Dhakal *et al.*, 1999); however, such derived weighting factors are only as good as the databases from which they are developed, as well as the errors in the cause–effect relationship implicit in such generalisations (Sidle and Ochiai, 2006).

In a simple GIS-based landslide hazard analysis, Gupta and Joshi (1990) mapped recent and old landslides on aerial photos in the Lower Himalayas and overlaid this information on geological maps, remotely sensed maps of land use and maps of major faults and thrust zones. Distance of existing slope failures from major tectonic features and slope aspect were used as surrogates for susceptibility to earthquake-triggered landslides. Other parameters used in the hazard analysis included lithology and land use. Each of the four geoenvironmental parameters (lithology, land use, distance from tectonic features and slope aspect) were equally weighted in this analysis, and the percentages of landslides in each geoenvironmental subcategory were computed and compared against the average landslide frequency. If subcategory values constituted >33% of the overall average value, they were weighted as high risk (2); values <33% lower than the average were weighted as low risk (0); and values in the range of ±33% of the mean were weighted as moderate risk (1) (Gupta and Joshi, 1990). While this methodology apparently focussed on earthquake-triggered landslides, this region also experiences rainfall-induced landslides. This exposes a major problem in such statistical analysis. This is the inability of such methods to distinguish between landslide-triggering mechanisms and different types of landslides that are sensitive to different triggering conditions. The inclusion of some type of precipitation data or threshold indices in such analyses would better link landslide occurrence to climate initiation processes and future climate change scenarios. In a retrospective study that assessed the effects of climate change on landslide reactivation, Buma (2000) found that a semi-empirical model of net precipitation successfully predicted episodes of landslide movement, based on a threshold of 3-month net antecedent precipitation. Such parameters could be included on a seasonal basis in landslide models and modified according to plausible climate change scenarios to assess the potential impacts of climate change on larger landslide reactivation. Likewise, triggering mechanisms for shallow landslides (e.g. rainfall and snowmelt) could be incorporated into empirical landslide hazard analyses, based on regional estimates of probabilities of rainfall intensity and total precipitation amounts, and regional snowpack data. Expected changes in these precipitation inputs due to climate change could then be incorporated into such analyses. More detailed hazard analysis should consider the potential of weighting each of the criteria based on local knowledge and relations to landslide intensity.

More sophisticated multivariate approaches to empirical landslide hazard analysis consider the interrelations amongst factors in terms of selection and weighting. Once all important parameters have been inventoried at appropriate scales, the presence or absence of landslides is then determined. Multiple regression or discriminant analysis is then typically used to analyse the resulting matrix (Mulder and van Asch, 1988; Carrara *et al.*, 1991; Rollerson *et al.*, 1997; Dhakal *et al.*, 2000). In some recent cases, neural network methods have been applied to weight causative factors (e.g. Lee *et al.*, 2004; Yesilnacar and Topal, 2005). While these methods employ increasingly sophisticated GIS, remote sensing and statistical/analytical tools, there appears to be a tendency to focus more on new methods rather than on trying to understand causal linkages for specific types of landslides (e.g. Varnum *et al.*, 1991; Guillande *et al.*, 1995; Lee *et al.*, 2004). In addition, one of the clear advantages of these analytical methods (unbiased factor selection and weighting) can also be a disadvantage because it may ignore field-based geomorphic and geotechnical expertise in such assessments (Rollerson *et al.*, 1997). To significantly improve multifactor landslide hazard assessments, three major issues need to be overcome: (1) methods need to be developed that can be applied in broader geographical regions or areas which experience different types of landslides (using different statistical analyses for different landslide types); (2) a clear focus needs to be placed on the underlying processes that relate to slope failure (e.g. rainfall–pore water response versus earthquakes); and (3) temporal as well as spatial attributes of landslide susceptibility need to be incorporated into the analysis (Sidle and Ochiai, 2006; van Westen *et al.*, 2006). Additionally, a major challenge is the need to better link specific land management activities into empirical landslide analyses. At present few multifactor empirical procedures address land management issues in a meaningful way, except for the inclusion of very general land cover classes (e.g. Kienholz *et al.*, 1984; Anbalagan, 1992). Finally, to incorporate climate change into empirical landslide models quantitatively, much better spatially explicit precipitation forecasts are necessary. These forecasts need to account for topographic complexity, as well as the effects of changes in precipitation patterns, on triggering thresholds for different types of landslides (van Westen *et al.*, 2006). Without such advances, only qualitative precipitation scenarios can be considered that have little relevance for landslide hazard prediction except at the broadest regional scale.

12.6.2 Physically based models

Distributed, physically based landslide models have been typically used to assess shallow, rapid landslides in relatively steep terrain based on a factor of safety analysis. Distributed, physically based landslide models have two unique requirements: (1) spatially and, in some cases, temporally (e.g. rooting strength) distributed model parameters are necessary; and (2) the model output must be spatially and temporally explicit because of the need to know the locations and timing of landslides (Sidle and Ochiai, 2006). The two models described here (SHALSTAB and dSLAM/IDSSM) have both been developed for, and applied to, shallow landslide problems in steep terrain occupied by temperate forests.

Montgomery and Dietrich (1994) developed a distributed, physically based landslide model (SHALSTAB) which couples digital terrain data with near-surface throughflow (i.e. TOPOG: O'Loughlin, 1986) and slope stability models. For simplicity, the model generally assumes that soils are cohesionless, slope-parallel subsurface flow occurs, unit weights of soils in the saturated and unsaturated zones are equal, and ignores the effects of vegetation root strength (Dietrich *et al.*, 2001). As such, conditionally unstable slopes are designated as those where the slope gradient equals the internal angle of friction of the soil. As soil mantles begin to saturate, the critical angle for failure decreases. An underlying assumption of SHALSTAB is that sites with the lowest ratios of effective precipitation to soil transmissivity are the least stable; this relationship holds well in many areas where SHALSTAB has been applied: in northern California, Washington and Oregon (Montgomery and Dietrich, 1994; Dietrich *et al.*, 2001). Further applications reveal that SHALSTAB frequently overpredicts landslides and performs best in steep catchments underlain by shallow bedrock and worst in less steep catchments underlain by thick glacial deposits (Montgomery *et al.*, 1998; Borga *et al.*, 2002; Fernandes *et al.*, 2004). This finding underlines the importNt of accurate representation of soil depth in such models. Because of the necessity to use steady-state rainfall inputs, SHALSTAB has not been tested for conditions where actual landslides are triggered during actual rainfall events.

A distributed, physically based shallow landslide model (dSLAM, later revised as IDSSM) that can assess the spatial and temporal effects of timber harvesting on slope stability incorporates: (1) infinite slope analysis; (2) continuous temporal changes in root cohesion and vegetation surcharge; and (3) stochastic influence of actual rainfall patterns on pore water pressure (Wu and Sidle, 1995; Sidle and Wu, 1999; Dhakal and Sidle, 2003). A root strength model developed by Sidle (1991) which simulates root decay and regrowth following timber harvest is used together with a vegetation surcharge model (Sidle, 1992) to simulate removal of tree weight and subsequent regrowth. The TAPES-C model was adapted in the topographic analysis to partition the catchment into relatively homogeneous elements because the 'stream-tubes' (TOPOTUBES) in this model are consistent with subsurface hydrologic and geomorphic processes (Moore *et al.*, 1988; Dhakal and Sidle, 2004a). Rainfall is applied as synthetic sequences or individual events. Only average values of input parameters are used for various spatially distributed parameters. The model was successfully tested in two steep, forested basins (1.18 km^2 and 1.12 km^2) in the Cedar Creek drainage of the Oregon Coast Ranges where a large storm in November 1975 caused widespread landsliding in the region (Wu and Sidle, 1995; Sidle and Wu, 1999). Simulated volumes (733 m^3 and 801 m^3) and numbers (four and seven) of landslides in the two basins agreed closely with values (734 m^3 and 749 m^3 and three and six respectively) measured in the field after the 1975 storm (Wu and Sidle, 1995, 1997). More recent applications of IDSSM in Carnation Creek, British Columbia, showed that partial cutting reduced landslides by 1.4–1.6-fold compared to clearcutting (Dhakal and Sidle, 2003) and that landslide occurrence was influenced by storm characteristics, including mean and maximum hourly intensity, duration, total rainfall and the temporal distribution of short-term intensity (Dhakal and Sidle, 2004b).

Although distributed, physically based models represent the most powerful landslide hazard analysis tools, their widespread application remains limited because of the high distributed data requirements (including digital elevation models: DEMs), expertise with GIS, and computer modelling. While some input data can be obtained from remote images and extracted from DEMs, these geotechnically based models require accurate distributed data on soil depth and other critical soil properties to be effective. Such data are typically not readily available. However, distributed, physically based landslide models have two major advantages over empirical approaches to landslide analysis: (1) they can directly incorporate rainfall–pore pressure dynamics, so that they have the potential to be incorporated into real-time warning systems for landslides; and (2) they can be used to evaluate long-term scenarios of vegetation cover and forecasts of climate change, related to spatial and temporal distribution of landslides. Notwithstanding the spatial problems related to current rainfall prediction capability in mountainous terrain (IPCC, 2007a), this method holds promise for long-term landslide erosion estimates under changing climates. Empirical analyses, on the other hand, are useful for landslide susceptibility mapping in data-poor regions of the world.

12.6.3 Do existing technologies and models still apply in a changing environment?

Both widespread land use activities and more concentrated disturbances affect the significance of severe surface erosion as well as the magnitude, frequency and type of landslides that occur in many mid-latitude temperate forests and steppes throughout the world. These anthropogenic activities can alter the thresholds for both initiation and acceleration of certain landslide types as well as gully erosion. Thus, empirical assessments and predictive models must be able to capture the manifestation of these anthropogenic and climate changes on the mechanisms that initiate and propagate extreme surface erosion and landslides. For example, an indirect, but important, consequence of climate change is the likely impact on the species composition of one-third of the world's existing forest (Acosta *et al.*, 1999). It is predicted that in some areas entire forest types may disappear and new ecosystems may become established. Vegetation species may be forced to migrate to higher elevations in response to predicted warming trends, although such changes in western North America may be very subtle (Peterson, 1998). Shifts in temperature and precipitation may reshape the boundaries between grasslands, shrublands, forests and other ecosystems. Such changes will affect the potential for severe erosion and landslides in mid-latitude regions by altering protective ground cover and deep rooting strength. Thus, it is imperative that empirical assessments and models capture such processes in simulations of extreme surface erosion and landslides under changing climatic and anthropogenic scenarios. Other dynamic processes that need to be represented in such predictive tools include the ability to represent changes in magnitude and frequency of precipitation (especially rainfall intensity) (Dhakal and Sidle, 2004b), changes in channel conditions that affect routing of sediment (due to changes in riparian vegetation, allochthonous inputs and channel structure) (Gomi *et al.*, 2002), changes in freeze–thaw and wetting–drying cycles that influence surface erosion and ravel (Sidle and Ochiai, 2006), and changes in weathering of soils and parent material that affect regolith strength (Chigira, 2002). Finally, downstream sedimentation resulting from increases in surface and landslide erosion reduces channel conveyance capacity and, together with the removal of frictional resistance in headwater channels impacted by debris flows, has strong implications for increased flooding (Lu and Higgitt, 1998; Thomas and Megahan, 1998; MacFarlane and Wohl, 2003). Thus, it is important to consider the spatial and temporal linkages between terrestrial and fluvial geomorphic processes in temperate ecosystems (e.g. Gomi *et al.*, 2002; Benda *et al.*, 2004).

12.7 Summary and conclusions

The likely impacts of possible climate change scenarios on landslide erosion outlined in Table 12.2 are in stark contrast to the forecasts inferred by the IPCC in their Fourth Assessment Report (e.g. Cruz *et al.*, 2007; IPCC, 2007b). Based on an understanding of landslide-triggering processes, it is estimated that six of the climate change scenarios would increase susceptibility to shallow landslides, while four would decrease susceptibility and two would exert no substantial change (Table 12.2). For slow, deep-seated landslides, only four climate change scenarios would increase susceptibility or movement rates, while six would decrease susceptibility and two would impose no significant change. Although some of the scenarios (particularly with combined increases in annual and extreme precipitation) are likely to cause larger increases in both shallow and deep landslides (compared to scenarios that decrease these landslides: Table 12.2), it is obvious that a uniform prediction of landslide increases, as implied by the IPCC, cannot be assumed. Furthermore, it is obvious from a summary of previous research that certain widespread land management practices that can be modified have a much larger and very predictable effect on landslide erosion. More sustainable solutions to this element of environmental change (i.e. anthropogenic practices) are much more manageable in the context of current knowledge and practical policy-making decisions than the uncertain impacts of global warming in mid-latitude temperate forest and rangelands. It is clear that anthropogenic practices like roads, forest conversion, timber harvesting, mining and recreation can all lead to sizeable increases in landslide erosion if practised incorrectly or in steep, unstable terrain. Thus, the most efficient course of action to reduce landslide erosion is to improve the planning and implementation of such human practices with the ultimate objective of sustainable land management. Such an approach will require close collaboration amongst government and local planning agencies, policy-makers, land managers, geoscientists and engineers. Many of the past failures in planning actions, policies and implementation have occurred not because of a lack of knowledge, but rather because of the lack of coordination of efforts and expertise.

Temperate mid-latitude forests and rangelands occupy a large portion of the Earth (Plate 36) and provide essential goods and services for global populations. By focussing on improved planning and management within many of these forests, it will be possible to keep landslide erosion within tolerable levels even in a warming climate. Building fewer roads, more carefully and in stable locations can go a long way in reducing landslide erosion in mountainous terrain (Sidle and Ochiai, 2006). Developing better zoning strategies for residential construction and recreation activities in steep terrain will also provide benefits. Likewise, utilising spatially distributed landslide prediction techniques to assist with planning of timber harvesting, including the designation of leave areas where trees should not be harvested and stable locations for roads, will substantially reduce landslide erosion. Such careful and long-term planning of forest operations is undoubtedly one of the most effective methods of minimising sediment delivery to streams and hazards to humans and property in temperate forests. Although sustainable forestry is possible on government-owned lands, many developed nations endowed with some of the most productive temperate forest resources have exhibited less than wise and realistic long-term planning during the past few decades (Sidle and Ochiai, 2006). Because of the continuing high demands for timber and increasing environmental pressure to reduce harvesting in temperate forests, more logging has occurred in the developing nations of Latin America and Southeast Asia where impacts are substantially greater and controls and enforcement of forest practices are poor (Sidle *et al.*, 2006). Natural resource agencies must implement long-term, practical perspectives of timber demands, weigh their relative environmental benefits and costs, and decide how to sustainably manage these resources.

To improve our knowledge of the interactions of environmental change (both climate and human-induced) with landslide erosion it is critical that government agencies invest in continuing landslide inventories or implement new long-term inventories that measure actual erosion fluxes for various land use practices, and ensure that detailed precipitation data are available for such areas. Granting agencies typically do not fund these important long-term investigations at the necessary spatial scales yet this information is at the cornerstone of articulating future policy and planning of mid-latitude forest and rangelands.

References

Abe, K. and Ziemer, R. R. (1991). Effect of tree roots on a shear zone: modeling reinforced shear stress. *Canadian Journal of Forest Research*, **21**, 1012, 1019.

Acosta, R. *et al.* (1999). *Climate Change Information Sheets*. Châtelaine, Switzerland: United Nations Environment Programme.

Amaranthus, M. P. R. *et al.* (1985). Logging and forest roads related to increased debris slides in southwest Oregon. *Journal of Forestry*, **83**, 229–233.

Anbalagan, R. (1992). Landslide hazard evaluation and zonation mapping in mountainous terrain. *Engineering Geology*, **32**, 269–277.

Arnaez-Vadillo, J. and Larrea, V. (1994). Erosion models and hydrogeomorphological function on hill-roads (Iberian system, La Rioja, Spain). *Zeitschrift für Geomorphologie (NF)*, **38**, 343–354.

Bansal, R. C. and Mathur, H. N. (1976). Landslides: the nightmare of hill roads. *Soil Conservation Digest*, **4**, 36–37.

Benda, L., Veldhuisen, C. and Black, J. (2003). Debris flows as agents of morphological heterogeneity at low-order confluences. *Bulletin of the Geological Society of America*, **115**, 1110–1121.

Benda, L. *et al.* (2004). Confluence effects in rivers: interactions of basin scale, network geometry, and disturbance regimes. *Water Resources Research*, **40**, W05402, doi:10.1029/2003WR002583.

Bishop, D. M. and Stevens, M. E. (1964). *Landslides on Logged Areas, Southeast Alaska*, Research Report NOR-1. Juneau: Forestry Service of the US Department of Agriculture.

Blanco-Canqui, H. and Lal, R. (2008). *Principles of Soil Conservation and Management*. Berlin: Springer-Verlag.

Blaschke, P. M., Trustrum, N. A. and DeRose, R. C. (1992). Ecosystem processes and sustainable land use in New Zealand steeplands. *Agriculture, Ecosystems and Environment*, **41**, 153–178.

Boardman, J. and Poesen, J., eds. (2006). *Soil Erosion in Europe*. Chichester: John Wiley.

Boer, M. M., Koster, E. A. and Lundberg, H. (1990). Greenhouse impacts in Fennoscandia: preliminary findings of a European workshop on the effects of climate change. *Ambio*, **19**, 2–10.

Borga, M. *et al.* (2002). Assessment of shallow landsliding by using a physically based model of hillslope stability. *Hydrological Processes*, **16**, 2833–2851.

Brardinoni, F., Slaymaker, O. and Hassan, M. A. (2003). Landslide inventory in a rugged forested watershed: a comparison between air-photo and field survey data. *Geomorphology*, **54**, 179–196.

Buma, J. (2000). Finding the most suitable slope stability model for the assessment of the impact of climate change on a landslide in southeast France. *Earth Surface Processes and Landforms*, **25**, 565–582.

Burgess, S. M. (2005). 2004–05 Tropical cyclone season summary. *Island Climate Update*, **57**, 6.

Burroughs, E. R. and Thomas, B. R. (1977). *Declining Root Strength in Douglas-Fir after Felling as a Factor in Slope Stability*, Research Paper No. INT-190. Ogden: Forestry Service of the US Department of Agriculture.

Burt, T. P. and Swank, W. T. (1992). Flow frequency responses to hardwood-to-grass conversion and subsequent succession. *Hydrological Processes*, **6**, 179–188.

Calder, I. R. (1990). *Evaporation in the Uplands*. Chichester: John Wiley.

Campy, M., Buoncristiani, J. F. and Bichet, V. (1998). Sediment yield from glacio-lacustrine calcareous deposits during the postglacial period in the Combe D'Ain (Jura, France). *Earth Surface Processes and Landforms*, **23**, 429–444.

Carrara, A. *et al.* (1991). GIS technique and statistical models in evaluating landslide hazard. *Earth Surface Processes and Landforms*, **16**, 427–445.

Chigira, M. (2002). The effects of environmental changes on weathering, gravitational rock deformation and landslides. In R. C. Sidle, ed., *Environmental Change and Geomorphic Hazards in Forests*. Wallingford: CAB International, pp. 101–121.

Cushman, S. A. and Wallin, D. O. (2000). Rates and patterns of landscape change in the Central Sikhote-alin Mountains, Russian Far East. *Landscape Ecology*, **15**, 643–659.

Dehn, M. and Buma, J. (1999). Modelling future landslide activity based on general circulation models. *Geomorphology*, **30**, 175–187.

Déqué, M. (2007). Frequency of precipitation and temperature extremes over France in an anthropogenic scenario: model results and statistical correction according to observed values. *Global and Planetary Change*, **57**, 16–26.

Dhakal, A. S. and Sidle, R. C. (2003). Long-term modeling of landslides for different forest management practices. *Earth Surface Processes and Landforms*, **28**, 853–868.

Dhakal, A. S. and Sidle, R. C. (2004a). Pore water pressure assessment in a forest watershed: simulations and distributed field measurements related to forest practices. *Water Resources Research*, **40**, W02405, doi:1029/2003WR002017.

Dhakal, A. S. and Sidle, R. C. (2004b). Distributed simulations of landslides for different rainfall conditions. *Hydrological Processes*, **18**, 757–776.

Dhakal, A. S., Amada, T. and Aniya, M. (2000). Landslide hazard mapping and its evaluation using GIS: an investigation of sampling scheme for grid-cell based quantitative method. *Photogrammetric Engineering and Remote Sensing*, **66**, 981–989.

Diamond, H. (2006). Review of recent tropical cyclone climatological research. *Island Climate Update*, **72**, 6.

Dietrich, W. E., Bullugi, D. and Real de Asua, R. (2001). Validation of the shallow landslide model, SHALSTAB, for forest management. In M. S. Wigmosta and S. J. Burges, eds., *Land Use and Watersheds: Human Influences on Hydrology and Geomorphology in Urban and Forest Areas*. Washington, DC: American Geophysical Union, pp. 195–227.

Douglass, J. E. and Hoover, M. D. (1988). History of Coweeta. In W. T. Swank and D. A. Crossley, eds., *Forest Hydrology and Ecology at Coweeta*. New York: Springer-Verlag, pp. 17–31.

Dunne, T. (1998). Critical data requirements for prediction of erosion and sedimentation in mountain drainage basins. *Journal of the American Water Resources Association*, **34**, 795–808.

Easterling, D. R. *et al.* (1999). Long-term observations for monitoring extremes in the Americas. *Climatic Change*, **42**, 285–308.

Ericksen, P. J., McSweeney, K. and Madison, F. W. (2002). Assessing linkages and sustainable land management for hillside agroecosystems in Central Honduras: analysis of intermediate and catchment scale indicators. *Agriculture, Ecosystems and Environment*, **91**, 295–311.

Evans, S. G. and Clague, J. J. (1994). Recent climate change and catastrophic geomorphic processes in mountain environments. *Geomorphology*, **10**, 107–128.

Ewers, R. M. *et al.* (2006). Past and future trajectories of forest loss in New Zealand. *Biological Conservation*, **133**, 312–325.

Eyles, R. J. (1971). Mass movement in Tangoio conservation reserve, northern Hawkes Bay. *Earth Sciences Journal*, **5**, 79–91.

Fairbairn, W. A. (1967). Erosion in the River Findhorn Valley. *Scottish Geographical Magazine*, **83**, 46–52.

Fernandes, N. F. *et al.* (2004). Topographic controls of landslides in Rio de Janeiro: field evidence and modelling. *Catena*, **55**, 163–181.

Fiksdal, A. J. (1974). A landslide survey of the Stequaleho Creek watershed. In C. J. Cedarholm and L. C. Lestelle, eds., *Observations of the Effects of Landslide Siltation on Salmon and Trout Resources of the Clearwater Basin, Jefferson County, Washington, 1972–73*. Seattle: Washington State Department of Natural Resources.

Food and Agricultural Organization (FAO) (2001). *Global Ecological Zoning for the Global Forest Resources Assessment 2000*, Final Report, Forest Resources Assessment, Working Paper No. 56. Rome: FAO.

Fowler, H. J. *et al.* (2005). New estimates of future changes in extreme rainfall across the UK using regional climate model integrations: 1. Assessment of control climate. *Journal of Hydrology*, **300**, 212–233.

Fransen, P. and Brownlie, R. (1995). Historical slip erosion in catchments under pasture and radiata pine forest, Hawke's Bay hill country. *New Zealand Journal of Forestry*, **40**, 29–33.

Friedman, S. M., and Friedman, K. A. (1988). *Reporting on the Environment: A Handbook for Journalists*. Bangkok: Asian Forum of Environmental Journalists.

Fu, C. and Wen, G. (1999). Variation of ecosystems over East Asia in association with seasonal, interannual and decadal monsoon climate variability. *Climatic Change*, **43**, 477–494.

Garrett, J. (1980). Catchment authority work in the Rangitikei area. *Aokautere Science Centre, New Zealand Ministry of Works and Development, Internal Report*, **21**, 23–26.

Gomi, T., Sidle, R. C. and Richardson, J. S. (2002). Understanding processes and downstream linkages of headwater systems. *BioScience*, **52**, 905–916.

Gresswell, S., Heller, D. and Swanston, D. N. (1979). *Mass Movement Response to Forest Management in the Central Oregon Coast Ranges*, Resources Bulletin No. PNW-84. Portland: Forestry Service of the US Department of Agriculture.

Guillande, R. P. *et al.* (1995). Automated mapping of the landslide hazard on the island of Tahiti based on digital satellite data. *Mapping Science and Remote Sensing*, **32**, 59–70.

Gupta, R. P. and Joshi, B. C. (1990). Landslide hazard zoning using the GIS approach: a case study from the Ramganga Catchment, Himalayas. *Engineering Geology*, **28**, 119–131.

Guthrie, R. H. (2002). The effects of logging on frequency and distribution of landslides in three watersheds on Vancouver Island, British Columbia. *Geomorphology*, **43**, 273–292.

Haeberli, W. and Burn, C. R. (2002). Natural hazards in forests: glacial and permafrost effects as related to climate change. In R. C. Sidle, ed., *Environmental Change and Geomorphic Hazards in Forests*. Wallingford: CAB International, pp. 167–202.

Haigh, M. J. (1984). Landslide prediction and highway maintenance in the Lesser Himalayas, India. *Zeitschrift für Geomorphologie (NF)*, **51**, 17–37.

Hawley, J. G. (1980). Introduction to workshop on the influence of soil slip erosion on hill country pastoral productivity. *Aokautere Science Centre, New Zealand Ministry of Works and Development, Internal Report*, **21**, 4–6.

Hewlett, J. D. and Helvey, J. D. (1970). Effects of forest clear-felling on the storm hydrograph. *Water Resources Research*, **6**, 768–782.

Hursh, C. R. and Brater, E. F. (1941). Separating storm hydrographs into surface- and subsurface-flow. *EOS, Transactions of the American Geophysical Union*, **22**, 863–871.

IPCC (2007a). *Climate Change 2007: The Physical Science Basis. Contribution of Working Group I to the Fourth Assessment Report of the Intergovernmental Panel on Climate Change*, S. Solomon *et al.*, eds. Cambridge: Cambridge University Press.

IPCC (2007b). *Climate Change 2007: Impacts, Adaptation and Vulnerability. Contribution of Working Group II to the Fourth Assessment Report of the Intergovernmental Panel on Climate Change*. M. L. Parry *et al.*, eds. Cambridge: Cambridge University Press.

Imaizumi, F. *et al.* (2006) Hydrogeomorphic processes in a steep debris flow initiation channel. *Geophysical Research Letters*, **33**, L10404, doi:10.1029/2006GL026250.

Innes, J. L. (1997). Historical debris-flow activity and climate in Scotland. *Paleoklimaforschung*, **19**, 233–240.

Iverson, R. M. and Major, J. J. (1987). Rainfall, ground-water flow and seasonal movement at Minor Creek landslide, northwestern California: physical interpretation of empirical relations. *Bulletin of the Geological Society of America*, **99**, 579–594.

Jakob, M. (2000). The impacts of logging on landslide activity at Clayoquot Sound, British Columbia. *Catena*, **38**, 279–300.

Kasai, M. *et al.* (2005). Impacts of land use change on patterns of sediment flux in Weraamaia catchment, New Zealand. *Catena*, **64**, 27–60.

Kelliher, F. M. *et al.* (1995). Estimating the risk of landsliding using historical extreme river flood data. *Journal of Hydrology (New Zealand)*, **33**, 123–129.

Ketcheson, G. L., and Froehlich, H. A. (1978). *Hydrology Factors and Environmental Impacts of Mass Soil Movements in the Oregon Coast Range*, Water Resources Research Institute Report. Corvallis: Oregon State University.

Khon, V. Ch. *et al.* (2007). Regional changes of precipitation characteristics in Northern Eurasia from simulations with global climate model. *Global and Planetary Change*, **57**, 118–123.

Kienholz, H. *et al.* (1984). Mapping of mountain hazards and slope stability. *Mountain Research and Development*, **4**, 247–266.

Kirkby, M. J. and Cox, N. J. (1995). A climatic index for soil erosion potential (CSEP) including seasonal and vegetation factors. *Catena*, **25**, 333–352.

Krongkaew, M. (2004). The development of the Greater Mekong Subregion (GMS): real promise or false hope? *Journal of Asian Economics*, **15**, 977–998.

Kuruppuarachchi, T. and Wyrwoll, K. H. (1992). The role of vegetation clearing in the mass failure of hillslopes: Moresby Ranges, Western Australia. *Catena*, **19**, 193–208.

Lambert, M. G. (1980). Pastoral production on eroded and uneroded slopes in the dry Wararapa hill country: an interim (12 month) report on the joint Grasslands Division and Aokautere Science Centre Project. *Aokautere Science Centre, New Zealand Ministry of Works and Development, Internal Report*, **21**, 7–12.

Landsea, C. W. *et al.* (1999). Atlantic Basin hurricanes: indices of climatic changes. *Climatic Change*, **42**, 89–129.

Law, F. (1956). The effect of afforestation upon the yield of catchment areas. *Journal of the British Waterworks Association*, **38**, 484–494.

Leathwick, J., McClone, M. S. and Walker, S. (2004). *New Zealand's Potential Vegetation Pattern*. Lincoln: Manaaki Whenua Press.

Lee, J. J., Phillips, D. L. and Dodson. R. F. (1996). Sensitivity of the US corn belt to climate change and elevated CO_2: 2. Soil erosion and organic carbon. *Agricultural Systems*, **52**, 503–521.

Lee, S. *et al.* (2004). Determination and application of the weights for landslide susceptibility mapping using an artificial neural network. *Engineering Geology*, **71**, 289–302.

Leung, L. R. *et al.* (2004). Mid-century ensemble regional climate change scenarios for the western United States. *Climatic Change*, **62**, 75–113.

Loaiciga, H. A. *et al.* (1996) Global warming and the hydrologic cycle. *Journal of Hydrology*, **174**, 83–127.

Lu, X. X. and Higgitt, D. L. (1998). Recent changes of sediment yield in the Upper Yangtze, China. *Environmental. Management*, **22**, 697–709.

Luckman, P. G., Gibson, R. D. and DeRose, R. C. (1999). Landslide erosion risk to New Zealand pastoral steeplands productivity. *Land Degradation and Development*, **10**, 49–65.

MacFarlane, W. A. and Wohl, E. (2003). Influence of step composition on step geometry and flow resistance in step-pool streams of the Washington Cascades. *Water Resources Research*, **39**, 1037, doi: 10.1029/2001WR001238.

Marden, M. and Rowan, D. (1993). Protective value of vegetation on Tertiary terrain before and during Cyclone Bola, East Coast, North Island, New Zealand. *New Zealand Journal of Forestry Science*, **23**, 255–263.

Marden, M. *et al.* (2005). Pre- and post-reforestation gully development in Mangatu Forest, East Coast, North Island, New Zealand. *River Research and Applications*, **21**, 757–771.

Mason, S. J. *et al.* (1999). Changes in extreme rainfall events in South Africa. *Climatic Change*, **41**, 249–257.

McGlone, M. S. (1989). The Polynesian settlement of New Zealand in relation to environmental and biotic changes. *New Zealand Journal of Ecology*, **12**, 115–130.

McGlone, M. S. (2002). A Holocene and latest Pleistocene pollen record from Lake Poukawa, Hawke's Bay, New Zealand. *Global and Planetary Change*, **33**, 283–299.

Megahan, W. F. and Ketcheson, G. L. (1996). Predicting downslope travel of granitic sediments from forest roads in Idaho. *Water Resources Bulletin*, **32**, 371–382.

Megahan, W. F. and Kidd, W. J. (1972). *Effect of Logging Roads on Sediment Production Rates in the Idaho Batholith*, Research Paper No. INT-123. Ogden: US Department of Agriculture Forest Service.

Megahan, W. F., Day, N. F. and Bliss, T. M. (1978). Landslide occurrence in the western and central Northern Rocky Mountain physiographic province in Idaho. *Proceedings of the 5th North American Forest Soils Conference*, Fort Collins, pp. 116–139.

Montgomery, D. R. (2007). *Dirt: The Erosion of Civilizations*. Berkeley: University of California Press.

Montgomery, D. R. and Dietrich, W. E. (1994). A physically based model for the topographic control on shallow landsliding, *Water Resources Research*, **30**, 1153–1171.

Montgomery, D. R., Sullivan, K. and Greenberg, H. M. (1998). Regional test of a model for shallow landsliding. *Hydrological Processes*, **12**, 943–955.

Montgomery, D. R. *et al.* (2000). Forest clearing and regional landsliding. *Geology*, **28**, 311–314.

Moore, I. D., O'Loughlin, E. M. and Burch, G. J. (1988). A contour based topographic model and its hydrologic and ecological applications. *Earth Surface Processes and Landforms*, **13**, 305–320.

Mulder, H. F. H. M. and van Asch, Th. W. J. (1988). Quantitative approaches in landslide hazard analysis. *Travaux de l'Institut de Géographie de Reims*, **69–72**, 43–53.

Nearing, M. A. *et al.* (2005). Modeling response of soil erosion and runoff to changes in precipitation and cover. *Catena*, **61**, 131–154.

Newman, E. B., Paradis, A. R. and Brabb, E. E. (1978). Feasibility and cost of using a computer to prepare landslide susceptibility maps of the San Francisco Bay Region, California. *US Geological Survey Bulletin*, **1443**.

New Zealand Ministry of Works and Development (1970). *Wise Land Use and Community Development*, Report of Technical Committee of Inquiry into the Problems of the Poverty Bay–East Cape District of New Zealand. Wellington: New Zealand Ministry of Works and Development

Nilsen, T. H. *et al.* (1979). Relative slope stability and land-use planning in the San Francisco Bay region, California. *US Geological Survey Professional Paper*, **944**.

Nogués-Bravo, D. *et al.* (2007). Exposure of global mountain systems to climate warming during the twenty-first century. *Global Environmental Change*, **17**, 420–428.

Nyaupane, G. P., Morais, D. B. and Dowler, L. (2006). The role of community involvement and number/type of visitors on tourism impacts: a controlled comparison of Annapurna, Nepal and Northwest Yunnan, China. *Tourism Management*, **27**, 1373–1385.

O'Loughlin, C. L. and Pearce, A. J. (1976). Influence of Cenozoic geology on mass movement and sediment yield response to forest removal, North Westland, New Zealand. *Bulletin of the International Association of Engineering Geology*, **14**, 41–46.

O'Loughlin, C. L. and Watson, A. J. (1979). Root-wood strength determination in radiata pine after clearfelling. *New Zealand Journal of Forestry Science*, **9**, 284–293.

O'Loughlin, C. L. and Watson, A. J. (1981). Root-wood strength deterioration in beech (*Nothofagus fusca* and *N. truncata*) after clearfelling. *New Zealand Journal of Forestry Science*, **11**, 183–185.

O'Loughlin, C. L. and Ziemer, R. R. (1982). The importance of root strength and deterioration rates upon edaphic stability in steepland forests. In *Carbon Uptake and Allocation in Subalpine Ecosystems as a Key to Management*, Proceedings of an IUFRO Workshop. Corvallis: Oregon State University, pp. 70–78.

O'Loughlin, C. L., Rowe, L. K. and Pearce, A. J. (1982). Exceptional storm influences on slope erosion and sediment yield in small forest catchments, North Westland, New Zealand. In *National Symposium on Forest Hydrology*, Melbourne, Institution of Engineers, pp. 84–91.

O'Loughlin, E. M. (1986). Prediction of surface saturation zones in natural catchments by topographic analysis. *Water Resources Research*, **22**, 794–804.

Page, M. J. and Trustrum, N. A. (1997). A late Holocene lake sediment record of the erosion response to land use change in a steepland catchment, New Zealand. *Zeitschrift für Geomorphologie (NF)*, **41**, 369–392.

Parkner, T. *et al.* (2006). Development and controlling factors of gullies and gully complexes, East Coast, New Zealand. *Earth Surface Processes and Landforms*, **31**, 187–199.

Peterson, D. L. (1998). Climate, limiting factors and environmental change in high-altitude forests of Western North America. In M. Beniston and J. L. Innes, eds., *The Impacts of Climate Variability on Forests*. Berlin: Springer-Verlag, pp. 191–208.

Pezza, A. B., Simmonds, I. and Renwick, J. A. (2007). Southern hemisphere cyclones and anticyclones: recent trends and links with decadal variability in the Pacific Ocean. *International Journal of Climatology*, **27**, 1403–1419.

Pfister, L., Humbert, J. and Hoffmann, L. (2000). Recent trends in rainfall–runoff characteristics in the Alzette River basin, Luxembourg. *Climatic Change*, **45**, 323–337.

Poesen, J. *et al.* (2003.) Gully erosion and environmental change: importance and research needs. *Catena*, **50**, 91–133.

Portela, A. H. and Aguirre, B. E. (2000). Environmental degradation and vulnerability in Cuba. *Natural Hazards Reviews*, **1**, 171–179.

Preston, N. J. and Crozier, M. J. (1999). Resistance to shallow landslide failure through root-derived cohesion in East Coast hill country soils, North Island, New Zealand. *Earth Surface Processes and Landforms*, **24**, 665–675.

Rice, R. M., Corbett, E. S. and Bailey, R. G. (1969). Soil slips related to vegetation, topography, and soil in southern California. *Water Resources Research*, **5**, 647–659.

Robison, E. G. *et al.* (1999). *Storm Impacts and Landslide of 1996*, Final Report, Forest Practices Technical Report No. 4. Salem: Oregon Department of Forestry.

Rollerson, T. P., Thomson, B. and Millard, T. H. (1997). Identification of coastal British Columbia terrain susceptible to debris flows. In *Debris-Flow Hazards Mitigation: Mechanics, Prediction and Assessment*. San Francisco: American Society of Civil Engineers, pp. 484–495.

Rowntree, P. R. (1993). Climatic models: changes in physical environmental conditions. In D. Atkinson, ed., *Global Climate Change: Its Implications for Crop Protection*. Farnham: British Crop Protection Council, pp. 13–32.

Salanger, M. J. and Griffiths, G. M. (2001). Trends in New Zealand daily temperature and rainfall extremes. *International Journal of Climatology*, **21**, 1437–1452.

Sánchez, E. C. *et al.* (2004). Future climate extreme events in the Mediterranean simulated by a regional climate model: a first approach. *Global and Planetary Change*, **44**, 163–180.

Schlyter, P. *et al.* (1993) Geomorphic process studies related to climate change in Karkevagge, Northern Sweden: status of current research. *Geografiska Annaler*, **75**A, 55–60.

Schwab, J. W. (1983). *Mass Wasting: October–November 1978 Storm, Rennel Sound, Queen Charlotte Islands, British Columbia*, Ministry of Forests Publication No. 91. Victoria: Ministry of Forests.

Schwab, J. W. (1988). Mass wasting impacts to forest land: forest management implications, Queen Charlotte timber supply area. In J. D. Louisier and G. W. Still, eds., *Degradation of Forested Land: Forest Soils and Risk*. Victoria: British Columbia Forest Service, pp. 104–115.

Sidle, R. C. (1991). A conceptual model of changes in root cohesion in response to vegetation management. *Journal of Environmental Quality*, **20**, 43–52.

Sidle, R. C. (1992). A theoretical model of the effects of timber harvesting on slope stability. *Water Resources Research*, **28**, 1897–1910.

Sidle, R. C., ed. (2002). *Environmental Change and Geomorphic Hazards in Forests*. Wallingford: CAB International.

Sidle, R. C. (2007). Dark clouds over Shangri-La. Opinion article. *The Japan Times*, March 15, 2007, p. 14.

Sidle, R. C. and Dhakal, A. S. (2002). Potential effects of environmental change on landslide hazards in forest environments. In R. C. Sidle, ed., *Environmental Change and Geomorphic Hazards in Forests*. Wallingford: CAB International, pp. 123–165.

Sidle, R. C. and Hornbeck, J. W. (1991). Cumulative effects: a broader approach to water quality research. *Journal of Soil and Water Conservation*, **46**, 268–271.

Sidle, R. C. and Ochiai, H. (2006). *Landslides: Processes, Prediction, and Land Use*. Washington, DC: American Geophysical Union.

Sidle, R. C. and Wu, W. (1999). Simulating effects of timber harvesting on the temporal and spatial distribution of shallow landslides. *Zeitschrift für Geomorphologie (NF)*, **43**, 185–201.

Sidle, R. C., Pearce, A. J. and O'Loughlin, C. L. (1985). *Hillslope Stability and Land Use*. Washington, DC: American Geophysical Union.

Sidle, R. C. *et al.* (2004). Interactions of natural hazards and humans: evidence in historical and recent records. *Quaternary International*, **118–119**, 181–203.

Sidle, R. C. *et al.* (2006). Erosion processes in steep terrain: truths, myths, and uncertainties related to forest management in Southeast Asia. *Forest Ecology and Management*, **224**, 199–225.

Singh, R. B. (1998). Land use/cover changes, extreme events and ecohydrological responses in the Himalayan region. *Hydrological Process*, **12**, 2043–2055.

Slaymaker, O. (2000). Assessment of the geomorphic impacts of forestry in British Columbia. *Ambio*, **29**, 381–387.

Slaymaker, O. (2001). Why so much concern about climate change and so little attention to land use change? *Canadian Geographer*, **45**, 71–78.

Soeters, R. and van Westen, C. J. (1996). Slope instability recognition, analysis, and zonation. In A. K. Turner and R. L. Schuster, eds., *Landslides: Investigation and Mitigation*, Special Report No. 247. Trans. Res. Board, National Res. Council, Washington DC: National Academic Press, pp. 129–177.

Soil and Water Conservation Society (2003). *Conservation Implications of Climate Change: Soil Erosion and Runoff from Cropland*. Ankeny: Soil and Water Conservation Society.

Sorensen, J. H. (2000). Hazard warning systems: review of 20 years of progress. *Natural Hazards Review*, **1**, 119–25.

Swank, W. T., Swift, L. W. and Douglass, J. E. (1988). Streamflow changes associated with forest cutting, species conversions and natural disturbance. In W. T. Swank and D. A. Crossley, eds., *Forest Hydrology and Ecology at Coweeta*. New York: Springer-Verlag, pp. 297–312.

Swanson, F. J. and Dyrness, C. T. (1975). Impact of clearcutting and road construction on soil erosion by landslides in the western Cascade Range, Oregon. *Geology*, **3**, 393–396.

Swanson, F. J. and Grant, G. (1982). *Rates of Soil Erosion by Surface and Mass Erosion Processes in the Willamette National Forest*. Final Report to Willamette National Forest. Corvallis: Forest Science Laboratory.

Swanson, F. J. and Swanston, D. N. (1977). Complex mass movement terrains in the western Cascade Range, Oregon. In *Reviews in Engineering Geology (Landslides)*, vol. 3. Boulder: Geological Society of America. pp. 113–124.

Swanston, D. N. (1981). Creep and earthflow erosion from undisturbed and management impacted slopes in the Coast and Cascade Ranges of the Pacific Northwest, USA. *IAHS Publication*, **132**, 76–94.

Swanston, D. N. and Marion, D. A. (1991). Landslide response to timber harvest in Southeast Alaska. *Proceedings of the 5th Federal Interagency Sedimentation Conference*, Las Vegas, pp. 10–49.

Swanston, D. N., Ziemer, R. R. and Janda, R. J. (1995). Rate and mechanics of progressive hillslope failure in the Redwood

Creek Basin, northwestern California. In *Geomorphic Processes and Aquatic Habitat in the Redwood Creek Basin, Northwestern California*. Professional Paper No. 1451-E, E1-E16. Washington, DC: US Geological Survey.

Thomas, R. B. and Megahan, W. F. (1998). Peakflow responses to clearcutting and roads in small and large basins, western Cascades, Oregon: a second opinion. *Water Resources Research*, **34**, 3393–3403.

Trenberth, K. E. and Owen, T. W. (1999). Workshop on indices and indicators for climatic extremes, Asheville, NC, USA, 3–6 June 1997: Breakout Group A – Storms. *Climatic Change*, **42**, 9–21.

Trustrum, N. A. and DeRose, R. C. (1988). Soil depth–age relationship of landslides on deforested hillslopes, Taranaki, New Zealand. *Geomorphology*, **1**, 143–160.

Trustrum, N. A. and Stephens, P. R. (1981). Selection of hill country pasture measurement sites by interpretation of sequential aerial photographs. *New Zealand Journal of Experimental Agriculture*, **9**, 31–34.

Trustrum, N. A., Lambert, M. G. and Thomas, V. J. (1983). The impact of soil slip erosion on hill country pasture production in New Zealand. *Proceedings of the 2nd International Conference on Soil Erosion and Conservation*, Honolulu.

van Westen, C. J., van Asch, T. W. J. and Soeters, R. (2006). Landslide hazard and risk zonation – why is it still so difficult? *Bulletin of the Engineering Geological Environment*, **65**, 167–184.

Varnum, N. C., Tueller, P. T. and Skau, C. M. (1991). A geographical information system to assess natural hazards in the east-central Sierra Nevada. *Journal of Imaging Technology*, **17** (2), 57–61.

Wasson, R. J. and Hall, G. (1982). A long record of mudslide movement at Waerenga-O-Kuri, New Zealand. *Zeitschrift für Geomorphologie (N.F.)*, **26**, 73–85.

Wemple, B. C., Swanson, F. J. and Jones, J. A. (2001). Forest roads and geomorphic process interactions, Cascade Range, Oregon. *Earth Surface Processes and Landforms*, **26**, 191–204.

Wolfe, M. D. and Williams, J. W. (1986). Rates of landsliding as impacted by timber management activities in northwestern California. *Bulletin of the Association of Engineering Geology*, **23**, 53–60.

Wu, W. and Sidle, R. C. (1995). A distributed slope stability model for steep forested hillslopes. *Water Resources Research*, **31**, 2097–2110.

Wu, W. and Sidle, R. C. (1997). Application of a distributed shallow landslide analysis model (dSLAM) to managed forested catchments in coastal Oregon. *IAHS Publication*, **245**, 213–21.

Wyss, W. and Yim, S. (1996). Vulnerability and adaptability of Hong Kong to hazards under climatic change conditions. *Water Air and Soil Pollution*, **92**, 181–190.

Xu, J. and Wilkes, A. (2002). People and ecosystems in mountain landscape of Northwest Yunnan, southwest China: causes of biodiversity loss and ecosystem degradation. *Global Environmental Research*, **6**, 103–110.

Yesilnacar, E. and Topal, T. (2005). Landslide susceptibility mapping: comparison of logistic regression and neural networks methods in a medium scale study, Hendek region (Turkey). *Engineering Geology*, **79**, 251–66.

Zhai, P. et al. (1999). Changes in climate extremes in China. *Climatic Change*, **42**, 203–18.

13 Tundra and permafrost-dominated taiga

Marie-Françoise André and Oleg Anisimov

13.1 Permafrost regions: a global change 'hotspot'

Permafrost is defined as any subsurface material that remains frozen continuously for at least two consecutive years. The seasonally thawed layer above is known as the 'active layer'. The permafrost regions occupy about 25% of the Earth's land area (Zhang *et al.*, 2000), mostly situated in the tundra and taiga zones of the northern hemisphere. The distribution of permafrost is usually represented as a series of concentric zones in which permafrost is continuous (underlies more than 90% of surface), discontinuous (50–90%) and sporadic (less than 50%) (Brown *et al.*, 1997) (Fig. 13.1). Permafrost occupies 85% of Alaska, more than 60% of Russia and 50% of Canada. Four million people are living in circumarctic/subantarctic regions, including 10% belonging to indigenous communities such as the Inuit, Sami (Lapps) and Chukchi (Bogoyavlenskiy and Siggner, 2004). Human concentrations are particularly important in northern Russia, with cities of over 100 000 inhabitants and large river ports (Fig. 13.2).

In circumarctic regions, most of the current and potential environmental and socioeconomic impacts of global warming are associated with permafrost thawing though surprisingly, as stressed by Nelson *et al.* (2002), much of the literature treating geohazards, social science and policy issues in polar regions fails to address issues related to permafrost adequately. However, an unprecedented research effort is being carried out among the multidisciplinary community involved in permafrost science under the auspices of the International Permafrost Association (IPA). Permafrost observations are internationally coordinated within the Global Terrestrial Network for Permafrost (Brown *et al.*, 2000b; Romanovsky *et al.*, 2002). Predictive modelling of permafrost behaviour is being developed, based on climatic scenarios from general circulation models (GCMs), and results from modelling are used in IPCC and ACIA assessment reports (IPCC, 2001a, 2001b, 2007a; ACIA, 2005). Frozen ground was listed by the International Union of Geological Sciences (IUGS) on its list of 'geoindicators' to be used to detect and assess environmental changes over relatively short periods (Berger and Iams, 1996). In this context, landform and landscape changes due to permafrost thaw are of special interest. For this reason, the international community of periglacial geomorphologists is actively involved in monitoring and remote sensing studies, and in palaeoenvironmental reconstructions. These studies aim to document and understand the past and present responses of arctic and subarctic landscapes to climate changes, and they provide data sets for model input and validation. In this chapter we focus on the results from the most recent studies relating climate change in permafrost regions to geomorphology. Analysis of the fundamental results obtained in earlier studies is beyond the scope of this volume; such a retrospective is, however, given in the numerous supporting publications cited in the text.

13.1.1 The Arctic: a climate change 'hotspot'

The Arctic is known as a climate change 'hotspot', where average temperatures have increased at almost twice the global average rate in the past 100 years (McBean *et al.*, 2005). Hotspots have been observed in parts of Alaska and Siberia, with a warming trend at 5 °C for the last century. Climate models predict that warming will be amplified in polar regions (Serreze *et al.*, 2000), particularly in the continental parts of North America and Eurasia (ACIA, 2005; IPCC, 2007a.). An additional Arctic warming of 4–7 °C is expected to occur over the next 100 years (ACIA, 2005), as well as increasing precipitation, mainly falling as

Geomorphology and Global Environmental Change, eds. Olav Slaymaker, Thomas Spencer and Christine Embleton-Hamann. Published by Cambridge University Press.

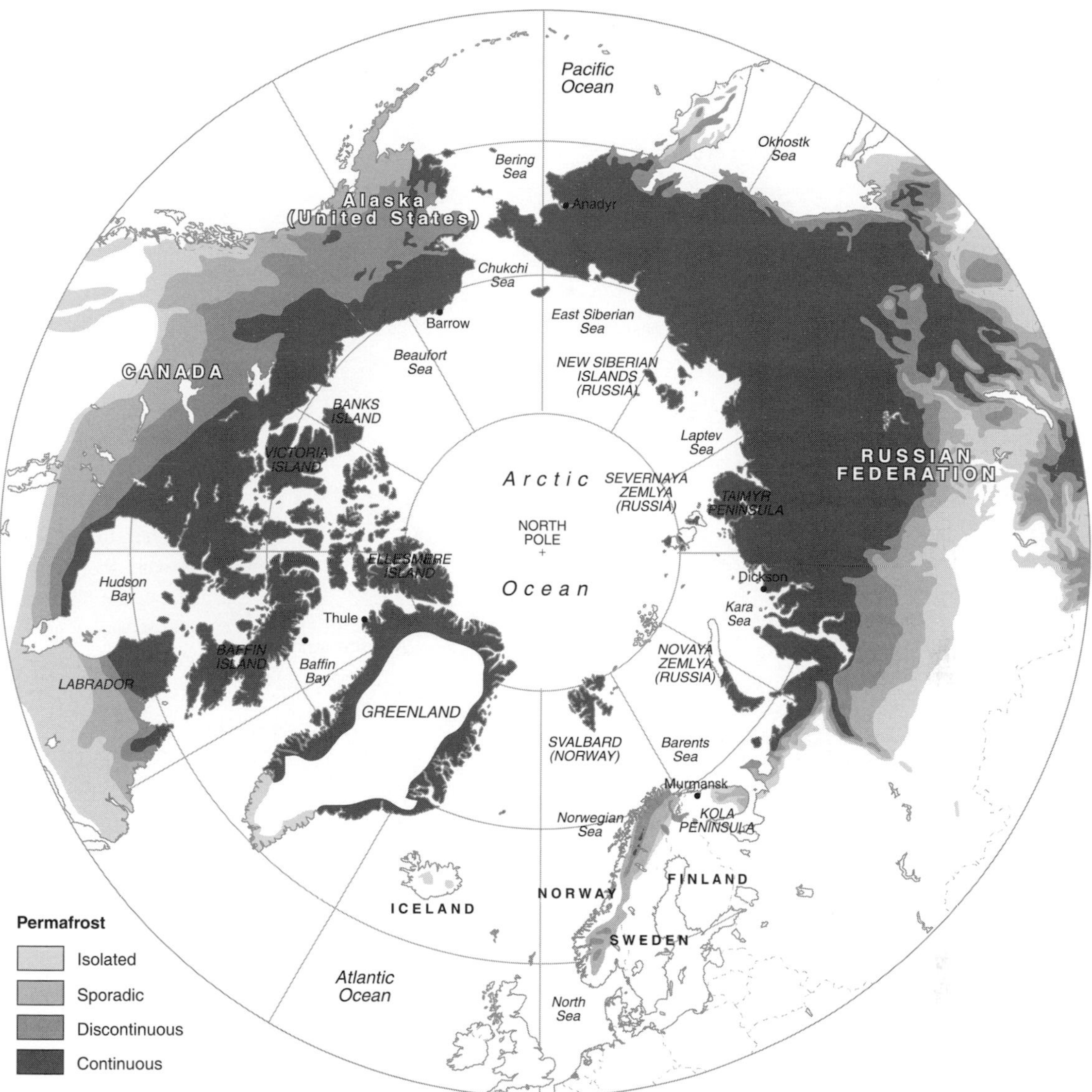

FIGURE 13.1. Permafrost zonation in the northern hemisphere (modified from Brown *et al.*, 1997). Values of 90–100%, 50–90% and less than 50% have been used to delineate the southern margins of the continuous, discontinuous and sporadic permafrost zones. Mean annual ground temperatures range from –8 °C/–13 °C in the northernmost zone to 0 °C/–2 °C in the southern sporadic zone, average permafrost thickness from 800 m to 10 m, and active layer thickness from 10 cm to over 3 m.

rain, following the increases in precipitation observed since at least the 1950s. Substantial decrease in snow cover is also projected.[1]

There is substantial evidence throughout the circumboreal region (Alaska, Canada, Russia) confirming the recent northward and upslope migration of the treeline in response to climate change. Two percent of Alaskan tundra on the Seward Peninsula has been already replaced by forest in the past 50 years (Lloyd *et al.*, 2003) and 70% of 200 investigated Alaskan locations show increased shrub abundance (Sturm *et al.*, 2001; Tape *et al.*, 2006). In central Siberia, forests are predicted to occupy current tundra positions (Soja *et al.*, 2007; Tchebakova *et al.*, 2007). In subarctic Sweden, the altitude of the treeline has risen by about 60 m

[1] More detailed documentation of the GCM-based climatic projections under various greenhouse emission scenarios may be found on the following IPCC websites: http://ipcc-ddc.cru.uea.ac.uk/ and http://igloo.atmos.uiuc.edu/IPCC/ (accessed in July 2008).

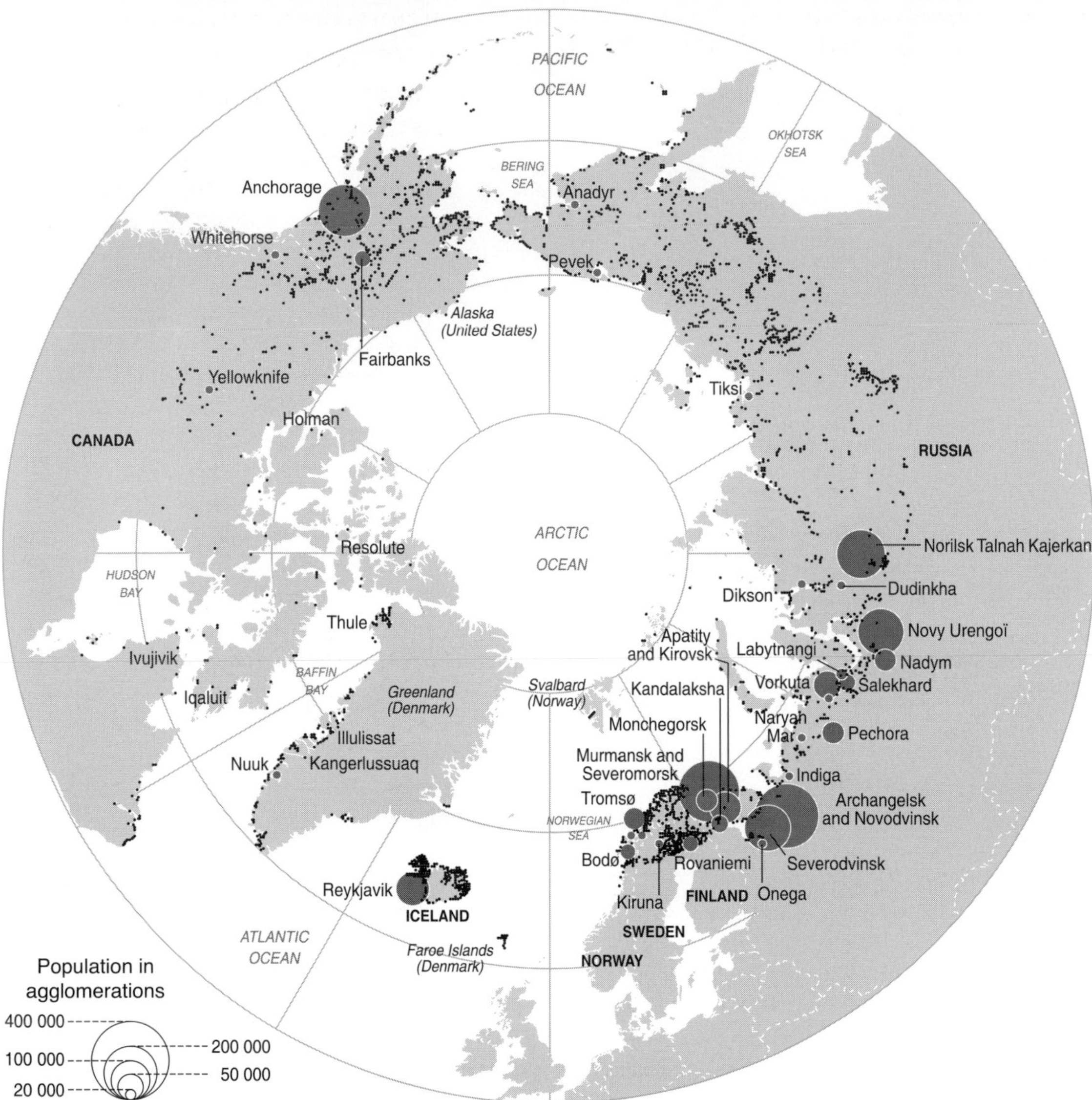

FIGURE 13.2. Human concentrations and scattered settlements in permafrost regions (from ACIA, 2005).

in the twentieth century and dry habitat vegetation has been replaced in places by wet habitat vegetation, whereas moist tussock tundra is drying up as permafrost degrades. On the whole, most subarctic alpine communities in Sweden are becoming drier (Callaghan, 2004; Christensen *et al.*, 2004; Malmer *et al.*, 2005; Truong *et al.*, 2006).

13.1.2 Socioeconomic changes associated with globalisation

Permafrost regions have been severely affected by globalisation and accompanying cultural, political, economic and land use changes, and by population growth. Although native communities have been sufficiently resilient to cope with past climate changes (e.g. the transition from Dorset to Thule cultures in arctic North America), the combined impacts of climate change and globalisation create new and unexpected challenges, due to drastic changes in lifestyle and social links (IPCC, 2001b). The contemporary economic development of the circumarctic region has brought expansion of infrastructures. Extraction of fossil fuels and mineral resources, transportation networks, industrial and civil facilities, engineering maintenance systems and tourism activities have increased substantially since 1950. The oil and gas fields of northern Alaska and western Siberia provide spectacular examples of this recent development. In Russia, 93% of natural gas and 75% of oil are produced in permafrost areas. Overall, the Russian permafrost regions contribute up to 70% of total Russian exports (Il'ichev *et al.*, 2003). Predictive assessment of permafrost-related risks indicates that in many areas residential buildings and industrial infrastructure could become unstable by the middle of the twenty-first century (Nelson *et al.*, 2001, 2002).

13.1.3 Permafrost thaw as a major consequence of global change and impacts on northern environments and societies

Discontinuous and sporadic permafrost is particularly sensitive to climate change because much of it is within 1 or 2 °C of thawing. Whereas few stability problems occur where permafrost is 'dry' (Bockheim and Tarnocai, 1998), thawing of ice-rich permafrost is often accompanied by subsidence and mass movements causing damage to infrastructure and landscape changes from forest to wetland ecosystems (e.g. Osterkamp *et al.*, 1997; Nelson *et al.*, 2001, 2002; Instanes *et al.*, 2005; IPCC, 2007b). As the Russian North is industrially well developed, the infrastructure of the oil and gas industry is of particular concern. Of the 35 000 failures affecting the 350 000-km-long network of pipelines in West Siberia, more than 20% are most probably due to deformation and weakening of foundations induced by permafrost thaw (Anisimov and Reneva, 2006). On the arctic plains of Alaska and Canada, coastline retreat in ice-rich permafrost results in land and habitat loss affecting ecosystems and human communities; some communities and activities are forced to relocate, while others face increasing risks and costs (ACIA, 2005; Rachold *et al.*, 2005). Last, permafrost thaw and associated vegetation changes have potential to impact the global climate through the release of additional carbon to the atmosphere. One recent study indicates that such an impact from thawing arctic wetlands will not be significant (Anisimov, 2007), although the impact from other mechanisms in terrestrial permafrost regions needs to be quantified better. Changes in subsea permafrost may also contribute to warming because significant amounts of methane hydrates are contained in the near-bottom layer of sediments on arctic continental shelves. Under warmer climate hydrates may become unstable leading to methane release to the atmosphere (Sloan, 2003; Maslin, 2004).

13.1.4 International research framework and geomorphological services

Periglacial geomorphologists have been increasingly involved in research programmes and working groups on landscape responses to climate change and permafrost thaw (e.g. IPA WG; IPA/IASC-ACD; ESF-PACE; IGU-GERTEC; ESF-SEDIFLUX; IAG-SEDIBUD). To obtain a more comprehensive picture of the spatial and temporal changes in permafrost, the Global Terrestrial Network for Permafrost (GTN-P) was developed in the 1990s under the auspices of the IPA (Brown *et al.*, 2000b; Burgess *et al.*, 2000). During the Fourth International Polar Year (2007–09), 21 countries are participating in the internationally coordinated 'Thermal State of Permafrost' programme (Brown and Christiansen, 2006).

Of special interest to geomorphologists is the Circumpolar Active Layer Monitoring (CALM) Program, which is an observational network involving more than 160 sites (Nelson and Brown, 1997; Brown *et al.*, 2000a). The active-layer monitoring and associated soil temperature recording are prerequisites for better understanding both of the inter-decadal variation of the thermal permafrost regime and of the linkage between ground temperatures and environmental factors including snow, vegetation and soil cover. They are crucial in so far as they provide invaluable data sets for model input and validation.

Active layer and landscape monitoring is also a useful tool to detect, characterise and quantify ongoing changes with direct impact on northern ecosystems, communities and infrastructure. Periglacial features such as frost mounds are particularly sensitive geoindicators, because they react rapidly to climate changes by collapsing. Monitoring and remote sensing studies of these small-scale landforms can therefore help to detect hotspots of recent and ongoing environmental changes. Moreover, geoecological monitoring of these landforms can contribute to a better understanding of the incidence of permafrost thaw on the carbon budget, which is useful to refine the emission scenarios. Geomorphologically based segmentation of arctic coastlines provides a useful basis for quantifying erosion rates and carbon and mineral sediment fluxes into the Arctic Ocean and for dynamical simulation of sediment flux intensity. Contributions from geomorphologists can also help to develop empirical models to assess the sensitivity of arctic coasts to environmental variability and human impacts, and to establish models to predict the future behaviour of the circumarctic coastal region in response to global change. By integrating maps and databases on geomorphology, landscapes, lithology and soil properties, geographical information systems (GIS) technology is one of the most effective bases for assessing the present-day state, trend and dynamics of environmental changes in permafrost regions. For example, databases of coastal change measurements can be used to identify hazard zones and to produce sensitivity maps of interest to land and resource managers.

13.2 Permafrost indicators: current trends and projections

13.2.1 Permafrost indicators: observational data and predictive modelling

Permafrost responds to climate change through changing ground temperature, depth of seasonal thawing and changing areal extent. The progressive increase in the active layer thickness could be a relatively fast response of permafrost to climatic warming. The increase in soil temperature and melting of the excess ground ice common in upper layers of permafrost in the high latitudes will favour development of potentially disruptive geomorphological processes. Due to latent heat involved in phase changes of water, these processes are relatively slow. Estimates show that under sustained warming complete thawing of 10-m-thick permafrost in the southernmost zone may take several decades, although the timing depends on the local climatic and soil conditions, and particularly on ground ice content. Subsidence at the surface induced by melting of ground ice (thermokarst) and mass movements (e.g. active layer detachment slides) may lead to widespread ecological disturbances and disruption of existing infrastructure (Nelson *et al.*, 2001, 2002).

Direct permafrost observations are limited and observational networks do not capture the full range of variability resulting from the differences in soil, landforms, vegetation and climatic conditions. This is why mathematical models of different complexity are most common tools for predicting the current and future state of permafrost over large, i.e. continental-scale, landscapes.

The past two decades have seen a dramatic rise in the number of permafrost models and in the variety of their applications. Review of literature indicates an increasing interest in spatial modelling of permafrost distribution as well as permafrost-related phenomena that may follow from global climate change. Simultaneously, major efforts are under way via the Global Terrestrial Network for Permafrost (GTN-P) (Brown *et al.*, 2000b; Burgess *et al.*, 2000; Romanovsky *et al.*, 2002) to create large geocryological databases useful for empirical modelling, analysis of observational data and tools for validation of modelling studies.

Results from permafrost models forced by several climatic scenarios (Table 13.1) indicate that by the middle of the twenty-first century the total area occupied by permafrost in the northern hemisphere may shrink by 15–30%, largely due to the thawing of the southern zone of sporadic and discontinuous permafrost but also due to the decrease in the areal continuity of the frozen ground in other zones. Predicted changes of the permafrost temperature have complex geographical pattern and range from 1–2 °C in the southern zone to 3 °C and more along the arctic continental slopes. The predicted seasonal thaw depth is also spatially variable, on average it may increase by 20–30%, and by more than 50% in the northernmost locations (Plate 39).

During the last decade models for predicting changes in permafrost at specific locations have been improved. At the same time projections at continental and hemispheric scales are obtained using climatic and soil forcing data with

TABLE 13.1. *Projected reduction of the near-surface permafrost area in the northern hemisphere under five general circulation models (GCMs)*

	Total permafrost area (M km^2) and % from modern			Continuous permafrost area (M km^2) and % from modern		
GCM	2030	2050	2080	2030	2050	2080
CCC	23.72	21.94	20.66	9.83	8.19	6.93
	87%	81%	76%	79%	66%	56%
ECHAM	22.30	19.31	17.64	9.37	7.25	5.88
	82%	71%	65%	75%	58%	47%
GFDL	24.11	22.38	20.85	10.19	8.85	7.28
	89%	82%	77%	82%	71%	59%
HadCM3	24.45	23.07	21.36	10.47	9.44	7.71
	90%	85%	78%	84%	76%	62%
NCAR	24.24	23.64	21.99	10.69	10.06	9.14
	89%	87%	81%	86%	81%	74%

relatively coarse spatial resolution, typically 0.5° of latitude or longitude or less. The results are thus characteristic of some 'typical' or 'average' permafrost conditions and do not capture the whole range of variability. Many of the processes that have important geomorphological implications, such as abrupt landsliding or thermokarst development, are governed by the complex interplay of various local stochastic factors, and climatic thresholds play an important role in initiating such processes. Intrinsic determinism of the currently existing permafrost models is a serious reason for interpreting model-based projections with caution. Ultimately, all such projections have to be consistent with observations.

A brief summary of observations indicating recent changes in permafrost parameters is given in the following sections.

13.2.2 Permafrost warming

Long-term temperature observations indicate that positive soil temperature trends were observed at several Siberian stations, including increases up to 1 °C during the last decade (Gilichinsky *et al.*, 1998; Pavlov and Moskalenko, 2002). In Alaska, since the mid-1970s, permafrost has warmed at most sites north of the Brooks Range from the Chukchi Sea to the Alaska–Canada border. Maximum warming of 3–4 °C for the arctic coastal plain, suggests a total permafrost warming of >6 °C at Prudhoe Bay during the last century, with most of the warming occurring in winter (Lachenbruch and Marshall, 1986; Osterkamp and Romanovsky, 1999; Osterkamp, 2007). In northwestern Canada, temperatures in the upper 30 m of permafrost have increased by up to 2 °C over the past 20 years (Nelson, 2003). Although cooling of permafrost in northeastern Canada has been often cited as an exception (Allard *et al.*, 1995), recent increases up to nearly 2 °C have occurred since the mid-1990s (Brown *et al.*, 2000a.; Beaulieu and Allard, 2003). There are pronounced regional variations in the temperature changes at the top of the permafrost layer; at many locations in the Arctic permafrost temperature has increased since the 1980s, by up to 3 °C (IPCC, 2007a). A recent model study of circumarctic soil temperatures over a 22-year period (1980–2001) confirmed this warming trend. A maximum of 0.035 °C a^{-1} warming was found in the continuous permafrost zone, and immediate thawing was either already occurring or expected to occur soon in the warmer sporadic permafrost zone (Oelke and Zhang, 2004). An overview of recent trends in permafrost temperature is provided in Table 13.2.

13.2.3 Reduction of near-surface permafrost areal extent

The maximum area covered by seasonally frozen ground has decreased by about 7% in the northern hemisphere since 1900, with a decrease in spring of up to 15% (IPCC, 2007a). Due to the thermal proximity to thawing conditions, permafrost degradation is particularly severe near its southern margin, and by the end of the twenty-first century this process may lead to northward shift of this boundary by a few hundred kilometres throughout much of northern North America and Eurasia (Nelson *et al.*, 2002). Sequential analysis of permafrost maps of western Siberia indicates two contrasting tendencies for the period 1950–2000: firstly, a southward shift of the southern boundary from 63° to 60° N, due to the cooling of the 1960s–1970s; secondly, a northward shift from 60° to 62° N in the last warm decades of the twentieth century (Anisimov *et al.*, 2002). In the Hudson Bay coastal plains, the permafrost started to degrade during the warm periods of fast forest expansion in the 1930s and 1940s, and in parts of the subarctic Québec, permafrost loss accelerated in the mid-1990s from 2.8% a^{-1} to 5.3% a^{-1} due to the combined effects of temperature rise and precipitation increase in the form of snow (Payette *et al.*, 2004).

13.2.4 Increased depth of thawing

In Alaska, active layer thickness did not increase, due to little summer temperature change, but 0.1 m a^{-1} thawing at the permafrost surface occurred at tundra and forest sites (Osterkamp, 2007). In Russia, despite pronounced warming, no general increase in the depth of seasonal thawing was noted, due to the influence of inter-annual variations in snow cover and other random factors like vegetation changes, which can mitigate the impact of warming on permafrost (Anisimov *et al.*, 2002; Stieglitz *et al.*, 2003; Anisimov and Belolutskaia, 2004; Shur *et al.*, 2005). Soil temperatures in snow bed habitats are warm during winter, depending mainly on the actual soil temperature in September at the day of first snow accumulation (Björk and Molau, 2007). Monitoring studies indicate that the snow cover is among the key drivers to the ground thermal regime that controls permafrost dynamics. Field experiments on snow fences at Barrow (Alaska) and Schefferville (Québec–Labrador) have shown that increasing snow depths can induce active layer thickening within a few years. Widespread increases in thaw depth are projected over most permafrost regions (ACIA, 2005).

TABLE 13.2. *Recent trends in permafrost temperature*

Region	Depth (m)	Period of record	Permafrost (°C change)	Reference
United States				
Trans-Alaska pipeline route	20	1983–2000	+0.6 to +1.5	Osterkamp, 2003; Osterkamp and Romanovsky, 1999
Barrow Permafrost Observatory	15	1950–2000	+1	Romanovsky *et al.*, 2002
Russia				
East Siberia	1.6–3.2	1960–1992	$+0.03\ a^{-1}$	V. E. Romanovsky, pers. comm., 2003
Northwest Siberia	10	1980–1990	+0.3 to +0.7	Pavlov, 1994
European North of Russia (continuous permafrost zone)	6	1973–1992	+1.6 to +2.8	Pavlov, 1994
European North of Russia (discontinuous permafrost zone)	6	1970–1995	up to 1.2	Oberman and Mazhitova, 2001
Canada				
Alert, Nunavut	15–30	1995–2000	$+0.15\ a^{-1}$	Smith *et al.*, 2003
Northern Mackenzie Basin, N.W.T.	28	1990–2000	$+0.1\ a^{-1}$	Couture *et al.*, 2003
Central Mackenzie Basin, N.W.T	15	1985–2000	$+0.03\ a^{-1}$	Couture *et al.*, 2003
Northern Québec	10	Late 1980s–mid-1990s	$-0.1\ a^{-1}$	Allard *et al.*, 1995
	10	1992–2005	$+0.2\ a^{-1}$	M. Allard, pers. comm., 2008
	20	1992–2005	$+0.1\ a^{-1}$	M. Allard, pers. comm., 2008
Norway				
Juvvasshøe, southern Norway	20	1999–2008	$+0.045\ a^{-1}$	Isaksen *et al.*, 2007
Svalbard				
Janssonhaugen	20	1998–2008	$+0.055\ a^{-1}$	Isaksen *et al.*, 2007

Source: Modified from Romanovsky *et al.* (2002).

13.3 Permafrost thaw as a driving force of landscape change in tundra/taiga areas

Ground ice is a common component of permafrost. If the volume of ground ice exceeds the total pore volume, 'excess ice' forms, which can occur in various forms including segregation ice, ice-wedge ice and massive tabular ice (Mackay, 1972). The term 'massive ice' refers to an ice body with an ice content of at least 250% on an ice-to-dry soil weight basis (Mackay, 1971). As to ice-rich permafrost, it usually comprises >75% ice. Both massive ice and ice-rich permafrost are especially sensitive to thaw induced by ground warming and the ice content is one of the key drivers of geomorphic response of permafrost to environmental change.

Degradation of permafrost is associated with the development of destructive geomorphological processes, such as coastal, fluvial and hillslope erosion. For instance, abrupt landsliding and slow mass movement (gelifluction) will be favoured by increased water content in the soil and at the gliding surfaces. Thawing-induced ground settlement will ultimately change the northern landscapes into thermokarst-affected terrain. In this respect, estimation of the sediment-dependent ice content of the upper part of the permafrost is highly important. Snow thickness and vegetation cover changes affect river activity, resulting in changes of channel morphology and erosion/deposition rates. Many of these processes are relatively well studied and may be predicted using process-oriented geomorphological and permafrost models coupled with scenarios of climate change. A quantitative index may be used to evaluate the potential development of potentially dramatic geomorphological processes under the projected future climatic conditions. The basic assumption behind a so-called 'settlement index' is that the intensity of such processes increases with the depth of seasonal thawing and with the ground ice content. These two parameters characterise the volume of the uppermost thawed material involved in the processes of coastal erosion, mass movement and sediment removal by surface runoff, and the rate of potential ground settlement due to ice thawing.

Such index partition the circumpolar region into areas with 'low', 'moderate' and 'high' susceptibility to climate-induced geomorphological changes. A zone in the high-susceptibility category extends discontinuously around the Arctic Ocean, indicating high potential for coastal erosion. Large portions of central Siberia, particularly the Sakha Republic (Yakutia), and the Russian Far East show moderate or high susceptibility. Areas of lower susceptibility are associated with mountainous terrain, landscapes in which bedrock is at or near the surface, and permafrost with low ice content.

13.3.1 Geomorphic responses to global change: thermokarst subsidence and thermoerosion

Thawing of ice-rich sediments leads to ground subsidence and often results in an irregular surface known as 'thermokarst terrain' by analogy with landscapes due to limestone solution. Thermokarst is the process by which specific landscape features result from the thawing of ice-rich permafrost or the melting of massive ice (van Everdingen, 2002). It includes two components: thermokarst subsidence (downwearing) and thermoerosion (backwearing).

Thermokarst subsidence

Thermokarst subsidence induces the formation of thaw depressions such as alases, numerous in Yakutia (Czudek and Demek, 1970; Soloviev, 1973), and thaw lakes, abundant in Alaska. Remote sensing reveals the spatial importance of thermokarst depressions in the arctic coastal plains such as the Lena Delta region (Section 13.3.2). In subarctic Alaska, over 40% of permafrost has been affected by thermokarst subsidence since the end of the LIA, and new thermokarst terrain is currently forming (Osterkamp *et al.*, 2000; Osterkamp, 2007). At study sites in northern Quebec, 76% of the present-day thermokarst area has formed since the late 1950s (Vallée and Payette, 2007).

Thermoerosion

Thermoerosion is particularly active in ice-rich unconsolidated marine deposits and is one of the key drivers of coastal retreat. It predominates in clayey sediments of the coasts of the Barents and Kara seas, with thermokarst cirques developing on the cliffs. Their size depends on the extent and thickness of ice beds (Vasiliev *et al.*, 2005). Active layer detachment slides and retrogressive thaw slumps are particularly abundant along the coasts of the northern Canadian archipelago (Lantuit and Pollard, 2005). Thermoerosion niches are widespread in the alluvial material of the Siberian riverbanks, where they contribute to rapid lateral erosion (Czudek and Demek, 1970). A 35-year diachronic GIS analysis (1967–2002) of the middle Lena River (Costard *et al.*, 2007) demonstrated that the highest erosional impact is found on vegetated islands, with mean values of 15 m a^{-1} (against 2 m a^{-1} for the channel banks). The comparison of the island head retreat before the temperature increase (1980–92) and since 1992 clearly highlights a strong acceleration of erosion (+24%). Recent modelling applied to the Lena River showed that thermal erosion is mainly driven by the water stream temperature increase during the flood season, which is four times more efficient than the discharge increase (Costard *et al.*, 2007; Randriamazaoro *et al.*, 2007).

Fluvial response

There are indications that the fluvial geomorphology in the northern lands has already been affected by changes in climate, permafrost and vegetation. Changes in the fluvial regimes are likely to continue into the future, ultimately leading to the transformation of the channel types. Anisimov *et al.* (2008) examined fluvial regime using data from 16 selected river gauges in north European Russia and applied a geomorphological model to study the potential transformation of the channel types under current and projected climatic conditions. According to the results of this study, river channels at four of 16 sites are potentially unstable even under the current conditions.

Quantification of thermokarst-affected terrain types in the Siberian coastal plains

The ice-rich 'ice complex' deposits of northeast Siberia are particularly sensitive to climate warming and have been submitted to extensive thermokarst processes since the early Holocene. Key sites in the Laptev Sea coastal lowlands near the Lena Delta were investigated within the frame of the multidisciplinary joint German–Russian research projects 'Laptev Sea 2000' (1998–2002) and 'Dynamics of Permafrost' (2003–05). Techniques using CORONA and Landsat-7 satellite images, and digital elevation models

FIGURE 13.3. Gullying from thermokarst activity along ice wedge polygons, Bylot Island, North Canadian Archipelago (from Fortier *et al.*, 2007).

(DEMs) were developed for upscaling field data in order to quantify permafrost landscape units. There was a special focus on thermokarst features, due to their role in the release of organic carbon into the ocean or the atmosphere (Grosse *et al.*, 2005, 2006, 2007). It appears that 50–80% of the study areas are affected by permafrost degradation and display a variety of associated landscape features (Plate 40): thermokarst basins, lakes and lagoons, thermoerosional cirques, gullies and valleys. This regionally focussed procedure can be extended to other areas to quantify terrain affected by permafrost degradation on a large scale and in high resolution.

13.3.2 Landform changes as geoindicators of global change

Of the multiple landscape changes induced by permafrost degradation, two are of particular interest in so far as they cover extensive areas and can be traced through remote sensing. The first one is the drastic change from networks of ice-wedge polygons into groups of hills called *baydzherakh*, a Yakutian term used to describe silty or peaty mounds, sometimes called 'graveyard mounds'. These mounds are separated by gullies following the collapsing ice wedges, which are actively forming in the Canadian Arctic due to permafrost degradation (Fortier *et al.*, 2007) (Fig. 13.3); their coalescence can result in the formation of badland thermokarst relief (French, 2007). Recent degradation of massive ice wedges has also been documented in arctic Alaska (Jorgenson *et al.*, 2006). The second landscape change, from frost mounds called *palsas* to thaw ponds, is widespread in more southern subarctic regions where they have been extensively studied.

Palsas as geoindicators of climate changes in subarctic regions

Originally used by the Sami People and Finns, the term *palsa* means a perennial frost mound with a core made of alternating layers of segregation ice and mineral soil material, and a superficial peat covering (Seppäla, 1988; Pissart, 2002). When the peat cover is very thin or absent, mounds are referred to as *lithalsas* (Harris, 1993). Mounds are usually less than 100 m in diameter and 5–10 m in height. Palsas and lithalsas occur in groups or 'fields' within subarctic bogs and mires, and represent one of the most marginal permafrost features at the outer limit of the discontinuous and sporadic permafrost zone, which makes them particularly sensitive to climatic fluctuations (Seppälä, 1988). They are widespread in northern Fennoscandia, western Siberia and Québec–Labrador. The cyclic development of palsas from frozen mounds to ponds surrounded by a rim or rampart (Plate 41) seems to be accelerated by climate warming which enhances thermokarst phenomena (Laberge and Payette, 1995; Osterkamp and Romanovsky, 1999; Osterkamp *et al.*, 2000; Nelson *et al.*, 2001). Innovative tools such as tomodensitometric analysis provide high-resolution images of the internal structure of palsas/lithalsas at various stages of their growth and decay (Calmels *et al.*, 2008).

Remote sensing and monitoring studies of palsas and derived thermokarst ponds allow the detection of hotspots of recent environmental changes. In Norway, some palsa mires have totally degraded in recent times (Sollid and Sørbel, 1998). In the southernmost palsa mire of Sweden, palsa extent has decreased by about 50% between 1960 and 1997 (Zuidhoff and Kolstrup, 2000). In northern Sweden and Finnish Lapland, warm and humid summers combined with increased snowfall favour rapid decay of palsa complexes, with almost complete collapse of individual palsas within 5–10 years (Seppälä, 1994; Zuidhoff, 2002; Luoto and Seppälä, 2003). In northern Québec, the key driver of palsa decay over the last 50 years has been reduction of frost penetration due to increased precipitation in the form of snow, and since the mid-1990s, accelerated thawing has been facilitated by the additional temperature rise (Payette *et al.*, 2004). In western Siberia, thermokarst subsidences develop so swiftly that lichens and dwarf shrubs growing on palsa summits simply settle down under the water. It also happens that late-stage palsas are colonised by trees which trap winter snow and inhibit further palsa growth through increased insulation from winter frost. On the whole, the recent increase in thermokarst development from palsas/lithalsas indicate the high sensitivity of these frost mounds to changes in temperature and precipitation, and predictive models of palsa distribution in subarctic Fennoscandia are currently being developed (Fronzek *et al.*, 2006).

13.3.3 Interactions between permafrost degradation, morphodynamics, vegetation and snow cover

It is well known that permafrost (in)stability depends not only on climate trends but also on various biophysical factors including vegetation cover and associated organic layers. It is the reason why Shur and Jorgenson (2007) developed a new permafrost classification system to describe the complex interaction of climatic and ecological processes in permafrost formation and degradation. This classification is of interest to predict the response of permafrost to climate changes and surface disturbances. For example, climate-driven, ecosystem modified permafrost can experience thermokarst even under cold conditions because of its ice-rich layer formed during ecosystem development.

In arctic tundra regions, thermokarst landform evolution and vegetational successions are closely linked as illustrated by the 'thaw lake cycle' from incipient small ponds at the intersection of ice wedge troughs to the lake formation, expansion, drainage and final colonisation by peat bog vegetation (Billings and Peterson, 1980; Hinkel *et al.*, 2003). In subarctic forested regions, thermokarst landforms are particularly abundant (Jorgenson and Osterkamp, 2005) (Fig. 13.4) and geoecological combinations vary according to ground ice types (e.g. ice wedges and massive ice), presence of water (e.g. rivers and lakes), pre-existing periglacial features (e.g. frost mounds and polygonal networks) and vegetation types (e.g. forest and bog).

Thermokarst phenomena create disturbances in the boreal forests, from the alteration of the forest physiognomy to drastic vegetation changes. Impacts of thermokarst on the boreal forest depend primarily on the ice content of the permafrost and on drainage conditions (Osterkamp *et al.*, 2000). At Alaskan sites underlain by ice-rich permafrost, thaw subsidence up to 6 m leads to the formation of a 'drunken forest' with black spruce trunks tilting in all directions. Finally, forest ecosystems can be completely destroyed and replaced by wet sedge meadows, bogs, thermokarst ponds and lakes. Interactions between thermokarst development, ecological successions and permafrost dynamics are complex both in space and time due to their combined influences on the insulating snow and organic layers. Vegetation changes may have positive feedback effects where there is a tendency for the replacement of insulating layers made of peat, lichens and mosses, by shrub vegetation with higher thermal conductivity and better ability to trap snow (Cornelissen *et al.*, 2001; Van Wijk *et al.*, 2003). During the second half of the twentieth century, increased forest densification associated with warming summers and formation of thermokarst ponds created conditions increasingly favourable to snow retention and groundwater circulation (Beaulieu and Allard, 2003). This chain of environmental impacts created a positive feedback loop that accelerated permafrost degradation over a 50-year period of gradual change in seasonal climate regime. But as forests are predicted to ultimately occupy current tundra positions of the continuous permafrost zone (Tchebakova *et al.*, 2007), this might increase the thickness of the insulating organic layer and mitigate the effects of climate warming on permafrost thaw.

Southwards, in the present boreal forests of the discontinuous permafrost zone, the increasing thaw depth favours the complete burning off of the insulating organic layer during fire episodes, which in turn leads to an increase of the active layer thickness and thawing of the top of permafrost (Viereck, 1982). In 2004–05, 4.5 m ha were consumed by summer fires in Alaska, of which 90% were in the discontinuous permafrost zone. In Yukon, Burn (1998) showed a lowering of 3.8 m in 39 years of the permafrost table due to burning of a spruce forest. To assess the long-term impacts of fire on permafrost, it is necessary to evaluate not only the permafrost sensitivity to fire, but also the capacity of permafrost to recover after fire. In continental Alaska, redevelopment of ecosystem-driven permafrost after fire is restricted to sites with poor drainage and fine-grained soil, mainly due to the effects of moisture and soil texture on the thermal conductivity of soils (Shur and Jorgenson, 2007). Even under the most conservative scenarios of climate change, which predict a 3 °C increase by 2100 (ACIA, 2005), the climate of the discontinuous permafrost zone will become unfavourable to permafrost, and in most of the areas that have been affected by fire, permafrost will not recover (Shur and Jorgenson, 2007).

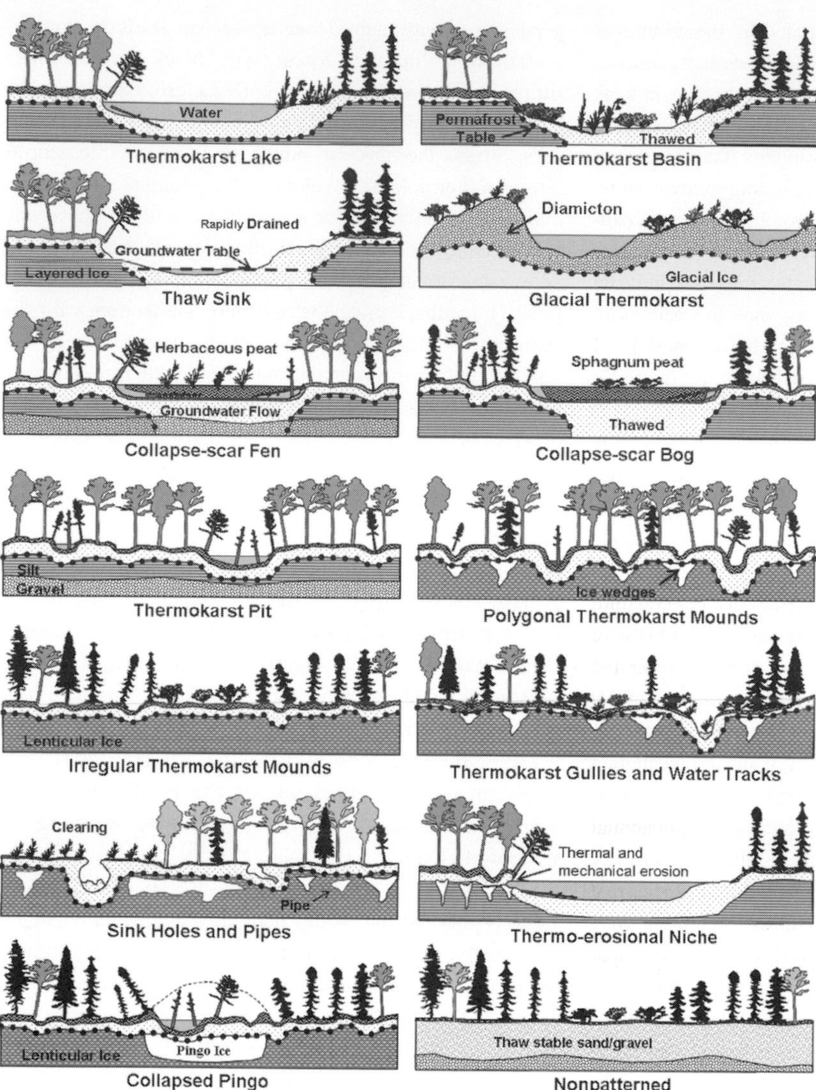

FIGURE 13.4. Schematic cross-sections illustrating the hydrologic, pedologic and vegetative characteristics of various modes of permafrost degradation in boreal Alaska (from Jorgenson and Osterkamp, 2005, their Figure 1).

13.4 Impact of landscape change on greenhouse gas release

13.4.1 Context and ongoing studies

According to the earlier studies, arctic soils contain approximately 455 Gt C, or 14% of the global soil carbon, of which about 50 Gt C are accumulated in the arctic wetlands (Anisimov and Reneva, 2006). Recent work has shown permafrost soil carbon pools to be much larger at depth than previously recognised because of cryogenic (freeze–thaw) mixing (Bockheim, 2007; Bockheim and Hinkel, 2007) and sediment deposition (Schirrmeister *et al.*, 2002). Available sparse data indicate that the entire northern circumpolar permafrost region may contain 1024 Gt of soil C in the surface 0 3 m depth (277 Gt of that in peatlands), with an additional 648 Gt of carbon locked in deep layers (~25 m thick) of aeolian and alluvial *yedoma* sediments (407 Gt), and deltaic deposits (241 Gt) of large arctic rivers (Zimov *et al.*, 2006; Schuur *et al.*, 2008). Several recent studies indicated high spatial variability with near-zero balance between the sink (photosynthetic uptake) and source (release due to soil decomposition) of carbon in the entire Arctic (Callaghan, 2004; Chapin *et al.*, 2005; Corradi *et al.*, 2005). The carbon turnover in the Arctic is projected to increase under the warmer climate; however the timing of the processes that determine the status of the Arctic as net sink or source varies. Increased trace gas emissions due to soil warming is likely to be the short-term response to climate change. In the longer-term warmer climate, more protracted growing periods, and northward movement of productive vegetation may increase photosynthetic carbon uptake.

The effect that the increase in the rate of soil carbon decomposition in the next few decades may have on the radiative forcing depends on the balance between the amounts of carbon emitted as CO_2 and CH_4. Methane has more than 20 times stronger greenhouse effect than an equal amount of CO_2. A few ecosystems in the Arctic, including wetlands, convert part of carbon that has been photosynthetically captured from the atmosphere as CO_2 to methane, which is further released as the product of organic soil decomposition. Because of this, even the areas and ecosystems that have net carbon sink status, such as tundra, may enhance the global radiative forcing if sufficient fraction of carbon is emitted as CH_4 (Friborg *et al.*, 2003; Callaghan, 2004).

Organic materials that are deposited in the arctic wetlands below the depth of seasonal thawing are currently not involved in the carbon cycle and may become available under warmer climatic conditions. Observations indicate that methane emissions in northern mires and peatlands are responsive to climatic variations. A detailed study of one mire shows that the climatic warming, deeper permafrost thawing and subsequent vegetation changes have been associated with increases in landscape-scale methane emissions in the range of 22–66% over the period 1970–2000 (Christensen *et al.*, 2004). Observations in northern Sweden indicate that the temperature and microbial substrate availability combined explain almost 100% of the variations in mean annual methane emissions (Christensen *et al.*, 2004).

Results from coupled carbon/permafrost models suggest that the flux of methane from Russian permafrost regions may increase by 6–8 Mt by 2050. The projected increase is compatible with the current annual net source of *c.* 20 Mt resulting from the balance between the much larger global source (*c.* 550 Mt) and sink (*c.* 530 Mt) of methane. However the effect of such changes on global climate will be small. If other sinks and sources remain unchanged, the projected increase in methane flux may raise the overall amount of atmospheric methane by *c.* 100 MT, or 0.04 ppm. Given that the sensitivity of the global temperature to 1 ppm of atmospheric methane is approximately 0.3 °C (IPCC, 2001a), additional radiative forcing resulting from such an increase may raise the global mean annual air temperature by 0.012 °C. This result indicates that many of the recent publications, both scientific and in the mass media, overstate the concerns associated with thawing wetlands in permafrost regions and the effect this process may have on the global climate system (Anisimov, 2007).

13.4.2 Geomorphological and geoecological services

The input of geoecological and geomorphological research in the carbon debate is threefold:

(a) spatial variability of terrestrial landscape changes involved in the carbon balance;
(b) evaluation of the impact of geomorphic processes on redistribution of soil organic carbon; and
(c) quantification of the carbon input within the Arctic Ocean.

In recent years, much attention has been paid to methane bubbling from thermokarst ponds and lakes as a positive feedback to climate warming (e.g. Zimov *et al.*, 1997). Walter *et al.* (2006) showed that upwelling accounts for 95% of methane emissions from Siberian thaw lakes and that the expansion of such lakes between 1974 and 2000 increased emissions by 58%. As modelling tends to minimise the effects of such increases on global climate (see Section 13.4.1), it is of interest to further investigate ongoing landscape trajectories, with special attention being paid to lake and pond area dynamics, both at local and zonal scales. Long-term monitoring of subarctic palsa plateaus located on the east coast of Hudson Bay, northern Québec (Payette *et al.*, 2004) (Fig. 13.5) indicate that the accelerated thawing of permafrost during the period 1957–2003 has induced the concurrent development of thermokarst ponds (carbon source) and peat accumulation through natural successional processes of terrestrialisation (carbon sink). These compensatory mechanisms have been also observed in other subarctic regions (Christensen *et al.*, 2004) and should be taken into account in emission scenarios. For the entire circumarctic region, projected losses of lake area due to further permafrost degradation imply a possible reduction by approximately 12% in methane emissions from northern lakes by 2100 (Smith *et al.*, 2007; Walter *et al.*, 2007).

Another key question deals with the effects of sustained warming on redistribution of soil organic carbon (SOC) in permafrost-affected soils. Recent soil studies in northern Alaska indicate that 55% of the SOC density of the active layer and near-surface permafrost can be attributed to redistribution from cryoturbation (Bockheim, 2007). As continued warming might accelerate cryoturbation, it will increase the incorporation of dense, high-molecular-weight SOC at depth, thereby enabling the soil to store more SOC than at present, and mitigating the loss of carbon dioxide to the atmosphere from increased soil respiration. The magnitude of cryoturbation will depend on geomorphological contexts, such as the occurrence of frost boils and ice wedge polygons.

The carbon input to the Arctic Ocean depends on both coastal and fluvial erosional activity and recent studies indicate that coastal erosion is a major source of the total organic carbon input (TOC). Based on a review of the existing literature (Rachold *et al.*, 2003) and on detailed field studies carried out in the Laptev and East Siberian seas

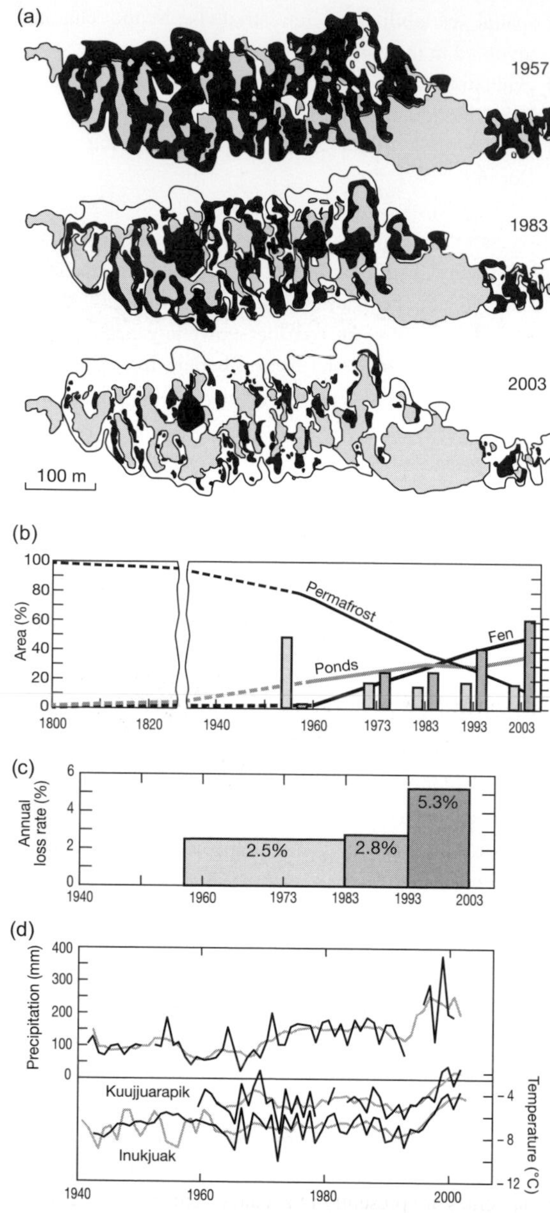

FIGURE 13.5. Peatland changes associated with permafrost thawing between 1957 and 2003 on the eastern coast of Hudson Bay, northern Québec, Canada (modified from Payette *et al.*, 2004). (a) Changing patterns of permafrost (black), thermokarst ponds (grey) and fen vegetation (white). (b) Changing cover (%) of permafrost, thermokarst ponds and fen vegetation; vertical bars correspond to number of thermokarst ponds (light grey) and palsa mounds (dark grey), the latter increasing due to the progressive fragmentation of the palsa plateau inherited from the Little Ice Age. (c) Annual rates of permafrost loss (%). (d) Precipitation data (Inukjuak weather station) and temperature data (Inukjuak and Kuujjuarapik weather stations).

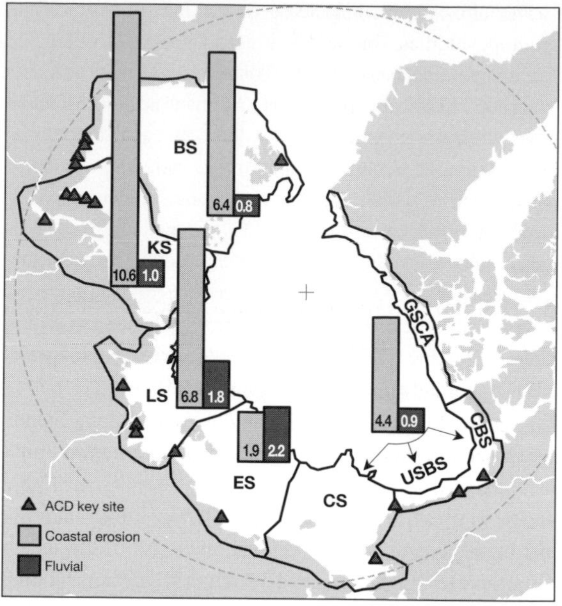

FIGURE 13.6. Comparison between riverine and coastal total organic carbon (TOC) input to the arctic seas (modified from Rachold *et al.*, 2005,). Arctic Coastal Dynamics (ACD) key sites are marked by triangles, and the ACD subdivision of the arctic coastline by major seas reads as follows: BS, Barents Sea; KS, Kara Sea; LS, Laptev Sea; ES, East Siberian Sea; CS, Chukchi Sea; USBS, US Beaufort Sea; CBS, Canadian Beaufort Sea; GSCA, Greenland Sea/Canadian Archipelago.

(Grigoriev and Rachold, 2003), the comparison between riverine and coastal TOC input is shown in Fig. 13.6. These are the best currently available estimates, but they may include errors ranging from *c.* 30% for the Laptev and East Siberian seas to one order of magnitude for the other seas (Rachold *et al.*, 2005). Estimated carbon input from eroding shorelines averages 149 Mg km^{-1} a^{-1} and totals 1.8×10^5 Mg a^{-1} for the entire Alaskan Beaufort Sea coast (Jorgenson and Brown, 2005). Coastal deposits of the Kara and Barents seas, where peat bog soils occupy relatively small areas, have a low organic carbon content: about 1% in clay sediments and not more than 0.7% in sands (Vasiliev *et al.*, 2005).

13.5 Socioeconomic impact and hazard implications of thermokarst activity

Circumarctic/subarctic permafrost regions are inhabited by 4 million people scattered in 370 settlements in tundra regions and several thousand settlements in the boreal forest. Human concentrations are particularly important in northern Russia, with cities with over 100 000 inhabitants (Yakutsk,

Noril'sk, Vorkuta) and large river ports (Salekhard on the Ob, Igarka and Dudinka on the Yenisei, and Tiksi on the Lena). The two main corridors of development associated with oil and gas are the Beaufort–Mackenzie–North Slope in North America and the Barents Sea–Pechora Basin in Russia. Both projects concern vulnerable regions with traditional caribou hunting or reindeer herding, and many sensitive coastal and marine habitats.

13.5.1 Damage to human infrastructure due to thaw-induced settling

Recent thaw subsidence associated with thermokarst has been reported in areas of Siberia and North America. Such subsidence has detrimental impacts on infrastructure built on permafrost. Ice-bonded sediments can have considerable bearing capacity and are often an integral part of engineering design in cold regions. The bearing capacity of permafrost decreases with warming, resulting in failure of pilings for buildings and pipelines. In Alaska, the runway serving the Prudhoe Bay oil fields has been reconstructed due to settling from melting permafrost (Hinzman *et al.*, 2005). In Nunavik, permafrost degradation is threatening the integrity of roads and airfields in 13 Inuit communities (Doré and Beaulac, 2007). In the Yakutsk region, a 2 °C rise in soil temperature has led to a decrease of 50% in the bearing capacity of frozen ground under buildings. The damage provoked by differential settlement affected hundreds of residential buildings, the Yakutsk airport and a power generating station. In 1998, the city was declared a natural disaster area. Russia is the most severely affected due to the extent of cities and river ports built on permafrost.

Accelerated frost thawing leads to costly increases in road damage and maintenance. In Alaska, it costs US$1.5 million to replace 1 km of road system (Weller and Lange, 1999). Increased economic costs are expected to affect infrastructure in permafrost regions, and engineers are already working on new design for permafrost terrain under changing conditions.

The impact of permafrost thaw on Russian cities and infrastructure

A 1998 survey of infrastructure in industrially developed parts of the Russian Arctic indicates that impacts of warming and thawing permafrost on engineered structures are already taking place (Anisimov and Lavrov, 2004). Many buildings in Russia's northern cities are in a potentially dangerous state: 10% in Noril'sk, 22% in Tiksi, 55% in Dudinka, 35% in Dicson, 50% in Pevek and Amderma, 60% in Chita and 80% in Vorkuta. Analysis of related accidents indicates that they increased by 42% in the city of Noril'sk, 61% in Yakutsk and 90% in Amderma in the period 1990–2000 compared to previous decades. A potentially dangerous situation has also been observed with respect to transportation routes and facilities. According to 1998 data, 46% of the roadbed under the Baikal–Amur railroad has been deformed by thawing of frozen ground, a 20% increase over the early 1990s. Runways in Noril'sk, Yakutsk, Magadan and other major Siberian cities are presently in a state of emergency. Serious situations have been observed in gas and oil pipelines traversing the Russian North. In 2001, for example, 16 breaks were reported on the Messoyakha–Noril'sk pipeline, causing significant economic and environmental damage. Some examples are illustrated in Fig. 13.7.

(a)

(b)

FIGURE 13.7. Damages provoked by permafrost thaw in Russian cities. (a) Collapse of a section of a residential building in Cherskiy, East Siberia, due to thawing of permafrost (photo by V. E. Romanovsky). (b) Extensive thermokarst development in a car park lot in Yakutsk (photo by N. Shiklomanov).

13.5.2 Predictive hazard mapping

Nelson *et al.* (2001) used the settlement index described in Section 13.3 to evaluate the potential threats to engineered structures due to warming and thawing of permafrost. The more recent study by Anisimov and Lavrov (2004) uses a modified hazard index that also includes the salinity of soils. Soil salinity is particularly important in the vicinity of the arctic shoreline. Currently emerging areas were previously located below sea level, and salt has been deposited in the upper soil layer. Even slight temperature variation may shift the balance between the ground ice and unfrozen water in such soils, which are thus particularly sensitive to climatic changes.

A predictive map based on a modified hazard index was constructed using the results from a permafrost model forced with climatic projection for 2050 (Plate 42). Areas of greatest hazard potential include the arctic coastline and parts of Siberia in which substantial development has occurred in recent decades. Particular concerns are associated with Yamal Peninsula which falls into the highest risk zone, because of the ongoing expansion of oil and gas extraction and of the transportation industry into this region. Although temperatures there are relatively low, frozen ground is already very unstable, largely because of its high salinity, and thus even slight warming may cause extensive thawing of permafrost and ground settlement. Such changes may potentially have serious impacts on the infrastructure, although the ultimate effect is construction-specific and largely depends on maintenance. Some structures may be relatively insensitive or easily adaptable to the projected changes, while others may be highly susceptible to degradation of permafrost. The problem is complicated by the fact that it is often impossible to differentiate between the effects of permafrost temperature increase and other factors, such as inadequate management or engineering design, on the damage. As reported by Kronik (2001), a majority of the damage to structures in Russian permafrost regions in the period 1980–2000 resulted from poor maintenance rather than climatic change. Similarly, Instanes (2003) emphasises that, in Svalbard, most problems with infrastructure relate to poor construction techniques. An example is provided by the runway at Svalbard Airport, Longyearbyen. Since the opening of the airport in 1975, the runway has repeatedly suffered from alternating thaw settlement and frost heave, mainly due to the fact that it was cut directly into ice-rich, frost-susceptible sediments; moreover, salty marine sediments were used as fill material and no insulation measures were taken (Instanes and Instanes, 1998; Humlum *et al.*, 2003).

13.6 Vulnerability of arctic coastal regions exposed to accelerated erosion

13.6.1 Natural and human causes of particular sensitivity of coastal regions

Of some 370 settlements in the tundra regions, more than 80% are located on the coast. Coastal arctic regions concentrate industrial facilities associated with oil and gas such as the Prudhoe Bay region in northern Alaska and the Pechora Basin in Russia. For this reason, they are particularly sensitive to rapid erosion enhanced by the contemporary combination of sea level rise, sea ice retreat and thawing of exposed permafrost, as illustrated by Shishmaref in Alaska (see below). Global average sea level rose at an average rate of 1.8 mm a^{-1} in the period 1961–2003 and the rate was faster in the period 1993–2003, about 3.1 mm a^{-1} (IPCC, 2007a). The extent of ice in Nordic seas during April has decreased by about 33% since the 1860s (www.arctic.noaa.gov/reportcard/seaice.html) and the annual average arctic sea ice extent has shrunk by 2.7% per decade since 1978, with larger decreases in summer of 7.4% per decade (IPCC, 2007a). Partly due to a longer open water season (e.g. Belchansky *et al.*, 2004), the erosional impact of storms is less mitigated by the sea ice cover, and wave action in ice-rich permafrost areas can produce extremely high rates of coastal erosion.

Shishmaref: an Alaskan Inuit village at sea

The Alaskan Inuit village of Shishmaref (600 inhabitants) is located on Sarichef Island which culminates at 6.5 m a.s.l. east of Bering Strait. In recent years, reduction of sea ice and permafrost thaw associated with climate warming (+4 °C in 30 years) and sea level rise induced accelerated coastal erosion. In 1997, a severe storm provoked a dramatic retreat of the coastline (–38 m), several houses were destroyed (Fig. 13.8) and 12 had to be removed. In 2001, huge waves threatened most of the village, and in 2002, the inhabitants decided to transfer the whole village to the mainland. Twenty more Alaskan villages will probably be forced to relocate in the coming years.

FIGURE 13.8. Shishmaref, an Alaskan village severely affected by coastal erosion, April 2007 (photo by D. Fortier).

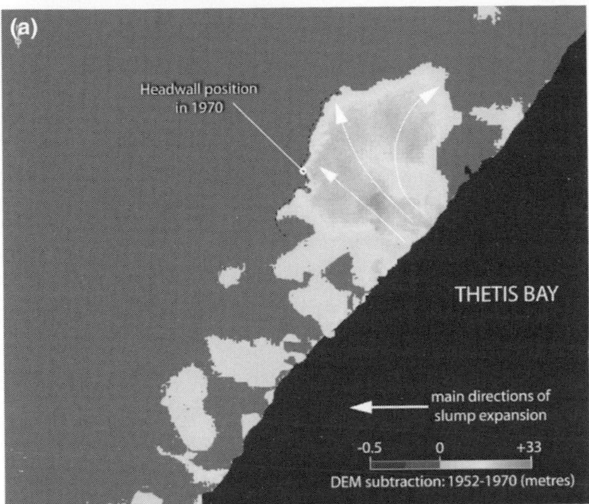

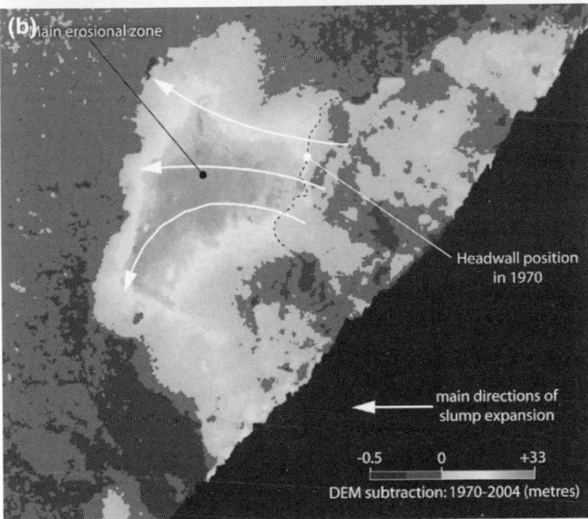

FIGURE 13.9. Volume loss maps associated with retrogressive thaw slump activity for the periods (a) 1952–70 and (b) 1970–2004 on Herschel Island, Yukon Territory (from Lantuit and Pollard, 2005, their Figures 10–11). Based on sequential DEMs and the associated three-dimensional geomorphic analysis, volume loss maps for 1952–70 and 1970–2004 show that the main zone of erosion has tripled in size between the two periods.

13.6.2 Accelerated coastal erosion in the Arctic: rates, processes, controls

Many arctic coasts are characterised by rapid erosion rates forced by sea level rise, periodic storms and permafrost thaw in ice-rich coastal deposits, and coastal erosion is investigated by researchers involved in the Arctic Coastal Dynamics (ACD) programme of the International Arctic Science Committee (IASC) and the International Permafrost Association (IPA) (Rachold *et al.*, 2005).

In recent decades, much of the coastline of the Canadian and Alaskan Beaufort Sea has been eroding at long-term rates of up to *c.* 20 m a^{-1} (e.g. Reimnitz *et al.*, 1988; Solomon, 2005). Average annual coastal erosion rates of 1 to 4 m a^{-1} are common in the Beaufort, Barents and Kara seas and are among the highest in the world. Recent trends consistent with climate change trends have not been clearly established to date.

Along the Beaufort Sea, coastal erosion has breached thermokarst lakes, causing draining of the lakes followed by marine flooding, and many freshwater lakes evolve into marine bays over time. In the National Petroleum Reserve in Alaska, quantitative remote sensing studies indicate that land loss from coastal erosion more than doubled between 1985 and 2005: the rate of erosion increased from 0.48 km^2 a^{-1} during 1955–85 to 1.08 km^2 a^{-1} during 1985–2005 (Mars and Houseknecht, 2007), most probably due to regional warming (Osterkamp, 2007). In the Yukon coastal plain, one of the most ice-rich and thaw-sensitive areas in the Canadian Arctic, retrogressive thaw slumps have recently increased in frequency and extent due to accelerated thawing of massive ground ice and coastal erosion (Lantuit and Pollard, 2005) (Fig. 13.9).

In Siberia, thermoterraces are widespread along the eroding coasts. They represent a unique feature of coastal geomorphology, for they progressively record quantitative information about the duration of their existence. This information can be used to calculate the average rate of the shoreline retreat over the lifespan of the terrace. The detailed method, including a geodetic survey of key points used for thermoterrace cross-profile measurements, is exposed by Are *et al.* (2005). Erosion rates calculated from thermoterrace dimensions are consistent with those derived by comparing geodetic survey results with aerial photographs.

The granulometry and ice content of coastal sediments are a major control on erosion rates and the associated sediment input in arctic seas. In the Laptev Sea, most of the sediment input comes from the so-called *Yedoma* or 'ice complex' (Fig. 13.10), which contain massive ice bodies. Subaerial erosion from this ice-rich deposit (ice content 40–70%) is 0.019 10^6 t a^{-1} km^{-1}, i.e. triple that from other Quaternary deposits with 10–30% ice content (Rachold *et al.*, 2000). For the mainland coast of the Alaskan Beaufort Sea, average erosion rates are 4 to 10 times higher in ice-rich muddy sediments than in ice-poor sandy to gravelly sediments (e.g. Jorgenson and Brown, 2005) (Fig. 13.11). Where sand-rich permafrost bluffs are exposed at the shoreline, sandy beaches protect permafrost bluffs from wave action (e.g. Reimnitz *et al.*, 1985). On the Alaskan coastal plain, once sandy barrier islands are eroded, erosion accelerates along mud-rich permafrost shorelines where wave undercutting triggers block collapse

FIGURE 13.10. 'Ice Complex' with large wedges, Bolshoy Lyahkhovsky Island, East Siberian Sea, 73° 33′ N (©ACD/AWI/ G. Grosse).

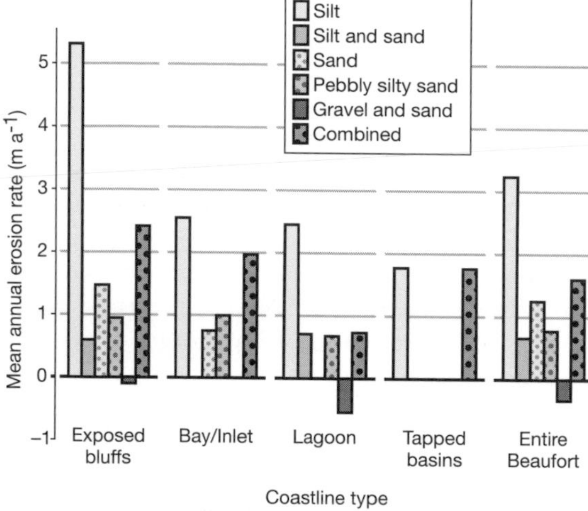

FIGURE. 13.11. Erosion rates by coastline type and soil texture for the mainland coast of the Alaskan Beaufort Sea (from Jorgenson and Brown, 2005, their Figure 4).

from permafrost bluffs (Mars and Houseknecht, 2007). Similar erosional processes can be observed along the Canadian and Russian arctic coasts. Once collapsed, permafrost blocks rapidly melt in seawater and wave action transports muddy sediments offshore.

On the whole, the sediment input to the Laptev Sea from coastal erosion is twice as large as the river input and in parts of the Alaskan coast such as the Colville River area, coastal erosion supplies seven times as much sediment to the Beaufort Sea as rivers (Reimnitz *et al.*, 1988; Rachold *et al.*, 2000; Grigoriev and Rachold, 2003). In contrast, the Canadian Beaufort Sea is dominated by the riverine sediment discharge mainly of the Mackenzie River, which exceeds coastal sediment input by a factor of *c.* 10 (Rachold *et al.*, 2000).

13.7 Discriminating the climate, sea level and land use components of global change

Many environmental changes occur over extensive areas in the absence of human pressure, for example, gullying along ice wedge polygons in tundra environments, and collapse of palsa fields in subarctic areas. In such cases, climate changes are the main control on landscape change. The accelerated coastal erosion in remote high arctic areas is due to the combination of sea level rise, sea ice retreat and thermokarst backwearing associated with warming. Lastly, permafrost degradation due to forest fires is accelerated by climate warming and direct human causes are rare.

In contrast, anthropogenic disturbances are responsible for striking changes in permafrost areas over relatively short timescales (see review in Brown, 1997). Construction of airstrips and roads can trigger thermokarst and associated subsidence (French, 1975; Nelson and Outcalt, 1982). Removal of the insulating vegetation cover can induce over 6 m of subsidence over a 26-year period, as demonstrated by experiments carried out near Fairbanks, Alaska (Péwé, 1954; Linell, 1973). Similarly, along the Kiruna–Narvik road opened in 1980, a permafrost lens in peatland terrain was disturbed by the construction work, provoking an overall 2.5 m subsidence (C. Jonasson, personal communication, 2008). Even trampling and overgrazing can trigger thermokarst and induce widespread thaw settlement (Mackay, 1970). Agencies regulating tourism in the Arctic are increasingly aware of the disrupting effects of trampling on permafrost terrain and have started developing new strategies either by strictly concentrating traffic along selected trails or by encouraging more diffuse patterns of use.

Discriminating and evaluating the respective roles of the natural and human components of environmental change is possible based on comparative studies like the one recently carried out in the Pechora Sea region (Ogorodov, 2004).

Evaluating the human impact on coastal stability in Varandei industrial complex, Pechora Sea

In the West Siberian Pechora Sea region, comparative monitoring of shoreline recession was carried out in areas affected and unaffected by economic activity, with special reference to the Varandei industrial zone (Ogorodov, 2004). The study area comprises two main geomorphological complexes:

(a) a 3–5 m high Holocene marine terrace, formed of ice-poor sand (5–10%) and partly covered by a dune belt reaching up to 12 m a.s.l.; under natural conditions, most of the terrace is being eroded at a rate of 0.5–2.5 m a^{-1} due to the deficit of beach material and the high gradient of the foredune slope;

(b) a 5–15 m high lacustrine–alluvial plain with numerous lakes, bogs and polygonal nets, mainly formed of loams and clays including ice wedges and massive ice beds; a thermoabrasion cliff is cut in these sediments, which is retreating at 1.8–2.0 m a^{-1}.

Active exploitation started in the 1970s and Varandei Island was subjected to the strongest human impact, with the oil terminal and a settlement with 3500 inhabitants being established on the well-drained dune of a marine terrace, which seemed more stable than the surrounding swampy plain. Construction of the industrial zone practically at the edge of the abrasion cliff was accompanied by sand removal from the foredune and beach. During exploitation, the coastal zone was submitted to severe mechanical disturbance, causing degradation of plant and soil cover of the whole dune belt of Varandei Island. Sands were subject to rapid deflation and thermoerosion inducing up to 3 m surface lowering within two decades. The overall cliff retreat reduced the amount of sediments supplied to the coastal zone, and as a result of human impact, the coastal erosion rate increased considerably from 0.5–2 to 7–10 m a^{-1}. Whereas coastal protection provoked the gradual decrease of coastal retreat rates to *c.* 2 m a^{-1} in places, it reduced sediment supply to the coast and enhanced erosion of the adjacent areas. Measurements taken in the period 1987–2000 indicate that the rate of coastal retreat around the settlement increased and reached 3–4 m a^{-1} (i.e. twice as high as in unaffected regions). Gullies up to 4 m deep are currently forming in the coastal bluff in one season. Several industrial and residential buildings were destroyed, and the oil terminal, airport and additional buildings and infrastructures are endangered, because the distance to the sea is only a few metres (Fig. 13.12). At Varandei, human impact clearly activates destructive coastal processes that considerably complicate industrial development of the coastal zone.

FIGURE 13.12. Wave-cut cliff near the Varandei oil terminal, Pechora Sea (© ACD/V. Tumskoi).

13.8 Lessons from the past

Whereas analysis of aerial photographs allows the assessment of rates of recent landscape change, palaeoenvironmental studies, including soil sampling, sedimentological analyses and radiocarbon dating, provide insights into the long-term response of geoecosystems to climate warming. In this context, identifying thermokarst sediments and sedimentary structures in the geological record is crucial (Murton, 2001), and particular attention should be drawn to previous warm periods such as the Holocene climatic optimum and the Medieval Warm Period, still poorly documented.

In unglaciated lowland regions of Siberia and northwest North America, the last deglaciation period, characterised by a warm and moist climate, saw the widespread initiation of thermokarst (Edwards *et al.*, 2007). The onset of thermokarst lake development dates back to *c.* 14 ka BP, and many lake initiation dates fall in the period *c.* 11–18 ka BP. A later major pulse of lake formation coincided with the Holocene thermal optimum in, for example, northeast Siberia where thermokarst lakes reached their maximum extent around *c.* 7–15 ka BP (Grosse *et al.*, 2007). Based on radiocarbon ages of wood fragments, Katamura *et al.* (2006) also suggest that most if not all thermokarst depressions (alases) in central Yakutia formed during the early Holocene under moist and warm conditions. The antiquity of Yakutian alases is also emphasised by Brouchkov *et al.* (2004), who did not find any evidence of noticeable increase of thermokarst terrain associated with recent climate warming. Similarly, in northern Alaska, radiocarbon dating of thaw lake sediments show that thermokarst development has been operating for at least 5500 a (Hinkel *et al.*, 2003; see below). On the Tanana Flats in central Alaska, it appears that 83% of thermokarst development has occurred before 1949, with a peak during the relatively warm mid to late eighteenth century (Jorgenson *et al.*, 2001).

Whereas contemporary development of thermokarst has been primarily described in subarctic margins, thaw unconformities observed in sediments of the western arctic coast of Canada show that climatic warming can trigger thermokarst disturbances farther north (Burn, 1997). It means that the continuous permafrost zone should not be considered as insensitive to climate warming, as also suggested by

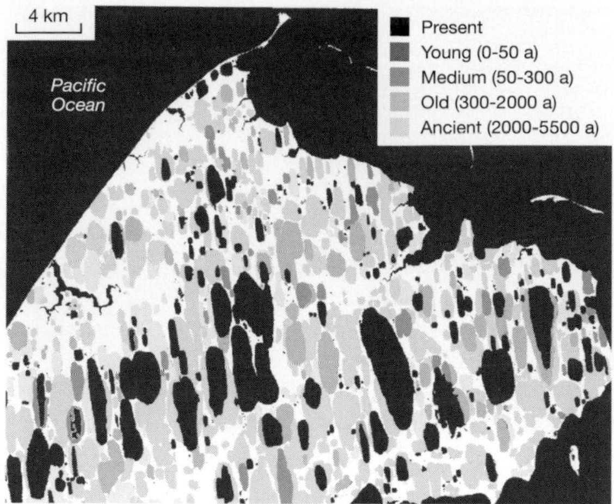

FIGURE 13.13. Age-classified Landsat-7 image of drained thaw lake basins on the Barrow Peninsula, Alaska (modified from K. Hinkel *et al.*, 2003) (www.geography.uc.edu/~kenhinke/DTLB/barrow_class.jpg, accessed 4 November 2008).

models and associated hazard zonation maps (Anisimov *et al.*, 1997; Nelson *et al.*, 2001, 2002).

The thaw lakes of the Barrow Peninsula and the geological record of thermokarst activity

About 20% of the arctic Alaskan coastal plain contains elliptical thaw lakes and drained thaw lake basins developed in ice-rich permafrost, with the longest axis oriented nearly perpendicular to the prevailing summer wind direction. Up to 75% of the Barrow Peninsula is covered with such lakes and basins. The 'thaw lake cycle', as described by previous authors (e.g. Hopkins, 1949; Carson and Hussey, 1962; Billings and Peterson, 1980) begins with the development of ponds at the intersection of ice wedge troughs or in low-centre polygons, that coalesce to form lakes that grow due to thermoerosion along margins, and finally drain, giving rise to a vegetated basin. However, lake evolution can be more complex and less cyclic than previously theorised, as suggested by detailed terrain and photogrammetric analyses (Jorgenson and Shur, 2007). Whatever the scenario, estimating the age of these thaw lake basins is of interest to put in context the current thermokarst activity. The relative age of 77 basins on the Barrow Peninsula was estimated by Hinkel *et al.* (2003) using the degree of plant community succession and verified by radiocarbon dating material collected from the base of the organic layer in 21 basins (Fig. 13.13). Using Landsat-7 imagery, a neural network classifying algorithm was developed from basin age-dependent spectra and texture. It appears that thermokarst activity has been operating on the Barrow Peninsula for at least 5500 a. Based on remote sensing and ^{14}C dating, basins can be arrayed into four age classes: 5500–2000 BP (ancient basins); 2000–300 BP (old basins); 300–50 BP (medium age basins); 50–0 BP (young basins). Soil studies indicate that the following properties of the organic layer increase with basin age: thickness of the organic layer, morphology of the organic horizons (i.e. degree of decomposition), and spatial variability in SOC in the organic layer (Bockheim *et al.*, 2004). Profile quantities of SOC average 48 kg m^{-3}, which is less than values reported from the arctic coastal plain (62 kg m^{-3}). The average long-term net accumulation of carbon in basins more than 100 years old is 13 g m^{-2} a^{-1}, which is comparable with values reported in the Russian Arctic.

13.9 Geomorphological services and recommendations for future management of permafrost regions

Large-scale predictive permafrost hazard maps are being developed (see Section 13.5.2) and Nelson *et al.* (2002) have emphasised the need for detailed mapping of hazards at local and regional scales. When combined with geophysical investigations, detailed geomorphological surveys and mapping are a crucial tool to meet the requirements of land management in large northern cities as in remote Native settlements.

Salluit (Nunavik): how to cope with demographic growth in sensitive permafrost terrain

Located on the southern shore of Hudson Strait, the village of Salluit, home to 1100 people, is built mainly on ice-rich marine clays on the floor of a steep-walled valley. The region lies in the continuous permafrost zone and the village was built in the 1950–70s, when the climate was cooling. The climate trend switched to fast warming during the 1990s and active layer detachment failures occurred in summer 1998 close to a residential area. These events prompted the launching of a study aiming to obtain a high-precision spatial and in-depth characterisation of permafrost conditions. As population growth is rapid (+30% between 1991 and 2001), 80 new dwellings are needed by 2026, as well as public and business services. Therefore, the ultimate goal of the study carried out by the Centre of Nordic Studies of Laval University (Allard *et al.*, 2004) was to make recommendations to select safe building areas. Twenty-two cores, geophysical soundings (GPR, electrical resistivity and reflection seismics) and cone penetration tests revealed that most of the area contains over 200%

(per dry weight) ice in the first 2 m below permafrost table. All landforms indicative of terrain instability (such as tension cracks, active layer failures, and thermoerosional gullies) were mapped (Plate 43). This integrated zoning approach allowed the delineation of favourable ground for construction in nearby bedrock outcrops and ice-poor coarse-grained soils (Plate 44).

Due to recent advancement of remote sensing technologies many surface features of permafrost terrain and typical periglacial landforms are now observable with a variety of sensors ranging from conventional aerial photography and high to medium resolution satellite imagery. Together with *in situ* measurements and modelling, data coming from remote sensing increase our understanding of the geomorphological changes in the northern lands.

Acknowledgements

This chapter benefited from scientific exchanges with many permafrost scientists. Special thanks are due to Michel Allard, Jim Bockheim, Fabrice Calmels, Hanne Christiansen, Daniel Fortier, Hugh French, Stefan Fronzek, Guido Grosse, Charles Harris, Ole Humlum, Torre Jorgenson, Hugues Lantuit, Frederick E. Nelson, Tom Osterkamp, Albert Pissart, Dan Riseborough, Matti Seppälä, Amber J. Soja, Johan Ludvig Sollid, Charles Tarnocai and Katey Walter. Oleg Anisimov's permafrost research is supported by the Russian Foundation for Basic Research, grants 07-05-00209 and 07605-13527. Jean-Pierre Magnier and Eric Leinberger's graphical assistance was much appreciated.

References

ACIA (2005). *Arctic Climate Impact Assessment*. Cambridge: Cambridge University Press. Available at www.acia.uaf.edu/

Allard, M., Wang, B. and Pilon, J. A. (1995). Recent cooling along the southern shore of Hudson Strait, Quebec, Canada, documented from permafrost temperature measurements. *Arctic and Alpine Research*, **28**, 157–166.

Allard, M. *et al.* (2004). *Salluit: une communauté en croissance sur un terrain sensible au changement climatique – Problématique du développement du village de Salluit, Nunavik*, Rapport final, Centre d'Etudes Nordiques. Quebec: Université Laval.

Anisimov, O. (2007). Potential feedback of thawing permafrost to the global climate system through methane emission. *Environmental Research Letters*, **2**, 14–21.

Anisimov, O. A. and Belolutskaiya, M. A. (2004). Predictive modelling of climate change impacts on permafrost: effects of vegetation. *Meteorology and Hydrology*, **11**, 73–81.

Anisimov, O. A. and Lavrov, C. A. (2004). Global warming and permafrost degradation: risk assessment for the infrastructure of the oil and gas industry. *Technologies of Oil and Gas Industry*, **3**, 78–83 (in Russian).

Anisimov, O. A. and Nelson, F. E. (1997a). Influence of climate change on continental permafrost in the northern hemisphere. *Meteorologiya I Gidrologiya*, **5**, 71–80 (in Russian – English translation appears in *Russian Meteorology and Hydrology* 1997/5).

Anisimov, O. A. and Nelson, F. E. (1997b). Permafrost zonation and climate change: results from transient general circulation models. *Climatic Change*, **35**, 241–258.

Anisimov, O. A. and Poljakov, A. V. (1999). Predicting changes of the air temperature in the first quarter of the twenty-first century. *Meteorology and Hydrology*, **2**, 25–31 (in Russian).

Anisimov, O. and Reneva, S. (2006). Permafrost and changing climate: the Russian perspective. *Ambio*, **35**, 169–175.

Anisimov, O. A., Shiklomanov, N. I. and Nelson, F. E. (1997). Effects of global warming on permafrost and active layer thickness: results from transient general circulation models. *Global and Planetary Change*, **15**, 61–77.

Anisimov, O. A. *et al.* (2002). Effect of climate change on permafrost in the past, present and future. *Izvestiya, Atmospheric and Oceanic Physics*, **38** (Suppl. 1), S25–S39.

Anisimov, O. A. *et al.* (2008). Predicting changes in alluvial channel patterns in North-European Russia under conditions of global warming. *Geomorphology*, **98**, 262–274.

Are, F. E. *et al.* (2005). Using thermoterrace dimensions to calculate the coastal erosion rate. *Geo-Marine Letters*, **25**, 121–126.

Beaulieu, N. and Allard, M. (2003). The impact of climate change on an emerging coastline affected by discontinuous permafrost: Manitounuk Strait, northern Quebec. *Canadian Journal of Earth Sciences*, **40**, 1393–1404.

Belchansky, G. I., Douglas, D. C., Platonov, N. G. (2004). Duration of the arctic sea ice melt season: regional and interannual variability, 1979–2001. *Journal of Climate*, **17**, 67–80.

Berger, A. R. and Iams, W. J., eds. (1996). *Geoindicators: Assessing Rapid Environmental Changes in the Earth System*. Rotterdam: Balkema.

Billings, W. D. and Peterson, K. M. (1980). Vegetational change and ice-wedge polygons through the thaw-lake cycle in arctic Alaska. *Arctic and Alpine Research*, **12**, 413–432.

Björk, R. G. and Molau, U. (2007). Ecology of alpine snowbeds and the impact of global change. *Arctic, Antarctic, and Alpine Research*, **39**, 34–43.

Bockheim, J. G. (2007). Importance of cryoturbation in redistributing organic carbon in permafrost-affected soils. *Soil Science Society of America Journal*, **71**, 1335–1342.

Bockheim, J. G. and Hinkel, K. M. (2007). The importance of 'deep' organic carbon in permafrost-affected soils of arctic Alaska. *Soil Science Society of America Journal*, **71**, 1889–1892.

Bockheim, J. G. and Tarnocai, C. (1998). Recognition of cryoturbation for classifying permafrost-affected soils. *Geoderma*, **81**, 281–293.

Bockheim, J. G. *et al.* (2004). Carbon pools and accumulation rates in an age-series of soils in drained thaw-lake basins,

Arctic Alaska. *Soil Science Society of America Journal*, **68**, 697–704.

Bogoyavlenskiy, D. and Siggner, A. (2004). Arctic demography. In N. Einarsson *et al.*, eds., *Arctic Human Development Report*. Akureyri, Iceland: Steffanson Arctic Institute, pp. 27–41.

Brouchkov, A. *et al.* (2004). Thermokarst as a short-term permafrost disturbance, central Yakutia. *Permafrost and Periglacial Processes*, **15**, 81–87.

Brown, J. (1997). Disturbance and recovery of permafrost terrain. In R. M. M. Crawford, ed., *Disturbance and Recovery in Arctic Lands: An Ecological Perspective*. Dordrecht: Kluwer, pp. 167–178.

Brown, J. and Christensen, H. H. (2006). Report of the International Permafrost Association. *Permafrost and Periglacial Processes*, **17**, 377–379.

Brown, J. *et al.* (1997). *International Permafrost Association Circum-Arctic Map of Permafrost and Ground Ice Conditions*, scale 1:10 000 000, Circum-Pacific Map Series, Map CP-45. Digital version available at www.geodata.soton.ac.uk/ipa/ (accessed July 2008).

Brown, J., Hinkel, K. M. and Nelson, F. E. (2000a). The Circumpolar Active Layer Monitoring (CALM) program: research designs and initial results. *Polar Geography*, **24**, 165–258.

Brown, J. *et al.* (2000b). The Global Terrestrial Network for Permafrost (GTN-P): a progress report. In *Rhythms of Natural Processes in the Earth Cryosphere*. Pushchino, Russia: Russian Academy of Sciences, pp. 203–204.

Burgess, M. M. *et al.* (2000). *The Global Terrestrial Network for Permafrost (GTN-P): Permafrost Monitoring Contributing to Global Climate Observations*, Geological Survey of Canada Current Research Paper No. 2000-E14 Available at www.nrcan.gc.ca/gsc/bookstore (accessed July 2008).

Burn, C. R. (1997). Cryostratigraphy, paleogeography, and climate change during the early Holocene warm interval, western Arctic coast, Canada. *Canadian Journal of Earth Sciences*, **34**, 912–935.

Burn, C. R. (1998). The response (1958–1997) of permafrost and near-surface ground temperatures to forest fire, Takhini River valley, southern Yukon Territory. *Canadian Journal of Earth Sciences*, **35**, 184–199.

Callaghan, T. V. ed. (2004). Climate change and UV-B impacts on Arctic tundra and polar desert ecosystems. *Ambio*, **33**, 385–479.

Calmels, F. C. and Allard, M. (2004). Ice-segregation and gas distribution in permafrost using tomodensitometric analysis. *Permafrost and Periglacial Processes*, **15**, 367–378.

Calmels, F., Allard, M. and Delisle, G. (2008). Development and decay of a lithalsa in Northern Québec: a geomorphological history. *Geomorphology*, **97**, 287–299.

Carson, C. E. and Hussey, K. M. (1962). The oriented lakes of arctic Alaska. *Journal of Geology*, **70**, 417–439.

Chapin, F. S. III *et al.* (2005). Polar systems. In H. Hassan, R. Scholes and N. Ash, eds., *Ecosystems and Human Well-Being: Current State and Trends*. Washington, DC: Island Press, pp. 717–743.

Christensen, T. R. *et al.* (2004) Thawing sub-arctic permafrost: effects on vegetation and methane emissions. *Geophysical Research Letters*, **31**, L04501.

Cornelissen, J. H. C. *et al.* (2001). Global change and Arctic ecosystems: is lichen decline a function of increases in vascular plant biomass? *Journal of Ecology*, **89**, 984–994.

Corradi, C. *et al.* (2005). Carbone dioxide and methane exchange of a northeast Siberian tussock tundra. *Global Change Biology*, **11**, 1910–1925.

Costard, F. *et al.* (2007). Impact of the global warming on the fluvial thermal erosion over the Lena River in central Siberia. *Geophysical Research Letters*, **34**, L14501, doi:10.1029/2007GL030212.

Couture, R. *et al.* (2003). On the hazards to infrastructure in the Canadian North associated with thawing of permafrost. *Proceedings of Geohazards 2003, 3rd Canadian Conference on Geotechnique and Natural Hazards*, Ottawa, pp. 97–104.

Czudek, T. and Demek, J. (1970). Thermokarst in Siberia and its influence on the development of lowland relief. *Quaternary Research*, **1**, 103–120.

Doré, D. and Beaulac, I. (2007). *Impact de la fonte du pergélisol sur les infrastructures de transport aérien et routier au Nunavik et adaptation*, Rapport No. GCT-2007-14. Quebec: Ministère des Transports du Québec.

Edwards, M. E. *et al.* (2007). Late-Quaternary dynamics of arctic thermokarst lakes contribute feedbacks to deglacial climate warming. In *Proceedings of the 17th INQUA Congress*, Sydney, July 2007.

Fortier, D., Allard, M. and Shur, Y. (2007). Observation of rapid drainage system development by thermal erosion of ice wedge on Bylot Island, Canadian Arctic Archipelago. *Permafrost and Periglacial Processes*, **18**, 229–243.

French, H. M. (1975). Man-induced thermokarst, Sachs Harbour airstrip, Banks Island, N.W.T. *Canadian Journal of Earth Sciences*, **12**, 132–144.

French, H. M. (2007). *The Periglacial Environment*. Chichester: John Wiley.

Friborg, T. *et al.* (2003). Siberian wetlands: where a sink is a source. *Geophysical Research Letters*, **30**, 2129, doi:10.1029/2003GL017797.

Fronzek, S., Luoto, M. and Carter, T. R. (2006). Potential effect of climate change on the distribution of palsa mires in subarctic Fennoscandia. *Climate Research*, **32**, 1–12.

Georgievskii, V. Y. (1998). On global climate warming effects on water resources. In *Water: A Looming Crisis*? Technical Documents in Hydrology No. 18. Paris: UNESCO, pp. 37–46.

Gilichinsky, D. A. *et al.* (1998). A century of temperature observations on soil climate: methods of analysis and long-term trends. In *Proceedings of the 7th International Conference on Permafrost*, Centre d'Etudes Nordiques, Université Laval, pp. 313–317.

Grigoriev, M. N. and Rachold, V. (2003). The degradation of coastal permafrost and the organic carbon balance of the Laptev and East Siberian Seas. In *Proceedings of the 8th International Conference on Permafrost*, Zürich, pp. 319–24.

Grosse, G. *et al.* (2005). The use of CORONA images in remote sensing of periglacial geomorphology: an illustration from the NE Siberian coast. *Permafrost and Periglacial Processes*, **16**, 163–172.

Grosse, G., Schirrmeister, L. and Malthus, T. J. (2006). Application of Landsat-7 satellite data and a DEM for the quantification of thermokarst-affected terrain types in the periglacial Lena-Anabar coastal lowland. *Polar Research*, **25**, 51–67.

Grosse, G. *et al.* (2007). Geological and geomorphological evolution of a sedimentary periglacial landscape in Northeast Siberia during the Late Quaternary. *Geomorphology*, **86**, 25–51.

Harris, S. A. (1993). Palsa-like mounds developed in a mineral substrate, Fox Lake, Yukon Territory. In *Proceedings of the 6th International Conference on Permafrost Proceedings*, Beijing, 1, pp. 238–243.

Hinkel, K. M. *et al.* (2003). Spatial extent, age, and carbon stocks in drained thaw lake basins on the Barrow Peninsula, Alaska. *Arctic, Antarctic and Alpine Research*, **35**, 291–300.

Hinzman, L. D. *et al.* (2005). Evidence and implications of recent climate change in northern Alaska and other Arctic regions. *Climatic Change*, **72**, 251–298.

Hopkins, D. M. (1949). Thaw lakes and thaw sinks in the Imuruk Lake area, Seward Peninsula. *Journal of Geology*, **57**, 119–131.

Humlum, O., Instanes, A. and Sollid, J. L. (2003). Permafrost in Svalbard: a review of research history, climatic background and engineering challenges. *Polar Research*, **22**, 191–215.

Il'ichev, V. A. *et al.* (2003). *Perspectives of the Development of the Modern Northern Settlements*. Moscow: Russian Academy of Architectural Science.

Instanes, D. and Instanes, A. (1998). Frozen ground temperature profiles at Svalbard Airport, Spitsbergen. In *Proceedings of the International Conference of Natural and Artificial Cooling Cold Regions Engineering*, Paris, pp. 229–237.

Instanes, A. (2003). Climate change and possible impact on Arctic infrastructure. In *Proceedings of the 8th International Permafrost Conference*, Zürich, pp. 461–466.

Instanes, A. *et al.* (2005). Infrastructure: buildings, support systems and industrial facilities. In C. Symon, L. Arris and B. Heal, eds., *Arctic Climate Impact Assessment (ACIA)*. Cambridge: Cambridge University Press, pp. 907–944.

IPCC (2001a). *Climate Change 2001: The Scientific Basis. Contribution of Working Group I to the Third Assessment Report of the Intergovernmental Panel on Climate Change*. Houghton, J. *et al.*, eds. Cambridge: Cambridge University Press. (pp. 349–416, Ramaswany, V. *et al.*, eds.)

IPCC (2001b). *Climate Change 2001: Impacts, Adaptation and Vulnerability. Contribution of Working Group II to the Third Assessment Report of the Intergovernmental Panel on Climate Change*. McCarthy, J. J. *et al.*, eds. Cambridge: Cambridge University Press. (pp. 801–841, Anisimov, O. and Fitzharris, B., eds.)

IPCC (2007a). *Climate Change 2007: The Physical Science Basis. Contribution of Working Group I to the Fourth Assessment Report of the Intergovernmental Panel on Climate Change*. Solomon, S. *et al.*, eds. Cambridge: Cambridge University Press. (pp. 1–18, Alley, R. B. *et al.*, eds.)

IPCC (2007b). *Climate Change 2007: Impacts, Adaptation, and Vulnerability. Contribution of Working Group II to the Fourth Assessment Report of the Intergovernmental Panel on Climate Change*. Parry, M. L. *et al.*, eds. Cambridge: Cambridge University Press. (pp. 653–685, Anisimov, O. A. *et al.*, eds.)

Isaksen, K. *et al.* (2007). Recent warming of mountain permafrost in Svalbard and Scandinavia. *Journal of Geophysical Research*, **112**, F02S04, doi:10.1029/2006JF000522.

Jorgenson, M. T. and Brown, J. (2005). Classification of the Alaskan Beaufort Sea coast and estimation of carbon and sediment inputs from coastal erosion. *Geo-Marine Letters*, **25**, 69–80.

Jorgenson, M. T. and Osterkamp, T. E. (2005). Response of boreal ecosystems to varying modes of permafrost degradation. *Canadian Journal of Forestry Research*, **35**, 2100–2111.

Jorgenson, M. T. and Shur, J. (2007). Evolution of lakes and basins in northern Alaska and discussion of the thaw lake cycle. *Journal of Geophysical Research*, **112**, F02S17, doi:10.1029/2006JF000531.

Jorgenson, M. T. *et al.* (2001). Permafrost degradation and ecological changes associated with a warming climate in central Alaska. *Climatic Change*, **48**, 551–579.

Jorgenson, M. T., Shur, J. L. and Pullman, E. R. (2006). Abrupt increase in permafrost degradation in Arctic Alaska. *Geophysical Research Letters*, **33**, L02503, doi:10.1029/2005GL024960.

Katamura, F. *et al.* (2006). Thermokarst formation and vegetation dynamics inferred from a palynological study in central Yakutia, eastern Siberia, Russia. *Arctic, Antarctic and Alpine Research*, **38**, 561–570.

Kronik, Y. A. (2001). Accident rate and safety of natural-anthropogenic systems in the permafrost zone. In *Proceedings of the 2nd Conference of Russian Geocryologists*, Moscow, **4**, pp. 138–146.

Laberge, M. J. and Payette, S. (1995). Long-term monitoring of permafrost change in a palsa peatland in northern Québec, Canada: 1983–1993. *Arctic, Antarctic and Alpine Research*, **27**, 167–171.

Lachenbruch, A. H. and Marshall, B. V. (1986). Changing climate: geothermal evidence from permafrost in the Alaskan Arctic. *Science*, **234**, 689–696.

Lantuit, H. and Pollard, W. H. (2005). Temporal stereophotogrammetric analysis of retrogressive thaw slumps on Herschel Island, Yukon Territory. *Natural Hazards and Earth System Sciences*, **5**, 413–423.

Linell, K. A. (1973). Long-term effects of vegetative cover on permafrost stability in an area of discontinuous permafrost. In *North American Contribution to the 2nd International Conference on Permafrost*. Washington, DC: National Academy Press, pp. 688–93.

Lloyd, A. H. *et al.* (2003). Patterns and dynamics of treeline advance on the Seward Peninsula, Alaska. *Journal of*

Geophysical Research, **108**, D2, 8161, doi:10.1029/2001JD000852.

Luoto, M. and Seppälä, M. (2003). Thermokarst ponds as indicators of the former distribution of palsas in Finnish Lapland. *Permafrost and Periglacial Processes*, **14**, 19–27.

Mackay, J. R. (1970). Disturbances to the tundra and forest tundra environment of the western Arctic. *Canadian Geotechnical Journal*, **7**, 420–432.

Mackay, J. R. (1971). The origin of massive icy beds in permafrost, western arctic coast, Canada. *Canadian Journal of Earth Sciences*, **8**, 397–422.

Mackay, J. R. (1972). The world of underground ice. *Annals of the American Association of Geographers*, **62**, 1–22.

Malmer, N. *et al.* (2005). Vegetation, climatic changes and net carbon sequestration. *Global Change Biology*, **11**, 1895–1909.

Mars, J. C. and Houseknecht, D. W. (2007). Quantitative remote sensing study indicates doubling of coastal erosion rate in the past 50 years along a segment of the Arctic coast of Alaska. *Geology*, **35**, 583–586.

Maslin, M. (2004). *Gas Hydrates: A Hazard for the Twenty-First Century*. London: Benfield Hazard Research Centre.

McBean, G. *et al.* (2005). Arctic climate: past and present. In C. Symon, L. Arris and B. Heal, eds., *Arctic Climate Impacts Assessment (ACIA)*. Cambridge: Cambridge University Press, pp. 21–60.

Murton, J. B. (2001). Thermokarst sediments and sedimentary structures, Tuktoyaktuk coastlands, western Arctic Canada. *Global and Planetary Change*, **28**, 175–192.

Nelson, F. E. (2003). Geocryology: (un)frozen in time. *Science*, **299**, 1673–1675.

Nelson, F. E. and Brown, J. (1997). Global change and permafrost. *Frozen Ground*, **21**, 21–24.

Nelson, F. E. and Outcalt, S. I. (1982). Anthropogenic geomorphology in northern Alaska. *Physical Geography*, **3**, 17–48.

Nelson, F. E., Anisimov, O. A. and Shiklomanov, N. I. (2001). Subsidence risk from thawing permafrost. *Nature*, **410**, 889–890.

Nelson, F. E., Anisimov, O. A. and Shiklomanov, N. I. (2002). Climate change and hazard zonation in the circum-arctic permafrost regions. *Natural Hazards*, **26**, 203–225.

Oberman, N. G. and Mazhitova, G. G. (2001). Permafrost dynamics in the north-east of European Russia at the end of the twentieth century. *Norwegian Journal of Geography*, **55**, 241–244.

Oelke, C. and Zhang, T. (2004). A model study of circum-arctic soil temperatures. *Permafrost and Periglacial Processes*, **15**, 103–121.

Ogorodov, S. A. (2004). Human impacts on coastal stability in the Pechora Sea. *Geo-Marine Letters*, **25**, 190–195.

Osterkamp, T. E. (2003). A thermal history of permafrost in Alaska. *Proceedings of the 8th International Conference on Permafrost*, Zurich, pp. 863–868.

Osterkamp, T. E. (2007). Characteristics of the recent warming of permafrost in Alaska. *Journal of Geophysical Research*, **112**, F02S02, doi:10.1029/2006JF000578.

Osterkamp, T. E. and Romanovsky, V. E. (1999). Evidence for warming and thawing of discontinuous permafrost in Alaska. *Permafrost and Periglacial Processes*, **10**, 17–37.

Osterkamp, T. E., Esch, D. C. and Romanovsky, V. E. (1997). Infrastructure: effects of climatic warming on planning, construction and maintenance. In *Proceedings of the BESIS Workshop: Implications of Global Change in Alaska and the Bering Sea Region*, Center for Global Change and Arctic System Research, Fairbanks, Alaska, pp. 115–127.

Osterkamp, T. E. *et al.* (2000). Observations of thermokarst and its impact on boreal forests in Alaska, USA. *Arctic, Antarctic and Alpine Research*, **32**, 303–315.

Pavlov, A. V. (1994). Current changes of climate and permafrost in the Arctic and Sub-Arctic of Russia. *Permafrost and Periglacial Processes*, **5**, 101–110.

Pavlov, A. V. and Moskalenko, N. G. (2002). The thermal regime of soils in the north of western Siberia. *Permafrost and Periglacial Processes*, **13**, 43–51.

Payette, S. *et al.* (2004). Accelerated thawing of subarctic peatland permafrost over the last 50 years. *Geophysical Research Letters*, **31**, L18208, doi:10.1029/2004GL020358.

Péwé, T. L. (1954). *Effect of Permafrost upon Cultivated Fields*, US Geological Survey Bulletin No. 989F. Washington, DC: US Government Printing Office.

Pissart, A. (2002). Palsas, lithalsas and remnants of these periglacial mounds: a progress report. *Progress in Physical Geography*, **26**, 605–621.

Rachold, V. *et al.* (2000). Coastal erosion vs. riverine sediment discharge in the Arctic shelf seas. *International Journal of Earth Sciences*, **89**, 450–460.

Rachold, V. *et al.* (2003). Modern terrigenous organic carbon input to the Arctic Ocean. In R. Stein and R. W. Macdonald, eds., *Organic Carbon Cycle in the Arctic Ocean: Present and Past*. Berlin: Springer-Verlag, pp. 33–55.

Rachold, V. *et al.* (2005). Arctic Coastal Dynamics (ACD): an introduction. *Geo-Marine Letters*, **25**, 63–68.

Randriamazaoro, R. *et al.* (2007). Fluvial thermal erosion: heat balance integral method. *Earth Surface Processes and Landforms*, **32**, 1828–1840.

Reimnitz, E., Graves, S. M. and Barnes, P. W. (1985). *Beaufort Sea Coastal Erosion, Shoreline Evolution, and Sediment Flux*, US Geological Survey Open-File Report No. 85–380. Washington, DC: US Government Printing Office.

Reimnitz, Z. E., Graves, S. M. and Barnes, P. W. (1988). *Beaufort Sea Coastal Erosion, Sediment Flux, Shoreline Evolution and the Erosional Shelf Profile*, US Geological Survey Map No. I-1182-G and text. Available at www.awi-potsdam.de/acd (accessed July 2008).

Romanovsky, V. E. *et al.* (2002). Permafrost temperature records: indicators of climate change. *EOS, Transactions of the American Geophysical Union*, **83**, 589–594.

Schirrmeister, L. *et al.* (2002). Paleoenvironmental and paleoclimatic records from permafrost deposits in the Arctic region of Northern Siberia. *Quaternary International*, **89**, 97–118.

Schuur, E. A. G. *et al.* (2008). Vulnerability of permafrost carbon to climate change: implications for the global carbon cycle. *BioScience*, **58**, 701–714.

Seppälä, M., 1988. Palsas and related forms. In M. J. Clark, ed., *Advances in Periglacial Geomorphology*. Chichester: John Wiley, pp. 247–278.

Seppälä, M. (1994). Snow depth controls palsa growth. *Permafrost and Periglacial Processes*, **5**, 283–288.

Serreze, M. C. *et al.* (2000). Observational evidence of recent change in the northern high-latitude environment. *Climatic Change*, **46**, 159–207.

Shur, Y. L. and Jorgenson, M. T. (2007). Patterns of permafrost formation and degradation in relation to climate and ecosystems. *Permafrost and Periglacial Processes*, **18**, 7–19.

Shur, Y. L., Hinkel, K. M. and Nelson, F. E. (2005). The transient layer: implications for geocryology and climate-change science. *Permafrost and Periglacial Processes*, **16**, 5–17.

Sloan, E. D. Jr (2003). Fundamental principles and applications of natural gas hydrates. *Nature*, **426**, 353–359.

Smith, L. C., Sheng, Y. and McDonald, G. M. (2007). A first pan-Arctic assessment of the influence of glaciation, permafrost, topography and peatlands on northern hemisphere lake distribution. *Permafrost and Periglacial Processes*, **18**, 201–208.

Smith, S. L., Burgess, M. M. and Taylor, A. E. (2003). High Arctic permafrost observatory at Alert, Nunavut: analysis of a 23-year data set. *Proceedings of the 8th International Conference on Permafrost*, Zurich, pp. 1073–1078.

Soja, A. J. *et al.* (2007). Climate-induced boreal forest change: predictions versus observations. In *Conference Program and Abstracts, Climate Change Impacts on Boreal Forest Disturbance Regimes*, Fairbanks, Alaska, p. 39.

Sollid, J. L. and Sorbel, L. (1998). Palsa bogs as a climate indicator: examples from Dovrefjell, southern Norway. *Ambio*, **27**, 287–291.

Solomon, S. M. (2005). Spatial and temporal variability of shoreline change in the Beaufort-Mackenzie region, North-West Territories, Canada. *Geo-Marine Letters*, **25**, 127–137.

Soloviev, P. A. (1973). Alas of thermokarst relief of Central Yakutia, Guidebook, *2nd International Permafrost Conference*, Yakutsk, USSR.

Stieglitz, M. *et al.* (2003). The role of snow cover in the warming of Arctic permafrost. *Geophysical Research Letters*, **30**, 1721.

Sturm, M., Racine, C. and Tape, K. (2001). Increasing shrub abundance in the Arctic. *Nature*, **411**, 546–547.

Tape, K., Sturm, M. and Racine, C. (2006). The evidence for shrub expansion in Northern Alaska and the Pan-Arctic. *Global Change Biology*, **12**, 686–702.

Tchebakova, N. M., Parfenova, E. I. and Soja, A. J. (2007). Potential climate induced vegetation change in central Siberia in the century 1960–2050. In *Conference Program and Abstracts, Climate Change Impacts on Boreal Forest Disturbance Regimes*, Fairbanks, Alaska, p. 26.

Truong, G., Palmé, A. E. and F. F. (2006). Recent invasion of the mountain birch *Betula pubescens* ssp. *tortuosa* above the treeline due to climate change: genetic and ecological study in northern Sweden. *European Society for Evolutionary Biology*, **20**, 369–380.

Vallée, S. and Payette, S. (2007). Collapse of permafrost mounds along a subarctic river over the last 100 years (northern Québec). *Geomorphology*, **90**, 162–170.

Van Everdingen, R. O. (2002). *Multi-Language Glossary of Permafrost and Related Ground Ice Terms*. Boulder: National Snow and Ice Data Center/World Data Center for Glaciology.

Van Wijk, M. T. *et al.* (2003). Long-term ecosystem level experiments at Toolik Lake, Alaska and at Abisko, Northern Sweden: generalizations and differences in ecosystem and plant type responses to global change. *Global Change Biology*, **10**, 105–123.

Vasiliev, A. *et al.* (2005). Coastal dynamics at the Barents and Kara Sea key sites. *Geo-Marine Letters*, **25**, 110–120.

Viereck, L. A. (1982). Effects of fire and firelines on active layer thickness and soil temperatures in interior Alaska. In *Proceedings of the 4th International Conference on Permafrost*, Ottawa, pp. 123–135.

Walsh, J. E. *et al.* (2005). Cryosphere and hydrology. In C. Symon, L. Arris and B. Heal, eds., *Arctic Climate Impacts Assessment (ACIA)*. Cambridge: Cambridge University Press, pp. 183–242.

Walter, K. M. *et al.* (2006). Methane bubbling from Siberian thaw lakes as a positive feedback to climate warming. *Nature*, **443**, 71–75.

Walter, K. M., Smith, L. C. and Chapin, F. S. III (2007). Methane bubbling from northern lakes: present and future contributions to the global methane budget. *Philosophical Transactions of the Royal Society of London A*, **365**, 1657–1676.

Weller, G. and Lange, M. (1999). *Impacts of Global Climate Change in the Arctic Regions*. Fairbanks: Center for Global Change and Arctic System Research, University of Alaska.

Zhang, T. *et al.* (2000). Further statistics on the distribution of permafrost and ground ice in the Northern Hemisphere. *Polar Geography*, **24**, 126–131.

Zimov, S. A., Schuur, E. A. G. and Chapin, E. S., III. (2006). Permafrost and the global carbon budget. *Science*, **312**, 1612–1613.

Zimov, S. A. *et al.* (1997). North Siberian lakes: a methane source fuelled by Pleistocene carbon. *Science*, **277**, 800–802.

Zuidhoff, F. S. (2002). Recent decay of a single palsa in relation to weather conditions between 1996 and 2000 in Laivadalen, northern Sweden. *Geografiska Annaler*, **84**A, 103–111.

Zuidhoff, F. S. and Kolstrup, E. (2000). Changes in palsa distribution in relation to climate change in Laivadalen, northern Sweden, especially 1960–1997. *Permafrost and Periglacial Processes*, **11**, 55–69.

14 Ice sheets and ice caps

David Sugden

14.1 Introduction

Astronauts who first orbited the Earth in the Space Shuttle described the Antarctic Ice Sheet as the most spectacular of the world's landscapes seen from space. The sight of the white snow surface, the surrounding Southern Ocean, and the sheer extent of the ice covering a continent one and a half times the size of the USA was dramatic and humbling (Fig. 14.1; Plate 45). The global scene would have been all the more dramatic about 20 ka BP at the height of the last ice age. An orbiting observer at the time would have seen the Laurentide Ice Sheet, an equally large ice mass located over Canada and the northern USA, a Greenland Ice Sheet more expanded than that of today, and a large ice sheet in northwest Europe extending from southern Britain across Scandinavia to Svalbard and northern Russia. At such a time global sea level was 120 m lower than it is today and the world's climatic and vegetation zones were compressed towards the equator (Plate 1). The contrast between the two scenes is powerful testament to the scale of the environmental changes that occur in response to natural cycles of insolation received by the Earth.

Ice sheets play a fundamental role in modulating global climate. The last few ice age cycles have lasted about 100 ka and display a sawtooth pattern with a long and irregular period of cooling as the ice sheets grow to their maximum, followed by an abrupt warming and a return to an interglacial climate similar to that of the present. Such a pattern is of a world hunting for a cooler equilibrium state only to cross a threshold which switches back to its warmer state, only for the cycle to repeat itself again. At present we do not know the feedbacks and links that explain this pattern, but the ice sheets are likely to be involved in amplifying the relatively minor changes in solar radiation received by the Earth as a result of Croll–Milankovitch orbital cycles. The substitution of a white reflective ice surface in place of darker rock and vegetation is one feedback process that would accentuate an initial cooling. The realisation that ice sheets are a dynamic feature of the Earth's climate with a history of large and sudden threshold changes acts as a warning from the past as we change the greenhouse gas content of the atmosphere to levels far beyond the natural envelope. Experience of the past suggests we should expect abrupt changes.

The present-day ice sheets influence life and the environment around the world. Present-day sea level is controlled by the volume of ice on land. The Antarctic Ice Sheet holds sufficient fresh water to raise global sea level by nearly 60 m were it to melt. The melting of the Greenland Ice Sheet would add a further 7 m. The Antarctic Ice Sheet is the polar hub of the southern hemisphere flywheel with westerly atmospheric circulation and ocean currents sweeping around it. It is this circulation pattern that has created our present global climate zones and, indeed, underlies the structure of this book. This is illustrated by the discovery that the stepped growth of the Antarctic Ice Sheet 34 Ma BP and its further expansion 14 Ma BP was linked to global cooling of the atmosphere and oceans as the Earth shifted from a greenhouse state with trees and animals near both poles to the present icehouse state with cold poles (Kennett, 1977; DeConto and Pollard, 2003; Miller *et al.*, 2008). Ocean circulation and productivity are also directly affected by ice sheets, as in the case of the production of cold antarctic bottom water which flows from beneath floating ice shelves and contributes to ocean overturning. Meltwater and ice from the Greenland Ice Sheet flows into the North Atlantic and contributes to the surface water which sinks and becomes part of the ocean conveyor that brings warm Atlantic water to the coasts of Europe from the southern hemisphere in exchange for a return flow of deep cold water (Crowley, 1992). Perhaps the ice sheets are also involved in the evolution of humankind through the modulation of global climate over the last 2 Ma.

Geomorphology and Global Environmental Change, eds. Olav Slaymaker, Thomas Spencer and Christine Embleton-Hamann. Published by Cambridge University Press.

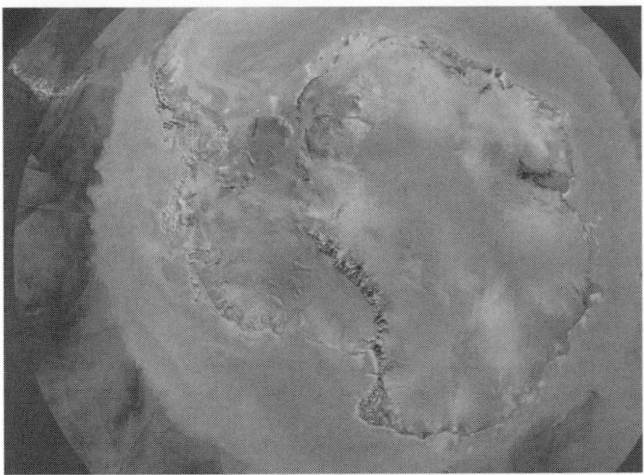

FIGURE 14.1. Composite satellite image of Antarctica showing the main ice domes of East and West Antarctica, the ice shelves of the Ross Sea and Weddell Sea embayments and the Antarctic Peninsula extending towards southernmost South America. The dark areas are mountains protruding above the ice and coastal oases. The sea ice extent represents winter conditions. The land ice is from AVHRR imagery and sea ice from Seawinds imagery. (Courtesy NASA/Goddard Space Flight Center, Scientific Visualization Studio.) (See also Plate 45 for colour version.)

It is clear that ice sheets are important players on the world scene and that it is important to understand how they work and to predict how they will respond to climate change forced by rising levels of greenhouse gases. One successful approach, as used by the scientists of the International Panel for Climate Change (IPCC, 2007), is (a) to identify the main factors affecting ice sheet health, in this case the interaction with the atmosphere that underlies the mass balance (balance between snow received and ice lost), (b) to employ theory to predict the relations between mass balance and glacier flow and (c) to model the future health of the ice sheet numerically under different scenarios. It is difficult to measure the surface processes and calibrate the models over such inhospitable domains directly in the field, so the models are tested by monitoring changes in ice surface elevation using satellite radar remote sensing (Zwally *et al.*, 2002; Wingham *et al.*, 2006a). As successful as the approach has been, many uncertainties remain. Firstly, the lack of surface process studies means that it is difficult to measure surface elevation remotely and thus mass balance. This is because the radar return is strongly related to the density of the upper layers and thus the returns are influenced by conditions in the snowpack. Secondly, the approach, at least in the 2007 IPCC report, could not consider the interface between glaciers and the ocean in detail because of the lack of understanding. This has become important in the case of Antarctica since several sea-terminating glaciers are thinning at rates far higher than expected, probably as a dynamic response to long-term sea level rise and rising sea temperatures accelerating the melting of bounding ice shelves.

Geomorphology has the potential to improve the predictions of ice sheet behaviour significantly. Geomorphologists specialise in process studies of glacier mass balance and are well positioned to refine the understanding of surface mass balance and glacier flow that is crucial to glacier modelling and prediction. Further, there is an additional dimension in that it is important to link ice sheet behaviour at different spatial and temporal scales if we are to have confidence in predictive models. Focus on longer-term ice sheet trajectories brings several benefits. Firstly, reconstructions of past ice sheet history provide a body of evidence with which to test and refine ice sheet models. A model that can reproduce the changes in ice extent and volume during an ice age cycle will have far more credibility than one that is unable to complete this task satisfactorily. Secondly, it is important to establish the longer-term trajectory of an ice sheet if one is to interpret present-day trends. For example, thinning of an ice sheet that has occurred continuously for the last 10 ka, as in the case of the Sarnoff Mountains in Marie Byrd Land region of West Antarctica (Stone *et al.*, 2003), might be a response to sea level rise since the Last Glacial Maximum (LGM) (Fig. 14.2); in contrast, thinning that has begun only in the last century could be related to human impact on climate. Thirdly, a longer-term study may be useful in identifying the processes and thresholds that are critical to the behaviour of an ice sheet. One example is the evidence of rapid retreat that has accompanied ice sheet deglaciation in the northern hemisphere when ice is grounded below sea level, as in the case of the collapse of marine parts of the North American and Eurasian ice sheets and certain fjord glaciers. It is the link between glacier flow and underlying topography that dominates patterns of ice retreat. John Mercer flagged this issue in the case of West Antarctica as long ago as 1978.

14.2 Distribution of ice sheets and ice caps

In this chapter the difference between an ice sheet and ice cap follows a long-held convention that features larger than 50 000 km^2 are called ice sheets and those smaller than this are ice caps (Armstrong *et al.*, 1973). Thus the present-day ice centred on uplands in Iceland are ice caps, while the Pleistocene ice that covered the whole of the island qualifies as a small ice sheet. The overwhelming mass of ice lies in Antarctica (88%) with 11.5% in the Greenland Ice Sheet and 0.5% in the world's remaining ice caps lying mostly in the Arctic basin and Patagonia (Table 14.1). The Antarctic Ice Sheet with an area of

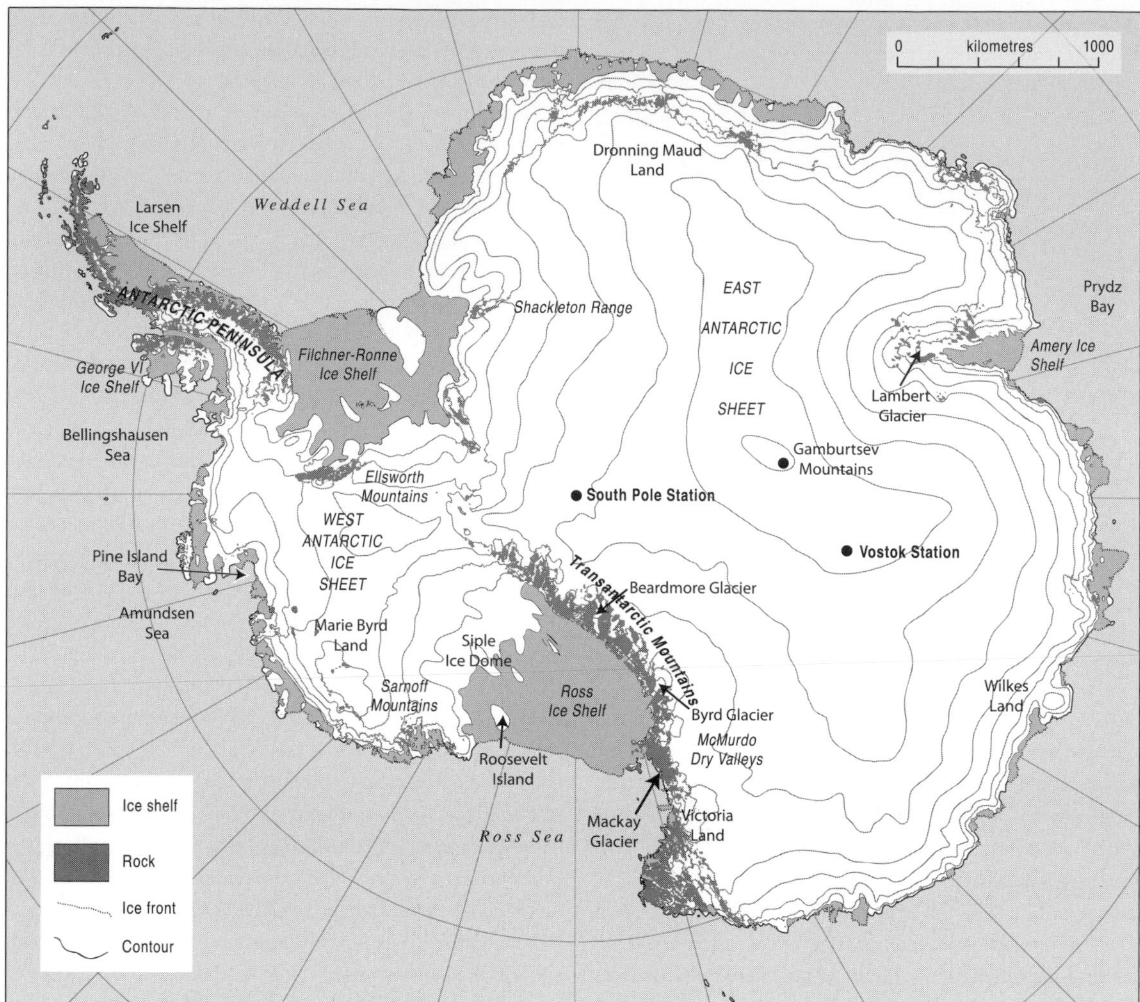

FIGURE 14.2. Location map of Antarctica showing the places mentioned in the text.

around 12 M km^2 consists of two main parts, named East Antarctica and West Antarctica after the hemispheres in which they lie (Fig. 14.1). The East Antarctic Ice Sheet is by far the biggest component and covers an area of 10 M km^2, while the West Antarctic Ice Sheet covers approximately 2 M km^2 (Fig. 14.2). The ice sheets are based on quite different bedrock bases (Fig. 14.3; Plate 46). East Antarctica is a coherent continent, while West Antarctica consists of continental fragments. The bedrock topography interpolated from radio-echo sounding is of a continent depressed isostatically by the weight of the overlying ice sheet. For comparison, Fig. 14.4 shows the bedrock base assuming isostatic recovery and the absence of any ice. In this latter reconstruction a hydrological model has been used to reconstruct possible preglacial river patterns on the terrestrial part.

The East Antarctic Ice Sheet is based on a continental remnant of Gondwana and is a broad dome measuring 4400–2600 km across and rising to altitudes in excess of 4200 m. In most places the ice ends at or near the continental margin. Here are the steepest slopes with ice surface altitudes typically rising 450 m within a distance of 10 km from the margin. The ice is over 4300 m thick in parts of the interior. The volume of ice in the ice sheet is calculated to be the equivalent of a global sea level rise of 52 m. Much of the subglacial bedrock base of the continent lies at altitudes that are currently close to or slightly below sea level. There are three main mountainous areas with present-day altitudes of 3000–4000 m (Fig. 14.2). One of the most spectacular is the uplifted rim of the Transantarctic Mountains which bounds the margin of the East Antarctic Ice Sheet continuously from

TABLE 14.1. *Ice sheets and ice caps: areas and volumes expressed as global sea level equivalent*

Ice sheet/ice cap	Area (1000 km^2)	Volume (m sea level equivalent)
East Antarctic	10 000	52
Greenland	1 700	7
West Antarctic	2 000	
Antarctic Peninsula	500	6
Ice caps		
Svalbard, Arctic Norway	37	
Franz Josef Land, Russia	14	
Novaya Zemlya, Russia	24	
Severnaya Zemlya, Russia	18	
Ellesmere Island, Canada	18	
Axel Heiberg Island, Canada	12	
Devon Island, Canada	16	
Baffin Island and other, Arctic Canada	43	
Vatnajökull, Iceland	8	
Greenland peripheries	70	
Patagonian ice caps	17	
	277	0.25
Last Glacial Maximum expansion		−120

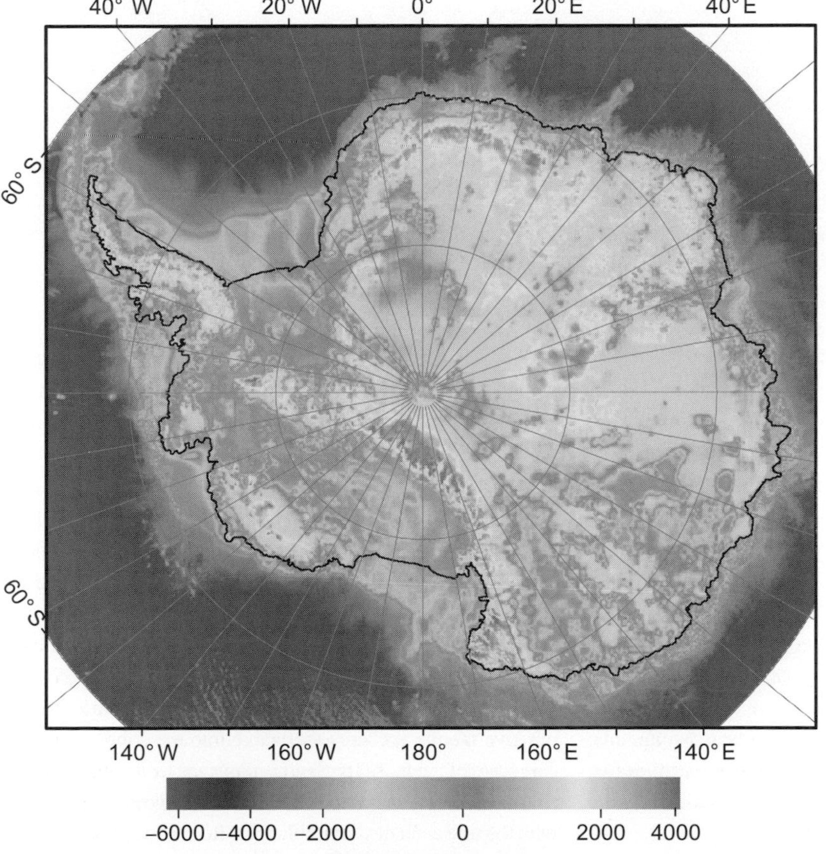

FIGURE 14.3. The present subglacial topography of Antarctica showing the continent and mountains of East Antarctica and the archipelago of West Antarctica. This is the idealised view of Antarctica if you remove the ice, maintain present sea level and ignore any isostatic adjustment (from Lythe *et al.*, 2001). (See also Plate 46 for colour version.)

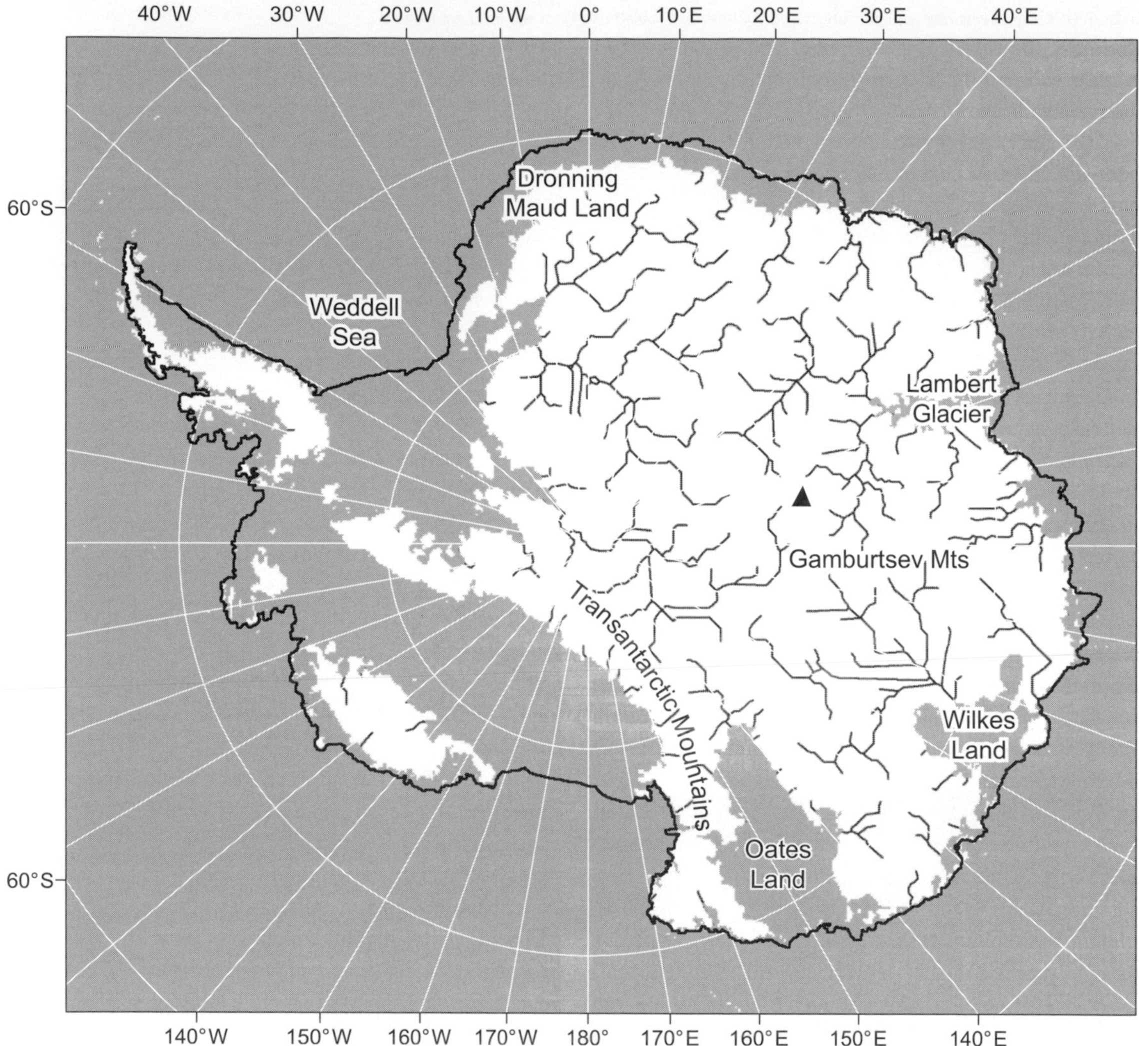

FIGURE 14.4. A reconstruction of the land area of Antarctica assuming isostatic adjustment for the loss of the overlying ice load. A hydrological model has been used to plot possible river drainage basins. The reconstruction simulates a possible antarctic fluvial environment that might have existed before glaciation (from Jamieson and Sugden, 2008).

the vicinity of the South Pole to northern Victoria Land. The mountain summits rise above the ice sheet surface and are breached by outlet glaciers draining the ice sheet. The Transantarctic Mountains continue into the Weddell Sea sector, but here they are broken up into discontinuous mountain groups, such as the Shackleton Range. A second mountainous area is in Dronning Maud Land. Here mountain summits, breached by outlet glaciers, form an imposing backdrop to the coastal lowlands and back onto a high subglacial plateau extending inland. The third mountain group, the Gamburtsev Mountains, is wholly subglacial. These mountains rise to altitudes of over 3000 m and are located in the middle of the continent.

The West Antarctic Ice Sheet is centred on an archipelago of mountainous continental remnants (Siddoway, 2008). Much of the ice sheet between the archipelago is grounded below sea level. The ice reaches an altitude of around 3000 m at its highest point in the approximate centre; its volume is equivalent to a global sea level rise of 6 m. The ice sheet consists of three main sectors, each draining into a different ocean, namely the Weddell Sea, Ross Sea and Amundsen Sea, respectively. In the Ross Sea and Weddell Sea embayments are two floating ice shelves that are the birthplace of massive tabular icebergs that calve into the southern ocean. The scale of the Ross Ice Shelf is remarkable; it covers an area of 850 000 km^2, larger than France, and yet goes up and down with the tide each day. The Filchner–Ronne Ice Shelf flanking the Weddell Sea is smaller, but still vast and covers an area of 360 000 km^2. The ice that flows into the Amundsen Sea is fringed by several much smaller ice shelves. The topography beneath the West Antarctic Ice Sheet is remarkable for its amplitude. There are mountains such as Mount Vinson in the Ellsworth Mountains that protrude through the ice and at 4897 m are the highest in the whole of Antarctica. And yet within 100 km there is a trench 2555 m below present sea level covered by the grounded ice sheet. The Antarctic Peninsula, which is not normally included as part of the West Antarctic Ice Sheet, adds another 0.5 M km^2 to the ice-covered area of Antarctica.

The Greenland Ice Sheet lies between latitudes 60°–80° N and occupies an area of some 1.7 M km^2 with a volume equivalent to a sea level rise of 7 m (Fig. 14.5). Its centre comprises two domes, that in the north reaching an altitude of around 3200 m and that in the south of 2850 m; the domes are linked by a saddle at around 2500 m. The ice sheet has a mean thickness of 1600 m and approaches a thickness of 3000 m in places. The underlying topography is of a central lowland shield with mountainous or upland rims bordering all but the mid-western coasts. In the south the mountain rim is dissected while in the north and east the uplands consist of plateaus.

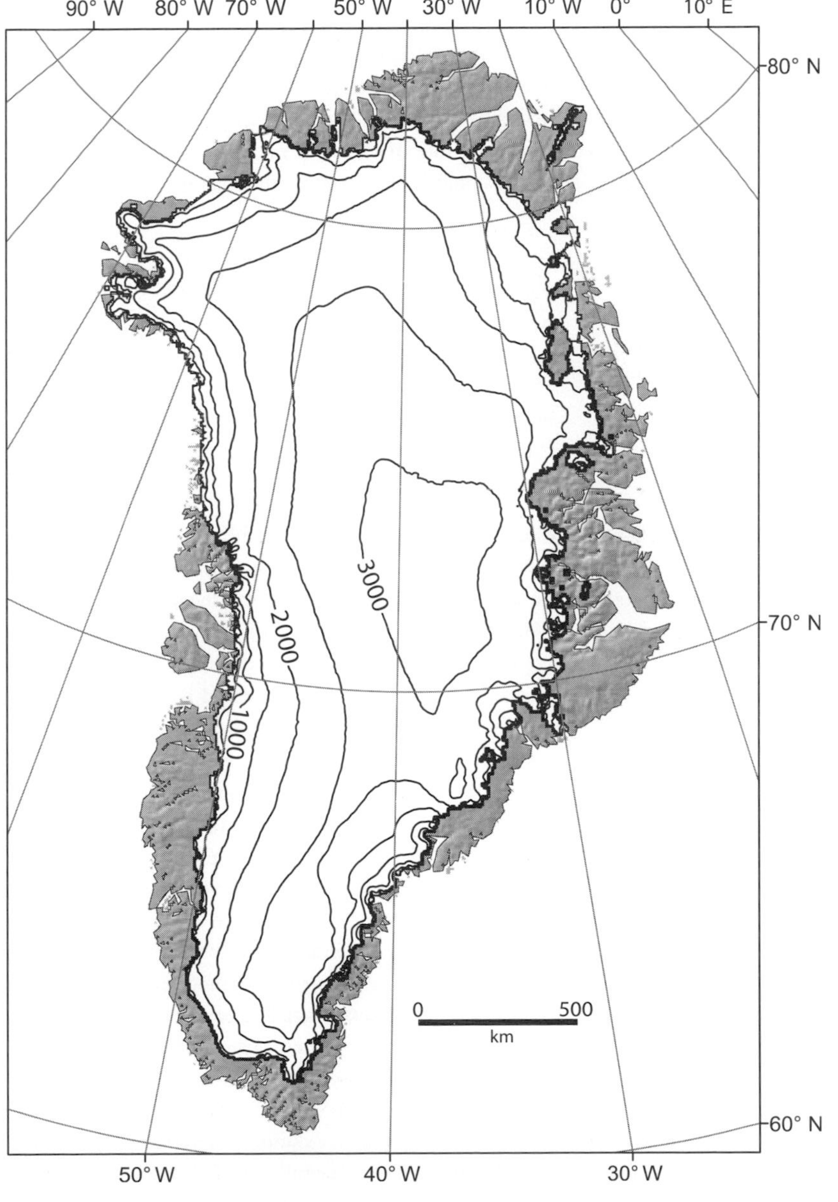

FIGURE 14.5. The morphology of the Greenland Ice Sheet. The summit of the ice sheet has an altitude of 3278 m and surface contours are shown at 500-m intervals. The map is based on data from the coherent ice sheet and ignores local glaciation in the coastal peripheries, especially the mountainous coast of southern and southeastern Greenland and small ice caps in the west and northeast (data from Bamber *et al*., 2001).

The remaining ice caps of the world contain ice equivalent to about 0.25 m of global sea level rise. This excludes mid-latitude mountain glaciers that are considered in Chapter 2 of this volume. About 40% of this total, accounting for an ice extent of 277 000 km^2 consists of shallow ice caps on the Arctic Islands of Canada, northern Greenland and the Eurasian Arctic. Individual ice caps vary in size from the largest at 17 500 km^2 in the case of Agassiz Ice Cap in Ellesmere Island to sizes of 2000–8000 km^2 in the case of various ice caps in Svalbard and on islands off northern Russia. Vatnajökull in southern Iceland is one of the most accessible and has an area of 8100 km^2. It was described scientifically in 1794–95 in perhaps the world's first comprehensive glaciological treatise, by S. Pálsson (Williams and Sigurðsson, 2004). The two largest ice caps elsewhere are in the South American Andes. The larger South Patagonian Icefield, extending for 350 km, is 13 000 km^2 in area while the smaller North Patagonian Icefield, about 100 km in length and 50 km across, is 4200 km^2 in area (Fig. 14.6). There is a question as to whether the Patagonian ice masses should be termed as 'ice fields', where the surface morphology is dominated by the underlying topography, or as 'ice caps' where a coherent ice dome dominates the underlying topography. In view of the coherence of the ice domes, the radial pattern of outlet glaciers, and the evidence that the ice expanded into a larger ice sheet during the LGM, the North Patagonian and South Patagonian ice caps are viewed as ice caps and discussed in this chapter.

During the LGM two major ice sheets developed on the temperate and polar land masses either side of the Atlantic Ocean. The Laurentide Ice Sheet in North America extended from the northern Canadian Arctic archipelago as far south as New York and the Midwest states of the USA and offshore to the southeast and north (Fig. 14.7a). There has been controversy about its former altitude depending on the criteria employed, but reconstructions based on the morphology of present ice sheets imply that it had a surface altitude in the centre of the order of 3500 m and a central thickness of over 4000 m (Sugden, 1977). The Eurasian Ice Sheet was centred on Scandinavia and the Gulf of Bothnia in the Baltic Sea and extended northwards over the Barents Sea to Svalbard and southwestwards over the North Sea to the British Isles (Fig. 14.7b). It incorporated the present Eurasian Arctic island ice caps near the Atlantic and flowed across the mountain axis of Scandinavia; the central dome rose to a central altitude in excess of 3000 m. The margins of the Greenland Ice Sheet thickened and extended offshore, especially in the west and more controversially in the east. Cordilleran ice sheets expanded in North America and Patagonia to form coherent wholes aligned along the main mountain axis. The main North American Cordilleran Ice Sheet was centred on the mountain ranges of western Canada and was some 1900 km long and 850 km across. Smaller ice caps extended along the Alaskan mountain ranges for a further 1500 km. The Patagonian Ice Sheet built up over a distance of 1800 km along the axis of the Andes, and in the main it extended some 200 km from coast in the west to the foothills of the Andes in the east (Fig. 14.6). Icelandic ice caps coalesced to form one ice sheet that extended beyond the coast onto the offshore shelf (Hubbard *et al.*, 2006).

14.3 Ice sheet and ice cap landscapes

The outstanding feature of ice sheets and ice caps is their dome morphology. The shape of the dome is fundamentally due to the strength of ice. The stronger the material the steeper and higher the dome will be; thus a pile of refrigerated butter will be higher than one of butter at room temperature. The rate of ice deformation is described by Glen's flow law and is expressed by the relation

$$\varepsilon = A\tau^n \tag{14.1}$$

where ε is the rate of deformation, A is a constant related to the temperature of the ice, τ is the shear stress and n is an exponent with a value of around 3 (Glen, 1955). The significance is that ice flow is highly sensitive to shear stress and this in turn is linked to ice thickness and surface slope. The latter relation can be expressed as:

$$\tau = \rho g h \sin\alpha \tag{14.2}$$

where τ is the shear stress, ρgh is the weight of overlying ice (ρ is the density of ice, g is the acceleration of gravity, h is the thickness of the glacier) and α is the surface slope (Nye, 1952). The important implication is that the thicker the ice the less the surface slope needs to be to permit flow. Moreover, the relation is exponential because of the high sensitivity of ice deformation to shear stress. Thus surface slopes on glaciers are progressively steeper near the margins as the ice thins and shallower in the interior where thick ice can accumulate. Such a broad pattern can be seen on a continental scale in Antarctica and Greenland and at a regional scale on any ice cap. The relationship of surface slope to ice thickness helps explain why on satellite images it is often possible to see the underlying topography of ice sheets expressed in the surface morphology by subtle changes in surface slope invisible on the ground.

The outer regions of ice sheet domes are typically marked by two landscape types: ice streams and ice shelves. Ice streams are linear zones of fast-flowing ice with velocities of hundreds of metres per year. The surface gradients are shallower than those of the adjacent ice dome

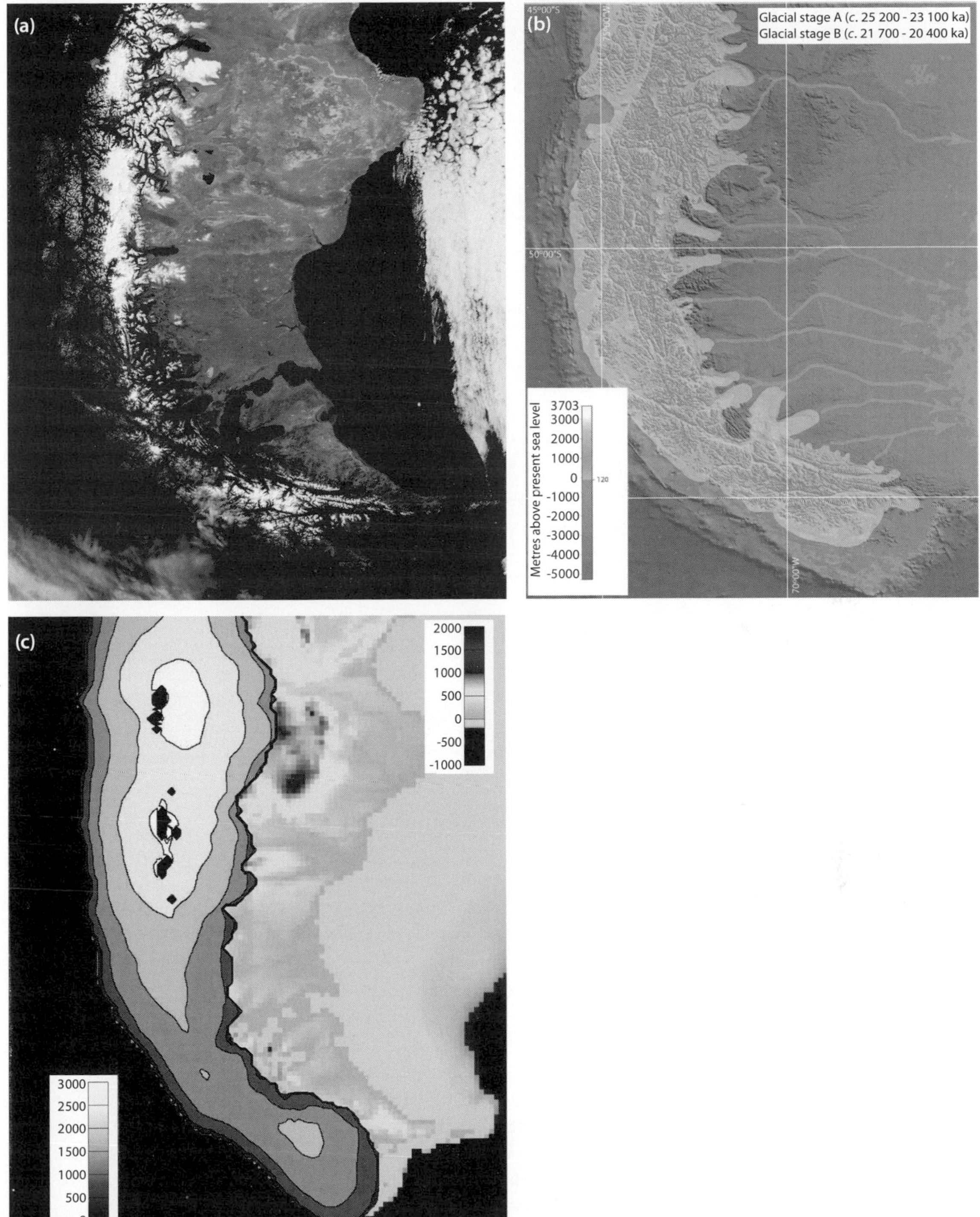

FIGURE 14.6. The Patagonian ice caps present and past. (a) Satellite image showing the present ice caps and adjacent snow-covered mountains (courtesy of NASA). (b) Reconstruction of the Patagonian ice sheet during the LGM based on field evidence (from McCulloch *et al.*, 2000). (c) Modelled reconstruction of the Patagonian ice sheet during the LGM (from Hulton *et al.*, 2002). Elevation (on) in the model is shown for the ice surface (bottom left) and for ice-free land and bathymetry (upper right).

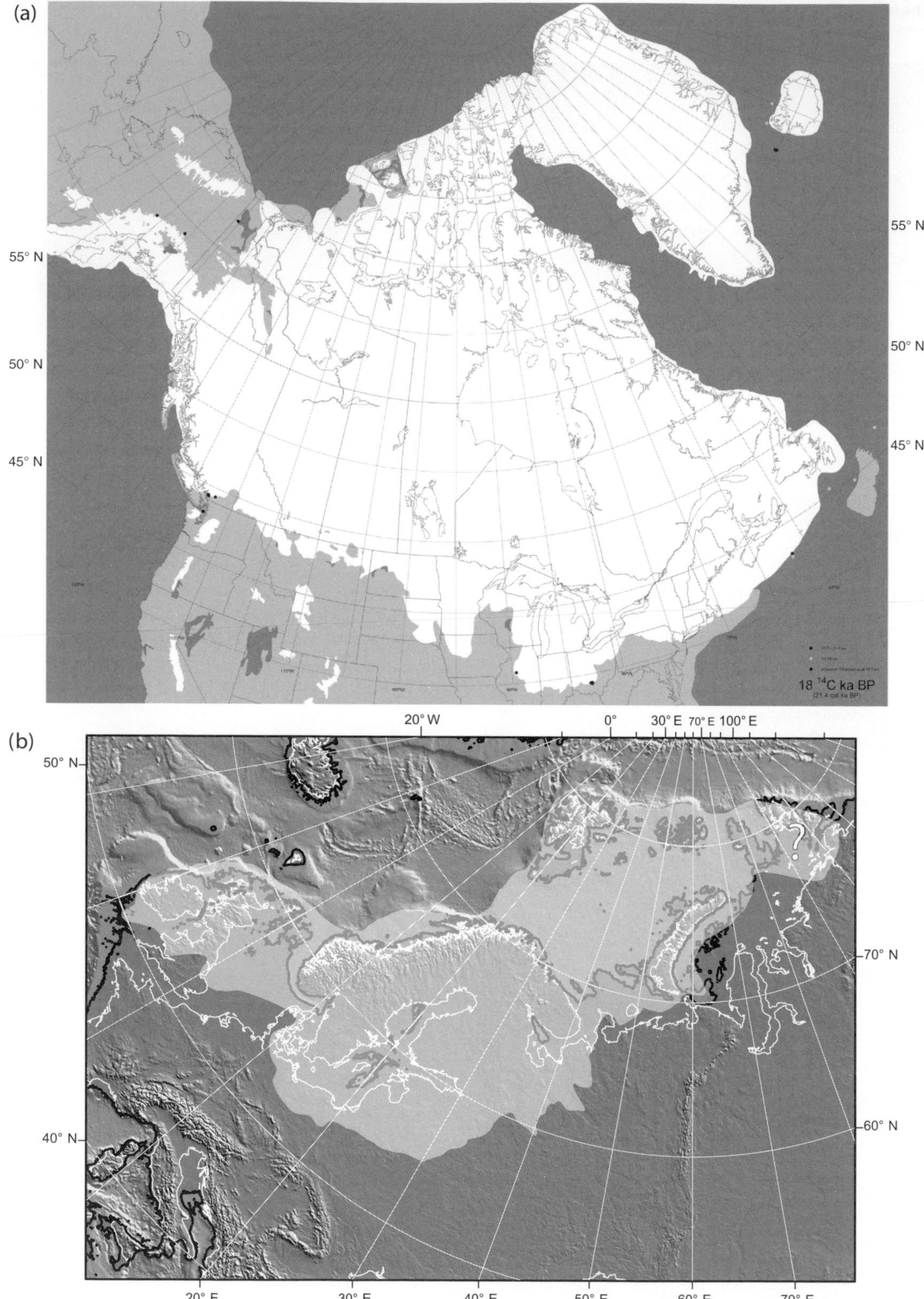

FIGURE 14.7. The northern hemisphere ice sheets at the LGM. (a) The Laurentide, Cordilleran, Greenland and Iceland ice sheets. Sea level lowering of 120 m exposes offshore shelves in the vicinity of Bering Strait (from Dyke, 2004). (b) Extent of the Eurasian ice sheet showing the linking ice masses of the British Isles, Scandinavia and the Barents Sea. Sea level lowering of 120 m, indicated by the black coastline, exposes the southern North Sea, the English Channel and the eastern Siberian offshore shelf. Ice limits after Sejrup *et al.* (2005). (Map compiled by Anthony Newton and Chris Place.)

because of enhanced sliding at their base and thus they form the axis of a depression in the ice sheet surface. Several ice streams may converge into one main trunk ice stream. The result of this convergence of flow is that ice streams discharge the bulk of ice in an ice sheet. This streaming of ice flow is well displayed in reconstructions of balance ice velocities for the Antarctic Ice Sheet (Fig. 14.8a). This is particularly so in the case of West Antarctica where much of the ice catchment is drained by ice streams flowing into the Ross and Filchner-Ronne ice shelves. A fine series of ice streams with sharply defined and crevassed boundaries flows into the Ross Ice Shelf (Fig. 14.9). These ice streams have unusually low surface gradients and remarkably their flow velocity at the junction with the ice shelf varies diurnally with ocean tides (Bindschadler *et al.*, 2003). Some of the ice streams in the area have shut down in recent times, for example the Siple ice stream some 450 years ago (Conway *et al.*, 2002). Commonly ice streams may lead into outlet glaciers that breach a mountain rim, as for example in the Transantarctic Mountains. Here for example is Beardmore Glacier, well known since the epic journeys of Robert Scott, which is 200 km long and 23 km wide and flows at 1 m d^{-1}. It is one of the largest of a series of outlet glaciers breaching the mountains every 60–100 km. The largest ice stream/outlet glacier complex is that of the Lambert Glacier system which is some 700 km long and 50 km wide. Here ice streams from a large catchment in the East Antarctic Ice Sheet converge on the head of a glacial trough before thinning out to form the Amery Ice Shelf, which is itself three times the size of Wales.

Ice streams and outlet glaciers are important in Greenland. Jakobshavn Isbrae, which drains about 6% of the Greenland Ice Sheet in the west, flows at the remarkable rate of 8–12 km a^{-1} (Fig. 14.10a). Its surface is broken up and an observer nearby can hear the cracks and groans as the ice flows. Sometimes it is possible to see movement with the naked eye, rather like the minute hand on a church clock. The glacier ends in a 30-m-high ice cliff from which in summer there is the constant and spectacular view of iceberg calving. The Northeast Greenland ice stream was only discovered in the mid-1990s and is 700 km long. It begins about 100 km from the ice divide and is 15–60 km wide and flows at velocities of 50–120 m a^{-1}. Perhaps the latter ice stream is an episodic flow feature. At a glaciological meeting in Edinburgh which first drew attention to this ice stream, Hal Lister, a member of the British North Greenland Expedition which traversed this part of the ice sheet in 1953–54, commented that he remembered no crevasses or other surface features indicating an ice stream. A possible implication is that it was not streaming at the time. As is common in Greenland such ice streams lead into outlet glaciers near the coast.

Ice shelves are most common in Antarctica where they comprise floating sheets of ice occupying embayments and accounting for about one-third of the coastline (Fig. 14.11). Their surfaces are flat and at their inland margin they may exceed 1000 m in thickness while at their coastal margins they are typically 100–200 m thick. At the inland border of an ice shelf there is often a zone of crevasses known as the strandcrack which marks the junction between grounded and floating ice and is subjected to groaning in response to

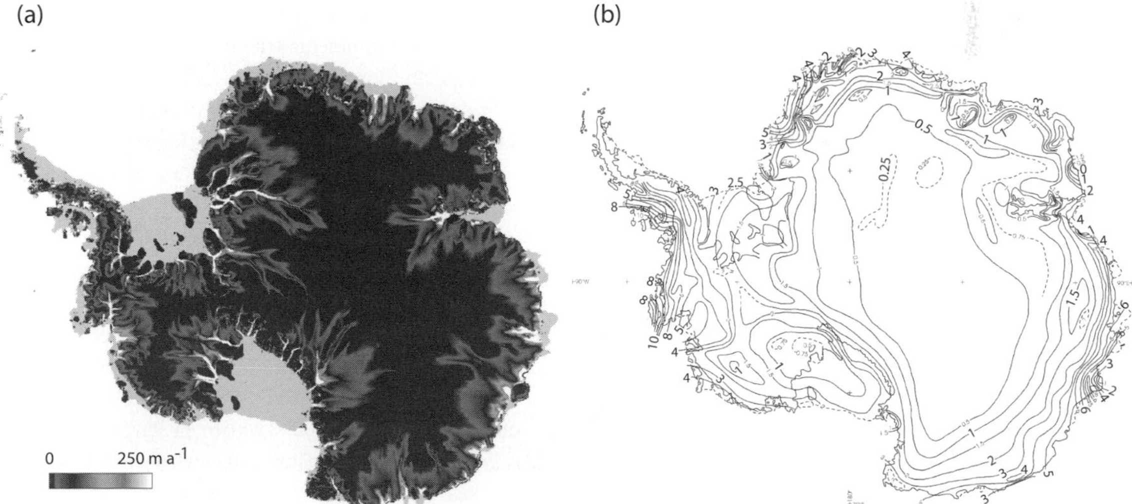

FIGURE 14.8. (a) Balance velocities calculated for the Antarctic Ice Sheet, showing the pattern of streaming flow within the ice sheet. The balance velocities are calculated by assuming equilibrium conditions and that all the ice accumulating upstream of a notional cross section flows through it (from Bamber *et al.*, 2000). (b) Accumulation on the Antarctic Ice Sheet showing the way totals fall from the steeper maritime margins towards the interior where values are desert like (from Giovinetti and Zwally, 2000).

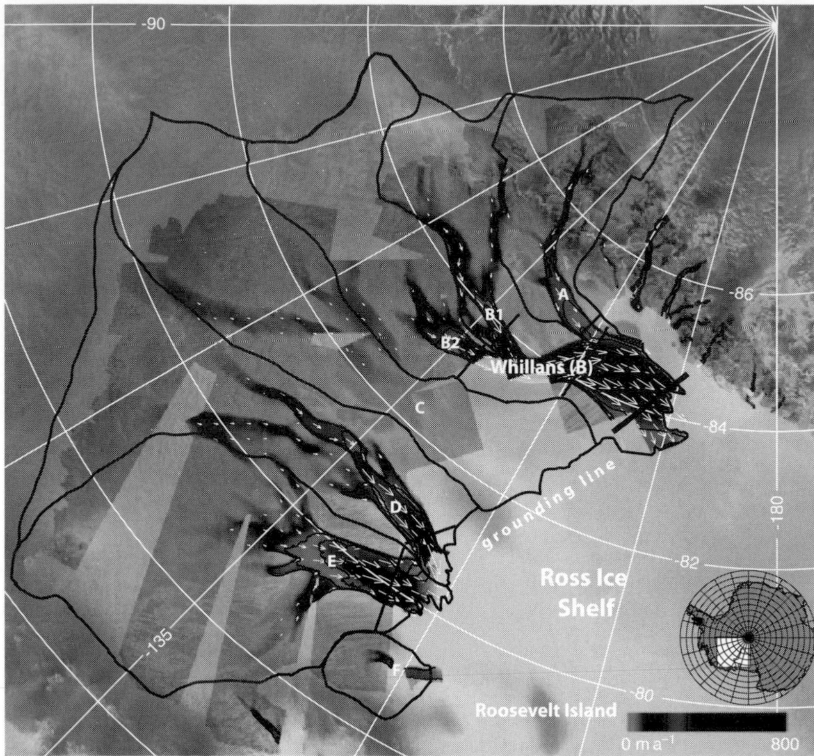

FIGURE 14.9. The pattern of high-velocity ice streams flowing into the Ross Ice Shelf from West Antarctica (left) and through the Transantarctic Mountains from East Antarctica (right). Ice velocities are calculated using radar interferometry on satellite imagery. Although not shown, ice velocities remain high after the ice crosses the grounding line and joins the ice shelf (from Joughin and Tulaczyk, 2002).

tidal flexure. Explorers pioneering routes towards the South Pole early in the twentieth century puzzled over such noises in the ice which on one occasion seemed to accompany their getting up in the morning. The front of an ice shelf appears as a cliff of ice. It was the vertical freeboard front of this ice shelf that blocked Captain Ross on his voyage south in 1841. So dominating is the landform that Ross himself wrote that he might 'with equal chance of success try to sail through the cliffs of Dover' (Kirwan, 1962, p. 171). The Ross Ice Shelf extends some 900 km inland and is about 800 km across at the coast. Smaller ice shelves include the Amery Ice Shelf in front of the Lambert outlet glacier, those flanking the coasts of Marie Byrd Land and Dronning Maud Land and those on both coasts of the Antarctic Peninsula. The last have hit the headlines because some, such as parts of the Larsen Ice Shelf, have disappeared sometimes catastrophically in the last few decades. The only ice shelves in the northern hemisphere are smaller and mainly in northern Greenland and Ellesmere Island.

14.4 Ice sheets and ice caps: mass balance

Ice sheets transfer ice from areas of accumulation to areas of loss or ablation. Ice accumulates where the climate is sufficiently cold to allow snowfall to survive year on year. The ice builds up until the surface slope is steep enough to cause ice to flow following the principles of Glen's flow law. The flow carries ice to an area where it can be disposed of. In warmer climes this may be a land margin where summer temperatures are sufficiently high to melt the ice and the resultant meltwater flows as seasonal rivers to the coast. In cold environments with minimal summer melting the ice typically flows into the sea as icebergs which then melt. In dry polar environments sublimation may be sufficient to achieve ablation without involving significant melting, as for example in the case of local glaciers in the McMurdo Dry Valleys of Antarctica (Fig. 14.10b).

It is helpful to distinguish an accumulation zone of an ice sheet, which sees a net accumulation of snow each year, from an ablation zone which experiences net loss each year. The boundary between the two is known as the equilibrium line. The altitude of the equilibrium line (ELA) varies according to the climate. In the case of the Greenland Ice Sheet it is at around 1100–1200 m. In Antarctica where mean annual temperatures are everywhere below zero, it is near sea level. It is important to stress that the ELA describes the net mass balance and thus it is possible to have melting in summer extending above the ELA. In such cases the summer meltwater may sink into the surface snowpack and refreeze and thus its mass is retained in the

FIGURE 14.10. (a) Jakobshavn Isbrae, an ice stream on the western margin of the Greenland Ice Sheet. The photograph, taken in 1989, is towards the interior of the ice sheet. The intensely broken up surface is the result of velocities of 8–12 km a^{-1}. The lineations are parallel to the direction of flow (photograph by David Sugden). (b) Hart Glacier is one of a series of local glaciers that terminate on land in the McMurdo Dry Valleys, Antarctica. The glacier loses mass by sublimation supplemented in warm summers by a brief period of melting. The ice temperature is around −17 °C. The moraines beyond the present margin, derived from rock debris falling onto the glacier from surrounding slopes, represent millions of years of transport and deposition (photograph by David Sugden).

accumulation zone. This is common in the case of the Greenland Ice Sheet and the ice caps of Arctic Canada.

In a perfectly symmetrical ice sheet the snow that accumulates at the centre sinks vertically as it is covered by successive accumulation amounting to a few centimetres of water equivalent each year. As it sinks it is stretched by lateral flow towards the ice sheet margin and thus annual layers become thinner and thinner until, near the base, 100 ka of accumulation may be represented only by a few metres of ice. In cold environments, such as the interior of Antarctica, where there is no surface melting, the transition from snow with interconnected air spaces to glacier ice with air bubbles may take place at a depth of 160 m over a period of 3 ka. A long and continuous record of accumulation in the absence of surface melting explains the value of ice cores in reconstructing climate. Such cores are taken from central ice domes where there has been minimum lateral flow. Experience has shown that the isotopic ratios in the ice tell of past temperatures while the gas in the air bubbles can hold a record of hundreds of thousands of years of atmospheric change (Petit *et al.*, 1999; EPICA, 2004) (see Fig. 1.8 in Chapter 1). Air flow through the surface snow of the ice sheet for over 1000 years or so adds an inherent uncertainty to the accurate dating of changes in atmospheric gas content, one that has been well debated in the case of whether carbon dioxide and methane changes follow, accompany or precede global temperature changes.

The pattern and magnitude of accumulation and ablation varies with climate and scale. A large ice sheet such as that in Antarctica controls its own climate to a large degree. The cold, high, dry continental interior escapes most storms and its low accumulation is nourished by 'diamond dust', small ice crystals such as those found in cirrus clouds (Fig. 14.8b). Coastal Antarctica feels the brunt of storms nourished in the westerlies that circle the continent and bring snow to the outer steeper slopes of the ice sheet. Air above the ice sheet is cooled by contact with the surface of the ice and becomes denser and flows radially outwards to the coast, with a tendency to divert to the left as the result of the Coriolis effect. These katabatic winds are channelled into topographic lows. Since such winds blow sea ice away from the coast, the early explorers, arriving by ship, often inadvertently located their bases at such particularly windy spots. Douglas Mawson in *The Home of the Blizzard* (1915) recorded gales on 340 days in one year! Once such winds reach the open coast and are freed from topographic constraints, they swing to the left as a result of the Coriolis effect to form the East Wind drift of currents and icebergs close to the coast of Antarctica.

The Greenland Ice Sheet is ten times smaller than Antarctica and is affected by the presence of the Gulf Stream to the south and the North Atlantic to the east. The ice sheet as a whole receives a higher proportion of precipitation from storms than Antarctica and the average accumulation rate of 30 cm a^{-1} is more than double that of Antarctica (Fig. 14.12). In summer almost half its surface area experiences melting and that produced below the runoff line flows off as meltwater. The variability displayed in Greenland ice cores shows that the climate in Greenland can change markedly and rapidly in response to changes in ocean conditions in the North Atlantic. Smaller ice caps are more at the mercy of climate in that they do not have the inertia to resist external changes. The Patagonian ice caps are the result of the westerlies

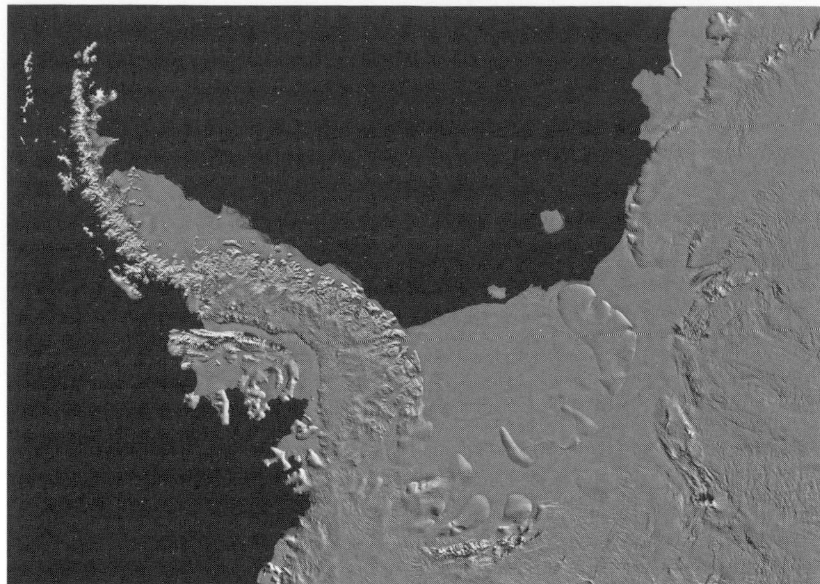

FIG 14.11. MODIS image showing the Filchner-Ronne Ice Shelf at the head of the Weddell Sea embayment, the Larsen Ice Shelf on the western flank of the Antarctic Peninsula and George VI Ice Shelf in the sound between the peninsula and Alexander Island to the east. Major ice streams flow into the Filchner-Ronne Ice Shelf and domes of grounded ice islands rise above its flat surface. The largest such ice island is Berkner Island. The mountains of the Ellsworth Massif lie inboard of the ice shelf to the south and mountain blocks of East Antarctica to the west of the embayment. Two large icebergs have calved into the Weddell Sea (courtesy of NASA/Goddard Space Flight Center).

crossing the mountain range of the Andes and depositing snow on a mountains axis sufficiently high to nourish ice. There is an extraordinary sharp precipitation gradient across the ice caps with western slopes experiencing annual snowfall totals in excess of 10 m water equivalent and eastern slopes ten times less. The mass balance gradient is enhanced by the orographic effect of the mountains with increased precipitation as the air is lifted on the upwind flank and dry föhn winds on the lee side that increase ablation.

The mass balance of an ice shelf involves interaction with the ocean. Ice accumulates from ice streams from the interior and in the case of Antarctica increasingly from surface accumulation near the coast. The flow of ice is unimpeded by bottom friction and thus the ice thins and flows to the coast with high velocities of up to 3 km a^{-1}. Often the ice shelf is pinned by topography where it abuts on land and thus there is a buttressing effect on glaciers that flow into the ice shelf from inland. Ablation is achieved by both iceberg calving and bottom melting. Icebergs, often the size of an English county, calve periodically, perhaps as infrequently as once every few decades. This normal behaviour sometimes seems alarming and a major event attracts global headlines. Bottom melting can occur from the intrusion of warm water from the ocean from ocean currents or tidal pumping. Rates of melting are usually of the order of a few metres each year but in exceptional cases have been recorded at over 20 m a^{-1}. Ice shelves are also sensitive to the effects of surface melting and occur mainly in environments with mean annual temperatures below − 10 °C. This explains their preponderance in Antarctica.

14.5 Ice flow and ice temperature

The mechanisms of glacier flow include creep and sliding. As follows from the flow law, most deformation takes place in the lowest layers of the ice. Creep normally accounts for all glacier flow in glaciers with basal ice below the pressure melting point. Theoretical predictions that there can be no sliding at the actual ice–bedrock interface have been supported by field observations of the protective capacity of glacier ice under cold-based conditions. In such situations most deformation takes place in the ice immediately above the base. As with other rocks, ice temperatures rise with depth as a result of geothermal heat flux. The rate of increase is around 1 °C per 55 m depth, but is affected by surface temperature, accumulation patterns and flow characteristics (Sugden, 1977). The role of ice velocity is especially important since a rate of 20 m a^{-1} generates as much heat as the average geothermal heat flux and this heat is concentrated at the base (Paterson, 1969). The implication is that cold-based ice is favoured in climates with cold mean annual temperatures. It also occurs in situations where topography causes the ice to be thin or to diverge and flow slowly.

Sliding at the ice–rock interface takes place when the ice is at the pressure melting point. The pressure melting point falls with increasing ice thickness and, for example, ice temperatures of −1.6 °C were discovered, along with meltwater, at the base of the Byrd ice core beneath 2164 m of ice in Marie Byrd Land. Once the basal ice is at the pressure melting point a feedback loop comes into play whereby all geothermal heat and that generated, for example by increased sliding, is used to melt basal ice. The amount of

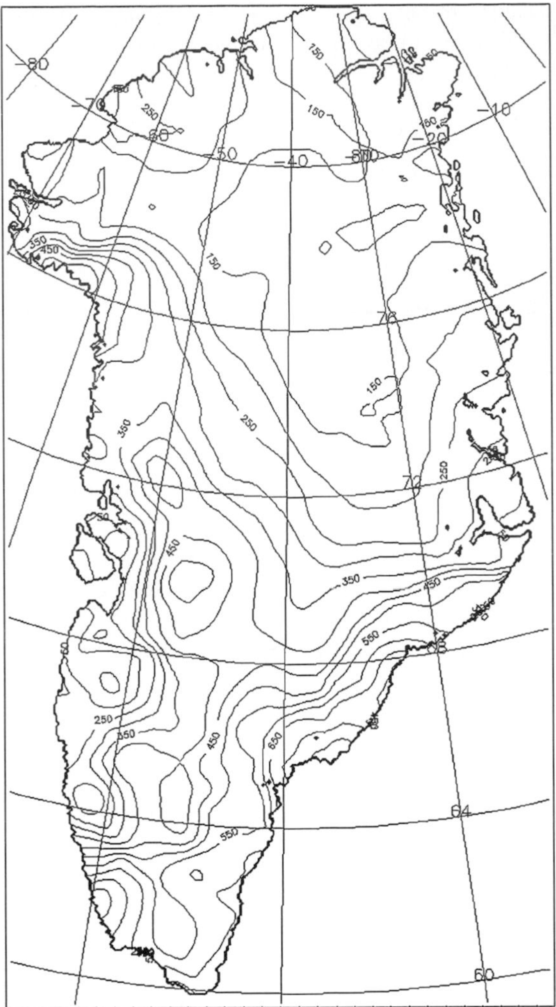

FIGURE 14.12. The accumulation rate (in g cm^{-2} a^{-1}) on the Greenland Ice Sheet, showing the high rates on the margins flanking the ice-free Atlantic Ocean and the low rates in the north and centre (from Bales *et al.*, 2001).

melting beneath warm-based ice, using the average value of the geothermal heat flux, is 6 mm a^{-1} (Paterson, 1969). In fast-flowing ice streams, basal sliding may account for 90% of glacier flow, with only 10% attributed to creep. In such situations and assuming a velocity of 1 km a^{-1}, the sliding may generate enough additional heat to melt a further 30 cm of basal ice, an important source of subglacial meltwater. Warm-based ice is favoured in environments with relatively warm mean annual temperatures, beneath thick ice and especially in areas where the underlying topography causes convergent ice flow and high ice velocities.

The presence of meltwater is especially significant when a warm-based glacier is flowing on soft sediments (Boulton, 1979). If, as is frequently the case, the pore water pressures in the sediment are sufficiently high, the sediment itself beneath the glacier deforms and contributes to glacier sliding. The most susceptible sediments are those with a mix of clay and silt-sized particles, such as those formed by abrasion beneath glaciers or those glacial products dropped into the sea or a lake. Sediment deformation means that sliding at or near the glacier base occurs at lower shear stresses than normal and thus the glacier surface slope is lower than normal. Deforming sediments have been observed beneath ice streams draining the West Antarctic Ice Sheet, especially those flowing into the Ross Ice Shelf, and this is likely to explain the unusually gentle surface gradients as the streams flow into the ice shelf.

Finally, basal meltwater creates a series of subglacial lakes. One of the exciting discoveries in Antarctica is the sheer number and size of subglacial lakes (Siegert, 2000) and the realisation that their presence can affect the way an ice sheet flows (Bell, 2008). Lake Vostok (beneath Vostok Station) is the most famous subglacial lake and is comparable in size to the Great Lakes in North America. It measures 260 km by 80 km and has a depth of over 500 m. Another, Lake Ellsworth, is attracting interest in the middle of West Antarctica (Siegert *et al.*, 2004). Such lakes build up in topographic depressions beneath basal ice at the pressure melting point and where the ice surface gradient is insufficient to drive the water down glacier. These conditions are most common beneath the central domes of deep ice sheets. Subglacial lakes reduce the friction at the glacier bed and in the case of the larger lakes the ice deforms easily and is picked out at the surface by a flat ice surface. It has been suspected that such lakes can drain catastrophically, as for example demonstrated by the distribution and scale of landforms visible in the ice-free margins of Antarctica (Fig. 14.13). In parts of the McMurdo Dry Valleys in the Transantarctic Mountains there are complex channel systems up to 300 m deep with staircases of potholes associated with water discharging over escarpments. They reflect the sudden drainage of huge volumes of water over the Transantarctic mountain rim at a time when there was a thicker, more extended Antarctic Ice Sheet (Denton and Sugden, 2005). Recently, the process of subglacial lake drainage was inferred from satellite remote sensing in East Antarctica by the sudden depression of one part of the ice surface and the consequent rise of another part of the ice surface some hundreds of kilometres apart (Wingham *et al.*, 2006b).

14.6 External controls and feedbacks

The main external variables affecting the behaviour of ice sheets and ice caps are climate, topography and sea level. In every case there is feedback between the ice and the

FIGURE 14.13. View of Wright Valley, one of the McMurdo Dry Valleys of Antarctica, showing meltwater channels and potholes of the Labyrinth in the foreground leading into the glacial trough of middle Wright Valley in the background. The forms were carved by several subglacial lake outbursts in the mid-Miocene (14–12 Ma BP). Hart Glacier shown in Fig. 14.10b is one of the glaciers on the flanks of the valley. Ice-covered Lake Vanda is in an overdeepening in the trough (photograph by David Sugden).

controlling variables. It is important to assess these effects if we are to interpret current trends confidently.

The climatic variability of ice sheets and ice caps affects their relative sensitivity to change. Thus the higher accumulation of ice in Greenland compared to Antarctica means that there is a higher turnover. In other words for a given unit area more ice comes in and more ice melts every year. An important implication is that although it is a far smaller ice sheet, its annual exchange of water with the oceans is 30% that of the larger Antarctic Ice Sheet. Thus it punches above its weight in terms of its response to climate change. Bearing in mind the abrupt changes in temperature indicated by the Greenland ice cores that can be correlated to changes in ocean circulation in the North Atlantic, one can state with some confidence that the Greenland Ice Sheet will be quick to respond to climate change. Such a conclusion is reinforced by feedback effects that would follow an initial warming. Wet surface snow has a higher albedo than dry snow and will absorb more heat. Further, surface melting in summer creates shallow lakes on slight irregularities on the surface of the ice sheet. These also lower the albedo of the ice surface. Thus increased surface melting introduces a feedback loop that enhances the effects of the initial warming. Similar arguments apply to the smaller ice caps with high accumulation rates, such as those in Patagonia. Although their volume is small they will be the first to respond to climatic change.

The Antarctic Ice Sheet, with the exception of the Antarctic Peninsula area, is too cold to respond dramatically to changes in surface climate. It is sufficiently far below the temperature threshold for surface melting to enhance ablation, at least under likely future scenarios. Thus the main effect of warming will be to increase the moisture content in the air and to increase the amount of snowfall. This would lead to a modest increase in ice thickness and increased ice flow. Interestingly, such predictions are borne out by analysis of atmospheric gases in ice cores suggesting that during full glacials when temperatures declined, the centre of the Antarctic Ice Sheet declined in altitude, presumably in response to reduced snowfall. Also, at times of warming, such as during the Pliocene (4–3 Ma BP), local glaciers in coastal Antarctica did indeed thicken and deposit moraines at altitudes above their present margins (Marchant *et al.*, 1994). In this case theory and field observations agree.

Sea level is an important control on ice sheet and ice cap extent because it introduces an additional and highly effective mechanism of ablation. Many glaciers in subpolar environments end in cliffs on the shore because wave sapping is an effective way of eroding and removing ice. Such a situation is common, for example in the South Shetland Islands off the Antarctic Peninsula, where it is possible to walk for long distances along the shore at low tide with an ice cliff rising vertically on the inland side at the high water mark. Where ice velocities are high and the mass of ice larger, grounded ice can move offshore. Observations on present glaciers suggest that glaciers flowing at velocities of 1 km a^{-1} can maintain a snout in water depths of 100 m while velocities of 10 km a^{-1} are necessary to survive in water depths of 300 m (Brown *et al.*, 1982). At water depths greater than this calving of icebergs and brash ice is able to remove any additional ice.

The sea level fluctuations of the Quaternary had an intimate two-way relationship with ice sheet growth and decay. The build-up of ice mainly in the northern hemisphere during the LGM lowered global sea level by around 120 m. If one goes along with the view that the Laurentide ice sheet was of similar volume to the Antarctic Ice Sheet, then the bulk of this water was stored in North America. What is interesting about the sea level curve is not only its broad sawtooth profile reflecting the slow build-up and rapid termination of each glaciation, but the sheer variability of the curve on shorter timescales. The variability relates partly to the cycles of climate change but partly to internal response mechanisms affecting the behaviour of ice sheets and ice caps. Some of the more dramatic changes on millennial timescales, Heinrich events, saw the abrupt collapse of parts of northern hemisphere ice sheets and the discharge of fleets of icebergs into the North Atlantic (Heinrich, 1988). The dynamics of marine-based ice streams discharging through Hudson Strait were important in triggering such abrupt changes (Andrews, 1998).

It seems likely that most sea level changes were driven by climatic changes in the northern hemisphere. As argued above, the Antarctic Ice Sheet is relatively immune to changes in surface climate. And yet it responds to sea level fall in several ways. The Ross Ice Shelf grounded and ice thickened sufficiently to flow and extend over the shallow offshore shelf. Outlet glaciers thickened near the coast as they adjusted to a more distant calving line or a thickened ice shelf. Thus there was a modest increase in ice volume in Antarctica triggered by changes in the northern hemisphere. It is another example of the way in which a positive feedback loop can accentuate an initial change.

It is interesting to note that of the three northern hemisphere ice sheets that were in existence at the LGM, the two that extended over sea basins have disappeared, namely the Laurentide Ice Sheet centred on Hudson Bay and the Eurasian Ice Sheet centred over the Barents, Baltic and North seas. In certain locations ice retreat into deepening water can introduce a feedback process whereby the ice retreats increasingly rapidly as the water deepens. In the case of the Laurentide Ice Sheet the final stages of collapse over Hudson Bay about 8 ka BP were close to catastrophic and caused global sea level to rise over 10 m in <1 ka. Perhaps the Greenland Ice Sheet has escaped such a fate so far because its base was sufficiently elevated to escape sea sapping and calving during retreat. Having said this, there is evidence from biomolecules at a depth of 2 km in the basal ice of the Greenland Ice Sheet of a forested landscape in the last million years that shows that the ice sheet was dramatically smaller than present during at least some interglacial periods (Willerslev *et al.*, 2007).

Topography influences the morphology and behaviour of ice sheets and ice caps at a variety of scales. At a continental scale, the geometry of the landmass helps determine the span and shape of an ice sheet. Thus the large East Antarctic and former Laurentide ice sheets are or were the highest in altitude at 3500–4000 m, while the narrower Greenland and West Antarctic ice sheets reached altitudes slightly exceeding 3000 m. In all cases the ice sheets are or were able to extend over the continental shelf at some stage but not into the deep water beyond the outer edge of the shelf. Topography is important at a regional scale in providing the upland massifs necessary for ice sheet seeding. It has long been argued that the uplands of eastern Canada were the birthplace of the Laurentide Ice Sheet (Andrews, 2006) while the mountain spine of Scandinavia fulfilled the same role in the case of the Scandinavian ice sheet (Kleman *et al.*, 2008). Modelling the initial growth of the Antarctic Ice Sheet suggests that it too began in the uplands of Dronning Maud Land, the Transantarctic Mountains and the subglacial Gamburtsev Mountains (Jamieson and Sugden, 2008). In such cases it is the presence of mountains close to a wet maritime coast that provide the optimum conditions for growth. In the case of the interior Gamburtsev Mountains it is the sheer altitude that permits growth in a continental climate.

There is a growing body of evidence to suggest that there may be a powerful feedback loop resulting from the modification of the topography by ice sheets, whereby ice sheets self limit their extent. The excavation of rock troughs tens of kilometres across and several kilometres deep by outlet glaciers crossing the uplifted margins of continents such as in eastern Canada, East Greenland and Antarctica has the effect of unloading the crust. Allowing for the substitution of the less dense ice for rock, it has been calculated that the effect is to accentuate the crustal uplift of the mountainous rim (Kerr and Gilchrist, 1996; Stern *et al.*, 2005). Under such circumstances the ice sheet is progressively isolated behind a rising mountain rim and increasingly decouples itself from climate. The higher the mountain rim the more stable the ice sheet becomes. Another example concerns the erosion and deepening of offshore continental shelves. It is possible to conceive of situations whereby an ice sheet oscillating in response to climate and sea level cycles progressively deepens the offshore shelf and troughs cut into the shelf. This could increase depths to the point when ice is no longer able to extend over the deepened parts of the shelf. Perhaps this is one explanation of why the East Antarctic Ice Sheet was unable to extend over its offshore shelf in the LGM (Anderson *et al.*, 2002).

The argument also applies at a local scale and especially to fluctuating outlet glaciers discharging into fjords. It may be that the fjord has been excavated to such a depth that a subsequent glacier advance is no longer possible because all the glacier's mass is lost through calving. The Lambert Glacier trough is one such example. During the initial stages of glaciation prior to the mid-Miocene, the ice extended out to the continental edge off Prydz Bay, but for the last *c.* 14 Ma it has been restricted to its trough because overdeepening of the trough has prevented grounding and thickening (Taylor *et al.*, 2004). One way in which glaciers can advance down deep troughs, postulated originally in Alaska and subsequently in South Georgia, is for a glacier to create a moraine bank at its snout and advance down the fjord on the back of a sediment bank (Clapperton *et al.*, 1989). In all of the above the role of relative sea level is crucial and an ice sheet can depress the crust near its margin isostatically. Indeed, the lagged isostatic depression during the course of a glacial cycle might itself deepen the offshore water and limit glacier growth.

14.7 Landscapes of glacial erosion and deposition

One way of looking at glacial erosion and deposition is as a means by which an ice sheet or ice cap transforms its bed to discharge ice in an efficient manner. The relations of ice flow to the geometry of the bed, the permeability of the underlying rocks and the distribution of deformable sediments will all affect ice sheet behaviour. The topic has been the concern of those studying landscapes both of former ice sheet beds and beneath existing ice sheets. Commonly, the two fields of expertise have been pursued by different groups of scientists and yet it seems worthwhile to link the two. After all, the northern hemisphere ice sheet beds are exposed for study, while the process responsible, the ice sheets, can be studied in action in Greenland and Antarctica. In particular, understanding the beds of former ice sheets helps interpret the behaviour and stability of the present ice sheets, especially the role of basal thermal regime and its influence on glacial erosion and deposition.

Ever since the glacial theory was accepted in the 1840s there has been discussion as to the relative efficacy of ice as an agent of erosion. The reason is that it is possible to see landscapes eroded by ice immediately adjacent to those with no sign of erosion by ice. Indeed, so sharp is the distinction, that in several fjord landscapes of the world you can sit on an unmodified plateau surface and look down into a trough eroded by ice to depths of over 1000 m. Another vivid example of the selectivity of glacial erosion is near Aberdeen in Scotland where there is a conspicuous case of a subglacial meltwater channel and adjacent tor silhouetted against the skyline on Clach na Ben. In this case cold-based ice protected the tor on the summit from erosion while subglacial meltwater carved a channel over the adjacent saddle. Such selectivity occurred beneath an overriding ice mass. Some readers may prefer the alternative explanation that the Devil took a bite out of the hill, found it too heavy, and dropped it nearby to form the tor.

An early suggestion that this selectivity in Scotland can be attributed to contrasts in basal thermal regime (Sugden, 1968) seems to be borne out by cosmogenic isotope analyses in Arctic Canada, Scandinavia and Antarctica (Kleman and Stroeven, 1997; Bierman *et al.*, 1999; Fabel *et al.*, 2002; Briner *et al.*, 2003; Sugden *et al.*, 2005). In such studies, tors, weathering regoliths, individual boulders and soils have survived inundation by ice sheets, sometimes demonstratably on several occasions. Models of former ice sheets show that such preserved landscapes were indeed covered by cold-based ice. The converse case is that when the ice is at the pressure melting point there is sliding between ice and rock and erosion takes place. In many places the result is a landscape of areal scouring where structural weaknesses have been exploited to form depressions and lake-filled hollows. The intervening bumps have been moulded by abrasion and plucking by overriding ice. The latter landscapes are widespread in the northern hemisphere on the shields of Canada and Scandinavia. In other places where glacial erosion has been selective the result is concentrated erosion that produces a trough overlooked by little-modified upland or plateau.

The patchwork of areal scouring and unmodified landscape is recognisable at a range of scales. At a subcontinental scale the lowlands beneath the thick centres of the Laurentide and Scandinavian ice sheets and on the western margin of the Greenland Ice Sheet were covered by warm-based ice and are characterised by areal scouring. In contrast the northern Canadian Arctic islands, northern Scandinavia and northeast Greenland were overridden by cold-based ice which has left little sign of glacial erosion. At local or regional scales it is typical to find areal scouring more common in low-lying valleys and at sites of converging ice flow. These are the locations where ice is thicker and faster flowing and thus the base is more likely to be at the pressure melting point. Conversely upland massifs commonly bear surfaces with subaerial weathering features and little evidence of glacial erosion. In extensive upland areas ice may carve troughs selectively through an otherwise little-modified surface, as for example in Baffin Island where John Andrews marvelled at the survival on the plateau of a centimetres-thick Tertiary deposit some 60 Ma old (Andrews, 2006). In these cases the thin ice over the plateau remains cold-based while temperatures at the pressure melting point are confined to troughs.

The pattern of troughs eroded by ice sheets and ice caps is also best described at two scales: subcontinental and regional/local. The former describes the effects of ice sheets such as those of the ice maximum in North America, Greenland and Scandinavia, while the latter describe the effects of smaller ice caps. This is important because it is a reminder that climate and ice sheet extent have fluctuated on many occasions. Reference to ice sheet cores shows that although ice maximum conditions represent the coldest episodes of the Pleistocene, intermediate levels of glaciation applied for longer periods of time.

The beds of mid-latitude northern hemisphere ice sheets demonstrate a continental-scale radial pattern of troughs, ice streams and meltwater routes. Troughs punch through the uplifted continental margins of Labrador, Baffin Island, East Greenland and Norway and form networks of fjords. In Baffin Island such troughs are carved through the main topographic divide. Many troughs have sinuous forms reminiscent of river valleys. Areas of low elevation also bear witness to former ice streams. In the latter case ice sheet drainage has been by means of a radial pattern of major ice streams, many of them only partially guided by the main outline of the underlying topography. Four massive ice streams up to 1100 km long and 100–300 km across flow radially from the northern flank of the Laurentide Ice Sheet and partially exploit the marine straits (and soft sediments) in Arctic Canada (Kleman *et al.*, 2006). Figure 14.14 shows an example from the former Scottish ice sheet where ice in the maritime west of Scotland converged from the western isles and the northwest mainland to flow offshore on soft sediments to the shelf edge where it deposited a huge 'delta' of till. In the convergent zone the ice transformed different bedrock lithologies into distinctive patterns of areal scouring, while in areas of faster flow are elongated streamlined ridges of bedrock and sediment. Many drumlin fields are related to ice streams with the more elongated forms reflecting faster flow in the centre of the stream and the sharp outer boundary of the field marking the ice stream boundary (Clark, 2006). One important observation is that ice streams demonstrating flow in one direction may be overprinted by another ice stream with a different direction of flow (Boulton and Clark, 1990). The implication is that the pattern of ice flow within an ice sheet can change character and direction abruptly as basal thermal conditions and ice morphology evolve (Kleman *et al.*, 2006). Finally, it is worth recording the subcontinental scale of meltwater flow radiating out from ice sheet centres in North America and Scandinavia. Eskers and linking rock channels can be traced for hundreds of kilometres. Typically such meltwater routes are marked by channels excavated in convexities in the underlying relief and esker ridges in depressions. The direction of flow is driven by ambient pressures in the ice and reflects primarily the ice surface gradient and secondarily the shape of the underlying topography (Shreve, 1972).

Local and regional patterns of glacial erosion are demonstrated both by corries and troughs. Corries are present in many upland massifs in mid-latitudes and reflect marginal glacial conditions where snow collects in sites that collect

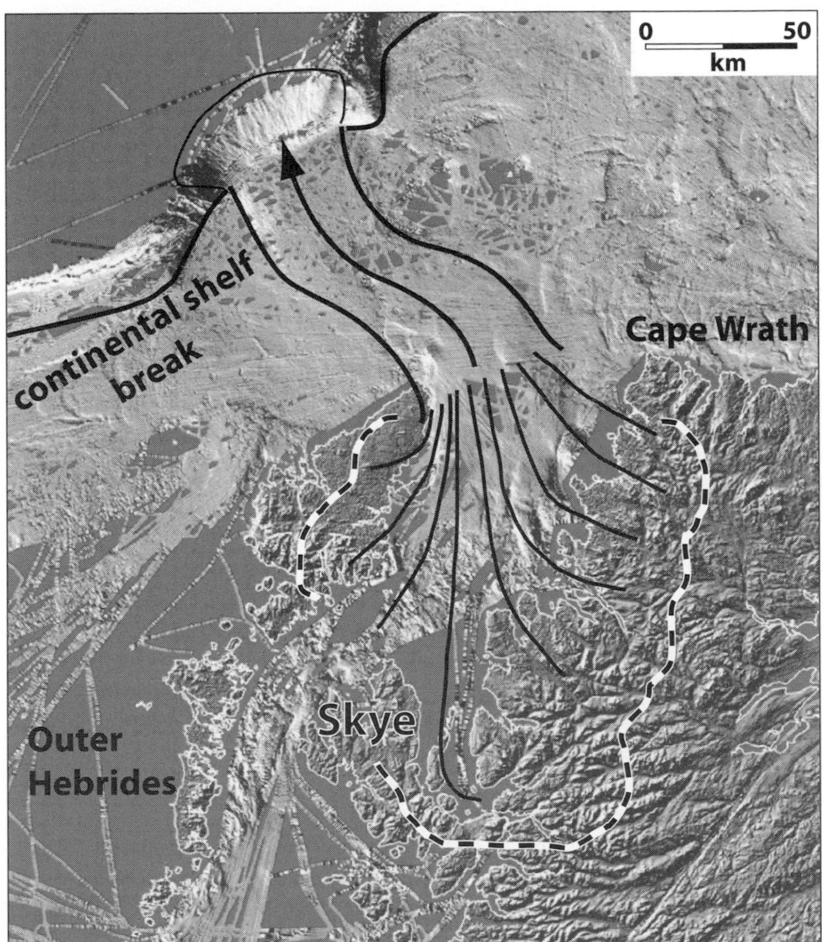

FIGURE 14.14. An image off the northwest coast of Scotland showing the large ice stream that concentrated ice from the Outer Hebrides and the mainland and flowed across the offshore shelf to deep water at the shelf edge. The landforms in both bedrock and deposits are streamlined to varying degrees by the overriding ice. Such ice stream beds are analogies for the beds beneath antarctic ice streams (from Bradwell *et al.*, 2007; © British Geomorphological Society and the Natural Environment Research Council).

drifted snow and are protected from melting by shading from the Sun. In the northern mid-latitudes the dominance of westerly winds and insolation favours a northeast orientation (Evans, 1969). Radial arrays of troughs radiating from mountain massifs reflect regional ice cap glaciations. Excellent examples are the troughs on both sides of the mountain axis of Scandinavia (Kleman *et al.*, 2008). Those on the eastern Swedish side are filled by lakes 10–100 km long and sufficiently continuous to be exploited by steamships for access in the early twentieth century. During ice age maxima the ice divide shifted towards the Baltic and the flow of ice reversed. Similar radial patterns of troughs occur in western Scotland, the Lake District in England and on island massifs in arctic Canada. The depth of the troughs and the clarity of their pattern imply that these phases of glaciation achieved some state of stability, probably related to topography. The implication is that topography exerted a control on ice sheet morphology such that the ice was able to maintain a regional radial flow under a range of different intermediate climates.

There are subcontinental patterns of glacial deposition. In North America the southern zone of the Laurentide Ice Sheet is marked by an arc of till deposition which surrounds the central zone of areal scouring. The deposits are made up of a succession of till sheets and reflect the progressive build-up of sediment by repeated warm-based glaciations. Many of the till sheets may have comprised deforming beds. A similar pattern occurs in Europe whereby the products of the glacial erosion of Scandinavia have been deposited in northern Germany, Poland and the southern Baltic states. In the case of maritime margins such as western Norway and Labrador much debris has been swept across the offshore shelf by ice streams flowing down troughs, but sediment can also accumulate on the ridges in between (Ottesen *et al.*, 2001). In all such cases it is the flow of ice at its maximum extent that drives the material outwards at a continental scale. In zones of complex and changing ice directions the eroded debris may move to and fro and remain in the central ice sheet area. This seems to be the situation in Scandinavia whereby the debris was initially shifted by regional ice caps towards the east, then survived as a thick sheet of drift beneath maximum ice sheets that were cold-based (Kleman *et al.*, 2008).

Armed with these insights from the northern hemisphere, it is useful to predict the pattern of erosion and deposition beneath the Antarctic Ice Sheet. Better knowledge of the bed will aid interpretation of the surface pattern of ice velocities, the location and behaviour of subglacial lakes and other aspects of ice sheet dynamics. At present information about the Antarctic Ice Sheet is sparse, especially since there are certain sectors where there is only a handful of observations in an area the size of western Europe. And yet even with such reservations the exercise can help identify interesting geomorphological issues for further study.

Firstly, it is helpful to highlight the long history of Antarctic glaciation and the similarities and differences with northern hemisphere ice sheets. The Antarctic has experienced ice sheet glaciation for 34 Ma. There was a remarkable period of *c.* 20 Ma from 34 to 14 Ma BP when the ice sheet was fluctuating in response to Croll–Milankovitch fluctuations in a manner analogous to that of the Laurentide Ice Sheet in the Pleistocene. Over 46 glacial cycles have been identified in sediment cores from Cape Roberts off the coast of the McMurdo Dry Valleys. The sediments reveal advances and retreats, and associated sea level fluctuations. Also over the period they show a progressive reduction in meltwater deposition, a change of vegetation from southern beech forest to tundra and a change in clay minerals from temperate forest soils to those forming in a periglacial environment (Barrett, 2007). Such features reflect the progressive cooling of Antarctica and of the wider world until, around 14 Ma BP, the ice sheet first advanced to the outermost offshore shelf. It then retreated to its present size which it has maintained for some 13.6 Ma or so. Thus in the initial stages of Antarctic glaciation an ice sheet similar in size to the Laurentide Ice Sheet fluctuated in a similar way but for an order of magnitude longer.

As a basis for prediction Fig. 14.15 shows different configurations of ice extent in Antarctica that are in equilibrium with a range of polar climates ranging from cool temperate to polar (Jamieson and Sugden, 2008). In addition, the model predicts the erosive potential of the ice at different stages. The model uses a three-dimensional ice flow routine to calculate basal ice temperatures. Erosive potential is related to ice velocity and basal drag and only comes into play when the basal ice is at the pressure melting point. The ice sheet model is forced by a degree-day model which calculates the amount of snow falling on the ice sheet in a year. We specify the climate on the basis that Antarctic glaciation began in a Patagonian-type climate, as demonstrated by fauna and flora in the earliest glacial deposits (Barrett, 2007). For the purposes of the experiment we allowed the model to change from the initial cool temperate climate to the current polar climate in a series of steps each 50 ka long. This means that the glaciers reach thermal equilibrium at each stage. The broad shape of the topography is assumed to be the same as that of today. Models such as this are simplified and include a number of uncertainties and yet they are a useful means of trying to capture the main features of ice sheet behaviour.

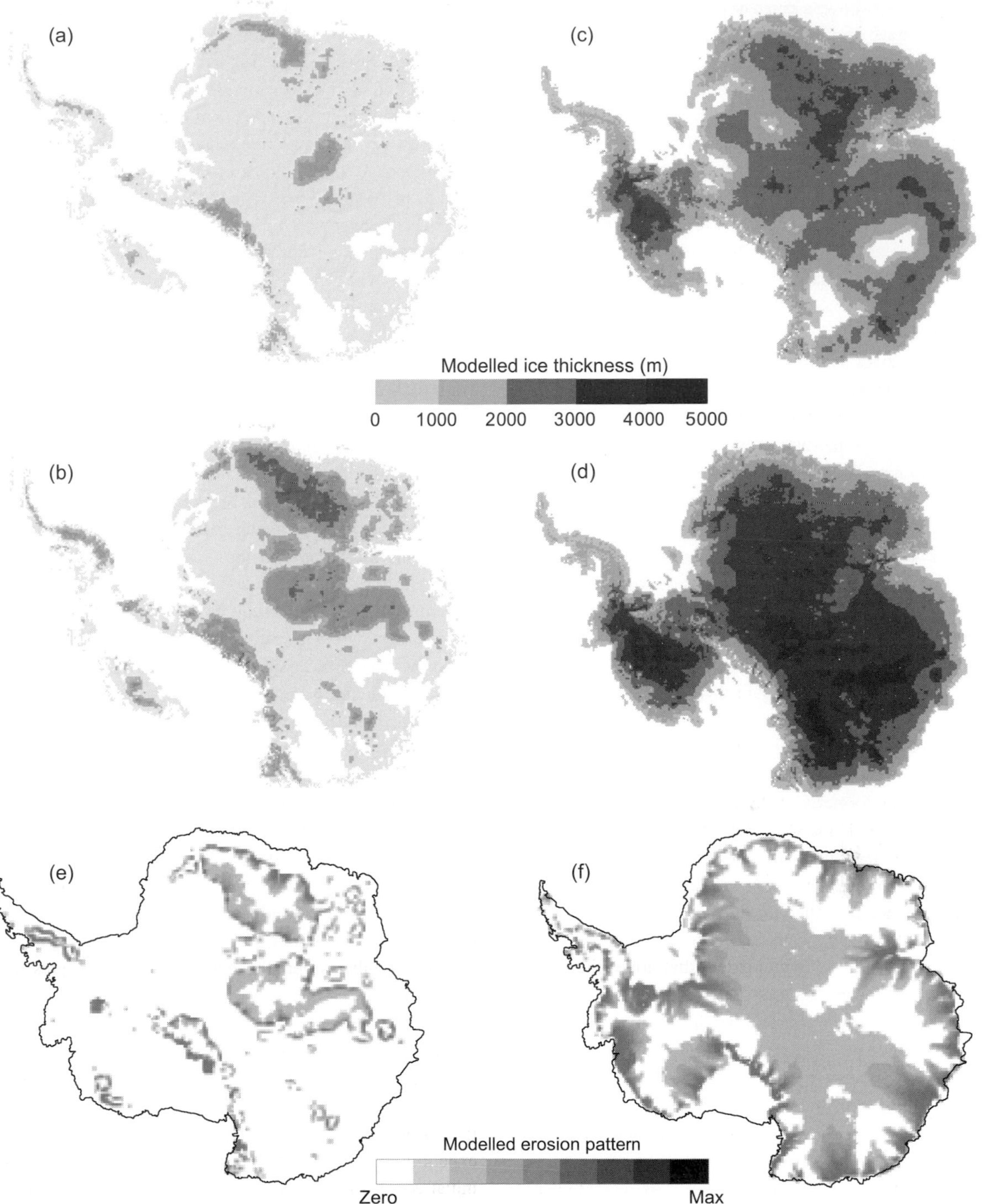

FIGURE 14.15. Modelled phases of glaciation in Antarctica under a range of climates from (a) Patagonian type, through intermediate states (b) and (c) to the present polar climate (d). The range illustrates a few of the possible ice configurations that would have characterised the first 20 Ma of Antarctic glaciation forced by Croll–Milankovitch fluctuations. (e) Shows the erosive potential of the ice sheet at an intermediate stage and (f) shows the erosive potential under the present ice sheet (modified from Jamieson and Sugden, 2008).

The model shows a range of possible stages of glaciation that may have been reached on many occasions in the initial 20 Ma of glaciation. There are several points of interest. First the model predicts that the main centres of glaciation are the upland massifs of the Transantarctic Mountains, Dronning Maud Land and the Gamburtsev Mountains. Also ice caps built up on the upland archipelago in West Antarctica. Ice radiated out from these mountain massifs, presumably on many occasions. The broad pattern of ice centres agrees with the predictions of another model driven by a general circulation model (GCM) set for preglacial times (Deconto and Pollard, 2003). This agreement between two approaches suggests that the models are capturing the main features of Antarctic Ice Sheet initiation in the main mountain groups. By analogy with the northern hemisphere one can predict that there will be corries in the mountains, regional patterns of troughs radiating out from the main massifs, areal scouring in the lowlands and minimal erosion on upland plateaus and summits. Since the mountain massifs are so distinctive and high, it is likely that a regional ice cover of this nature will be a common, even stable, feature of the various glacial cycles. Over time the products of erosion will be transferred radially to the surrounding lowlands or, in the case of West Antarctica, into any adjacent basin below sea level.

The model shows how further growth causes the ice sheets to merge and impose a continental radial pattern. The zone of maximum erosion advances as a wave towards the coast with particular focus on the main topographic depressions, formerly large river basins (Figs. 14.3 and 14.4). Finally the ice sheet stabilises with the margin near the coast for millions of years under a climate regime dominated by the ice sheet and similar to that of today. The implication is that for the last 13.6 Ma or so glacial erosion has been concentrated in broad areas in the upper hinterlands of the main ice streams and along the course of ice streams themselves. In locations where such streams flow through the marginal mountain rim the erosion will be of hard rock, while in the case of ice streams on former glaciomarine deposits, such as in West Antarctica, glacial erosion will be achieved by bed deformation. Much of the higher inland topography is covered by cold-based ice.

Using the model as a basis for comparison with the ice sheets of the northern hemisphere raises several observations and predictions. The first observation is that the radial pattern of troughs and ice streams at the scale of the ice sheet as a whole is clearly visible in the model and in reality at the margins of the ice sheet. As might be expected from such a large and long-lived ice sheet, troughs like that of the Lambert Glacier are large by the standards of the northern hemisphere. A second observation is that the ice streams in the straits of the Canadian Arctic archipelago are similar in scale to those in West Antarctica. In both cases the ice streaming may be enhanced by rift structures favouring the accumulation of glaciomarine sediments on sea floor depressions during lesser regional glaciations.

The model can be used to make some predictions about the subglacial morphology of Antarctica related to the pattern of basal thermal regime. There will be two types of landscape of areal scouring. In places where the flow is in the same direction during both regional and continental glaciation, one would expect maximum lineation and smoothing in the direction of ice flow. In places where the direction of flow changes at each stage of glaciation, then the landscape will be irregular and less smoothed. There is a contrast between the coastward and inland flanks of the mountain massifs of Dronning Maud Land and the Transantarctic Mountains in that the coastal flanks maintain a constant direction of flow at all stages of glaciation, while the inland flanks experience a change of flow direction as ice sheets wax and wane. Although the Gamburtsev Mountains are located in the interior, the same contrast applies to the surrounding lowlands. Ice flowing directly towards the sea has a constant flow direction at most stages of glaciation, while flow directions towards the interior are more complex and vary with ice sheet growth and decay. A second prediction is that upland areas will tend to be little modified by overrunning ice. In this case the uplands have thin ice covers under both regional and continental glaciation and are likely to have been overridden by cold-based ice at all stages of glaciation.

Finally, it is possible to comment on the possible debris deposition in and around Antarctica. The implication of a wave of erosion near the margins of ice sheets that build up on individual massifs and coalesce to form one major ice sheet is that the eroded material may be transported by complex routes but will eventually end up in offshore zones at the mouths of the main topographic basins. These include four basins, namely those draining into the southeastern Weddell Sea, the basin of the Lambert Glacier, and basins beneath the ice terminating at the coast both in Wilkes Land and northern Victoria Land. One possible interesting case of possible capture of an ice catchment comes from the case of Byrd Glacier (Fig. 14.2). This is the biggest trough cutting through the Transantarctic Mountains and it drains a substantial part of the ice sheet inboard of the mountains (Fig. 14.8a). Presumably it is transporting much material to the Ross Sea via the Ross Ice Shelf. Is it possible that this outlet glacier has carved a trough sufficiently large to capture much of the ice and debris that would have initially drained elsewhere? The potential diversion is indicated by comparison of the Byrd catchment with the underlying topography (Figs. 14.3 and 14.4). If such a diversion occurred, there could be a sedimentary record of the event in the offshore sediments. Having said this, uncertainties about the subglacial topography and indeed on the possible existence of a preglacial river valley mean that such a view is highly speculative.

One can predict that there will be zones of deposition in the depressions between the Gamburtsev Mountains and Dronning Maud Land on the one hand, and between the Gamburtsev Mountains and the Transantarctic Mountains on the other hand. Erosion by regional ice flow will have transported debris to the lowland from both mountain flanks. Such material may have survived if, during subsequent more expansive glaciations, the deposits were covered by cold-based ice, as occurred in parts of Sweden. The model implies that there are such areas, especially at intermediate elevations. Depending on how long regional ice conditions applied during the 14 Ma of fluctuating ice masses, the volume of sediment could be considerable.

14.8 How will ice sheets and ice caps respond to global warming?

The remainder of this chapter examines the future of the world's ice sheets and ice caps. It builds on the review in the previous pages of the factors at different scales that determine how such ice masses behave. In the pages that follow the focus is on how these various factors hang together in specific situations, namely Antarctica, with separate treatment of the West and East Antarctic ice sheets, Greenland Ice Sheet and smaller ice caps.

14.8.1 Antarctic Ice Sheet

The Antarctic Ice Sheet is by far the most important player in that it alone would account for a *c.* 60-m rise in sea level, were it to melt. On average the ice sheet exchanges a volume of ice and snow each year which is the equivalent to 5.1 mm of global sea level. The input is from snowfall from the atmosphere and the pattern of high accumulation on the marine peripheries and minimal accumulation in the centre is shown in Fig. 14.8b. Output is by means of calving into the ocean. The scale of the ice sheet and the magnitude of this exchange means that a relatively minor change in the balance due to climate change could have far-reaching effects.

In recent years most attention has been focussed on the possible response to warming and the issue has been of great concern to the IPCC community. The conclusion has been that although there is considerable uncertainty, the effects of warming are likely to be close to neutral in that different processes balance out (IPCC, 2007). The uncertainty is mainly due to the lack of firm data on the optimum spatial and time scales. The various attempts to measure the mass balance of the ice sheet are plagued by the lack of data on snow accumulation and iceberg calving and estimates vary in magnitude and even in the direction of change. Satellite measurements hold the prospect of improving the accuracy and coverage of data but it will be many years before it is possible to distinguish the overall climate signal from local variability. The two main sets of processes are those involved with the atmosphere and those that link with the sea.

Warming will increase the water content of the atmosphere and lead to an increase in snowfall. Since surface temperatures in Antarctica are everywhere below zero, there will be no surface melting to compensate. Modelling suggests that a temperature rise of over 5 °C would increase accumulation in Antarctica while a rise of over 8 °C would be needed for melting to begin to balance accumulation (Huybrechts, 1993, 2004). Such estimates must be treated with caution since they take no account of other climatic changes such as storm tracks which could affect the figures either way. Nonetheless they indicate that under likely climatic scenarios warming of the atmosphere will cause accumulation to rise and, at least in the short term, for ice thicknesses to increase with the overall effect of abstracting water from the ocean.

Much less is known about the interface with the ocean but it seems likely that warming of the oceans is likely to lead to more bottom melting and thinning of ice shelves and that this in turn could lead to the thinning of ground-based ice streams and outlet glaciers flowing into the ice shelves. Ice shelf sensitivity to ocean temperature changes is indicated by the loss of George VI Ice Shelf in the early Holocene climatic optimum when deep-ocean foraminifera existed in what is now a lake impounded by the current ice shelf (Bentley *et al.*, 2005). It is easy to believe that such ocean water flowing beneath the Ross and Filchner-Ronne ice shelves could lead to thinning and eventually ice shelf loss. However, the expectation is counteracted by another feedback process which is related to sea ice formation. Freezing of the sea surface increases the density of ocean water which then sinks beneath the ice shelf and accounts for melting at the grounding line (Bentley, 2004). With global warming and less sea ice forming, this process could be suppressed. Here is another example of the complexities of the Earth system that add uncertainty to our predictions.

There are warning signs that, in spite of the uncertainty, things are beginning to change. The Antarctic Peninsula has seen the reduction and sometimes catastrophic collapse of several ice shelves in the twentieth century (Vaughan and

Doake, 1996). Glaciers flowing into these ice shelves have thinned abruptly as a response (De Angelis and Skvarca, 2003). These losses have accompanied documented atmospheric warming. The Antarctic Peninsula climate is much warmer than that of the main West Antarctic ice shelves and indeed the losses all occur where there is surface melting and mean annual temperatures are above −10 °C. But is this an early warning of what could occur? A second observation is that certain ice streams in the Amundsen Sea sector of Antarctica that are grounded below sea level are thinning and lowering at rates far higher than predicted by ice sheet models (Fig. 14.16). Are we seeing a positive feedback loop whereby thinning reduces the ice freeboard and that this further enhances flow and thinning? A third observation would come from those studying ice sheet deglaciation in the northern hemisphere. Here, there is evidence of abrupt episodes of ice loss associated with calving, as for example in the dying throes of the Laurentide ice sheet (Dyke, 2004). How different are current observations in West Antarctica from those that would have been seen by an observer on the shores of Hudson Bay in Canada 8 ka BP?

East Antarctic Ice Sheet

The proposition that the East Antarctic Ice Sheet has survived intact for the last 13.6 Ma, discussed above, overlooks an intense debate about the stability of the ice sheet

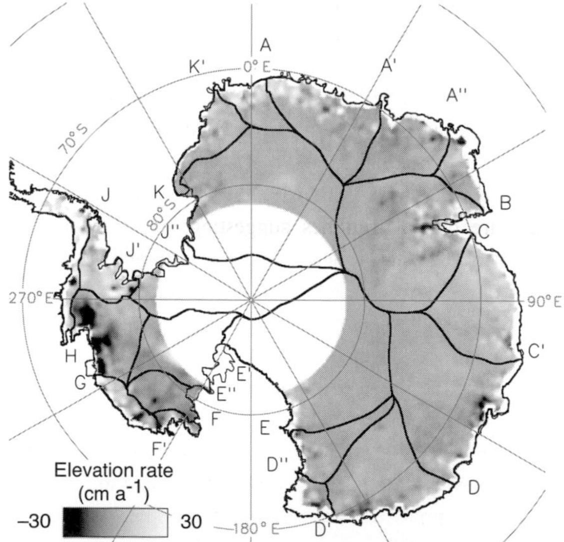

FIGURE 14.16. Rate of elevation change of the Antarctic ice sheet, 1992–2003, as measured by radar from satellites. There is no coverage near the South Pole. The changes show modest thickening in much of East Antarctica and thinning in West Antarctica inland of the Amundsen Sea, an area where the ice sheet is grounded below sea level (from Shepherd *et al.*, 2001).

and whether it is susceptible to warming. It is worth reviewing some of the arguments used in the debate since they illustrate the way in which geomorphology can contribute to our understanding of the working of the Earth system.

The background to the debate is that study of oxygen isotopes on carbonates in marine cores suggested that the Antarctic Ice Sheet grew step by step in response to the changing distribution of land and sea accompanying the break-up of Gondwana and to positive feedback as the growth of the ice sheet led to progressive cooling (Kennett, 1977). But then marine diatoms of Pliocene age were found in the Sirius Group deposits, glacial tills found at high elevations in the Transantarctic Mountains (Webb *et al.*, 1984). The diatoms were mixed with vegetation remains indicating temperature conditions some 20–25 °C higher than today. Assuming the diatoms were transported by glaciers, the most likely source for them was marine basins in interior Antarctica. It followed that the continent must have been largely deglaciated in the Pliocene to allow the diatoms to grow in inland seas and that there was a subsequent ice sheet advance that entrained them and deposited them in the Transantarctic Mountains.

The hypothesis attracted support from adjacent fields of science, for example from those seeking higher Pliocene sea levels in order to explain shorelines in the eastern USA (Dowsett and Cronin, 1990). Barrett *et al.* (1992) pointed out that if this deglaciation in the Pliocene was true, then the East Antarctic Ice Sheet must be regarded as sensitive to global warming of a few degrees and could be at risk in the future, leading to sea level rise of tens of metres. This was a challenging hypothesis of great importance to the world and its effect was to stimulate further work.

One line of attack was to look at the Sirius Group deposit in the light of landscape evolution. In a series of papers George Denton and colleagues showed that the Sirius Group deposits formed early in the history of the Transantarctic Mountains and that the present landscape was essentially a relict from the mid Miocene, some 14 Ma BP (Denton *et al.*, 1993; Sugden *et al.*, 1995). The preservation of buried glacier ice 8 Ma old, argon/argon dates on volcanic ash trapped in tills and avalanche deposits, cosmogenic isotope analysis of bedrock and surface boulders, and the presence of undisturbed Miocene volcanic cones all testify to the persistence for millions of years of a cold polar climate. None of the above could have survived had the climate been even a few degrees warmer, since surficial features would have been removed by weathering, melting or erosion by running water. Since the present climate is dominated by the presence of the ice sheet, the implication is that the East Antarctic Ice Sheet has survived intact for at least the last 13.6 Ma, including during the warmer Pliocene. Indeed, moraines of Pliocene age record that the glaciers thickened and advanced modestly in response to warming at the time (Marchant *et al.*, 1994). These arguments for stability did not explain the presence of the marine diatoms and it was only subsequently that they were explained as wind blown. Under such circumstances the Sirius Group deposits with overridden tree remains are likely to date back to the initial glaciation of Antarctica.

There is evidence of thickening of peripheral parts of the East Antarctic Ice Sheet in response to sea level lowering during the LGM. Typically the lower reaches of glaciers thickened. For example in the case of the Mackay Glacier flowing through the Transantarctic Mountains in southern Victoria Land the glacier thickened by 150 m 20 km from the coast and progressively less up glacier, until at a distance of 85 km up glacier the thickening had fallen to 40 m. This pattern can be seen on other Transantarctic outlet glaciers (Denton *et al.*, 1989) and the Lambert Glacier, which remained confined to its trough throughout the LGM. Cosmogenic isotope analysis on nunataks protruding above the ice shows that ice surface elevations on the ice sheet dome in MacRobertson Land west of the Lambert Glacier thickened modestly in response to a minor advance at the coast (Mackintosh *et al.*, 2007). The interior of the ice sheet was colder during the LGM and analysis of the ice cores suggest that the central elevation fell modestly at the time.

This record of the past behaviour of the East Antarctic Ice Sheet suggests that it is a quasi-stable feature of the Earth's environment, at least while the Antarctic continent is situated over the South Pole. The ice sheet controls its climate and responds to warming by thickening slightly as a result of increased snowfall and thinning in response to cooling and a reduction of snowfall. The main response is a result of sea level changes instigated by ice sheet fluctuations in the northern hemisphere. Sea level lowering causes outlet glaciers to thicken near the coast.

These generalisations are borne out by recent satellite observations of current changes in ice surface elevation (Fig. 14.16). Although there is no information from the vicinity of the South Pole, there is evidence of widespread thickening. It is tempting to attribute this to temperature or circulation changes associated with global warming. The only exceptions are marginal areas where the ice is grounded below sea level as in part of Wilkes Land. Possibly thinning in these latter locations is a response to sea level rise following retreat of northern hemisphere ice sheets after the LGM.

The overriding conclusion, backed up by field observations and theory on a range of timescales, is that the East Antarctic Ice Sheet is a stable feature of our world and that its 52 m of sea level equivalent is likely to survive any foreseeable future climatic scenario. Indeed, the ice sheet will thicken and help counteract the rise of sea level caused by melting of other ice sheets.

West Antarctic Ice Sheet

The dynamics of the West Antarctic Ice Sheet are dominated by ice streams, many of them grounded below sea level, flowing into the Ross, Weddell and Amundsen/Bellingshausen seas. Since the ice streams drain the bulk of the ice and since they flow fast, the implication is that the ice sheet will be supersensitive to sea level change and will respond to change rapidly. The discovery of a tidal signal to flow tens of kilometres inland of the grounding line is a vivid demonstration of the sensitivity of ice stream flow to changing ocean conditions (Murray *et al.*, 2007).

The sensitivity of the West Antarctic Ice Sheet is borne out by its history during the LGM. Everywhere it thickened and, unlike much of East Antarctica, it may have reached to the outer edge of the offshore shelf in some sectors. The Ross Ice Shelf grounded and extended to the vicinity of northern Victoria Land. The expanded and grounded ice mass dammed lakes in the McMurdo Dry Valleys and caused Transantarctic Mountain outlet glaciers like the Mackay Glacier to be diverted north, as revealed by bends in their offshore troughs. In the Amundsen Sea area and Antarctic Peninsula there are troughs, meltwater channels, ice-scoured landscapes and till sheets all showing that ice extended up to hundreds of kilometres offshore to the shelf edge (Anderson, 1999; Ó Cofaigh *et al.*, 2005). Ice advanced offshore in the Weddell Sea sector, but there is less certainty about how extensive it was (Bentley, 1999). It is assumed that the centre of the ice sheet thickened in response to its increased size, but by how much is unclear. There are glacial trimlines suggesting ice thickened by about 1000 m in the Ellsworth Mountains, but this probably relates to an early stage of glaciation in Antarctica, when surface ice temperatures were close to the melting point. Meanwhile, ice core studies and recent cosmogenic studies point to more limited thickening, perhaps of the order of a few hundred metres.

Retreat from the LGM extent has occurred in different styles and at different times, partly reflecting different subglacial topographies. The western shelf of the Antarctic Peninsula was deglaciated by 10 ka BP and the continuity and pattern of landforms as revealed by sonar suggests that the retreat was punctuated by phases of rapid collapse of hundreds of kilometres of ice in a few decades. On the other eastern side of the peninsula, retreat was steadier, with stillstands marked by distinct moraines. In the Ross Sea sector,

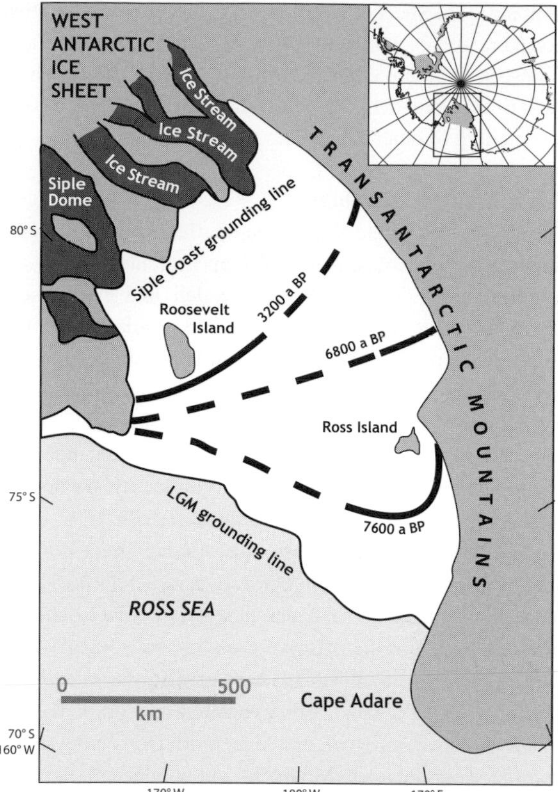

FIGURE 14.17. The swinging door pattern of retreat of the grounding line in the Ross Sea embayment during the Holocene, based on glacial geological evidence in the Transantarctic Mountains and ice stratigraphy on Roosevelt Island (from Conway et al., 1999).

there is evidence of steady retreat of the grounded Ross Ice Shelf throughout the Holocene (Fig. 14.17). The grounded ice margin of the ice shelf has retreated at a rate of some 100 m a^{-1} for the last 8 ka (Conway *et al.*, 1999). The reconstruction is based on the age of abandoned moraines leading into the ice shelf from the Transantarctic Mountains and the age of structures in the ice grounded on Roosevelt Island. A similar conclusion was obtained from cosmogenic isotope work on the adjacent mountains of Marie Byrd Land where there has been steady thinning of over 800 m from 12 ka BP to the present (Stone *et al.*, 2003). At the time of writing there is no comparable evidence of ice sheet trajectories in the Holocene in any other sector of West Antarctica.

The twentieth century saw the retreat of several ice shelves in the Antarctic Peninsula, especially after the 1980s. During this period the area has experienced higher than average regional warming which has caused collapse through the effects of increased surface meltwater (Doake and Vaughan, 1991) and through bottom melting by ocean water (Shepherd *et al.*, 2004). Similar collapses have been documented in the past, especially in the early Holocene around 9 ka BP when George VI Ice Shelf disappeared and marine foraminifera show that the coastal seas of Antarctica were warmer than at present (Bentley *et al.*, 2005). Larsen A Ice Shelf collapsed in 1995, while the collapse of Larsen B in 2002 saw an ice shelf 3250 km^2 in area collapse in 3 weeks. A floating ice shelf does not of itself affect sea level, but the glaciers flowing into the ice shelf then thin and maintain enhanced flow velocities for at least decades and thus do contribute to sea level rise (Rignot, 2006).

Figure 14.16 shows the rate of change in ice surface elevation in the 11-year period from 1992 to 2003, as measured by satellite altimeters. What is striking is the extent of thinning in the Amundsen Sea sector and the high rate of thinning inland of Pine Island Bay. The observations confirm that ice is continuing to thin in those parts of the mountains of Marie Byrd Land that have experienced thinning for the last 12 ka. But the most striking discovery is the catchment of Pine Island Bay. Here, further glaciological study has revealed average bottom melting rates of ice shelves of 24 m a^{-1}, retreat of the grounding line by 1 km a^{-1} in the 1990s (Rignot, 1998) and evidence that thinning extends over 150 km inland along ice streams (Shepherd *et al.*, 2001). There are several surprises. Firstly, the rates of bottom melting are far higher than predicted. Secondly, the effects of bottom melting and thinning have been transmitted up glacier far faster than expected and suggest that ice streams can respond dynamically to change in decades rather than millennia. Thirdly, the loss of a buttressing effect by ice shelf thinning, and not necessarily collapse, is sufficient to provide the initial trigger for thinning inland.

These are all profoundly important observations and show that the coupling between ice sheets and the ocean involves processes that are not yet understood. Moreover, such detailed processes are not yet included in ice sheet models that have been used to predict the response of an ice sheet to climate change. Payne (1999) employed a high-resolution model including longitudinal stresses to show how such ice streams can respond within decades to a wave of enhanced sliding that is transmitted up glacier from the grounding line. The trigger in this case is likely to be the thinning of the ice shelf at the grounding line caused by the influx of abnormally warm ocean water.

The West Antarctic Ice Sheet emerges as a major point of sensitivity in a warming world. It is dominated by ice streams that are sensitive to sea level and to ocean temperatures and it seems that some ice streams can respond to change in decades rather than millennia. This change of view marks a sea change in our perspective on the response of the Antarctic Ice Sheet to warming. Previously, the prevailing view, as argued in the IPCC reports, is that the response of the Antarctic Ice Sheet to warming would be neutral. In other words, the increased snowfall and thickening in East Antarctica would balance marine melting in West Antarctica. Also, the current generation of ice sheet models do not incorporate rapid ice dynamic effects and thus significant changes to ice sheet volume are predicted to take millennia rather than decades (Huybrechts and De Wolde, 1999). But the new observations show that it is the rapid response of ice streams that will dominate the response on human time-spans. Depending on the size of the ice streams involved and how much of the drainage basin thins, we should not be surprised to experience global sea level changes of up to 1 m within a century. Loss of the Pine Island catchment alone could raise global sea level by the equivalent of 0.3 m. This is a particularly important point to make since, as a matter of principle, and justifiably, the 2007 IPCC report plays down the role of poorly understood science, such as that of marine collapse. Such an approach could mistakenly be taken to indicate a reduced risk of significant sea level rise within a century.

14.8.2 Greenland Ice Sheet

The Greenland Ice Sheet contains the equivalent of around 7 m of sea level (Fig. 14.18). It differs from Antarctic ice sheets in that it has a zone of surface and marginal melting that is sensitive to temperature changes. Melting can affect approximately half of the ice sheet surface in any single summer. Moreover, the altitudinal distribution of the ice sheet enhances its sensitivity to temperature change. This is because the melting zone reaches onto the upper gentle slopes of the ice sheet and thus a relatively small temperature change can greatly change the area subjected to melting. Accumulation is through snowfall on the ice sheet and the pattern is typical of ice sheets in that totals are highest around the maritime peripheries and least in the centre and the continental north. Ablation is roughly equally divided between melting and calving. The melting occurs as meltwater streams discharging into fjords around the coast and onto outwash plains in central western Greenland. In addition there is bottom melting of floating glacier tongues in northern Greenland. Calving is from tidewater glaciers. The largest is Jakobshavn Isbrae which can lose *c.* 8 km of its length in a single summer season through calving. But other active tidewater glaciers end in the fjords of southeastern and eastern Greenland. The meltwater and ice that reaches the sea contributes to the mix of cold surface fresh water with the North Atlantic Drift that drives the oceanic thermohaline overturning of the North Atlantic.

The Greenland Ice Sheet withdrew from its LGM extent offshore in line with the other mid-latitude ice sheets. By the time of the Younger Dryas cold episode (12.5 ka BP), all three ice sheets had withdrawn onto land but were still

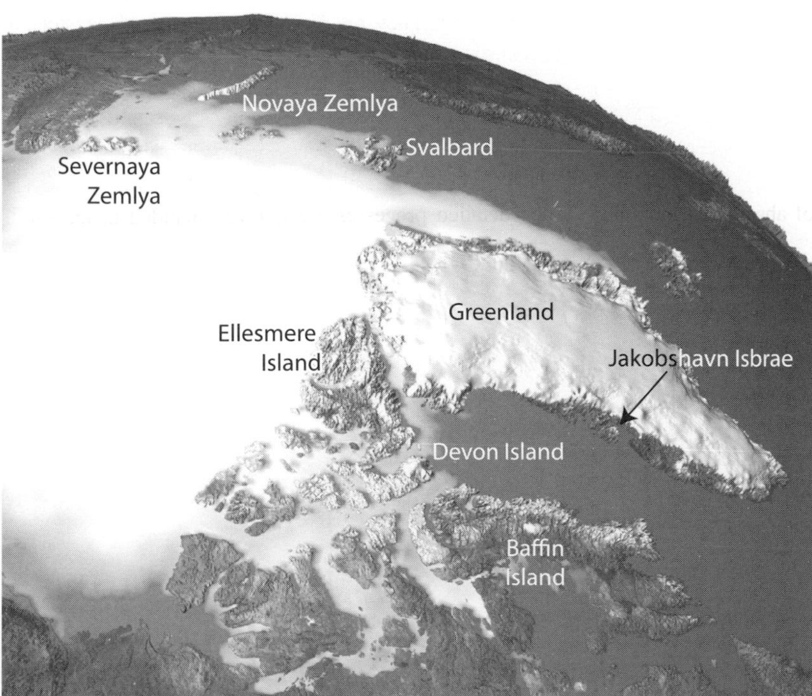

FIGURE 14.18 The Greenland Ice Sheet and arctic ice caps that contribute meltwater to the North Atlantic and Arctic oceans. The ice caps on Baffin Island (Barnes Ice Cap), Devon Island and Ellesmere Island are on high plateaus. The Eurasian island ice caps are at lower elevations (courtesy of NASA).

extensive. In the early Holocene the Greenland Ice Sheet retreated behind its present margin, perhaps by hundreds of kilometres in the west (Weidick, 1993). At this point the fortunes of the various ice sheets diverged and the Laurentide and Scandinavian ice sheets continued to retreat and disappear, while the Greenland Ice Sheet advanced to its present extent, depressing the Earth crust isostatically as a result. Interestingly, analysis of basal ice in Greenland ice cores and sea level reconstructions suggest that the Greenland Ice Sheet was smaller than at present during previous interglacials and that it may have contributed an additional 2–3 m towards sea level rise at such times (Cuffey and Marshall, 2000; Willerslev *et al.*, 2007). Perhaps the climatic deterioration in the mid-Holocene was just sufficient to tip the balance and reverse the trajectory of decline.

Estimates of the current mass balance of Greenland vary with the approach used and at present it is not possible to know with certainty whether the ice mass is decreasing or increasing. It is especially difficult to interpret recent satellite-derived changes in surface elevation because of the complicating effect of surface melting on snow density and because of isostatic depression of the crust as a result of ice sheet expansion late in the Holocene. Further, it is difficult to estimate calving volumes when basic data about calving depths is missing from about one-third of the calving outlets. Also, there is an internal control on calving speeds that may be independent of climate. For example, Jakobshavn Isbrae, which drains some 6% of the ice sheet, slowed down in 1985–92 but has since doubled its velocity and caused the ice to thin for a distance of 30 km inland.

In spite of the uncertainty, a campaign of flights and radio-echo sounding of the ice sheet in the 1990s has identified a consistent pattern. The interior above an altitude of *c.* 2000 m seems to be in balance and has maintained its surface altitude with little change. The marginal zones of the ice sheet experienced thinning with rates exceeding 1 m a^{-1} close to the coast, while outlet glaciers thinned in their lower reaches (Krabill *et al.*, 2000; Alley *et al.*, 2005). Overall the project concluded that there was a net loss of ice and that Greenland could be contributing 0.13 mm a^{-1} to global sea level rise (Thomas, 2004). The thinning at lower altitudes coincides with a period from 1979 to 2002 when the extent of summer melting increased by 16% (Abdalati and Steffen, 2001). It is tempting to link the two and to suggest that the ice sheet is losing mass through the effects of high summer temperatures.

There are two possible mechanisms to explain a correlation between increased surface melting and ice sheet thinning. One process, limited to tidewater glaciers, is increased melting by warm seawater leading to thinning, reduced basal friction and faster flow. As in Antarctica there are processes by which this accelerated ice flow extends up glacier into the wider catchment. Another process is the routing of surface meltwater to the glacier base which then has the same effect of enhancing sliding (Nienow *et al.*, 2005). One particular feature of Greenland is that shallow surface lakes of meltwater accumulate on the ice (Fig. 14.19). They are sufficiently large to be visible in

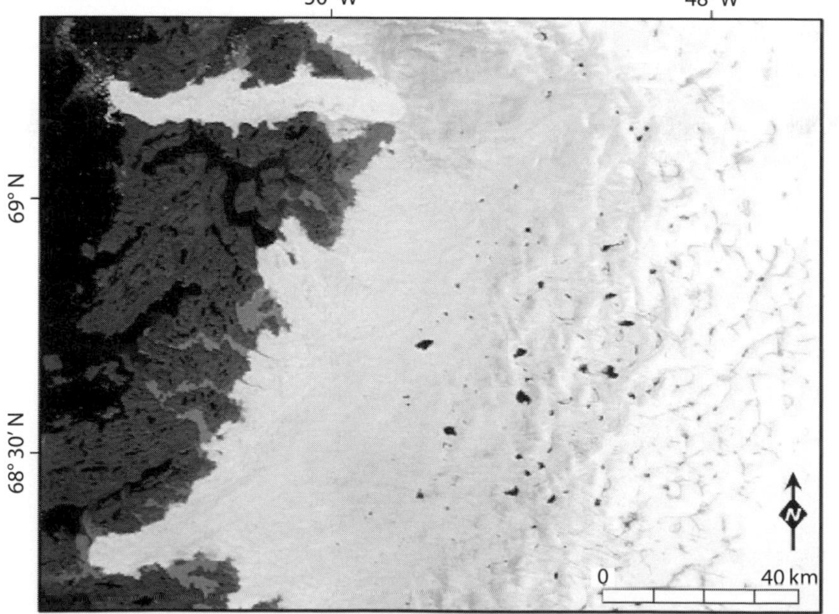

FIG. 14.19. MODIS-image of the western edge of the Greenland Ice Sheet showing surface lakes in the ablation zone on 29 July 2006 (250 m resolution). The lakes build up in depressions on the ice surface and at least some are known to drain suddenly into the ice sheet. Jakobshavn Isbrae and its frontal fjord full of brash ice and icebergs is at the top of the image (courtesy of MODIS Rapid Response Project at NASA/GSFC; image prepared by Aud Sundal).

summer from overflying commercial airliners on their way between Europe and North America. Such lakes are known to drain suddenly into the ice by means of tunnels (Das *et al.*, 2008). The latter are sufficiently large and continuous to be subsequently explored by speleologists. Such englacial drainage is a surprise because there is a thick layer of ice below the pressure melting point that would normally freeze any penetrating water. Nonetheless, the lakes drain and it remains to be discovered how the meltwater from these lakes finds its way to the base and affects glacier sliding and ice sheet thinning. Perhaps earlier discoveries of thick layers of basal ice exposed at the ice edge, with isotopic signatures indicating freezing inland of the ice margin, represent the wholesale freezing on of such water bodies at the base (Sugden *et al.*, 1987).

So the Greenland Ice Sheet emerges as another sensitive ice sheet with the potential to respond to global warming over decades to centuries (Witze, 2008). Models consistently show that ice in western Greenland disappears under warmer climatic scenarios. This is illustrated in Fig. 14.20 by an ice sheet model which is coupled to an atmospheric GCM. In this case the model is run with average temperatures *c.* 2 °C warmer than those that are in equilibrium with the present ice sheet extent. The simulation shows a reduction in volume that is equivalent to a rise in sea level of 1.9 m and that it could be achieved in 2 ka (Fig. 14.20c). In this scenario modest atmospheric warming increases surface melting, modifies accumulation totals (increase on the high centre and reduction at lower altitudes), and causes significant ice loss in the west. In models such as this there is uncertainty about the amplitude and rate of change for two main reasons. Firstly, the critical processes operating at the ice sheet surface and at the base of the main outlet glaciers are imperfectly understood and, if they are incorporated into the model, are likely to speed up the rate of change by as much as ten times. Secondly, it is difficult to judge when the model simulates present-day conditions effectively and thus the starting point is uncertain. What is important is the direction of change which is a reduction in volume over time-spans of centuries.

When assessing the risks of such a reduction in the volume of the Greenland Ice Sheet, it is worth recalling that it may have been smaller than present during previous interglacials when climate was similar to that predicted to occur in the future. Moreover, sea level may have been 2–3 m higher than at present. It is also important that the

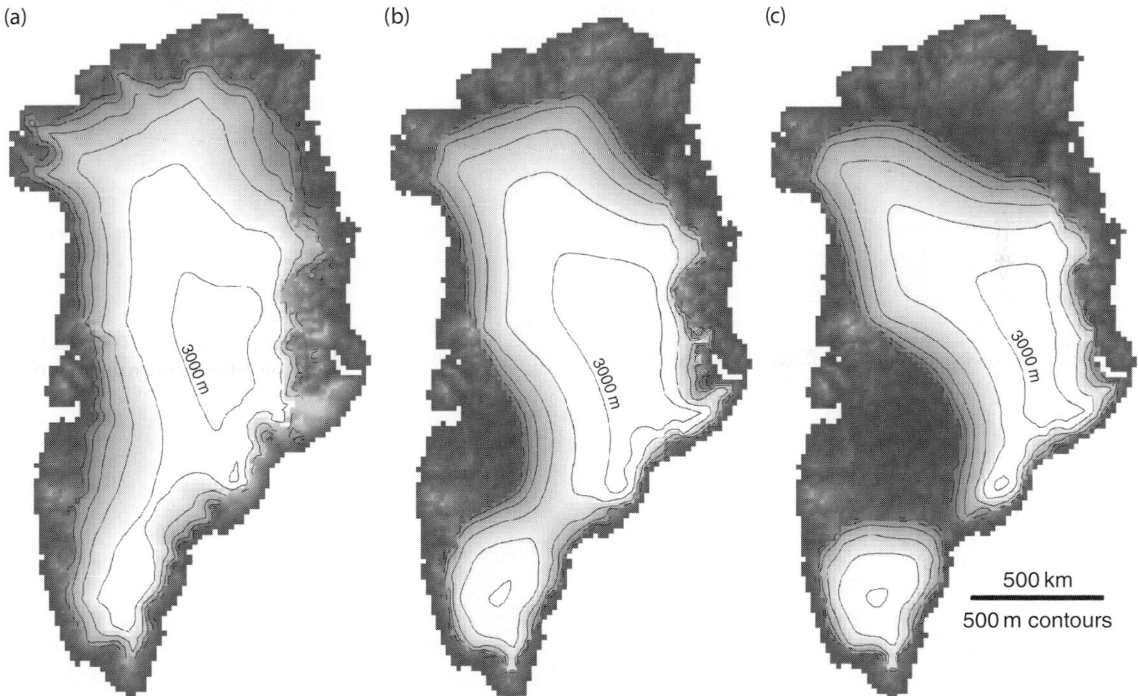

FIGURE 14.20. Modelling the response of the Greenland Ice Sheet to atmospheric warming. In these experiments the GLIMMER ice sheet model is coupled to a GCM. (a) The model starts with the present ice extent and (b) and (c) simulate ice extent following mean annual warming of *c.* 2 °C degrees. The simulations show a global sea level rise of (b) 0.2 m after 1 ka and (c) 1.9 m after 2 ka. The model does not take account of changes in ice dynamics that could shorten the response time dramatically (M. Karatay and N. Hulton, personal communication, 2008).

meltwater from the southern part of the ice sheet plays a role in the operation of the Atlantic thermohaline circulation. The predicted meltwater volumes (up to 0.11 M m^3 s^{-1}) have the potential to slow down or even close down the circulation according to certain oceanic models (Rahmstorf, 2000). The result of the latter would be to abruptly cool the North Atlantic arena. The Greenland Ice Sheet is readily accessible and a concentrated attack on processes of mass balance and flow could narrow the uncertainty. It is difficult to imagine a greater priority for study by the populations bordering the North Atlantic Ocean.

14.8.3 Ice caps

Arctic Islands

The ice caps in the Arctic Islands hold a small component of the world's fresh water stored on land (<0.25 m sea level equivalent), and yet they have the potential to respond to climate change within decades and thus on human timespans. The quick response is related to their small size. Mass balance turnover of ice is large in proportion to ice volume and loss of ice through a surge or a decade of warm summers may seriously lower the accumulation area of the ice cap and make it even more sensitive to loss. Another effect of small size is that the ice extent is too restricted to exert mitigating feedback effects on climate; rather, the difference in albedo between the surrounding land surfaces and the ice enhances summer melting. Finally, the limited altitudinal range makes ice caps especially sensitive to changes in climate; this is because the ELA is commonly located on the gentle upper slopes of the ice cap and thus a modest change can have a major effect on the mass balance.

The Arctic Island ice caps are children of the Holocene (Fig. 14.18). During the last glacial most ice caps were part of larger ice masses, namely the continental Laurentide, Greenland, Scandinavian and Icelandic ice sheets. Following the mid to early Holocene warmth that saw most of these small ice caps disappear, they reformed because of their elevation and proximity to open sea which provides the necessary moisture for snowfall. Thus in the dry continental climate of northern Canada and Greenland, the ice caps of Devon and Ellesmere islands are centred on high plateaus adjacent to the sea. In the more maritime climate of the Barents Sea, ice caps, such as those on Svalbard, can survive at lower elevations close to sea level. Further east in Siberia, the climate is too dry to nourish ice caps.

One characteristic of the ice caps is that they experience surface melting over much of their area in summer and low temperatures in winter. Summer meltwater may refreeze on the surface to form superimposed ice and, indeed, in some situations in arctic Canada all the accumulation is in the form of superimposed ice (Koerner, 1970). The effect of summer melting and refreezing has restricted the value of ice cores as a means of establishing past trends in accumulation and ablation. Also there is a complex relation between climate trends in different parts of the Arctic which means that it is difficult to generalise across the area. Particularly important is the North Atlantic Oscillation (NAO) which links to the Arctic Oscillation (AO) and brings spatial and temporal variability on timescales of decades. When the NAO is positive, there is a pressure gradient between high pressure over the Azores and low pressure over Iceland. Under such conditions winter storms cross the Atlantic into the Eurasian Arctic bringing warmer temperatures and more snowfall, while northern Greenland and arctic Canada experience cold, dry conditions (Dowdeswell and Hagen, 2004). Thus one would expect variability in glacier response from time to time and from place to place.

Estimates of the response of arctic ice caps to climate change have been based on such methods as direct mass balance measurements and remote sensing of surface altitude and iceberg calving. With the exception of parts of northern Ellesmere Island, the ice caps have few floating ice tongues and thus the effect of bottom melting through ocean melting is reduced. The consensus is that, on the basis of measurements made in the latter part of the twentieth century, arctic ice caps have experienced a reduction in ice volume contributing to sea level rise of *c.* 0.05 mm a^{-1} (Dyurgerov and Meier, 1997). This is about 25% of the total produced worldwide by glaciers and ice caps, excluding the Antarctic and Greenland ice sheets.

Patagonian ice caps

The Patagonian ice caps are located in the warmest and wettest environment of any ice cap and therefore ice turnover is high and glacier fluctuations respond rapidly to any climate change. The ice caps have elevations that range from 800 m in the west to 1500 m on the eastern side. Several nunataks with elevations of >3000 m rise abruptly above the ice surface, for example Monte San Valentín in north and Volcán Lautaro in the south. The equilibrium line varies from glacier basin to glacier basin and is at altitudes of 900–1300 m (Aniya *et al.*, 1996). The ice caps experience contrasting climates. There is a north–south gradient as the climate becomes cooler and wetter to the south. But the most dramatic contrast is from east to west and reflects the position of the ice caps athwart the southern westerlies. The western flank of the ice caps is exposed to orographically enhanced snowfall, supplemented by rime ice from supercooled cloud, while the eastern flank experiences dry föhn winds. The contrast is revealed by measurements from the accumulation zone on each flank. The record accumulation rate of 15.4 m a^{-1} is found in the Tyndall Glacier catchment to the west, while the totals are a mere 0.3 m a^{-1} in the accumulation zone at Cerro Gorra Blanca in the east (Casassa *et al.*, 2006). On average the annual accumulation is thought to be 5–10 m a^{-1}.

During the LGM the ice sheet expanded to form a coherent ice sheet extending 1800 km along the Andes from the southern tip of the continent. Figure 14.6 compares ice sheet reconstructions based on field evidence with those based on modelling. The latter is based on a cooling of 6 °C and the northward migration of the westerlies. What is interesting is that it is only possible to simulate the actual morphology of the expanded ice sheet by reducing precipitation in southern latitudes and increasing it in the north. If the precipitation distribution remains the same as at the present day then the ice expands eastwards into a major ice sheet that extends over the lowlands towards the Falkland Islands (Hulton *et al.*, 1994). Thus the argument for the northward migration of the westerlies during the LGM is strong. Although it is likely that the resolution of the model overestimates the surface altitude and ice volume, it appears that the Patagonian Ice Sheet contributed to global sea level lowering of *c.* 1 m at the LGM (Hulton *et al.*, 2002).

The fortunes of Patagonian outlet glaciers over the last 50 years have been studied from air photographs and satellite imagery (Fig. 14.21). Overall there has been glacier retreat, but with remarkable variability from place to place and some strong advances. O'Higgins Glacier retreated 13 km in 41 years at a mean rate of 330 m a^{-1}, which is an order of magnitude faster than the mean rate of retreat for the outlet glaciers of the North Patagonian ice cap of 30 m a^{-1}. Overall, ice volume loss from Patagonia over the last 50 years has been estimated to be the equivalent of a mean sea level rise of 0.038 mm a^{-1} (Aniya, 1999). Such a trend correlates with temperature warming of 0.05 °C a^{-1} sustained for 40 years in southern South America after 1948 (Naruse, 2006).

However, it is the anomalous behaviour of Upsala and Pío XI glaciers that suggests caution in generalising about glacier trends on decadal timescales. The relative stability of Upsala Glacier from 1945 to 1978 coincides with its snout grounding on a rock bar. Its subsequent rapid retreat took place with the snout ending in deeper water. Moreover, the anomalous behaviour of Pío XI Glacier could relate to a surge or to ice/sediment relations at the snout terminating in the water of the terminal lake and fjord.

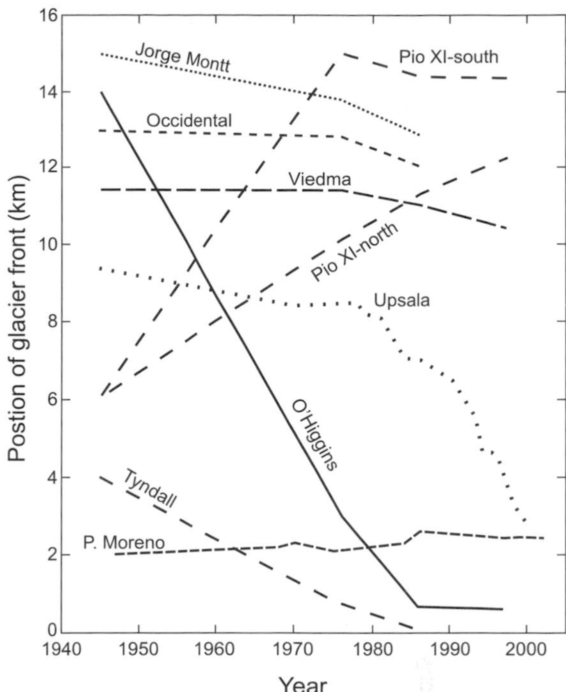

FIGURE 14.21. The varied response of individual Patagonian outlet glaciers to climate change in the last 50 years (from Naruse, 2006).

Patagonian experience shows that local factors, often topographically controlled, affect glacier fluctuations over decades. The irony is that the glaciers that should be the first to respond to climate change reveal the innate complexity of the way in which glaciers react to non-climatic factors.

14.8.4 Wider implications

Ice sheets and ice caps are complex systems with multiple feedbacks occurring on a range of spatial and time scales. Many of these feedbacks have been recognised but new unsuspected relations keep emerging. The spatial scales range from global to metres size, while the timescales involve processes operating over tens of millions of years to those on diurnal tidal timescales or less. There are examples of remarkable feedbacks across the scales. For example, global climate can create a continental-sized ice sheet, while the basal ice temperature at a single point at the grounding line of an ice sheet can trigger a regional thinning response with implications for global sea level. Again, the effect of millions of years of regional glaciation in West Antarctica that deposited glacial mud into the adjacent straits is now influencing the behaviour of ice streams flowing into the Ross Ice Shelf on a diurnal timescale; moreover these ice streams influence the future stability of the West Antarctic Ice Sheet. The implication is that it is necessary to understand the processes and evolution of ice sheets and ice caps at all scales if we are to be confident of any predictions about their future response to climate change (see Chapter 1, Section 1.3.2).

The sheer size of ice sheets and the long time-spans involved in their evolution means that modelling is the only realistic way of predicting future ice sheet behaviour. At the time of the IPCC report in 2007, sophisticated three-dimensional thermomechanical ice sheet models utilised the flow law and temperature of the ice and were driven by a generalised relation to ice surface slope. They utilised a coarse cellular grid of tens of kilometres and necessarily relied on poorly understood assumptions about the role of glacier sliding and the relations between calving speed and ocean conditions. Although state of the art at the time, such models tend to stress the stability of ice flow and also the long time-spans required for an ice sheet to respond to change. Also their relatively coarse resolution means that they struggle to capture the behaviour of ice streams that are increasingly recognised as the key to the dynamics of ice sheets. There is now a major interest in developing the necessary high-resolution models.

Such models will need geomorphological information on process. As seen in Fig. 14.16, satellites offer the ability to monitor ice surface altitude and mass balance over the vast continental areas of ice sheets and also on inaccessible ice caps and outlet glaciers. The quality of the interpretation of such data will depend on how well we understand the seasonal processes of accumulation and ablation. In areas such as Greenland where there are strong regional climatic gradients, it would seem wise to invest in surface mass balance studies. The link between surface mass balance and basal hydrology is another unknown. It has long been known that an increase in subglacial meltwater pressure can lead to rapid glacier sliding. Could it be that the drainage of surface meltwater on the Greenland Ice Sheet is contributing to enhanced sliding and thinning? If so, how does surface meltwater penetrate ice that is below the pressure melting point? Why does it not freeze to the base? Another uncertainty is the magnitude of calving and the controls on calving both in Greenland and West Antarctica. It is important to know how much bottom melting occurs. It is also important to know the bed topography in detail in order to identify whether or not an outlet glacier is close to a threshold of retreat.

Two ice sheets stand out for further study of glaciological processes. Firstly, the Greenland Ice Sheet is close to Europe and North America and influences the turnover of the Atlantic Ocean conveyor. We know that Greenland is the ice sheet that has the potential to respond rapidly to climate warming and, moreover, has a history demonstrating that it melts in climates only slightly warmer than that of today and could contribute to a further sea level rise of 2–3 m. It is difficult to overestimate the urgency of glaciological process studies on the Greenland Ice Sheet. Secondly, the uncertainty about the future of the West Antarctic Ice Sheet revolves around the interaction with the ocean. The loss of ice from catchments such as those draining into Pine Island Bay could raise sea level by up to 1 m within a century. Here it is imaginative ways of studying the undersurfaces of floating ice shelves and the base of ice streams that will bring real benefits. Identifying the urgency of two ice sheet studies serves to stress the general importance of improving glaciological theory. Of course, this can and should be achieved by many types of study in glaciers all over the world; the choice will depend on the optimum location for a particular study. The important message here is to improve our predictions of ice sheet response while the ice sheets still exist!

Geomorphological study of ice sheet evolution also has a role to play in firming up predictions of ice sheet response to change. The long history of the East Antarctic Ice Sheet serves to demonstrate its resilience and control over its climate. Survival during the warmer Pliocene demonstrates that it can coexist in a warmer world than at present. Moreover, the evidence of Pliocene thickening of land-bound glaciers, such as Taylor Glacier, is a good test of predictions that the main effect of warming will be thickening in response to increased snowfall. In contrast, the decline of the Ross Sea sector of the West Antarctic Ice Sheet throughout the Holocene is demonstration of a system response lagging the main rise in sea level by *c.* 10 ka. It affects the interpretation of current satellite-derived trends and throws the role of ocean–ice feedbacks into sharp focus.

Finally, the reconstruction of ice sheet and ice cap history is vital as a means of testing and refining models. Can a model reproduce the glacier fluctuations of an ice sheet during a glacial cycle, bearing in mind changes in climate and sea level? Can we model the build-up and decay of mid-latitude ice sheets during a glacial cycle and their subsequent loss? Are we optimising our knowledge base and exchanging perspectives between those Earth scientists working on ice sheet beds in former mid-latitudes and those working on present ice sheets? Good progress has been made but one of the limitations is good empirical data from the geomorphological community. The arrival of cosmogenic isotope analysis is important in this regard since it has the potential to provide modellers with much of what they require in Antarctica and Greenland, namely point data from exposed rock outcrops quantifying rates of ice thinning, duration of burial and even insights into subglacial conditions of erosion, deposition and basal thermal regime. Moreover, in former mid-latitudes the technique is opening up the use of former beds to establish the changing dynamics and thermal regime of ice sheets during glacial cycles.

14.9 Conclusion and summary

Ice sheets and ice caps play a vital role in modulating global climate and sea level. Their present distribution influences the pattern of global climate, vegetation zones and ocean circulation and it is their non-linear response to modest changes in solar insolation cycles that brought us the ice ages of the last 2–3 Ma. At present ice sheets and ice caps store a volume of fresh water that, should it melt, would be equivalent to a rise in sea level of around 65 m. During the ice age the growth of ice sheets and ice caps caused average global sea level to fall around 120 m below present. This chapter shows how studies of glaciological processes, such as mass balance and glacier flow, combined with the history of past fluctuations and landscape change, are necessary to understand how ice sheets behave. Also, there is much to gain from bringing together the perspective of those who study former mid-latitude ice sheets in the northern hemisphere and those working on present-day ice sheets and ice caps; for example, the chapter uses knowledge of the beds of former mid-latitude ice sheets to model and predict the nature of the subglacial landscapes in Antarctica. Together this provides a wealth of information with which to refine and constrain glaciological models of ice sheet and ice cap behaviour. The priority is to develop such models at an appropriate resolution, to ensure that they can simulate past ice sheet and ice cap changes accurately, and then to predict their response to possible future environmental changes in a manner convincing to policy-makers.

The overriding message of this chapter is that ice sheets and ice caps represent the greatest risk to the world through their influence on sea level. Small ice caps, having rapid response times, are already diminishing in size in response to global warming and have the potential to contribute up to 0.25 m in a century. The Antarctic and Greenland ice sheets have longer response times. Existing ice sheet models and current satellite observations of the interaction with surface climate all point to a slow response over several centuries and this is the basis of the relatively modest predictions of sea level change contained, for example, in the 2007 IPCC reports. But studies of past behaviour point to instances when parts of ice sheets and ice caps can collapse abruptly, often in response to changes in sea level, ocean water temperature or the internal dynamics of glacier flow. Such instances are infrequent and difficult to predict but they carry the risk of a rapid rise in global sea level of the order of 1 m in a century, rates that have been exceeded in the past. Pine Island Glacier in West Antarctica is one such example of a glacier grounded on a bed below sea level that is thinning rapidly today; alone it carries the risk of raising global sea level by around 0.3 m within a century. We should be alert for this and other such instabilities, especially in those parts of Antarctica where ice is grounded below sea level, and in Greenland where surface melting and meltwater run off is so involved. So important and so much geomorphology to do!

Acknowledgements

This review owes much to experience gained in the course of research projects supported by the Natural Environment Research Council (UK), the National Science Foundation, Office of Polar Programs (US), the Royal Society, London, and the Carnegie Trust for the Universities of Scotland. I am most grateful to many colleagues for their help, but especially to Anthony Newton for assistance with the figures, and to John Andrews, Johan Kleman, Stewart Jamieson, Peter Nienow, Andrew Shepherd and an anonymous referee for their discerning comments on the text.

References

Abdalati, W. and Steffen, K. (2001). Greenland ice sheet melt extent: 1979–1999. *Journal of Geophysical Research*, **106**, D24, 33 983–33 988.

Alley, R. B. *et al.* (2005). Ice-sheet and sea-level changes. *Science*, **310**, 456–460.

Anderson, J. (1999). *Antarctic Marine Geology*. Cambridge: Cambridge University Press.

Anderson, J. B. *et al.* (2002). The Antarctic Ice Sheet during the Last Glacial Maximum and its subsequent retreat: a review. *Quaternary Science Reviews*, **21**, 49–70.

Andrews, J. T. (1998). Abrupt changes (Heinrich events) in late Quaternary North Atlantic marine environments: a history and review of data and concepts. *Journal of Quaternary Science*, **13**, 3–16.

Andrews, J. T. (2006). The Laurentide Ice Sheet: a review of history and processes. In P. G. Knight, ed., *Glacier Science and Environmental Change*. Oxford: Blackwell, pp. 201–207.

Aniya, M. (1999). Recent glacier variations of the Hielo Patagónicos, South America, and their contribution to sea-level change. *Arctic, Antarctic and Alpine Research*, **31**, 165–173.

Aniya, M., Sato, H. and Naruse, R. (1996). The use of satellite and airborne imagery to inventory outlet glaciers of the Southern Patagonian Icefield, South America. *Photogrammetric Engineering and Remote Sensing*, **62**, 1361–1369.

Armstrong T. E., Roberts, B. and Swithinbank, C. W. M. (1973). *Illustrated Glossary of Snow and Ice*. Cambridge: Scott Polar Research Institute.

Bales, R. C. *et al.* (2001). Accumulation over the Greenland ice sheet from historical and recent records. *Journal of Geophysical Research*, **106**, D24, 33 813–33 826.

Bamber, J. L., Vaughan, D. G. and Joughin, I. (2000). Widespread complex flow in the interior of the Antarctic ice sheet. *Science*, **287**, 1248–1250.

Bamber, J. L., Ekholm, S. and Krabill, W. B. (2001). A new, high-resolution digital elevation model of Greenland fully validated with airborne laser altimeter data. *Journal of Geophysical Research*, **106**, 6733–6745.

Barrett, P. J. (2007). Cenozoic climate and sea level history from glacimarine strata off the Victoria Land coast, Cape Roberts Project, Antarctica. In M. J. Hambrey *et al.*, eds., *Glacial Sedimentary Processes and Products*. Oxford: Blackwell, pp. 259–287.

Barrett, P. J. *et al.* (1992). Geochronological evidence supporting Antarctic deglaciation three million years ago. *Nature*, **359**, 816–818.

Bell, R. E. (2008). The role of subglacial water in ice-sheet mass balance. *Nature Geoscience*, **1**, 297–304.

Bentley, C. R. (2004). Mass balance of the Antarctic ice sheet: observational aspects. In J. L. Bamber and A. J. Payne, eds., *Mass Balance of the Cryosphere*. Cambridge: Cambridge University Press, pp. 459–489.

Bentley, M. J. (1999). Volume of Antarctic ice at the Last Glacial Maximum, and its influence on global sea level change. *Quaternary Science Reviews*, **18**, 1569–1595.

Bentley, M. J. *et al.* (2005). Early Holocene retreat of George VI Ice Shelf, Antarctic Peninsula. *Geology*, **33**, 173–176.

Bierman, P. R. *et al.* (1999). Mid-Pleistocene cosmogenic minimum-age limits for pre-Wisconsinan glacial surfaces in southwestern Minnesota and southern Baffin Island, a multiple nuclide approach. *Geomorphology*, **27**, 25–39.

Bindschadler, R. A. *et al.* (2003). Tidally controlled stick-slip discharge of a West Antarctic ice stream. *Science*, **301**, 1087–1089.

Boulton, G. S. (1979). Processes of glacier erosion on different substrata. *Journal of Glaciology*, **23**, 15–37.

Boulton, G. S. and Clark, C. D. (1990). A highly mobile Laurentide ice sheet revealed by satellite images of glacial lineations. *Nature*, **346**, 813–817.

Bradwell, T., Stoker, M. and Larter, R. (2007). Geomorphological signature and flow dynamics of the Minch palaeo-ice stream, NW Scotland. *Journal of Quaternary Science*, **22**, 609–617.

Briner, J. P. *et al.* (2003). Last glacial maximum ice sheet dynamics in Arctic Canada inferred from young erratics perched on ancient tors. *Quaternary Science Reviews*, **22**, 437–444.

Brown, C., Meier, M. and Post, A. (1982). *Calving Speed of Alaskan Calving Glaciers, with Application to Columbia Glacier*, US Geological Survey Professional Paper No. 1258-C. Washington, DC: US Government Printing Office.

Casassa, G., Rivera, A. and Schwikowski, M. (2006). Glacier mass-balance data for southern South America (30°–56° S). In P. G. Knight, ed., *Glacier Science and Environmental Change*. Oxford: Blackwell, pp. 239–241.

Clapperton, C. M., Sugden, D. E. and Pelto, M. (1989). Relationship of land terminating and fjord glaciers to Holocene climatic change, South Georgia, Antarctica. In J. Oerlemans, ed., *Glacier Fluctuations and Climatic Change*. Dordrecht: Kluwer, pp. 57–75.

Clark, C. D. (2006). Mega-scale glacial lineations and cross-cutting ice-flow landforms. *Earth Surface Processes and Landforms*, **18**, 1–29.

Conway, H., Hall, B. L. and Denton, G. H. (1999). Past and future grounding line retreat of the West Antarctic ice sheet. *Science*, **286**, 280–283.

Conway, H., Catania, G. and Raymond, C. F. (2002). Switch of flow direction in an Antarctic ice stream. *Nature*, **419**, 465–467.

Crowley, T. J. (1992). North Atlantic deep water cools the southern hemisphere. *Paleoceanography*, **7**, 489–497.

Cuffey, K. M. and Marshall, S. J. (2000). Substantial contribution of sea-level rise during the last interglacial from the Greenland ice sheet. *Nature*, **404**, 591–594.

Das, S. B. *et al.* (2008). Fracture propagation to the base of the Greenland Ice Sheet during supraglacial lake drainage. *Science*, **320**, 778–781.

De Angelis, H. and Skvarca, P. (2003). Glacier surge after ice shelf collapse. *Science*, **299**, 1560–1562.

DeConto, R. M. and Pollard, D. (2003). Rapid Cenozoic glaciation of Antarctica induced by declining atmospheric CO_2. *Nature*, **421**, 245–249.

Denton, G. H. and Sugden, D. E. (2005). Meltwater features that suggest Miocene ice sheet overriding of the Transantarctic Mountains in Victoria Land, Antarctica. *Geografiska Annaler*, **87**A, 1–19.

Denton, G. H. *et al.* (1989). Late Wisconsin and early Holocene glacial history, inner Ross embayment, Antarctica. *Quaternary Research*, **31**, 151–182.

Denton, G. H. *et al.* (1993). East Antarctic ice sheet sensitivity to Pliocene climatic change from a dry valleys perspective. *Geografiska Annaler*, **75**A, 155–204.

Doake, C. S. M. and Vaughan, D. G. (1991). Rapid disintegration of the Wordie Ice Shelf in response to atmospheric warming. *Nature*, **350**, 328–330.

Dowdeswell, J. A. and Hagen, J. O. (2004). Arctic ice caps and glaciers. In J. L. Bamber and A. J. Payne, eds., *Mass Balance of the Cryosphere*. Cambridge: Cambridge University Press, pp. 527–557.

Dowsett, H. J. and Cronin, T. M. (1990). High eustatic sea level during the middle Pliocene: evidence from the southeastern US Atlantic coastal plain. *Geology*, **18**, 435–438.

Dyke, A. S. (2004). An outline of North American deglaciation with emphasis on central and northern Canada. In J. Ehlers and P. L. Gibbard, eds., *Quaternary Glaciations: Extent and Chronology*, Part II. Amsterdam: Elsevier, pp. 373–424.

Dyurgerov, M. B. and Meier, M. F. (1997). Year-to-year fluctuations of global mass balance of small glaciers and their contribution to sea-level change. *Arctic and Alpine Research*, **29**, 392–402.

EPICA community members (2004). Eight glacial cycles from an Antarctic ice core. *Nature*, **429**, 623–628.

Evans, I. S. (1969). The geomorphology and morphometry of glacial and nival areas. In R. J. Chorley, ed., *Water, Earth and Man*. London: Methuen, pp. 369–380.

Fabel, D. *et al.* (2002). Landscape preservation under Fennoscandian ice sheets determined from in situ produced ^{10}Be and ^{26}Al. *Earth and Planetary Science Letters*, **201**, 397–406.

Giovinetti, M. B. and Zwally, H. J. (2000). Spatial distribution of net surface accumulation on the Antarctic ice sheet. *Annals of Glaciology*, **31**, 171–178.

Glen, J. W. (1955). The creep of polycrystalline ice. *Proceedings of the Royal Society of London A*, **1175**, 519–538.

Heinrich, H. (1988). Origin and consequences of cyclic ice rafting in the Northeast Atlantic Ocean during the past 130 000 years. *Quaternary Research*, **29**, 143–152.

Holtedahl, H. (1958). Some remarks on the geomorphology of continental shelves off Norway, Labrador and southeast Alaska. *Journal of Geology*, **66**, 461–471.

Hubbard, A. *et al.* (2006). A modelling insight into the Icelandic Last Glacial Maximum ice sheet. *Quaternary Science Reviews*, **25**, 2283–2296.

Hulton, N. R. J. *et al.* (1994). Glacier modeling and the climate of Patagonia during the Last Glacial Maximum. *Quaternary Research*, **42**, 1–19.

Hulton, N. R. J. *et al.* (2002). The Last Glacial Maximum and deglaciation in southern South America. *Quaternary Science Reviews*, **21**, 233–241.

Huybrechts, P. (1993). Glaciological modelling of the Cenozoic East Antarctic ice sheet: stability or dynamism? *Geografiska Annaler*, **75**A, 221–238.

Huybrechts, P. (2004). Antarctica: modelling. In J. L. Bamber and A. J. Payne, eds., *Mass Balance of the Cryosphere*. Cambridge: Cambridge University Press, pp. 491–523.

Huybrechts, P. and de Wolde, J. (1999). The dynamic response of the Greenland and Antarctic ice sheets to multiple-century climatic warming. *Journal of Climate*, **12**, 2169–2188.

IPCC (2007). *Climate Change 2007: The Physical Science Basis. Contribution of Working Group I to the Fourth Assessment Report of the Intergovernmental Panel on Climate Change*. Solomon, S. *et al.*, eds. Cambridge: Cambridge University Press.

Jamieson, S. S. R. and Sugden, D. E. (2008). Landscape evolution of Antarctica. In A. K. Cooper *et al.*, eds., *Antarctica: A Keystone to a Changing World, Proceedings of the 10th International Symposium on Antarctic Earth Science*. Washington, DC: National Academies Press, pp. 39–54.

Jamieson, S. S. R. *et al.* (2005). Cenozoic landscape evolution of the Lambert Basin, East Antarctica: the relative role of rivers and ice sheets. *Global and Planetary Change*, **45**, 35–49.

Joughin, I. and Tulaczyk, S. (2002). Positive mass balance of the Ross ice streams, Antarctica. *Science*, **295**, 476–480.

Kennett, J. P. (1977). Cenozoic evolution of Antarctic glaciation, the circum-Antarctic ocean, and their impact on global paleoceanography. *Journal of Geophysical Research*, **82**, 3843–3860.

Kerr, A. R. and Gilchrist, A. R. (1996). Glaciation, erosion and the evolution of the Transantarctic Mountains. *Annals of Glaciology*, **23**, 303–308.

Kirwan, L. P. (1962). *A History of Polar Exploration*. Harmondsworth: Penguin.

Kleman, J. and Stroeven, A. P. (1997). Preglacial surface remnants and Quaternary glacial regimes in northwestern Sweden. *Geomorphology*, **19**, 35–54.

Kleman, J. *et al.* (2006). Reconstruction of palaeo-ice sheets: inversion of their glacial geomorphological record. In P. G. Knight, ed., *Glacier Science and Environmental Change*. Oxford: Blackwell, pp. 192–198.

Kleman, J., Stroeven, A. P. and Lundqvist, J. (2008). Patterns of Quaternary ice sheet erosion and deposition in Fennoscandia and a theoretical framework for explanation. *Geomorphology*, **97**, 73–90.

Koerner, R. M. (1970). Some observations on superimposition of ice on the Devon Island ice cap, N. W. T., Canada. *Geografiska Annaler*, **52**A, 57–67.

Krabill, W. *et al.* (2000). Greenland ice sheet: high elevation balance and peripheral thinning. *Science*, **289**, 428–430.

Lythe, M., Vaughan, D. G. and the BEDMAP Consortium (2001). BEDMAP: a new ice thickness and subglacial topographic model of Antarctica. *Journal of Geophysical Research*, **106**, B6, 11 335–11 351.

Mackintosh, A. *et al.* (2007). Exposure ages from mountain dipsticks in Mac Robertson Land, East Antarctica, indicate little change in ice-sheet thickness since the Last Glacial Maximum. *Geology*, **35**, 551–554.

Marchant, D. R. *et al.* (1994). Quaternary changes in level of upper Taylor Glacier, Antarctica; implications for palaeoclimate and East Antarctic Ice Sheet dynamics. *Boreas*, **25**, 29–43.

Mawson, D. (1915). *The Home of the Blizzard*. London: Heinemann.

McCulloch, R. D. *et al.* (2000). Climatic inferences from glacial and palaeoecological evidence at the last glacial termination, southern South America. *Journal of Quaternary Science*, **15**, 409–417.

Mercer, J. H. (1978). West Antarctic ice sheet and CO_2 greenhouse effect: a threat of disaster. *Nature*, **271**, 321–325.

Miller, K. G. *et al.* (2008). A view of Antarctic ice-sheet evolution from sea-level and deep-sea isotope changes during the late Cretaceous-Cenozoic. In A. K. Cooper *et al.*, eds., *Antarctica: A Keystone to a Changing World, Proceedings of the 10th International Symposium on Antarctic Earth Science*. Washington, DC: National Academies Press, pp. 55–70.

Murray, T. *et al.* (2007). Ice flow modulated by tides at up to annual periods at Rutford ice stream, West Antarctica. *Geophysical Research Letters*, **34**, L18503, doi:10.1029/2007GL031207.

Naruse, R. (2006). The response of glaciers in South America to environmental change. In P. G. Knight, ed., *Glacier Science and Environmental Change*. Oxford: Blackwell, pp. 231–238.

Nienow, P. *et al.* (2005). Hydrological controls on diurnal ice flow variability in valley glaciers. *Journal of Geophysical Research*, **110**, F04002, doi:10.1029/2003JF000112.

Nye, J. F. (1952). The mechanics of glacier flow. *Journal of Glaciology*, **2**, 82–93.

Ó Cofaigh, C. *et al.* (2005). Flow of the West Antarctic Ice Sheet on the continental margin of the Bellingshausen Sea at the Last Glacial Maximum. *Journal of Geophysical Research*, **110**, B11103, doi:10.1029/2005JB003619.

Ottesen, D. *et al.* (2001). Glacial processes and large-scale morphology on the mid-Norwegian continental shelf. In O. J. Martinsen and T. Dreyer, eds., *Sedimentary Environments Offshore Norway: Palaeozoic to Recent*. Amsterdam: Elsevier, pp. 441–449.

Paterson, W. S. B. (1969). *The Physics of Glaciers*. Oxford: Pergamon.

Payne, A. J. (1999). A thermomechanical model of ice flow in West Antarctica. *Climate Dynamics*, **15**, 115–125.

Petit, J. R. *et al.* (1999). Climate and atmospheric history of the past 420 000 years from the Vostok ice core, Antarctica. *Nature*, **399**, 429–436.

Rahmstorf, S. (2000). The thermohaline circulation: a system with dangerous thresholds? *Climatic Change*, **46**, 247–256.

Rignot, E. J. (1998). Fast recession of a West Antarctic glacier. *Science*, **281**, 549–551.

Rignot, E. J. (2006). Changes in ice dynamics and mass balance of the Antarctic ice sheet. *Philosophical Transactions of the Royal Society of London A*, **364**, 1637–1655.

Shepherd, A. *et al.* (2001). Inland thinning of Pine Island glacier, West Antarctica. *Science*, **291**, 862–864.

Shepherd, A. Wingham, D. and Rignot, E. (2004). Warm ocean water is eroding West Antarctic Ice Sheet. *Geophysical Research Letters*, **31**, L23402, doi:10.1029/2004GL021106.

Shreve, R. L. (1972). Movement of water in glaciers. *Journal of Glaciology*, **11**, 205–214.

Siddoway, C. S. (2008). Tectonics of the West Antarctic rift system: new light on the history and dynamics of distributed intracontinental extension. In A. K. Cooper *et al.*, eds., *Antarctica: A Keystone to a Changing World, Proceedings of the 10th International Symposium on Antarctic Earth Science*. Washington, DC: National Academies Press, pp. 91–114.

Siegert, M. J. (2000). Antarctic subglacial lakes. *Earth Science Reviews*, **50**, 29–50.

Siegert M. J. *et al.* (2004). Subglacial Lake Ellsworth: a candidate for in situ exploration in West Antarctica. *Geophysical Research Letters*, **31**, L23403, doi:10.1029/2004GL021477.

Sjerup, H. P. *et al.* (2005). Pleistocene glacial history of the NW European continental margin. *Marine and Petroleum Geology*, **22**, 1111–1129.

Stern, T. A., Baxter, A. K. and Barrett, P. J. (2005). Isostatic rebound due to glacial erosion within the Transantarctic Mountains. *Geology*, **33**, 221–224.

Stone, J. O. *et al.* (2003). Holocene deglaciation of Marie Byrd Land, West Antarctica. *Science*, **299**, 99–102.

Sugden, D. E. (1968). The selectivity of glacial erosion in the Cairngorm Mountains, Scotland. *Transactions of the Institute of British Geographers*, **45**, 79–92.

Sugden, D. E. (1977). Reconstruction of the morphology, dynamics and thermal characteristics of the Laurentide ice sheet at its maximum. *Arctic and Alpine Research*, **9**, 21–47.

Sugden, D. E. and Denton, G. H. (2004). Cenozoic landscape evolution of the Convoy Range to Mackay glacier area, Transantarctic Mountains: onshore to offshore synthesis. *Bulletin of the Geological Society of America*, **116**, 840–857.

Sugden, D. E. *et al.* (1987). Evidence of two zones of entrainment beneath the Greenland Ice Sheet. *Nature*, **328**, 238–241.

Sugden, D. E. *et al.* (1995). Preservation of Miocene glacier ice in East Antarctica. *Nature*, **376**, 412–414.

Sugden, D. E. *et al.* (2005). Selective glacial erosion and weathering zones in the coastal mountains of Marie Byrd Land, Antarctica. *Geomorphology*, **67**, 317–334.

Sugden, D. E., Bentley, M. J. and Ó Cofaigh, C. (2006). Geological and geomorphological insights into Antarctic ice sheet evolution. *Philosophical Transactions of the Royal Society of London A*, **364**, 1607–1625.

Taylor, J. *et al.* (2004). Topographic controls on post-Oligocene changes in ice-sheet dynamics, Prydz Bay, East Antarctica. *Geology*, **32**, 197–200.

Thomas, R. H. (2004). Greenland: recent mass balance measurements. In J. L. Bamber and A. J. Payne, eds., *Mass Balance of the Cryosphere*. Cambridge: Cambridge University Press, pp. 393–436.

Vaughan, D. G. and Doake, C. S. M. (1996). Recent atmospheric warming and retreat of ice shelves on the Antarctic Peninsula. *Nature*, **379**, 328–331.

Webb, P. N. *et al.* (1984). Cenozoic marine sedimentation and ice volume variation on the east Antarctic craton. *Geology*, **12**, 287–291.

Weidick, A. (1993). Neoglacial change of ice cover and the related response of the Earth's crust in west Greenland. *Rapport Grønlands Geologiske Undersøgelse*, **159**, 121–126.

Willerslev, E. *et al.* (2007). Ancient biomolecules from deep ice cores reveal a forested southern Greenland. *Science*, **317**, 111–114.

Williams, R. S. Jr and Sigurðsson, O., eds. (2004). *Draft of a Physical, Geographical, and Historical Description of Icelandic Ice Mountains on the Basis of a Journey to the Most Prominent of Them in 1792–1794 by Sveinn Pálsson (1795)*. Cambridge: Icelandic Literary Society and International Glaciological Society.

Wingham, D. J. *et al.* (2006a). Mass balance of the Antarctic Ice Sheet. *Philosophical Transactions of the Royal Society of London A*, **364**, 1627–1635.

Wingham, D. J *et al.* (2006b). Rapid discharge connects subglacial lakes. *Nature*, **440**, 1033–1036.

Witze, A. (2008). Losing Greenland? *Nature*, **452**, 798–802.

Zwally, H. J. *et al.* (2002). Surface melt-induced acceleration of Greenland ice-sheet flow. *Science*, **297**, 218–222.

15 Landscape, landscape-scale processes and global environmental change: synthesis and new agendas for the twenty-first century

Thomas Spencer, Olav Slaymaker and Christine Embleton-Hamann

15.1 Introduction: beyond the IPCC Fourth Assessment Report

15.1.1 Changing structures for IPCC-type science

It is difficult to escape from the framing of the climate change debate by the IPCC; indeed, that is where we began at the start of this volume. In each of the Assessment Reports, including the Fourth Assessment Report in 2007, the IPCC methodology has been a strongly 'top–down' process, developing a complete set of emissions scenarios, leading to a complete set of climate change scenarios and finally resulting in a range of matching impact and adaptation analyses (Fig. 15.1a). Such an approach is robust and defensible (as it needs to be) because it is underpinned by not only a thorough exploration of the socioeconomic and political forces lying behind the emissions scenarios but also by their subsequent translation into outputs from sophisticated three-dimensional atmosphere–ocean general circulation models (AOGCMs), themselves subject to model intercomparison testing. Model runs in this framework are, however, time-consuming and expensive and the methodological flowchart is cumbersome. This is because of the need to work through each of these stages sequentially to arrive at new impacts from changed emissions scenarios. Within this context, and in beginning the discussion of a possible Fifth Assessment Report, an IPCC Expert Group (Moss *et al.*, 2008) has proposed a so-called parallel approach whereby research into climate projections and emissions and socioeconomic scenarios are decoupled, taking place alongside one another (Fig. 15.1b).

Alongside this model are proposed changes in the role of the IPCC itself; at its meeting in Mauritius in April 2006 the IPCC decided (not, it should be noted, without internal debate) (IISD Reporting Services, 2006), that rather than directly co-ordinating and approving new scenarios itself, it would seek 'to catalyse the timely production by others of new scenarios for a possible Fifth Assessment Report'. The combination of a new structure and a new approach (and see also Racs and Swart, 2007) thus places emphasis on how the linkages between the different boxes in Fig. 15.1b might be established and maintained.

15.1.2 From global to regional, from AD 2100 to AD 2035 and AD 2300

The nature of the IPCC is such that it is strongly driven by the needs of an 'end user' community of international conventions (e.g. United Nations Framework Convention on Climate Change (UNFCCC), Convention on Biodiversity), global public and intergovernmental organisations (FAO, WHO), sub-global organisations with regulatory powers (EU), governments (national, regional, local) and NGOs rather than answering to the needs of an 'intermediate user' community of researchers who use scenarios from another segment of the research community as input into their own work. Thus in a list of ten users the research community was ranked ninth with only 'other' below it (Moss *et al.*, 2008). End user interests in regional and national adaptation and mitigation strategies need to be driven not by the outputs of global models but rather by scenarios provided at regional, and even local, levels. Furthermore, these issues are immediate because end users are coming under pressure to respond to potentially increasingly stringent international controls on emissions. Thus there is a strong user-led demand for new scenarios for the 'near term' (to *c.* AD 2035). Such scenarios are complicated by the need to consider the interaction of shorter timescales of atmospheric and land surface change with the longer timescales

Geomorphology and Global Environmental Change, eds. Olav Slaymaker, Thomas Spencer and Christine Embleton-Hamann. Published by Cambridge University Press.

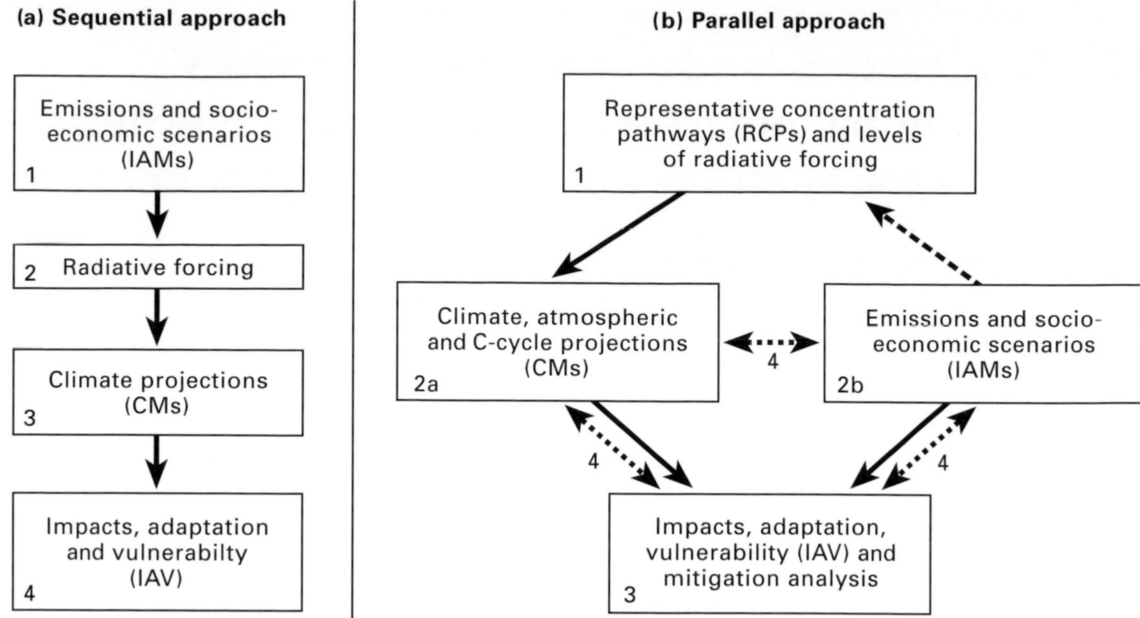

FIGURE 15.1. (a) Previous sequential IPCC approach (b) proposed parallel approach. Representative concentration pathways (RCPs) replace the earlier concept of benchmark emissions scenarios. Numbers indicate analytical steps (2a and 2b proceed concurrently). Arrows indicate transfers of information (solid lines), selection of RCPs (dashed) and integration of information, with feedbacks (dotted) (from Moss *et al.*, 2008).

of ongoing sea level and ice sheet response. Furthermore, because current policy seeks in the immediate future to simply slow the rate of increase in carbon dioxide emissions rather than to stabilise atmospheric carbon dioxide (Broecker, 2007), then a number of 'overshoot' and 'peak and decline' scenarios come into play, whereby atmospheric carbon dioxide concentrations, and associated temperature increases, exceed target levels before declining to stabilisation (Huntingdon and Lowe, 2007). Although they lie outside the remit of this text, it is worth noting that such scenarios, perhaps extending out to AD 2300, take environmental science into uncharted territory, raising many questions (such as the relation of the peak to final stabilised values and the expected length of time above stabilisation) and these include questions concerned with patterns of landform and ecosystem recovery once the decline starts towards stabilisation or, alternatively, the prospect of irreversible landscape change such that no recovery is possible.

Taken as a whole, with the restructuring of research effort and the refocussing of research effort towards smaller spatial scales and towards both shorter and longer timescales than the AD 2100 benchmark, these new ways of thinking about the best way to tackle the climate change problem provide both an opportunity and a challenge for geomorphology. But where does geomorphology stand now?

15.1.3 Placing geomorphology

In the highly segmented world of IPCC science, the science community is divided into three groups: the climate modelling community (Working Group I); the impacts, adaptation and vulnerability community (Working Group II) and the integrated assessment modelling community (Working Group III). In the post-2007 IPCC model it is clear that these three groups are going to need to talk to one another in ways that have not been the case in the past.

Working Group I is self-defined. Working Group III consists mainly of social scientists and those in the field of energy technology (IPCC, 2007c). Geomorphology's constituency is, therefore, Working Group II. From the IPCC perspective (Moss *et al.*, 2008), the Impacts, Adaptation and Vulnerability (IAV) group (a) has a small research base relative to climate science; (b) is loose, lacking coherence and structure; and (c) is not strongly scenario-based, being generally analytical in character rather than model-based and emphasising risks and vulnerabilities rather than projections of impacts. It is instructive to view the engagement of geomorphologists with Working

Group II's contribution to the Fourth Assessment Report. Whilst in no way failing to recognise the liberal sprinkling of geomorphologists at the level of contributing author, at the top two tiers of chapter authorship, of the 48 Coordinating Lead Authors and 130 Lead Authors (i.e. authors who feature in the official citation of an IPCC chapter), there are 6 and 11 authors respectively whose research might be termed geomorphological or hydrological in character. The inescapable conclusion is that there is much scope to increase geomorphology's engagement over the next 5 years of research towards a Fifth Assessment. What has geomorphology to offer and how might its contribution be framed?

15.1.4 Time and space scales in geomorphology, global environmental change and landscape change

In establishing geomorphology's intersection with the global environmental change debate, Slaymaker *et al.* (Chapter 1, and specifically Fig. 1.1) make the case for the range of spatial scale of analysis to be 1–100 000 km^2 (thus excluding individual landforms at the lower end of the scale and landscape belts at the upper end of the scale). The corresponding temporal scale is perhaps 1–100 years, the temporal scale relevant to human life and livelihood and the time period over which mitigation and adaptive strategies in response to environmental change might take place.

This raises a well-known difficulty for geomorphological studies because this space sits squarely between the two dominant modes of explanation in geomorphology, variously termed historical/configurational v. mechanistic, or evolutionary v. equilibrium, or landscape v. reductionist, in nature (e.g. Douglas, 1982; Smith *et al.*, 2002; Richards and Clifford, 2008; Thomas and Wiggs, 2008; and see Chapter 1). In reality, of course, no such stark dichotomy exists. Landscapes are the product of controls and defining processes at a range of scales which nest within each other; thus in southern South America, local features sit within a broader regional context where the controls are relief and ice cover and these controls themselves sit within an overarching plate tectonic framework (Plate 47). The relative importance of historical v. modern controls varies as a function of size and age of landforms and landscapes (Schumm, 1985; Chapter 1, Fig. 1.5). In particular, there is an important distinction to be made between landscapes that are glacierised (present glacier cover), glaciated in the past and never-glaciated landscapes. Observations in the previous paragraph apply primarily to never-glaciated landscapes. The effects of climate change on glacierised landscapes are direct, especially in generating greater natural hazards, such as jökulhlaups, glacial lake drainage and loss of water supply from glacial meltwaters (Chapter 2). Landscapes that were glaciated in the past are conditionally unstable and show delayed responses to climate change over millennia. The distinctive hydrologic response of paraglacial environments is illustrated in Fig. 2.13.

These observations, do not, however, make the scale linkage problem, the issue of transferring knowledge between systems of different magnitude, any less intransigent. To some extent, the first seven chapters in this volume, on azonal environments, take a more *process*, and perhaps smaller-scale, approach whereas the seven chapters that follow, organised around the world's biomes, look more at the *response* to those processes at larger spatial and longer temporal scales. In the latter chapters, therefore, the role of landscape inheritance has significance. Thus for savanna landscapes 'inheritance is – an important legacy in regard to the relationship between geomorphology and ecology and in terms of landscape sensitivity to extreme events such as droughts and floods' (Chapter 9, p. 256). In the context of geomorphology and global environmental change, and as the preceding chapters in this volume have shown, enormous challenges in scientific explanation are generated when uncertain near-future process regimes come up against uncertain landscape responses. This is a considerable challenge for geomorphology in expanding away from the knowledge gained in humid temperate catchment systems into environments where considerable climatic variability, including the role of extreme natural events, is the norm and where non-linear, threshold-controlled processes–form interactions are dominant. We consider these issues further below. What is clear, however, is that it is important to recognise the appropriate space-time scale of any particular case study and not to attempt explanations which draw information from different spatial scales across different time scales (e.g. lakes: Dearing and Jones in Chapter 3; coasts: Cowell *et al.* in Chapter 6).

15.2 Geomorphological processes and global environmental change

15.2.1 The two-dimensional, instantaneous modelled land surface

The last decade has been typified by a huge research effort into the changes in land use and land cover caused by human activity and the associated impacts on land surface processes, ecosystem services and biodiversity and their feedbacks on climate, perhaps best typified by the Millennium Ecosystem Assessment (see Chapter 1, Appendix 1.3). The

Assessment reports show that the degree of landscape modification by human agency has been considerable and global in scale over historical timeframes and shows no sign of abating; population growth and economic development may well drive a further 20% of terrestrial ecosystems into agricultural production within the next 40 years (Millennium Ecosystem Assessment, 2005). The challenge of understanding this anthropogenically led dynamic has been taken up incrementally by the IPCC scientific community, starting from the first clear sky and undifferentiated land surface global climate models of the 1970s (Le Treut *et al.*, in IPCC, 2007a). Thus the science has evolved to take into account the links between climate and biogeochemical systems: how changes in the land surface (vegetation, soils, water) under human activity affect climate through changes in radiative (e.g. albedo) and non-radiative (e.g. hydrological cycle) terms; and how a range of processes in terrestrial ecosystems (including vegetation cover, productivity and soil and plant respiration) influence the flux of carbon between land surfaces and the atmosphere (Denman *et al.*, in IPCC, 2007a). However, for climate modelling purposes, this information needs to be presented in grid cell form and amalgamations of cells then draped over land surfaces. The size of the cells may have been considerably revised downwards to progressively finer meshes but they remain grid cells all the same. The IPCC world is therefore a broadly two-dimensional world, not a three-dimensional geomorphological one. The implications of 'smoothing out' relief are considered in more detail below.

15.2.2 Climate–land-cover–landform linkages

Furthermore, even across such surfaces, the linkages between climate, land cover and landforms are neither simple nor instantaneous. Firstly, there is a tendency to see climate-driven landscape change as, at best, progressive, and at worst, accelerative. There is, however, no reason why climate cannot in certain circumstances drive reversals in historical trends or initiate quite new modes of landscape change (for an illustration for coasts see Chapter 6). This is because, secondly, changes in boundary conditions can occur not only as a consequence of extrinsic changes in the earth–ocean–atmosphere system but also from intrinsic system state changes. In estuarine (Chapter 5), coastal (Chapter 6) and coral reef (Chapter 7) environments, for example, a key non-climatic control is that of sediment supply. In secondary rainforest (Chapter 8), savanna (Chapter 9), Mediterranean landscapes (Chapter 11) and permafrost (Chapter 13) fire dynamics play an important role in internal ecosystem functioning and landscape change. Sometimes, but not always, these intrinsic changes intersect with anthropogenic activities. Thirdly, the magnitude, mode and timeframes of morphological adjustment (responsiveness) are likely to vary considerably between different geomorphic elements of the landscape. Fourthly, morphodynamic feedbacks exist that are not only temporally specific but also cascade across timescales to provide a degree of self-organisation to geomorphic development. Thus, for example, Chapter 7 points out that at centennial to millennial timescales, sea level change modulates reef growth. But this reef growth in turn influences wave, current and sedimentation processes that govern the short-term morphological development of the reef and the sedimentary landforms that sit on reef platforms. Fifthly, many landscape systems show lagged responses between their different components. This behaviour can colour the interpretation of system 'health'. Thus, for example, short-term, ecological scenarios of decline in reef coral cover over years to decades have typically been used to infer deleterious consequences for reef geomorphology. In fact the reef structure is determined by much longer-term, centennial to millennial processes, a context in which short-term changes in benthic cover may be of little significance (Perry *et al.*, 2008).

We should not forget the feedbacks between land use change, land cover change, changing albedo and climate change in some biomes. As Chapter 9 reports, if rangeland and cropland in Africa increase by more than 5.6 M km^2 by the year 2050, reductions in carbon flux and increases in albedo might lead to temperature increases by up to 1.5 °C (DeFries *et al.*, 2002). Furthermore, there are important linkages between geomorphic processes, landscape change and the global carbon cycle. Too little is known about the redistribution of soil carbon by geomorphic processes and the spatial variability of carbon sequestration in soil profiles under different climates and vegetation covers. Studies of carbon sequestration processes and storage patterns in salt marshes (e.g. Chmura *et al.*, 2003) and mangrove swamps (e.g. Twilley *et al.*, 1992) are still at an early stage. And in permafrost environments, a landscape change component now needs to be added to developing coupled carbon–permafrost models (Chapter 13).

Whilst there has been considerable research effort in establishing the linkages between climate change and changes in vegetation extent, structure and function, there have been relatively few attempts to extend such studies to climate-related changes in geomorphological attributes. Figure 15.2 provides some of the few conceptual biogeomorphological models which have been devised to look at the nature of climate–earth surface–ecosystem linkages. These models are highly simplified in that the extrinsic forcing factors are represented as simple step functions.

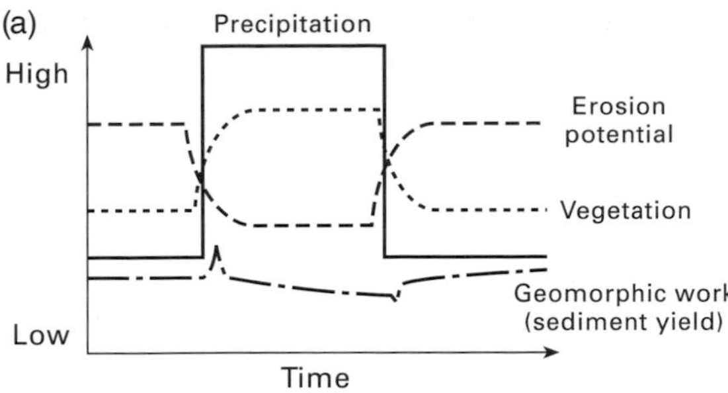

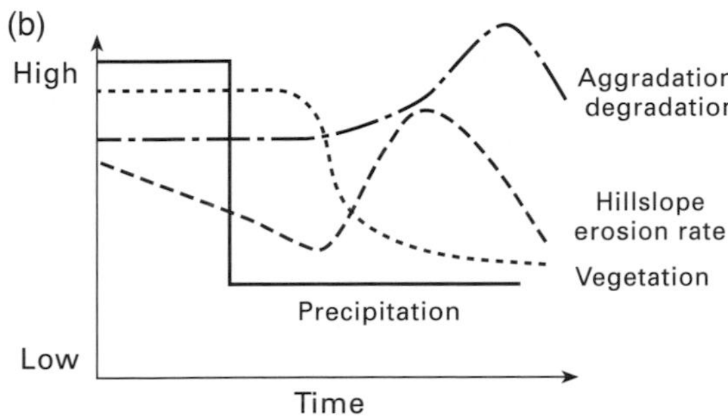

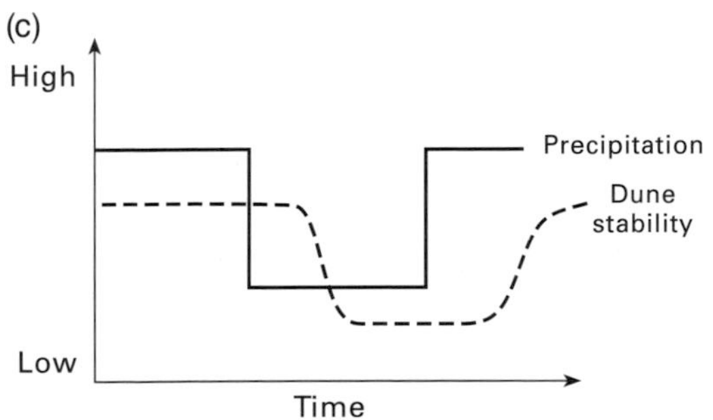

FIGURE 15.2. Conceptual models for the linkages between climate (precipitation) forcing and ecological and geomorphological responses for (a) drainage basin processes; (b) arid hillslopes; and (c) vegetated dunefields (from Viles *et al.*, 2008).

Vegetation responses may relate to community composition (C3 v. C4 plants) and the relative proportions of herbs and grasses, shrubs and trees (e.g. Chapter 5, Chapter 9). They also need to be visualised spatially, particularly given that many land-surface–vegetation systems are characterised by complex three-dimensional canopies in dynamic, and thus transient, clumps, patches and gaps.

Viles *et al.* (2008) identify a series of questions that will need to be answered if we are to move from these simple models to parameterised models aimed at landscape change prediction: how can the relevant parameters, and the linkages between them, be identified?; what are the most appropriate forms of response curve?; and what are the timescales involved, and to what degree do landscape histories influence such timescales?

15.3 Landscapes and global environmental change

15.3.1 Multiple drivers of landscape change

Scaling up from a consideration of geomorphological processes to the landscape scale, it is clear that there are many drivers of environmental change. Yet the interactive coupling between relief, land use and climate has thus far been poorly represented in climate change discussions (e.g. Osmond *et al.*, 2004). The IPCC (2007a) provides a thorough assessment of climate, and sea level, as drivers but the chapters in this volume show that there are in fact four broad groupings and only two of these are directly or indirectly climate-induced. The drivers are: hydroclimate and runoff; sea level; direct human activity; and topographic relief. The first three of these have both spatial and temporal components; on the timescale of climate change, and except in a few exceptional circumstances, relief acts as a spatial control only. These drivers are not all active in every azonal and zonal landscape considered in this volume and their relative importance varies between environments and biomes (Table 15.1). Furthermore, in this simple first stage of analysis, it is assumed that there are no interactions between the drivers and that relations are additive rather than multiplicative.

Within some of these biomes, such as the tropical rainforest (Chapter 8) and coral reefs (Chapter 7), it is particularly difficult to provide a single global score for the driver of human activity. Mountain environments show how complex the interaction between these drivers can be. Thus, in British Columbia mountain environments are controlled by historical legacy (the so-called paraglacial effect), relief and climatic variability but show little human disturbance, the Austrian Alps are a 'human enhanced' landscape, where differentiation comes from the complexity of socioeconomic processes, and in the Ethiopia Highlands there is an ongoing, contested debate about the relative importance of hydroclimate and runoff (in a location where 'climate' itself is difficult to define), topographic relief and a rapidly changing human impact from rapid population growth (Chapter 2). However, some patterns do emerge in a relative scaling across a three-point scale (Table 15.1) which to some extent mirrors the five-driver model of biodiversity change proposed by Sala *et al.* (2000). In mountain environments, topographic relief is clearly the major driver but in some locations, as discussed above, human activity has become dominant (and not necessarily destructively so). The relative importance of the four drivers is similar in lake environments, although here environmental degradation from human activities (i.e. eutrophication and acidification) in many cases makes humans the dominant force.

TABLE 15.1. *Importance of the four drivers of landscape change by environment and biome (qualitative scoring from 3 = highest to 1 = lowest; zero score = not significant)*

Environment or biome	Hydroclimate and runoff	Sea level	Human activity	Topographic relief
Azonal				
Mountains	*		**	**
Lakes and lake catchments	*		***	**
Rivers	**	*	***	*
Estuaries, coastal marshes, tidal flats and coastal dunes	*	**	***	
Beaches, cliffs and deltas	*	**	***	*
Coral reefs	*	**	*	
Zonal				
Tropical rainforests	*	*	*	*
Tropical savannas	**	*	***	*
Deserts	**	*	*	*
Mediterranean landscapes	*	*	***	*
Temperate forests and rangelands	*	*	***	*
Tundra and permafrost-dominated taiga	**	*	*	*
Ice sheets and ice caps	**	*		*

For rivers, the balance shifts from topographic relief towards hydroclimate and runoff. Clearly for coastal environments, the dominant interplay is between sea level controls and varying levels of human impact. For a general assessment of the tropical rainforest biome, the four drivers are of equal importance whereas for savanna landscapes human activity and variations in hydroclimate (particularly the role of different rainfall regimes) are the dominant drivers. Hydroclimate dominates the desert biome. Whilst human activity is the key control over Mediterranean and temperate biomes, this control is of minimal impact at the large scale in tundra and ice-covered biomes where hydroclimate is the main driver. Clearly projecting future climate and associated process change onto these varying backdrops is likely to lead to strongly differentiated environmental futures.

We now consider elements of each of these drivers in turn. As a linking illustrative device of the underlying complexity of these issues, we look at how the debate over potential changes in near-future storminess intersects with these drivers. Windstorms (and coupled storm surges) account for a greater proportion of damaging events than any other form of natural disaster worldwide and are thus of considerable societal concern, particularly given the assumption that they will increase their global impact as a result of climate change (e.g. Berz, 2005). And we choose this element of the climate change debate not because of its high profile and strongly politicised role (e.g. Curry *et al.*, 2006; Mooney, 2007) but because it illustrates both the rapidly evolving nature of much of the discussion on climate change mechanisms, and their deepening complexity, but also the challenge faced by geomorphology and ecology in incorporating critical, yet regional-scale, and hence poorly known and poorly modelled, processes into the better forecasting of near-future landscape change.

15.3.2 The (relatively) unambiguous role of climate

In certain instances, the role of climate is clear. Thus, for example, in the case of ice sheets (Chapter 14), the IPCC methodological chain of establishing the main factors that determine ice sheet 'health'; the use of theory to build models of mass balance and ice flow; the modelling of future health; and the testing of these models against remotely sensed data, whilst capable of considerable development and refinement using geomorphologically driven data collection and modelling, has proved to be a fruitful means of exploring the likely reaction of large-scale ice masses to global warming. Whilst the drivers of sea level and topographic relief are important, an approach to landscape change in such environments which is fundamentally based around scientific questions relating to climate and climate change is not an unreasonable one. Similarly, in the tundra and permafrost-dominated taiga biomes (Chapter 13), the strong controls on the areal extent of permafrost and depth of the active layer to temperature and the strong linkages, through thermokarst subsidence processes and thermoerosion, between climate change and landscape change, make establishing the role of climate relatively unambiguous; where there is human impact this can be assessed through comparison with adjacent areas of permafrost where human impacts are minimal. In both these biomes, therefore, it is relatively straightforward to identify key research questions (e.g. better monitoring of ice sheet surface processes and basal ice processes beneath outlet glaciers, monitoring of changes in the thickness of the active layer in permafrost) and key geographical locations for research (e.g. the West Antarctica Ice Sheet, the Greenland Ice Sheet, the discontinuous permafrost zone in Russia, Canada and Alaska) and to develop appropriate monitoring networks further (e.g. Chapter 13: Circumpolar Active Layer Monitoring (CALM) Program). Furthermore, the lack of human impact has allowed the use of Quaternary analogues to inform the likely impact of near-future climate changes on landscape change in a way that has been more problematic in other biomes.

15.3.3 Hydroclimate and runoff

Changing climate is intensifying the global hydrological cycle, leading to significant changes in precipitation, runoff and evapotranspiration, with a probable increase in hydrological extremes (Chapter 1; Chapter 4). However, it is well known that for the same emissions scenario, different general circulation models (GCMs) produce varying patterns of spatial environmental change and that inter-model agreement is much better for projected temperature changes than for precipitation, to the extent that for some regions precipitation changes can show a difference in sign (Murphy *et al.*, 2004; IPCC, 2007a). The deployment of meteorological satellites in the 1970s began to make possible the large-scale averaging of changes in precipitation and recently Wentz *et al.* (2007) have reported that global mean precipitation increased by $7.4 \pm 2.6\%$ per °C for the period 1987–2006. General circulation models predict much smaller increases of 1–3% per °C (Held and Soden, 2006), although this may be due to the difficulty of comparing the short-term (20-year) trends in the satellite observational record to the centennial trends derived from GCMs (Lambert *et al.*, 2008). If climate modelling is to be decoupled from the other activities (Fig. 15.1b), then it might be attractive to generate a very considerable number of runs from a single GCM (e.g. Stainforth *et al.*, 2005), or provide multi-ensemble runs from different GCMs (e.g. Murphy *et al.*, 2004), to arrive at a better idea of the envelope of possible climate states and a narrowing of the bands of prediction uncertainty. Yet these approaches have, for example, not thus far been successful in predicting climate change impacts on freshwater resources, largely because of the use of ensemble means and a disregard for inter-annual, seasonal and shorter-term variations in precipitation characteristics (IPCC, 2007b).

Many of the chapters in this volume show that the definition of climate itself is difficult. This is particularly true for those transitional biomes such as the savanna biome (Chapter 9) and the Mediterranean (Chapter 11); in both cases there is a gradation of climatic regimes with each of these regimes themselves being subject to considerable inter-annual variability. Furthermore, there is a considerable challenge in predicting landscape change under conditions of changing seasonality, when part of the year may be wetter and part of the year drier than at present (e.g. Mediterranean landscapes: Chapter 11). In particular, short-term rainfall is the most uncertain and most difficult variable to estimate at meaningful spatial and temporal resolutions (IPCC, 2007a), yet is the most important for many geomorphological processes and landscape-changing events. Furthermore, even at the present time, the linkage of such short-term events to climate change is obscured by the lack of widespread long-term precipitation records, variable data quality and inter-annual and decadal climatic variability (Viles and Goudie, 2003; Chapters 8 and 10). A particular difficulty is that historical data sets generally do not permit the resolution of the underlying tendencies in landscape change because the environmental variability signal overwhelms any trend on the decadal timescale. This is the case across the environments and biomes covered in this volume. Thus, for example, on the coast of southeastern Australia, mean-trend shoreline change becomes comparable to the decadal fluctuations only after the passage of about 180 years, yet the longest high-resolution data set now spans little more than 30 years (Chapter 6). For both the rainforest (Chapter 8) and savanna (Chapter 9) biomes too little long-term monitoring is available to observe possible transient and non-equilibrium responses to climate and land use change. And for ice sheets and ice caps, data on snow accumulation and iceberg calving rates derived from radar remote sensing are too recent to allow the separation of the climate signal from local variability (Chapter 14).

Scaling and the vexed question of near-future storminess

Chapter 14 gives a very telling example of the influence of scale on problem specification and understanding. Notwithstanding their considerable three-dimensional sophistication, the thermomechanical ice sheet models in the IPCC Fourth Assessment (IPCC, 2007a) utilised a coarse cellular grid of tens of kilometres which necessarily emphasised ice sheet stability and long response times, whereas in fact it has now become clear that it is the behaviour of spatially restricted, climate-sensitive and highly dynamic ice streams that most probably hold the key to the understanding of ice sheet dynamics – and to rates and magnitudes of global sea level change.

There is also a mismatch in spatial scales between GCMs and hydrological processes, and the inability to simulate realistic precipitation patterns that lead to extreme events. Various scaling techniques have been developed to address this problem, both in the form of dynamical downscaling which attempts to establish true physical linkages between the climate at large and regional scales and statistical methods that use empirical relations between large-scale climate variables and local weather variables and which assume that these relations are transferable with changes in near-future boundary conditions. Unfortunately, however, it is becoming clear that higher-resolution spatial models have a greater dependence upon local relief and on regional-scale feedbacks (such as land cover change, soil moisture change and the number of frost days) which are less likely to scale in a linear fashion with global radiative forcing (Mitchell *et al.* 1999; Ruosteenoja *et al.*, 2007).

In addition, such scaling exercises have great difficulty dealing with extreme events. Overall, the frequency and intensity of extreme weather events such as storms and hurricanes are expected to change with scenarios of climatic warming. Theoretical arguments (e.g. Holland, 1997) and modelling studies (e.g. Bengtsson *et al.*, 2007) indicate that tropical cyclone wind speeds should increase with increasing sea surface temperature. Direct observational verification for this relation has been established over the past 35–40 years for Atlantic tropical cyclones (Emanuel, 2005; Webster *et al.*, 2005; Saunders and Lea, 2008) and more globally since 1981 (Elsner *et al.*, 2008). Given these various lines of argument, it has not been difficult to assign the rising trend of hurricane activity in the Atlantic to greenhouse gas-induced ocean warming (e.g. Mann and Emanuel, 2006; Holland and Webster, 2007) and, given the pronounced Atlantic warming projected for the twenty-first century (IPCC, 2007a), to see dire predictions of increased tropical cyclone and hurricane magnitude and frequency in the near future. Such scenarios are of considerable concern given the physical and societal impacts that such events can cause on low-lying coasts (Plates 48 and 49). However, there has been an equally large number of critics who have argued that the data are not reliable enough to confirm such a relation, particularly in trying to marry data from the pre-satellite era with contemporary storm tracking (e.g. Landsea, 2005, 2007; Chang and Gao, 2007; Mann *et al.*, 2007). Over historical timescales, the average frequency of Atlantic hurricanes decreased from the 1760s until the early 1990s, reaching very low values in the 1970s and 1980s. Thus Nyberg *et al.* (2007) argue that the period of enhanced hurricane activity since 1995 is not unusual in the longer-term record and that the last decade merely represents a return to 'normal' hurricane activity rather than a response to increasing sea surface temperature. Furthermore, dating of storm overwash deposits in a coral reef lagoon in Puerto Rico further suggest that such decadal fluctuations in storminess nest within centennial- to millennial-scale fluctuations in hurricane incidence, perhaps modulated by ENSO and the West African monsoon (Donnelly and Woodruff, 2007). Finally, recent modelling that randomly 'seeds' embryonic hurricane cells into GCM models and studies the resulting patterns of hurricane intensity, frequency and location, suggests a reduction in global frequency but localised increases in intensity (Emanuel *et al.*, 2008). Similarly, for Atlantic hurricanes, Knutson *et al.* (2008) have predicted overall substantially fewer tropical storms (−27%) and hurricanes (−18% (and hurricanes making landfall by –30 %)) under warm climate runs, suggesting that sea surface temperature is not the primary controlling factor but rather circulation changes and/or moisture. This conclusion echoes the review of the changing spatial patterning of Caribbean hurricane incidence described in Chapter 8 of this volume.

Climate change, coasts and geomorphology

If the environmental forcing controls are uncertain, as the storminess debate clearly shows, then several chapters in this volume show that there are further layers of complexity. It is clear that the resolution of the debate on the possible changing magnitude, frequency and location of tropical storms with global environmental change has important implications for sandy coasts, coastal salt marshes and mangrove swamps and coral reefs, which lie within the tropical storm belts and on their margins if the present storm belts expand under global environmental change. However, the exact near-future nature and patterning of these impacts are difficult to resolve. Firstly, individual storm tracks are generally narrow (<30 km) and thus the chance of a particular location being hit in any one hurricane season is low. Whilst short-term prediction of storm tracks has become very precise (e.g. Hurricane Katrina (2005): McCallum and Heming, 2006), predictions of track position over a week ahead of landfall remains highly problematic. There is a strong interaction between storm track orientation, regional bathymetry and nearshore hydrodynamics. Thus, for example, the south-to-north track of Hurricane Donna (1960) produced a 4 m storm surge in the Middle Florida Keys as water was driven north across Florida Bay; by comparison, the east-to-west track of the similar strength Hurricane Betsy (1965) across the bay resulted in a lower storm surge at Key Largo (although subsequently much greater impact on Lake Pontchartrain and the city of New Orleans) (Perkins and Enos, 1968). Secondly, Chapter 7 shows that in coral reef environments there are non-linear relations between storm intensity, storm type and geomorphic damage. Whereas hurricanes with typical wind speeds of 120–150 km hr^{-1} result in a patchwork of impacted and non-impacted areas, determined by water depth, reef front aspect and reef topography in relation to storm direction, severe storms, with wind speeds in excess of 200 km hr^{-1}, may overcome the structural resistance of the reef as a whole, reducing three-dimensional complexity to an unstable rubble plain, unconducive to coral re-establishment, and producing a hiatus to reef recovery lasting for up to 50 years. Hurricanes with slow forward speeds result in prolonged coastal wave attack and beach scouring which is felt for hours to days in advance of the wind stress; these systems can be large in extent and be accompanied by high rainfall and extensive flooding of low-lying areas. By comparison, compact, fast-moving and intense hurricanes (such as Hurricane Andrew (1992)) generate little wave scour but rather brief but extremely strong unidirectional currents and onshore surges, particularly where enhanced by wave shoaling or funnelled through narrow passes between islands. Such storms leave beaches and coastal barriers relatively intact but do great damage to subtidal seagrass meadows (Tedesco *et al.*, 1995). Thirdly, hurricanes are often only the trigger for ongoing biogeomorphological response: thus Knowlton *et al.* (1990) documented ongoing coral mortality following Hurricane Allen (1980), an order of magnitude more severe than the impact of the storm itself; and more than 2 years after Hurricane Donna, storm-damaged mangrove was still dying in the Florida Keys (Craighead and Gilbert, 1962). And over longer timescales, there are important interactions between ecosystem and landform recovery rates, the intervals between storms and landscape change (for reef environments see Chapter 7).

Climate change, hydrology and geomorphology

The uncertainties of landscape-scale response are also considerable in environments other than those associated with land–sea interactions. Thus, for example, Chapter 12 and Chapter 8, for temperate and tropical environments respectively, both argue that regions experiencing increases in annual or seasonal precipitation may well see increased periods of deep-seated mass wasting processes and more shallow, rapid landslides with greater precipitation variability which includes shorter-term precipitation inputs. These impacts might be offset by changes in evapotranspiration in a warming climate – or not, because rising atmospheric carbon dioxide levels mean that less water loss is needed in transpiration for the same carbon gain. These changes come down to the level of the relative frequency of sunny versus rainy days, as transpiration losses are generally much higher on sunny days, and the availability of soil water for transpiration, which will vary with soil characteristics and soil depths as well as rainfall factors (Chapter 8). Changes in evapotranspiration demands will affect rates and styles of vegetation growth, including rooting depth, and these controls will also feed back into changing landslide and deep-seated failure dynamics. Such a landsliding phase would have important influences on the other parts of geomorphological systems, notably in terms of large, sudden increases in supply of bedload and suspended sediment to river systems, enhancing subsequent slopewash rates through bare areas of landslide scars, and potentially providing a mechanism for drainage network extension via conversion of linear landslide scars into valley-side ephemeral channels. Further consequences will include increased channel size (particularly if river systems were previously supply-limited) and higher suspended sediment, bedload and solute yields (Chapter 8).

15.3.4 Sea level

Slaymaker *et al.* (Chapter 1) document the record of sea level change over the twentieth century, documenting its inter-decadal variability and spatial patterning in different decades, and review estimates for near-future change, highlighting the uncertainties generated by the difficulty of incorporating a term for accelerated ice flow (and see also Alley *et al.*, 2008). In spite of these caveats, there is a general feeling that the reporting of widespread erosion of sandy beaches can be explained by sea level rise (Chapter 6). There is concern over accelerated sea level rise; thus, for example, Reed *et al.* (2008) argue that whilst many coastal marshes in the US mid-Atlantic region will most probably survive under future sea level rise rates 2 mm a^{-1} greater than current rates, many will become vulnerable if the increase reaches 7 mm a^{-1}.

The assessment of the impact of sea level rise at the coast is, however, not entirely straightforward (Chapter 5); whilst sea level rise can be measured at an annual timescale, its influence on coastal landforms is really only demonstrable at the scale of decades (at least while sea level rise rates remain below ~1 cm a^{-1}). Most scenario modelling of regional sea level rise, in the absence of either climate or sea level models at the regional scale, is reduced to applying global mean sea level scenarios onto regional to local estimates of land movement. Furthermore, a more geomorphological perspective shows that there are multiple controls upon coastal landforms and that these controls operate differentially across a much wider range of timescales. The fundamental relation – for salt marsh surfaces, beaches and reef islands – is that between the change in accommodation space, caused by sea level movements, and sediment availability (Chapters 5–7). Changes in the direction and intensity of wave climates, and magnitude–frequency characteristics of storms, neither of which are, as yet, well specified in atmosphere–ocean models, impact on the spatial distribution of accommodation space (Chapter 6). Thus, for example, at 'normal' levels of wave energy, salt marshes are highly efficient dissipaters of wave energy, yet at some as yet poorly defined higher energy threshold they are eroded and can retreat rapidly, redefining the relative areal coverage of salt marsh versus fronting mudflat (Moller and Spencer, 2002). Sediment availability is influenced by both changes in climate and human activities over inter-annual to decadal timescales. In some locations and some coastal systems, sediment supply may be in the form of pulsed inputs related to short-lived flood events which may in turn be storm- or cyclone-generated. In estuaries, changes in freshwater runoff (in both overall magnitude and seasonal variation) and changes in salinity become important because these boundary conditions affect estuarine mixing characteristics, circulation patterns and sediment transport, deposition and entrainment processes. In many subtidal, intertidal and supratidal environments the relations between climate and substrate are mediated by either vegetation or benthic invertebrate communities. For these communities, alterations in air temperature, water temperature and precipitation, often over timescales below annual means, become significant. As community structure changes in response to these climatic shifts so earth surfaces can cross thresholds, from being biostabilised to being bioerosional, and vice versa (e.g. Widdows *et al.*, 2004).

15.3.5 Topographic relief

The notion that climate change drives land cover change recognises one of the controls on landscape change but an equally critical driver is topography. Thus, for example, the use of the climatically defined term 'desert' hides a fundamental distinction between the extensive sand seas and sand dunes of the low relief, ancient shield deserts of the Arabian Peninsula, Australia and southern Africa and the range fronts, alluvial fans and basins of the high relief, tectonically active deserts of South America, western North America and Central Asia (Chapter 10). Similarly, on Mt. Kilimanjaro, topographic relief provides a strong control on the distribution of biomes, from mountain summits to savanna landscapes at lower levels (Plate 50).

There have been numerous attempts to define this topographic control on geomorphological processes and landscape change, variously involving basin area and maximum elevation (e.g. Milliman and Syvitski, 1992), the ratio of maximum basin elevation to basin length (e.g. Summerfield and Hulton, 1994) or mean height (Ahnert, 1984). All these correlations are hampered by variable, and often poor, data quality resulting from difficulties relating to the spatial and temporal sampling of sediment load characteristics. As most recording stations for suspended sediment loads are in the lowlands, the down-catchment nature of the 'sediment cascade' becomes critical and how sediments are stored, for shorter or longer periods, in a whole variety of sediment sinks (lakes, alluvial fans, proglacial zones and floodplains) (Chapter 2, Chapter 3, Chapter 4). Even allowing for these complexities, a multivariate analysis of sediment yield data and associated parameters reveals that a combination of environmental and topographic factors can only explain about 60% of the variance in global sediment yield data; Hovius (1998) suggests that the additional control of tectonic activity must be factored into the sediment yield equation. Scaled for drainage basin area, sediment yields increase through almost five orders of magnitude from tectonic cratons (typical sediment yields of $100\,\mathrm{t\,km^{-2}\,a^{-1}}$) to contractional mountains (up to $10\,000\,\mathrm{t\,km^{-2}\,a^{-1}}$) (Fig. 15.3). Tectonically active mountain belts not only provide relief to drive erosion processes but they also combine high regolith loss with

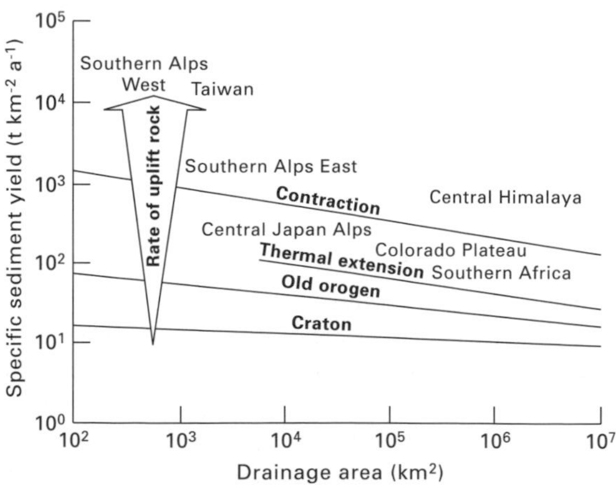

FIGURE 15.3. Specific sediment yield versus drainage area by tectonic setting (from Hovius, 1998).

the rapid uplift of new bedrock into the weathering zone to continually refresh these erosion processes and maintain high sediment yields.

Relief, tectonics and storminess

Recent research has further informed the nature of these relief controls on landscape change across more narrowly defined space and timescales. On the rapidly uplifting (5–7 mm a^{-1}) island of Taiwan, erosion appears driven by the interaction of erodible substrates, rapid deformation in the form of frequent earthquakes and typhoon-driven runoff variability. Earthquakes produce sediment by rock mass shattering and landsliding, and landslides and debris flows are also triggered by typhoon-generated storm runoff which flushes sediments from the mountains (Dadson *et al.*, 2003). Thus changes in the frequency, magnitude and track positioning of typhoons in the Philippine Sea, western North Pacific Ocean consequent upon climate change will have consequences for denudation rates; these processes are more important than simple relief and average precipitation controls (e.g. Andes: Montgomery *et al.*, 2001; Himalayas: Finlayson *et al.*, 2002). Furthermore, these sediment transfer processes also involve the transmission of large quantities of non-fossil particulate organic carbon. These organic sediments may be carried in turbid, hyperpycnal river plumes which trigger turbidity currents and bypass nearshore wetland depocentres (e.g. Mulder and Syvitski, 1995) for burial in deeper waters on continental shelves (e.g. Galy *et al.*, 2007). The transfer of organic carbon from the terrestrial biosphere to the oceans via erosion and fluvial transport is a major pathway in the global carbon cycle (Stallard, 1998; Schlunz and Schneider, 2000) and any increased efficiency of this pathway from storm-driven processes might be an important negative feedback on climate change (Hilton *et al.*, 2008).

15.3.6 Human activity

As of November 2008, the world's population is estimated to be about 6.72 billion (6 720 000 000); this absolute figure continues to grow, although the rate of growth has almost halved since its peak of 2.2% a^{-1}, which was reached in 1963. The world's population, on its current growth trajectory, is expected to reach nearly 9 billion by the year AD 2042. There is, therefore, considerable substance to the claim that agriculture has been the greatest force of land transformation on this planet (Lambin and Geist, 2006); nearly a third of the Earth's land surface is currently being used for growing crops or grazing cattle (FAO, 2007). Half of the global population now lives in cities, and within two decades, nearly 60% of the world's people will be urban dwellers. Urban growth is most rapid in the developing world, where cities gain an average of 5 million residents every month (UN-HABITAT, 2008); these demographics make enormous demands on environmental resources and have severe impacts on landscapes. A microcosm of this dynamic is illustrated by Plate 51. The population of the city of Las Vegas, Nevada, USA has risen from 125 000 in 1970 to 478 000 in 2000. However, it is the growth of the metropolitan area of Las Vegas that is most remarkable; this had (as of 2007) a population of 1.84 million inhabitants and was one of the fastest-growing regions of the United States. The implications for the desert landscape of southern Nevada is clear from Plate 51.

Not surprisingly, therefore, for many authors in this volume it is human activity which has set the character of the contemporary landscape. Thus, for example; Chapter 4 shows that rivers have been directly manipulated and incidentally affected by human activities for more than a thousand years, with particularly dramatic effect during the twentieth century. Climate change influences runoff from the land and the flow of rivers but, in all settled parts of the world, the effect of human manipulation of land surface conditions, regulation of flows and abstraction of water supplies will continue to be far more important determinants of change in river flow regimes. Similarly, sediment yield to streams and the consequent stability of stream channels will continue to be affected by the effect of human activity on the land surface. In the Mediterranean, forestry, agriculture and mining have altered the landscape in a series of settlement and landscape exploitation phases associated with different cultures from Neolithic times (Chapter 11). Elsewhere human impact, or at least the destructive phase of human occupancy, has been more recent. Thus, in tropical savannas:

> the modern distribution of savanna ecosystems is … something of an artificial view because the nature and extent of human activity, especially during the last 100 years or so, has markedly disturbed and even destroyed savanna vegetation in many of the regions indicated, while at the same time leading to its expansion into regions that were tropical rainforest during the altithermal. (Chapter 9, p. 249).

Many of these changes reflect the great rural to urban shifts in population during the twentieth century and the great fluxes of migrants around the developing world. Thus, for example, coastal impacts of climate change relevant to society can only be contemplated meaningfully in the context of likely changes to human settlement patterns and other coastal use projected over the same time scale (Nicholls *et al.*, 2008). Population growth in mountain environments (Chapter 2) and the expansion of human exploitation of tundra and permafrost-dominated taiga (Chapter 13) are more recent still. Finally,

> geomorphic effects of anthropogenic disturbance are potentially particularly high in the humid tropics and, given the scale of current and projected future forest clearance and land use change, will arguably continue to greatly outweigh those of climatic change. (Chapter 8, p. 226).

It is not difficult, therefore, to see human activity as the dominant driver in landscape change today, and into the near future.

As Chapter 1 argues, it has been usual to analyse the role of human activity as though it were outside the geosystem (e.g. see Fig. 1.6), a weakness that fails to recognise the accelerating interdependence of humankind and the geosystem. By contrast, the IPCC (2007b) specifically addresses landscape vulnerability, a function of the character, magnitude and rate of environmental change and variation to which a system is exposed, its sensitivity and its adaptive capacity. In general, those landscapes that have the least capacity to adjust are the most vulnerable; and geomorphology controls that ability to adjust. Furthermore, there have been several major efforts in the last decade to evaluate land cover and land use change and to use such dynamics to better understand the coupled human–environment system. Appendix 1.4 (Chapter 1) gives details of the Land Use and Land Cover Change (LUCC) Project; one might also mention the following Global Land Project (GLP). However, Liverman and Roman Cuesta (2008) conclude that progress has been limited, partly on a practical level as a result of the difficulties of scaling often patchy and partial socioeconomic data sets to satellite imagery detailing land cover changes and partly more generally because of the difficulty in explaining human actions and policy decisions. Even scenario-based approaches fail to identify the full range of possible futures, not least because of the inability to identify unexpected 'surprises' (e.g. AIDS, uptake of biofuel technologies) which have major socioeconomic and environmental consequences (Millennium Ecosystem Assessment, 2005).

Environmental hazards and storminess: the US Gulf and Atlantic coasts

This is not a text on environmental hazards *per se* but such hazards cannot be ignored. The continued growth in global population described above and its increasing concentration into urban centres (2 billion people are likely to be added to the cities of developing countries over the next 20 years) hugely increase human vulnerability to extreme natural events, regardless of whether or not they are set to increase as a result of climate change. It is likely in the future that the world will experience several disasters in a year that kill more than 10 000 people and that events which kill more than 1 million will soon be upon us (Huppert and Sparks, 2006). These frightening statistics are strongly influenced by the exposure of developing countries but Hurricane Katrina demonstrated that developed countries also lack the ability to properly prepare for, and respond to, extreme natural events (e.g. Independent Levee Investigation Team Final Report, 2006; American Society of Civil Engineers Hurricane Katrina External Review Panel, 2007). Pielke *et al.* (2008) estimate the damage that historical storms affecting the USA would have caused if they had made landfall under contemporary socioeconomic conditions, adjusting historical damages by changes in national inflation, growth in wealth and changes in population in the US counties affected by each storm (and for a similar approach, but based on housing units, see Collins and Lowe (2001)). In this approach there is no long-term trend in losses (Fig. 15.4); indeed, the damage in the period 1926–35 was nearly 15% higher than in 1996–2005. The 1926 Great Miami hurricane appears as the single most destructive storm (with normalised losses of US$157 billion), although Hurricane Katrina ranks second highest (US$81 billion) and the 2004 and 2005 Atlantic hurricane seasons are particularly extreme, containing 7 of the 30 most damaging storms over the period 1900–2005. Estimated property damage from Hurricane Ike (2008) was US$27 billion (2008 prices), making Ike the third costliest hurricane in US history, after Katrina (2005) and Andrew (1992) (Plates 48 and 49).

The current trend in the USA is for a doubling of losses every 10 years; thus a storm like the Great Miami hurricane could result in losses of US$500 billion as early as the 2020s. Figure 15.5 graphically illustrates this increased vulnerability. Development of Miami Beach began in the 1910s and 1920s (Fig. 15.5a); although Miami was badly impacted by the Great Miami hurricane of 1926 (Fig. 15.4), by 1930 its population had reached 6500. In the 2000 census the population of Miami Beach was over 87 000. The Miami urbanised area – by 2007 the fourth largest urbanised area in the USA (after New York, Los Angeles and Chicago) – is a narrow coastal strip, 180 km long but only 32 km at its widest, supporting a population of almost 5 million inhabitants at densities of 1700 individuals km^{-2}, rising to over 10 000 individuals km^{-2} in the city of Miami (UNPD, 2008). By 2025, it is estimated that Florida's population will exceed 27 million, compared to 5 million in 1960 (Barnes, 2007). Since 1926, the city of Miami has only suffered two direct hits from hurricanes (Hurricane King (1950) and Hurricane Cleo (1964)), although it has

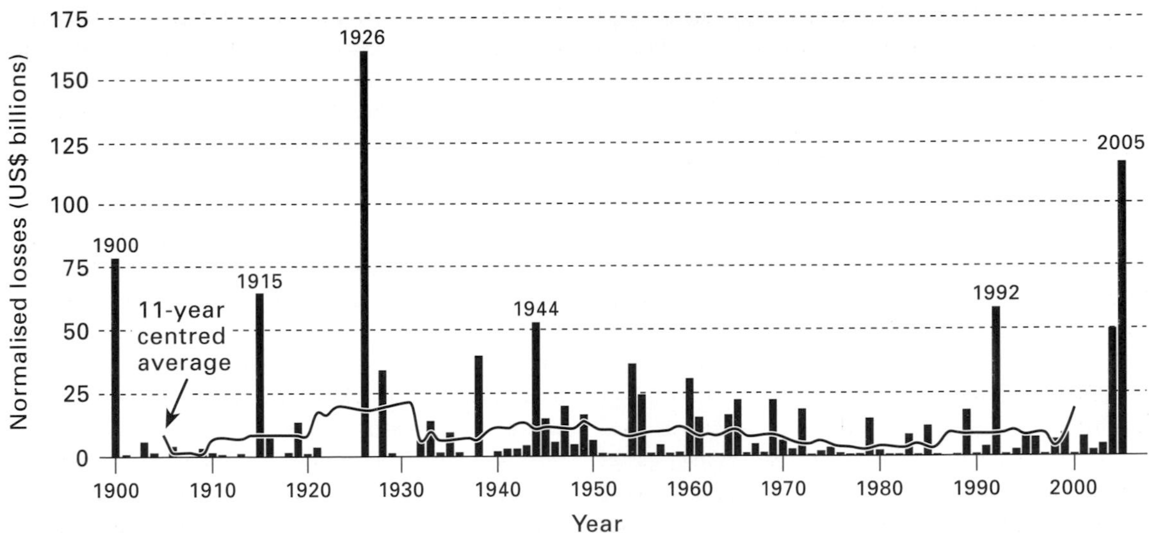

FIGURE 15.4. Normalised (adjusted for changes in inflation, wealth and population) US Gulf coast and Atlantic hurricanes, 1900–2005. Line represents 11-year centred moving average. Note: high losses in 2004 and 2005 hurricane seasons were exceptional in the long-term record (from Pielke *et al.*, 2008; with permission from ASCE).

been affected by flooding and wind damage from a series of hurricanes over the last 40 years (from Hurricane Betsy (1965) to Hurricane Wilma (2005)). Along with New Orleans and New York, it remains one of the most vulnerable large cities in the USA, and globally, to hurricane impact.

Environmental hazards and storminess: the UK east coast

In the UK, catastrophe loss modelling of a repeat of the southern North Sea storm surge of 31 January – 1 February 1953, but with current sea defences (including the Thames barrier) in place, predicts loss estimates of between £5 million, where the storm surge coincides with a neap tide, to more than £2 billion where the surge accompanies the highest astronomical spring tide and is associated with widespread defence breaching. Taking into account the probability of different surge-tide events, the 'expected loss' for a repeat of the 1953 event is £470 million, reflecting the inundation of 9500 properties (Muir Wood *et al.*, 2005). Although many of the northwest European flood defence schemes (e.g. the Rotterdam Barrier and Eastern Scheldt Barrier) have been designed for a lifetime of 200 years, accounting for 50 cm of sea level rise, many of the reinforcements and extensions to the line of flood defence put in place in the UK after 1953 are approaching the end of their design life (Select Committee for Agriculture, 1998). The potential impacts are not insignificant: modelling for eastern England (Nicholls and Wilson, 2001) suggests that under a 'high' climate change scenario (sea level rise of 71 cm by 2050 and + 107 cm by 2080, referenced to 1990) the 1-in-100-year defence standard could be reduced to 1 in 2–8 years by 2050, with many defences at or below the 1-in-1-year standard by 2080. At the same time, under the IPCC SRES A2 scenario, the current + 1.5 m surge event at Immingham, Humber Estuary, eastern England which is predicted to be exceeded once every 120 years on average under the present climate, will be exceeded once every 7 years on average by the 2080s, given modelled climate change and continued local land subsidence (Lowe and Gregory, 2005). Furthermore, the Thames Gateway project, a major urban regeneration plan for East London and the margins of the Thames estuary, envisages the construction of 120 000 new homes, all in areas designated as high flood risk (Office of the Deputy Prime Minister, 2003).

Pielke and Sarewitz (2005, p. 256) comment:

> two well documented aspects of the climate–society relationship are largely absent from the climate debate: (1) the awareness that, over time, societal changes – demographic, social, economic, and other changes in the characteristics of human populations – are primary factors in climate impacts on humans and human impacts on the environment and (2) viable strategies for responding to such changes lie predominantly in the area of societal governance, not in efforts to control the future behaviour of climate.

It is important that geomorphology contributes to these issues of 'societal governance'.

FIGURE 15.5. (a) Miami Beach, 1925 (from the Wendler Collection, State Library and Archives of Florida); (b) South Beach, Miami Beach and (background) downtown Miami, January 2006.

15.3.7 Summary

The discussion above shows that the controls on landscape change are not restricted to climate alone. Landscapes evolve at varying rates and in varying ways to four major drivers – hydroclimate and runoff, sea level, topographic relief and human activity – and the relative balance of these drivers varies between the major biomes on the Earth's surface. In many biomes, however, the role of human activity has increased markedly over the last 100 years to become the dominant driver, both now and into the foreseeable future. There is a danger in some localities that we place too much emphasis on policy-making and management for climate change whilst paying too little attention to the acceleration of landscape modification by growing human populations. The example of changes in storminess in the near future shows that the physical processes that lie behind the variability in storm magnitudes, frequencies and positioning remain uncertain and that the modelling of patterns of future storminess is currently constrained by the spatial resolution of the GCM models available for this purpose. However, it is encouraging that the simplistic linkages between storminess and global warming prompted by the extreme 2004 and 2005 Atlantic hurricane seasons, and culminating in some responses to the impact of Hurricane Katrina on coastal Louisiana and the city of New Orleans, have been tempered, in the scientific community at least, by a much deeper realisation of the complexity of hurricane–climate relations and, in some quarters, by a willingness to reassess formerly held positions. The above review also makes clear, however, that the uncertainties that lie behind atmosphere–ocean modelling are then propagated through to further sets of uncertainties that relate to the relations between climate and ecosystems, and ecosystems and landforms, and the assembly of landforms into landscapes. It is likely that all these uncertainties are additive (Fig. 15.6), resulting in low predictive capability and suggesting risk-based forecasting as a more fruitful way forward. Whereas deterministic predictions provide a single estimate of future landscape change, forecasts indicate the range of feasible impact projections, and probabilistic forecasts assign the likelihood of occurrence across the range of impact magnitudes (e.g. see Chapter 6 for 'best estimate' v. 'risk averse' forecasting in relation to shoreline retreat estimates).

15.4 Conclusions: new geomorphological agendas for the twenty-first century

In conclusion, we return now to questions posed near the beginning of this chapter: what has geomorphology to offer the global environmental change debate and how might this contribution be framed?

Fundamentally, it is important to now build a series of geomorphological scenarios for potential near-future environmental change for the azonal ecosystems and terrestrial biomes identified in this volume. This exercise needs to take into account the varying roles of the four drivers identified earlier in this chapter and to recognise the fundamental interaction between those drivers that are global and systemic (hydroclimate and sea level) in nature and those that are regional to local, cumulative (topographic relief, land cover and land use changes) in character. Some progress has been made in linking models of climate, vegetation and land use change and in predicting changes in global biodiversity (Sala *et al.*, 2000); a parallel exercise based on geomorphological principles and focussing on

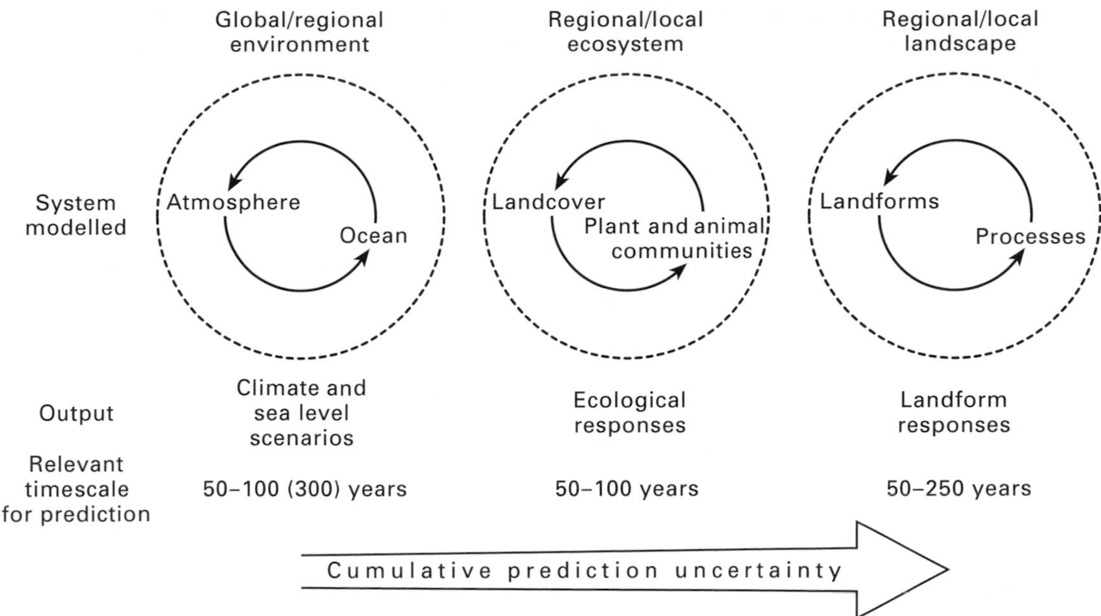

FIGURE 15.6. Trends in prediction uncertainty in climate–ecology–landform models.

geodiversity would now be of value. A particular challenge will be to distinguish those parts of landscapes that are close to threshold and which, because of their vulnerability, deserve priority attention.

15.4.1 Challenges to building geomorphological scenarios in response to global environmental change

There are, however, a number of challenges to scenario building in geomorphology. Underlying these approaches is a need to tackle the age-old problem that the key issues lie at those intermediate scales which are neither small-scale process based or long-term geological stratigraphy type questions (where geomorphology intersects with Quaternary science). Such difficulties are not restricted to geomorphology; they are very common in environmental science. Thus Brantley (2008) points out that soils are defined not only by rock particles but also by minerals, nutrients, organic matter, biota and water, and that each of these defining properties has a particular timescale, and a particular research community (geology, geochemistry, ecology, hydrology) associated with it. Environmental problems occur when attempts to remedy problems at one scale, such as the addition of fertilisers to compensate for long-term soil fertility decline at the nutrient cycling scale, lead to problems at another scale, such as the escape of nutrients and the potential eutrophication of water bodies at the timescale of soil water flow. She concludes that 'learning how soils will change in the future will require observations and models that cross time scales' (Brantley, 2008, p. 1455) and identifies the need for models that describe not only how soil components (sediment record, chronosequences and observations of modern-day fluxes) react alone but how they interact with each other in response to tectonic, climate and anthropogenic forcing.

A good example of these difficulties of scale concerns how rapidly the large amounts of carbon stored as soil organic matter will respond to global warming. On time-scales of months to years, the rates of accumulation and loss of soil carbon are estimated from observed rates of fresh plant litter addition to soils and mass loss during decomposition respectively, these rates being controlled by litter quality, soil faunal and microbial community composition and climate. On millennial timescales, changes in carbon stocks cannot be observed directly but are calculated from inferences based on the age of organic matter as measured by radiocarbon. Here the amount and age of soil carbon are controlled by changes in mineral surfaces related to weathering. Perhaps not surprisingly, the variation in soil carbon storage demonstrated by these different methodologies, applied over different timescales, with differing process controls, gives markedly different rates of change: ~2 to 10 Mg C ha^{-1} a^{-1} from measurements of short-term litter dynamics and ~0.02 Mg C ha^{-1} a^{-1} for geologic time estimates. Furthermore, neither of these methodologies can address the fundamental problem of interest which has a different timescale. This is the soil carbon response to global change involving organic carbon stocks that change over decades to centuries. Where measurements at this timescale are available – from soil carbon dynamics after known disturbance or the use of radiometric markers in soil carbon pools from well-dated events – they show that substantial stores of soil carbon can accumulate or be lost at intermediate rates (~0.1 to 10 Mg C ha^{-1} a^{-1}). Here the controlling processes are not only climatic but also involve complex interactions with soil properties, soil fauna and microbial communities, and vegetation community dynamics (Trumbore and Czimczik, 2008). Furthermore, most detailed studies of soil carbon age come from small plot experiments undertaken at best over a few years, whereas processes operating at larger spatial scales over decades to centuries (such as erosion, fire or vegetation change) may ultimately determine the impact of changing soil states on atmospheric carbon dioxide. For example, fire-dominated Mediterranean (Chapter 11) and boreal (Chapter 13) ecosystems accumulate surface litter between burning events. Increasing burned area in a given year can return carbon faster to the atmosphere than it accumulates in unburned areas, making the region a net carbon source. Rapidly changing land use patterns, as observed in the ever-wet and seasonal tropics (Chapters 8 and 9), can be more important for evaluating soil carbon balance than the factors causing variable rates of carbon loss or gain in a small plot experiment. Such landscape-scale processes are crucial for the global carbon budget and geomorphologists should bring their expertise to bear alongside soil geochemists and ecologists as both more sophisticated field studies and ecosystem carbon models are developed. Furthermore, it must be recognised that as timescales change, so the nature of earth–atmosphere interactions may change. Thus the dominance of sink (photosynthetic uptake) and source (release due to soil decomposition) of carbon is likely to vary with timescale: increased trace gas emissions due to soil warming is likely to be the short-term response to climate change but over the longer term warmer climates, extended growing periods, and northward movement of productive vegetation may increase photosynthetic carbon uptake (Chapter 13).

Some of these issues are related to the fundamental character of geomorphological process–response systems (see Section 15.2 above). However, over the last 40 years geomorphology has made considerable progress in formulating a wide range of models that cover such issues as characteristic form, threshold exceedence, complex response and landscapes of transition (see Chapter 1 for a fuller discussion of these models). There is a strong modelling strand in geomorphology, from Kirkby (1971) and Ahnert (1976) to the development of 'reduced complexity' models (Brasington and Richards, 2007), such as the cellular automaton models of channel dynamics and landscape evolution (e.g. CHILD: Tucker *et al.*, 2001; CAESAR: Coulthard *et al.*, 2007) which addresses these intermediate-scale issues and which suggests that geomorphology is well placed to tackle these scale issues.

A further advantage is the linkage between geomorphology and Quaternary science, although there is a need to be mindful of the 'no past analogue' problem. A Quaternary perspective can provide the long-term framework within which to test models of landscape response to near-future climate change properly. A longer time-frame also gives the possibility of establishing longer-term trajectories to inform process understanding to identify where thresholds to landscape change might lie from past behaviour. Thus Chapter 14 points out that knowledge of the past behaviour of ice sheets and ice caps may identify those ice-covered areas that can collapse abruptly, often in response to changes in sea level, ocean water temperature or the internal dynamics of glacier flow.

15.4.2 The intrinsic value of geomorphology and geodiversity

IPCC science, not only from its established structure but also through its Working Group III, has very successfully linked climate science, and climate scientists, to issues of international policies and agreements, mitigation and adaptive strategies and even issues of sustainable development on climate change. Yet the focus is dominantly on climate. In a different way, the international biological community has been able to establish international conventions that seek to minimise biodiversity losses on a global scale. There have, however, been no comparable initiatives at the international level over the loss of geodiversity (a measure of the variety and uniqueness of landforms, landscapes and geological formations; see Chapter 1 for a fuller discussion) and this in spite of the clearly close linkages between biodiversity and geodiversity. An important role for the geomorphological community, therefore, is to promote the importance of incorporating notions of geodiversity into arguments for the protection and preservation of landforms and earth surface processes. In the same way that biodiversity 'hotspots' have been identified, so attention should be given to the identification of geomorphological hotspots as sites or regions of special value in terms of geodiversity and high vulnerability to environmental change.

15.4.3 Geomorphological services, sustainability and vulnerability

There is a considerable literature on the concept of ecological services, including the monetary valuation of those services (e.g. Costanza, 1997). Chapter 7, for example, shows that coral reefs possess benefits under four categories of ecosystem services (Millennium Ecosystem Assessment, 2005): *regulation* of incident oceanographic swell conditions to control reef and lagoon circulation, reduce shoreline erosion, protect beaches and coastlines from storm surges, and control beach and island formation; *provision* of aggregates for building (coral and sand), as well as the provision of land surface area and associated subsurface water resources, especially through reef island construction; *supporting* nutrient cycling and active carbonate production to build reef and reef island structures; and *cultural* benefits that include spiritual identity for indigenous communities and potential for tourism and recreation-based income. As this list shows, and as Chapter 5 points out for coastal marshes, tidal flats and sand dunes, many of those services identified as ecological could equally be termed geomorphological; indeed many seem more geomorphological than ecological. The idea of ecosystem services has not been without its critics (e.g. McCauley, 2006) but the geomorphological community should make more of the notion of geomorphological services than is currently the case.

The concept of sustainability is highly contested in a field like geomorphology because the drivers of change are themselves constantly changing and landscapes and their soils are, over century timescales, frequently collapsing, due to overexploitation. While the value of sustainability has achieved wide currency in principle, the implementation has proved difficult. Nevertheless, geomorphologists have to ask the question 'in what sense can landscapes be sustained over century or millennial timescales in the face of constantly changing human activities, sea level changes and climate change?' It is apparent that, as Diamond (2005) and Montgomery (2007) have pointed out, the removal of soil cover will reduce livelihood options for people and agriculture. The careful management of land, and its biogeochemical and aesthetic properties, enhances long-term human security and should be a priority for the global environmental agenda. Improving sustainability may include a component of environmental restoration. However, as Chapter 4 makes clear, we should not fall into the trap of believing that the world is getting better because we see apparent improvements in some environments in the developed world. There is a need to look critically at the actual geomorphological and ecological effectiveness of these projects, how these improvements compare with the scale of historical degradation, and how progress in the developed world scales in the light of continued degradation in the developing world. As geomorphologists we should be sufficiently confident of our understanding of both the natural and human-modified world to put forward large-scale proposals for environmental improvement, based on sound physical and ecological principles. A fine example of such an approach are the radical, strongly geomorphologically driven plans to extensively re-engineer the lower Mississippi River, including the abandonment of the present bird's-foot delta, to create more sustainable communities and economies on the US Gulf coast (Technical Group Envisioning the Future of the Gulf Coast Conference, 2006).

Debates over the relative merits of climate change mitigation versus adaptation point inexorably in the short to medium term towards a need to focus on adaptation. It is important, therefore, that geomorphologists investigate the role of geomorphology in promoting the development of adaptive systems. Reliance on reactive, autonomous adaptation to the cumulative effects of environmental change is likely to prove ecologically and socioeconomically costly. By contrast, planned and anticipatory adaptation strategies can provide multiple benefits, although it is important to recognise that there are limits on their implementation and effectiveness.

There is a need to couple socioeconomic as well as geomorphological vulnerability and to integrate the multiple impacts of land use changes on society and landscapes. For example, not only socioeconomic well-being, food security and health but also water resources, the carbon cycle and the functioning of geomorphic systems should be considered together. The linkage between land and water use needs to be better understood and incorporated into vulnerability studies. Chapter 3 alerts the reader to the extreme vulnerability of small lake catchments and wetlands for this precise reason. Water impacts on land use change are an important issue, as illustrated by irrigation farming in drylands (Chapters 9 and 10) and city expansion in deserts (Chapter 10; Plate 51). One of the most important trade-offs facing many societies engaged in intensive agriculture is between water quality, agricultural development and urban expansion. Emerging results from complexity research on patterns of geodiversity at multiple scales show strong linkages between landscape conservation and livelihood security.

While geomorphological research has tended to focus on so-called 'slow variables', a big challenge is to better integrate extreme events of all kinds: not only hydroclimate events (e.g. at ENSO-type scales, decadal scales, etc.; see Chapters 3, 8 and 9 for example) but also human-caused events (e.g. wars, conflicts, socioeconomic shocks, discussed in Chapters 1 and 2 and earlier in this chapter). These so-called 'fast' variables are often decisive in determining the resilience and collapse of geosystems (Holling, 2001). Surprises happen but integration of surprises into landscape change research has not developed rapidly enough. The concept of resilience establishes a clear connection between risks from extreme events and socioeconomic well-being (Lambin and Geist, 2006).

Throughout this book we have illustrated the impacts of human activities through analyses of large numbers of case studies. A methodological challenge to geomorphology is to move beyond *a posteriori* analyses of results towards comparative analyses of case studies. But such comparative analyses require standard data collection systems, which are rarely available for land use and landscape changes. Geomorphologists should consider seriously the need to expand the portfolio of analytical methods beyond multiple regressions to include narratives, system-based approaches, network analysis and complexity theory in order to address these highly complex and interrelated cause and effect relations. This is the essential burden of Holling's (2001) panarchy metaphor.

15.4.4 Closing statement

Continued changes in climate will ultimately tell us how landscapes will respond to global environmental change, perhaps rather sooner than was envisaged a decade ago. However, forecasting possible changes will be a safer path to follow, particularly given the importance of earth surface processes in sustaining societies. A geomorphology for the twenty-first century should have a strong underlying focus on making communities more resilient to the effects of climate change, particularly in helping those who are the most vulnerable and least able to cope with a changing environment. Scientists have a range of choices in interfacing with decision-makers. One of the most important roles is in helping to expand, or at least clarify, the scope of options available for responding to global environmental change. That should include the geomorphological viewpoint.

References

Ahnert, F. (1976). Brief description of a comprehensive three-dimensional process–response model of landform development. *Zeitschrift für Geomorphologie, Supplementband*, **25**, 29–49.

Ahnert, F. (1984). Local relief and the height limits of mountain ranges. *American Journal of Science*, **284**, 1035–1055.

Alley, R. B. *et al.* (2008). Understanding glacier flow in changing times. *Science*, **322**, 1061–1062.

American Society of Civil Engineers Hurricane Katrina External Review Panel (2007). *The New Orleans Hurricane Protection System: What Went Wrong and Why*. Available at www.asce.org/files/pdf/ERPreport.pdf.

Barnes, J. (2007). *Florida's Hurricane History*, 2nd edn. Chapel Hill: University of North Carolina Press.

Bengtsson, L. *et al.* (2007). How may tropical cyclones change in a warmer climate. *Tellus A*, **59**, 539–561.

Berz, G. (2005). Windstorm and storm surges in Europe: loss trends and possible counter-actions from the viewpoint of an international insurer. *Philosophical Transactions of the Royal Society of London A*, **363** ,1431–1440.

Brantley, S. L. (2008). Understanding soil time. *Science*, **321**, 1454–1455.

Brasington, J. and Richards, K. S. (2007). Reduced-complexity, physically based geomorphological modelling for catchment and river management. *Geomorphology*, **90**, 171–177.

Broecker, W. S. (2007). Climate change: CO_2 arithmetic. *Science*, **315**, 1371.

Chang, E. K. M. and Guo, Y. (2007). Is the number of North Atlantic tropical cyclones significantly underestimated prior to the availability of satellite observations? *Geophysical Research Letters*, **34**, L14801.

Chmura G. L. *et al.* (2003) Global carbon sequestration in tidal, wetland soils. *Global Biogeochemical Cycles*, **17**, 1111, doi:10.1029/2002GB001917.

Collins, D. J. and Lowe, S. P. (2001). *A Macro Validation Dataset for US Hurricane Models*. Arlington: Casualty Actuarial Society. Available at www.casact.org/pubs/forum/01wforum/01wf217.pdf.

Costanza, R. (1997). The value of the world's ecosystem services and natural capital. *Nature*, **387**, 259.

Coulthard, T. J., Hicks, M. D. and Van De Wiel, M. J. (2007). Cellular modelling of river catchments and reaches: advantages, limitations and prospects. *Geomorphology*, **90**, 192–207.

Craighead, F. C. and Gilbert, V. C. (1962). The effects of Hurricane Donna on the vegetation of southern Florida. *Quarterly Journal of the Florida Academy of Sciences*, **25**, 1–28.

Curry, J. A., Webster, P. J. and Holland, G. J. (2006). Mixing politics and science in testing the hypothesis that greenhouse warming is causing a global increase in hurricane intensity. *Bulletin of the American Meteorological Society*, **87**, 1025–1038.

Dadson, S. J. *et al.* (2003). Links between erosion, runoff variability and seismicity in the Taiwan orogen. *Nature*, **426**, 648–651.

DeFries, R. S., Bounoua, L. and Collatz, C. J. (2002). Human modification of the landscape and surface climate in the next fifty years. *Global Change Biology*, **8**, 438–458.

Diamond, J. (2005). *Collapse: How Societies Choose to Fail or Succeed*. London: Penguin.

Donnelly, J. P. and Woodruff, J. D. (2007). Intense hurricane activity over the past 5000 years controlled by El Niño and the West African monsoon. *Nature*, **447**, 465–468.

Douglas, I. (1982). The unfulfilled promise: earth surface processes as a key to landform evolution. *Earth Surface Processes and Landforms*, **7**, 101.

Elsner, J. B., Kossin, J. P. and Jagger, T. H. (2008). The increasing intensity of the strongest tropical cyclone. *Nature*, **455**, 92–95.

Emanuel, K. (2005). Increasing destructiveness of tropical cyclones over the past 30 years. *Nature*, **436**, 686–688.

Emanuel, K., Sundararajan, R. and Williams, J. (2008). Hurricanes and global warming: results from downscaling IPCC AR4 simulations. *Bulletin of the American Meteorological Society*, **89**, 347–367.

FAO (2007). *State of the World's Forests Report 2007*. Rome: Food and Agriculture Organization of the United Nations.

Finlayson, P., Montgomery, D. R. and Hallet, B. (2002). Spatial coincidence of rapid inferred erosion with young metamorphic massifs in the Himalayas. *Geology*, **30**, 219–222.

Galy, V. *et al.* (2007). Efficient organic carbon burial in the Bengal Fan sustained by the Himalayan erosional system. *Nature*, **450**, 407–410.

Held, I. M. and Soden, B. J. (2006). Robust responses of the hydrological cycle to global warming. *Journal of Climate*, **19**, 5686–5699.

Hilton, R. G. *et al.* (2008). Tropical-cyclone-driven erosion of the terrestrial biosphere from mountains. *Nature Geoscience*, **1**, 759–762.

Holland, G. J. (1997). The maximum potential intensity of tropical cyclones. *Journal of Atmospheric Science*, **54**, 2519–2541.

Holland, G. J. and Webster, P. J. (2007). Heightened tropical cyclone activity in the North Atlantic: natural variability or climate trend? *Philosophical Transactions of the Royal Society of London A*, doi:10.1008/rsta.2007.2083.

Holling, C. S. (2001) Understanding the complexity of economic, ecologic and social systems. *Ecosystems*, **4**, 390–405.

Hovius, N. (1998). Controls on sediment supply by large rivers. In K. W. Shanley and P. J. McCabe, eds., *Relative Role of Eustasy, Climate and Tectonics in Continental Rocks*. Tucson: Society of Economic Paleontologists and Mineralogists, pp. 3–16.

Huntingford, C. and Lowe, J. (2007). 'Overshoot' scenarios and climate change. *Science*, **316**, 829.

Huppert, H. E. and Sparks, R. S. J. (2006). Extreme natural hazards: population growth, globalisation and environmental change. *Philosophical Transactions of the Royal Society of London A*, **364**, 1875–1888.

IISD Reporting Services (2006). Summary of the 25th session of the Intergovernmental Panel on Climate Change. *Earth Negotiations Bulletin* **12**. Available at www.iisd.ca/meetings/2006.htm (accessed 20 October 2008).

Independent Levee Investigation Team Final Report (2006). *Investigation of the Performance of the New Orleans Flood Protection Systems in Hurricane Katrina on August 29, 2005*. Available at www.ce.berkeley.edu/~new_orleans/

IPCC (2007a). *Climate Change 2007: The Physical Science Basis. Contribution of Working Group I to the Fourth Assessment Report of the Intergovernmental Panel on Climate Change*. Solomon, S. *et al.*, eds. Cambridge: Cambridge University Press.

IPCC (2007b). *Climate Change 2007: Impacts, Adaptation and Vulnerability. Contribution of Working Group II to the Fourth Assessment Report of the Intergovernmental Panel on Climate Change*. Parry, M. L. *et al.*, eds. Cambridge: Cambridge University Press.

IPCC (2007c). *Climate Change 2007: Mitigation of Climate Change. Contribution of Working Group III to the Fourth Assessment Report of the Intergovernmental Panel on Climate Change*. Metz, B. *et al.*, Cambridge: Cambridge University Press.

Kirkby, M. J. (1971). Hillslope process–response models based on the continuity equation. *Institute of British Geographers Special Publication*, **3**, 15–30.

Knowlton, N., Lang, J. C. and Keller, B. D. (1990). Case study of natural population collapse: post-hurricane predation on Jamaican staghorn corals. *Smithsonian Contributions to the Marine Sciences*, **31**, 1–25.

Knutson, T. R. *et al.* (2008). Simulated reduction in Atlantic hurricane frequency under twenty-first-century warming conditions. *Nature Geoscience*, **1**, 359–364.

Lambert, F. H. *et al.* (2008). How much will precipitation increase with global warming? *EOS, Transactions of the American Geophysical Union*, **89**, 193–194.

Lambin, E. F. and Geist, H. J., eds. (2006). *Land-Use and Land-Cover Change: Local Processes and Global Impacts*. Berlin: Springer-Verlag.

Landsea, C. W. (2005). Hurricanes and global warming. *Nature*, **438**, E11–E13.

Landsea, C. W. (2007). Counting Atlantic tropical cyclones back to 1900. *EOS, Transactions of the American Geophysical Union*, **88**, 197, 202.

Liverman, D. M. and Roman Cuesta, R. M. (2008). Human interactions with the Earth system: people and pixels revisited. *Earth Surface Processes and Landforms*, **33**, 1458–1471.

Lowe, J. A. and Gregory, J. M. (2005). The effects of climate change on storm surges around the United Kingdom. *Philosophical Transactions of the Royal Society of London A*, **363**, 1313–1328.

Mann, M. E. and Emanuel, K. A. (2006). Atlantic hurricane trends linked to climate change. *EOS, Transactions of the American Geophysical Union*, **87**, 233, 238, 241.

Mann, M. E., Sabbatelli, T. A. and Neu, U. (2007). Evidence for a modest undercount bias in early historical Atlantic tropical cyclone counts. *Geophysical Research Letters*, **34**, L22707.

McCallum, E. and Heming, J. (2006). Hurricane Katrina: an environmental perspective. *Philosophical Transactions of the Royal Society of London A*, **364**, 2099–2115.

McCauley, D. J. (2006). Selling out on nature. *Nature*, **443**, 27–28.

Millennium Ecosystem Assessment (2005). *Ecosystems and Human Well-Being*, vol. 2, *Scenarios*. Washington, DC: Island Press.

Milliman, J. D. and Syvitski, J. P. M. (1992). Geomorphic/tectonic control of sediment discharge to the ocean: the importance of small mountainous streams. *Journal of Geology*, **100**, 525–544.

Mitchell, J. F. B. *et al.* (1999). Towards the construction of climate change scenarios. *Climatic Change*, **41**, 547–581.

Möller, I. and Spencer, T. (2002). Wave dissipation over macro-tidal saltmarshes: effects of marsh edge typology and vegetation change. *Journal of Coastal Research, Special Issue*, **36**, 506–521.

Montgomery, D. R. (2007). *Dirt: The Erosion of Civilizations*. Berkeley: University of California Press.

Montgomery, D. R., Balco, G. and Willett, D. (2001). Climate, tectonics and morphology of the Andes. *Geology*, **27**, 579–582.

Mooney, C. (2007). *Storm World: Hurricanes, Politics and the Battle over Global Warming*. Orlando: Harcourt.

Moss, R. *et al.* (2008). *Towards New Scenarios for Analysis of Emissions, Climate Change, Impacts, and Response Strategies*. Geneva: Intergovernmental Panel on Climate Change.

Muir Wood, R. *et al.* (2005). Catastrophe loss modelling of storm-surge flood risk in eastern England. *Philosophical Transactions of the Royal Society of London A*, **363**, 1407–1422.

Mulder, T. and Syvitski, J. P. M. (1995). Turbidity currents generated at river mouths during exceptional discharges to the world's oceans. *Journal of Geology*, **103**, 285–299.

Murphy, J. M. *et al.* (2004). Quantification of modelling uncertainties in a large ensemble of climate change simulations. *Nature*, **430**, 768–772.

Nicholls, R. J. and Wilson, T. (2001). Integrated impacts on coastal areas and river flooding. In I. P. Holman and P. J. Loveland, eds., *REGIS: Regional Climate Change Impact Response Studies in East Anglia and North West England*. Oxford: UKCIP, pp. 54–101.

Nicholls, R. J. *et al.* (2008). Climate change and coastal vulnerability assessment: scenarios for integrated assessment. *Sustainability Science*, **3**, 89–102.

Nyberg, J. *et al.* (2007). Low Atlantic hurricane activity in the 1970s and 1980s compared to the past 270 years. *Nature*, **447**, 698–701.

Office of the Deputy Prime Minister (2003). *Sustainable Communities: Building for the Future*. Available at www.odpm.gov.uk/stellent/groups/odpm_communities/documents/sectionhomepage/odpm_communities_page.hcsp.

Osmond, B. *et al.* (2004). Changing the way we think about global change research: scaling up in experimental ecosystem science. *Global Change Biology*, **10**, 393–407.

Perkins, R. D. and Enos, P. (1968) Hurricane Betsy in the Florida–Bahama area: geologic effects and comparison with Hurricane Donna. *Journal of Geology*, **76**, 710–717.

Perry, C. T., Spencer, T. and Kench, P. K. (2008) Carbonate budgets and reef production states: a geomorphic perspective on the ecological phase-shift concept. *Coral Reefs*, **27**, 853–866.

Pielke, R. A. Jr, and Sarewitz, D. (2005). Bringing society back into the climate debate. *Population and Environment*, **26**, 255–268.

Pielke, R. A. Jr *et al.* (2008). Normalised hurricane damage in the United States: 1900–2005. *Natural Hazards Review*, **9**, 29–42.

Raes, F. and Swart, R. (2007). Climate assessment: what's next? *Science*, **318**, 1386.

Reed, D. J. *et al.* (2008). Site-specific scenarios for wetlands accretion as sea level rises in the mid-Atlantic region. In J. G. Titus and E. M. Strange, eds., *Background Documents Supporting Climate Change Science Program Synthesis and Assessment Product 4.1*, EPA Report No. 430R07004. Washington, DC: US Environmental Protection Agency, pp. 133–174.

Richards, K. S. and Clifford, N. J. (2008). Science, systems and geomorphologies: Why LESS may be more. *Earth Surface Processes and Landforms*, **33**, 1323–1340.

Ruosteenoja, K., Tuomenvirta, H. and Jylhäs, K. (2007). GCM-based regional temperature and precipitation change estimates for Europe under four SRES scenarios applying a super-ensemble pattern-scaling method. *Climatic Change*, **80**, 193–208.

Sala, O. E. *et al.* (2000). Global biodiversity scenarios for the year 2100. *Science*, **287**, 1770–1774.

Saunders, M. A. and Lea, A. S. (2008). Large contribution of sea surface warming to recent increase in hurricane activity. *Nature*, **451**, 557–560.

Schlünz, B. and Schneider, R. R. (2000). Transport of terrestrial organic carbon to the oceans by rivers: re-estimating flux and burial rates. *International Journal of Earth Sciences*, **88**, 599–606.

Schumm, S. A. (1985). Explanation and extrapolation in geomorphology: seven reasons for geologic uncertainty. *Transactions of the Japanese Geomorphological Union*, **6**, 1–18.

Select Committee on Agriculture (1998). *Sixth Report: Flood and Coastal Defence*. London: HMSO.

Smith, B. J., Warke, P. A. and Whalley, W. B. (2002). Landscape development, collective amnesia and the need for integration in geomorphological research. *Area*, **33**, 409–418.

Stainforth, D. A. *et al.* (2005). Uncertainty in predictions of the climate response to rising levels of greenhouse gases. *Nature*, **433**, 403–406.

Stallard, R. F. (1998). Terrestrial sedimentation and the carbon cycle: coupling weathering and erosion to carbon burial. *Global Biogeochemical Cycles*, **12**, 231–257.

Summerfield, M. A. and Hulton, N. J. (1994). Natural controls of fluvial denudation rates in world drainage basins. *Journal of Geophysical Research*, **99**(B7), 13 871–13 883.

Technical Group Envisioning the Future of the Gulf Coast Conference (2006). *Envisioning the Future of the Gulf Coast: Final Report and Findings*. Available at www.futureofthegulfcoast.org/files/finalreport.pdf.

Tedesco, L. P. *et al.* (1995). Impact of Hurricane Andrew on South Florida's sandy coastline. *Journal of Coastal Research, Special Issue*, **18**, 59–82.

Thomas, D. S. G. and Wiggs, G. F. S. (2008). Aeolian responses to global change: challenges of scale, process and temporal integration. *Earth Surface Processes and Landforms*, **33**, 1396–1418.

Trumbore, S. E. and Czimczik, C. I. (2008). An uncertain future for soil carbon. *Science*, **321**, 1455–1456.

Tucker, G. E. *et al.* (2001). The Channel-Hillslope Integrated Landscape Development (CHILD) model. In R. S. Harmon and W. D. Doe, eds., *Landscape Erosion and Sedimentation Modeling*. Norwell: Kluwer, pp. 349–388.

Twilley, R. R., Chen, R. H. and Hargis, T. (1992). Carbon sinks in mangroves and their implications to carbon budget of tropical coastal ecosystems. *Water, Air and Soil Pollution* **64**: 265–288.

UNPD (2008). *World Urbanization Prospects: The 2007 Revision Population Database*. Available at http://esa.un.org/unup/.

UN-HABITAT (2008). *State of the World's Cities 2008/2009: Harmonious Cities*. Nairobi: UN-HABITAT.

Viles, H. A. and Goudie, A. S. (2003). Interannual, decadal and multidecadal scale climatic variability and geomorphology. *Earth Science Reviews*, **61**, 105–131.

Viles, H. A. *et al.* (2008). Biogeomorphological disturbance regimes: progress in linking ecological and geomorphological systems. *Earth Surface Processes and Landforms*, **33**, 1419–1435.

Webster, P. J. *et al.* (2005). Changes in tropical cyclone number, duration and intensity in a warming environment. *Science*, **309**, 1844–1846.

Wentz, F. J. *et al.* (2007). How much more rain will global warming bring? *Science*, **317**, 233–235.

Widdows, J. *et al.* (2004). Role of physical and biological processes in sediment dynamics of a tidal flat in Westerschelde Estuary, SW Netherlands. *Marine Ecology Progress Series*, **274**, 41–56.

Index

Thomas Spencer

Page numbers in italics refer to figures and tables, in bold to plates

abstraction of water 11, 98, 413
accelerated erosion 87–89
  in mountains 37, 52
  of Arctic coastal regions 358
  of Nile coastline following dam construction 119
  of soil in tropical savannas 257, 258, 261
accommodation space 15, 164, *168*, 197, 412
acidification 19, 89, 94, 407
  lake experiments 80
  post-industrial, of lakes 83, 91, 92, 94
adaptive capacity 7, 414
  in mountains 61–62
  options for coastal areas 175, 176
adaptive systems 27–28, 419
aeolian systems in deserts 286–291
afforestation 13, 56, 61, 92, 102, 123, 308, 322
  in mountains 46, *47*
agriculture 10, 20, 21, 25, 92
  agricultural drainage impacts on runoff production 101
  and runoff 100–101
  as greatest force of land transformation 413
  canals for agricultural irrigation 108
  development in the Mediterranean and land degradation 309
  erosion rates under pastoral agriculture in Africa 231
  in mountain biomes 37, 46, 56
  intensification in the coastal zone 150
  paddy agriculture 94
  urban agriculture 91
alluvial fans 42, 91, 280, 283, 285, 287, 313, 412
Antarctica *369*, *370*, **15**
  Hart Glacier *379*, *382*
  McMurdo Dry Valleys *382*
  present subglacial topography *371*, **15**
  pre-glacial fluvial environment *372*
Antarctic Ice Sheet 8, 10, 16, 17, 368, 369, 382, 383, 386
  East Antarctic Ice Sheet 16, 370–372, 383, 398
  Equilibrium line altitude 378
  Filchner–Ronne Ice Shelf 372, *380*
  Ross Ice Shelf 372, 377, 378, 383, 388, 391, 392, 397
  subglacial lakes 381
  West Antarctic Ice Sheet 16, 372–373, 397, 409
Antarctic Peninsula 2, 16, 373, 378, 382, 391, 392
Anthropocene 8, 10, 89
Aral Sea 9, 71, 72, 86–87, *88*, 285, 286, *286*, 287
Arctic as climate change hotspot *52*, 344–346
Arctic ice caps *393*
arroyos and the 'arroyo problem' 279, 285
Atlantic Reef Province 15
Atlantic Thermohaline Circulation, role of Greenland Ice Sheet meltwater 396, 398
atmospheric carbon dioxide increasing concentration 1, 11
  and changes in Mediterranean vegetation 313
  and rainforest tree growth and forest extent 237
  and water-use efficiency of plants 12, 240

bajadas **16**
bauxite 258, 310
baydzherakh 352
biodiversity 22–23, 37, 58, 94, 107, 405, 407, 416
  Convention on Biodiversity 31
  'hotspots' 418
  losses 418
biogeomorphological models 406, *407*
biomes 2, 7, 8, 20, 26, 28, 407, 408, 409, **1**
Black Sea 151
boundary layer meteorology and presence of a tree canopy 322
Bruun Rule 149, 165–166

C3 photosynthetic pathway 12, 140, 217, 248, 252, 262, 264, 407, 419
C4 photosynthetic pathway 99, 140, 217, 248, 251, 254, 264, 281, 407
calcrete 258
campos cerrados 251, 264
canals 108–109
  Jonglei Canal, Sudan 120
carbon cycle 355, 406, 407, 419
  and carbon sequestration 406
  and carbon sink status of the Amazon rainforest 238
  and chemical weathering as a carbon sink 257
  carbon input to Arctic Ocean 355, *356*
  on coral reefs 180

transfer of organic carbon to continental shelves 413
turnover in Arctic soils 354
carbonate production in coral reef environments 182–188
and bioerosion 184, 187
and carbonate sediment producers 183
environmental controls on 183
framework-building corals 182
generation of coral rubble 184
secondary framework production of calcareous encrusters 184
secondary framework production by precipitation of cements 184
Caspian Sea 9, 71, 72, 79, *81*
catastrophe loss modelling 415
cellular automaton models 418
characteristic form and landform evolution 25, *25*, 418
in tropical savannas 263
chemical weathering in the ever-wet tropics 221, 256
clearcutting and landslide erosion rates 331
and lag time to maximum landsliding 331
clearfelling in contemporary Mediterranean landscapes 311
climate – land cover – landform linkages 406–407
climate modelling for tropical areas 243
climatic geomorphology 4–5, *4*
morphoclimatic/morphogenetic regions 4
utility and validity in the tropics 217
coastal classification 159–162, *160*, *161*, 176
advancing coasts 161
based on relative sea level trends 160
by Finkl 161
by Shepard 160, 168
by Valentin 161
Curray model of long-term coastal change 165
emergent coasts 160
geotectonic classification 160
retreating coasts 161
submergent coasts 160
typology of coastal forms 159
coastal evolution of the coast of the Netherlands 168–169, *173*, **7**
coastal geomorphic change, drivers and scales of change 15, 132–133, *132*, *163*
acceleration in rates of change 158
and acute erosion hazards 162
and chronic erosion hazards 162
barrier progradation **7**
coastal morphodynamics 132
coastal populations 130
rates on Arctic coasts 359, 361
risk-based prediction and adaptation 174–176
socioeconomic changes on coasts 158, *416*
coastal geomorphology, history of scientific thought 131
coastal marshes and tidal flats 136–142
carbon fixation 140
economic value assessment 141, *142*
future sea level rise 411
geomorphic settings 136
infilling of the tidal frame *138*
loss from 'coastal squeeze' 141, *141*
Mississippi Deltaic Plain 140
process regime 136
rate of marsh formation 138
response to increased salinity 140
sediment stability in the intertidal zone *139*
storm surge impacts 139
surface elevation changes 140
coastal sand dune systems 142–150, *144*
aeolian sand transport 143–144
and increases in global temperature 148–149
destabilisation of vegetated dune blowouts 146
dune–beach interaction 143–145, *143*, **6**
dunefields 146–147, *146*
embryo dunes 143
foredune system 142–145
increased frequency/intensity of storms 149
overwash fan 143
overwash terrace 145
parabolic dunes 146
plant species and dune morphology 148
plant species and dune stability 148
plant zonation on dunes 147–148
role of beach morphodynamics 144
sand supply 144, 147
transgressive dunefields 147, **6**
vegetation–dune interaction 147–148
warming in the Arctic 149
coastal tract cascade 19, 158, 159, 162–174, *162*, *167*, *170*
barrier rollover 172, *174*
overwash 172
parabolic dunes 172
shoreface slope *166*
shoreline fluctuations 163–164
simulation modelling and probability of shoreline recession *175*
the quantitative coastal tract 166–167
tidal inlets 173–174
transgressive dune sheets 172
coasts and soft cliffs 170–172, **7**
complex response of landscapes 26, 418
continental runoff, historical trends 12
coral bleaching 187
acclimation and adaptation 193
and local hydrodynamics 193
and 'time to extinction' models 193
'mass bleaching' 193
relation to ocean temperatures 193
susceptibility between species 193
coral reef cycling of calcium carbonate 184–185
carbonate budgets 184–186, *184*, 187
reef budgetary states 185, *185*
reef production status 186
the 'carbonate factory' 181, 199, 203
coral reef distribution *7*
coral reef growth – sea level relations 189–191, *190*
'catch-up' reefs 189, 198
contemporary growth and responses to near-future sea level rise 192
'give-up' reefs and drowned carbonate banks 189–190
'keep-up' reefs 189, 198
reef accretion in the Holocene 191, *191*

coral reef growth – sea level relations (cont.)
  standards of metabolic performance 192, *192*
  thicknesses of Indo-Pacific province Holocene reefs *189*
coral reefs and ocean acidification 194–195
coral reef landforms 180–182, *181*, *196*, **8**
  and eco-geomorphic units 181, *182*
coral reef sedimentary landforms 181, 195–202, *195*, *197*
  and reef ecological condition 208
  anthropogenic effects 202–204
  constructional and erosional impacts of cyclones and hurricanes 200
  evolution 197–199
  Holocene high energy window 198
  morphodynamics 199–200
  remobilisation of beaches in next century 205
  sea level change, reef growth and landform relations 197–198, *201*
  sediment supply 198–199, *200*
  Shoreface Translation Model 201
coral reefs and trajectories of response to global environmental change 205–208
  with increased sea surface temperatures 192–193
coral reefs in Discovery Bay, Jamaica 187–188, *188*
coral reefs, interaction between biological and physical processes 181
  coral diseases 203
  ecological decline 180
  limits to coral growth *183*
  overfishing 187
  'phase shift' dynamics 186, 187, *187*, 203
  storm impacts on reefs 193–194, 411
  wave energy and reef morphology 197
cosmogenic isotope analysis 384, 392, 398
Coweeta Experimental Forest North Carolina USA 322
Croll–Milankovitch orbital cycles 368, 386
cropland 20, 30, 406, 413
  abandonment of irrigated cropland 281
  and source areas of storm runoff 99, 407
  erosion rates for the contiguous United States *21*
  future increases in Africa 254
  in Austria 59
Cyclone Nargis 2

dambos 258–259, *258*, 270
dams *59*, 107–108, *115*
  and alteration of streamflow characteristics in desert regions 282
  Aswan Dams, River Nile 108, 118–120, *119*
  construction in mountains 47, **3**
  impacts on the Danube Delta 151
  impacts on the Ebro Delta 170
  in the Indus basin 117–118, *117*
  on the Colorado River 115–117, *116*
  on the Mekong River 120
  on the River Amazon 237
  on the River Yangtze 114
  reduced sediment flux to coast *48*, 176, 315
Dead Sea 71, 79, 81, 285
debris avalanches 17, 53, 57
debris dams on rainforest rivers 224, 228, 230
debris flows 24, *24*, 26, 28–29, 41, 53, 57, *58*, 322
  and decreased rooting strength in the soil mantle 328
debris slides 53
deep weathering in the tropics 218, 221, 223, 225, 250, 256–257, *256*
deforestation 13, 20, 21, 31, 37, 56, 63, 90, 92, 237, 307, 322
  and forest hydrology 102
  and tropical hydrology 103
  in the Amazonian rainforest 239
degradation of arid ecosystems *282*
deltas 2, 76, 105
  Danube Delta 151–152, *151*, **6**
  Ebro Delta 168, 169–170, *171*, *172*
  in the mountain biome 45
  Irrawaddy Delta 2
  loss in the Mediterranean as a result of sea level rise 315
  Mississippi Delta 2
  Po Delta 168, 169
  Sacramento–San Joaquin Delta 125, 135, 136
  sedimentation in lake deltas *77*, 79
dendrochronology, in mountains 39
denudation rates estimated for the contiguous United States *18*
desert climates 278
  and climate change scenarios *277*
  defined by the aridity index 277
desert rivers 283–285
  aggradation and incision of channels 283
  changing water balances 283
  nature of flood events 283
  response to climate change *279*, **10**
desertification 21, 63, 281
  in China 281
  with climate change in Mediterranean landscapes 313, 315, 317
deserts 276, *277*
  difficulty of predicting near-future environmental change 276
  increasing populations, especially urban 281, **10**
  landform types *278*
  process-based models of biophysical systems 291
  rainfall 279
  regional, geologic and tectonic environment 278
  surface disturbance by offroad vehicles 282
  surface subsidence following excessive groundwater withdrawal 281
Digital Elevation Models (DEMs) 18, 65, 336, 351
disturbance regimes 6, 53, 61, 64, 65
dry ravel 322, 326, *327*
duricrusts 257–258, 263
dust storms 287–288, *288*, **10**
  dune activity and dust deposition in adjacent oceans *281*
  dust generation and condition of surface sediments 287
  dust production and human impacts on land cover 287
  emissions from deserts correlated with ENSO cycles 279
  future changes in dust emissions 288
  inter-annual frequency and magnitude 287
  major dust source areas 285, 287
  pans as sources of global dust 260

Saharan dust plume **10**
threshold wind velocity required for entrainment and emission 287

earthflows 17, 325
earthquakes 2, 6, 57, 90, 325, 335 (see also seismic hazards)
quake lakes, Szechwan 2, 57, 62
ecological footprint 25
ecosystem and ecological services 30, 405, 419
and coastal wetlands 130, 131
and coral reefs 180, 208
emergent properties of geomorphological systems 6
El Niño Southern Oscillation (ENSO) dynamics 11, 14, 16, 410, 420
and coral bleaching 186, 193
and future climate change Mediterranean landscapes 313
and global climate shift in 1976–77 232
and groundwater recharge in deserts 280
and rainfall variability in deserts 280
and shoreline fluctuations 163, 200
and tropical savanna dynamics 251
ENSO droughts and rainforest dynamics 238
ENSO events and increased incidence of extreme precipitation events 326
landscape responsiveness in relation to ENSO 264
signal in lakes 85, 86
environmental hazards and storminess on the UK east coast 415
environmental hazards and storminess on US coasts 414–415
environmental refugees from coral atolls 180
estuaries 133–136, *133*
changes in runoff regime 135
classification 133
climate forcing 134–135
estuarine processes *134*, 412
future salinities 135
increased temperatures 135
interaction between fresh and saline waters 133
salinity penetration under climate change 136
sediment cascade *134*, 135
Ethiopian Highlands 62–64, 407, **4**
environmental rehabilitation in Tigray 63
improved landscapes **4**
land degradation 63, *64*
Eurasian ice sheet 9, 374
eutrophication 19, 79, 91, 93, 94, 407, 417
artificial eutrophication 90
in lakes 79, 89, 90–91, 94
incipient eutrophication 90
industrial eutrophication 90
evapotranspiration 11, 12, 98, 237, 303, 409, 411–412
demands of vegetation cover and deep-seated landsliding 325

ferricrete 258
fire dynamics and landscape change 406
and land degradation in the Mediterranean *305*, 307, 316
and origin and evolution of tropical savannas 251
in discontinuous permafrost zone 353
increased hazard from higher temperatures 330
increased risk in Mediterranean landscapes 308
increased threat to rainforest 237
role of aboriginal hunter–gatherers in Australia 252, 308
floodplains 105, 106–107
forest clearance and hydrological change 321
forest conversion to pasture and accelerated landslide and gully erosion 328–330
forest harvesting and increased likelihood of landslide initiation 330
forest species composition and climate change 337
forestry 20
forest management and runoff 102–103
in mountains 46, 56
management in Mediterranean 315–317
frequency and intensity of extreme weather events 325, 326, 410
frequency and magnitude of geomorphic events 26
frozen ground 49–50
seasonal variations 54
fynbos 307

General Circulation Models (GCMs) 13, 44, 409, 410
geoconservation 23
geodiversity and geomorphology 3, 22–23, 418
in mountains 37, 41
potential losses in lake catchments and wetlands 94
geoecological monitoring of periglacial landforms 347
geoindicators 23
palsas in subarctic regions 352–353
periglacial features 347
permafrost 344
geomorphic hotspots 7, 23, 270
geomorphic services in coastal systems 153
coastal sand dune systems 150
coastal wetlands 130, 133
on coral reefs and modification with landform change 208
geomorphic thresholds 7, 26, 418
and arroyo cutting 285
and landform and ecosystem dynamics in deserts 279
annual rainfall threshold for active dunes 265
climatic thresholds to thermokarst development 349
for sand movement and dune reactivation 263
to gully formation in tropical savannas 261
to landsliding 328
vegetation cover threshold for dune mobility 265
geomorphological changes with climate change rainforest 240–242
geomorphological inheritance *6*
in savanna landscapes 255–256
geomorphological services 419
in permafrost regions 362–363
geomorphological thresholds in rainforests with climate change 244
geomorphology scale linkage problem 405
glacial–interglacial cycles 6, *6*, 8
glacial erosion and deposition landscapes 384–389
arc of till deposition 386
areal scouring 384

glacial erosion and deposition landscapes (cont.)
corries 385
drumlin fields 385
eskers 385
troughs eroded by ice 384, 386
glacier extent in the tropics *52*
glacier flow mechanisms 380–381
glacier ice cores 41
glacier lake outburst floods 53, 57, 61, 62, 65, 405
glacier surging 2, 57
glacier–runoff–sediment transport relations 51
glaciers and ice caps 51
historical shrinkage of tropical glaciers 51
mass balance changes in last fifty years 51
reductions in glacier length since Little Ice Age 51
thinning, mass loss and retreat of mountain glaciers 51
total mass balance and contribution to sea level rise *51*, *52*
upward shift in equilibrium line with climate change 51
Global Environmental Outlook (GEO) scenarios 29
global mean precipitation rates and trends 12, 409
and rainforest tree growth 237, 238
and twenty-first-century regional variations 325
in Mediterranean landscapes 312–313
global mean surface temperature trends 11
global population growth 19, 20, 413, 414
global sediment flux 19, 21
global sediment yield and glacier meltwaters 51
Grand Canyon *23*, 25
grazing 100, 413
and destruction of biological soil crusts in deserts 282
and land drainage in peatlands 101–102
in the mountain biome 46
Great Salt Lake 79, 81
greenhouse gas emissions 1
methane emissions in northern mires and peatlands 355
Greenland 8
Greenland ice cores 379, 382, 394
Greenland Ice Sheet 16, 17, 368, *373*, 382, 383, *393*, 409
equilibrium line altitude 378
Greenland Ice Sheet shallow surface meltwater lakes 394, *394*, 398
ice sheet model 395
ice stream of Jacobshavn Isbrae *379*, *394*
modelling responses to warming with GLIMMER ice sheet model *395*
grounding line dynamics during Holocene, Ross Sea embayment *392*
gullies as long-term chronic sources of sediment to streams 329
gullies in the seasonal tropics 261
gypcrete 258

Heinrich events 10, 382
hillslope hydrological cycle 98–100, *100*
partial source area model with infiltration-excess overland flow 99
theory of runoff production from Horton and Hursh 99
variable source area model with saturation-excess overland flow 99–100
Holocene Epoch 8, 10, 14, 38
African humid period 280, 281
and rainforest climates 215
dune activity episodes 289
glacier fluctuations 39
in lakes 78, 81, 83
Holocene Optimum 9, 10, **1**
in the Arctic 361
in Antarctica 389
Huanghe (Yellow) River 19, 121
South-to-North Water Transfer Project 108, 121
human activity and climate change 19–21, 413–415
human activity, population and land use in mountains 45–49
human footprint on Earth **2**
human impacts in Mediterranean landscapes 307–310
Humboldt, Alexander von 1
hurricane activity 15, *415*
Great Miami Hurricane of 1926 414
Hurricane Allen 187, 411
Hurricane Andrew 140, 411, 414
Hurricane Bebe 200
Hurricane Betsy 411, 415
Hurricane Cleo 414
Hurricane Donna 411
Hurricane Hattie 200
Hurricane Hugo 140, 224
Hurricane Ike 414, **16**
Hurricane Katrina 2, 115, 140, 410, 414, 416
Hurricane King 414
Hurricane Rita 140
hydroclimate and runoff as systemic drivers of environmental change 10–14
hydro-isostatic adjustments 14
hydrological cycle 11, 40–41, 71, 409
in mountains 44
and rivers 98
water balance studies in Lancashire UK 322
hydrological responses to climate change in rainforest 240
hypoxic zones in the Gulf of Mexico 21, *22*

ice age cycles 368
'ice complex' deposits northeast Siberia 351, 359, *360*
ice core records 8, *8*, 379
ice sheet and ice cap landscapes 374–378
Antarctic outlet glaciers 377, 383, 391
Greenland outlet glaciers 377
ice dome morphology 374
ice shelves 374, 377, 380, 392
ice streams 374, 377, *378*, *385*, 391, 392
ice sheet and ice cap responses to global warming 389–399
Antarctic Ice Sheet 389–393
Antarctic Peninsula 389
East Antarctic Ice Sheet 390–391
Greenland Ice Sheet 393–396
ice caps in Arctic islands 396
Patagonian ice caps 396–397
Patagonian Ice Sheet *15*, 374
Patagonian outlet glaciers 397, *397*

West Antarctic Ice Sheet 391–393
ice sheet models 398
for Antarctica 386, *387*
ice sheets and ice caps distribution 369–374
ice sheets and ice caps mass balance 378–380
accumulation on the Antarctic Ice Sheet *377*
accumulation on the Greenland Ice Sheet *381*
balance velocities on the Antarctic Ice Sheet *377*
for Antarctica 379
for the Greenland Ice Sheet 379
ice surface elevation monitoring by radar from satellites 369, *390*, 391, 392
ice-wedge polygons 352
Indian Ocean tsunami on reef islands of the Maldives 202
Indo-Pacific reef province 14, 205
inselbergs *132*, 226, 256, 263, 270
IPCC storylines and scenario family 24, 28–29, *29*, 62, 235, 403

jökulhlaups 57, 405

Kalahari regions *266*
barchan dunes 266
blowouts 266
climate change, land use and land cover dynamics 269
dunefields *267*
GCM simulations of future dune activity 268, *268*
impacts of climate change on land use management 269
linear/longitudinal dunes 267, *267*, 268
longitudinal dune reactivation with climate change 270, **9**
marked increases in modelled twenty-first century dune activity 268
optically stimulated thermoluminescence (OSL) dating of dunes 267
parabolic dunes 266
radiocarbon dating of dunes 267
sand dune activity in the Holocene 267
seif dunes in the Kalahari 267
sensitivity of landscapes 262, 267
transverse dune ridges 267

lake catchment systems 74–75, *76*, 93
catchment erosion factor *89*
Bussjösjön, southern Sweden 89, *89*
Havgårdssjön, southern Sweden 89
Lake Baikal 72, 75, 82, *82*, 90
Lake Biwa 82
Lake Chad 79, 86–87, 285
Lake Eyre 79, 81, 285, 286, *286*, 291
Lake Patzcuaro, Mexico 89
Lake Taihu, China 91
Lake Tutira, North Island, New Zealand 90
Lake Waikopiro, North Island, New Zealand 90
land use and erosion rates in lake catchments *92*
Schwarzsee, Swiss Alps 90
Seebergsee, Swiss Alps 90
lake geochemistry 83
lake hydrology 71
and surface warming since the 1960s 92
lake internal processes 78–80
biological activity in lakes 79
chemical activity in lakes 79
density contrasts and lake sedimentation processes 79
physical mixing in lakes 78–79
sedimentation processes in lakes 79–80
lake level fluctuations 81, 92
lake sediments 8, *75*, *77*, *80*, *84*, *86*, *87*
and records of climatic variability 85–86
and role of basin area 77
coupling of time and space scales in sedimentation 78
diatoms in lake sediments 83
experimental model of sedimentation 84
hydroclimatic changes from lake sediments 85–86
in mountains 41, 44, 56, *73*
land clearance and lake sedimentation 89–90, 328
particle size and stratigraphy in lakes 81–83, *82*
pastoral land use and lake sedimentation 90
pollen in African lake cores 264
proxy data in lake sediments 81–83, *81*, *85*
sediment laminations and rhythmites in lakes 83
sedimentation and seasonal rainfall model for lakes 84–85, *85*
lake types 72–74, *72*, *73*, 76, *78*, *80*, 94
land cover change 19, 21, 25, 30–31, 55
uncertainty of change under climate change in tropical savannas 254
land degradation
following forest clearance 321
in Mediterranean landscapes 310, 312
land ethic 25
Land Use and Land Cover Change (LUCC) project 30, 414
land use change 19, 21, 30–31, 55, 62
and soil carbon balance 418
and surface runoff *100*
changing ideas on understanding of land use change in savannas *255*
prediction of future change in tropical savannas 254
land use in mountains 46–48, *60*
landscape as an intermediate-scale region 1
landscape change 8–9, 25–26, *408*
landscape inheritance 248, 405
landscapes of transition 26–27, 53, 61, 418
landslides 26, 41, 53, 112, 223, 243, 321, 411
and enhanced slopewash on landslide scars 411
and root strength deterioration following tree removal 328, 330, *331*
and sediment delivery to channels 326, *327*, 329, 411, **12**
climate induced v. anthropogenically induced 334
control by soil moisture 325, 326, 332
depth of failure and rainstorm characteristics *241*
distributed, physically based models 335–336
frequency in clearcut v. natural forest 331–332, *331*
GIS-based landslide hazard analysis 334
multivariate approaches to landslide hazard analysis 335
neural network methods for landslide hazard analysis 335
role of roads, mining and residential development 47, 328, *333*, 334, **12**
snowmelt-triggered 325, *327*

Last Glacial Maximum (LGM) 9, 15, *376*, **1**
and evolution of tropical savannas 252
and extent of global deserts 280
and low lake levels in Africa 264
and rainforest climates 215
and reduced aeolian accumulation in the Kalahari 268
and the East Antarctic Ice Sheet 391
and the Greenland Ice sheet 393
and the Patagonian ice caps 397
and trade wind intensity and persistence 280
and vegetation belt lowering in Africa 264
in Australia 265
in Madagascar 264
tropical ocean temperatures *9*
laterite 258
Laurentide Ice Sheet 9, 373, 374, 382, 383
Little Ice Age 39, 83
and Mediterranean rainfall variability 307
in British Columbia 53
loess 2, 9, 56

Medieval Warm Period 39
Mediterranean agro-sylvo-pastoral system 309–310, 311
dehesa 308, 309, 311, **11**
montado 308, 309
Mediterranean climate characteristics 297, *298*, 299
Malda Anomaly in climate records 306
'*pro pluvia*' rogation ceremonies in Mediterranean historical archives 307
rainfall intensities 299
Mediterranean forest fires 303, **11**
intensified runoff 303
sediment yields 303
Mediterranean geology and soils 297–299, *298*
soil salinisation **11**
terra rossa 299
Mediterranean hydrology and fluvial geomorphology 300–301
and forest management 308
fiumara 300, 301
floods 300–301, *301*, 306, 314
impact of mass urbanisation 315
increased runoff with vegetation removal in California 314
jumara 300
Quaternary terrace levels 305
rambla 300, 314, **11**
reductions in water availability with climate change 313
Mediterranean land subsidence 302
Mediterranean vegetation *298*, 302–303
Quaternary vegetation change 305
Mediterranean sea level change 306, 314–315
Mediterranean slope processes 301–302
gullying 301
mass movements 301, 302
Quaternary periglacial activity 305
sheet wash 301
Mediterranean wind erosion 302
mid-latitude temperate forests and rangelands 321, 322, 323–325, *324*, **12**
Millennium Ecosystem Assessment 7, 25, 30, 406, 414
miombo woodland 248, 251
mitigation and adaptation to climate change by inhabitants of arid regions 291
modelling ice sheet and ice cap histories 398
models of ice sheet mass balance and ice flow 408
moraine dam failure 53
morphodynamics 5
feedbacks and landscape change 406
of coral reef structures 181
Moruya Beach, SE Australia 163
mountain ecological and geoecological zonation 40–41, *40*, *41*
mountain geomorphic process zones 41
mountain Holocene climate change 38–40, *42*
glacier advances and historical records 39, *39*
ice cores 39
lake sediments 39
palaeoecology 39
mountain hydroclimate and runoff 44–45
Mountain Protected Areas (MPAs) 48
mountain regions 4, *55*, **3**
population growth 37, 56, 407
of Austria 58–61, **4**
of British Columbia, Canada 53–55, **4**
of Tajikistan 57, **4**
temperate regions with high population density 46
temperate regions with low population density 46
mountain roads and the destabilisation of hillsides 332–334
mountain sediment cascade 42–44, *43*
coarse debris system 43
cryosphere system 43
fine grained system 43–44
geochemical system 44
mountain types 38, *45*, *46*, *47*, 65
high 38
low 38
mid-elevation 38
polar 46
tropical 46
very high 38
Mt Fujiyama, Japan 24
Mt Kilimanjaro *52*, 61, **16**
multi-model projected climate changes **2**

Narrabeen Beach, SE Australia 163
natural hazards 23–24
North Atlantic Oscillation (NAO) 11, 13, 16
and future climate change Mediterranean landscapes 313
and Arctic Island ice cap dynamics 396
North Sea, storm surge of 1953 415

ocean acidification 180, 194
Oetztal ice man 39
outlet glaciers 383, 409

over-fishing on reefs 202
overgrazing 62, 87, 92
and land degradation in the Mediterranean 308
and reactivation of stabilised dunes 282
and triggering of thermokarst *348*, 360
overwash 201, 410

Pacific Decadal Oscillation and shoreline fluctuations 16, 163, 200
signal in lakes 86
paired catchment experiments in forest hydrology 321
on Plynlimon, Wales 13, 322
palsas 352, 355, **13**
cyclic development 352
decay within last fifty years 352
remote sensing and monitoring 352
Pan European Soil Erosion Risk Assessment (PESERA) **11**
panarchy metaphor 27, *27*, 64, 420
paraglacial landscapes 53, 405
and sedimentation in lakes 83
sediment pulses in river systems 111
sediment yields *53*
paramo 40
Patagonian ice caps *375*
periglacial activity 54, 305
lakes and ponds *74*, *362*, **5**
monitoring and remote sensing 347
permafrost 2, 9, 10, 49, 50, *50*, *52*, 54, 61, *345*, 408
and human settlements *346*
application of remote sensing to permafrost terrain and landforms 363
classification 353
coupled carbon/permafrost modelling 355, 406
definition 344
degradation 57, *354*
global distribution 344
ground ice as a component 350
hazard mapping **14**
settlement index 351, 357
subsea instability of methane hydrates 347
thaw impacts on infrastructure and vegetation communities 347, *357*
permafrost indicators of climate change 348–349, 352
accelerated coastal erosion *358*, *359*, *360*, *361*
increased depth of thawing 349
peatland changes *356*
reduction of near-surface permafrost extent 349
snow cover control of ground thermal regime 349
warming of permafrost in China 50
permafrost models 348–349, 350, 358
predictive modelling of behaviour 344, *348*, 360, **13**
physical-based models in geomorphology and hydrology 98
prediction of geomorphological effects of climate change 234
plant introductions into desert regions 282
plantation forests 329
and extensive shallow landsliding *329*
erosion rates under tropical plantations 231, **9**
plate tectonics 5
and climate change from plate mobility 251
control of mountain relief 42, **16**
playa lakes 79, 278, 279
Pleistocene Epoch 8
pollutant aerosols 11
process geomorphology 5
process–response systems 5, *5*, 279
punctuated equilibrium 25, 26

Ramsar Convention 93
reafforestation 56
and reduced sediment supply to channels 329, 333
'reduced complexity' models 418
regulation of river flows 413
decreased sediment load in the River Nile 131
reduced sediment flux to coast 176
reservoirs 19, 21, 47, 59, 60, 71, 91–92, 98, 232, 237
resilience 6
of coral islands 202
of reef systems 186, 205
of wet tropics to disturbance by climate change and human impacts 234
responsiveness of geomorphic elements in the landscape 181, 406
river and lake ice 49
river channel functioning 103–109
control by dykes 98
downstream hydraulic geometry 104, *104*, 109
form of river channels 103–106
upland v. trunk channels 105
river morphology and effects of human activities 106–109
diversions and canals 108–109
Mississippi River channel redesign 107
modification of channel form 107
modifying the flow regime 107–108
reinforcement of river banks 106–107, 313
river restoration 121–125
of the Colorado River 123
river sediment transport and sedimentation 109–114
aggrading systems 111
bed-material transport in rainforest zone 224
degrading systems 111
drainage network, channel type domains and sediment yields *106*, *111*
principles of sediment transport *110*
sediment cycle 109–112
sediment reservoirs and storage times *111*
sediment supply to channels and rock weathering 110
transit times of sediments within drainage basins *105*, 110, 135
water quality 109
within-channel sediment storage modes 110
river sedimentation and effects of human activities 112–114
and agricultural land use 112
and forest land use 112–113
and mining and quarrying 113
and urban land conversion 114

rock avalanches 53, 57, 61
rock glaciers 50
rockslides 17, 57
rockfalls 17, 53, 57, 61
runoff 12, 409, 413
glacier melt 51

salt and saline lakes 71, 278
sand dune systems 2, *287*, 288–291, *291*
atmospheric erosivity and dune activity 268
classification as active, dormant or relict 288, *289*, **10**
crescentic dunes 288
dune activity and climate *280*, 288, 289, *291*
dune mobility index 268, 290–291, *290*
dune surface erodibility 268
linear dunes 288
lunette dunes associated with pans 259, 260
sediment supply and dune activity 288–289
star dunes 288
transverse dunes 288
variations in wind energy and dune mobility 288
vegetated and stabilised in savannas 255
vegetation cover and dune dynamics 289
scaling in geomorphology 5, *6*, 160, 162, 409–410
sea ice recent shrinkage of extent in the Arctic 358
sea level *14*, 15–16, 411
and ice sheet ablation 382
and ice sheet growth and decay 382
sea level rise 14–17, *15*, *16*, *17*
and coastal sand dune systems 149–150
and erosion of sandy beaches 164
Holocene meltwater pulses 190
loss of reef islands 195
rapid rise from ice sheet and ice cap collapse 399
recent rates on arctic coasts 358
response of coastal marshes 131, 140–141
with climate change in the Mediterranean basin 315
sea surface temperatures (SSTs) and hurricane development 237
sediment availability 412
in deserts 289
on coasts 2, 164
sediment budgets 18–19, 75, 172, 202
sediment cascade 17–18, *18*, 412
in mountains 41
on coasts 131
sediment supply and landscape change 406, 412
to coral reefs 188, 197
to the coast 20, 169, 226
sediment yields 63, 75, 412, *412*, 413
from grazing land in East Africa 231
in rainforest catchments 223
model of Langbein and Schumm 263, 269
specific sediment yields in mountains 42
specific sediment yields in river systems 111
specific sediment yields to lakes 77
seismic hazards 57
self-organisation in landforms 181, 406
sensitivity of landscapes 6, 7, 28, 414
in tropical savannas 262–263
of arctic coasts to environmental variability and human impacts 347
of coasts 159
of coral reef sedimentary landforms to sea level change *203*, 205
of desert environments 276
of mountain ecosystems to changing climate 40, 61
of the Greenland Ice Sheet 395
of the West Antarctic Ice Sheet 391, 392
shifting agriculture in the humid tropics 226–228
silcrete 258
snow 49
artificial 56, 61, 62
northern hemisphere extent *52*
water equivalent 49
snow avalanches 49, 54, 57
snow gliding 60
snowline 40, *54*
snowmelt 49
and triggering of slow, deep-seated landslides 326, 327, 332
snowpack decline predictions 327
'societal governance' and geomorphology 415
soil carbon timescales of accumulation and loss 417
redistribution of soil carbon 355, 406
soil conservation in the American Midwest 112
soil creep 322
in the humid tropics 223
rates and forest logging 332
soil degradation 31
soil erosion 20, 56
by runoff 112
in the Mediterranean 301–302
Stern Review Report on the Economics of Climate Change 25
storm hydrograph changes with conversion of vegetation cover 323
streamflow 12–13
sustainability and geomorphology 24, 419
sustainability in mountain systems 65
sustainability of coastal geomorphic systems 151
Sustainable Urban Drainage Systems (SUDS) 102
Svalbard 2, 46, 64

taiga 4
terminal lake basins in deserts 285–286
termites, influence of global climate change on distribution and activity 257
and soil nutrient contents 251
bioturbation by termites 257
terraces and terracing 56
in mountains 48
in rainforest 231, 243
terrestrial wetlands 71, 92–93
of the Danube Delta 152
Quaternary pollen sequences in African wetlands 264
thermoerosion 351, 409
thermokarst 'thaw lake cycle' 353, 362, **13**
thermokarst 409
development from early Holocene 361, 362
gullying *352*

impacts on the boreal forest 353
subsidence 351
thermokarst socioeconomic impact and hazard implications 356–358
impacts on Russian cities 357
predictive hazard mapping 357, 362
thaw subsidence and human infrastructure 357
thermoterraces on eroding Arctic coasts 359
Thornthwaite–Mather water balance method 44
Three Gorges Dam and the Li-Jiang (Yangtze) River 23, 92, 108
Tibetan plateau 38, 65, 91
Tidal flats 137–140
and wetland restoration 125
biostabilisation 138–139, *139*
colonisation by emergent macrophytes 137–138
ice rafting 139
timber harvesting and slope stability 46, 308, 336
timber line 40, 54, *54*
in Austria 58
time and space scales in geomorphology *2*, *132*, 181, 208, 279, 405
and geomorphic system response to rapid change *263*
tipping points 2
and rainforest survival 238
topographic relief 17–19, 412–413
relief, tectonics and storminess 413
treeline 39, **16**
recent northward and upslope migration in circumboreal region 345
tropical cyclone magnitude, track and frequency changes 233–234, *234*
historical records of Atlantic tropical cyclones 410
hurricane frequency since 1995 in the North Atlantic 131
tropical rainforest biome and long-term history 215–217, *216*
early Holocene rainforest extent in Africa 216
Last Glacial Maximum rainforest extent in Africa 216
lowering of sea level and rainforest extent in Malesia 217
refugia in the Amazon Basin 217
tropical rainforest climates 214–215, *220*
boundary with humid subtropical forests 215
convectional rainfall 215
ENSO event changes and the magnitude–frequency of dry periods 232
high rainfall intensities and totals 218, 224
high water vapour capacity of tropical atmospheres 218
perhumidity index 214
predictions of changes in temperature and rainfall 235–237
rainstorm magnitude and frequency changes 232–233, *233*
recent trends in rainfall 231–232
recent upward shifts in mean temperature 231
tropical rainforest logging and timber removal 228–231, *229*
clear-felling 228
erosional impacts of selective logging 228
high lead logging 228
impacts enhanced or modified by climate change 242
Reduced Impact Logging (RIL) 230, 242
sediment supply to reef systems 202
skid trails 228
tractor logging 228
tropical rainforest replacement by savannas 239
tropical rainforest response to climate change *235*, 238
changes in drainage density 241
higher suspended sediment, bedload and solute yields 241
increased channel size 241
increased landsliding 241
increases in tropical cyclones and heavy rainstorms 237
rates of weathering and chemical denudation 240
response to higher temperatures 238
runoff increases with increased rainfall 240
slopewash rates 240
tropical rainforest soils and vegetation 215
edaphic variations and forest formations 215
soil groups 215
tropical rainforest zone, geomorphological characteristics 217–231
drainage densities in the humid tropics 225, *225*
tropical rainforest zone, geomorphological processes 219–224
suspended sediment transport 224
bed-material transport 224
channel cross-sectional area and land use *228*
slope and catchment erosion rates *227*
tropical rainforest zone, hydrological processes 220–221, *221*
chemical denudation and runoff *222*
evapotranspiration 220
hillslope runoff 221
Hortonian overland flow 221
interception 220
pipeflow in the seasonal tropics 261
pipeflow on tropical rainforest hillslopes 221, 222
river flow 220
saturation overland flow 221, 224
sediment yield, runoff and land use *223*
slopewash 221–222, *222*
throughflow 221
tropical savannas 249, *249*, *250*, 413, **16**
classification 250
climatic characteristics 249
colluvium 257
ecological basis for description 250
fire–climate feedbacks *254*
interrelationships between grazing and fire 252
manipulation of grazing and fire regime by people 252
nature and depth of weathering front 256
of southern Africa *251*
pans in arid tropical savanna 259–260, *260*, *261*
rangeland, future increases in Africa 254
role of obligate grazers, browsers and mixed feeders 252
role of termites 252
stone lines 257, 258
types controlled by precipitation amount and seasonality 250
uncertainty in the nature and rate of climate change 253–254
uncertainty of future fire climate and vegetation linkages 253
uncertainty of future land use management responses to climate change 254
wind erosion and deposition 262
tundra 4, 37, 54, 345, 355

uncertainty and geomorphological change 29, 409, *417*
additive uncertainities in the climate–ecosystem–landforms chain 416
and near-future landslides 330
annual rainfall and ENSO cycle intensity in rainforest areas 244
in the Mediterranean 306
in the response of Antarctic Ice Sheet to global warming 389
interaction of West Antarctic Ice Sheet with the ocean 398
landscape-scale response to climate change 411
of mass balance of Greenland Ice Sheet 394
of near-future process regimes 405
on coasts 174
with regional climate models in complex terrain 332
Universal Soil Loss Equation (USLE) 63
applied to the Mediterranean 301
urbanisation 20, 56, 92, 231, 237, 314, 413
and expansion into desert landscapes 413, **16**
and increased surface runoff in Mediterranean landscapes 310, 313, *314*
and runoff 102
sewering and phosphate detergents in lakes and their eutrophication 90, **5**

varves 45, 83, 86
vulnerability of landscapes 2–3, 7, 27, 29, 414, 419
high levels in Mediterranean landscapes 315
in the mountains 61
of arctic coastal regions exposed to accelerated erosion 358–360
of glacier extent and behaviour to rising temperatures 2
of island states 131
of landscapes in the wet tropics 234
of terrestrial wetlands and lakes 93–94
of tropical rainforest landscapes 231
to flood impacts in Mediterranean landscapes 313

Wadden Sea 169
water control by dams and diversions 114–120, 282, 285, 309
water yield changes with conversion of vegetation cover 322–323
World Heritage Convention 22
World Heritage List 22, 31, 46, 48

Younger Dryas *8*, 10, 393

zebra mussel (*Dreissena polymorpha*) 91

PROPERTY OF
SENECA COLLEGE
LIBRARIES
NEWNHAM CAMPUS